T0180420

Lecture Notes in Computer Science

Lecture Notes in Artificial Intelligence   14177

Founding Editor

Jörg Siekmann

Series Editors

Randy Goebel, *University of Alberta, Edmonton, Canada*
Wolfgang Wahlster, *DFKI, Berlin, Germany*
Zhi-Hua Zhou, *Nanjing University, Nanjing, China*

The series Lecture Notes in Artificial Intelligence (LNAI) was established in 1988 as a topical subseries of LNCS devoted to artificial intelligence.

The series publishes state-of-the-art research results at a high level. As with the LNCS mother series, the mission of the series is to serve the international R & D community by providing an invaluable service, mainly focused on the publication of conference and workshop proceedings and postproceedings.

Xiaochun Yang · Heru Suhartanto ·
Guoren Wang · Bin Wang · Jing Jiang · Bing Li ·
Huaijie Zhu · Ningning Cui

Editors

Advanced Data Mining and Applications

19th International Conference, ADMA 2023
Shenyang, China, August 21–23, 2023
Proceedings, Part II

Springer

Editors
Xiaochun Yang
Northeastern University
Shenyang, China

Guoren Wang
Beijing Institute of Technology
Beijing, China

Jing Jiang
University of Technology Sydney
Sydney, NSW, Australia

Huaijie Zhu
Sun Yat-sen University
Guangzhou, China

Heru Suhartanto
The University of Indonesia
Depok, Indonesia

Bin Wang
Northeastern University
Shenyang, China

Bing Li
Agency for Science, Technology
and Research (A*STAR)
Singapore, Singapore

Ningning Cui
Anhui University
Hefei, China

ISSN 0302-9743   ISSN 1611-3349 (electronic)
Lecture Notes in Artificial Intelligence
ISBN 978-3-031-46663-2   ISBN 978-3-031-46664-9 (eBook)
https://doi.org/10.1007/978-3-031-46664-9

LNCS Sublibrary: SL7 – Artificial Intelligence

This Springer imprint is published by the registered company Springer Nature Switzerland AG
The registered company address is: Gewerbestrasse 11, 6330 Cham, Switzerland

Paper in this product is recyclable.

Preface

The 19th International Conference on Advanced Data Mining and Applications (ADMA 2023) was held in Shenyang, China, during August 21–23, 2023. Researchers and practitioners from around the world came together at this leading international forum to share innovative ideas, original research findings, case study results, and experienced insights into advanced data mining and its applications. With the ever-growing importance of appropriate methods in these data-rich times, ADMA has become a flagship conference in this field. ADMA 2023 received a total of 503 submissions from 22 countries across five continents. After a rigorous double-blind review process involving 318 reviewers, 216 regular papers were accepted to be published in the proceedings, 123 were selected to be delivered as oral presentations at the conference, 85 were selected as poster presentations, and 8 were selected as industry papers. This corresponds to a full oral paper acceptance rate of 24.4%. The Program Committee (PC), composed of international experts in relevant fields, did a thorough and professional job of reviewing the papers submitted to ADMA 2023, and each paper was reviewed by an average of 2.97 PC members. With the growing importance of data in this digital age, papers accepted at ADMA 2023 covered a wide range of research topics in the field of data mining, including pattern mining, graph mining, classification, clustering and recommendation, multi-objective, optimization, augmentation, and database, data mining theory, image, multimedia and time series data mining, text mining, web and IoT applications, finance and healthcare. It is worth mentioning that ADMA 2023 was organized as a physical-only event, allowing for in-person gatherings and networking. We thank the PC members for completing the review process and providing valuable comments within tight schedules. The high-quality program would not have been possible without the expertise and dedication of our PC members. Moreover, we would like to take this valuable opportunity to thank all authors who submitted technical papers and contributed to the tradition of excellence at ADMA. We firmly believe that many colleagues will find the papers in these proceedings exciting and beneficial for advancing their research. We would like to thank Microsoft for providing the CMT system, which is free to use for conference organization, Springer for their long-term support, the host institution, Northeastern University, for their hospitality and support, Niu Translation and Shuangzhi Bo for their sponsorship. We are grateful for the guidance of the steering committee members, Osmar R. Zaiane, Chengqi Zhang, Michael Sheng, Guodong Long, Xue Li, Jianxin Li, and Weitong Chen. With their leadership and support, the conference ran smoothly. We also would like to acknowledge the support of the other members of the organizing committee. All of them helped to make ADMA 2023 a success. We appreciate the local arrangements, registration and finance management from the local arrangement chairs, registration management chairs and finance chairs Kui Di, Baoyan Song, Junchang Xin, Donghong Han, Guoqiang Ma, Yuanguo Bi, and Baiyou Qiao, the time and effort of the proceedings chairs, Bing Li, Huaijie Zhu, and Ningning Cui, the effort in advertising the conference by the publicity chairs and social network and social media coordination chairs, Xin Wang, Yongxin

Tong, Lina Wang, and Sen Wang, and the effort of managing the Tutorial sessions by the tutorial chairs, Zheng Zhang and Shuihua Wang, We would like to give very special thanks to the web chair, industry chairs, and PhD school chairs Faming Li, Chi Man Pun, Sen Wang, Linlin Ding, M. Emre Celebi, and Zheng Zhang, for creating a successful and memorable event. We also thank sponsorship chair Hua Shao for his sponsorship. Finally, we would like to thank all the other co-chairs who have contributed to the conference.

August 2023   Xiaochun Yang
                                                                        Bin Wang
                                                                        Jing Jiang

Organization

Chair of the Steering Committee

Xue Li   University of Queensland, Australia

Steering Committee

Osmar R. Zaiane	University of Alberta, Canada
Chengqi Zhang	Sydney University of Technology, Australia
Michael Sheng	Macquarie University, Australia
Guodong Long	Sydney University of Technology, Australia
Xue Li	University of Queensland, Australia
Jianxin Li	Deakin University, Australia
Weitong Chen	Adelaide University, Australia

Honor Chairs

Xingwei Wang	Northeastern University, China
Xuemin Lin	Shanghai Jiao Tong University, China
Ge Yu	Northeastern University, China

General Chairs

Xiaochun Yang	Northeastern University, China
Heru Suhartanto	The University of Indonesia, Indonesia
Guoren Wang	Beijing Institute of Technology, China

Program Chairs

Bin Wang	Northeastern University, China
Jing Jiang	University of Technology Sydney, Australia

Local Arrangement Chairs

Kui Di	Northeastern University, China
Baoyan Song	Liaoning University, China
Junchang Xin	Northeastern University, China

Registration Management Chairs

Donghong Han	Northeastern University, China
Guoqiang Ma	Northeastern University, China
Yuanguo Bi	Northeastern University, China

Finance Chair

Baiyou Qiao	Northeastern University, China

Sponsorship Chair

Hua Shao	Shenyang Huaruibo Information Technology Co., Ltd., China

Publicity Chairs

Xin Wang	Tianjin University, China
Yongxin Tong	Beihang University, China
Lina Wang	Wuhan University, China

Social Network and Social Media Coordination Chair

Sen Wang	University of Queensland, Australia

Proceeding Chairs

Bing Li	Agency for Science, Technology and Research (A*STAR), Singapore
Huaijie Zhu	Sun Yat-sen University, China
Ningning Cui	Anhui University, China

Tutorial Chairs

Zheng Zhang	Harbin Institute of Technology, Shenzhen, China
Shuihua Wang	University of Leicester, UK

Web Chair

Faming Li	Northeastern University, China

Industry Chairs

Chi Man Pun	University of Macau, China
Sen Wang	University of Queensland, Australia
Linlin Ding	Liaoning University, China

PhD School Chairs

M. Emre Celebi	University of Central Arkansas, USA
Zheng Zhang	Harbin Institute of Technology, Shenzhen, China

Program Committee

Meta Reviewers

Bohan Li	Nanjing University of Aeronautics and Astronautics, China
Can Wang	Griffith University, Australia
Chaokun Wang	Tsinghua University, China
Cheqing Jin	East China Normal University, China
Guodong Long	University of Technology Sydney, Australia
Hongzhi Wang	Harbin Institute of Technology, China
Huaijie Zhu	Sun Yat-sen University, China
Jianxin Li	Deakin University, Australia
Jun Gao	Peking University, China
Lianhua Chi	La Trobe University, Australia
Lin Yue	University of Newcastle, Australia
Tao Shen	University of Technology Sydney, Australia

Wei Emma Zhang	University of Adelaide, Australia
Weitong Chen	Adelaide University, Australia
Xiang Lian	Kent State University, USA
Xiaoling Wang	East China Normal University, China
Xueping Peng	University of Technology Sydney, Australia
Xuyun Zhang	Macquarie University, Australia
Yanjun Zhang	Deakin University, Australia
Zheng Zhang	Harbin Institute of Technology, Shenzhen, China

Reviewers

Abdulwahab Aljubairy	Macquarie University, Australia
Adita Kulkarni	SUNY Brockport, USA
Ahoud Alhazmi	Macquarie University, Australia
Akshay Peshave	GE Research, USA
Alex Delis	Univ. of Athens, Greece
Alexander Zhou	Hong Kong University of Science and Technology, China
Baoling Ning	Heilongjiang University, China
Bin Zhao	Nanjing Normal University, China
Bing Li	Institute of High Performance Computing, A*STAR, Singapore
Bo Tang	Southern University of Science and Technology, China
Carson Leung	University of Manitoba, Canada
Changdong Wang	Sun Yat-sen University, China
Chao Zhang	Tsinghua University, China
Chaokun Wang	Tsinghua University, China
Chaoran Huang	University of New South Wales, Australia
Chen Wang	Chongqing University, China
Chengcheng Yang	East China Normal University, China
Chenhao Zhang	University of Queensland, Australia
Cheqing Jin	East China Normal University, China
Chuan Ma	Zhejiang Lab, China
Chuan Xiao	Osaka University and Nagoya University, Japan
Chuanyu Zong	Shenyang Aerospace University, China
Congbo Ma	University of Adelaide, Australia
Dan He	University of Queensland, Australia
David Broneske	German Centre for Higher Education Research and Science Studies, Germany

Dechang Pi	Nanjing University of Aeronautics and Astronautics, China
Derong Shen	Northeastern University, China
Dima Alhadidi	University of New Brunswick, Canada
Dimitris Kotzinos	ETIS, France
Dong Huang	South China Agricultural University, China
Dong Li	Liaoning University, China
Dong Wen	University of New South Wales, Australia
Dongxiang Zhang	Zhejiang University, China
Dongyuan Tian	Jilin University, China
Dunlu Peng	University of Shanghai for Science and Technology, China
Eiji Uchino	Yamaguchi University, Japan
Ellouze Mourad	University of Sfax, Tunisia
Elsa Negre	LAMSADE, Paris-Dauphine University, France
Faming Li	Northeastern University, China
Farid Nouioua	Université Mohamed El Bachir El Ibrahimi de Bordj Bou Arréridj, Algeria
Genoveva Vargas-Solar	CNRS, France
Gong Cheng	Nanjing University, China
Guanfeng Liu	Macquarie University, Australia
Guangquan Lu	Guangxi Normal University, China
Guangyan Huang	Deakin University, Australia
Guannan Dong	University of Macau, China
Guillaume Guerard	ESILV, France
Guodong Long	University of Technology Sydney, Australia
Haïfa Nakouri	ISG Tunis, Tunisia
Hailong Liu	Northwestern Polytechnical University, China
Haojie Zhuang	University of Adelaide, Australia
Haoran Yang	University of Technology Sydney, Australia
Haoyang Luo	Harbin Institute of Technology (Shenzhen), China
Hongzhi Wang	Harbin Institute of Technology, China
Huaijie Zhu	Sun Yat-sen University, China
Hui Yin	Deakin University, Australia
Indika Priyantha Kumara Dewage	Tilburg University, The Netherlands
Ioannis Konstantinou	University of Thessaly, Greece
Jagat Challa	BITS Pilani, India
Jerry Chun-Wei Lin	Western Norway University of Applied Sciences, Norway
Jiabao Han	NUDT, China
Jiajie Xu	Soochow University, China
Jiali Mao	East China Normal University, China

Jianbin Qin	Shenzhen University, China
Jianhua Lu	Southeast University, China
Jianqiu Xu	Nanjing University of Aeronautics and Astronautics, China
Jianxin Li	Deakin University, Australia
Jianxing Yu	Sun Yat-sen University, China
Jiaxin Jiang	National University of Singapore, Singapore
Jiazun Chen	Peking University, China
Jie Shao	University of Electronic Science and Technology of China, China
Jie Wang	Indiana University, USA
Jilian Zhang	Jinan University, China, China
Jingang Yu	Shenyang Institute of Computing Technology, Chinese Academy of Sciences
Jing Du	University of New South Wales, Australia
Jules-Raymond Tapamo	University of KwaZulu-Natal, South Africa
Jun Gao	Peking University, China
Junchang Xin	Northeastern University, China
Junhu Wang	Griffith University, Australia
Junshuai Song	Peking University, China
Kai Wang	Shanghai Jiao Tong University, China
Ke Deng	RMIT University, Australia
Kun Han	University of Queensland, Australia
Kun Yue	Yunnan University, China
Ladjel Bellatreche	ISAE-ENSMA, France
Lei Duan	Sichuan University, China
Lei Guo	Shandong Normal University, China
Lei Li	Hong Kong University of Science and Technology (Guangzhou), China
Li Li	Southwest University, China
Lin Guo	Changchun University of Science and Technology, China
Lin Mu	Anhui University, China
Linlin Ding	Liaoning University, China
Lizhen Cui	Shandong University, China
Long Yuan	Nanjing University of Science and Technology, China
Lu Chen	Swinburne University of Technology, Australia
Lu Jiang	Northeast Normal University, China
Lukui Shi	Hebei University of Technology, China
Maneet Singh	IIT Ropar, India
Manqing Dong	Macquarie University, Australia

Mariusz Bajger	Flinders University, Australia
Markus Endres	University of Applied Sciences Munich, Germany
Mehmet Ali Kaygusuz	Middle East Technical University, Turkey
Meng-Fen Chiang	University of Auckland, New Zealand
Ming Zhong	Wuhan University, China
Minghe Yu	Northeastern University, China
Mingzhe Zhang	University of Queensland, Australia
Mirco Nanni	CNR-ISTI Pisa, Italy
Misuk Kim	Sejong University, South Korea
Mo Li	Liaoning University, China
Mohammad Alipour Vaezi	Virginia Tech, USA
Mourad Nouioua	Mohamed El Bachir El Ibrahimi University, Bordj Bou Arreridj, Algeria
Munazza Zaib	Macquarie University, Australia
Nabil Neggaz	Université des Sciences et de la Technologie d'Oran Mohamed Boudiaf, Algeria
Nicolas Travers	Léonard de Vinci Pôle Universitaire, Research Center, France
Ningning Cui	Anhui University, China
Paul Grant	Charles Sturt University, Australia
Peiquan Jin	University of Science and Technology of China, China
Peng Cheng	East China Normal University, China
Peng Peng	Hunan University, China
Peng Wang	Fudan University, China
Pengpeng Zhao	Soochow University, China
Philippe Fournier-Viger	Shenzhen University, China
Ping Lu	Beihang University, China
Pinghui Wang	Xi'an Jiaotong University, China
Qiang Yin	Shanghai Jiao Tong University, China
Qing Liao	Harbin Institute of Technology (Shenzhen), China
Qing Liu	Data61, CSIRO, Australia
Qing Xie	Wuhan University of Technology, China
Quan Chen	Guangdong University of Technology, China
Quan Z. Sheng	Macquarie University, Australia
Quoc Viet Hung Nguyen	Griffith University, Australia
Rania Boukhriss	University of Sfax, Tunisia
Riccardo Cantini	University of Calabria, Italy
Rogério Luís Costa	Polytechnic of Leiria, Portugal
Rong Zhu	Alibaba Group, China
Ronghua Li	Beijing Institute of Technology, China
Rui Zhou	Swinburne University of Technology, Australia

Rui Zhu	Shenyang Aerospace University, China
Sadeq Darrab	Otto von Guericke University Magdeburg, Germany
Saiful Islam	Griffith University, Australia
Sayan Unankard	Maejo University, Thailand
Senzhang Wang	Central South University, China
Shan Xue	University of Wollongong, Australia
Shaofei Shen	University of Queensland, Australia
Shi Feng	Northeastern University, China
Shiting Wen	Zhejiang University, China
Shiyu Yang	Guangzhou University, China
Shouhong Wan	University of Science and Technology of China, China
Shuhao Zhang	Singapore University of Technology and Design, Singapore
Shuiqiao Yang	UNSW, Australia
Shuyuan Li	Beihang University, China
Silvestro Roberto Poccia	University of Turin, Italy
Sonia Djebali	Léonard de Vinci Pôle Universitaire, Research Center, France
Suman Banerjee	IIT Jammu, India
Tao Qiu	Shenyang Aerospace University, China
Tao Zhao	National University of Defense Technology, China
Tarique Anwar	University of York, UK
Thanh Tam Nguyen	Griffith University, Australia
Theodoros Chondrogiannis	University of Konstanz, Germany
Tianrui Li	Southwest Jiaotong University, China
Tianyi Chen	Peking University, China
Tieke He	Nanjing University, China
Tiexin Wang	Nanjing University of Aeronautics and Astronautics, China
Tiezheng Nie	Northeastern University, China
Uno Fang	Deakin University, Australia
Wei Chen	University of Auckland, New Zealand
Wei Deng	Southwestern University of Finance and Economics, China
Wei Hu	Nanjing University, China
Wei Li	Harbin Engineering University, China
Wei Liu	University of Macau, Sun Yat-sen University, China
Wei Shen	Nankai University, China
Wei Song	Wuhan University, China

Weijia Zhang	University of Newcastle, Australia
Weiwei Ni	Southeast University, China
Weixiong Rao	Tongji University, China
Wen Zhang	Wuhan University, China
Wentao Li	Hong Kong University of Science and Technology (Guangzhou), China
Wenyun Li	Harbin Institute of Technology (Shenzhen), China
Xi Guo	University of Science and Technology Beijing, China
Xiang Lian	Kent State University, USA
Xiangguo Sun	Chinese University of Hong Kong, China
Xiangmin Zhou	RMIT University, Australia
Xiangyu Song	Swinburne University of Technology, Australia
Xianmin Liu	Harbin Institute of Technology, China
Xianzhi Wang	University of Technology Sydney, Australia
Xiao Pan	Shijiazhuang Tiedao University, China
Xiaocong Chen	University of New South Wales, Australia
Xiaofeng Gao	Shanghai Jiaotong University, China
Xiaoguo Li	Singapore Management University, Singapore
Xiaohui (Daniel) Tao	University of Southern Queensland, Australia
Xiaoling Wang	East China Normal University, China
Xiaowang Zhang	Tianjin University, China
Xiaoyang Wang	University of New South Wales, Australia
Xiaojun Xie	Nanjing Agricultural University, China
Xin Cao	University of New South Wales, Australia
Xin Wang	Southwest Petroleum University, China
Xinqiang Xie	Neusoft, China
Xiuhua Li	Chongqing University, China
Xiujuan Xu	Dalian University of Technology, China
Xu Yuan	Harbin Institute of Technology, Shenzhen, China
Xu Zhou	Hunan University, China
Xupeng Miao	Carnegie Mellon University, USA
Xuyun Zhang	Macquarie University, Australia
Yajun Yang	Tianjin University, China
Yanda Wang	Nanjing University of Aeronautics and Astronautics, China
Yanfeng Zhang	Northeastern University, China
Yang Cao	Hokkaido University, China
Yang-Sae Moon	Kangwon National University, South Korea
Yanhui Gu	Nanjing Normal University, China
Yanjun Shu	Harbin Institute of Technology, China
Yanlong Wen	Nankai University, China

Yanmei Hu	Chengdu University of Technology, China
Yao Liu	University of New South Wales, Australia
Yawen Zhao	University of Queensland, Australia
Ye Zhu	Deakin University, Australia
Yexuan Shi	Beihang University, China
Yicong Li	University of Technology Sydney, Australia
Yijia Zhang	Jilin University, China
Ying Zhang	Nankai University, China
Yingjian Li	Harbin Institute of Technology, Shenzhen, China
Yingxia Shao	BUPT, China
Yishu Liu	Harbin Institute of Technology, Shenzhen, China
Yishu Wang	Northeastern University, China
Yixiang Fang	Chinese University of Hong Kong, Shenzhen, China
Yixuan Qiu	The University of Queensland, Australia
Yong Zhang	Tsinghua University, China
Yongchao Liu	Ant Group, China
Yongpan Sheng	Southwest University, China
Yongqing Zhang	Chengdu University of Information Technology, China
Youwen Zhu	Nanjing University of Aeronautics and Astronautics, China
Yu Gu	Northeastern University, China
Yu Liu	Huazhong University of Science and Technology, China
Yu Yang	Hong Kong Polytechnic University, China
Yuanbo Xu	Jilin University, China
Yucheng Zhou	University of Technology Sydney, Australia
Yue Tan	University of Technology Sydney, Australia
Yunjun Gao	Zhejiang University, China
Yunzhang Huo	Hong Kong Polytechnic University, China
Yurong Cheng	Beijing Institute of Technology, China
Yutong Han	Dalian Minzu University, China
Yutong Qu	University of Adelaide, Australia
Yuwei Peng	Wuhan University, China
Yuxiang Zeng	Hong Kong University of Science and Technology, China
Zesheng Ye	University of New South Wales, Sydney, Australia
Zhang Anzhen	Shenyang Aerospace University, China
Zhaojing Luo	National University of Singapore, Singapore
Zhaonian Zou	Harbin Institute of Technology, China
Zheng Liu	Nanjing University of Posts and Telecommunications, China

Zhengyi Yang	University of New South Wales, Australia
Zhenying He	Fudan University, China
Zhihui Wang	Fudan University, China
Zhiwei Zhang	Beijing Institute of Technology, China
Zhixin Li	Guangxi Normal University, China
Zhongnan Zhang	Xiamen University, China
Ziyang Liu	Tsinghua University, China

Contents – Part II

Knowledge Graph

CKGE: Improving Distance Based Knowledge Graph Embedding
via Contrastive Learning 3
    Yafei Liu, Shuaishuai Zu, and Li Li

Knowledge Graph Completion with Information Adaptation
and Refinement 16
    Yifan Liu, Bin Shang, Chenxin Wang, and Yinliang Zhao

Duet Representation Learning with Entity Multi-attribute Information
in Knowledge Graphs 32
    Yuanbo Xu, Yuanbo Zhang, Yongjian Yang, Hangtong Xu, and Lin Yue

TKGAT: Temporal Knowledge Graph Representation Learning Using
Attention Network 46
    Shaowei Zhang, Zhao Li, Xin Wang, Zirui Chen, and WenBin Guo

Separate-and-Aggregate: A Transformer-Based Patch Refinement Model
for Knowledge Graph Completion 62
    Chen Chen, Yufei Wang, Yang Zhang, Quan Z. Sheng, and Kwok-Yan Lam

Two Birds with One Stone: A Link Prediction Model for Knowledge
Hypergraph Based on Fully-Connected Tensor Decomposition 78
    Jun Pang, Hong-Chao Qin, Yan Liu, and Xiao-Qi Liu

HEM: An Improved Parametric Link Prediction Algorithm Based
on Hybrid Network Evolution Mechanism 91
    Dejing Ke and Jiansu Pu

Joint Embedding of Local Structures and Evolutionary Patterns
for Temporal Link Prediction 107
    *Tingxuan Chen, Jun Long, Liu Yang, Guohui Li, Shuai Luo,
    and Meihong Xiao*

Multimedia

DQN-Based Stitching Algorithm for Unmanned Aerial Vehicle Images 125
    Ji Ma, Wenci Liu, and Tingwei Chen

HM-QCNN: Hybrid Multi-branches Quantum-Classical Neural Network
for Image Classification 139
  Haowen Liu, Yufei Gao, Lei Shi, Lin Wei, Zheng Shan, and Bo Zhao

DetOH: An Anchor-Free Object Detector with Only Heatmaps 152
  Ruohao Wu, Xi Xiao, Guangwu Hu, Hanqing Zhao, Han Zhang,
  and Yongqing Peng

LAANet: An Efficient Automatic Modulation Recognition Model Based
on LSTM-Autoencoder and Attention Mechanism 168
  Qing Li and Xin Zhou

A Compact Phoneme-To-Audio Aligner for Singing Voice 183
  Meizhen Zheng, Peng Bai, and Xiaodong Shi

Song-to-Video Translation: Writing a Video from Song Lyrics Based
on Multimodal Pre-training 198
  Feifei Fu, Zelong Sun, Guoxing Yang, Xiaolong He, and Zhiwu Lu

Medical Image Analysis

Breast Cancer Histopathology Image Classification Using Frequency
Attention Convolution Network 217
  Ruidong Lu, Qiule Sun, Xueyan Ding, and Jianxin Zhang

Wavelet-SVDD: Anomaly Detection and Segmentation with Frequency
Domain Attention 230
  Linhui Zhou, Weiyu Guo, Jing Cao, Xinyue Zhang, and Yue Wang

Anatomical-Functional Fusion Network for Lesion Segmentation Using
Dual-View CEUS 244
  Peng Wan, Chunrui Liu, and Daoqiang Zhang

SpMVNet: Spatial Multi-view Network for Head and Neck Organs at Risk
Segmentation 257
  Hongzhi Liu, Shiyu Zhu, Qianjin Feng, and Yang Chen

SNN-BS: A Clinical Terminology Standardization Method Using Siamese
Networks with Batch Sampling Strategy 272
  Xiao Wei, Xiaoxin Wang, and Nengjun Zhu

Natural Language

Read Then Respond: Multi-granularity Grounding Prediction
for Knowledge-Grounded Dialogue Generation 291
   Yiyang Du, Shiwei Zhang, Xianjie Wu, Zhao Yan, Yunbo Cao,
   and Zhoujun Li

SUMOPE: Enhanced Hierarchical Summarization Model for Long Texts 307
   Chao Chang, Junming Zhou, Xiangwei Zeng, and Yong Tang

An Extractive Automatic Summarization Method for Chinese Long Text 320
   Jizhao Zhu, Wenyu Duan, Naitong Yu, Xinlong Pan, and Chunlong Fan

A Likelihood Probability-Based Online Summarization Ranking Model 334
   Shuhao Yue, Dunhui Yu, and Di Xie

Spatial Commonsense Reasoning for Machine Reading Comprehension 347
   Miaopei Lin, Meng-xiang Wang, Jianxing Yu, Shiqi Wang,
   Hanjiang Lai, Wei Liu, and Jian Yin

Multimodal Learning for Automatic Summarization: A Survey 362
   Zhicheng Zhang, Yibo Sun, and Shiyan Su

Deep Knowledge Tracing with Concept Trees 377
   Yupei Zhang, Rui An, Wenxin Zhang, Shuhui Liu, and Xuequn Shang

Privacy and Security

Cryptography-Inspired Federated Learning for Generative Adversarial
Networks and Meta Learning 393
   Yu Zheng, Wei Song, Minxin Du, Sherman S. M. Chow, Qian Lou,
   Yongjun Zhao, and Xiuhua Wang

A Hessian-Based Federated Learning Approach to Tackle Statistical
Heterogeneity 408
   Adnan Ahmad, Wei Luo, and Antonio Robles-Kelly

Analyzing the Convergence of Federated Learning with Biased Client
Participation 423
   Lei Tan, Miao Hu, Yipeng Zhou, and Di Wu

Privacy Lost in Online Education: Analysis of Web Tracking Evolution 440
   Zhan Su, Rasmus Helles, Ali Al-Laith, Antti Veilahti, Akrati Saxena,
   and Jakob Grue Simonsen

DANAA: Towards Transferable Attacks with Double Adversarial Neuron
Attribution 456
   Zhibo Jin, Zhiyu Zhu, Xinyi Wang, Jiayu Zhang, Jun Shen,
   and Huaming Chen

CRNN-SA: A Network Intrusion Detection Method Based on Deep
Learning 471
   Wanxiao Liu, Jue Chen, and Xihe Qiu

Applications

Conspiracy Spoofing Orders Detection with Transformer-Based Deep
Graph Learning 489
   Le Kang, Tai-Jiang Mu, and Xiaodong Ning

A Novel Explainable Rumor Detection Model with Fusing Objective
Information 504
   Junlong Wang, Dechang Pi, Mingtian Ping, and Zhiwei Chen

VLS: A Reinforcement Learning-Based Value Lookahead Strategy
for Multi-product Order Fulfillment 518
   Ryan Wickman, Junxuan Li, and Xiaofei Zhang

Deep Reinforcement Learning for Stock Trading with Behavioral Finance
Strategy 535
   Shilong Deng, Zetao Zheng, Hongcai He, and Jie Shao

Ensemble Learning Based Employment Recommendation Under
Interaction Sparsity for College Students 550
   Haiping Zhu, Yifei Zhao, Yuchen Wu, Yan Chen, Wenhao Li,
   Qinghua Zheng, and Feng Tian

How Does ChatGPT Affect Fake News Detection Systems? 565
   Bo Li, Jiaxin Ju, Can Wang, and Shirui Pan

CNGT: Co-attention Networks with Graph Transformer for Fact
Verification 581
   Jing Yuan, Chen Chen, Chunyan Hou, and Xiaojie Yuan

Multi-modal

Supervised Discriminative Discrete Hashing for Cross-Modal Retrieval 599
   Xingyu Lu and Chi-Man Pun

A Knowledge-Enhanced Inferential Network for Cross-Modality
Multi-hop VQA 614
Shiqi Wang, Jianxing Yu, Miaopei Lin, Shuang Qiu, Xiaofeng Luo,
and Jian Yin

Multi-head Similarity Feature Representation and Filtration for Image-Text
Matching 629
Mengqi Jiang, Shichao Zhang, Debo Cheng, Leyuan Zhang,
and Guixian Zhang

Multimodal Conditional VAE for Zero-Shot Real-World Event Discovery 644
Zhuopan Yang, Di Luo, Jiuxiang You, Zhiwei Guo, and Zhenguo Yang

MAMRP: Multi-modal Data Aware Movie Rating Prediction 660
Mingfu Qin, Qian Zhou, Wei Chen, and Lei Zhao

MFMGC: A Multi-modal Data Fusion Model for Movie Genre
Classification 676
Xiaorui Yang, Qian Zhou, Wei Chen, and Lei Zhao

TED-CS: Textual Enhanced Sensitive Video Detection with Common
Sense Knowledge 692
Bihui Yu, Linzhuang Sun, Jingxuan Wei, Shuyue Tan, Yiman Zhao,
and Liping Bu

Author Index 709

Knowledge Graph

CKGE: Improving Distance Based Knowledge Graph Embedding via Contrastive Learning

Yafei Liu, Shuaishuai Zu, and Li Li[✉]

School of Computer and Information Science, Southwest University,
Chongqing, China
lily@swu.edu.cn

Abstract. Knowledge graph embedding (KGE) is critical in various downstream applications as it represents entities and relations in a knowledge graph as low-dimensional vectors. The embeddings of the entities and relations denote their semantics on the knowledge graph, which affects the effectiveness of the model. Recently, distance-based (DB) models have demonstrated great explanatory power in KGE. However, most existing DB models focus solely on single triples to independently optimize the scoring function, disregarding the interconnections among different triples. To address this issue, we propose CKGE, a novel contrastive learning approach that enhances the performance of DB models while remaining versatile enough to apply to different DB models. Specifically, CKGE improves the alignment and uniformity of DB models, meaning that the embedding of the same semantic entities should remain close under different relations, and embeddings for random entities should scatter on the hypersphere. Additionally, we present a supervised contrastive learning approach to optimize in-batch negative methods, thereby improving the learning of semantic entities. Extensive experiments on four benchmark datasets demonstrate that CKGE yields significant improvements in link prediction, especially for large-scale datasets such as ogbl-wikikg2.

Keywords: Knowledge graph embedding · Distance based model · Contrastive Learning

1 Introduction

Knowledge graphs usually represent structured human knowledge in the form of (head entity, relation, tail entity). Although knowledge graphs usually contain billions of triples, they still suffer from the incompleteness problem due to a lot of factual triples missing, which needs knowledge graph completion (KGC). Knowledge graph embedding (KGE) has been proposed for this problem, which embeds all entities and relations into a low dimensional space and aims to predict missing links between entities.

© The Author(s), under exclusive license to Springer Nature Switzerland AG 2023
X. Yang et al. (Eds.): ADMA 2023, LNAI 14177, pp. 3–15, 2023.
https://doi.org/10.1007/978-3-031-46664-9_1

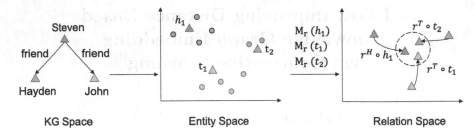

Fig. 1. A toy example showing how DB models can exhibit entities and relations representation on KGs.

Recently, some distance based (DB) models, which use the spatial distance of two entities after the transformation of the relation to judge whether two entities have a certain relation or not, have shown great explanation and power in KGE. In general, according to the way of manipulating relations, we divide the majority of DB models roughly into two groups: translation families and rotation families. In knowledge graph embeddings, translation models such as TransE [1], TransR [10], and TransD [8] primarily focus on addressing relation mappings including 1-to-N, N-to-1, and N-to-N relationships. Meanwhile, the newer rotation models, exemplified by RotatE [17] and PairRE [3], have expanded their scope to cater to a range of relation patterns including symmetry/antisymmetry, inverse, composition, and subrelation. For both categories, the underlying assumption for a valid triple (h, r, t) is that after undergoing a relational transformation r, the head entity h should be proximate to the tail entity t. Intuitively, the smaller the spatial distance between two entities post-transformation, the higher their likelihood of representing a valid relationship in reality.

However, most existing DB models only focus on single triples to independently optimize the scoring function while ignoring the interconnection among different triples. Motivated by the sentence embedding representation [6], alignment and uniformity are also observed by KGE. As shown in Fig. 1, suppose that some triples have different tail entities but share the same head entity and relation like *(Steven, friend, Hayden)* and *(Steven, friend, John)*. For alignment, the tail entities *Hayden* and *John* should be as close as possible to the head entity *Steven* after relation *friend* transformation respectively. While the entities *Steven, Hayden,* and *John* should also maintain distance uniformity with other entities in the knowledge graph after relation transformation, which is beneficial to the link prediction task as a measure of the quality of KGE.

Here we propose CKGE, a novel method that effectively constrains entities to improve the performance of KGE via contrastive learning, especially for the DB models. Specifically, Our motivation is based on the observation that the same semantic entities should keep alignment as shown in Fig. 1 and maintain distance uniformity with other entities in the knowledge graph. First, we abstract key procedures from mainstream DB models and present a unified DB model paradigm. Secondly, based on the above paradigm and analysis, we design an in-batch division method for positive and negative samples without extra data.

Thirdly, we further conduct a supervised contrastive learning method to optimize the in-batch negative method using semantic labels, which is able to learn the same semantic entities better. CKGE is widely applicable to various distance based models, including TransE, TransH, PairRE, etc. Experiments show that CKGE yields consistent and significant improvements in datasets for the knowledge graph completion task.

In summary, our main contributions are as follows:

- As far as we know, we are the first to propose the DB model framework with contrastive learning. The proposed CKGE constrains the representation of the same semantic entities in different triples.
- We present a unified DB model paradigm by abstracting diverse DB models, and theoretically prove that CKGE is widely applicable to various DB models.
- Experiments show that CKGE yields consistent and significant improvements on four benchmark datasets for link prediction tasks. It is worth noting that CKGE improves nearly 6% on the large scale knowledge graph (ogbl-wikikg2).

2 Related Work

2.1 KGE Model

Knowledge graph embedding models can be broadly classified into three categories [15]: *distance based models* (DB), *tensor decomposition models* (TDB), and *neural network based Models* (NN).

Distance Models project entities on the knowledge graph into space. Generally, the closer the spatial distance between two entities after the transformation of relation, the greater the probability of validity in the real world. And the score function have the formulation of $s(h_i, r_j, t_k) = -\|\Gamma(h_i, r_j, t_k)\|$, where Γ is a model-specific function. We divide the majority of DB models roughly into two families according to the way they manipulate relations. Moreover, although many research attempts to design more complicated scoring function [2, 22], we think that the aforementioned DB models are powerful enough and our proposed unified DB model paradigm is based on these.

Tensor Decomposition Based Models formulate the KGC task as a triadic binary tensor completion challenge. Within the framework of RESCAL [14], each relationship is depicted using a matrix of full rank, with its scoring mechanism defined through a bilinear approach, which is $f_r(\mathbf{h}, \mathbf{t}) = \mathbf{h}^\top \mathbf{M}_r \mathbf{t}$. However, full-rank matrices are prone to overfitting, DistMult [21] defines $\mathbf{M}_r$ as a diagonal matrix to solve it. ComplEx [19] emerged to tackle DistMult's limitations in handling antisymmetric relations, integrating complex-valued embeddings. However, its ability to handle the composition pattern remains limited, and both its spatial and temporal complexities have grown significantly.

Neural Network Based Models leveraging neural architectures have also made good progress in recent years. ConvE [4], R-GCN [16], and KBGAT [12] have incorporated convolutional neural networks, graph convolutional networks, and graph attention networks into KGE, respectively. Yet, due to the opaque nature of NNs, they often lack clear interpretability.

2.2   Contrastive Learning

Contrastive learning seeks to optimize representations by drawing positive pairs closer and distancing negative pairs. This process ensures that similar samples cluster together while dissimilar ones remain distant. This approach has found broad applications in both computer vision and natural language processing. Acting akin to regularization techniques, contrastive learning leverages negative samples to stabilize the loss function within mini-batches. In the realm of knowledge graphs, [20] employs this method for efficient training with an expansive set of negative samples. Yet, their model relies on textual data and pre-trained models, overlooking semantically similar entities.

3   Methods

In this section, we present CKGE. Section 3.1 introduces a consolidated DB model paradigm, encapsulating essential processes of prevalent translation and rotation models. Subsequently, Sects. 3.2 and 3.3 detail unsupervised and supervised CKGE, respectively. The supervised approach utilizes semantically similar entities as labels, enhancing the unsupervised version.

3.1   A Unified DB Model Paradigm

We provide a unified view of several DB models, by showing that they are restricted versions under our paradigm. We first propose a unified version of the distance scoring function:

$$f_r(\boldsymbol{h}, \boldsymbol{t}) = \left\| \boldsymbol{r^1} \circ \mathbf{M_r}\left(\boldsymbol{h}\right) + \boldsymbol{b_r} - \boldsymbol{r^2} \circ \mathbf{M_r}\left(\boldsymbol{t}\right) \right\|_{1/2} \tag{1}$$

where $\boldsymbol{h}, \boldsymbol{t}$ represent entities embedding. $\boldsymbol{M_r}(\cdot)$ represents the relation-specific projecting matrix of entity vectors. Inspired by PairRE [3], each relation is characterized as two weight vectors ($\boldsymbol{r^1}$ and $\boldsymbol{r^2}$), corresponding to the transformations towards the head and tail entities, respectively. The symbol $\circ$ illustrates the functional transition induced by relation r, understood as a rotation maneuver within a complex space in RotatE. Next, we show that our unified DB Model Paradigm can cover the ideas of most mainstream DB models.

TransE is the most classic translation based model, Given a triple (h, r, t), in which entities and relations are projected in Euclidean Space. There is no entity mapping and relation translation. Compared to Eq. 1, the score function of TransE is trivial to be rewritten as removing $\boldsymbol{r^1}$, $\boldsymbol{r^2}$ and $\boldsymbol{M_r}$. The distance score is defined as:

$$f_{\mathbf{TransE}}(\boldsymbol{h}, \boldsymbol{t}) = \left\| \boldsymbol{h} + \boldsymbol{r} - \boldsymbol{t} \right\|_{1/2} \tag{2}$$

TransX includes TransH, TransR, and TransD, which represent different projection operations for entities. For example, for TransH, $M_r(\cdot)$ is supposed to be projected to the hyperplane with $M_r(h) = h - w_r^\top h w_r$. Compared to Eq. 1, the score function of TransX is trivial to be rewritten as removing r^1 and r^2. The distance score is defined as:

$$f_{\text{TransX}}(h, t) = \|M_r(h) + b_r - M_r(t)\|_{1/2} \qquad (3)$$

RotatE and PairRE study more relationships pattern compared by TransE and TransX. For a triple (h, r, t), the relation means the rotation of the entity, which is different from the translation in TransE. Compared to Eq. 1, the score function of RotatE/PairRE is trivial to verify as the score function can be rewritten as removing M_r. PairRE is the complete model of RotatE with paired vectors for each relation representation, which rotates the head and tail entities separately to better model the sub-relationship. The distance score is defined as:

$$f_{\text{RotatE/PairRE}}(h, t) = \|r^1 \circ h + b_r - r^2 \circ t\|_{1/2} \qquad (4)$$

3.2 Unsupervised CKGE

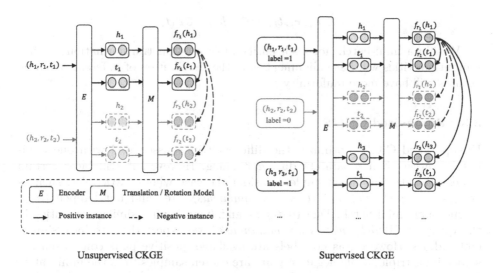

Unsupervised CKGE   Supervised CKGE

Fig. 2. Illustrations of unsupervised CKGE and supervised CKGE. The supervised CKGE considers similar semantics entities as labels to improve the unsupervised CKGE.

Now we introduce the unsupervised CKGE, and Fig. 2 shows the details. The basic idea of our approach is to treat each triple itself as a positive sample pair and different triples as negative sample pairs. Thus our method generates more negative samples to provide to the model for learning and is able to represent alignment and uniformity better for sparse entities.

For the unified DB Model framework, we would split the distance scoring function into two parts for (h, r, t): $r^1 \circ \mathbf{M_r}(h) + b_r$ and $r^2 \circ \mathbf{M_r}(t)$ (or $r^1 \circ \mathbf{M_r}(h)$ and $b_r + r^2 \circ \mathbf{M_r}(t)$). Note that we use $f_r(h)$ and $f_r(t)$ to replace $r^1 \circ \mathbf{M_r}(h) + b_r$ and $r^2 \circ \mathbf{M_r}(t)$ respectively. For DB models, the score function factually uses the distance between $f_r(h)$ and $f_r(t)$ as a basis to judge whether (h, r, t) is valid. For each triple, $f_r(h)$ and $f_r(t)$, the head entity and tail entity after the mapping matrix and relation transformation, are used as positive sample pairs. For the remaining triples in the same mini-batch, the head/tail entity pairs are used as negative sample pairs. For unsupervised CKGE, triples within the same batch are widely used as negative samples. With a mini-batch of N pairs, we adopt the InfoNCE loss function to calculate a sample $f_{r_i}(h_i)$ and its positive sample $f_{r_i}(t_i)$ as the contrastive loss:

$$\mathcal{CL}(h_i) = -\log \frac{e^{\text{sim}(f_{r_i}(h_i), f_{r_i}(t_i))/\tau}}{\sum_{j=1}^{N}(e^{\text{sim}(f_{r_i}(h_i), f_{r_j}(h_j))/\tau} + e^{\text{sim}(f_{r_i}(h_i), f_{r_j}(t_j))/\tau})}, \quad (5)$$

where τ is a temperature hyperparameter and $\text{sim}(f_r(h), f_r(t))$ is the cosine similarity. As shown in Fig. 2, it illustrates unsupervised contrastive learning in CKGE. For head entity and tail entity in the triple(h_i, r_i, t_i), we have:

$$\mathcal{CL}(h_i, r_i, t_i) = \mathcal{CL}(h_i) + \mathcal{CL}(t_i) \quad (6)$$

Within a mini-batch, for each entity after relation transformation, similar semantic entities will have alignment, and the distribution of different semantic entities will have more uniformity.

3.3   Supervised CKGE

Unsupervised CKGE optimizes the different triples associated with one mini-batch, rather than separately during training. However, it still faces certain issues. To take this for example, consider the case where two triples, such as *(tigers, is, mammals)* and *(lions, is, mammals)*, are valid but happen to be in the same mini-batch. Due to *tigers* and *lions* having similar semantics in the query of *"Which entities are mammals?"*, we expect them to have similar embeddings. However, as no labels are available, positive pairs come from the same single triple, while negative pairs are chosen samples from the mini-batch. This causes tigers and lions to be pushed apart as negative pairs due to their presence in different triples, resulting in a negative gain.

To address the aforementioned issue, we draw inspiration from supervised contrastive learning [9] and employ a supervised contrastive learning loss (Eq. 7) to train the model. Specifically, we define the positive samples of those triples that share the same relation and head entity/tail entity as the label. The triples sharing the head or tail entities are treated as positive sample pairs like *(tigers, is, mammals)* and *(lions, is, mammals)*. For a given triple, we use all its positive samples in the same mini-batch, and define the improvement loss as follows:

$$CL(\boldsymbol{h_i}) = -\log \frac{\sum\limits_{a \in P} \left(e^{\text{sim}(f_{r_i}(h_i), f_{r_a}(h_a))/\tau} + e^{\text{sim}(f_{r_i}(h_i), f_{r_a}(t_a))/\tau}\right)}{\sum_{j=1}^{N} \left(e^{\text{sim}(f_{r_i}(h_i), f_{r_j}(h_j))/\tau} + e^{\text{sim}(f_{r_i}(h_i), f_{r_j}(t_j))/\tau}\right)}, \tag{7}$$

where P is defined as the set in which triples have the same label in the same mini-batch.

3.4 Training Objective

For a given training triple (h_i, r_i, t_i) in a KG, our framework computes the joint loss as follows:

$$\mathcal{L}(h_i, r_i, t_i) = \mathcal{L}_s + \lambda \mathcal{L}_{cl} \tag{8}$$

where $\mathcal{L}_s$ signifies the scoring function of the DB model. $\mathcal{L}_{cl}$ represents the weighted contrastive loss discussed in Sects. 3.2 and 3.3. λ is a balanced hyper-parameter.

For the scoring function $\mathcal{L}_s$, the self-adversarial negative sampling loss [17] is typically employed for training purposes:

$$\mathcal{L}_s = -\log \sigma(\gamma - f_r(h, t)) - \sum_{i=1}^{n} p(h_i', r, t_i') \log \sigma(f_r(h_i', t_i') - \gamma) \tag{9}$$

where γ stands for a set margin, while σ denotes the sigmoid function. And we introduce a weighted contrastive loss that assigns λ as a hyper-parameter. (h_i', r, t_i') refers to the ith negative triple and $p(h_i', r, t_i')$ indicates the weight assigned to this negative sample. $p(h_i', r, t_i')$ is defined as follows:

$$p\left((h_i', r, t_i') \mid (h, r, t)\right) - \frac{\exp f_r(h_i', t_i')}{\sum_j \exp f_r(h_j', t_j')} \tag{10}$$

4 Experiments

4.1 Experimental Setting

Dataset Setting. Our model's effectiveness is assessed through link prediction across four benchmark knowledge graphs: FB15k-237 [18], WN18RR [4], YAGO3-10 [11], and ogbl-wikikg2 [7]. Table 1 offers a summary of these datasets' statistical data. Compared to FB15k and WN18, FB15k-237 and WN18RR have their inverse relations excluded, emphasizing primarily symmetry/antisymmetry and composition patterns. Ogbl-wikikg2, derived from the Wikidata knowledge base, surpasses the scale of other benchmarks by a considerable margin. Navigating complex relation mappings becomes an added challenge besides the standard relation patterns.

Evaluation Protocol. For link prediction evaluation, we employ MR, MRR, and Hit@N as metrics. Given a test triple (h, r, t), the objective is to replace a missing head or tail entity, resulting in pairs like $(h, r, ?)$ or $(?, r, t)$. KGE models rank these triples based on their scores. MR represents the mean rank of accurate entities; a lower value is preferable. MRR calculates the average inverse rank of these entities, while Hit@N determines the fraction of correct entities among the top n. For both MRR and Hit@N, higher values indicate better performance.

Table 1. Entity, relation, and triple counts for the datasets utilized in our experiments.

Dataset	#Entity	#Relation	#Train	#Valid	#Test
FB15k-237	14,541	237	272,115	17,535	20,466
WN18RR	40,943	11	86,835	3,034	3,134
YAGO3-10	123,182	37	1,079,040	5,000	5,000
ogbl-wikikg2	2,500,604	535	16,109,182	429,456	598,543

Implementation and Baseline. For the main experiments, we implement our models based on the implementation of PairRE [3], which is the recent state-of-the-art. In order to maintain a controlled test, all hyperparameters are kept the same with origin experiments except the hyperparameters related to comparative learning like t and λ. And We employ grid search, optimizing hyperparameters based on the validation datasets' performance. Specifically, we search temperature hyperparameter in { 0.5, 0.1, 0.05, 0.01, 0.005 }, and search contrastive learning loss coefficients in { 0.05, 0.1, 1 }. Our proposed models CKGE are implemented by PyTorch 1.12.0 and trained on a Linux server with GTX 3090. We compare the performance of CKGE against several KGE models with different families, including RotatE [17], RotatE3D [5], QuatE [22], DisMult [21], ComplEx [19], ConvE [4], R-GCN [16], and ConvKB [13].

4.2   Main Results

Comparisons for FB15k-237, WN18RR, and YAGO3-10 datasets are shown in Table 2. We can see that our model's performance yields consistent metrics improvements using different datasets compared to other models. Since our model shares the same hyper-parameter settings and implementation with PairRE, comparing it with this state-of-the-art model is fair to show the advantages and disadvantages of the proposed model. It is noteworthy that CKGE works very well on the WN18RR dataset at Hit@1 than PairRE, but is lower than RotatE. Compared to other datasets, the knowledge graph for the WN18RR dataset is denser because the number of relations has only 11. As we will demonstrate in the following experiments, CKGE can theoretically be integrated with any model that fits this paradigm in Sect. 3.1.

Table 2. Link prediction results on FB15k-237, WN18RR, and YAGO3-10. While CKGE's results are from our experiments, data for other models are sourced from their respective papers.

Model	FB5k-237				WN18RR				YAGO3-10		
	Hits		MR	MRR	Hits		MR	MRR	Hits		MRR
	@10	@1			@10	@1			@10	@1	
ConvE	0.491	0.239	246	0.316	0.520	0.400	4187	0.430	0.620	0.350	0.440
R-GCN	0.417	0.153	–	0.248	–	–	–	-	–	–	–
ConvKB	0.483	–	196	0.302	0.508	–	<u>2741</u>	0.220	-	–	–
DisMult	0.485	0.225	301	0.311	0.490	0.390	5100	0.430	0.540	0.240	0.340
ComplEx	0.486	0.227	376	0.313	0.510	0.410	5261	0.440	0.550	0.260	0.360
TransE	0.527	0.231	173	0.329	0.529	0.013	3414	0.223	0.673	0.391	0.492
TransH	0.534	0.236	171	0.335	0.499	0.010	3937	0.214	0.645	0.357	0.357
RotatE	0.533	0.241	177	0.338	0.552	0.417	2923	0.462	0.670	0.402	0.495
RotatE3D	0.543	0.250	165	0.347	**0.579**	**0.442**	3328	**0.489**	–	–	–
QuatE	0.495	0.221	176	0.311	<u>0.564</u>	<u>0.436</u>	3472	<u>**0.481**</u>	–	–	–
PairRE	<u>0.544</u>	<u>0.256</u>	<u>160</u>	<u>0.351</u>	0.522	0.400	2867	0.440	<u>0.675</u>	<u>0.436</u>	<u>0.522</u>
PairRE-CKGE	**0.550**	**0.260**	**155**	**0.355**	0.554	0.426	**2651**	0.463	**0.687**	**0.446**	**0.531**

Additionally, we run CKGE on many different relation mapping types, including 1-1, 1-N, N-1, and N-N. The results of CKGE on different relation categories on FB15k and ogbl-wikikg2 are shown in Table 3. We have observed that our model exhibits excellent performance in heterogeneous relationships such as 1-N, N-1, and N-N, particularly on the ogbl-wikikg2 dataset. This demonstrates that CKGE improves the alignment and uniformity of the DB models.

Table 3. Experimental results on FB15k and ogbl-wikikg2 by relation mapping.

Model	FB15k				ogbl-wikikg2			
	1-1	1-N	N-1	N-N	1-1	1-N	N-1	N-N
TransE	0.887	0.822	0.766	0.895	0.074	0.063	0.400	0.220
ComplEx	**0.939**	<u>0.896</u>	0.822	0.902	<u>0.394</u>	<u>0.278</u>	0.483	0.504
RotatE	<u>0.923</u>	0.840	0.782	0.908	0.164	0.144	0.431	0.261
PairRE	0.785	**0.899**	<u>0.872</u>	**0.940**	0.262	0.270	<u>0.594</u>	<u>0.587</u>
PairRE-CKGE	0.919	0.846	**0.964**	<u>0.935</u>	**0.589**	**0.549**	**0.696**	**0.759**

4.3   Model Analysis

To demonstrate the generality of our approach, we applied CKGE to TransE, TransH, and PairRE in the ogbl-wikikg2 dataset respectively. Table 4 shows the effectiveness of CKGE. We think CKGE will play a huge potential role in the large-scale model in the future. It is worth noting that TransE-CKGE gets an MRR score of 0.355, but TransH-CKGE only gets an MRR score of 0.337. Incorporated with CKGE, TransE gets a 9.2% improvement on MRR, which outperforms the improvement 7.3% score of TransH. Because for TransE, CKGE can constrain more samples, including triples with identical head entities and those with identical head entities and relations. But for TransH, the conditions for CKGE to be effective are more rigorous, only including the samples with the same head entity and relation.

Table 4. Added CKGE to TransE, TransH, and PairRE. Experiment results show that metrics MRR and Hit are significantly improved.

Model	ogbl-wikikg2			
	MRR	Hit@10	Hit@3	Hit@1
TransE	0.263	0.360	0.286	0.206
TranE-CKGE	**0.355**	**0.395**	**0.360**	**0.329**
TransH	0.264	0.360	0.287	0.208
TransH-CKGE	**0.337**	**0.388**	**0.347**	**0.304**
PairRE	0.522	0.621	0.539	0.469
PairRE-CKGE	**0.582**	**0.699**	**0.607**	**0.520**

We compare unsupervised CKGE and supervised CKGE in Table 5. The difference between unsupervised CKGE and supervised CKGE is used unsupervised contrastive learning and supervised contrastive learning. We add them to TransH and PairRE in the FB15k-237 dataset. For TransH-CKGE and PairRE-CKGE used the unsupervised contrastive learning method, the version that used the supervised contrastive learning method brings consistent improvements. Therefore, we adopt supervised CKGE with training as much as possible.

Table 5. Results on unsupervised CKGE and supervised CKGE for FB15k-237 dataset.

Model		FB5k-237				
		MR	MRR	Hit@1	Hit@3	Hit@10
TransH	+unsupervised CKGE	170	0.335	0.236	0.374	0.531
	+supervised CKGE	**168**	**0.336**	**0.238**	**0.375**	**0.533**
PairRE	+unsupervised CKGE	156	0.353	0.258	0.389	0.548
	+supervised CKGE	**155**	**0.355**	**0.260**	**0.394**	**0.550**

4.4 Visualization

We employed T-SNE to visualize triples, highlighting how CKGE promotes consistency and uniformity by aligning entities with analogous semantics post-relation transformation.

Given a query, $(h_i, r_j, ?)$ - with h_i and r_j as the head entities and relations respectively - the goal of link prediction is to determine the valid t_k. From ogbl-wikikg2, we randomly picked 10 queries having a 1-to-N relation mapping type. The entity embeddings generated by TransE are visualized using T-SNE. In Fig. 3, every entity is depicted as a 2D point; points of the same color and number represent different $(h_i, r_j, ?)$ contexts. As evident in Fig. 3, CKGE ensures that entities within the same (h_i, r_j) context have closely aligned and compact representations.

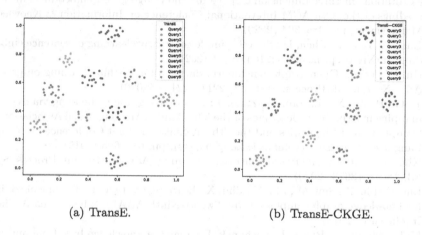

(a) TransE.   (b) TransE-CKGE.

Fig. 3. We visualized the embeddings of tail entities using T-SNE. Points sharing a color and number correspond to the same (h, r) context.

5 Conclusion and Future Work

In this study, we introduce CKGE, a contrastive learning framework tailored for distance-based knowledge graph embedding models. We noted that in such models, positive pairs often diverge into subsets: one closely tied to the head entity and the other to the tail entity following a relationship shift. Experimental results reveal that CKGE enhances the efficiency of distance-based models on standard datasets, notably in large-scale graphs. The efficacy of contrastive learning suggests its potential applicability in various other domains, warranting future exploration. Given the power of contrastive learning, a potential avenue for future research is to extend this approach to models in other fields.

References

1. Bordes, A., Usunier, N., Garcia-Duran, A., Weston, J., Yakhnenko, O.: Translating embeddings for modeling multi-relational data. In: Advances in Neural Information Processing Systems 26 (2013)
2. Cao, Z., Xu, Q., Yang, Z., Cao, X., Huang, Q.: Dual quaternion knowledge graph embeddings. In: Proceedings of the AAAI Conference on Artificial Intelligence. vol. 35, pp. 6894–6902 (2021)
3. Chao, L., He, J., Wang, T., Chu, W.: Pairre: knowledge graph embeddings via paired relation vectors. arXiv preprint arXiv:2011.03798 (2020)
4. Dettmers, T., Minervini, P., Stenetorp, P., Riedel, S.: Convolutional 2d knowledge graph embeddings. In: Proceedings of the AAAI Conference on Artificial Intelligence, vol. 32 (2018)
5. Gao, C., Sun, C., Shan, L., Lin, L., Wang, M.: Rotate3d: representing relations as rotations in three-dimensional space for knowledge graph embedding. In: Proceedings of the 29th ACM International Conference on Information & Knowledge Management, pp. 385–394 (2020)
6. Gao, T., Yao, X., Chen, D.: Simcse: simple contrastive learning of sentence embeddings. arXiv preprint arXiv:2104.08821 (2021)
7. Hu, W., et al.: Open graph benchmark: datasets for machine learning on graphs. Adv. Neural. Inf. Process. Syst. **33**, 22118–22133 (2020)
8. Ji, G., He, S., Xu, L., Liu, K., Zhao, J.: Knowledge graph embedding via dynamic mapping matrix. In: Proceedings of the 53rd Annual Meeting of the Association for Computational Linguistics and the 7th International Joint Conference on Natural Language Processing (volume 1: Long papers), pp. 687–696 (2015)
9. Khosla, P., et al.: Supervised contrastive learning. Adv. Neural. Inf. Process. Syst. **33**, 18661–18673 (2020)
10. Lin, Y., Liu, Z., Sun, M., Liu, Y., Zhu, X.: Learning entity and relation embeddings for knowledge graph completion. In: Twenty-Ninth AAAI Conference on Artificial Intelligence (2015)
11. Mahdisoltani, F., Biega, J., Suchanek, F.: Yago3: a knowledge base from multilingual wikipedias. In: 7th Biennial Conference On Innovative Data Systems Research. CIDR Conference (2014)
12. Nathani, D., Chauhan, J., Sharma, C., Kaul, M.: Learning attention-based embeddings for relation prediction in knowledge graphs. In: Proceedings of the 57th Annual Meeting of the Association for Computational Linguistics, pp. 4710–4723 (2019)
13. Nguyen, D.Q., Nguyen, T.D., Nguyen, D.Q., Phung, D.: A novel embedding model for knowledge base completion based on convolutional neural network. arXiv preprint arXiv:1712.02121 (2017)
14. Nickel, M., Tresp, V., Kriegel, H.P.: A three-way model for collective learning on multi-relational data. In: ICML (2011)
15. Rossi, A., Barbosa, D., Firmani, D., Matinata, A., Merialdo, P.: Knowledge graph embedding for link prediction: a comparative analysis. ACM Trans. Knowl. Dis. Data (TKDD) **15**(2), 1–49 (2021)
16. Schlichtkrull, M., Kipf, T.N., Bloem, P., van den Berg, R., Titov, I., Welling, M.: Modeling relational data with graph convolutional networks. In: Gangemi, A., et al. (eds.) ESWC 2018. LNCS, vol. 10843, pp. 593–607. Springer, Cham (2018). https://doi.org/10.1007/978-3-319-93417-4_38

17. Sun, Z., Deng, Z.H., Nie, J.Y., Tang, J.: Rotate: Knowledge graph embedding by relational rotation in complex space. arXiv preprint arXiv:1902.10197 (2019)
18. Toutanova, K., Chen, D.: Observed versus latent features for knowledge base and text inference. In: Proceedings of the 3rd Workshop on Continuous Vector Space Models and Their Compositionality, pp. 57–66 (2015)
19. Trouillon, T., Welbl, J., Riedel, S., Gaussier, É., Bouchard, G.: Complex embeddings for simple link prediction. In: International Conference on Machine Learning, pp. 2071–2080. PMLR (2016)
20. Wang, L., Zhao, W., Wei, Z., Liu, J.: Simkgc: simple contrastive knowledge graph completion with pre-trained language models. arXiv preprint arXiv:2203.02167 (2022)
21. Yang, B., Yih, W.t., He, X., Gao, J., Deng, L.: Embedding entities and relations for learning and inference in knowledge bases. arXiv preprint arXiv:1412.6575 (2014)
22. Zhang, S., Tay, Y., Yao, L., Liu, Q.: Quaternion knowledge graph embeddings. In: Advances in Neural Information Processing Systems 32 (2019)

Knowledge Graph Completion with Information Adaptation and Refinement

Yifan Liu, Bin Shang, Chenxin Wang, and Yinliang Zhao[✉]

Xi'an Jiaotong University, Xi'an, China
zhaoy@xjtu.edu.cn

Abstract. At present, knowledge graph completion (KGC) is mainly divided into structure-based methods and language-based methods, which characterize the structural information and semantic information of knowledge graphs, respectively. Though existing works have developed methods to integrate both information, we argue their end-to-end training manner suffers discrepancy, compatibility, resources redundancy issues. Therefore, we propose a novel two-stage training paradigm for tackling KGC task, *i.e.* information adaptation and refinement (KGCIAR). Specifically, KGCIAR has two stages, 1) adaptation and 2) refinement. In the adaptation stage, we fine-tune the PLM with the input of descriptive information and supervised by the KG structural information. In the second refinement stage, we freeze the adapted PLM model and infer the description embeddings of entities and relations. Then, those embeddings are leveraged as the entity/relation initial embeddings. Finally, we train a lightweight KGC model. Moreover, we devise two novel objectives for knowledge adaptation, which are self-supervised adaptation and structure-aware contrastive adaptation. Furthermore, we systematically compare the performance of different lightweight KGC models for information refinement. The experiments on KGC task and various variants analyses demonstrate that KGCIAR is effective in harnessing both structure and language information in KG.

Keywords: Knowledge graph completion · Pre-trained Language Model · Knowledge Adaptation

1 Introduction

In recent years, knowledge graphs massively contribute to the rapid developments of artificial intelligence, digitization and big data [18,38]. A knolwedge graph (KG) is a semantic structure [28] for storing, representing and reasoning knowledge, which enables computers to understand human knowledge, and improves the organization of complex data. Generally, a knowledge graph consists of entities, attributes, and relations. By modeling these entities, attributes, and relations as semantic networks, knowledge graphs can enhance the performance in question answering [16], recommendation systems [35], language understanding [18], and etc.

© The Author(s), under exclusive license to Springer Nature Switzerland AG 2023
X. Yang et al. (Eds.): ADMA 2023, LNAI 14177, pp. 16–31, 2023.
https://doi.org/10.1007/978-3-031-46664-9_2

Knowledge graphs are usually implemented manually or semi-automatically. However, due to the sparsity issue, some hidden relations are unable to be fully captured [35], which leads to the incompleteness of the knowledge graph. Moreover, expansion of the knowledge graph increases its size and complexity. This induces more missing and incomplete information, which impairs the quality of knowledge graphs. Therefore, automatically discovering and completing missing information [32, 33, 44], *i.e.*, *knowledge graph completion* (KGC), is necessary.

Existing KGC methods lay in two directions [33]: 1) structure-based KGC and 2) language-based KGC. Structure-based methods learn representation of entities and relations via their IDs and *structure information* [7, 12, 17, 29, 39, 42]. Then they predict missing entities and relations using various scoring functions. In those language-based approach [11, 32, 33, 37, 41, 44], textual *descriptive information* of entities and relations are incorporated to supplement the semantic information for the KGC task. Structure-based methods is effective in capturing the connectivity information among entities and relations. For example, distance-based models, such as TransE [7], TransH [39], TransD [17] and RotatE [29], define different metric spaces for learning embeddings. However, they are unable to leverage the descriptive semantics. To cope with this problem, language-based KGC method has emerged [11, 32, 41]. The descriptive information are encoded by language models [44] and optimized with the distance objective functions. Recently, the incredible successes of pre-trained language foundation models [6] provides off-the-shelf powerful tools to represent those descriptive information. KG-BERT [44] and SimKGC [33] directly optimizes the pre-trained BERT models with KGC objectives. StAR [32], BLP [11] and KEPLER [37] add the loss function of the embedding-based model in order to simultaneous harness structural information and descriptive information.

Most existing methods integrate the structure and descriptive information in an end-to-end training manner, which are however problematic. Firstly, there is *discrepancy* between structure and descriptive information. Two entities may be described similarly, while they share few common structure semantics. In this sense, directly integrating the structural and descriptive information leads to a contradicting optimization. Consequently, the KGC performance is hindered. Secondly, there is *compatibility* issue if simultaneously characterizing structural information and descriptive information. Existing works [7, 17, 39] demonstrates a simple model with only trainable embedding layers for entities and relations is effectively in modeling structural information. In contrast, a large-size language model [13, 22] is required to well characterize the descriptive information. Therefore, it is incompatible to harness both information with one single model. Finally, the *resources redundancy* issue is ignored in existing text-based KGC models. Recent works [32, 33, 44] fine-tune a large pre-trained language model (PLM) with loss functions designed for KGC tasks. Since those PLMs are pre-trained with massive language corpus and various tasks [25], most parameters are irrelevant to a specific KG and KGC task. During inference stage, those large language models are required to load into memory, which results in unnecessary resources consumption.

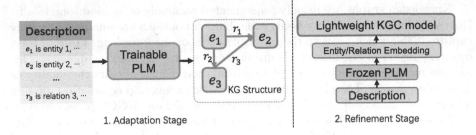

1. Adaptation Stage   2. Refinement Stage

Fig. 1. The training paradigm of KGCIAR. In the adaptation stage, we fine-tune the PLM with the input of descriptive information and supervised by the KG structural information. In the refinement stage, we freeze the adapted PLM and train a lightweighted KGC model.

To this end, this paper investigates a new KGC training paradigm, *i.e.* information adaptation and refinement, which is named as **KGCIAR**. As illustrated in Fig. 1, KGCIAR has two stages, 1) adaptation and 2) refinement. In the adaptation stage, we fine-tune the PLM with the input of descriptive information and supervised by the KG structural information. In this way, we adapt the PLM to be aware of the KG-based information. In the second refinement stage, we first freeze the adapted PLM model and infer the description embeddings for entities and relations. Then, those embeddings are leveraged as the entity/relation initial embeddings. Finally, we train a lightweight KGC model. KGCIAR is an effective training paradigm. Firstly, the adaptation stage fuses the descriptive and structural information via the structure-guided fine-tuning of the PLM, which alleviates the discrepancy problem. Additionally, the refinement stage refines the knowledge from adapted PLM by completing KGC task. Since we freeze the parameter of PLM and only train a lightweight KGC model, there is no compatibility issue. Moreover, we only use the lightweight KGC model for completing knowledge, which is rather resources efficient.

In this paper, we investigate the potentials of adapting different PLMs to KGC tasks. We identify it is still challenging in knowledge adaptation stage. Firstly, if enforcing the PLM to optimize over the KG structure, there is a knowledge collapse problem [19]. Secondly, the critical sparsity issue leads to the lack of adaptation supervision signals. Therefore, we devise two objectives for adaptation, which are Self-Supervised Adaptation (SSA) and Structure-aware Contrastive Adaptation (SCA). By simultaneously optimizing both targets, the PLM is well adapted to the information in KG. Moreover, regarding the refinement stage we systematically compare the performance of different lightweight KGC models in refining the knowledge for KGC tasks, including translation-based methods, convolution-based methods, and etc. Overall, the main contributions of this paper are as follows:

– We identify three critical issues when incorporating both structural and descriptive information for KGC task in an end-to-end training manner. And

we are the first work proposing a novel two-stage information adaptation and refinement training paradigm to simultaneously resolve all three issues.
- We systematically study the novel two-stage training paradigm, which includes information adaptation and knowledge refinement. We propose two novel objectives for the first adaptation stage, which are the self-supervised adaptation loss and structure-aware contrastive adaptation loss.
- We conduct extensive experiments on KGC task, including different variants of the proposed two-stage KGCIAR. The results demonstrate the effectiveness of the two-stage training paradigm for KGC.

2 Related Work

2.1 Pre-trained Language Foundation Models

Most recently, pre-trained foundation models [2,8,13,25] attract widespread attention due to its efficacy in resolving various tasks. Foundation models are widely used in various fields. For example, in the medical field, foundation models based on multimodal data in the healthcare ecosystem can effectively diagnose diseases [6]. In the field of vision, multi-modal foundation models such as Flamingo [2] can generate realistic images and videos, which promote the development of computer vision technology. Those large pre-trained language models (PLMs) such as BERT [13], T5 [25], GPT-3 [8] exhibit tramandous ability in solving languge tasks. BERT [13] is pre-trained in an unsupervised manner and then fine-tuned on specific tasks for optimal performance. T5 [25] firstly investigates the unifying format of language input to solve arbitrary tasks. GPT-3 [8] devises a much larger model size and uses more training data to autoregressive generate text. Most recently, ChatGPT is released, which is developed based on the GPT-3 [8]/GPT-3.5/GPT-4 architecture, demonstrating amazing ability in interacting with humans, This paper investigates how to effectively adapting the knowledge of PLM to KG and harness those knowledge for KG completion task.

2.2 Knowledge Graph Completion

Knowledge graph completion (KGC) refers to filling in missing information by predicting the relation between entities or related entities given another entity and the relation in an existing knowledge graph [7,12,33,44]. We can obtain the embedding vector representation of entities and relations by optimizing with the KG structures. For example, translation-based KGC models [7,17,39] learn entity and relation embeddings via minimizing the distance between the relation-wise head entity embedding and tail entity embedding. RotatE [29] maps entity and relation vectors to a complex space, defining the relation as the rotation from the head entity to the tail entity. DistMult [42] is a tensor decomposition-based model that represents each relation using a diagonal matrix, and calculates the similarity between entity pairs using dot products. ConvE [12] uses convolutional and pooling layers to calculate scores between entity pairs. Recent research leverage both structural and descriptive information of entities and relations in KG to perform KGC. DKRL [41] takes advantage of the entity description information by encoding the description information into the entity representation using two encoders, namely, Continuous Bag-of-Words model (CBOW) and Convolutional Neural Network (CNN), to encode the textual information of entity descriptions. Then, DKRL [41] jointly learns the knowledge representation using both triples and description information. KG-BERT [44] uses the pre-trained BERT model for KGC, representing entities and relations as their names or descriptions. Triples are packed as a name/description sequence and fine-tuned as the input sentence for the BERT model. StAR [32] splits the triples into two groups, with the head entity and relation in one group and the tail entity in another group. It uses a twin model to asymmetrically encode text in two groups and combines loss functions from both representation learning and spatial structure. BLP [11] and KEPLER [37] convert entity semantic information into entity embedding vectors through pre-trained foundation models, and train entity embedding vectors and relational embedding vectors through knowledge graph embedding models. In this way, the structural information of knowledge graphs is fused. KEPLER-Rel [37] is a sub-model of KEPLER [37], which also converts the semantic information of relations into relation embedding vectors. SimKGC [33] proposes a new contrastive learning method based on the twin model, which can use more negative samples and improves the loss function using InfoNCE as the loss function. At present, the text-based KGC method has become the mainstream method for KGC research because of its more information input and full utilization of the semantic information of the knowledge graph. How to better combine it with knowledge graph embedding to make full use of the advantages of both, we think it is one of the most urgent problems to be solved in current research.

3   Preliminary

A knowledge graph $\mathcal{G}$ is a directed graph, in which the set of all nodes is the entity set $\mathcal{E}$, and the set of all edges is the relation set $\mathcal{R}$. The sets of entities $\mathcal{E}$ and relations $\mathcal{R}$ are usually of various types, such as people, locations, events, and etc. We denote $\mathcal{G} = \{(h, r, t)\}$, because the knowledge graph is constructed from triples. In the data structure, each triple represents that there is a relationship edge r between the entity node h and the entity node t in the knowledge graph. Triples (h, r, t) capture the relationships between entities in the graph. For instance, a triple $(Barack\ Obama, born\_in, Hawaii)$ represents the fact that *Barack Obama was born in Hawaii*.

In this paper, we use bold text $\mathbf{h}$, $\mathbf{r}$, and $\mathbf{t}$ represent the embedding for the h, r, and t, respectively. To train the model to distinguish between true and false triples, negative triples are generated by randomly replacing either the head or tail entity of a true triple with a different entity from $\mathcal{E}$, or by replacing the relation with a different relation from $\mathcal{R}$. These negative triples represent false statements and are used as negative examples during training. The set of negative triples is denoted as $\mathcal{N}$, and the model is optimized to maximize the scores of true triples while minimizing the scores of negative ones.

KGC task is formulated as two types of prediction task, *i.e.*, head entity prediction and tail entity prediction, denoted as $(h, r, ?)$ and $(?, r, t)$, respectively. We predict ? by ranking all entities in the KG. In addition, r^{-1} represents the inverse relationship of the relation r. If there is a triplet (h, r, t), then the triplet (t, r^{-1}, h) must also exist. Therefore, we added an inverse relation during training, and trained the inverse triplet corresponding to each triplet.

4 Proposed Framework

4.1 Overall Framework of KGCIAR

Hereafter, we introduce the KGCIAR (KG completion via Information Adaptation and Refinement) framework, which is a novel two-stage training paradigm for KG completion task. The overall framework of KGCIAR is shown in Fig. 1, which contains adaptation stage and refinement stages. In the adaptation stage, we first input the entity description and relation description into a PLM model, such as BERT [13] and S-BERT [27], and generate the embedding for entities and relations. Then, we use the KG structure information to calculate the training loss. After adaptation training stage, we have a *KG adapted language model* (KGA-LM), which has the ability to infer embeddings of entities and relations. During the second refinement stage, we freeze the parameter of this KGA-LM such that it only has the inference ability. Then, we use this KGA-LM to to infer the embedding of entity and relations, and send those embeddings to a lightweight KGC model. During refinement stage, we train a KGC model with the initial embedding from KGA-LM model and supervised by the KGC targets, *i.e.* predict head or tail entities given an entity and a relation. After refinement stage, we can directly use the lightweight KGC model to complete the KGC task. Next, we introduce these two stages in details.

4.2 Knowledge Adaptation for PLM

As aforementioned, directly inferring the embedding of descriptive information from PLM is unable to align their semantics information with the KG structural information. Therefore, we propose to conduct knowledge adaptation of the PLM. However, it is challenging tasks due to the following reasons: Firstly, PLMs are generally trained from a large corpus, which has rather different knowledge distribution from the KG. If enforcing the PLM to optimize over the KG structure, there is a knowledge collapse problem [19]. Secondly, KG structure is usually of critical sparsity issue, which leads to lack of adaptation supervision signals. To tackle both challenges, we propose two objectives for adaptation, which are self-supervised adaptation (SSA) loss and structure-aware contrastive adaptation (SCA) loss.

Self-supervised Adaptation. By employing this SSA loss, we adapt the PLM towards the distribution of descriptive information in the KG. Specifically, we adopt the similar idea in [14,33] by passing the embedding from PLM model to the dropout layers twice. There are many dropout layers in the internal structure of the pre-trained foundation model BERT. The dropout layer randomly masks the embedding in one forward pass. Hence, for the same input, we have two embeddings. For example, given a tail entity descriptive information as denoted as D_t and PLM generate its embedding two times as $\mathbf{t}$ and $\mathbf{t}^+$. The $\mathcal{L}_{SSA}$ loss is formulated as follows:

$$\mathcal{L}_{SSA} = -\log \frac{e^{(\phi(\mathbf{t},\mathbf{t}^+)-\gamma)/\tau}}{e^{(\phi(\mathbf{t},\mathbf{t}^+)-\gamma)/\tau} + \sum_{i=1}^{|\mathcal{N}|} e^{\phi(\mathbf{t},\mathbf{t}_i^-)/\tau}}, \quad (1)$$

where $\phi(\cdot,\cdot)$ is used to calculate the similarity between two vectors. $\mathcal{N}$ denotes the number of negative entities in the batch, where $\mathbf{t}_i^-$ represents their embeddings. In this paper, we adopt the cosine similarity as a score between the two vectors. The reason is that in the knowledge prediction period, we also use the cosine similarity as the metric space for ranking entities. γ is the additive margin, which improves the separation between positive samples and nearby negative samples by introducing a margin around the positive samples, resulting in higher scores for positive samples. τ is the temperature hyper-parameter for contrastive learning.

Structure-aware Contrastive Adaptation. With SSA, the PLM is adapted to the distribution of entity and relation descriptive information in KG. However, those embeddings are not aligned with the structure information. Therefore, we propose another structure-aware contrastive learning (SCA) loss. To be specific, given input text description (D_h, D_r, D_t) for head entity, relation and tail entity, PLM generates their embeddings as $\mathbf{h}$, $\mathbf{r}$ and $\mathbf{t}$, respectively. In order to incorporate the structural information of knowledge graphs into training, we combine $\mathbf{h}$ and $\mathbf{r}$ using common embedding models and normalize the combined vector using L_2 normalization to obtain the embedding vector $\mathbf{e}_{hr}$, which can represent the combined semantic information of the head entity and the relation. We study three combination methods to obtain the embedding vector $\mathbf{e}_{hr}$: addition, element-wise product, and convolution.

$$\mathbf{e}_{hr} = \begin{cases} \mathbf{h} + \mathbf{r} & \text{if using addition} \\ \mathbf{h} \circ \mathbf{r} & \text{if using element-wise product} \\ \mathbf{h} * \mathbf{r} & \text{if using convolution,} \end{cases} \quad (2)$$

where $*$ denotes the convolution operation of embeddings as in ConvE [12]. In the experiment, we will discuss how the combination method is selected.

Hereafter, we adopt a contrastive learning for optimization. Specifically, in the same batch, we pull the combined head relation embedding $\mathbf{e}_{hr}$ closer to the embedding vectors of tail entities in KG, *i.e.* the positive samples, and push away from those entities in the same batch but not existing in the knowledge graph as the tail entities, *i.e.* the negative samples. The SCA function is

$$\mathcal{L}_{\text{SCA}} = -\log \frac{e^{\phi(\mathbf{e}_{hr},\mathbf{t})/\tau}}{e^{\phi(\mathbf{e}_{hr},\mathbf{t})/\tau} + \sum_{i=1}^{|\mathcal{N}|} e^{\phi(\mathbf{e}_{hr},\mathbf{t}^-)/\tau}}. \tag{3}$$

This SCA loss optimizes the embeddings by connecting the head and tail entities via relations, which reveals the structural information in KG. The idea of contrastive learning is to bring similar samples closer and push away dissimilar samples, and the goal is to learn a good semantic representation space from samples. Therefore, the PLM is adapted to the structural information.

Final Adaptation Loss. Finally, we combine the SSA loss and SCA loss with a balance weight α, where $\alpha \in [0,1]$, to obtain the overall loss function of the adaptation stage.

$$\mathcal{L} = \alpha\mathcal{L}_{\text{SSA}} + (1-\alpha)\mathcal{L}_{\text{SCA}}. \tag{4}$$

Next, we introduce the second knowledge refinement stage.

4.3   Knowledge Refinement

Though in the adaptation stage, the KGA-LM is able to generate entity and relation embeddings and those embeddings preserve both the descriptive and structural information, it is still sub-optimal due to the incompatibility issue. Therefore, in this section, we introduce how to refine the embedding from adapted PLM for entities and relations. During this stage, we freeze the KGA-LM and inferring the embedding of entities and relations by using the descriptive information as the initial embeddings. Then, we pass the initial embeddings to a lightweight KGC model and train this model with objective.

The KGCIAR paradigm has no constraints for the KGC model in the refinement stage. For example, we can use translation-based models, convolution-based model, encoder-based model, and etc. In this paper, we observe that those embedding-based KGC model, such as TransE [7] performs better than other KGC model. As such, we use the embedding from adapted PLM as the initialization of the entity and relation emebedding, and fine-tuning those with a distance measurement loss. Speficially, if setting the KGC model to be TransE, the optimization loss is formulated as follows:

$$\mathcal{L}_{\text{KGC}} = \frac{1}{N} \sum_{i=1}^{N} \max\left(0, \lambda + \psi(\mathbf{h}, \mathbf{r}, \mathbf{t}) - \psi\left(\mathbf{h}, \mathbf{r}, \mathbf{t}_i^-\right)\right), \tag{5}$$

where N denotes the total number of negative samples for training. λ represents the margin value which controls the degree of separation between positive and negative samples. $\mathbf{t}_i^-$ is the negative tail entity embeddings. The scoring function ψ for the triplet is defined as the L_1 distance between the $\mathbf{h}+\mathbf{r}$ embedding vector and the $\mathbf{t}$ embedding vector as

$$\psi(\mathbf{h}, \mathbf{r}, \mathbf{t}) = \|\mathbf{h} + \mathbf{r} - \mathbf{t}\|_1. \tag{6}$$

Note that the embeddings of entity and relation are initialized from adapted PLM and are optimized with the KGC loss $\mathcal{L}_{KGC}$.

4.4 Discussion of KGCIAR

In this section, we discuss different variants of KGCIAR. In the adaptation stage, we can choose different PLMs, we propose three combination methods of head and relation embedding vectors. We put those different combination methods as the superscript. For example, if using addition combination, we denote it as KGCIAR^{add}. As such, we also have KGCIAR^{mult}, and KGCIAR^{conv}. According to experimental results, we observe that addition combination generally performs better. Therefore, KGCIAR is the KGCIAR^{add} model unless otherwise specified.

During the refinement stage, in addition to using the basic embedding model TransE [7], we can also use other KGC models to fine-tune entity and relation embeddings. In this paper, we study several KGC models such as TransH [39], TransD [17], ConvE [12], and DistMult [42]. We denote KGC model in the subscript. For example, KGCIAR$_{transe}$ adopts the KGC model to be transE. As such, we also have KGCIAR$_{transh}$, KGCIAR$_{transd}$, KGCIAR$_{conve}$ and KGCIAR$_{mult}$ variants, respectively. Since we observe better performance of using TransE, by default we use the TransE model as the lightweight KGC model for the KGCIAR framework.

Furthermore, we design another type of KGC model that only has fully-connected mapping layers as KGC model. It is designed to compare with those embedding-based models. We name this paradigm as KGCIAR-MLP model. By comparing KGCIAR-MLP model, we should clearly verify the better performance of embedding-based models. Specifically, we modified the triple score function of the embedding model by adding an MLP layer to it. By training the parameters of the MLP layer, we transform the embeddings from KGA-LM to be refined in KGC task. In the KGCIAR-MLP model, the trainable parameters in the second stage are those MLP layers. And we treat the embedding from KGA-LM as the input feature.

5 Experiments

5.1 Setups

Datasets. We conduct the experiments on two datasets, FB15k-237 [5] and Wikidata5M [31]. FB15k-237 is a subset of Freebase that contains 14,541 entities and 237 relations. Wikidata5M, is a subset of Wikidata that contains about 5 million entities and 822 relations. The scale of the dataset is much larger than that of FB15k-237, and the number of training sets is about 20 times that of FB15k-237. We believe that on a large-scale dataset, it is more challenging and can better demonstrate the performance of the model. Both datasets have description information for entities. We follow previous work [33] to preprocess the datasets for training, validation and test.

Evaluation Metrics. The evaluation metrics for KGC are adopted in commonly used metrics, *i.e.* the Mean Reciprocal Rank (MRR) and the Hit Rate (H@k). For the KGC task, we calculate the prediction metrics of the head entity and the prediction metrics of the tail entity. For the final metrics, we use the average of the prediction metrics with respective to all triplets.

5.2 Baselines and Hyper-parameters

In terms of baseline selection, we compared with both structure-based models and language-based models. For the structure-based models, due to the limited benchmarks on the large dataset Wikidata5M, we selected three representative models, TransE [7], DistMult [42], and RotatE [29]. For the text-based models, we chose DKRL [41], KG-BERT [44], StAR [32], KEPLER [37], and SimKGC [33]. Among them, DKRL [41], KG-BERT [44], and SimKGC [33] only use pre-trained foundation models to train semantic information. StAR [32] and KEPLER [37] combine with embedding-based models and also learn the structural information of the knowledge graph during the training process.

Regarding the hyper-parameter, in the first step of the experiment, we set the initial temperature τ to 0.05, the additive margin γ to 0.02 and the batch size to 1024. In the second step, we set the number of negative samples $\mathcal{N}$ to 25 for each positive sample, set the margin λ to 4.0 for the dataset FB15k-237, and set the margin λ to 12.0 for the dataset Wikidata5M. For the selection of negative samples $\mathcal{N}$ and the weight α of two losses, we experiment and select the optimal values for each dataset. In this paper, we investigated different PLMs, including *bert-base-uncased, roberta-base, sentence-transformers/all-MiniLM-L6-v2* in the model zoo on Huggingface[1]. We found that *bert-base-uncased* is generally the best one for KGCIAR. Therefore, all experiments are conducted based on it.

5.3 Overall Knowledge Completion Performance

For the KGCIAR model, we choose the combination method of addition in the first step, and use TransE [7] as the KGC in the second step to conduct experiments. We compare the results of the experiment with the baseline models, which are shown in Table 1. The reported values of those baseline model follow SimKGC [33] and BLP [11].

According to the experimental results, compared with the language-based model, our proposed model KGCIAR has achieved better performance on both datasets, both exceeding the best baseline SimKGC [33]. And compared with structure-based models, our proposed model surpasses the embedding model RotatE [29], which suggests that our KGCIAR method is effective in harnessing the structural information. Overall, we should observe that RotatE is the best baseline in FB15k-237 while SimKGC is the best baseline in Wikidata. This indicates that existing work has the limitation in combining both structural information and descriptive information, due to our aforementioned three issues.

---

[1] https://huggingface.co/models.

Table 1. Main results for FB15k-237 and Wikidata5M datasets.

Method	FB15k-237				Wikidata5M			
	MRR	H@1	H@3	H@10	MRR	H@1	H@3	H@10
structure-based methods								
TransE [7]	27.9	19.8	37.6	44.1	25.3	17.0	31.1	39.2
DistMult [42]	28.1	19.9	30.1	44.6	25.3	20.8	27.8	33.4
RotatE [29]	33.8	24.1	37.5	53.3	29.0	23.4	32.2	39.0
language-based methods								
DKRL [41]	21.5	13.5	23.1	37.9	16.0	12.0	18.1	22.9
KG-BERT [44]	23.6	14.5	25.8	42.0	–	–	–	–
StAR [32]	29.6	20.5	32.2	48.2	–	–	–	–
KEPLER [37]	–	–	–	–	21.0	17.3	22.4	27.7
SimKGC [33]	33.6	24.9	36.2	51.1	35.8	31.3	37.6	44.1
KGCIAR	**34.8**	**25.3**	**38.4**	**53.7**	**37.0**	**31.8**	**39.5**	**46.1**

5.4 Variant Analysis

As we discussed in Sect. 4.4, there are a series of different variants of KGCIAR. Therefore, in this section, we conduct systematic experiments on the FB15k-237 dataset for those variant models. We first study the relationship between the combination methods of head entity $\mathbf{h}$ and relation $\mathbf{r}$ in the adaptation stage and KGC models in the refinement stage. Due to space limitation, we only report the results of $\text{KGCIAR}_{\text{mult}}^{\text{add}}$, $\text{KGCIAR}_{\text{mult}}^{\text{mult}}$, $\text{KGCIAR}_{\text{conve}}^{\text{add}}$ and $\text{KGCIAR}_{\text{conve}}^{\text{conv}}$ in Table 2. We also add the results of DistMult [42] and ConvE [12] for comparison.

Table 2. Adaptation combination methods comparison on FB15k-237 dataset.

	DistMult	$\text{KGCIAR}_{\text{mult}}^{\text{add}}$	$\text{KGCIAR}_{\text{mult}}^{\text{mult}}$	ConvE	$\text{KGCIAR}_{\text{conve}}^{\text{add}}$	$\text{KGCIAR}_{\text{conve}}^{\text{conv}}$
MRR	28.1	31.0	**32.0**	31.7	32.9	**33.1**
H@1	19.9	22.7	**23.3**	23.2	24.1	**24.2**
H@3	30.1	34.1	**35.1**	34.7	36.0	**36.4**
H@10	44.6	47.4	**49.8**	49.2	50.8	**51.3**

The experimental results are twofold. Firstly, we observe that our variants model perform better than those KGC model, which demonstrate the effectiveness of our proposed KGCIAR framework. The adaptation stage is effectively adapt the language knowledge into the KGC model to improve the performance. Secondly, the choice of combination method in the first stage should be align with the KGC model in the second stage. For example, if we use the DistMult as the base KGC model in the second stage, then the element-wise product should be employed in the adaptation stage, *i.e.* the $KGCIAR_{mult}^{mult}$. In this we, we can ensure that the PLM is aware of the KG-based information in the first stage and fuses the descriptive and structural information well. Next, We fix the combination of **h** and **r** in the first stage to be addition and compare the performance among different KGC models in the refinement stage. We adopt $KGCIAR_{transe}^{add}$, $KGCIAR_{transh}^{add}$, $KGCIAR_{transd}^{add}$ and KGCIAR-MLP to conduct experiments. The experimental results are shown in Table 3.

Table 3. Variant results using different KGC models on FB15k-237 dataset.

	KGCIAR-MLP	$KGCIAR_{transe}^{add}$	$KGCIAR_{transh}^{add}$	$KGCIAR_{transd}^{add}$
MRR	31.9	34.8	**35.0**	**35.0**
H@1	22.4	25.3	**25.5**	25.4
H@3	35.4	38.4	**38.5**	**38.5**
H@10	50.6	53.7	54.0	**54.1**

According to Table 3, we have the following observations. Firstly, KGC IAR_{transh}^{add}, $KGCIAR_{transd}^{add}$ and $KGCIAR_{transe}^{add}$ are generally comparable. In their original papers, TransH and TransD are more complex than TranE, and they usually performs better. However, since our KGCIAR framework already adapt the language knowledge into the KGC model and refine the entity/relation embeddings, a simple TransE model achieves the comparable performance. This indicates the effectiveness of our KGCIAR framework in improving the performance with a lightweight KGC model. In addition, we also find that the performance of KGCIAR-MLP is worse than other variants. KGCIAR-MLP differs from other variant in its refinement stage. It directly use the embedding from KGA-LM as the input for MLP, without updating the entity/relation embeddings. This demontrastes that refining of entity and relation embeddings as the KGC model is necessary to achieve the optimal performance. We hypothesis that an MLP layer is not sufficient for maintain the structure information for KG. By analyzing the experimental results of the above variant models, we believe that selecting a lightweight KGC model and matching it with a optimal combination method can retain the superiorty of KGCIAR framework.

5.5   Optimization Analysis

In this section, we investigate the training process of the KGCIAR framework. We adopt the H@10 and loss with respect to the epochs in the refinement stage as evaluation metrics. We conduct the experiments on the dataset FB15k-237, we select $KGCIAR_{mult}^{mult}$ and $KGCIAR_{transe}^{add}$ to compare with their corresponding KGC models *i.e.* DistMult and TransE the experimental results are shown in Fig. 2a. Firstly, according to the loss trends, we find that both KGCIAR variant converge much faster than the original KGC model. It suggests that after adaptation, the language knowledge are well adapted to the KG structure information. Therefore, refinement of KGC model converges faster than directly train a KGC model. Secondly, both KGCIAR variants performs better than the KGC model, which indicates the superiority of the KGCIAR framework in integrating both the language knowledge and structural knowledge. Thirdly, KGCIAR$_{transe}^{add}$ performs better than all other, which justify the efficacy of a lightweight KGC model in the refinement stage.

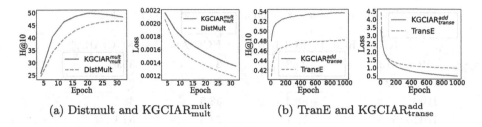

(a) Distmult and KGCIAR$_{mult}^{mult}$ (b) TranE and KGCIAR$_{transe}^{add}$

Fig. 2. Comparison for H@10 and Loss w.r.t. Epochs.

5.6   Ablation Study

To validate the necessity of our two-stage model KGCIAR, we demonstrate the effectiveness of the first adaptation stage and the second refinement stage in improving model performance through ablation experiments. We conduct experiments on the FB15k-237 and Wikidata5M datasets. The performance of KGCIAR, the model with only the adaptation stage, and the model with only the refinement stage on each dataset are shown in Table 4.

Table 4. Ablation study on FB15k-237 and Wikidata5M datasets with KGCIAR.

	FB15k-237				Wikidata5M			
	MRR	H@1	H@3	H@10	MRR	H@1	H@3	H@10
KGCIAR	**34.8**	**25.3**	**38.4**	**53.7**	**37.0**	**31.8**	**39.5**	**46.1**
w/o adaptation	27.9	19.8	37.6	44.1	25.3	17.0	31.1	39.2
w/o refinement	31.2	22.3	34.0	49.2	28.5	21.6	32.2	39.8

It can be seen from Table 4 that when the first adaptation stage or the second refinement stage is used alone on the FB15k-237 dataset and the Wikidata5M dataset, the results obtained by the KGCIAR model training are far lower than the model effect when the two stages are combined. This shows that after the structural and descriptive information in KG are fused through the adaptation stage, further refinement is required to fully learn the structural and descriptive information in KG. And the performance of the model drops off even more when only the refinement stage is used. This shows that without knowledge adaptation, the model lacks the learning of the descriptive information of entities and relations, which leads to a significant decline in model performance.

6 Conclusion

This paper discuss three critical issues in exiting end-to-end training paradigm for integrating both structural knowledge and language knowledge to tackle knowledge graph completion task, *i.e.* discrepancy, compatibility and resources-redundancy. We propose a novel two-stage training framework KGCIAR, which includes the adaptation stage and the refinement stages. It trains a KGA-LM for adapting the language knowledge to KG structure. To overcome the optimization problems, we devise a self-supervised adaptation loss and a structure-aware contrastive adaptation loss. During the refinement stage, we systematically investigates the relationship between the adaptation methods and the selection of lightweight KGC models. By conducting experiments, we observe that the combination methods in adapation stage should match the choice of KGC model. Overall, the proposed framework has achieved the best performance on both FB15k-237 and Wikidata5M datasets.

References

1. Alammary, A.S.: Bert models for arabic text classification: a systematic review. Appl. Sci. **12**(11), 5720 (2022)
2. Alayrac, J.B., et al.: Flamingo: a visual language model for few-shot learning. Adv. Neural. Inf. Process. Syst. **35**, 23716–23736 (2022)
3. Amit, S.: Introducing the knowledge graph. America: Official Blog of Google (2012)
4. Amit, S.: Introducing the knowledge graph: Things, not strings. Official Blog (of Google) (2012)
5. Bollacker, K., Evans, C., Paritosh, P., Sturge, T., Taylor, J.: Freebase: a collaboratively created graph database for structuring human knowledge. In: Proceedings of the 2008 ACM SIGMOD International Conference on Management of Data, pp. 1247–1250 (2008)
6. Bommasani, R., et al.: On the opportunities and risks of foundation models. arXiv preprint arXiv:2108.07258 (2021)
7. Bordes, A., Usunier, N., Garcia-Duran, A., Weston, J., Yakhnenko, O.: Translating embeddings for modeling multi-relational data. In: Advances in Neural Information Processing Systems 26 (2013)
8. Brown, T., et al.: Language models are few-shot learners. Adv. Neural. Inf. Process. Syst. **33**, 1877–1901 (2020)
9. Chen, T., Kornblith, S., Norouzi, M., Hinton, G.: A simple framework for contrastive learning of visual representations. In: International Conference on Machine Learning, pp. 1597–1607. PMLR (2020)
10. Chen, Z., Wang, Y., Zhao, B., Cheng, J., Zhao, X., Duan, Z.: Knowledge graph completion: a review. IEEE Access **8**, 192435–192456 (2020)
11. Daza, D., Cochez, M., Groth, P.: Inductive entity representations from text via link prediction. In: Proceedings of the Web Conference 2021, pp. 798–808 (2021)

12. Dettmers, T., Minervini, P., Stenetorp, P., Riedel, S.: Convolutional 2d knowledge graph embeddings. In: Proceedings of the AAAI Conference on Artificial Intelligence, vol. 32 (2018)
13. Devlin, J., Chang, M.W., Lee, K., Toutanova, K.: Bert: pre-training of deep bidirectional transformers for language understanding. arXiv preprint arXiv:1810.04805 (2018)
14. Gao, T., Yao, X., Chen, D.: Simcse: simple contrastive learning of sentence embeddings. arXiv preprint arXiv:2104.08821 (2021)
15. Hogan, A., et al.: Knowledge graphs. ACM Comput. Surv. (CSUR) 54(4), 1–37 (2021)
16. Huang, X., Zhang, J., Li, D., Li, P.: Knowledge graph embedding based question answering. In: Proceedings of the Twelfth ACM International Conference on Web Search and Data Mining, pp. 105–113 (2019)
17. Ji, G., He, S., Xu, L., Liu, K., Zhao, J.: Knowledge graph embedding via dynamic mapping matrix. In: Proceedings of the 53rd Annual Meeting of the Association for Computational Linguistics and the 7th International Joint Conference on Natural Language Processing (volume 1: Long papers), pp. 687–696 (2015)
18. Ji, S., Pan, S., Cambria, E., Marttinen, P., Philip, S.Y.: A survey on knowledge graphs: representation, acquisition, and applications. IEEE Trans. Neural Netw. Learn. Syst. 33(2), 494–514 (2021)
19. Jing, L., Vincent, P., LeCun, Y., Tian, Y.: Understanding dimensional collapse in contrastive self-supervised learning. arXiv preprint arXiv:2110.09348 (2021)
20. Lin, Y., Han, X., Xie, R., Liu, Z., Sun, M.: Knowledge representation learning: a quantitative review. arXiv preprint arXiv:1812.10901 (2018)
21. Lin, Y., Liu, Z., Sun, M., Liu, Y., Zhu, X.: Learning entity and relation embeddings for knowledge graph completion. In: Proceedings of the AAAI Conference on Artificial Intelligence, vol. 29 (2015)
22. Liu, Yet al.: Roberta: a robustly optimized bert pretraining approach. arXiv preprint arXiv:1907.11692 (2019)
23. Nguyen, D.Q., Nguyen, T.D., Nguyen, D.Q., Phung, D.: A novel embedding model for knowledge base completion based on convolutional neural network. arXiv preprint arXiv:1712.02121 (2017)
24. Nickel, M., Tresp, V., Kriegel, H.P., et al.: A three-way model for collective learning on multi-relational data. In: ICML, vol. 11, pp. 3104482–3104584 (2011)
25. Raffel, C., et al.: Exploring the limits of transfer learning with a unified text-to-text transformer. J. Mach. Learn. Res. 21(1), 5485–5551 (2020)
26. Ramesh, A., et al.: Zero-shot text-to-image generation. In: International Conference on Machine Learning, pp. 8821–8831. PMLR (2021)
27. Reimers, N., Gurevych, I.: Sentence-bert: sentence embeddings using siamese bert-networks. arXiv preprint arXiv:1908.10084 (2019)
28. Sowa, J.F.: Semantic networks. Encyclopedia Artifi. Intell. 2, 1493–1511 (1992)
29. Sun, Z., Deng, Z.H., Nie, J.Y., Tang, J.: Rotate: knowledge graph embedding by relational rotation in complex space. arXiv preprint arXiv:1902.10197 (2019)
30. Vaswani, A., et al.: Attention is all you need. In: Advances in Neural Information Processing Systems 30 (2017)
31. Vrandečić, D., Krötzsch, M.: Wikidata: a free collaborative knowledgebase. Commun. ACM 57(10), 78–85 (2014)
32. Wang, B., Shen, T., Long, G., Zhou, T., Wang, Y., Chang, Y.: Structure-augmented text representation learning for efficient knowledge graph completion. In: Proceedings of the Web Conference 2021, pp. 1737–1748 (2021)

33. Wang, L., Zhao, W., Wei, Z., Liu, J.: Simkgc: simple contrastive knowledge graph completion with pre-trained language models. arXiv preprint arXiv:2203.02167 (2022)

34. Wang, Q., Mao, Z., Wang, B., Guo, L.: Knowledge graph embedding: a survey of approaches and applications. IEEE Trans. Knowl. Data Eng. **29**(12), 2724–2743 (2017)

35. Wang, S., et al.: Metakrec: collaborative meta-knowledge enhanced recommender system. In: 2022 IEEE International Conference on Big Data (Big Data), pp. 665–674. IEEE (2022)

36. Wang, T., Isola, P.: Understanding contrastive representation learning through alignment and uniformity on the hypersphere. In: International Conference on Machine Learning, pp. 9929–9939. PMLR (2020)

37. Wang, X., et al.: Kepler: a unified model for knowledge embedding and pre-trained language representation. Trans. Assoc. Comput. Ling. **9**, 176–194 (2021)

38. Wang, Y., Liu, Z., Fan, Z., Sun, L., Yu, P.S.: Dskreg: differentiable sampling on knowledge graph for recommendation with relational gnn. In: Proceedings of the 30th ACM International Conference on Information & Knowledge Management, pp. 3513–3517 (2021)

39. Wang, Z., Zhang, J., Feng, J., Chen, Z.: Knowledge graph embedding by translating on hyperplanes. In: Proceedings of the AAAI Conference on Artificial Intelligence, vol. 28 (2014)

40. Wang, Z., Ng, P., Ma, X., Nallapati, R., Xiang, B.: Multi-passage bert: a globally normalized bert model for open-domain question answering. arXiv preprint arXiv:1908.08167 (2019)

41. Xie, R., Liu, Z., Jia, J., Luan, H., Sun, M.: Representation learning of knowledge graphs with entity descriptions. In: Proceedings of the AAAI Conference on Artificial Intelligence, vol. 30 (2016)

42. Yang, B., Yih, W.t., He, X., Gao, J., Deng, L.: Embedding entities and relations for learning and inference in knowledge bases. arXiv preprint arXiv:1412.6575 (2014)

43. Yang, Y., et al.: Improving multilingual sentence embedding using bi-directional dual encoder with additive margin softmax. arXiv preprint arXiv:1902.08564 (2019)

44. Yao, L., Mao, C., Luo, Y.: Kg-bert: bert for knowledge graph completion. arXiv preprint arXiv:1909.03193 (2019)

45. Yuan, L., et al.: Florence: A new foundation model for computer vision. arXiv preprint arXiv:2111.11432 (2021)

46. Zhang, Y., Yao, Q., Kwok, J.T.: Bilinear scoring function search for knowledge graph learning. IEEE Trans. Pattern Anal. Mach. Intell. (2022)

Duet Representation Learning with Entity Multi-attribute Information in Knowledge Graphs

Yuanbo Xu[1], Yuanbo Zhang[1], Yongjian Yang[1], Hangtong Xu[1(✉)], and Lin Yue[2]

[1] MIC Lab, College of Computer Science and Technology, Jilin University, Changchun, China
{yuanbox,yyj}@jlu.edu.cn, {zhangyb19,xuht21}@mails.jlu.edu.cn
[2] The University of Newcastle, Callaghan, Australia
Lin.Yue@newcastle.edu.au

Abstract. Representation Learning (RL) of knowledge graphs aims to project both entities and relations into a continuous low-dimension space. Most methods concentrate on learning entities' representations with structure information indicating the relations between entities (Trans- methods), while the utilization of entity multi-attribute information is insufficient for some scenarios, such as cold start issues or zero-shot problems. How to utilize the complex and diverse multi-attribute information for RL is still a challenging problem for enhancing knowledge graph embedding research. In this paper, we propose a novel RL model *Duet Entity Representation Learning* (DERL) for knowledge graphs, which takes advantage of entity multi-attribute information. Specifically, we devise a novel encoder *Entity Attribute Encoder* (EAE), which encodes both entity attribute types and values to generate the entities' attribute-based representations. We further learn the entities' representations with both structure information and multi-attribute information in DERL. We evaluate our method on two tasks: the knowledge graph completion task and the zero-shot task. Experimental results on real-world datasets show that our method outperforms other baselines on two downstream tasks by building effective representations for entities from their multi-attribute information. The source code of this paper can be obtained from https://anonymous.4open.science/r/DUET-adma2023/.

Keywords: Multi-attribute · Representation Learning · Knowledge graphs

1 Introduction

Knowledge graphs (KGs) provide a massive amount of structure information for entities and relations, which have been successfully utilized in various fields such as knowledge inference [21] and question answering [23]. Typical KGs like Freebase [1], or YAGO3 [8] usually model the multi-relational information with many structure triples represented as (*head entity*, relation, *tail entity*), which is also abridged as (h, r, t).

Currently, most RL methods focus on structure information but ignore attribute information in KGs. For example, in Fig. 1, we show two entity multi-attribute information in a structure triple sampled from DWY100K [13]. Although some works have realized the importance of multi-attribute information such as DT-GCN [11], they haven't

X. Yang et al. (Eds.): ADMA 2023, LNAI 14177, pp. 32–45, 2023.
https://doi.org/10.1007/978-3-031-46664-9_3

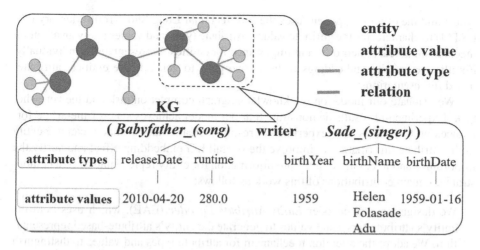

Fig. 1. Example of entity multi-attribute information in DWY100K.

fully used the rich semantic information. First, they didn't embed the attribute types and attribute values jointly and applied them to improve the entity representation semantic accuracy directly. Second, they didn't consider using both structure information and multi-attribute information to improve the overall RL effect. In addition, entity multi-attribute information is generally stored in KGs in the form of triples and the attribute triples can not tackle successive attribute values and might suffer from issues like one-to-multi or multi-to-one relations in KGs. Furthermore, entity multi-attribute information is usually diverse and complex: different entities may have multiple attribute types in KGs, and even different attribute values may have various data structures and value granularities. For example, in Fig. 1, $Babyfather\_(song)$ and $Sade\_(singer)$ have different attribute types; the three attribute types of $Sade\_(singer)$ that correspond to attribute values have different forms and structures. Entity multi-attribute information is too complex to use for learning embeddings directly. In the meanwhile, intuitively, different attribute types and values play different degrees of importance in the entities' representations. If entity multi-attribute information cannot be used reasonably, the entities' representations will lose a large amount of accurate semantic information thus reducing their semantic accuracy.

To address those problems, we first design a novel encoder EAE, which can encode the complex and diverse entity multi-attribute information to generate the entities' attribute-based representations. Moreover, we propose a novel RL model DERL for KGs, combining structure and multi-attribute information to improve KG embedding. In the DERL model, an entity's representation is responsible for jointly modeling the corresponding structure information and multi-attribute information.

For learning structure information, we follow a typical RL method TransE [2] and regard the relation in each structure triple as a translation from the head entity to the tail entity. For learning multi-attribute information, we use the EAE to learn the entities' attribute-based representations. In our EAE, we set up a training model that contains two embedding components. One component embeds the entity's different attribute types, and the other component uses the bi-directional Long Short-Term Memory (Bi-LSTM) to characterize the attribute values. Attribute types and values apply an attention mechanism to learn their different importance for entities' representations individually. Finally, we use the embeddings of these two parts to generate the entities' attribute-based representations.

We evaluate our model on the knowledge graph completion task and the zero-shot task. Experimental results demonstrate that our model achieves state-of-the-art performances on both tasks. Our experimental results indicate that our model can use entity multi-attribute information to improve the overall KG embedding effect and verify the importance and necessity of attribute information for entity representation. We demonstrate the main contributions of this work as follows:

- We design a novel encoder *Entity Attribute Encoder* (EAE), which uses both the entity's attribute types and values to generate the entity's attribute-based representation. We adopt the attention mechanism for attribute types and values to distinguish the importance of different attribute information to the entity's representation.
- We propose a novel RL model *Duet Entity Representation Learning* (DERL), which utilizes both entity structure information and multi-attribute information for enhancing RL's effect.
- We evaluate the DERL model's effectiveness on the knowledge graph completion and zero-shot tasks. Experimental results on real-world datasets illustrate that the DERL model consistently outperforms other baselines on these two tasks. To the best of our knowledge, this is the first work attempt to use entity multi-attribute information to solve the zero-shot problem.

2   Problem Formulation

We first introduce the symbols used in this paper. Given a structural triple $(h, r, t) \in T$, while $h, t \in E$ stand for entities, $r \in R$ stands for the relation. Respectively, h and t are the head entity and the tail entity. $a \in A$ stands for the attribute type and $v \in V$ stands for the attribute value. $c \in v$ stands for the attribute value character. T stands for the whole training set of structural triples. E is the set of entities, R is the set of relations, A is the set of attribute types, and V is the set of attribute values. We propose two kinds of representations for each entity to utilize structure information and multi-attribute information in DERL.

Definition 1. Structure-Based Representations: e_s represents the entity's structure-based representation. e_{sh} and e_{st} are the structure-based representations based on the head entity and the tail entity. $\mathbf{r}$ represents the relation's representation. These representations could be learned through existing translation-based models.

Definition 2. Attribute-Based Representations: e_a represents the entity's attribute-based representation. e_{ah} and e_{at} are the attribute-based representations based on the head entity and the tail entity. We will propose an encoder to construct this kind of representation in the following section.

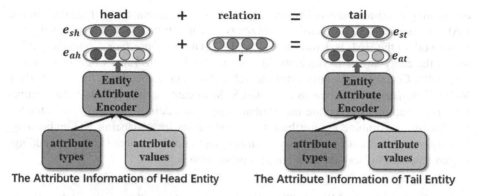

Fig. 2. The Overall Architecture of DERL Model

3   Methodology

3.1   Overall Architecture

We attempt to utilize entity structure information as well as multi-attribute information in the DERL model. Following the framework of translation-based methods, we define the overall energy function as follows:

$$S(h, r, t) = S_s + S_a, \tag{1}$$

where $S_s = ||e_{sh} + \mathbf{r} - e_{st}||$. S_s is an energy function based on structure-based representations, which is the same as the translation-based methods. S_a is an energy function based on attribute-based representations and structure-based representations. To make the learning process of S_a compatible with S_s. We define S_a as:

$$S_a = S_{as} + S_{sa} + S_{aa}, \tag{2}$$

where $S_{as} = ||e_{ah} + \mathbf{r} - e_{st}||$ and $S_{sa} = ||e_{sh} + \mathbf{r} - e_{at}||$, in which one of the head entity or the tail entity is the structure-based representation, and the other is the attribute-based representation. $S_{aa} = ||e_{ah} + \mathbf{r} - e_{at}||$, the head entity and the tail entity are both attribute-based representations. According to the overall energy function, the overall architecture of the DERL is demonstrated in Fig. 2. We learn the entities' structure-based representations and relations' representations from TransE. And we learn the entities' attribute-based representations from EAE. Under the overall energy function, we can get the attribute-based representations and the structure-based representations simultaneously. The overall energy function will project these two types of entities' representations into the same vector space with relation representations shared by all four energy functions, which will be promoted between two types of representations.

3.2   Entity Attribute Encoder

Entity multi-attribute information is difficult to use due to its complexity, heterogeneity, and different levels of importance. These problems directly lead to the difficulty of learning multi-attribute information into entities' representations. Therefore in our EAE, we consider the encoding of attribute types and attribute values respectively. The framework of the EAE is demonstrated in Fig. 3. The *Attribute Type Embedding (ATE)* learns the entity attribute type embeddings, and the *Attribute Value Character Embedding (AVCE)* learns the entity attribute value character embeddings. In the *Attribute Value Embedding (AVE)*, we use the Bi-LSTM to capture the attribute value characters' information and generate the attribute value embeddings. We apply the attention mechanism to combine the attribute types and values for enhancing KG embedding. Finally, we combine the attribute type embeddings and the attribute value embeddings to generate the entities' attribute-based representations.

Attribute Type Embedding (ATE). We first count the attribute types and randomly generate an embedding for each attribute type. Because each entity has a different number of attribute types and values, we adopt the zero-filling strategy to unify the numbers. To prevent the zero-filling strategy from affecting the model's training, we separately generate the same embedding for all zeros to prevent problems such as vanishing gradients. Given the entity's M attribute types: $A = (a_0, a_1, ..., a_M)$, we obtain the following embeddings of the entity's M attribute types:

$$\mathbf{A} = (\boldsymbol{a}_0, \boldsymbol{a}_1, ..., \boldsymbol{a}_M). \tag{3}$$

Attribute Value Character Embedding (AVCE). We first count the characters that appear in the attribute values. Then we randomly generate an embedding for each attribute value character. Because the numbers of characters in each attribute value are different, we also utilize the zero-filling strategy and generate the same embedding for all zeros. Given the attribute value N characters: $v_i = (c_0, c_1, ..., c_N)$, we get the following attribute value character embeddings:

$$\boldsymbol{v}_i = (\boldsymbol{c}_0, \boldsymbol{c}_1, ..., \boldsymbol{c}_N). \tag{4}$$

Attribute Value Embedding (AVE). We observe that the different attribute values might appear differently in KGs. For example: "2012-12-12" and "180 cm" represent a person's birthday and height respectively. In mono-lingual KGs, the attribute value can be considered as a sequence of characters with the same vocabulary. [15] proves that the LSTM can effectively capture the sequence information between characters. Therefore we choose the Bi-LSTM to learn the sequence information between characters from beginning to end. The following equations define the Bi-LSTM cell:

$$f_t = \sigma(\mathbf{W}_f[h_{t-1}, c_t] + \mathbf{b}_f), \tag{5}$$

$$i_t = \sigma(\mathbf{W}_i[h_{t-1}, c_t] + \mathbf{b}_i), \tag{6}$$

$$\tilde{H}_t = \tanh(\mathbf{W}_H[h_{t-1}, c_t] + \mathbf{b}_H), \tag{7}$$

$$H_t = f_t \odot H_{t-1} + i_t \odot \tilde{H}_t, \tag{8}$$

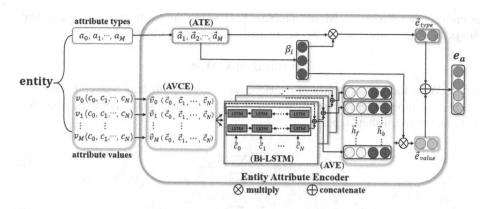

Fig. 3. The Framework of Entity Attribute Encoder

$$o_t = \sigma(\mathbf{W}_o[h_{t-1}, c_t] + \mathbf{b}_o), \tag{9}$$

$$h_t = o_t \odot \tanh(H_t), \tag{10}$$

where $\odot$ denotes a vector multiplication, f_t, i_t, o_t are the forget gate, input gate, and out gate of the Bi-LSTM cells. $\mathbf{W}_f$, $\mathbf{W}_i$, $\mathbf{W}_H$, $\mathbf{W}_o$ are weight matrices. σ is the sigmoid function. $\mathbf{b}_f$, $\mathbf{b}_i$, $\mathbf{b}_H$, $\mathbf{b}_o$ are biases. Bi-LSTM is divided into the forward LSTM (F-LSTM) and the backward LSTM (B-LSTM). The F-LSTM reads the input character embeddings. For example, the F-LSTM reads the attribute value character embeddings $v_i = (c_0, c_1, ..., c_N)$ from left to right. The B-LSTM reads the attribute value character embeddings reversely. The outputs of the F-LSTM and B-LSTM are:

$$h_f = \text{F-LSTM}(c_N, h_{f-1}), \tag{11}$$

$$h_b = \text{B-LSTM}(c_0, h_{b+1}). \tag{12}$$

The initial states of the Bi-LSTM are set to zero vectors. After reading the embedding of all characters contained in an attribute value, we concatenate the final hidden states of the two-direction LSTM outputs to generate the attribute value embedding:

$$v_i = [h_f; h_b]. \tag{13}$$

Given the entity's M attribute values: $V = (v_0, v_1, ..., v_M)$, we get the following the attribute value embeddings:

$$\mathbf{V} = (v_0, v_1, ..., v_M). \tag{14}$$

Attention for Attribute Types and Attribute Values. An entity's attribute-based representation assembles all the entity attribute information, but not all attribute information is equally important to an entity's representation. To learn the importance of different attribute types and attribute values for an entity's representation, we adopt the attention mechanism to solve this problem [22]. Given the entity attribute type embeddings: $\mathbf{A} = (a_0, a_1, ..., a_M)$, we calculate their attention weights:

$$\beta_i = \text{softmax}(\mathbf{A}^\mathrm{T}\mathbf{W}_t a_i), \tag{15}$$

where $\mathbf{W}_t$ is the weight matrix of a_i. Here we utilize the attribute type embeddings to get the attention weights. The attention weights of attribute value should be consistent with that of its attribute type:

$$e_{type} = \sum_{i=0}^{M} \beta_i a_i, \tag{16}$$

$$e_{value} = \sum_{i=0}^{M} \beta_i v_i, \tag{17}$$

we concatenate e_{type} and e_{value} to get the entity's attribute-based representation:

$$e_a = [e_{type}; e_{value}]. \tag{18}$$

3.3 Objective Formalization

We utilize a margin-based score function as our training objective, which is defined as follows:

$$L = \sum_{(h,r,t)\in T} \sum_{(h',r',t')\in T'} \max(\gamma + S(h,r,t) - S(h',r',t'), 0), \tag{19}$$

where margin γ means the artificially defined minimum distance between positive and negative examples. $S(h, r, t)$ is the overall energy function, in which both head and tail entities have two kinds of representations: structure-based representations and attribute-based representations. The above energy functions are defined as the $L1$-norm. It is verified by experiments that the DERL's effects based on $L1$-norm are better than the DERL's effects based on $L2$-norm. T' is the negative sample set of T, which we define as follows:

$$T' = (h',r,t)|h' \in E \cup (h,r',t)|r' \in R \cup (h,r,t')|t' \in E, \tag{20}$$

which means one of the entities or relations in a triple can be randomly replaced by another one. Since we have two entities' representations, if a triple already exists T, it will not treat it as a negative sample because the entity can be either a structure-based representation or an attribute-based representation.

3.4 Optimization and Implementation Details

DERL model can be defined as a parameter set $\theta = (\mathbf{E}, \mathbf{R}, \mathbf{A}, \mathbf{C}, \mathbf{W}, \mathbf{B})$. $\mathbf{E}$ stands for the embedding set of entities and $\mathbf{R}$ stands for the embedding set of relations. They can be randomly initialized or trained by previous translation-based methods such as TransH

[18] and TransR [7]. **A** stands for the embedding set of attribute types and **C** stands for the embedding set of attribute value characters and they are initialized randomly. **W** and **B** represent the weight set and bias set of Bi-LSTM and attention mechanism in EAE, which can be initialized randomly. We utilize the mini-batch stochastic gradient descent (SGD) to optimize our model, where chain rules are applied to update the variables and parameters. We use GPU to accelerate training.

4   Experiments

4.1   Datasets and Experiment Settings

Datasets. In our experiments, we use the DWY100K [13] to evaluate our models' knowledge graph completion effect. For the zero-shot task, we build a new dataset FB24K-New based on FB24K [6] to simulate a zero-shot scenario. We select 12,789 entities as In-KG entities in FB24K and select 5,179 entities in FB24K that are related to In-KG entities as Out-of-KG entities. We extract the structure triples which contain In-KG entities and Out-of-KG entities and add them to the test set. Our test set is split into 4 types: (I - I), (O - I), (I - O), and (O - O). I represent an In-KG entity, and O represents an Out-of-KG entity. The DWY100K, FB24K, and FB24K-New details are listed in Table 1 and Table 2.

Table 1. Statistics of DWY100K

Datasets	#Ent	#Rel	#Attr	#Attr tr	#Rel tr
DBP-WD-Dbpedia	100,000	330	351	381,166	463,294
DBP-WD-Wikidata	100,000	220	729	789,815	448,774
DBP-YG-Dbpedia	100,000	302	334	451,646	428,952
DBP-YG-Wikidata	100,000	31	23	118,373	502,563

Experiment Settings. In the DERL model, the margin γ set among $\{1.0, 2.0, 3.0\}$. The learning rate λ set among $\{0.0005, 0.0003, 0.001\}$. We set different learning rates for different representation type combinations. The optimal configurations of the DERL are: $\lambda = 0.001$, $\gamma = 1.0$. We set the size of character embedding and attribute type embedding to 32. We set the attention weight size to 64 and the size of the hidden layer of Bi-LSTMs to 16. The dimensions of the attribute-based representation and structure-based representation are set to 64. The dimension of the relation's representation is set to 64. We set two evaluation settings named "Raw" and "Filter": "Filter' drops the repeated triples in the training stage (when we alternate the entities and relations, the reconstructed triple has a chance to be an existing triple), while "Raw" does not.

Table 2. Statistics of FB24K and FB24K-New

Dataset	#Ent	#Rel	#Attr	#Attr tr	#Rel tr
FB24K	23,634	673	314	207,610	216,409
Dataset	#Ent	#I - I	#O - I	#I - O	#O - O
FB24K-New	17,968	100,249	12,699	400	135

4.2 Knowledge Graph Completion Task

Due to the incompleteness and complexity of KGs, many KGs are missing triples, and a large number of potential relations between entities in the KGs are not discovered. Knowledge graph completion aims to learn appropriate entities' and relations' representations to discover the latent, correct triples. In addition, the knowledge graph completion task has been widely used to evaluate the quality of knowledge representations [24].

Evaluation Protocol. We will report four prediction results based on our models. The DERL(Structure) only utilizes structure-based representations for all entities when predicting the missing ones. While DERL(Attribute) only utilizes attribute-based representations. The DERL(Union) is a simple joint method considering the weighted concatenation of both entities' representations. The DERL(Ablation) only uses attribute information for training. We use three measures as our evaluation indicators: Mean Rank, Hits@10 and Hits@1 [19,20]. In our experiment, we select TransE [2], ComplEx [16], SimplE [5], RotatE [14], QuatRE [9], ParamE [3], TransRHS [24], DT-GCN [11], and HittER [4] as baselines, which will be discussed in the Related Work.

Experimental Results. Table 3 and Table 4 present the entity and relation prediction results respectively. Our analysis draws the following conclusions: (1) most DERL models outperform all baselines on both Mean Rank, Hits@10, and Hit@1. It indicates that the entities' representations with multi-attribute information perform better in knowledge graph completion, which not only proves that EAE can effectively encode attribute information but also shows that DERL model can learn an accurate entity's representation. (2) DERL(Structure) shows good performance, although it is inferior to some experimental results. After the mutual promotion of two kinds of information, compared with some models (such as TransE, ComplEx, DT-GCN) performance effects have been improved. The results indicate that two entities' representations can learn and share the same vector space. This proves that two kinds of information can be jointly trained to improve the RL's overall effect. (3) The DERL models' results outperform baselines on Mean Rank. The Mean Rank can well reflect the overall quality of knowledge representation and determine the prediction results. In this paper, we use entity multi-attribute information as semantics information to improve the entity representation semantic precision. Therefore, the DERL models' results are much better than the baselines' results on Mean Rank. The case studies indicate that we may not know the

Table 3. Entity Prediction Results in Knowledge Graph Completion Task

Model	DBP-WD-Dbpedia				DBP-WD-Wikidata				DBP-YG-Dbpedia				DBP-YG-Wikidata			
	Mean Rank		Hits@10(%)		Mean Rank		Hits@10(%)		Mean Rank		Hits@10(%)		Mean Rank		Hits@10(%)	
	Raw	Filter	Raw	Filter	Raw	Filter	Raw	Filter	Raw	Filter	Raw	Filter	Raw	Filter	Raw	Filter
TransE	343	242	30.1	41.9	341	254	30.1	41.4	354	258	28.9	40.3	445	379	14.4	27.8
ComplEx	341	240	30.3	41.7	338	259	30.3	42.3	349	261	29.3	39.6	443	381	13.9	27.6
SimplE	332	251	29.8	40.5	356	287	28.4	40.3	402	301	26.4	39.1	487	392	11.9	25.6
RotatE	314	229	33.4	45.2	322	244	31.8	46.9	341	249	30.9	45.2	432	377	14.6	30.1
QuatRE	322	261	27.3	38.1	354	287	26.7	38.5	371	270	24.3	36.5	447	379	14.8	30.2
ParamE	437	311	22.3	31.1	367	298	22.7	32.4	381	331	20.3	30.5	533	401	10.7	21.2
TransRHS	310	237	33.5	45.9	327	245	31.2	47.2	345	251	29.9	44.2	431	379	14.1	29.7
DT-GCN	351	245	32.1	33.2	519	388	31.7	33.3	354	287	30.3	45.3	455	367	14.0	22.6
HittER	401	302	27.8	44.7	444	363	30.7	45.1	368	311	26.4	46.9	384	357	13.8	28.6
DERL(Ablation)	310	241	33.9	49.8	320	251	33.2	44.1	345	251	29.9	44.2	422	370	14.5	29.5
DERL (Structure)	330	245	31.9	41.4	341	251	30.5	43.1	342	249	29.5	43.7	443	382	14.3	27.1
DERL (Attribute)	311	236	34.9	50.8	317	239	34.3	47.6	321	247	34.1	46.8	429	371	14.9	30.5
DERL (Union)	307	221	35.1	51.7	318	245	34.1	50.5	312	240	34.4	47.1	381	352	14.8	30.1
Improv.	1.0%	3.5%	4.8%	12.6%	1.6%	2.1%	7.0%	7.8%	8.5%	3.6%	8.5%	0.4%	0.8%	1.4%	0.7%	1.0%

entities' details only by using the structure information, but we may know the entity better by learning rich potential information from entity multi-attribute information.

Table 4. Relation Prediction Results in Knowledge Graph Completion Task

Model	DBP-WD-Dbpedia				DBP-WD-Wikidata				DBP-YG-Dbpedia				DBP-YG-Wikidata			
	Mean Rank		Hits@1(%)		Mean Rank		Hits@1(%)		Mean Rank		Hits@1(%)		Mean Rank		Hits@1(%)	
	Raw	Filter	Raw	Filter	Raw	Filter	Raw	Filter	Raw	Filter	Raw	Filter	Raw	Filter	Raw	Filter
TransE	4.01	3.53	37.3	41.1	3.97	3.58	32.2	43.5	4.22	3.52	30.4	41.7	6.13	5.51	18.8	29.2
ComplEx	3.98	3.55	36.8	46.8	3.92	3.54	36.4	45.8	3.91	3.56	35.6	42.7	6.19	5.42	18.4	26.4
SimplE	4.45	3.98	31.8	35.8	4.91	4.53	33.4	34.6	4.78	4.56	32.3	43.6	7.19	6.58	10.4	23.1
RotatE	3.54	3.21	41.4	50.8	3.51	3.09	44.9	54.1	3.76	3.32	43.9	50.1	6.02	5.52	18.6	33.4
QuatRE	4.21	3.93	38.6	47.2	3.87	3.51	41.4	50.5	4.02	3.88	36.6	44.3	7.14	6.11	16.9	30.9
ParamE	4.54	3.99	36.4	43.2	3.57	3.14	44.1	54.8	4.54	3.91	34.3	41.2	8.24	6.31	13.9	25.9
TransRHS	3.72	3.51	42.8	51.4	4.55	4.05	35.1	46.7	3.79	3.36	39.2	51.3	6.17	5.55	16.7	31.6
DT-GCN	5.28	5.05	21.3	29.4	6.91	5.53	23.4	34.1	6.98	6.56	20.1	23.4	7.48	6.52	11.7	17.8
HittER	4.07	3.43	43.5	50.3	5.48	4.56	39.6	54.9	5.48	5.01	37.7	49.8	5.58	5.33	14.5	33.6
DERL(Ablation)	3.54	3.31	41.8	50.4	3.61	3.19	44.1	50.7	3.51	2.99	47.2	54.6	5.82	5.46	18.2	30.9
DERL (Structure)	3.91	3.55	37.4	48.2	3.96	3.51	40.5	45.4	3.97	3.64	38.1	44.1	6.21	5.57	17.8	25.5
DERL (Attribute)	3.61	3.22	46.5	53.3	3.41	3.02	46.1	56.6	3.51	3.07	44.1	54.8	5.94	5.69	18.8	32.9
DERL (Union)	3.53	3.09	47.7	51.9	3.57	2.98	45.7	55.3	3.34	2.89	47.8	56.1	5.51	5.17	18.9	34.3
Improv.	0.3%	3.7%	9.7%	4.0%	2.8%	3.6%	2.7%	3.1%	11.2%	13.0%	8.9%	9.4%	1.3%	3.0%	0.5%	2.9%

4.3 Knowledge Graph Completion in Zero-Shot Task

How to embed the new entities in the KGs and apply them is the main purpose of the zero-shot task. However, it is difficult to embed the Out-of-KG entities directly, and efficiently finding the latent relations between Out-of-KG and In-KG entities is difficult.

In this paper, we use multi-attribute information to learn the Out-of-KG entities' representations, which solves the problems that the Out-of-KG entities can't embed directly and the knowledge graph completion in zero-shot tasks.

Evaluation Protocol. We select DKRL [19], ConMask [12], and OWE [10] as our baselines which will be discussed in the Related Work. We utilize Hits@10, and Hits@1 [19] for entity and relation prediction. We only present the results on the "Filter" setting. We present four results in the experiment, and the (O - I), (I - O), and (O - O) have been explained above; the Total is the combined result of these three test sets.

Experimental Results. Fig. 4 shows the experimental results of (O - I), (I - O), (O - O), and Total. We can observe that: (1) In most cases, DERL significantly outperforms other models on all four types of test sets. DERL achieves about 16.2% improvement in entity prediction and 5.7% improvement in relation prediction. It demonstrates that DERL can effectively utilize the Out-of-KG entity multi-attribute information into the entity's representation to handle the zero-shot problem. (2) The entity description information and multi-attribute information belong to the text information of the entity, but the DERL model performs better in entity prediction, relation prediction, and Mean Rank. It not only shows the effectiveness of the DERL model in embedding text information and capturing entity semantic information but also explains the advantages of using entity attribute information to solve the zero-shot problem. (3) From Fig. 4, we can see that some DERL's results are not ideal, which may be because two entities belong to two entity spaces. Therefore, the connections between In-KG and Out-of-KG entities are still in need of enhancement.

5   Related Work

5.1   Knowledge Graph Embedding

In recent years, knowledge graph embedding methods have achieved great success and promotion. TransE [2] follows the rule ($h + \mathbf{r} \approx t$) to embed the entities and relations. SimplE [5] not only uses the Polyadia-Score but also utilizes the inverse of the relation. ParamE [3] extends current embedding methods by combining the nonlinear-fitting ability of neural networks and translational properties. ComplEx [16] first introduces the Complex-Spaces to capture symmetric and antisymmetric relations. RotatE [14] treats the relation as a rotation from the head entity to the tail entity. QuatRE [9] defines the Quaternion-Space with Hamilton-Product to enhance correlations between head and tail entities. TransRHS [24] utilizes the relative positions between vectors and spheres to enhance the generalization between relations. HittER [4] proposes a Transformer-based RL model to enhance the effects of entities and relations. DT-GCN [11] makes full use of the advantages of multiple-types entity's attribute values to explore the expressiveness of the entity's representation.

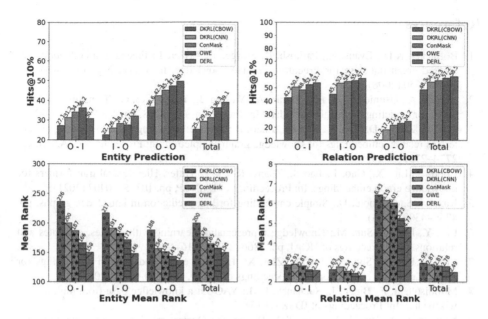

Fig. 4. Entity and Relation Prediction Results in Zero-Shot Task

5.2 Zero-Shot Problem

Zero-shot problem is a key issue in Knowledge Graph Completion because of the data sparsity (including entity and relations). Currently, few models are in a position to solve the zero-shot problem by using ancillary information. DKRL [19] proposes to use entity description information to generate entities' representations to solve the zero-shot problem. ConMask [12] comprehensively utilizes entities' names and textual information to deal with zero-shot situations. OWE [10] combines the entities' names and description information in the Transformation Space to improve open-world link prediction. To benefit the zero-shot problem in KGs, we utilize ancillary information directly to learn attribute-based representation and structure-based representation jointly, thus enriching the sparse information hidden in knowledge graphs.

6   Conclusion

In this paper, we propose a novel RL model (DERL) that utilizes both structure and multi-attribute information to improve the RL's effect in KGs. To effectively encode entity multi-attribute information, we also design an attribute information encoder EAE. Experimental results on real-world datasets demonstrate that the DERL model consistently outperforms other baselines on the knowledge graph completion task and zero-shot task. [17]

References

1. Bollacker, K.D., Evans, C., Paritosh, P., Sturge, T., Taylor, J.: Freebase: a collaboratively created graph database for structuring human knowledge. In: Proceedings of ICMD, pp. 1247–1250 (2008)
2. Bordes, A., Usunier, N., García-Durán, A., Weston, J., Yakhnenko, O.: Translating embeddings for modeling multi-relational data. In: Proceedings of NIPS, pp. 2787–2795 (2013)
3. Che, F., Zhang, D., Tao, J., Niu, M., Zhao, B.: Parame: regarding neural network parameters as relation embeddings for knowledge graph completion. In: Proceedings of AAAI, pp. 2774–2781 (2020)
4. Chen, S., Liu, X., Gao, J., Jiao, J., Zhang, R., Ji, Y.: Hitter: Hierarchical transformers for knowledge graph embeddings. In: Proceedings of EMNLP, pp. 10395–10407 (2021)
5. Kazemi, S.M., Poole, D.: Simple embedding for link prediction in knowledge graphs, pp. 4289–4300 (2018)
6. Lin, Y., Liu, Z., Sun, M.: Knowledge representation learning with entities, attributes and relations. In: Proceedings of IJCAI, pp. 2866–2872 (2016)
7. Lin, Y., Liu, Z., Sun, M., Liu, Y., Zhu, X.: Learning entity and relation embeddings for knowledge graph completion. In: Proceedings of AAAI, pp. 2181–2187 (2015)
8. Mahdisoltani, F., Biega, J., Suchanek, F.M.: YAGO3: a knowledge base from multilingual wikipedias. In: Proceedings of IDSR (2015)
9. Nguyen, D.Q., Vu, T., Nguyen, T.D., Phung, D.: Quatre: relation-aware quaternions for knowledge graph embeddings. CoRR abs/ arXiv: 2009.12517 (2020)
10. Shah, H., Villmow, J., Ulges, A., Schwanecke, U., Shafait, F.: An open-world extension to knowledge graph completion models. In: Proceedings of the AAAI, pp. 3044–3051 (2019)
11. Shen, Y., Li, Z., Wang, X., Li, J., Zhang, X.: Datatype-aware knowledge graph representation learning in hyperbolic space. In: Proceedings of CIKM, pp. 1630–1639 (2021)
12. Shi, B., Weninger, T.: Open-world knowledge graph completion. CoRR abs/ arXiv: 1711.03438 (2017)
13. Sun, Z., Hu, W., Zhang, Q., Qu, Y.: Bootstrapping entity alignment with knowledge graph embedding. In: Proceedings of IJCAI, pp. 4396–4402 (2018)
14. Sun, Z., Deng, Z., Nie, J., Tang, J.: Rotate: knowledge graph embedding by relational rotation in complex space. In: Proceedings of ICLR (2019)
15. Trisedya, B.D., Qi, J., Zhang, R.: Entity alignment between knowledge graphs using attribute embeddings. In: Proceedings of AAAI, pp. 297–304 (2019)
16. Trouillon, T., Welbl, J., Riedel, S., Gaussier, É., Bouchard, G.: Complex embeddings for simple link prediction. In: Proceedings of ICML, pp. 2071–2080 (2016)
17. Wang, S., et al.: Knowledge graph representation via hierarchical hyperbolic neural graph embedding. In: 2021 IEEE International Conference on Big Data (Big Data), Orlando, FL, USA, 15–18 December 2021, pp. 540–549. IEEE (2021). https://doi.org/10.1109/BigData52589.2021.9671651
18. Wang, Z., Zhang, J., Feng, J., Chen, Z.: Knowledge graph embedding by translating on hyperplanes. In: Proceedings of AAAI, pp. 1112–1119 (2014)
19. Xie, R., Liu, Z., Jia, J., Luan, H., Sun, M.: Representation learning of knowledge graphs with entity descriptions. In: Proceedings of AAAI, pp. 2659–2665 (2016)
20. Xie, R., Liu, Z., Luan, H., Sun, M.: Image-embodied knowledge representation learning. In: Proceedings of IJCAI, pp. 3140–3146 (2017)
21. Yang, B., Yih, W., He, X., Gao, J., Deng, L.: Embedding entities and relations for learning and inference in knowledge bases. In: Bengio, Y., LeCun, Y. (eds.) In Proceedings of ICLR (2015)

22. Yang, K., Liu, S., Zhao, J., Wang, Y., Xie, B.: COTSAE: co-training of structure and attribute embeddings for entity alignment. In: Proceedings of AAAI, pp. 3025–3032 (2020)
23. Yin, J., Jiang, X., Lu, Z., Shang, L., Li, H., Li, X.: Neural generative question answering. In: Proceedings of IJCAI, pp. 2972–2978 (2016)
24. Zhang, F., Wang, X., Li, Z., Li, J.: Transrhs: A representation learning method for knowledge graphs with relation hierarchical structure. In: Proceedings of IJCAI, pp. 2987–2993 (2020)

TKGAT: Temporal Knowledge Graph Representation Learning Using Attention Network

Shaowei Zhang, Zhao Li, Xin Wang$^{(\boxtimes)}$, Zirui Chen, and WenBin Guo

College of Intelligence and Computing, Tianjin University, Tianjin, China
{zhangsw,lizh,wangx,zrchen,wenff}@tju.edu.cn

Abstract. Temporal knowledge graph representation learning models can capture more comprehensive semantic information, which has higher practical application value and gradually attracts wide attention. However, the existing temporal knowledge graph representation learning models usually have challenges in encoding temporal information and capturing rich structural information. In this paper, we propose a novel temporal knowledge graph representation learning model, named TKGAT, which is based on graph neural networks using Bochner's theorem to design time encoding function that can flexibly learn relative time information. Furthermore, attention network is adopted to model different relations features and the self-attention mechanism is optimized by the decoupled attention method, so that the attention weight matrix incorporates more extensive temporal and structural information and learns the correlations between entity and temporal features. The extensive experiments have shown that the proposed model can consistently outperform state-of-the-art models over all benchmark datasets.

Keywords: temporal knowledge graph · representation learning · decoupled attention

1 Introduction

A great amount of data generated in daily life often takes the form of graph structure, such as social networks, financial transactions and literature citations. Researchers have adopted the form of triple $(subject, relation, object)$ to represent semantic information in data, and construct large-scale knowledge graphs (KG) such as DBpedia, FreeBase, and WordNet [25]. However, the KGs are usually incomplete due to data sparsity, which makes knowledge graph completion (KGC) a priority task. Knowledge graph representation learning expresses underlying semantic information by mapping the triples into continuous low-dimensional vector spaces, which is proved to be an efficient method for KGC [11].

S. Zhang, Z. Li—Contributed equally to this research.

© The Author(s), under exclusive license to Springer Nature Switzerland AG 2023
X. Yang et al. (Eds.): ADMA 2023, LNAI 14177, pp. 46–61, 2023.
https://doi.org/10.1007/978-3-031-46664-9_4

Static KG representation learning models that ignore the temporal information, which can lead to an inaccurate semantic representation. As depicted in Fig. 1 (a), there are three relations *Praise or endorse, Make optimistic comment* and *Criticize or denounce* between *Barack Obama* and *Iran*, such knowledge

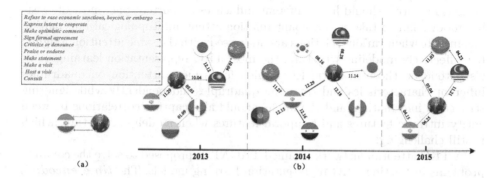

Fig. 1. Example of the temporal knowledge graph

can cause confusion when temporal information is neglected since these three relations are in conflict. Figure 1 (b) depicts a sample of the temporal knowledge graph (TKG), the relations between *Barack Obama* and *Iran* made clarity as the temporal information has been added. We can also observe that *Iran* has an *Express intent to cooperate* with *China*, *Consult* with *Afghanistan* and *Host a visit* with *Syria*, these three relation types will have various impacts on *Iran*, and the topology of countries and relations around *Iran* also determines the character of *Iran*. Therefore, effectively modeling the topological features of KG is essential for KG representation learning. Besides, capturing temporal features in TKG is also crucial. As shown in Fig. 1 (b), the relation *Make a visit* between *Barack Obama* and *South Korea* occurred at time *2014-08-16*, however, *Barack Obama* has an relation *Make optimistic comment* with *Iran* at time *2014-12-29*, since the long time interval between the two events, the former will have less influence on the latter as time passes, which also reveals that more significant temporal characteristics are typically provided by the relative time. Our model aims to well capture the topological and temporal features in TKG, in contrast to the static KG representation learning models, which ignore temporal information and process the TKG directly in a static manner, resulting in incomplete and inaccurate expression of semantic information.

In recent years, TKG representation learning has received extensive attention from both academia and industry [7], which incorporates the corresponding temporal features when expressing the semantic information in data. However, most of the current TKG representation learning models usually face many challenges. (1) The sensible time encoding, since the TKG topology is dynamic, entities should have various features at different times. Besides, time encoding should satisfy the inherent properties of time, such as the relative time can usually carry more meaningful information than absolute time, for example, when a user buys a product on the internet, the temporal information of browsing and staying on a certain product is more important than the order of browsing the products. However, the previous models mostly used simple feed-forward neural networks or recurrent neural networks to capture temporal features, which lack of in-depth theory; (2) Modeling relations appropriately, distinct relations around an entity should have different influences on current one, most of existing models fail to take into account relation attention. Topological information incomplete when various relations are addressed with the same attention weights; (3) Effectively modeling structure, the most TKG representation learning models extend on those in static KG, which focus more attention on quadruples inherent characteristics and treat the quadruples independently while ignoring structural information, and the model should also capture correlations between entity intrinsic features and temporal features when modeling structure, which is still challenging.

A TKG attention networks, named TKGAT, is proposed to solve the common problems in existing TKG representation learning models. The *time encoding function* based on the Bochner's theorem [23] has been adopted to capture temporal features, which is well suited to model the properties of relative time and has a deep theoretical foundation. The weights of the different relation types are constructed by the attention network to reflect the relevant to central entity. The self-attention mechanism [19] has proved its powerful ability in various tasks, the position encoding is replaced by time encoding and *decoupled attention* [6] is applied to optimize self-attention, which can incorporate more extensive knowledge graph features and effectively capture the correlations between entity and time. Our contributions in this paper can be summarized as follows.

(1) We propose a novel temporal knowledge graph representation learning model, TKGAT, which encodes temporal information based on Bochner's theorem and uses attention networks to capture different relations weight in order to efficiently model relational information and improve model performance.
(2) By separating structure and time encoding to optimize the traditional self-attention mechanism, a decoupled attention approach is designed, which combines graph neural networks to efficiently capture correlations between entity and temporal features.
(3) The model proposed in this paper achieves the best experimental results on three public datasets, further demonstrating the effectiveness of the model and outperforming baseline methods.

The rest of this paper is organized as follows. Section 2 presents related works. We introduce preliminaries in Sect. 3. We describe the proposed model in detail in Sect. 4. Section 5 reports the experimental results, and we conclude in Sect. 6.

2 Related Work

In this section, the traditional static KG representation learning models and the TKG representation learning models are introduced.

2.1   Static Knowledge Graph Representation Learning

At present, most of the existing knowledge graph representation learning models are suitable for static KG, which can be classified into three categories. The first category is the translation-based model, which makes the head and tail entities satisfy the translation constraints of the relation, and measure the truth of the triples by calculating the Euclidean distance between the head and tail entity vectors after the translation. TransE [1], TransH [20], and TransR [13] are the most representative models, since the simple and efficient nature of TransE, there are a series of subsequent works that extended on TransE. The second category is the semantic matching based model, which evaluates the plausibility of a fact by matching the underlying semantic information of entities and relations in the vector space. RESCAL [15], DistMult [24], ComplEx [18], and SimplE [9] are the simplest and most widely used models. The third category is neural network-based model, which mainly takes advantage of the excellence of neural networks in feature extraction and non-linear fitting to model KG features, representative models include ConvE [3], ConvKB [14], and RGCN [16]. However, all these models ignore the temporal information and fail to reflect the real-world change properties, resulting in lower accuracy in TKG.

2.2   Temporal Knowledge Graph Representation Learning

In recent years, temporal knowledge graph representation learning has gradually become a hot research topic. Most existing models primarily focus on extending static KG representation learning to TKG. TTransE [7] adds temporal information to the score function of the TransE and makes it satisfy the temporal information based translation constraint. HyTE [2] extends the TransH model, which projects entities and relations to a time-specific hyperplane to realize the embedding of temporal information. TA-TransE [4] represents the relation type and temporal information as a sequence of characters, then uses the LSTM to learn the time-aware representation of relation types. TComplEx [10] extends ComplEx and considers the score of each quadruple as fourth-order tensor decomposition. TeRo [21] borrows ideas from TransE and RotatE [17], which defines the temporal evolution of entity embedding as a rotation and regards relation as translation. ATiSE [22] incorporates temporal information into entity and relation representations by using additive time series decomposition and uses a multi-dimensional Gaussian distribution to represent temporal uncertainty. Inspired by diachronic word embedding, DE-SimplE [5] incorporates temporal information into diachronic entity embedding and has the capability of modeling various relation patterns. Compared to our model, these models fail to capture the rich structural information and the correlations between entity and temporal features. Another line of work on TKG representation learning employs neural networks, RE-NET [8] adopts a R-GCN based aggregator and recurrent event encoder to model the historical information. RE-GCN [12] learns the evolutional representations of entities and relations by capturing the structural dependencies and sequential patterns. However, those models focus on TKGC extrapolation task, i.e., inferring the feature facts in a sequence, which are fundamentally different from our work.

3   Preliminaries

In this section, we present the preliminaries of our work, including the definition of temporal knowledge graph and graph neural network.

3.1   Temporal Knowledge Graph

In this paper, we represent a temporal knowledge graph as $\mathcal{G} = \{(s, r, o, t)\} \subseteq \mathcal{V} \times \mathcal{R} \times \mathcal{V} \times \mathcal{T}$, where $\mathcal{V}$, $\mathcal{R}$ and $\mathcal{T}$ indicate the sets of nodes, edges, and timestamps, respectively. Temporal knowledge graph completion (TKGC) is to solve the problem of incompleteness in TKG. Assume that the whole true facts set is $\mathcal{F} \subseteq \mathcal{V} \times \mathcal{R} \times \mathcal{V} \times \mathcal{T}$, TKG should be a subset of the whole true facts set since the incompleteness of TKG, i.e., $\mathcal{G} \subseteq \mathcal{F}$. TKGC is the reasoning from $\mathcal{G}$ to $\mathcal{F}$. According to the time range, TKGC has two settings, interpolation and extrapolation. Given a temporal knowledge graph $\mathcal{G}$ with timestamps t range from t_1 to t_T, for the interpolation setting, TKGC predicts missing fatcs with $t_1 < t < t_T$; In contrast, for the extrapolation setting, TKGC predicts missing fatcs with $t > t_T$, i.e., predicting future facts based on past ones. More formally, the purpose of TKGC is to predict either the subject in a given query $(?, r, o, t)$ or the object in a given query $(s, r, ?, t)$. Our work is focus on the TKGC for the interpolation settings.

3.2   Graph Neural Network

Graph neural network (GNN) enjoys several advantages such as the ability to effectively handle non-Euclidean data, which makes it a great success in processing graph data. The core idea of GNN is the message propagation mechanism, i.e., the central node features are constructed by aggregating information from neighbors. In order to obtain the features of the central node i through multiple layers of GNN, each GNN layer will implement the following two steps: (1) Message Propagation, get messages from all neighbors of node i; (2) Message Aggregation, aggregate messages from all neighbor nodes then combines with the features of node i in the previous layer to obtain the features in the current layer. The above processes are defined as follows:

$$\mathbf{h}_{\mathcal{N}_i^k}^l \leftarrow AGG \left(\{ \mathbf{h}_j^{l-1}, \forall j \in \mathcal{N}_i^k \} \right) \tag{1}$$

$$\mathbf{h}_i^l \leftarrow \sigma \mathbf{W}^l \left(\mathbf{h}_i^{l-1} \| \mathbf{h}_{\mathcal{N}_i^k}^l \right) \tag{2}$$

Steps (1) and (2) correspond to the Eqs. 1 and 2, respectively. Where $\mathcal{N}_i^k$ denotes the k neighbors of node i, h_i^l denotes the hidden layer state of node i at l-th layer, and AGG is a specific function for aggregating the features of neighbors, which can be implemented using long short term memory (LSTM), self-attention mechanisms, etc. In this paper, we use a decoupled attention approach to implement *AGG*, which is able to capture more extensive features. The representative GNN models include graph convolutional networks (GCN) and graph attention networks (GAT), both of which assign weights to neighbors explicitly or implicitly during the aggregating features.

4 Our Approach

The Fig. 2 depicts the architecture of our model. Overall, the model is based on the encoder-decoder architecture. The encoder module maps entities into a continuous low-dimensional vector space and incorporates structural and temporal features simultaneously. In view of the fact that the relations are usually irrelevant to the temporal information, the temporal features are integrated into the vector of the entity in our model. Since the different relation types have different impacts on subject, the encoder module first integrates the relation features into the objects according to the type attention weights, then employs a decoupled attention method to learn the interactions between the subjects and objects in terms of structure and time. Finally, the quadruple based (s, r, o, t) is converted into the triple (s_t, r, o_t), decoder module can directly evaluate triples using the static KG embedding methods.

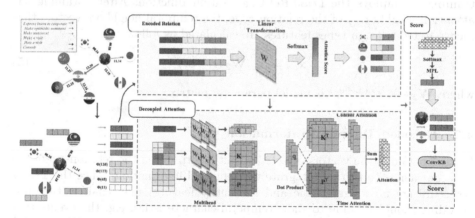

Fig. 2. The architecture of the TKGAT model. In this figure, in order to evaluate the truth of the quadruple (*Barack Obama, Make optimistic comment, Iran, 2014-12-29*). Firstly, we find the temporal neighbors where the interaction time with *Barack Obama* before *2014-12-29*, encoded relation module combines the vectors of the subject *Barack Obama*, relations and temporal neighbors together to calculate attention weights and integrates the relation features into the temporal neighbors. Secondly, time encoding function based on Bochner's Theorem is applied to capture relative time features. Thirdly, decoupled attention module learns vector of *Barack Obama* by capturing the structural and temporal feature, an analogous approach is used for *Iran*. Finally, static KGs embedding model ConvKB is adopted to evaluate score of triple that integrated temporal features.

4.1   Encoded Relation Information

Assume that there are $|\mathcal{R}|$ relation types and $|\mathcal{V}|$ entities in the temporal knowledge graph $\mathcal{G}$, the initial vectors of all entities and relations are represented as sets $\mathrm{E} = \{\mathbf{e}_i\}_{i=1}^{|\mathcal{V}|}$ and $\mathrm{R} = \{\mathbf{r}_i\}_{i=1}^{|\mathcal{R}|}$ respectively, where $\mathbf{e}_i \in \mathbb{R}^{d_e}$ represents the initial vector of i-th entity and $\mathbf{r}_i \in \mathbb{R}^{d_r}$ represents the initial vector of i-th relation, d_e and d_e represent the initial vectors dimension of entity and relation respectively. Given a quadruple (s, r, o, t), according to the inherent characteristics of time, i.e., information about future events cannot influence the ones of the present moment, the temporal neighbors of subject s are denoted as $\mathcal{N}_s^{t_k < t} = \{(r_i, o_j, t_k) | (s, r_i, o_j, t_k) \in \mathcal{G}, t_k < t\}$. Since various relation types have different effects on the subject, we combine the subject vector $\mathbf{e}_s$, relation vector $\mathbf{r}_i$, and the object vector $\mathbf{e}_j$ together and calculate the attention weights by the sofmax function. Finally, the relation feature is incorporated into the corresponding object vector, where the attention weights are calculated as follows.

$$\mathbf{u}_{r_i,o_j} = \mathbf{W}_1 \left(\mathbf{e}_s \,\|\, \mathbf{r}_i \,\|\, \mathbf{e}_j\right) \tag{3}$$

$$\alpha_{i,j} = \mathrm{softmax}\left(\mathbf{u}_{r_i,o_j}\right) = \frac{\exp\left(\sigma\left(\mathbf{p} \cdot \mathbf{u}_{r_i,o_j}\right)\right)}{\sum_{(r_m,o_n,t_i)\in\mathcal{N}_s^{t_i<t}}\exp\left(\sigma\left(\mathbf{p} \cdot \mathbf{u}_{r_m,o_n}\right)\right)} \tag{4}$$

where $\mathbf{W}_1 \in \mathbb{R}^{d_e \times (2d_e + d_r)}$, $\mathbf{p} \in \mathbb{R}^{d_e}$ are parameters learned during the model training, σ employs the LeakyReLU activation function. After obtaining the attention weights $\alpha_{i,j}$ of the relation type, the temporal neighbors vectors that incorporated relation types features are calculated as follows:

$$\mathbf{x}_{i,j} = \alpha_{i,j}\mathbf{W}_2 \left(\mathbf{r}_i \,\|\, \mathbf{e}_j\right) \tag{5}$$

where $\mathbf{W}_2 \in \mathbb{R}^{d_e \times (d_e + d_r)}$ is model parameter matrix.

4.2   Encoded Temporal Information

Having obtained the vectors of entitits that incorporated the relations information, our aim is to further integrate the temporal information. Since the TKG's structure are no longer static and the entity features may change, the time encoding should be able to show temporal characteristics, e.g. the events that happened a long time ago have less impact on the current events. We employ the time encoding function mapping from the time domain to the continuous differentiable functional domain proposed by literature [23], which is based on Bochner's Theorem and can be compatible with gradient descent in model training, we denoted it as $\Phi(t)$ and the definition as follows:

$$t \to \Phi(t) := \sqrt{\frac{1}{d_t}} \left[\cos(\omega_1 t), \sin(\omega_1 t), ..., \cos(\omega_n t), \sin(\omega_n t)\right] \tag{6}$$

where $\omega = [\omega_1, ..., \omega_{d_t}]^T$ are learnable parameters.

4.3   Encoded Structural Information

Since the topology of the TKG contains important information, we borrow the core idea of GNN, i.e., using message propagation mechanism to capture the structural information. In order to aggregate the messages from neighbors coupled with attention weights, we adopt the decoupled attention method based on self-attention mechanism.

Given a quadruple (s, r, o, t), the temporal neighbors of subject s are $\mathcal{N}_s^{t_k < t}$. At time t, the vector of the subject s at layer l-th is represented as $\mathbf{h}^l$, when $l = 1$, $\mathbf{h}^l = \mathbf{e}_s$, i.e., the initial vector of s. The subject s corresponding object under relation r_j is o_i, and its vector at lth layer is represented as $\mathbf{h}_i^l$, when $l = 1$, $\mathbf{h}_i^l = \mathbf{x}_{i,j}$, which is obtained by the encoded relation module. Since the relative time, rather than absolute time, usually reveals critical temporal information, we directly encode the relative time $\{t - t_1, t - t_2, .., t - t_k\}$ using the time encoding function, then we obtain the temporal encoding of neighbors $\{\Phi(t - t_1), \Phi(t - t_2), ..., \Phi(t - t_k)\}$, where k denotes the number of neighbors of s at time t.

The traditional self-attention mechanism are used to process sequence structure, which add or combine the two vectors that are used to represent the content and position information of the token to construct its feature. However, this approach can't effectively capture the correlation between content and position features. Inspired by DeBERTa [6], we apply time encoding to replace position encoding and calculate the weights by decoupled attention method.

The query vector at layer l is $\mathbf{q} = \mathbf{W}_q \mathbf{h}^{l-1}, \mathbf{W}_q \in \mathbb{R}^{d_h \times d_e}$ is the model parameter matrix, the vector of temporal neighbours and temporal encoding are constructed as matrices $\mathbf{Z}_E$ and $\mathbf{Z}_T$ respectively, which are represented at the $l - 1$ layer as:

$$\mathbf{Z}_E = \left[\mathbf{h}_1^{(l-1)}, \mathbf{h}_2^{(l-1)}, ..., \mathbf{h}_k^{(l-1)} \right] \in \mathbb{R}^{d_e \times k} \tag{7}$$

$$\mathbf{Z}_T = [\Phi(t - t_1), \Phi(t - t_2), ..., \Phi(t - t_k)] \in \mathbb{R}^{d_t \times k} \tag{8}$$

Applying linear transformation on matrices $\mathbf{Z}_E$ and $\mathbf{Z}_T$:

$$\mathbf{K} = \mathbf{W}_K \mathbf{Z}_E, \mathbf{P} = \mathbf{W}_T \mathbf{Z}_T, \mathbf{V} = \mathbf{W}_V \mathbf{Z}_E \tag{9}$$

where $\mathbf{W}_K, \mathbf{W}_V \in \mathbb{R}^{d_h \times d_e}$, $\mathbf{W}_T \in \mathbb{R}^{d_h \times d_t}$ are model parameters, the attention matrix obtained by the decoupled attention approach as following:

$$\widetilde{\mathbf{A}}_{0,j} = [\mathbf{q}]^\top \mathbf{K}_j + [\mathbf{q}]^\top \mathbf{P}_j \tag{10}$$

the attention matrix $\widetilde{\mathbf{A}} \in \mathbb{R}^{1 \times k}$, where $\mathbf{K}_j$ and $\mathbf{P}_j$ denote the j-th column of the matrix $\mathbf{K}$ and $\mathbf{P}$ respectively. In the process of calculating attention, $[\mathbf{q}]^\top \mathbf{K}_j$ is used to capture the correlation between the subject s and the j-th neighbour object in terms of structure, and $[\mathbf{q}]^\top \mathbf{P}_j$ is used to capture the correlation between the subject s and the j-th neighbour object in terms of time, the final attention matrix is obtained by adding the two above. We apply the softmax function to get the weights, then the final feature vector of temporal neighbors is obtained by weighted sum.

$$h_{\mathcal{N}_s^{<t}}^l = \text{softmax}\left(\frac{\widetilde{\mathbf{A}}_{0,j}}{\sqrt{2d_h}}\right)\mathbf{V} \tag{11}$$

In order to maintain the original features of the subject s, we concatenate the final feature vector of temporal neighbors with the s hidden vector at $(l-1)$-th layer, then pass it to a multilayer perceptron to capture non-linear interactions.

$$\mathbf{h}^l = \text{MPL}\left(\mathbf{h}_{\mathcal{N}_s^{\tau_k<\tau}}^l \,||\, \mathbf{h}^{l-1}\right) = \text{ReLU}\left(\left[\mathbf{h}_{\mathcal{N}_s^{\tau_k<\tau}}^l \,||\, \mathbf{h}^{l-1}\right]\mathbf{W}_0^l + \mathbf{b}_0^l\right)\mathbf{W}_1^l + \mathbf{b}_1^l \tag{12}$$

$$\mathbf{W}_0^l \in \mathbb{R}^{2d_h \times d_h}, \mathbf{b}_0^l \in \mathbb{R}^{d_h}, \mathbf{W}_1^l \in \mathbb{R}^{d_h \times d_o}, \mathbf{b}_1^l \in \mathbb{R}^{d_o}$$

where $\mathbf{W}_0^l$, $\mathbf{b}_0^l$, $\mathbf{W}_1^l$ and $\mathbf{b}_1^l$ are model parameters, d_o denotes the dimension of the final output vector. We also show that the proposed model can be easily extended to the multi-head setting which can improve performance and stability. Suppose there are m different head, and $\text{head}^{(i)} = \mathbf{h}_{\mathcal{N}_s^{\tau_k<t}}^{l(i)}$, we concatenate the m head outputs with s and then carry out the same procedure as Eq. 12.

$$\widetilde{\mathbf{h}}^l = \text{MPL}\left(\text{head}^{(1)} \,||\, ..., \,||\, \text{head}^{(m)} \,||\, \mathbf{h}^{l-1}\right) \tag{13}$$

4.4   Decoder and Training

Given a quadruple $\eta = (s, r, o, t)$, the encoder module of the TKGAT provides vectors with temporal information $(\tilde{\mathbf{s}}_t, \mathbf{r}, \tilde{\mathbf{o}}_t)$. Since the temporal information has been incorporated into the entity vector, the static KG model score function can be used to evaluate the triples. Among the currently existing methods, TKGAT adopts ConvKB as the decoder, the score function defined as following:

$$f(\eta) = \left(\overset{|\Omega|}{\underset{n=1}{||}} g\left([\mathbf{s}_t, \mathbf{r}, \mathbf{o}_t] * \omega^n\right)\right)\mathbf{W} \tag{14}$$

where Ω denotes the set of convolution kernels, ω^n denotes the n-th convolution kernel, and $\omega \in \Omega$. $\mathbf{W}_c$ denotes the parameters matrix of the linear transformation, Ω and $\mathbf{W}_c$ share parameters during the model training, the activation function $g(\cdot)$ employs ReLU, $*$ denotes the convolution operation. The output vectors of the $|\Omega|$ convolution operations are concatenated into a single vector, then linear transformation is applied to obtain the final score.

During the model training, the parameters of are learned using gradient-based optimization in mini-batches. For each quadruple $\eta = (s, r, o, t) \in \mathcal{G}$, we sample a negative set of entities $S = \{o'|(s, r, o', t) \notin \mathcal{G}\}$, then the cross-entropy loss function is used to train the model, which defined as follows:

$$\mathcal{L} = -\sum_{\eta \in \mathcal{G}} \frac{\exp\left(f(s, r, o, t)\right)}{\exp\left(\sum_{o' \notin \mathcal{G}} f(s, r, o', t)\right)} \tag{15}$$

Note that, without losing generality, we used the above loss and negative samples for subject queries. The algorithm 1 shows the training process in detail.

Algorithm 1: TKGAT training algorithm

Input: Temporal knowledge graph $\mathcal{G}$, initialization vector dimension for entity, relation, and timestamp d_e, d_r, and d_t, number of negative samples n, number of iterative rounds n_{iter}, number of batches n_b; batch size m_b

Output: Vector representation of entities, vector representation of relations

Initialize the vector of entity $\mathbf{e}_i$ with $N\left(0, \frac{1}{d_e}\right)$;

Initialize the vector of relation $\mathbf{r}_i$ with $N\left(0, \frac{1}{d_r}\right)$;

for $n = 1, ..., n_{iter}$ **do**

    for $i = 1, ..., n_b$ **do**

        $\mathcal{D}_{batch} \leftarrow Sample(\mathcal{D}_{train}, m_b)$;

        `// Sample` m_b `instances from training set`

        for $(s, r, o, t) \in \mathcal{D}_{batch}$ **do**

            $\mathcal{D}_{batch} \leftarrow \mathcal{D}'_{train} \cup \{s', r, o', t\}$;

            `// Negative samples by replacing the subject and object`

            $\mathbf{x}_{i,j} \leftarrow \alpha_{i,j} \mathbf{W}_2 \left(\mathbf{r}_i \,\|\, \mathbf{e}_j\right)$;

            `// Encoded relation information according to Equation 5`

            $\Phi(t - t_i) \leftarrow$ relative time encoding according to Equation 6;

            $\widetilde{\mathbf{h}} \leftarrow$ vector of entity according to Equation 10, 11, 13 ;

        end

        Training the model according to the Equation 14, 15 ;

    end

end

---

5 Experiments

In this section, to verify the effectiveness of the proposed model, we conduct experiments on link prediction tasks on three public datasets. We first introduce the experimental setup, including datasets, evaluation metrics, baselines, and implementation, and then analyze the experimental results. Furthermore, we perform several ablation studies to demonstrate the effectiveness of each main component of the proposed model.

5.1 Experimental Setup

Datasets. We evaluate our proposed models on the link prediction tasks, and three public TKGs datasets are used in our experiments. The statistics of the datasets are summarised in Table 1. For the Integrated Crisis Early Warning System (ICEWS) dataset, we use two subsets provided by [4]: ICEWS14, corresponding to facts in 2014, and ICEWS05-15, corresponding to facts between 2005 and 2015. For the Global Database of Events, Language, and Tone (GDELT) dataset, we use subsets which corresponding to facts from 1 April 2015 to 31 March 2016, each piece of data has a corresponding timestamp. We use the same splits of training, validation, and testing sets as provided by [5].

Evaluation Metrics. For each quadruple $(s, r, o, t) \in \mathcal{D}_{test}$, where $\mathcal{D}_{test}$ represents the test dataset, we generate two queries: $(s, r, ?, t)$ and $(?, r, o, t)$. For the first query, the model evaluates all entities and obtains scores $f(s, r, o', t)$, $\forall o' \in \mathcal{E}$, with an analogous approach used for the second query. According to the final scores, the rank of the given quadruple is obtained, and we report *mean reciprocal rank* (MRR) which is defined as:

$$MRR = \frac{1}{2\,|\mathcal{D}_{test}|} \sum_{\eta \in \mathcal{D}_{test}} \left(\frac{1}{rank\,(o|s, r, t)} + \frac{1}{rank\,(s|r, o, t)} \right) \qquad (16)$$

where $\eta = (s, r, o, t)$, $|\mathcal{D}_{test}|$ denotes the size of the test dataset. We also report $Hits@1$, $Hits@3$, and $Hits@10$ measures where $Hits@k$ represents the percentage of correct quadruple in the k highest ranked predictions, $Hits@k$ defined as:

$$Hit@k = \frac{1}{2\,|\mathcal{D}_{test}|} \sum_{\eta \in \mathcal{D}_{test}} \mathbb{I}_{(rank(o|s,r,t) \leq k)} + \mathbb{I}_{(rank(s|r,o,t) \leq k)} \qquad (17)$$

where $\mathbb{I}_{(.)}$ is an indicator function, $\mathbb{I}_{(cond)}$ is 1 if *cond* holds and 0 otherwise.

Table 1. Statistics of datasets.

Dataset	Entities	Relations	Training	Validation	Test
ICEWS14	6,869	230	72,826	8,941	8,963
ICEWS05-15	10,094	251	368,962	46,275	46,092
GDELT	500	20	2,735,685	341,961	341,961

Baselines. We test the performance of the proposed model against a variety of strong baselines, including static KG representation learning models and TKG representation learning models. Note that all these static models are applied without considering the time information in the input, including: TransE [1], DistMult [24], ComplEx [18], and SimplE [9]. The other TKG representation learning baselines models include: TTransE [7], HyTE [2], TA-TransE [4], DE-SimplE [5], ATiSE [22], and TeRo [21]. As TGAT [23] is specifically designed to handle dynamic network graphs not TKG, we have not compared with it.

Implementation. We implemented our model and the baselines in PyTorch and conducted the experiments on an NVIDIA Tesla V100 GPU. The vectors dimension of the entity, relation, and time are fixed to 128. We also tried to use different score functions to train the model, finally, we chose the ConvKB model as our decoder. The number of temporal neighbors samples is set to 20 for ICEWS14 and ICEWS05-15 datasets, 50 for the GDELT dataset. Theoretically, the information from multi-hop neighbors can be aggregated in our model, to

Table 2. Evaluation results on link prediction. The best results are in bold and the second-best results are underlined.

Dataset	ICEWS14				ICEWS05-15				GDELT			
Metrics	MRR	Hits@1	Hits@3	Hits@10	MRR	Hits@1	Hits@3	Hits@10	MRR	Hits@1	Hits@3	Hits@10
TransE	0.280	0.094	-	0.637	0.294	0.090	-	0.663	0.155	0.060	0.178	0.335
DistMult	0.439	0.323	-	0.672	0.456	0.337	-	0.691	0.210	0.133	0.224	0.365
ComplEx	0.474	0.370	0.523	0.689	0.485	0.377	0.531	0.702	0.213	0.132	0.234	0.374
Simple	0.478	0.373	0.530	0.689	0.486	0.376	0.535	0.705	0.211	0.128	0.231	0.382
TTransE	0.255	0.074	-	0.601	0.271	0.084	-	0.616	0.115	0.0	0.160	0.318
HyTE	0.297	0.108	0.416	0.655	0.316	0.116	0.445	0.681	0.188	0.0	0.165	0.326
TA-TransE	0.275	0.095	-	0.625	0.299	0.096	-	0.668	-	-	-	-
TA-DistMult	0.477	0.363	-	0.686	0.474	0.346	-	0.728	0.206	0.124	0.219	0.365
DE-TransE	0.326	0.124	0.467	0.686	0.314	0.108	0.453	0.685	0.126	0.0	0.181	0.350
DE-DisMult	0.501	0.392	0.569	0.708	0.484	0.366	0.546	0.718	0.213	0.130	0.228	0.376
DE-SimplE	0.526	0.418	0.592	0.725	0.513	0.392	0.578	0.748	_0.230_	_0.141_	_0.248_	_0.403_
ATiSE	0.550	0.436	_0.629_	_0.750_	0.519	0.378	0.606	0.794	-	-	-	-
TeRo	_0.562_	_0.468_	0.621	0.732	_0.586_	_0.469_	_0.668_	_0.795_	-	-	-	-
TKGAT (ours)	**0.574**	**0.502**	**0.655**	**0.752**	**0.607**	**0.504**	**0.676**	**0.813**	**0.256**	**0.154**	**0.290**	**0.441**

speed up training, only the information about the 2-hop neighbors is aggregated. The number of attention heads and negative samples is set to 4 and 200 respectively, and the Adam SGD optimizer is applied to train model, we set 0.001 as the learning rate for all datasets.

5.2 Results and Analysis

Table 2 shows the experimental results of link prediction on ICEWS14, ICEWS05-15, and GDELT datasets. From the result, we can observe that the static KG representation learning models fell behind TKG models in most cases. The primary reason is static KG models only learned one representation for each entity or relation, without taking into account the temporal information.

The results also demonstrate the state-of-the-art performance of our approach for link prediction tasks. As we can see, the TKGAT model significantly improves on the suboptimal TeRo model for most metrics. The typical TKG representation learning models DE-SimplE, ATiSE, and TeRo, which pay more attention to model temporal information while ignoring to capture of the TKG topology structural information. In contrast, our model is based on the GNN framework, which has the advantage of building structural features. Besides, our model adopted attention networks to model relation weights and decoupled attention is applied to incorporate more extensive TKG structural features, which allowed our model accurately to describe entities and relations characteristics. TKGAT obtained central entity features by aggregating temporal neighbours, a large number of network parameters were used to learn the features, which increased a little model complexity but improved the accuracy. Meanwhile, time encoding function based on Bochner's theorem was employed to model relative time features, which further improved the model performance.

The experimental results also exhibt that the improvement in ICEWS05-15 and GDELT is greater than ICEWS14 dataset. the main reason is the comparatively small scale of the ICEWS14 dataset, in order to achieve the best prediction results, a large amount of training data is required. In addition, the results show that the model performance on the ICEWS14 and ICEWS05-15 datasets are better than those on the GDELT datasets, the major reason is the quite small scale of the entities and relation types in the GDELT dataset, however, the interactions between entities are extremely complex, which makes challenging to extract effective information from the extremely complex interactions. Furthermore, the quality of the GDELT dataset is slightly lower, resulting in a relatively lower accuracy.

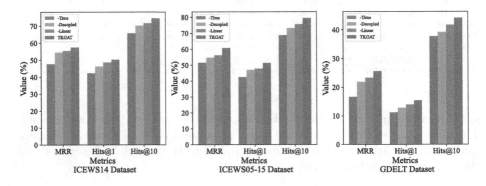

Fig. 3. Ablation study on three datasets

5.3   Ablation Study

To verify the effectiveness of each component in TKGAT, firstly, we implemented a version of TKGAT with all temporal attention weights set to the same value (-Time) to prove the validity of the time encoding function based on Bochner's theorem. Secondly, we removed the decoupled attention module (-Decoupled) and adopted the traditional self-attention mechanism directly to calculate attention scores between different entities. Finally, we incorporated relations information directly into the object using a linear transformation (-Linear) to verify the effectiveness of modeling relation weights.

As shown in Fig. 3, the TKGAT-Time model significantly reduced on MRR metric in all datasets, which proved the effectiveness of the time encoding function, and we can also notice that building temporal features in TKG is essential. In addition, the results show that the TKGAT-Decoupled model performed worse than the TKGAT model, which proved that the decoupled attention method is beneficial for improving the performance of the attention mechanism, and the correlations between entity and temporal features captured by decoupled attention are effective for TKG representation learning. We can also observe that the TKGAT-Linear model worked slightly worse than the TKGAT model, which indicates the effectiveness of capturing relations weights.

6 Conclusion

In this paper, we present a novel model, called TKGAT, for temporal knowledge graph representation learning. Specifically, time encoding function based on Bochner's theorem was applied to efficiently model relative time information, decoupled attention was adopted to capture the correlations between entity and temporal features, and the different relations influences were learned by attention network. Experimental results show that the TKGAT can effectively model temporal knowledge graph features. The ablation study also demonstrates the effectiveness of each component of TKGAT. For future work, the generation of time-aware discriminative negative samples is worth exploring.

Acknowledgment. This work is supported by the National Key R&D Program of China (2020AAA01 08504), the Key Research and Development Program of Ningxia Hui Autonomous Region (2023ZDYF0574), and the National Natural Science Foundation of China (61972275).

References

1. Bordes, A., Usunier, N., Garcia-Duran, A., Weston, J., Yakhnenko, O.: Translating embeddings for modeling multi-relational data. In: Proceedings of the 26th International Conference on Neural Information Processing Systems, pp. 2787–2795. NIPS'13, Curran Associates Inc. (2013)
2. Dasgupta, S.S., Ray, S.N., Talukdar, P.: Hyte: Hyperplane-based temporally aware knowledge graph embedding. In: Proceedings of the 2018 Conference on Empirical Methods in Natural Language Processing, pp. 2001–2011. EMNLP'18, Association for Computational Linguistics (2018)
3. Dettmers, T., Minervini, P., Stenetorp, P., Riedel, S.: Convolutional 2D knowledge graph embeddings. In: Proceedings of the Thirty-Second AAAI Conference on Artificial Intelligence, pp. 1811–1818. AAAI'18, AAAI Press (2018)
4. García-Durán, A., Dumančić, S., Niepert, M.: Learning sequence encoders for temporal knowledge graph completion. In: Proceedings of the 2018 Conference on Empirical Methods in Natural Language Processing, pp. 4816–4821. EMNLP'18, Association for Computational Linguistics (2018)
5. Goel, R., Kazemi, S.M., Brubaker, M., Poupart, P.: Diachronic embedding for temporal knowledge graph completion. In: Proceedings of the Thirty-Fourth AAAI Conference on Artificial Intelligence, pp. 3988–3995. AAAI'20, AAAI Press (2020)
6. He, P., Liu, X., Gao, J., Chen, W.: Deberta: Decoding-enhanced bert with disentangled attention. arXiv preprint arXiv:2006.03654 (2020)
7. Jiang, T., et al.: Encoding temporal information for time-aware link prediction. In: Proceedings of the 2016 Conference on Empirical Methods in Natural Language Processing, pp. 2350–2354. EMNLP'16, Association for Computational Linguistics (2016)
8. Jin, W., Qu, M., Jin, X., Ren, X.: Recurrent event network: Autoregressive structure inferenceover temporal knowledge graphs. In: Proceedings of the 2020 Conference on Empirical Methods in Natural Language Processing (EMNLP), pp. 6669–6683. Association for Computational Linguistics, Online (2020)

9. Kazemi, S.M., Poole, D.: Simple embedding for link prediction in knowledge graphs. In: Proceedings of the 32nd International Conference on Neural Information Processing Systems, pp. 4284–4295. NeurIPS'18, Curran Associates, Inc. (2018)

10. Lacroix, T., Obozinski, G., Usunier, N.: Tensor decompositions for temporal knowledge base completion. In: International Conference on Learning Representations, pp. 1–12. ICLR'20 (2020)

11. Li, Z., Liu, X., Wang, X., Liu, P., Shen, Y.: Transo: a knowledge-driven representation learning method with ontology information constraints. World Wide Web, pp. 1–23 (2022)

12. Li, Z., et al.: Temporal knowledge graph reasoning based on evolutional representation learning. In: Proceedings of the 44th International ACM SIGIR Conference on Research and Development in Information Retrieval, pp. 408–417 (2021)

13. Lin, Y., Liu, Z., Sun, M., Liu, Y., Zhu, X.: Learning entity and relation embeddings for knowledge graph completion. In: Proceedings of the Twenty-Ninth AAAI Conference on Artificial Intelligence, pp. 2181–2187. AAAI'15, AAAI Press (2015)

14. Nguyen, D.Q., Nguyen, T.D., Nguyen, D.Q., Phung, D.: A novel embedding model for knowledge base completion based on convolutional neural network. In: Proceedings of the 2018 Conference of the North American Chapter of the Association for Computational Linguistics: Human Language Technologies, pp. 327–333. NAACL'18, Association for Computational Linguistics (2018)

15. Nickel, M., Tresp, V., Kriegel, H.P.: A three-way model for collective learning on multi-relational data. In: Proceedings of the 28th International Conference on Machine Learning, pp. 809–816. ICML'11, Omnipress (2011)

16. Schlichtkrull, M., Kipf, T.N., Bloem, P., van den Berg, R., Titov, I., Welling, M.: Modeling relational data with graph convolutional networks. In: Gangemi, A., Navigli, R., Vidal, M.-E., Hitzler, P., Troncy, R., Hollink, L., Tordai, A., Alam, M. (eds.) The Semantic Web: 15th International Conference, ESWC 2018, Heraklion, Crete, Greece, June 3–7, 2018, Proceedings, pp. 593–607. Springer, Cham (2018). https://doi.org/10.1007/978-3-319-93417-4_38

17. Sun, Z., Deng, Z.H., Nie, J.Y., Tang, J.: Rotate: Knowledge graph embedding by relational rotation in complex space. In: Proceedings of the Seventh International Conference on Learning Representations, pp. 328–337. ICLR'19 (2019)

18. Trouillon, T., Welbl, J., Riedel, S., Gaussier, E., Bouchard, G.: Complex embeddings for simple link prediction. In: Proceedings of the 33rd International Conference on International Conference on Machine Learning, pp. 2071–2080. ICML'16, JMLR.org (2016)

19. Vaswani, A., et al.: Attention is all you need. In: Advances in Neural Information Processing Systems 30 (2017)

20. Wang, Z., Zhang, J., Feng, J., Chen, Z.: Knowledge graph embedding by translating on hyperplanes. In: Proceedings of the Twenty-Eighth AAAI Conference on Artificial Intelligence, pp. 1112–1119. AAAI'14, AAAI Press (2014)

21. Xu, C., Nayyeri, M., Alkhoury, F., Shariat Yazdi, H., Lehmann, J.: Tero: A time-aware knowledge graph embedding via temporal rotation. In: Proceedings of the 28th International Conference on Computational Linguistics, pp. 1583–1593. COLING'20, International Committee on Computational Linguistics (2020)

22. Xu, C., Nayyeri, M., Alkhoury, F., Yazdi, H.S., Lehmann, J.: Temporal knowledge graph embedding model based on additive time series decomposition. arXiv preprint arXiv:1911.07893 (2019)

23. Xu, D., Ruan, C., Körpeoglu, E., Kumar, S., Achan, K.: Inductive representation learning on temporal graphs. In: 8th International Conference on Learning Representations. ICLR'20 (2020)
24. Yang, B., Yih, S.W.t., He, X., Gao, J., Deng, L.: Embedding entities and relations for learning and inference in knowledge bases. In: Proceedings of the third International Conference on Learning Representations, pp. 809–816. ICLR'15 (2015)
25. Zhang, F., Wang, X., Li, Z., Li, J.: Transrhs: A representation learning method for knowledge graphs with relation hierarchical structure. In: Proceedings of the Twenty-Ninth International Joint Conference on Artificial Intelligence, pp. 2987–2993. IJCAI'20, International Joint Conferences on Artificial Intelligence Organization (2020)

Separate-and-Aggregate: A Transformer-Based Patch Refinement Model for Knowledge Graph Completion

Chen Chen[1], Yufei Wang[2], Yang Zhang[2], Quan Z. Sheng[2], and Kwok-Yan Lam[1(✉)]

[1] Nanyang Technological University, Singapore, Singapore
{s190009,kwokyan.lam}@ntu.edu.sg
[2] Macquarie University, Sydney, Australia
yufei.wang@students.mq.edu.au, yang.zhang21@hdr.mq.edu.au,
michael.sheng@mq.edu.au

Abstract. Knowledge graph completion (KGC) is the task of inferencing missing facts from any given knowledge graphs (KG). Previous KGC methods typically represent knowledge graph entities and relations as trainable continuous embeddings and fuse the embeddings of the entity h (or t) and relation r into hidden representations of query $(h, r, ?)$ (or $(?, r, t)$) to approximate the missing entities. To achieve this, they either use shallow linear transformations or deep convolutional modules. However, the linear transformations suffer from the expressiveness issue while the deep convolutional modules introduce unnecessary inductive bias, which could potentially degrade the model performance. Thus, we propose a novel Transformer-based Patch Refinement Model (**PatReFormer**) for KGC. **PatReFormer** first segments the embedding into a sequence of patches and then employs cross-attention modules to allow bi-directional embedding feature interaction between the entities and relations, leading to a better understanding of the underlying KG. We conduct experiments on four popular KGC benchmarks, WN18RR, FB15k-237, YAGO37 and DB100K. The experimental results show significant performance improvement from existing KGC methods on standard KGC evaluation metrics, e.g., MRR and H@n. Our analysis first verifies the effectiveness of our model design choices in **PatReFormer**. We then find that **PatReFormer** can better capture KG information from a large relation embedding dimension. Finally, we demonstrate that the strength of **PatReFormer** is at complex relation types, compared to other KGC models.

Keywords: Knowledge Graph Completion · Transformer · Cross-Attention

1 Introduction

Knowledge graphs (KGs) have emerged as a powerful tool for representing structured knowledge in a wide range of applications, including information retrieval, question answering and recommendation systems. A typical KG is represented as a large collection of triples $(head\ entity, relation, tail\ entity)$, denoted as (h, r, t). Despite having large amount of KG triples, many real-world KGs still suffer from incompleteness, e.g., massive valid triples are missing. To alleviate this issue, the task of knowledge graph

© The Author(s), under exclusive license to Springer Nature Switzerland AG 2023
X. Yang et al. (Eds.): ADMA 2023, LNAI 14177, pp. 62–77, 2023.
https://doi.org/10.1007/978-3-031-46664-9_5

completion (KGC) is proposed [1, 7, 8, 13], which is to predict the missing entity given the query $(h, r, ?)$ or $(?, r, t)$.

Existing methods for KGC generally learn continuous embeddings for entities and relations, with the goal of capturing the inherent structure and semantics of the knowledge graph. They define various scoring functions to aggregate the embeddings of the entity and relation, forming a hidden representation of query $(h, r, ?)$ (or $(?, r, t)$) and determine the plausibility between the query representation and missing entity embedding. Essentially, these scoring functions are a set of computation operations on interactive features of the head entity, relation and tail entity. Early KGC models like TransE [1], DistMult [2] and ComplEx [3] use simple linear operations, such as addition, subtraction and multiplication. Despite the computational efficiency, these simple and shallow architectures are incapable of capturing complicated features, e.g., poor expressiveness. To improve the model expressiveness, some recent KGC models integrate the deep neural operations into the scoring function. ConvE [8], as the start of this trend, applies standard convolutional filters over reshaped embeddings of input entities and relations, and subsequent models [6, 9] follow this trend to further improve the expressiveness of the feature interaction between entities and relations. Although these convolution-based KGC models have achieved significant empirical success, they impose unnecessary image-specific inductive bias (i.e., locality and translation equivariance) to the KGC embedding models, potentially degrading the model performance.

To combat these limitations, in this paper, we propose a novel Transformer-based Patch Refinement Model (**PatReFormer**) for the KGC task. The Transformer model is first proposed to handle Natural Language Processing (NLP) tasks [24] and demonstrates superior capability in other visual tasks [37]. More recently, with the recent progress of Vision Transformer (ViT) [28], attention-based modules achieve comparable or even better performances than their CNN counterparts on many vision tasks. Through attention mechanism, ViT-based models could dynamically focus on different embedding regions to obtain high-level informative features. What is more, ViT-based models do not impose any image-specific inductive bias, allowing them to handle a wider range of input data. Motivated by this, **PatReFormer** follows a "Separate-and-Aggregate" framework. In the separation stage, **PatReFormer** segments the input entity and relation embeddings into several patches. We explore three different separation schemes: *1)* directly folding the embedding vector into several small patches; *2)* employing several trainable mapping matrices to obtain patches; and *3)* using randomly initialized, but orthogonal mapping matrix to obtain patches. In the aggregation stage, unlike [28, 32] which use standard Transformer architecture, **PatReFormer** uses a cross-attentive architecture that deploys two separate attention modules to model the bi-directional interaction between the head entities and relations.

To evaluate our proposed approach, we conduct experiments on several benchmark datasets, including WN18RR, FB15k-237, YAGO37, and DB100K, for the KGC tasks. Our experiments show that **PatReFormer** successfully outperforms both non-Transformer-based and Transformer-based KGC methods, demonstrating the effectiveness of our approach. Our analysis shows the effectiveness of our cross-attention module design, patch-based position design, and embedding segmentation design. We find that **PatReFormer** is capable to learn useful KG knowledge using a large embedding dimension, while previous KGC models cannot. Finally, we demonstrate the advantages of **PatReFormer** in complex relation types, compared to previous KGC methods.

2   Related Work

Non-Neural-Based Methods. A variety of non-neural based models are proposed for KGC leveraging simple vector space operations, such as dot product and matrix multiplication, to compute scoring function. TransE [1] and its subsequent extensions [33,34] learn embeddings by representing relations as additive translations from head to tail entities. DistMult [2] uses multi-linear dot product to characterize three-way interactions among entities and relations. ComplEx [3] represents entities and relations as complex-valued vectors, achieving an optimal balance between accuracy and efficiency. HolE [15] utilizes cross-correlation, the inverse of circular convolution, for matching entity embeddings. More recently, SEEK [17] proposes a framework for modeling segmented knowledge graph embeddings and demonstrates that several existing models, including DistMult, ComplEx, and HolE, can be considered special cases within this framework.

Neural-Based Methods. Neural network (NN) based methods have also been explored. Approaches such as [35,36] employ a Multi-Layer Perceptron (MLP) to model the scoring function. Moreover, Convolutional Neural Networks (CNN) have been utilized for KGC tasks. ConvE [8] applies convolutional filters over reshaped head and relation embeddings to compute an output vector, which is then compared with all other entities in the knowledge graph. Subsequent works, including ConvR [6] and InteractE [9] enhance ConvE by fostering interactions between head and relation embeddings.

Transformer-Based Methods. The Transformer model known for employing self-attention to process token sequences has achieved remarkable success in NLP tasks. This success is attributed not only to its capacity for handling long-range dependencies but also to its tokenization concept. Recently, this concept has been extended to other domains, such as computer vision through Vision Transformers [28] and multi-modality with Two-stream Transformers [31]. These approaches have a common thread: they decompose the data (text or images) into smaller patches and process them using attention mechanisms. In the field of KGC, recent works have incorporated textual information and viewed entity and relation as the corresponding discrete descriptions. These methods often utilize pre-trained Transformers for encoding. However, high-quality textual KG data is not always accessible. As a result, our proposed method eschews additional textual information, instead integrating the tokenization concept into KGC to enhance performance.

3   Method

3.1   Knowledge Graph Completion

A *Knowledge Graph* can be represented as $(\mathcal{E}, \mathcal{R}, \mathcal{T})$ where $\mathcal{E}$ and $\mathcal{R}$ denote the sets of entities and relations respectively. $\mathcal{T}$ is a collection of tuples $[(h, r, t)_i]$ where head

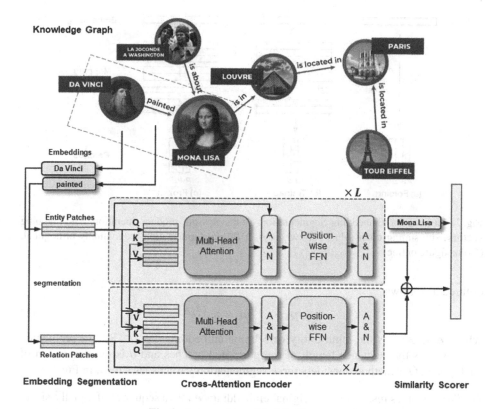

Fig. 1. An overview of `PatReFormer`

and tail entity $h, t \in \mathcal{E}$ and relation $r \in \mathcal{R}$. The task of *Knowledge Graph Completion* includes the head-to-tail prediction (e.g., predicting the head entity h in the query $(?, r, t)$) and the tail-to-head prediction (e.g., predicting the tail entity t in the query $(h, r, ?)$).

In this paper, following previous works [1,6,8], we represent head and tail entities h and t as e_h and $e_t \in \mathbb{R}^{d_e}$ and relation r as $e_r \in \mathbb{R}^{d_r}$. Our objective is to learn a function $\mathrm{F} : \mathbb{R}^{d_e} \times \mathbb{R}^{d_r} \to \mathbb{R}^{d_e}$ such that given tuple (h, r, t), the output of $\mathrm{F}(e_h, e_r)$ closely approximates e_t. For tail-to-head prediction, we additionally generate the reversed tuple (t, r^{-1}, h) and train the output of $\mathrm{F}(e_t, e_{r^{-1}})$ to be closed to e_h.

3.2 `PatReFormer`

In this section, we will introduce the details of `PatReFormer`. Figure 1 shows the overview of our `PatReFormer` model, which comprises three components: *Embedding Segmentation*, *Cross-Attention Encoder*, and *Similarity Scorer*.

Embedding Segmentation. At this stage, `PatReFormer` converts entity and relation embeddings into sequences of patches. Formally, a segmentation function $pat(\cdot)$ is

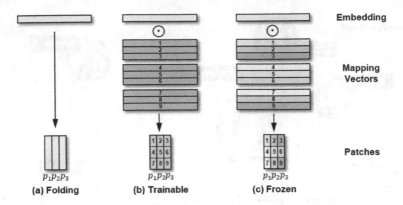

Fig. 2. Variants for Embedding Segmentation. $\odot$ denotes dot product operation. The mapping vectors with similar color (blue, yellow, grey) of frozen segmentation are mutually orthogonal. (Color figure online)

defined as follows:

$$p_0, p_1, \cdots, p_k = pat(e) \tag{1}$$

where $e \in \mathbb{R}^{k \cdot d}$ is the input entity or relation embeddings. $p_i \in \mathbb{R}^d$ are segmented patches. k is the sequence length of the generated patches and d is the dimension of each patch. Our method considers three segmentation variants, as shown in Fig. 2:

Folding involves reshaping the original embeddings e into a sequence of equally-sized, smaller patches. Formally,

$$pat(\cdot) : p_{i,j} = e_{i*d+j} \tag{2}$$

Trainable Segmentation employs a set of mapping vectors v with adaptable parameters, enabling the model to learn and optimize the mapping function during training. This function can be written as:

$$pat(\cdot) : p_{i,j} = u_{i,j} \odot e \tag{3}$$

where $u_{i,j}$ are trainable vectors.

Frozen Segmentation utilizes the function with fixed parameters, precluding updates during the training process. Notably, the frozen Segmentation function comprises a set of matrices populated with mutually orthogonal vectors. This design choice aims to facilitate the generation of embedding patches that capture distinct aspects of an entity or relation, thereby enhancing the model's ability to represent diverse features. The patches are generated by:

$$pat(\cdot) : p_{i,j} = u_{i,j} \odot e \text{ , where } u_{i,j} \odot u_{i,k} = 0 \text{ for all } j, k \tag{4}$$

The value of $u_{i,j}$ is obtained from the orthogonal matrix U_i, which is generated through singular value decomposition (SVD) of a randomly initialized matrix M_i. i.e.,

$$M_i = U_i \Sigma V_i^\top \tag{5}$$

Cross-Attention Patch Encoder. After segmenting entity and relation embedding into patches, we then aggregate these patches together via *Cross-Attention Patch Encoder* which is based on a Siamese-Transformer architecture. We will discuss its details below.

Positional Embedding. The original Transformer model encodes them with either fixed or trainable positional encoding to preserve ordering information. However, unlike visual patches from images or words from the text, in the **PatReFormer** model, the patches from embeddings do not hold any much spatial information (i.e., the values in the first and last dimension alone do not carry particular semantic meaning). We thus remove the positional embedding in **PatReFormer**. We verify the effectiveness of this design in Sect. 5.

Cross-Attention Layer. Our proposed cross-attention layer process the entity and relation patches interactively with two separated attention modules:

$$h_h^i = \begin{cases} \mathrm{MHA}_{\mathrm{ER}}^i(h_h^{i-1}, h_r^{i-1}, h_r^{i-1}) & i > 0 \\ pat(e_h) & i = 0 \end{cases} \tag{6}$$

$$h_r^i = \begin{cases} \mathrm{MHA}_{\mathrm{RE}}^i(h_r^{i-1}, h_h^{i-1}, h_h^{i-1}) & i > 0 \\ pat(e_r) & i = 0 \end{cases} \tag{7}$$

where h_h^i, h_r^i denote hidden representation of the i-th layer for head entity and relation respectively. $\mathrm{MHA}_{\mathrm{ER}}$ and $\mathrm{MHA}_{\mathrm{RE}}$ denotes Entity-to-Relation and Relation-to-Entity Attention module respectively. Both modules are based on the multi-head attention (MHA) mechanism, though they have different sets of parameters and inputs. The MHA module operates as follows:

$$\mathrm{MHA}(Q, K, V) = \mathrm{Concat}(\mathrm{head}_1, \mathrm{head}_2, \cdots, \mathrm{head}_H)W^o, \tag{8}$$

$$\text{where } \mathrm{head}_i = \mathrm{Attention}(QW_i^Q, KW_i^K, VW_i^V) \tag{9}$$

$W_i^Q \in \mathcal{R}^{d \times d_s}, W_i^K \in \mathcal{R}^{d \times d_s}, W_i^V \in \mathcal{R}^{d \times d_s}$ are projection matrix. $d_s - d/H$ where H is the predefined number of attention heads. Attention$(\cdot)$ is the scaled dot-product attention module:

$$\mathrm{Attention}(Q, K, V) = \mathrm{softmax}(\frac{QK^{\mathsf{T}}}{\sqrt{d}})V \tag{10}$$

where $Q \in \mathcal{R}^{N \times d}, K \in \mathcal{R}^{M \times d}, V \in \mathcal{R}^{M \times d}$, and N and M denote the lengths of queries and keys (or values).

Position-wise Feed-Forward Network Layer. The position-wise feed-forward network (FFN) refers to fully connected layers, which perform the same operation on each position of the input independently.

$$\mathrm{FFN}(X) = \mathrm{ReLU}(XW_1 + b_1)W_2 + b_2 \tag{11}$$

where X is the output of the *Cross-Attention Layer* i.e., h_h^i or h_r^i. $W_1 \in \mathcal{R}^{d \times d_f}$, $b_1 \in \mathcal{R}^{d_f}$, $W_2 \in \mathcal{R}^{d_f \times d}$, $b_2 \in \mathcal{R}^d$ are trainable weights and bias. To facilitate the optimization on deep networks, **PatReFormer** employs a residual connection [29] and Layer Normalization [30] on *Corss-Attention Layer* and *FFN*.

Similarity Scorer. We employ a scoring function to evaluate the relevance between the output from the Cross-Attention Encoder and the target entity embedding. Specifically, we concatenate the hidden representations obtained from the two Transformers sub-modules and project them back to the entity dimension using a linear layer.

$$e' = \text{Concat}(\overline{X_e}, \overline{X_r})W_o + b_o \tag{12}$$

In this context, $W_o \in \mathcal{R}^{(d_e+d_r)\times d_e}$, $b_o \in \mathcal{R}^{d_e}$ are weights and bias of the linear layer, respectively. $\overline{\cdot}$ is the operation to reshape Transformer output into a vector. Subsequently, we compute the dot product of the projected vector e' and the target entity embedding e_t. A sigmoid function is then applied to the result to ensure the final output falls within the $[0, 1]$ range. This scorer can be expressed as:

$$s = \text{Sigmoid}(e' \odot e_t) \tag{13}$$

Algorithm 1 provides a full procedure of our proposed **PatReFormer** method.

Algorithm 1 PatReFormer for Computing the Score of a Triple in a KG

Input: Embedding for entities and relations, E and R; head entity h, relation r and tail entity t; tokenization function $tok(\cdot)$
Output: the score of triple (h, r, t)
1: $e_h, e_r, e_t \leftarrow E.get(h), R.get(r), E.get(t)$ # get embeddings for h, r and t
2: $e_h \leftarrow tok(e_h)$
3: $e_r \leftarrow tok(e_r)$
4: **for** $i = 1$ to L **do**
5:     $e_h \leftarrow \text{LayerNorm}(\text{MHA}(e_h, e_r, e_r) + e_h)$
6:     $e_h = \text{LayerNorm}(\text{FFN}(e_h) + e_h)$
7:     $e_r \leftarrow \text{LayerNorm}(\text{MHA}(e_r, e_h, e_h) + e_r)$
8:     $e_r = \text{LayerNorm}(\text{FFN}(e_r) + e_r)$
9: **end for**
10: $e' \leftarrow \text{Concat}(\overline{e_h}, \overline{e_r})W_o + b_o$
11: $s \leftarrow \text{Sigmoid}(e' \odot e_t)$
12: **return** s

3.3 Training and Inference

For training, we leverage the standard binary cross entropy loss with label smoothing:

$$\mathcal{L}_{\text{BCE}} = -\frac{1}{N} \sum_{i=1}^{N} [y_i \log(s_i) + (1 - y_i) \log(1 - s_i)] \tag{14}$$

where p_i and y_i are the score and label of the i-th training instance respectively. $y_i \in [\epsilon, 1-\epsilon]$, where ϵ is the label smoothing value. For inference, **PatReFormer** computes the scores of the query $(h, r, ?)$ for every possible entities and rank them based on the corresponding scores. More details are presented in Sect. 4.1.

4 Experimental Results

In this section, we evaluate **PatReFormer** against various baselines in the KGC task on multiple benchmark KGs.

4.1 Experimental Setup

Dataset. Our proposed method is evaluated on four publicly available benchmark datasets: FB15K-237 [19], WN18RR [8], YAGO37 [20] and DB100K [21]. A summary of these datasets is provided in Table 1. FB15K-237 and WN18RR are widely-used benchmarks derived from FB15K and WN18 [1], respectively. They are free from the inverting triples issue. FB15K-237 and WN18RR were created by removing the inverse relations from FB15K and WN18 to address this issue. DB100K and YAGO37 are two large-scale datasets. DB100K was generated from the mapping-based objects of core DBpedia [22], while YAGO37 was extracted from the core facts of YAGO3 [23].

Table 1. Statistics of datasets.

Dataset	#Ent	#Rel	#Train	#Valid	#Test
WN18RR	40,943	11	86,835	3,034	3,134
FB15K-237	14,541	237	272,115	17,535	20,466
DB100K	99,604	470	597,572	50,000	50,000
YAGO37	123,189	37	989,132	50,000	50,000

Evaluation Protocol. Our experiment follows the filtered setting proposed in [1]. Specifically, for each test triple (h, r, t), two types of triple corruption are considered, i.e., tail corruption $(h, r, ?)$ and $(t, r^{-1}, ?)$. Every possible candidate in the knowledge graph is used to replace the entity, forming a set of valid and invalid triples. The goal is to rank the test triple among all the corrupted triples. In the filtered setting, any true triples observed in the train/validation/test set except the test triple (h, r, t) are excluded during evaluation. The evaluation metrics include the mean reciprocal rank (MRR) and the proportion of correct entities ranked in the top n (H@n) for $n = 1, 3, 10$. The evaluation is performed over all test triples on both types of triple corruption.

Table 2. Optimal hyperparameters for various KGC benckmarks

	η	L	d_e	d_r	p_1	p_2	p_3
WN18RR	1e-3	2	100	5000	0.1	0.1	0.4
FB15K-237	1e-3	12	100	3000	0.3	0.1	0.4
DB100K	5e-4	4	200	5000	0.1	0.1	0.4
YAGO37	1e-4	4	200	1000	0.1	0.1	0.1

Table 3. Experimental results of baseline models on FB15K-237, WN18RR.

	FB15K-237				WN18RR			
	MRR	H@1	H@3	H@10	MRR	H@1	H@3	H@10
Non-Transformer-Based Methods								
TransE [1]	.279	.198	.376	.441	.243	.043	.441	.532
DistMult [2]	.241	.155	.263	.419	.430	.390	.440	.490
ComplEx [3]	.247	.158	.275	.428	.440	.410	.460	.510
R-GCN [4]	.249	.151	.264	.417	-	-	-	-
SACN [5]	.350	.260	.390	.540	.470	.430	.480	.540
ConvR [6]	.350	.261	.385	.528	.475	.443	.489	.537
RotatE [7]	.338	.241	.375	.533	.476	.428	.492	.571
ConvE [8]	.325	.237	.356	.501	.430	.400	.440	.520
InteractE [9]	.354	.263	-	.535	.463	.430	-	.528
AcrE [10]	.358	.266	.393	.545	.459	.422	.473	.532
Transformer-based methods								
KG-BERT [11]	-	-	-	.420	.216	.041	.302	.524
MTL-KGC [12]	.267	.172	.298	.458	.331	.203	.383	.597
StAR [13]	.296	.205	.322	.482	.401	.243	.491	**.709**
GenKGC [14]	-	.192	.355	.439	-	.287	.403	.535
PatReFormer (ours)	**.364**	**.271**	**.400**	**.551**	**.480**	.439	**.499**	.558

Implementation Details. We implement **PatReFormer** in PyTorch[1]. In this experiment, we fix mini-batch size $\mathcal{B}$ to 256, Transformer dimensions d to 50, and label smoothing value ϵ to 0.1. The other hyper-parameters are tuned via grid search. Specifically, we select learning rate η from {1e-4, 5e-4, 1e-3}, number of layers L from {2, 4, 12}, entity embedding size d_e {100, 200}, relation embedding size d_r {1000, 3000, 5000}. All dropout ratios, i.e., p_1 on embedding patches, p_2 on Cross-Attention Encoder and p_3 on the linear layer in Similarity Scorer, are tuned in {0.1, 0.2, 0.3, 0.4}. We use Adam [18] to optimize our model. On each dataset, we select the optimal configuration according to the best MRR on the validation set within 2500 epochs. The optimal configurations of **PatReFormer** on the four datasets are listed in Table 2.

4.2 Experimental Results

Table 3 presents a comprehensive comparison of our proposed **PatReFormer** model, against the baseline models on two popular FB15K-237 and WN18RR benchmarks. Our experimental results indicate that **PatReFormer** is highly competitive against the state-of-the-art models. Specifically, **PatReFormer** achieves improvements of 0.009 in MRR and 0.6% in H@10 compared to the previous models used

---

[1] https://pytorch.org/.

Table 4. Experimental results of several models evaluated on DB100K, YAGO37.

	DB100K				YAGO37			
	MRR	H@1	H@3	H@10	MRR	H@1	H@3	H@10
TransE [1]	.111	.016	.164	.270	.303	.218	.336	.475
DistMult [2]	.233	.115	.301	.448	.365	.262	.411	.575
HolE [15]	.260	.182	.309	.411	.380	.288	.420	.551
ComplEx [3]	.242	.126	.312	.440	.417	.320	.471	.603
Analogy [16]	.252	.142	.323	.427	.387	.302	.426	.556
SEEK [17]	.338	.268	.370	.467	.454	.370	.498	.622
AcrE [10]	.413	.314	.472	.588	-	-	-	-
PatReFormer (ours)	**.436**	**.353**	**.479**	**.589**	**.523**	**.449**	**.567**	**.656**

on WN18RR. On WN18RR, **PatReFormer** obtains better results in terms of MRR (0.480 vs. 0.476) and H@3 (0.499 vs. 0.492) and is competitive in the H@10 and H@1 metrics. We attribute this discrepancy to the fact that the WN18RR dataset is a lexicon knowledge graph that relies heavily on textual information. As a result, the KGC models that incorporate pre-trained language models, such as StAR and MTL-KGC, achieve better performance than **PatReFormer** in those metrics.

To further verify the effectiveness of **PatReFormer** on larger KG, we evaluate our method on DB100K and YAGO37. Table 4 presents the performance comparison of **PatReFormer** with other baseline KGC models. On both benchmarks, **PatReFormer** outperforms existing methods on all evaluation criteria. In particular, **PatReFormer** demonstrates superiority on YAGO37 with a significant relative improvement of 15.2% (0.523 vs 0.454) and 5.5% (0.656 vs 0.622) in MRR and H@10 respectively. These findings indicate the feasibility and applicability of **PatReFormer** on real-world large-scale knowledge graphs.

5 Analysis

In this section, we investigate **PatReFormer** from various perspectives. In the first place, we show the effectiveness of the design choices in **PatReFormer**. We then show that **PatReFormer** is capable of capturing more knowledge via a large embedding dimension. Finally, we demonstrate the advantages of **PatReFormer** in complex knowledge relations. All experiments are conducted on FB15K-237.

5.1 Impact of Cross Attention

In this section, we aim to examine the effectiveness of cross-attention in our proposed model by comparing it with two variants: 1) full self-attention, in which entity and relation patches are combined together before being fed into the model, and full self-attention is applied on the combined input; and 2) separate self-attention, in which each Transformer conducts self-attention on entity and relation patches independently before concatenating their results in the Similarity Scorer. The experimental results demonstrate that our proposed cross-attention method outperforms both the full self-attention and separate self-attention variants. We hypothesize that the cross-attention mechanism only learns to connect patches from different embeddings (i.e., patches from the same embedding never interact with each other), avoiding unnecessary interference from a single embedding. This could be the primary reason why cross-attention outperforms the full self-attention variant. Furthermore, the separate self-attention variant lacks interaction between entities and relations, which could explain the significant performance drop (Tables 5 and 6).

Table 5. Analysis for model structure on FB15K-237. Att. denotes attention.

	MRR	H@1	H@3	H@10
PatReFormer	.3640	.2708	.3997	.5506
Full Self-Att	$.3599_{\downarrow.0041}$	$.2656_{\downarrow.0052}$	$.3966_{\downarrow.0031}$	$.5476_{\downarrow.0030}$
Sep. Self-Att	$.3387_{\downarrow.0253}$	$.2503_{\downarrow.0205}$	$.3698_{\downarrow.0299}$	$.5161_{\downarrow.0345}$

5.2 Impact of Positional Encoding

The original Transformer model [24] involves positional encoding to convey positional information of sequential tokens. To examine the impact of positional encoding on **PatReFormer**, we conduct an experiment with two variants: 1) trainable positional encoding (TPE) and 2) fixed positional encoding (FPE). Our experimental results demonstrate that the model without positional encoding (**PatReFormer**) outperforms the other two variants. We believe that this is due to the nature of embeddings patches, which inherently capture the features of entities or relations in a non-sequential manner. As a result, integrating positional encoding into the model introduces extraneous positional information, causing a decline in performance.

Table 6. Analysis for positional encoding (PE) on FB15K-237. Our proposed **PatReFormer** does not apply PE.

Models	MRR	H@1	H@3	H@10
PatReFormer	.3640	.2708	.3997	.5506
w/ TPE	$.3354_{\downarrow.0286}$	$.2474_{\downarrow.0234}$	$.3660_{\downarrow.0337}$	$.5107_{\downarrow.0399}$
w/ FPE	$.2580_{\downarrow.1060}$	$.1897_{\downarrow.0811}$	$.2789_{\downarrow.1208}$	$.3907_{\downarrow.1599}$

5.3 Impact of Segmentation

In this section, we explore the impact of segmentation on our proposed model, specifically examining the performance without using segmentation, and employing folding, trainable, and frozen segmentation. Our experimental results in Table 7 present that the utilization of segmentation yields a substantial performance improvement. With respect to the segmentation methods, frozen segmentation outperforms the other two variants. We believe this is due to the orthogonal vectors employed in frozen segmentation, which enhance the model's capacity to discern features of embeddings from distinct perspectives. Conversely, trainable segmentation, which allows parameters freely update during training, may face difficulties in achieving this. These findings emphasize the importance of selecting segmentation variants in the context of knowledge graph completion tasks. The superior performance of frozen segmentation suggests that these orthogonal vectors can be advantageous in extracting diverse features from entity and relation embeddings.

Table 7. Analysis for tokenization variants on FB15K-237.

	MRR	H@1	H@3	H@10
PatReFormer	.3640	.2708	.3997	.5506
w/o Seg	$.3501_{\downarrow.0139}$	$.2592_{\downarrow.0116}$	$.3850_{\downarrow.0147}$	$.5316_{\downarrow.0190}$
Folding Seg	$.3623_{\downarrow.0017}$	$.2695_{\downarrow.0013}$	$.3979_{\downarrow.0018}$	$.5488_{\downarrow.0018}$
Trainable Seg	$.3572_{\downarrow.0068}$	$.2642_{\downarrow.0066}$	$.3936_{\downarrow.0061}$	$.5433_{\downarrow.0073}$

5.4 Effectiveness of PatReFormer via a Large Relation Embedding Dimension

In a typical KG, the number of relations is much less than the number of entities. Thus, we hypothesize that the KGC models that can effectively handle a large relation embedding dimension should achieve superior KGC performance. We verify this hypothesis in this section. Figure 3 shows a clear performance increasing trend for PatReFormer as the length of relation embeddings increases. However, the other baseline KGC models, such as TransE, ConvE, and RotatE, do not deliver similar improvement; RotatE even suffers from performance delegations after the embedding dimension increases. This result shows that PatReFormer could capture more knowledge by using a large embedding dimension, while other methods cannot due to their insufficient modeling expressiveness. Such ability allows PatReFormer to capture more knowledge for relation embeddings and achieve better performance.

5.5 Analysis on Different Types of Relations

In this section, we analyze the performance of different types of relations for various models: TransE, ConvE, RotatE and PatReFormer. To categorize the relations, we

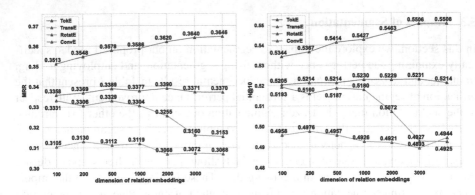

Fig. 3. Analysis of relation embedding size on FB15K-237

considered the average number of tails per head and heads per tail, grouping them into four distinct types: 1-1 (one-to-one), 1-N (one-to-many), N-1 (many-to-one), and N-N (many-to-many). The results presented in Table 8 demonstrate that our **PatReFormer** model outperforms the other models in handling more complex relation types, such as 1-N, N-1, and N-N. This indicates that the increased interaction in our model allows it to capture intricate relationships more effectively. We note that TransE and ConvE perform better for simpler one-to-one relations. We believe there could be two reasons behind this phenomenon: 1) TransE and ConvE are intrinsically adept at representing simple relations (i.e., one-to-one), and 2) the limited number of evaluation instances for this category might result in biased results. Despite this, this experiment verifies the strength of our proposed **PatReFormer** model in modeling complex relation types and highlights its potential applicability to a wide range of more complicated KGC tasks.

Table 8. Experimental results by relation categories for KGC methods on FB15K-237.

	#triples	TransE		ConvE		RotatE		PatReFormer	
		MRR	H@10	MRR	H@10	MRR	H@10	MRR	H@10
1-1	192	**.4708**	.5520	.4384	.5546	.3315	.5078	.3339	**.5625**
1-N	1,293	.2388	.3650	.2532	.3789	.2719	.4017	**.2828**	**.4203**
N-1	4,185	.3975	.4972	.4151	.5187	.4207	.5168	**.4647**	**.5698**
N-N	14,796	.2877	.5063	.3133	.5315	.3167	.5337	**.3432**	**.5564**

6   Conclusion

In this paper, motivated by the recent advances in Transformers, we propose a novel Transformer-based Patch Refinement model **PatReFormer** for knowledge graph completion. **PatReFormer** includes three main components: Embedding Segmentation, Cross-Attention Encoder, and Similarity Scorer. We first segment the knowledge graph embeddings into patches and then apply a Transformer-based cross-attention encoder to model interaction between entities and relations. Finally, the Similarity Scorer combines the encoded representations to compute the similarity between inputs and target entities. The experiments on four benchmark datasets (WN18RR, FB15k-237, DB100K and YAGO37) show that our proposed **PatReFormer** outperforms existing state-of-the-art knowledge graph completion (KGC) approaches. These results validate the effectiveness of our approach and highlight the potential advantages of incorporating patch-based embeddings and cross-attention mechanisms in such tasks.

Acknowledgement. This research is supported by the National Research Foundation, Singapore and Infocomm Media Development Authority under its Trust Tech Funding Initiative and Strategic Capability Research Centres Funding Initiative. Any opinions, findings and conclusions or recommendations expressed in this material are those of the author(s) and do not reflect the views of National Research Foundation, Singapore and Infocomm Media Development Authority.

References

1. Bordes, A., Usunier, N., Garcia-Duran, A., Weston, J., Yakhnenko, O.: Translating embeddings for modeling multi-relational data. In: NIPS, pp. 1–9 (2013)
2. Yang, B., Yih, W.T., He, X., Gao, J., Deng, L.: Embedding entities and relations for learning and inference in knowledge bases. In: ICLR, San Diego, CA, USA (2015)
3. Trouillon, T., Welbl, J., Riedel, S., Gaussier, È., Bouchard, G.: Complex embeddings for simple link prediction. In: ICML, pp. 2071–2080. PMLR (2016)
4. Schlichtkrull, M., Kipf, T.N., Bloem, P., van den Berg, R., Titov, I., Welling, M.: Modeling relational data with graph convolutional networks. In: Gangemi, A., et al. (eds.) ESWC 2018. LNCS, vol. 10843, pp. 593–607. Springer, Cham (2018). https://doi.org/10.1007/978-3-319-93417-4_38
5. Shang, C., Tang, Y., Huang, J., Bi, J., He, X., Zhou, B.: End-to-end structure-aware convolutional networks for knowledge base completion. Proc. AAAI **33**, 3060–3067 (2019)
6. Jiang, X., Wang, Q., Wang, B.: Adaptive convolution for multi-relational learning. In: Proceedings of the 2019 NAACL-HIT, pp. 978–987 (2019)
7. Sun, Z., Deng, Z.-H., Nie, J.-Y., Tang, J.: Rotate: Knowledge Graph Embedding by Relational Rotation In Complex Space, p. 2019. ICLR, New Orleans, LA, USA, May (2019)
8. Dettmers, T., Minervini, P., Stenetorp, P., Riedel, S.: Convolutional 2D knowledge graph embeddings. In: Proceedings of the AAAI, vol. 32 (2018)
9. Vashishth, S., Sanyal, S., Nitin, V., Agrawal, N., Talukdar, P.: Interacte: improving convolution-based knowledge graph embeddings by increasing feature interactions. In AAAI **34**, 3009–3016 (2020)
10. Ren, F., et al.: Knowledge graph embedding with atrous convolution and residual learning. In: Proceedings of COLING, Barcelona, Spain (Online), December 8–13, pp. 1532–1543 (2020)
11. Yao, L., Mao, C., Luo, Y.: KG-BERT: BERT for knowledge graph completion. CoRR, abs/1909.03193 (2019)
12. Kim, B., Hong, T., Ko, Y., Seo, J.: Multi-task learning for knowledge graph completion with pre-trained language models. In: Proceedings of the 28th COLING 2020, Barcelona, Spain (Online), December 8–13, 2020, pp. 1737–1743 (2020)

13. Wang, B., Shen, T., Long, G., Zhou, T., Wang, Y., Chang, Y.: Structure-augmented text representation learning for efficient knowledge graph completion. In WWW '21, Virtual Event / Ljubljana, Slovenia, April 19–23, 2021, pp. 1737–1748 (2021)
14. Xin Xie, X., et al.: From discrimination to generation: Knowledge graph completion with generative transformer. In: WWW'22, Lyon, France, April 2022, pp. 162–165. ACM (2022)
15. Nickel, M., Rosasco, L., Poggio, T.: Holographic embeddings of knowledge graphs. In: Proceedings of the AAAI, vol. 30 (2016)
16. Liu, H., Wu, Y., Yang, Y.: Analogical inference for multi-relational embeddings. In: ICML, pp. 2168–2178. PMLR (2017)
17. Xu, W., Zheng, S., He, L., Shao, B., Yin, J., Liu, T.Y.: SEEK: segmented embedding of knowledge graphs. In: Proceedings of the 58th ACL 2020, Online, July 5–10, 2020, pp. 3888–3897 (2020)
18. Kingma, D.P., Ba, J.: Adam: A method for stochastic optimization. In: 3rd ICLR 2015, San Diego, CA, USA, May 7–9, 2015, Conference Track Proceedings (2015)
19. Toutanova, K., Chen, D.: Observed versus latent features for knowledge base and text inference. In: Proceedings of the 3rd Workshop on Continuous Vector Space Models and their Compositionality, pp. 57–66 (2015)
20. Ding, B., Wang, Q., Wang, B., Guo, L.: Improving knowledge graph embedding using simple constraints. In: Proceedings of the 56th ACL 2018, Melbourne, Australia, July 15–20, 2018, Volume 1: Long Papers, pp. 110–121 (2018)
21. Guo, S., Wang, Q., Wang, L., Wang, B., Guo, L.: Knowledge graph embedding with iterative guidance from soft rules. In: Proceedings of the AAAI, vol. 32 (2018)
22. Bizer, C., et al.: Dbpedia-a crystallization point for the web of data. J. Web Semantics **7**(3), 154–165 (2009)
23. Mahdisoltani, F., Biega, J., Suchanek, F.M.: Yago3: A knowledge base from multilingual wikipedias. In: 7th CIDR Conference (2014)
24. Vaswani, A., et al.: Attention is all you need. In: Advances in Neural Information Processing Systems 30: Annual Conference on Neural Information Processing Systems 2017, December 4–9, 2017, Long Beach, CA, USA, pp. 5998–6008 (2017)
25. Devlin, J., Chang, M.W., Lee, K., Toutanova, K.: BERT: pre-training of deep bidirectional transformers for language understanding. In: Proceedings of the 2019 NAACL: Human Language Technologies, NAACL-HLT 2019, Minneapolis, MN, USA, June 2–7, 2019, Volume 1 (Long and Short Papers), pp. 4171–4186 (2019)
26. Liu, Y., et al.: Roberta: A robustly optimized BERT pretraining approach (2019). CoRR, abs/1907.11692
27. Raffel, C., et al.: Exploring the limits of transfer learning with a unified text-to-text transformer. J. Mach. Learn. Res., **21**, 140:1–140:67 (2020)
28. Dosovitskiy, A. et al.: An image is worth 16x16 words: Transformers for image recognition at scale. In: 9th ICLR 2021, Virtual Event, Austria, May 3–7 (2021)
29. He, K., Zhang, X., Ren, S., Sun, J.: Deep residual learning for image recognition. In: 2016 IEEE Conference on CVPR 2016, Las Vegas, NV, USA, June 27–30, 2016, pp. 770–778. IEEE Computer Society (2016)
30. Ba, J.L., Kiros, J.R., Hinton, G.E.: Layer normalization (2016). CoRR, abs/1607.06450
31. Lu, J., Batra, D., Parikh, D., Lee, S.: Vilbert: Pretraining task-agnostic visiolinguistic representations for vision-and-language tasks. In: NIPS 2019(December), pp. 8–14,: Vancouver, pp. 13–23. BC, Canada (2019)
32. Radford, A.: Learning transferable visual models from natural language supervision. In: ICML 2021 Jul 1 (pp. 8748–8763). PMLR (2021)
33. Lin, Y., Liu, Z., Sun, M., Liu, Y., Zhu, X.: Learning entity and relation embeddings for knowledge graph completion. In: Proceedings of AAAI, January 25–30, 2015, Austin, Texas, USA, pp. 2181–2187. AAAI Press (2015)

34. Wang, Z., Zhang, J., Feng, J., Chen, Z.: Knowledge graph embedding by translating on hyperplanes. In: Proceedings of AAAI, July 27–31, 2014, Québec City, Québec, Canada, pp. 1112–1119. AAAI Press

35. Dong, X.: Knowledge vault: A web-scale approach to probabilistic knowledge fusion. In: Proceedings of the 20th ACM SIGKDD, pp. 601–610 (2014)

36. Ravishankar, S., Talukdar, P.P: Revisiting simple neural networks for learning representations of knowledge graphs. In: 6th AKBC@NIPS 2017, Long Beach, California, USA, December 8 (2017)

37. Parmar, N., et al.: Image transformer. In: ICML, pp. 4055–4064. PMLR (2018)

Two Birds with One Stone: A Link Prediction Model for Knowledge Hypergraph Based on Fully-Connected Tensor Decomposition

Jun Pang[1,2], Hong-Chao Qin[3(✉)], Yan Liu[4], and Xiao-Qi Liu[1]

[1] Wuhan University of Science and Technology, Hubei 430065, China
[2] Hubei Key Laboratory of Intelligent Information Processing and Realtime Industrial System, Hubei 430065, China
[3] Beijing Institute of Technology, Beijing 100081, China
hcqin@bit.edu.cn
[4] Wuhan Jiangang Middle School, Hubei 430050, China

Abstract. Knowledge hypergraph link prediction aims to predict missing relationships in knowledge hypergraphs and is one of the effective methods for graph completion. The existing optimal knowledge hypergraph link method based on tensor decomposition, i.e., GETD (Generalized Model based on Tucker Decomposition and Tensor Ring Decomposition), has achieved good performance by extending Tucker decomposition, but there are still two main problems: (1)GETD does not establish operations or connections between any two tensor factors, resulting in limited representation of tensor correlation (referred to as finiteness); (2)The tensor decomposed by GETD is highly sensitive to the arrangement of tensor patterns (referred to as sensitivity). In response to the above issues, we propose a knowledge hypergraph link prediction model, called GETD⁺, based on fully-connected tensor decomposition(FCTN). By combining Tucker decomposition and FCTN, a multi-linear operation/connection is established for any two factor tensors obtained from tensor decomposition. This not only enhances the representation ability of tensors, but also eliminates sensitivity to tensor pattern arrangement. Finally, the superiority of the GETD⁺ model was verified through a large number of experiments on real knowledge hypergraph datasets and knowledge graph datasets.

Keywords: Knowledge hypergraph · Link prediction · Tensor Decomposition

1 Introduction

Knowledge hypergraph is a Semantic Web [1] that uses hyperedges to describe multiple relationships. It has a wide range of applications, including semantic search [2,3], knowledge question and answer [4], recommendation and decision [5,6], etc. Due to incomplete knowledge acquisition and other reasons, the construction of knowledge hypergraphs in real life is incomplete, which affects the use of knowledge hypergraphs. As one of the effective methods for supplementing the knowledge hypergraph, knowledge hypergraph linkage prediction aims to predict missing relationships, especially hyper relationships, based on known entities and relationships in the knowledge hypergraph.

At present, knowledge hypergraph link prediction methods can be mainly divided into three categories based on technical differences: methods based on distance models, methods based on tensor decomposition, and methods based on neural networks. Most methods based on distance models are relatively simple, but they do not have complete expressiveness; The methods based on tensor decomposition generally have complete expressiveness, but their computational complexity is relatively large; The methods based on neural networks have achieved good experimental results by utilizing the modeling ability of neural networks for nonlinear complex relationships, but this type of method is a difficult to understand black box form. Due to the complete expressiveness and strong explanatory power of the methods based on tensor decomposition, we study this type of methods.

To the best of our knowledge, the existing optimal method for the second type is GETD [7]. It is the first to apply tensor decomposition method to knowledge hypergraph link prediction, by extending Tucker decomposition [8] and combining Tucker decomposition with Tensor Ring decomposition [9]. The ability to decompose higher-order tensors into multiple third-order tensors not only fully expresses all relationship types, but also reduces the complexity of the model. However, the GETD model has two shortcomings: (1) Firstly, the GETD model only establishes operations or connections between adjacent factor tensors, rather than any two factor tensors, which leads to a finite representation of tensor correlation (i.e., finiteness). (2) Secondly, the tensor decomposed by GETD remains invariant only when the target tensor's pattern undergoes cyclic shift or reverse permutation. This means that this decomposition is highly sensitive (i.e., sensitive) to the arrangement of tensor patterns, resulting in inflexibility in decomposition and application.

To address the above issues, we introduce a fully connected tensor network FCTN decomposition [10], which decomposes an n-order tensor into a set of n-order factors and establishes multilinear operations or connections between any two factors. FCTN decomposition has the superior ability to directly characterize the intrinsic correlation between any two tensor modules, and is invariant to any arrangement of tensor modules. Therefore, we propose a knowledge hypergraph link prediction method based on fully connected tensor decomposition. By combining Tucker decomposition and FCTN decomposition, it improves the finite representation of tensor correlation and the high sensitivity to tensor pattern arrangement in the GETD model.

The main contributions of this paper are summarized below.

(1) To solve the finite problem of the GETD model, we introduce fully-connected tensor network decomposition and establish a connection between any two factors.

(2) To address the sensitivity issue of the GETD model, we propose an improved method based on the GETD model, i.e., GETD$^+$, which combines Tucker decomposition and FCTN decomposition to make the decomposition more flexible.

(3) We conduct extensive experiments on binary knowledge graph datasets and knowledge hypergraph datasets and the results validate the effectiveness and superiority of the proposed method.

The rest of this paper is organized as follows: Sect. 2 reviews related works, Sect. 3 overviews and discusses the proposed method, Sect. 4 conducts experiments and analysis, and Sect. 5 concludes the paper.

2 Related Works

The research works on knowledge hypergraph link prediction based on representation learning can be divided into three categories according to the different technologies used: methods based on distance models, methods based on tensor decomposition, methods based on neural networks.

Methods Based on Distance Models. The most typical method of using distance model to predict links in the knowledge graph is TransE [1], which embeds entities and binary relation in the knowledge graph into low-dimensional vectors. It believes that each relationship r in the knowledge graph is a translation transformation from entity h to tail entity t, that is, meeting the requirements of $h + r \approx t$. Later, many variants based on distance model are proposed, but most of them can only deal with binary relation. The idea of using distance model to predict hypergraph links is to first model the relationship as a certain transformation operation between entities in a multivariate relationship, then learn the embedded representation according to the relationship between entities and relationships, and then apply it to link prediction tasks. m-TransH [11] method maps entities to the knowledge hypergraph multivariate relationship hyperplane, and defines scoring functions with the weight of mapping results. M-TransH first used the distance model-based method to solve the problem of knowledge hypergraph link prediction, but there is a problem that does not have full expressiveness. RAE [12] further improved the relevance hypothesis on the basis of m-TransH. Considering the probability of two entities appearing in a multivariate relationship at the same time, this probability was introduced into the loss function and the full connected neural network was used to train the model. However, RAE involved star to cluster conversion in modeling, can cause permanent loss of certain attribute features. A common drawback of distance models is that most translational distance models do not have complete expressiveness [13], thus they have certain limitations in relationship modeling.

Methods Based on Tensor Decomposition. This type of methods represents relationships as high-order tensors, and then decomposes the high-order tensors into multiple low-order tensors to learn embedding representations. Due to the good performance of such methods in binary knowledge graphs, researchers have extended the tensor-decomposition based knowledge graph linkage prediction methods to knowledge hypergraphs. SimplE [14] and ComplEx [15] both use the constraints of binary relation for operations, which is difficult to extend to hyper relational data in an equivalent operation mode. The recently proposed GETD [7] model is an extension of the Tucker ER [8] model in dealing with knowledge hypergraph link prediction problems, combining Tucker decomposition [8] and Tensor Ring decomposition [9]. GETD first decomposes the higher-order tensor representation of multivariate relationships into a kernel tensor and several factor tensors. To solve the problem of having too many kernel tensor parameters, it continues to decompose the kernel tensor into multiple third-order tensors to reduce model complexity Although the GETD model can fully express all relationship types, it has two issues: finiteness and sensitivity. As mentioned above, tensor decomposition methods typically decompose high-order tensors into multiple low-order tensors and have a strong mathematical theoretical foundation.

Methods Based on Neural Networks. The models based on neural networks can learn the interactive information between entities, the structural information of knowledge graphs, etc., and improve the performance of representation learning in relation modeling, structural modeling, etc. Therefore, a large number of existing neural network methods have been used in knowledge hypergraph link prediction tasks. The model based on the traditional neural network learns the interactive information within the multiple relationships. For example, NaLP [13] and tNaLP+ [16] represent each multivariate relationship as a set of key value pairs (where keys are relationships and values are entities), and then use convolutional and fully connected neural networks to learn the multivariate relationship HINGE [17] and NeuInfer [18] consider the primacy and inferiority of structural information, and believe that using only key value pairs to represent multivariate relationships will result in suboptimal models. Therefore, multivariate relationships are represented as primary triples and a set of auxiliary key value pairs. The model based on graph neural networks combines graph structure information to complete the modeling of knowledge hypergraphs. HyperMLN [19] combines the knowledge hypergraph embedding model and Markov logic network to complete link prediction. The final prediction results can be explained by logical rules and weight values to explain the reasoning path of multiple relationships, achieving link prediction with interpretability of knowledge hypergraphs. StarE [12] is the first and currently the best method to use graph neural networks for knowledge hypergraph link prediction. Using graph convolutional neural networks to learn multi hop domain information of target entities can effectively learn graph structure information. QCKGE [20] is proposed to implement knowledge graph embedding based on quaternion transformation and convolutional neural network. GRA-GAT [21] propose a global relationship assisted graph attention network based on graph convolutional neural networks. But QCKGE and GRA-GAT mainly focus on link prediction of binary knowledge graphs, and there is currently no promotion of knowledge hypergraphs. H-AKRL [22] proposed a knowledge representation learning model based on hypergraph neural networks, which models the correlation between entities and attributes at a higher level. Although both H-AKRL and our method can complete embedding learning tasks, the research object of this paper is knowledge hypergraphs rather than knowledge graphs. The neural network method fully utilizes the modeling ability of neural networks for nonlinear complex relationships, and achieves effective prediction of missing elements by learning the structural and semantic features of the graph. However, it does not have interpretability and belongs to a black box model.

3   GETD$^+$ Model

In this section, we provide a detailed introduction to our knowledge hypergraph link prediction method, i.e., GETD$^+$, based on fully connected tensor network decomposition. Firstly, the design of the GETD$^+$ model is presented, and then the scoring function and training process of the GETD$^+$ model are described in detail.

3.1   GETD$^+$ Model Design

From the perspective of tensor completion, GETD$^+$ represents an n-ary knowledge hypergraph as a binary valued $(n+1)$ order tensor $\chi \in \{0,1\}^{n_r \times n_e \times n_e \times ... \times n_e}$ $(n_r = |R|, n_e = |E|)$, where the first order represents a relationship and the other orders represent entities. $\chi_{i_1 i_2 ... i_n} = 1$ indicates that the corresponding multivariate relationship fact is true, and $\chi_{i_1 i_2 ... i_n} = 0$ indicates that the corresponding multivariate relationship is false or does not exist. Correspondingly, given the relationships and any n-1 entities in n-ary relationships, the link prediction problem is simplified to predicting missing entities in multivariate relationships. For example, given $(i_r, ?, i_2, i_3, ..., i_n)$, predicting the first entity can be determined by the maximum score of the corresponding pattern vector for each entity. However, directly using high-order tensors to represent knowledge hypergraphs can lead to high complexity in subsequent training models, requiring the use of corresponding tensor decomposition methods to reduce the order of tensors. The key issue is to preserve useful information in higher-order tensors while reducing the order of the tensor. Our GETD$^+$ first uses Tucker decomposition to decompose the higher-order tensor representing the knowledge hypergraph into a set of factor matrices and a relatively small core tensor, as shown in Eq. 1. Due to the complete expressiveness of Tucker decomposition, it completely encodes the similarity between entities and relationships in the elements of the core tensor.

$$\phi(i_r, i_1, i_2, ..., i_n) = W X_1 r_{i_r} X_2 e_{i_1} X_3 e_{i_2} X_4 ... X_{n+1} e_{i_n}, \tag{1}$$

where $W \in \mathbb{R}^{d_r \times d_e \times d_e \times ... \times d_e}$ is tensor of order $n + 1$, r_{i_r} and rows of R and E, representing the embedding vectors of relationships and entities. But as the order of higher-order tensors increases, the order of this core tensor will also increase accordingly. Therefore, GETD$^+$ also introduces the FCTN decomposition method to decompose the reshaped core tensor into $n + 1$ factor tensors, as shown in Eq. 2, further reducing the number of parameters. Any two FCTN factor tensors G_{k_1} and G_{k_2} have equally sized patterns for tensor contraction operations, which enables FCTN decomposition to fully characterize the intrinsic correlation between any two patterns of the target tensor.

$$
\begin{aligned}
\widehat{W}_{j_1 j_2 \cdots j_{n+1}} = & \Sigma_{r_{1,2}=1}^{R_{1,2}} \Sigma_{r_{1,3}=1}^{R_{1,3}} \cdots \Sigma_{r_{1,n+1}=1}^{R_{1,n+1}} \Sigma_{r_{2,3}=1}^{R_{2,3}} \Sigma_{r_{2,n+1}=1}^{R_{2,n+1}} \cdots \Sigma_{r_{n,n+1}=1}^{R_{n,n+1}} \\
& \{ G_1(j1, r_{1,2}, r_{1,3}, \ldots, r_{1,n+1}) \\
& \ G_2(r_{1,2}, j_2, r_{2,3}, \ldots, r_{2,n+1}) \\
& \ \ldots \\
& \ G_{n+1}(r_{1,n+1}, r_{2,n+1}, \ldots, r_{n-1,n+1}, r_{n,n+1}, j_{n+1}) \}
\end{aligned}
\tag{2}
$$

Therefore, FCTN decomposition can be expressed as $\widehat{W} = FCTN(G_k{}_{k=1}^{n+1}) = FCTN(G_1, G_2, \ldots, G_{n+1})$. The main framework of GETD$^+$ is shown in Fig. 1 ($n = 2$). Specifically, on the left side of the Fig. 1 is the outer structure of Tucker decomposition, and on the right side is the FCTN decomposition of the core tensor.

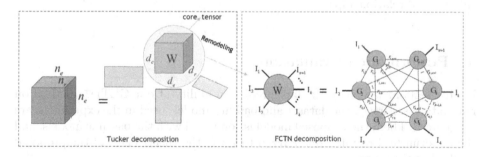

Fig. 1. Illustration diagram of the main framework for the GETD$^+$ method.

3.2 Scoring Function and Loss Function

According to Tucker decomposition and FCTN decomposition, the scoring function of GETD$^+$ is defined as Eq. 3.

$$
\begin{aligned}
Score = & \ \phi(i_r, i_1, i_2, \ldots, i_n) \\
= & \ FCTN(G_1, G_2, \ldots, G_{n+1}) X_1 r_{i_r} X_2 e_{i_2} X_3 \ldots X_{n+1} e_{i_n}
\end{aligned}
\tag{3}
$$

Use the negative sampling technique in paper [7] to obtain negative samples . The loss function adopts the commonly used logarithmic loss function, and the loss function of GETD$^+$ can be defined as Eq. 4.

$$L_{i_r,i_1,i_2,...,i_n} = \Sigma_{j=1}^n L_{(i_r,i_1,i_2,...,i_n)}^{(j)}$$
$$= \Sigma_{j=1}^n - \Phi(i_r,i_1,i_2,...,i_n) + log(e^{\Phi(i_r,i_1,...,i_n)} + \Sigma_{x \in N_{(i_r,i_1,...,i_n)}^{(j)}})$$
(4)

3.3   Model Training

GETD$^+$ is trained in small batches. All the multivariate relationships and each entity field in them are considered for training. Algorithm 1 gives the pseudocode of the training algorithm, with training set S, number of training rounds and entities, relationship embedding dimensions, etc. as inputs. The embedding of entities and relationships is randomly initialized in the first row of the algorithm. During the training process, the third row of the algorithm samples a small batch of size, where each observation value is considered for training in lines 4–11. Specifically, for each multivariate relationship in, the algorithm constructs a negative sample set in line 6. Then use Eq. 3 in line 7 to calculate the score, and further use Eq. 4 in line 8 to calculate the multi-class logarithmic loss. Finally, the algorithm updates the model parameters based on the loss gradient. The model complexity of GETD$^+$ is $O(n_e d_e + n_r d_r)$, where n_e and n_r are the number of entities and relationships.

4   Performance Evaluation

In this section, we verify the link prediction ability of our GETD$^+$ method on real datasets. Firstly, the dataset and baseline models used in the experiment are introduced. Then, the proposed model is compared with the baseline models, and the experimental results are analyzed to verify the effectiveness of the proposed method in this paper.

4.1   Experimental Setup

The experimental datasets, baseline models, and experimental parameter settings are introduced. We use two public real data sets of 3-ary and 4-ary and two common public data sets of binary relation to evaluate the proposed model, as follows.

(1) WikiPeople [13]: This is a public relations dataset extracted from Wikidata about human entities. Due to the significant sparsity of multivariate relationships after the number of elements exceeds 5, we only extract ternary and quaternary relationships, which are called WikiPeople-3 and WikiPeople-4, respectively.

(2) JF17K [12]: This is a public relations dataset developed based on Freebase. Similar to WikiPeople, the high element relationship data in JF17K is also

sparse, so we extract ternary and quaternary relationships called JF17K-3 and JF17K-4, respectively.

(3) WN18 [23]: This is a subset database based on the binary relation data set WordNet. It is a database about lexical relationships between words.

(4) FB15k [24]: This is a subset database based on the binary relation data set Freebase. It is a data stream about facts in the real world, including movies, sports, etc.

Table 1. Statistical information of the datasets.

Dataset	$|E|$	$|R|$	Train	Valid	Test
WikiPeople-3	12270	66	20656	2582	2582
WikiPeople-4	9528	50	12150	1519	1519
JF17K-3	11541	104	27635	3454	3455
JF17K-4	6536	23	7607	951	951
WN18	40943	18	141442	5000	5000
FB15k	14951	1345	483142	50000	59071

The above WikiPeople and JF17K datasets are randomly divided into a training set, a validation set, and a test set in an 8:1:1 ratio. WN18 and FB15k are tested based on the data provided in reference [22]. The specific statistical information is shown in Table 1, where $|E|$ and $|R|$ represent the number of entities and relationships, respectively, and $Train$, $Valid$, and $Test$ represent the number of facts in the training set, validation set, and test set, respectively. The proposed GETD$^+$ model mainly focuses on using tensor decomposition methods to solve the problem of knowledge hypergraph link prediction. Therefore, GETD$^+$ is only compared with the current optimal tensor decomposition class methods, such as n-CP, n-TuckER, and GETD. At the same time, to verify the link prediction effect of the GETD$^+$ model on binary knowledge graphs, the baseline models also includes link prediction methods for binary knowledge graphs, such as TransE, DistMult, ComplEx, and TuckER.

(1) N-CP [7] is an extended method of CP decomposition, which represents the same entity as different embedding vectors by modeling the role information of the entity at different positions.

(2) N-TuckER [7] is an extended method of TuckER decomposition, used in multivariate relational link prediction tasks.

(3) GETD [7] decomposes the core tensor into multiple third-order tensors by combining Tucker decomposition and Tensor Ring decomposition to achieve link prediction of knowledge hypergraphs.

(4) TransE [11] is the most primitive binary knowledge graph linking prediction method, treating each relationship in the knowledge graph as a translation transformation from the starting entity vector to the ending entity vector.

(5) DistMult [25] considers the spatial mapping matrix in the model as a diagonal matrix by introducing the modeling of relationship vectors.

(6) ComplEx [15] maps entity and relationship types to low-dimensional complex spaces and proposes a matrix decomposition method based on complex numerical vectors.

(7) TuckER [8] decomposes a large tensor into a relatively small tensor and multiplies three factor matrices, greatly reducing the complexity of the model.

We adopt the popular average reciprocal ranking MRR and hit rate Hits@k evaluate the performance of the hyperedge link prediction model for knowledge hypergraph. These two evaluation indicators are related to the ranking of positive cases in the test set among all negative cases. The ranking calculation method is as follows: given a set of multivariate relationships F, let f be any positive example in the test set. For an entity at any position in f, replace it with all other entities that are not this element to obtain a set of negative examples of the relationship. Then, remove the positive examples already included by F from these negative examples and obtain candidate samples. Next, input the candidate samples obtained in the previous step and the original positive case f into the prediction model to obtain a score, and rank based on the score to obtain the ranking of positive case f. Hits@k Represents the ratio of all positive samples in the top k of the test set, which is the number of positive relationships in the top k of the test set divided by the number of all relationships in the test set. The calculation formula is shown in the Eq. 5. The value range of Hits@k is $[0, 1]$, and a larger value indicates that the inference algorithm performs better.

$$Hit@k = \frac{\Sigma_{f\in F_{test}} rank(f) \leq n?1:0}{count(F_{test})},$$ (5)

where $randk(f) \leq n?1:0$ is the conditional expression that determines whether the ranking enters the top n. If true, the result is 1, otherwise it is 0. $count(F_{test})$ represents the total number of relationships included in the test set. Average reciprocal ranking (MRR) calculates the average value of the reciprocal of each positive example rank. The calculation formula is shown in the Eq. 6. The range of MRR values is $[0, 1]$, and the larger the value, the better the effect.

$$MRR = \frac{\Sigma_{f\in F_{test}} \frac{1}{rank(f)}}{count(F_{test})}$$ (6)

We take Hits@1, Hits@10, and MRR to evaluate the effectiveness of the proposed model as performance evaluation indicators. The experimental environment is Ubuntu 18.04, RTX3090, Anaconda, Python 3.6, CUDA11.0, and Python 1.7.1. For the fairness of the experiment, the baseline models all use the optimal experimental results. For datasets WikiPeople-3 and JF17K-3, set the entity and relationship embedding size to $d_e = d_r = 50$. For datasets WikiPeople-4 and JF17K-4, set the entity and relationship embedding size to $d_e = d_r = 25$. The learning rate is set to 0.0001. In addition, the model also uses batch standardization and dropout functions to prevent overfitting of model training.

4.2 Experimental Results and Analysis

In this section, we compare and analyze the performance of the proposed GETD⁺ method with the baseline methods, and investigate the predictive performance of the binary knowledge graph linkage of GETD⁺.

Table 2. Links prediction results on WikiPeople data set.

Method	WikiPeople-3				WikiPeople-4			
	MRR	Hits@10	Hits@3	Hit@1	MRR	Hits@10	Hits@3	Hits@1
n-CP	0.330	0.496	0.356	0.250	0.265	0.445	0.315	0.169
n-TuckER	0.365	0.548	0.400	0.274	0.362	0.570	0.432	0.246
GETD	0.373	0.558	0.401	0.284	0.386	0.596	0.462	0.265
GETD⁺	**0.402**	**0.587**	**0.437**	**0.301**	**0.415**	**0.618**	**0.498**	**0.286**

Multivariate Relationship Link Prediction. We compare Hits@1, Hits@3, Hits@10 and MRR of the proposed GETD⁺ model and baseline models respectively on WikiPeople-3, WikiPeople-4, JF17K-3 and JF17K-4 data sets. The experimental results are shown in Table 2 and Table 3, where the bolded data is the optimal result of each evaluation index. Table 2 and Table 3 show that the GETD⁺ model proposed in this paper achieves the best performance on all performance evaluations of all data sets. For WikiPeople-3 data set, Hits@1, Hits@3, Hits@10 and MRR have increased by 1.7–5.1%, 3.6–8.1%, 2.9–9.1% and 2.9–7.2%, respectively. For WikiPeople-4 dataset, Hits@1, Hits@3, Hits@10 and MRR increase by 2.1–11.7%, 3.6–18.3%, 2.2–17.3% and 2.9–15%, respectively. For JF17K-3 data set, Hits@1, Hits@3, Hits@10 and MRR are increased by 1.9–5.3%, 2.6–5.4%, 3.8–6.7% and 5.3–8.5%, respectively. For JF17K-4 data set, Hits@1, Hits@3, Hits@10, and MRR increased by 0.4–2.6%, 1.3–3.6%, 0.8–3.1%, and 2.6–4.9%, respectively. From the above experimental results, it can be seen that due to the different embeddings of different entity domains, n-CP is relatively weak, while GETD and n-Tucker capture the interaction between entities and the relationships with potential tensors/core tensors, with good performance. Moreover, GETD has solved the problem of excessive core tensor parameters in the n-TuckER model, and the experimental performance is superior to n-TuckER. The GETD⁺ model solves the randomness problem of tensor factor arrangement and the finite representation problem of tensor correlation on the basis of GETD, and achieves optimal performance in experiments.

Binary Relation Link Prediction To investigate the applicability of GETD⁺ in the link prediction of binary knowledge graph, we compare the Hits@1, Hits@3, Hits@10 and MRR of the proposed GETD⁺ model and baseline models on the WN18 and FB15K datasets, respectively. The embedding size in the

Table 3. Links prediction results on JF17K data set.

Method	JF17K-3				JF17K-4			
	MRR	Hits@10	Hits@3	Hit@1	MRR	Hits@10	Hits@3	Hits@1
n-CP	0.700	0.827	0.736	0.635	0.787	0.890	0.821	0.733
n-TuckER	0.727	0.852	0.761	0.664	0.804	0.902	0.841	0.748
GETD	0.732	0.856	0.764	0.669	0.810	0.913	0.844	0.755
GETD$^+$	**0.785**	**0.894**	**0.790**	**0.688**	**0.836**	**0.921**	**0.857**	**0.759**

experimental setup is set to $= 200$, and the other settings are the same as the prediction of multiple relationship links. The experimental results are shown in Table 4, where the bold data represent the optimal result for each evaluation indicator.

Table 4. Links prediction results on WN18 or FB15k data set.

Method	WN18				FB15k			
	MRR	Hits@10	Hits@3	Hit@1	MRR	Hits@10	Hits@3	Hits@1
TransE	0.454	0.934	0.823	0.089	0.380	0.641	0.472	0.231
DistMult	0.822	0.936	0.914	0.728	0.654	0.824	0.733	0.546
ComplEx	0.941	0.947	0.945	0.936	0.692	0.840	0.759	0.599
TuckER	0.953	0.958	0.955	**0.949**	0.795	**0.892**	0.833	0.741
GETD	0.948	0.954	0.950	0.944	0.824	0.888	0.847	0.787
GETD$^+$	**0.956**	**0.970**	**0.956**	0.946	**0.837**	0.890	**0.854**	**0.795**

5 Conclusion

In this paper, we propose a GETD$^+$ model to address the two shortcomings of the existing knowledge hypergraph link prediction model GETD, namely the limitation and sensitivity issues. By introducing fully-connected tensor network (FCTN) decomposition and combining Tucker decomposition and FCTN decomposition, the GETD$^+$ model has the superior ability to directly characterize the internal correlation between any two tensor modules, and is invariant to any arrangement of tensor modules, thus solving the finite and sensitivity problems of the GETD model. The superiority of the GETD$^+$ model is verified on four real knowledge hypergraph datasets. To verify the applicability of the model on binary knowledge graphs, experiments were conducted on two commonly used binary knowledge graph datasets, and the experimental results show that GETD$^+$ also has superiority in the link prediction of binary knowledge graph.

Acknowledgment. The work is supported by the National Natural Science Foundation of China (No.62372342, No. 61702381).

References

1. Wen, J.F., Li, J.X., Mao, Y.Y., et al.: On the representation and embedding of knowledge bases beyond binary relations. In: Twenty-Fifth International Joint Conference on Artificial Intelligence (IJCAI), pp. 1300–1307 (2016)
2. Almousa, M., Benlamri, R., Khoury, R.: A novel word sense disambiguation approach using wordnet knowledge graph. arXiv preprint arXiv:2101.02875 (2021)
3. Dai, Z., Li, L., Xu, W.: CFO: conditional focused neural question answering with largescale knowledge bases. In: 54th Annual Meeting of the Association for Computational Linguistics (ACL), pp. 800–810 (2016)
4. Chen, Y., Wu, L., Zaki, M.J.: Bidirectional attentive memory networks for question answering over knowledge bases. In: ACL, pp. 2913–2923 (2019)
5. Ji, S., Feng, Y., Ji, R., et al.: Dual channel hypergraph collaborative filtering. In: 26th ACM International Conference on Knowledge Discovery and Data Mining (SIGKDD), pp. 2020–2029 (2020)
6. Yu, W., Qin, Z.: Graph convolutional network for recommendation with low-pass collaborative filters. In: International Conference on Machine Learning (ICML), pp. 10936–10945 (2020)
7. Liu, Y., Yao, Q.M., Li, Y.: Generalizing tensor decomposition for N-ary relational knowledge bases. In: 29th International World Wide Web Conferences (WWW), pp. 1104–1114 (2020)
8. Balaevi, I., Allen, C., Hospedales, T.M.: TuckER: tensor factorization for knowledge graph completion. In: ICML, pp. 5184–5193 (2019)
9. Zhao, QB., Zhou, G.X., Xie, S.L., et al.: Tensor ring decomposition. arXiv preprint arXiv:1606.05535 (2016)
10. Zheng, Y.B., Huang, T.Z., Zhao, X.L., et al.: Fully-connected tensor network decomposition and its application to higher-order tensor completion. In: AAAI, vol. 35(12), pp. 11071–11078 (2021)
11. Bordes, A., Usunier, N., Garcia-Duran, A., et al.: Translating embeddings for modeling multi-relational data. In: 27th Annual Conference on Neural Information Processing Systems (NIPS), pp. 2787–2795 (2013)
12. Zhang, R.C., Li, J.P., Mei, J.J., et al.: Scalable instance reconstruction in knowledge bases via relatedness affiliated embedding. In: WWW, pp. 1185–1194 (2018)
13. Guan, S.P., Jin, X.L., Wang, Y.Z., et al.: Link prediction on N-ary relational data. In: WWW, pp. 583–593 (2019)
14. Kazemi, S.M., Poole, D.: Simple embedding for link prediction in knowledge graphs. In: NIPS, pp. 4289–4300 (2018)
15. Trouillon, T., Welbl, J., Riedel, S., et al.: Complex embeddings for simple link prediction. In: ICML, pp. 2071–2080 (2016)
16. Guan, S.P., Jin, X.L., Guo, J.F., et al.: Link prediction on n-ary relational data based on relatedness evaluation. IEEE Trans. Knowl. Data Eng. (TKDE) **35**(1), 672–685 (2023)
17. Nguyen, D.Q., Nguyen, T.D., Dat, Q.N., et al.: A novel embedding model for knowledge base completion based on convolutional neural network. In: ACL, pp. 327–333 (2018)
18. Dettmers, T., Minervini, P., Stenetorp, P., et al.: Convolutional 2D knowledge graph embeddings. In: AAAI, pp. 1811–1818 (2018)

19. Chen, Z., Wang, X., Wang, C., et al.: Explainable link prediction in knowledge hypergraphs. In: 31st ACM International Conference on Information and Knowledge Management (CIKM), pp. 262–271 (2020)
20. Gao, Y., Tian, X., Zhou, J., et al.: Knowledge graph embedding based on quaternion transformation and convolutional neural network. In: 17th International Conference on Advanced Data Mining and Applications (ADMA), pp. 128–136 (2021)
21. Hou, R., Zhu, W., Zhu, C.: Global relation auxiliary graph attention network for knowledge graph completion. In: 5th International Conference on Artificial Intelligence and Big Data (ICAIBD), pp. 532–538 (2022)
22. Xu, Y.W., Zhang, H.J., Cheng, K., et al.: Knowledge graph embedding with entity attributes using hypergraph neural networks. Intell. Data Anal. **26**(4), 959–975 (2022)
23. Bordes, A., Glorot, X., Weston, J., Bengio, Y.: A semantic matching energy function for learning with multi-relational data. Mach. Learn. **94**(2), 233–259 (2013). https://doi.org/10.1007/s10994-013-5363-6
24. Antoine, B., Nicolas, U., Alberto, G.D., et al.: Irreflexive and hierarchical relations as translations. arXiv preprint arXiv:1304.7158 (2013)
25. Yang, B.S., Yih, W.T., He, X.D., et al.: Embedding entities and relations for learning and inference in knowledge bases. In: International Conference on Learning Representations (ICLR) (2015)

HEM: An Improved Parametric Link Prediction Algorithm Based on Hybrid Network Evolution Mechanism

Dejing Ke[1] and Jiansu Pu[2(✉)]

[1] University of Electronic Science and Technology of China, Sichuan 611731, China
[2] Big Data Visual Analysis Lab, University of Electronic Science
and Technology of China, Sichuan 610000, China
`jiansu.pu@uestc.edu.cn`

Abstract. Link prediction plays an important role in the research of complex networks. Its task is to predict missing links or possible new links in the future via existing information in the network. In recent years, many powerful link prediction algorithms have emerged, which have good results in prediction accuracy and interpretability. However, the existing research still cannot clearly point out the relationship between the characteristics of the network and the mechanism of link generation, and the predictability of complex networks with different features remains to be further analyzed. In view of this, this article proposes the corresponding link prediction indices Reg, DFPA and LW on regular network, scale-free network and small-world network respectively, and studies their prediction properties on these three network models. At the same time, we propose a parametric hybrid index HEM and compare the prediction accuracy of HEM and many similarity-based indices on real-world networks. The experimental results show that HEM performs better than other indices. In addition, we study the factors that play a major role in the prediction of HEM and analyze their relationship with the characteristics of real-world networks. The results show that the predictive properties of factors are closely related to the features of networks.

Keywords: Link Prediction · Complex Networks · Network Evolution · Data Mining

1 Introduction

The network represents the relationship between entities in the form of connections, which is an effective and popular abstraction of the complex real world. Network science has been involved in biological, social, communication and economic fields and achieved fruitful achievement [1,2]. In network science, network evolution and link prediction are two most challenging and attractive directions.

Network evolution mechanism is one of the most important aspect of the research of complex networks. It aims to understand the root causes of changes in network structure and function. Currently there have been a lot of models to study network evolution mechanism. Such as ER, WS, BA and so on [3–6]. And

X. Yang et al. (Eds.): ADMA 2023, LNAI 14177, pp. 91–106, 2023.
https://doi.org/10.1007/978-3-031-46664-9_7

link prediction is an attracted and challenge task in complex network. Link prediction aims to predict missing links and new links in the network through existing structural information in the network. Link prediction can helps us to understand and infer the connection mechanism of complex networks. And Link prediction has been applied into all kinds of fields. There are a lot of efficient research of link prediction algorithms at present. No matter how a link prediction algorithm is expressed, it is essentially a guess of network evolution mechanism. A good link prediction algorithm can more accurately reveal the evolution behavior of a network [7].

The research of link prediction and complex networks is developing rapidly, but it also faces many challenges. Firstly, the existing similarity algorithms often perform well in the face of a few networks, but they are no longer effective when dealing with a wider range of real-world networks, including directed networks, weighted networks, heterogeneous edge networks and other complex situations [8–10]. Secondly, there is a strong correlation between the link prediction algorithm and the network structure characteristics and the link predictability of the network in theory [11,12]. However, how to describe and express the relationship between them is a challenging task. In addition, through link prediction, the evolution characteristics of the network can be reproduced to a certain extent, and the research on the evolution behavior of complex networks can be promoted, but the research on this aspect is still relatively lacking; on the other hand, link prediction needs to face large-scale real data at the application level, and our algorithm needs stronger adaptability and more efficient calculation [13].

Therefore, starting from these challenges, this paper attempts to study through the following aspects. Firstly, this paper studies the characteristics of regular networks, scale-free networks and small-world networks. According to these characteristics, we propose the corresponding link prediction indices Reg, DFPA and LW. Through these indices, we aim to verify: link prediction indices are often related to the characteristics of the network when predicting; a single index often cannot cope with many networks, and indices that fit a certain network characteristics will always be better for the network. After that, we propose a parametric hybrid index HEM. We hope that through this hybrid index, we can get a better generalization performance index that integrates the characteristics of different networks. This index has better adaptability and more accurate prediction effect on complex real-world networks.

In this article we first introduce some basic network evolution models, then introduce the evaluation metrics of link prediction and some representative similarity-based algorithms. Finally we introduce our proposed indices based on network evolution mechanism.

2   Related Work

At present, link prediction has been applied widely in recommendation systems [14,15], mining biological information [16,17], reconstructing network information [18,19], and evaluating network evolution models [20,21]. Current link prediction methods mainly include methods based on structural similarity, network embedding, matrix completion, ensemble learning and neural network methods, etc. [22–24].

Among all the link prediction algorithms, the similarity-based algorithms are favored in many fields because of its simplicity and good interpretability. The similarity-based algorithms compute the similarity of each pair of nodes. Then similarity is used for prediction. The similarity-based algorithms include local similarity-based and global similarity-based indices. The local ones often take "common neighbor" into mainly account, such as CN, Satlon, Jaccard, Sorensen, HPI, HDI, LHN1, etc. [25]. The global ones always take higher-order paths into consideration, like LP, Katz and LHN2 and LO [26–28]. And some indices predict links by randomly walking, like LRW and SRW [23,29]. And Some takes other global information [25]. The more information is considered, a better the performance there will be, but it also brings higher computational cost.

All the link prediction algorithms calculate the connection probability between nodes in the network and express the network connection mechanism to some extent. Through the study of network evolution mechanism, if we can deeply grasp the relationship between nodes in network evolution and deeply understand the basis of connections in the network, we are more likely to propose an excellent link prediction algorithm. Based on this idea, we proposes the link prediction algorithm via the evolution characteristics of the network.

So we firstly construct regular networks, scale-free networks and small-world networks and proposes our algorithms accordingly. We then perform link prediction on these networks to analyze the feature of indices.

Secondly, we propose an combined algorithm. The index sets two parameters for the prediction factors. We sample the parameters and perform predictions on some real-world networks. The results show that our index performs better than many classical similarity-based indices. We hope that through the combination of simple characteristic indices, we can conduct a more efficient and interpretable index.

Finally, we analyze the dominant factors of the hybrid index. Experiments show that the accuracy and the upper limit are determined by the main factors. In addition, we find that the main factors are always related to the characteristics of the network, which coincides with the prediction properties of individual index.

When performing predictions, we often pay attention to the best results, and parameter sampling should also be oriented to the upper limit of the index. Finding the main factor can help to optimize the sampling problem.

3   Network Model and Link Prediction

In this section, we will briefly introduce some network evolution models, link prediction evaluation metrics and similarity-based indices.

3.1   Network Evolution Model

The study of complex networks plays an increasingly important role in mathematics, statistical physics, computer science and other fields [30]. In order to study specific feature of networks, this article will focus on regular network, small world network and scale-free network. We choose them because they have the most common and basic characteristics of complex networks. And we hope to simulate the feature of complex network by their simple features.

(1) **Regular Network**. In the regular network each node has the same number of neighbors. Many crystal networks or protein networks in the field of chemistry can be regarded as regular networks.

(2) **Scale-Free Network**. Networks with power-law degree distribution are called scale-free networks [31]. The scale-free network always can be generate by preferential attachment, that is, new nodes tend to be connected to nodes with high degree.

(3) **Small-World Network**. The small-world network depicts the phenomenon of large clustering coefficient and small average short path length in the real world network. Social networks, protein networks, food chain networks, cultural networks and so on have been proved to have the characteristics of small-world networks. In small-world network the nodes tend to connect with their close neighbors.

3.2   Link Prediction Evaluation Metrics

Reference [23] proposed two methods to evaluate the accuracy of link prediction algorithms, namely AUC (area under the receiver operating characteristic curve) and Precision. The briefly review of them are below.

AUC. The AUC metric evaluates the accuracy of the algorithms by comparing the score of missing links and the nonexistent links. Suppose there are n independent comparisons in total. Among these comparisons, there are n1 times the missing link having a greater score and n2 times missing link and nonexistent link have the same score. Then the AUC value can be calculated as:

$$AUC = \frac{n_1 + 0.5n_2}{n} \tag{1}$$

When AUC is equal to 0.5, the prediction accuracy of the algorithm is equivalent to random prediction. The closer the AUC value is to 1, the better the prediction accuracy of the algorithm is.

Precision. The Precision metric sorts the scores of missing links and nonexistent links in descending order. We take the sorted top-L links as the predicted ones. Among these L links, N links belong to the test set. Then the Precision can be calculated as:

$$Precision = \frac{N}{L} \tag{2}$$

Compared with AUC, the Precision only focuses on whether the top L links are predicted accurately.

3.3   Link Prediction Similarity-Based Algorithms

The similarity-based algorithms for link prediction compute a similarity score S_{xy} for each pair of nodes x and y, which directly represent the link possibility between x and y. The algorithms can be classified into two categories: local similarity indices and global similarity indices. Here we choose some representative indices to introduce (These indices are similar to the indices proposed in this paper in terms of expression. So they are chosen to better analyze and explain the differences. We ignore some indices that are not comparable). The details are as follows.

3.4   Local Similarity Indices

(1) Common Neighbor (CN) [25]

$$S_{xy}^{CN} = |\Gamma(x) \cap \Gamma(y)| \tag{3}$$

$\Gamma(x)$ denotes the set of neighbors of the node x. In the CN index, the more common neighbors two nodes have, the more likely they are to connect.

(2) Salton Index [25]

$$S_{xy}^{Salton} = \frac{|\Gamma(x) \cap \Gamma(y)|}{\sqrt{k_x \times k_y}} \tag{4}$$

k_x and k_y denote the degree of nodes x and y, respectively.

(3) Resource Allocation Index (RA) [25]

$$S_{xy}^{RA} = \sum_{z \in \Gamma(x) \cap \Gamma(y)} \frac{1}{k_z} \tag{5}$$

The RA index defines the amount of resources x allocates to y.

(4) Cannistraci-Hebb index (CH) [32]

$$S_{xy}^{CH} = \sum_{z \in \Gamma(x) \cap \Gamma(y)} \frac{1 + k_z^i}{1 + k_z^e} \tag{6}$$

where k_z^i denotes the number of links of z with other common neighbors of x and y, and k_z^e denotes the number of links between z and nodes other than x and y or their common neighbors.

(5) Local Path Index (LP) [25]

$$S^{LP(n)} = A^2 + \epsilon A^3 + \epsilon^2 A^4 + \cdots + \epsilon^{n-2} A^n \tag{7}$$

where ϵ is a free parameter and n is the maximum order.

3.5   Global Similarity Indices

(1) Katz Index [26]

$$S_{xy}^{Katz} = (I - \beta A)^{-1} - I = \beta A_{xy} + \beta^2 A_{xy}^2 + \beta^3 A_{xy}^3 + \cdots \qquad (8)$$

β is the free parameter. I is the identity matrix. The contribution of higher order path can be controlled by adjusting β. This index considers all path sets. It calculates all the paths and assigns less weight to long paths in an exponential decay.

(2) Linear Optimization index (LO) [28]

$$S^{LO} = \alpha A(\alpha A^T A + I)^{-1} A^T A = \alpha A^3 - \alpha^2 A^5 + \alpha^3 A^7 - \alpha^4 A^9 + \cdots \qquad (9)$$

α is a free parameter. I is identity matrix and A is adjacency matrix. When α is small enough, LO degenerates to the index that calculates only the 3-hop paths A^3.

4   Link Prediction Based on Network Evolution Mechanism

According to the characteristics of regular networks, scale-free networks and small-world networks, this article proposes link prediction indices for these three networks, and proposes a hybrid indices for complex networks based on the three indices. Note that all the link prediction results in this article are obtained by using the 10-fold cross-validation method on test networks.

4.1   Index Based on Regular Networks

According to the characteristics of regular networks, this article proposes a link prediction index called Reg. Reg is expressed as follows:

$$S_{xy}^{Reg} = \frac{1}{\sqrt{k_x \times k_y}} \qquad (10)$$

k_x and k_y represent the degree of nodes x and y, respectively. In the formula, the nodes with larger degree are less likely to be connected. Small nodes are more likely to generate connections. By suppressing the connection probability of large degree nodes and promoting the connection probability of small degree nodes, the degree balance is achieved to a certain extent.

In order to study the performance of Reg index, we compared the link prediction accuracies of Reg index, CN index and Salton index on random regular network (see results in Table 1).

We can see that the Reg index is significantly better than other indices. Due to the randomness of the regular network, the CN index has an AUC value of only 0.5, while the Satlon index shows random results even with the same computational factor (i.e., $\frac{1}{\sqrt{k_x \times k_y}}$) as Reg index. As the degree of each node increases, the prediction performance of Reg index will gradually decrease.

Table 1. Accuracies on regular networks

Network	Reg_3	Reg_8	Reg_13	Reg_18	Reg_23	Reg_28	Reg_33
Cn	0.500	0.497	0.493	0.493	0.494	0.492	0.489
Salton	0.500	0.498	0.493	0.496	0.500	0.503	0.506
Reg	**0.942**	**0.839**	**0.784**	**0.752**	**0.729**	**0.712**	**0.698**

Accuracies are measured by the AUC value. The number of nodes of the network are all 2000. The results are calculated on random regular network whose each node has 3, 8, 13, 18, 23, 28 and 33 neighbors, respectively. And these 7 regular networks are denoted as Reg_3, Reg_8, Reg_13, Reg_18, Reg_23, Reg_28 and Reg_33, respectively.

4.2 Index Based on Scale-Free Networks

In reference to the article [25], a link prediction index PA corresponding to the preferential attachment principle is proposed. The expression of PA is as follows.

$$S_{xy}^{PA} = k_x \times k_y \qquad (11)$$

This article also proposes a link prediction algorithm called DFPA (Difference Preferential Attachment) for scale-free networks. The expression is as follows.

$$S_{xy}^{DFPA} = \frac{max(k_x, k_y)}{min(k_x, k_y)} \qquad (12)$$

Compared with PA index, DFPA index pays more attention to the connection between nodes with large degree and nodes with small degree. Nodes with similar degree are more stable and less likely to connect with each other. Therefore, small degree nodes and large degree nodes develop faster according to DFPA index. Besides, the connection probability between nodes with large degree is smaller than PA.

We compare the link prediction accuracies of PA and DFPA on scale-free networks constructed by BA model. The results are shown in Fig. 1. Note that accuracies are measured by the AUC value. The number of nodes of the networks are all 2000. Based on the BA model, each time the new nodes generate 1, 2, 4, 8, 16, 32 and 64 links, respectively. Thus there are 7 kinds of scale-free networks.

According to the prediction results of PA and DFPA in these scale-free networks, DFPA performs better when the network is sparse. As the degree of each node increase, the performance of PA gradually becomes better, while that of DFPA shows a downward trend. However, DFPA has a higher upper limit than PA in prediction.

There is a definition of degree assortativity in article [33], when it is greater than 0, nodes with similar degrees tend to connect with each other. When it is less than 0, nodes with different degrees are more likely to connect with each other. DFPA considers the latter case. In theory, the DFPA index also predicts accurately on disassortative networks.

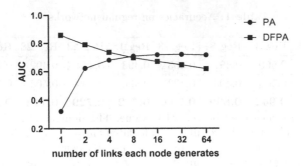

Fig. 1. Accuracies of PA and DFPA on scale-free networks

4.3   Index Based on Small-World Networks

In small-world network, each node is connected to the nearest k nodes. Based on that, this article proposes the LW (local world) index. The LW index considers that when two nodes have paths of length less than k or $k + 1$, the two nodes are possible to have connection. The expression of LW index is as follows.

$$S^{LW} = A^k + A^{k+1} \tag{13}$$

k is the free parameter. A is the adjacency matrix of the network. A^k calculates the number of paths with length k between each pair of nodes. The paths calculated by A^k may go back and forth on some edges. So in order to consider both odd-order paths and even-order paths, LW calculates the sum of A^k and A^{k+1}.

k in LW represents the breadth and scope of information, which is similar to n in LP index. Compared with LP and Katz index, LW index does not consider that the lower order path has a higher weight. The weight of the path is related to the size of k and network structure. And the LW index has a small computational complexity.

To facilitate the comparison of LP and LW indices, we define the LPK index as:

$$S^{LPK} = A^2 + A^3 + \cdots + A^k + A^{k+1} \tag{14}$$

LPK is the case where the ϵ parameter of LP is set to 1 and the order n of LP is set to $k + 1$.

For instance, we define LP2, LP4 and LP8 as the cases where the k value of LPK takes 2, 4 and 8 respectively. Similarly, define LW2, LW4, and LW8 as the cases where the k value of the LW index takes 2, 4 and 8, respectively.

We see that LPK and LW are basically equal. It is because $A^k + A^{k+1}$ are almost cover the information of A^i when i less than k.

4.4   Hybrid Index Based on Complex Network

Among the above three indices, Reg and DFPA are indices based on degree distribution, and LW is the index based on network topology. According to the three link prediction indices proposed by different network models, this article proposes a hybrid index called HEM (Hybrid Evolution Mechanism). The expression of HEM is as follows.

$$S_{xy}^{HEM} = S_{xy}^{Reg\,\alpha} \times S_{xy}^{DFPA^{1-\alpha}} \times S_{xy}^{LW} \tag{15}$$

According to equation (10), (12) and (13), the above formula can be expanded as:

$$S_{xy}^{HEM} = \frac{1}{\sqrt{k_x \times k_y}}^{\alpha} \times \frac{max(k_x, k_y)}{min(k_x, k_y)}^{1-\alpha} \times (A^k + A^{k+1})_{xy} \tag{16}$$

There are two free parameters α and k in the HEM index. The α parameter is used to balance the degree distribution. The role of the k parameter is the same as in LW, representing the range of paths included.

By adjusting the α parameter, we can achieve the optimal balance of the HEM index in the link prediction on the mixed networks of regular networks and disassortative networks. When α is close to 1, the HEM index tends to predict on regular networks; when α is close to 0, the HEM index tends to predict on disassortative networks. The k parameter represents the path range considered in the prediction of LW index. If the k value is set too small, some high-order paths may not be taken into account for prediction. If it is too large, the paths that should not be considered will be involved. Therefore, the α and k parameters need to be adjusted simultaneously during the experiment.

In order to test the link prediction accuracy of the HEM index, this article selects the following network data sets (see in Table 2). The multiple edges are regarded as one single edge, and the directed edge is regarded as an undirected edge. The self-connections are not taken into account. In addition, we only consider the giant component when one network is not well connected.

Table 2. The features of 11 real-world networks

Network	N	M	K	Δ	D	C	ρ
PPI	2375	11693	9.85	118	15	0.306	0.454
NS	1461	2742	3.75	34	17	0.694	0.462
Grid	4941	6594	2.67	19	46	0.08	0.003
INT	5022	6258	2.49	106	15	0.012	−0.138
PB	1222	16714	27.36	351	8	0.32	−0.221
Yeast	2361	6646	5.63	64	11	0.13	−0.099
FBC	4039	88234	43.69	1045	8	0.606	0.064
HSS	1858	12534	13.49	272	14	0.141	−0.085
GrQc	5242	14484	5.53	81	17	0.53	0.659
AS	6474	12572	3.88	1458	9	0.252	−0.182
ER	1174	1417	2.41	10	62	0.017	0.127

Where N and M denote the number of nodes and edges of the network, respectively; K denotes the average degree; Δ denotes the maximum degree; D denotes the network diameter; C denotes the clustering coefficient; ρ denotes the degree assortativity. PPI is a protein-protein interaction network [34]. NS is a network of co-authorships in the area of network science [35]. Grid contains information about the power grid of the Western States of the United States of America [4]. INT represents the router-level topology of the Internet [36]. PB is a network of hyperlinks between political blogs about politics in the United States of America [37]. Yeast is a protein-protein interaction network in budding yeast [38]. FB consists of "friends lists" from Facebook, whose data was collected from survey participants using this Facebook app [39]. HSS represents the network of friendships between users of the website hamsterster.com [40]. GrQc is the collaboration network from the e-print arXiv and covers scientific collaborations between authors papers submitted to General Relativity and Quantum Cosmology category [40]. AS is the network of autonomous systems of the Internet connected with each other [40]. ER is the international E-road network, a road network located mostly in Europe [40].

There are many similarity indices in link prediction. This paper only selects some indexes that are similar to the indexes proposed in this paper in terms of expression. On the one hand, it is better to control variables and understand the factors that cause the difference in accuracy between indexes. On the other hand, some indices are quite different from the indicators in this paper in terms of predictive properties and computational performance, so that the predictive differences of the indicators cannot be accurately grasped, and the interpretability is also poor.

So this article compares the prediction accuracies of the HEM index and other similarity-based indices like CN, Salton, PA, RA, CH, LPK, Katz and LO on these networks. In these 11 networks, we calculate the AUC value and Precision value of these link prediction algorithms (see results in Table 3 and Table 4). Where The L value of Precision is 100. The parameter values in both Katz and LO indices are set to 0.01. The values of k parameter in LPK are selected as 2, 4 and 8, respectively. In the HEM index, we simultaneously sampled the α parameter and the k parameter. The values of α are selected as 0.0, 0, 25, 0.5, 0.75 and 1.0, respectively; and the values of k are selected as 2, 4 and 8, respectively. Among the 15 results obtained by combining the two parameters, we take the best result of the HEM index and record the α and k parameters when the AUC value is maximized.

Table 3. Algorithms' accuracy quantified by AUC

Network	PPI	Grid	INT	PB	Yeast	FB	HSS	GrQc	AS	NS	ER
Cn	0.893	0.589	0.559	0.919	0.706	0.992	0.805	0.922	0.696	0.943	0.526
Salton	0.892	0.588	0.559	0.875	0.705	0.992	0.789	0.922	0.676	0.944	0.526
PA	0.823	0.442	0.472	0.902	0.788	0.831	0.866	0.740	0.738	0.631	0.338
RA	0.894	0.589	0.559	0.923	0.706	**0.995**	0.809	0.923	0.700	0.944	0.526
CH	0.866	0.698	0.569	0.856	0.522	0.992	0.589	**0.938**	0.606	**0.988**	**0.713**
LP2	**0.939**	0.638	**0.633**	**0.932**	**0.839**	0.984	0.936	0.930	0.762	0.946	0.555
LP4	0.906	0.708	0.572	0.915	0.818	0.962	0.878	0.921	0.660	0.943	0.627
LP8	0.825	**0.772**	0.378	0.897	0.770	0.911	0.830	0.846	0.623	0.934	0.692
Katz	0.920	0.660	0.378	0.925	0.821	0.611	0.915	0.914	0.690	0.945	0.629
LO	0.935	0.560	0.623	0.929	0.813	0.986	**0.952**	0.846	**0.787**	0.852	0.486
α	0.50	0.75	0.50	0.75	0.00	1.00	0.75	1.00	0.00	1.00	1.00
k	2	8	2	2	2	2	2	4	2	4	8
HEM	**0.958**	**0.902**	**0.922**	**0.936**	**0.869**	**0.989**	**0.953**	**0.961**	**0.944**	**0.987**	**0.858**

Table 4. Algorithms' accuracy quantified by Precision

Network	PPI	Grid	INT	PB	Yeast	FB	HSS	GrQc	AS	NS	ER
Cn	0.474	0.000	0.008	0.078	0.003	0.040	0.003	0.354	0.059	0.200	0.000
Salton	0.000	0.000	0.000	0.000	0.000	0.001	0.000	0.011	0.000	0.046	0.000
PA	0.409	0.000	0.014	0.082	0.009	0.033	0.089	0.222	0.131	0.005	0.000
RA	0.002	0.000	0.000	0.028	0.001	0.041	0.000	0.000	0.016	0.004	0.000
CH	0.267	0.005	0.000	0.010	0.008	0.006	0.000	0.140	0.026	0.229	0.000
LP2	0.548	0.037	0.280	0.412	0.144	0.661	0.297	0.629	**0.253**	0.252	0.000
LP4	0.531	**0.046**	0.243	0.391	0.117	0.689	0.186	0.641	0.227	**0.253**	0.000
LP8	0.523	0.035	0.218	0.349	0.099	**0.694**	0.161	**0.644**	0.213	0.251	**0.001**
Katz	0.533	0.001	0.009	0.261	0.003	0.612	0.015	0.522	0.099	0.201	0.000
LO	**0.603**	**0.046**	**0.379**	0.414	**0.198**	0.037	**0.964**	0.301	0.185	0.230	**0.001**
α	0.50	1.00	1.00	0.75	0.75	0.00	1.00	0.50	0.00	0.75	0.00
k	2	4	2	2	2	2	2	8	4	4	4
HEM	**0.978**	0.051	0.159	**0.524**	0.178	**0.993**	0.731	**0.759**	0.081	0.273	0.002

According to the results of AUC, HEM performs much better than other indices in Grid, INT, AS and ER networks. In PPI, PB, Yeast, HSS, GrQc and NS networks, the prediction accuracies of HEM is also higher than other indices. For FB network, HEM and many other indices perform very well, the prediction accuracies are basically reaching 100%.

According to the results of Precision, the performance of HEM index on PPI, FB, HSS networks is much better than other indices, especially on PPI and FB networks, the Precision values of the HEM index are almost 1. HEM also has a better improvement on PB and GrQC networks compared to the classic indices. In contrast, in the AUC results, the HEM index outperforms in Grid, INT, and AS networks, but underperform in Precision compared to other indices, which indicates that most of correct predictions from the HEM index for these networks come from the second half of the lists of links.

Also in the tables we see that the parameters of HEM index differ when taking the maximum AUC and Precision values. Therefore, we need to study the role of parameters in the HEM index and their relationship with network characteristics.

5   Analysis of HEM Index

In order to understand the influence of different parameters, study which factor, including Reg, DFPA and LW, plays a major role in the prediction. Here we propose two methods.

(1) Calculate the prediction accuracies of different factors separately, and choose two factors with the highest accuracy.
(2) Sample α and k, then choose the top 5 combinations of α and k parameters from where the HEM index has the highest prediction accuracy. Where α takes the average value, and k takes the mode. If α is equal to 0.5, we only consider the k. Or when α is close to 0, take the factor DFPA; when it close to 1, take Reg.

The first method discusses the performance of individual factors, and the second method calculates the parameters that have a greater impact on the prediction. In practical considerations, The second method is used as the main reference, and the results obtained by the first method can make us have a better understanding of the characteristics of the network.

Here we discuss the situation when the prediction accuracy measured by the AUC value. The results of two methods may be different when it measured by the Precision value, but it has the same way. In this article we consider 5 factors, they are Reg, DFBA, LW2, LW4 and LW8.

We compare the main factors of the 11 networks obtained by the two methods, results are shown in Table 5.

Table 5. The main factors of 11 networks obtained by the method 1 and method 2

Method	PPI	Grid	INT	PB	Yeast	FB	HSS	GrQc	AS	NS	ER
1	LW2	LW8	DFPA, LW2	LW2	LW2	LW2	LW2	LW2	DFPA, LW2	LW2	REG, LW8
2	LW2	LW8	DFPA, LW2	LW2	DFPA, NW2	LW2	LW2	REG, LW4	DFPA, LW2	REG, LW4	REG, LW8

It can be seen that the results obtained by the two methods are basically the same except for the three networks of Yeast, GrQc and NS. In Yeast, the main factors calculated by method 2 has DFPA. While in method 1, DFPA in yeast performs better than Reg. In GrQc and NS networks, the main factors obtained by method 2 has Reg, while according to method 1, Reg factor performs worse than DFPA factor. Therefore, the influencing factors cannot be simply determined by the individual prediction accuracy.

Observe the several networks with high clustering coefficient: NS, FB, PB and GrQc, they have LW2 as their main factors based on the first method. LW2 performs very well on these networks, especially on FB. The FB network is the dense network with high clustering coefficient, and the prediction accuracies of LW indices basically reaches 1. So we guess that the LW index may be related to the clustering coefficient of the network. Besides, we can also observe that the density of the network also has a certain influence on the prediction of LW. For example, although the NS network has the highest clustering coefficient, the average degree of the network is only 3.75, far sparser than the FB network, and the LW2 and LW4 indices perform less well than on the FB network. Moreover, the main factors in the NS network obtained by the second method are Reg and LW4, indicating that due to the sparsity, a wider k in LW and additional consideration of regularity are needed to have a better prediction performance on the NS network. In addition, although the clustering coefficient of HSS network is low, the network is denser, then the performance of the LW index on the network is as good as that on the PPI and PB networks, whose clustering coefficient are much larger.

Both Grid and ER networks are sparse, and the diameter of the two networks is very large compared to other networks. Therefore, LW index needs to consider wider paths to predict the links. The main factors obtained in method 1 and method 2 are both LW8. The degree assortativity of the INT and AS networks is observed to be negative, indicating that the networks have the tendency of differential connection. Thus in these two networks, DFPA as their main factor performs the best among all the factors.

Moreover, the maximum degree of network AS is 1485, indicating that the degree distribution is very unbalanced, and the preferential attachment is more obvious. So the prediction performance of DFPA factor alone on AS network is also better. The maximum degree of GrQc, ER and NS networks is relatively small, indicating that the degree distribution of the network is relatively balanced. So on these 3 networks, the corresponding results obtained in the second method, Reg are their main factors. Though the Yeast network also has a small maximum degree, the degree assortativity is negative, indicating that connections on the network are still difference preferential. Correspondingly in the second method, DFPA is the main factor on Yeast network.

In summary, the Reg factor often acts on networks with relatively balanced degree distribution, that is, when the maximum degree is relatively small, we can take the Reg index into account to predict links. The DFPA index is usually more effective on networks with negative degree assortativity. The prediction performance of LW index is determined by clustering coefficient, average degree and network diameter. When clustering coefficient is higher and the network is denser, the link prediction of LW index is always more accurate. The size of the k of LW index depends largely on the diameter and average distance of the network.

By arranging the above results, we compare the prediction results(measured by the AUC value) of individual factors and hybrid index by tabular statistic (see in Table 6).

Table 6. Results of individual factors and hybrid index

	PPI	Grid	INT	PB	Yeast	FB	HSS	GrQc	AS	NS	ER
best of Factors	0.939	0.772	0.849	0.932	0.839	0.984	0.936	0.930	0.929	0.946	0.693
best of HEM	**0.958**	**0.902**	**0.922**	**0.936**	**0.869**	**0.989**	**0.953**	**0.961**	**0.944**	**0.987**	**0.858**

So we can see that the main factor largely determines the upper limit of the prediction accuracy of the hybrid index.

In general, the hybrid index always has a better prediction performance than the single index. The prediction performance is mainly determined by the main factor, and other factors may have some influence to the prediction, which will help to improve the overall result.

If we can determine the factors that have a greater impact in the link prediction of different networks, then we can save the sampling on the parameters of the HEM index that have little impact and reduce the computational complexity. Depending on the upper limit of the main factors, we can also have some idea of the upper limit of the HEM index. Determining the main factors can also give us some insight into the characteristics of the network.

6　Conclusion and Future Work

The link prediction indices proposed in this article, are based on the idea of simulating evolution mechanism through simple rules.

Thus, we firstly proposes corresponding link prediction algorithms on regular networks, scale-free networks and small-world networks respectively and studies their prediction properties on these three network models. Then we propose a parametric hybrid index, which has higher prediction accuracy than many similarity-based indices on real-world complex networks. Finally we studies the main predictors in the hybrid index, and analyzes and summarizes their relationship with network features.

In the future work, we will further refine the link prediction algorithms according to the network evolution mechanism. Firstly, we need to consider more details of topology structure. After all, path information is not sufficient to define the existence of links. Secondly, we only considers the mixed degree distribution of the regular network and the disassortativitive network. Therefore, it is necessary to consider the degree distribution more exactly in future research.

Acknowledgment. This work was supported by the National Natural Science Foundation of China (Grant Nos. 62272088 and U19A2078).

References

1. Newman, M.: Networks. Oxford University Press, New York (2018)
2. Barabási, A.-L.: "Network Science." Network Science (2016)
3. Erdos, P.L., Rényi, A.: On the evolution of random graphs. Trans. Am. Math. Soc. **286**, 257–257 (1984)
4. Watts, D.J., Strogatz, S.H.: Collective dynamics of 'small-world' networks. Nature **393**(6684), 440–442 (1998)
5. Barabási, A.-L., Albert, R.: Emergence of scaling in random networks. Science **286**(5439), 509–512 (1999)
6. Clauset, A., Moore, C., Newman, M.E.J.: Hierarchical structure and the prediction of missing links in networks. Nature **453**(7191), 98–101 (2008)
7. Wang, L., Shang, C.: Research on link prediction problem in scale-free network. Comput. Eng. **38**(3), 67–70 (2012)
8. Lü, L., Zhou, T.: Link prediction in weighted networks: the role of weak ties. EPL (Europhys. Lett.) **89**(1), 18001 (2010)
9. Leskovec, J., Huttenlocher, D., Kleinberg, J.: Predicting positive and negative links in online social networks. In: Proceedings of the 19th International Conference on World Wide Web (2010)
10. Murata, T., Moriyasu, S.: Link prediction of social networks based on weighted proximity measures. In: IEEE/WIC/ACM International Conference on Web Intelligence (WI 2007). IEEE (2007)
11. Lü, L., et al.: Toward link predictability of complex networks. Proc. Natl. Acad. Sci. **112**(8), 2325–2330 (2015)
12. Tan, S.Y., et al.: Link predictability of complex network from spectrum perspective. Acta Physica Sinica Chinese Edition **69**(8), 088901 (2020)
13. Lin-Yuan, L.: Link prediction on complex networks. J. Univ. Electron. Sci. Technol. China (2010)
14. Lü, L., et al.: Recommender systems. Phys. Rep. **519**(1), 1–49 (2012)
15. Bagci, H., Karagoz, P.: Context-aware friend recommendation for location based social networks using random walk. In: Proceedings of the 25th International Conference Companion on World Wide Web (2016)
16. Fakhraei, S., et al.: Network-based drug-target interaction prediction with probabilistic soft logic. IEEE/ACM Trans. Comput. Biol. Bioinf. **11**(5), 775–787 (2014)
17. Sridhar, D., Fakhraei, S., Getoor, L.: A probabilistic approach for collective similarity-based drug-drug interaction prediction. Bioinformatics **32**(20), 3175–3182 (2016)
18. Squartini, T., et al.: Reconstruction methods for networks: the case of economic and financial systems. Phys. Rep. **757**, 1–47 (2018)
19. Peixoto, T.P.: Reconstructing networks with unknown and heterogeneous errors. Phys. Rev. X **8**(4), 041011 (2018)
20. Wang, W.-Q., Zhang, Q.-M., Zhou, T.: Evaluating network models: a likelihood analysis. EPL (Europhys. Lett.) **98**(2), 28004 (2012)
21. Zhang, Q.-M., et al.: Measuring multiple evolution mechanisms of complex networks. Sci. Rep. **5**(1), 1–11 (2015)
22. Zhou, T.: Progresses and challenges in link prediction. Iscience **24**(11), 103217 (2021)
23. Lü, L., Zhou, T.: Link prediction in complex networks: a survey. Phys. A **390**(6), 1150–1170 (2011)

24. Mutlu, E.C., Oghaz, T.A.: Review on graph feature learning and feature extraction techniques for link prediction. arXiv preprint arXiv:1901.03425 (2019)
25. Zhou, T., Lü, L., Zhang, Y.-C.: Predicting missing links via local information. Eur. Phys. J. B **71**(4), 623–630 (2009)
26. Lü, L., Jin, C.-H., Zhou, T.: Similarity index based on local paths for link prediction of complex networks. Phys. Rev. E **80**(4), 046122 (2009)
27. Leicht, E.A., Petter, H., Newman, M.E.J.: Vertex similarity in networks. Phys. Rev. E **73**(2), 026120 (2006)
28. Pech, R., et al.: Link prediction via linear optimization. ArXiv abs/1804.00124 (2018): n. pag
29. Liu, W., Lü, L.: Link prediction based on local random walk. EPL (Europhys. Lett.) **89**(5), 58007 (2010)
30. Hou, L., et al.: Recent progress in controllability of complex network. Wuli Xuebao/Acta Physica Sinica **64**(18), 0188901 (2015)
31. Caldarelli, G.: Scale-Free Networks: Complex Webs in Nature and Technology. Oxford University Press, Oxford (2007)
32. Muscoloni, A., Abdelhamid, I., Cannistraci, C.V.: Local-community network automata modelling based on length-three-paths for prediction of complex network structures in protein interactomes, food webs and more. BioRxiv, 346916 (2018)
33. Newman, M.E.J.: Assortative mixing in networks. Phys. Rev. Lett. **89**(20), 208701 (2002)
34. von Mering, C., et al.: Comparative assessment of large-scale data sets of protein-protein interactions. Nature **417**(6887), 399–403 (2002). https://doi.org/10.1038/nature750
35. Newman, M.E.J.: Finding community structure in networks using the eigenvectors of matrices. Phys. Rev. E Stat. Nonlinear Soft Mater. Phys. 74 3 Pt 2, 036104 (2006)
36. Mahajan, R., et al.: Inferring link weights using end-to-end measurements. In: International Memory Workshop (2002)
37. Adamic, L.A., Glance, N.S.: The political blogosphere and the 2004 U.S. election: divided they blog. In: LinkKDD 2005 (2005)
38. Jeong, H., et al.: Lethality and centrality in protein networks. Nature **411**(6833), 41–2 (2001). https://doi.org/10.1038/35075138
39. Mcauley, J.J., Leskovec, J.: Learning to discover social circles in ego networks. Neural Information Processing Systems Curran Associates Inc. (2012)
40. Leskovec, J., Kleinberg, J., Faloutsos, C.: Graph evolution: densification and shrinking diameters. ACM Trans. Knowl. Discov. Data **1**(1), 2 (2007)

Joint Embedding of Local Structures and Evolutionary Patterns for Temporal Link Prediction

Tingxuan Chen[1], Jun Long[2(✉)], Liu Yang[1(✉)], Guohui Li[1], Shuai Luo[1], and Meihong Xiao[1]

[1] School of Computer Science and Engineering, Central South University, Changsha, China
{chentingxuan,yangliu,guohuili,luoshuai06,xiaomeihong,junlong}@csu.edu.cn
[2] Big Data Institute, Central South University, Changsha, China

Abstract. Link prediction tackles the prediction of missing facts in an incomplete knowledge graph (KG) and has been widely explored in reasoning and information retrieval. The vast majority of existing methods perform link prediction on static KGs, with the assumption that the relational facts are generally correct. However, some facts may not be universally valid, as they tend to evolve. Despite the prevalence of temporal knowledge graphs (TKGs) with evolving facts, the studies on such data for temporal link prediction are still far from resolved. In this paper, we propose SiepNet, a novel graph neural network for temporal link prediction, driven by local **S**tructural **I**nformation and **E**volutionary **P**atterns. Specifically, SiepNet captures the local structural information based on a relation-aware GNN architecture, and incorporates temporal attention to model long- and short-range historical dependencies hidden in TKGs. Moreover, SiepNet integrates local structures and evolutionary patterns to enhance the semantic representation of evolving facts in TKGs. The extensive experiments on five real-world TKG datasets demonstrate the effectiveness of our approach SiepNet in temporal link prediction, compared with the state-of-the-art methods.

Keywords: Temporal knowledge graph · Graph embedding · Temporal link prediction · Representation learn · Evolutionary patterns

1 Introduction

Knowledge graphs (KGs) organize and store real-world facts, enabling multifarious downstream applications, such as knowledge retrieval, question answering, and recommender systems [12]. KGs encode factual knowledge in the form of triple (s, r, o) as directed graphs, where nodes correspond to the subject entity s or object entity o, and edges represent the relation r among them. Owing to the high cost of knowledge fusion and dynamics of facts, most KGs often suffer from incompleteness [31]. Thus, link prediction becomes a crucial task, which intends

X. Yang et al. (Eds.): ADMA 2023, LNAI 14177, pp. 107–121, 2023.
https://doi.org/10.1007/978-3-031-46664-9_8

to recover the most probable missing facts. Since real-world KGs contain millions of multi-relational facts, traditional symbolic and logic-based approaches cannot be extended to large-scale KGs for link prediction.

Recently, KG embedding has emerged as a promising method for link prediction. It attempts to learn multi-dimensional vectorial representations of entities and relations in KGs, while using a scoring function to evaluate the plausibility of a triplet. Represented by TransE [1], these translation-based approaches achieve a good trade-off between model complexity and link prediction performance by modelling relations as translation operations on entity embeddings. However, the vast majority of existing embedding methods perform link prediction on static KGs, with the assumption that the relational facts in KGs are generally correct.

Actually, facts always evolve over a specific period of time [3]. Therefore, researchers construct temporal knowledge graphs (TKGs) to store ever-growing temporal information either explicitly or implicitly, such as YAGO [24] and ICEWS [16]. Figure 1 shows an example of a temporal knowledge graph (TKG), where the fact (*Donald Trump, president of, USA*) was accurate only from 2017 to 2020. However, traditional KG embedding methods cannot address the issue of TKGs, where facts often show temporal dynamics. For example, they often confuse entities such as *Trump* and *Biden* when predicting (?, *president of, USA, 2021*). Additionally, TKG embeddings carrying temporal information are challenging due to the sparsity and irregularities of temporal expressions [5].

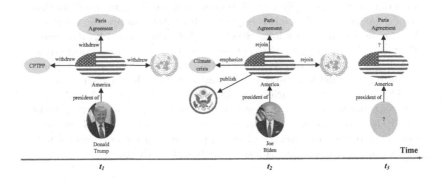

Fig. 1. Example of temporal knowledge subgraphs.

To solve the challenges, Know-Evolve [27] and its extension DyRep [28] predict future events based on ground truths of preceding events at inference time. As a result, these methods cannot predict missing events in future time-stamps without ground truths. To capture more information based on past facts, Jin proposed a novel autoregressive architecture RE-NET [14], which models facts as probability distributions over TKGs. However, RE-NET learns representations of entities and relations by implicitly exploiting temporal information without distinguishing dynamic dependencies across facts.

In this work, we observe that TKGs are dynamically heterogeneous graphs with multiple relationships, i.e., the local structures of graphs are always diverse under different time windows, and the facts evolve across time windows. As an example in Fig. 1, the local structure information of the entity *America* comes from 4 entities and 2 relations at t_1. While at t_2, the local structure of the entity *America* changes significantly, resulting in not only the emergence of new entities and relations but also the absence of some entities and relations at t_1. Moreover, the fact (*Donald Trump, president of, America*) at t_1 evolves into (*Joe Biden, president of, America*) at t_2.

To this end, we propose SiepNet, a novel graph neural network for temporal link prediction, driven by local **S**tructural **I**nformation and **E**volutionary **P**atterns. The main ideas of SiepNet are (1) capturing graph structure dependencies based on a relation-aware GNN architecture, (2) learning long-range and short-range evolutionary patterns of TKGs using an attention-based recurrent network, and (3) integrating local structures and evolutionary patterns to strengthen the representation learning of facts, which improves the performance of temporal link prediction. We summarize our main contributions as follows:

- We propose a representation learning model SiepNet for temporal link prediction, which simultaneously considers local structures and evolutionary patterns hidden in TKGs.
- We design an attention-based recurrent network to tackle dynamic dependencies across entities over time, which helps to distinguish the impact of different historical facts on future facts inference.
- To validate the effectiveness of our model, we conduct extensive experiments on five real-world TKGs containing millions of multi-relational facts with different time intervals, where our model consistently outperforms other baselines in terms of temporal link prediction.

2   Related Work

Towards temporal link prediction, we restrict our focus to recent works on TKG embedding methods, including geometric models and neural network models.

Geometric Models. These models attempt to minimize the distance between two entity vectors translated by geometric transformations of relations. TTransE [17] extends TransE [1] for static KGs to TKGs by adding temporal constraints. TA-TransE [5] embeds temporal information into relation types, which can be used with existing scoring functions for temporal link prediction in TKGs. HyTE [3] utilizes time-specific normal vectors directly to generate representations of entities and relations over different time-stamps. Nevertheless, these geometric models cannot infer future facts according to past facts and cannot be further extended to extrapolate settings.

Neural Network Models. These models use deep neural networks to learn underlying features of time-stamps for link prediction. RE-NET [14] combines a recurrent neural network and a neighborhood aggregator to model event sequences. CyGNet [34] predicts future facts by modelling observed facts with a copy-generation network. TITer [25] continuously transfers query nodes to new nodes through relevant temporal facts based on time-aware reinforcement learning strategies, and generates representation vectors of unseen entities using an IM module. CluSTeR [19] performs temporal reasoning on TKGs by joint reinforcement learning and a graph convolution network. RE-GCN [20] learns evolutionary representations of facts at each timestamp, by modelling KG sequences recurrently using a recurrent evolutionary network. However, the performance of these neural network models is limited by repetitive patterns.

3   Problem Definition

We consider a temporal knowledge graph as a sequence of graph snapshots, ordered ascending based on time-stamps, namely $G = \{G_1, G_2, \cdots, G_\tau\}$, where $G_t = (V_t, E_t)$ represents the snapshot at a particular time slice t ($t \in 1, 2, \cdots, \tau$) with an entity set V_t and a relation set E_t. V_t corresponds to the subject entity s or object entity o at a time slice t, and E_t represents the relation r between them. Thus, a fact in G_t is denoted by a quadruple (s, r, o, t) with a time slice t, in which $s \in V_t$, $o \in V_t$ and $r \in E_t$.

Given the preceding observed facts in G, the temporal link prediction aims to predict the missing facts of the current time slice t, i.e., to predict the unseen subject entity s given $(?, r, o, t)$ (object entity o given $(s, r, ?, t)$, and relation r given $(s, ?, o, t)$) at a particular time slice t.

4   Methodology

4.1   The Model Architecture

The proposed model SiepNet depicted in Fig. 2 consists of two main components: (1) Local Structural Information Aggregation, and (2) Evolutionary Patterns Aggregation. First of all, we design a relation-aware GNN to capture the local structural information from multi-relational and multi-hop neighbors of each single graph snapshot. Then, we explore long-range and short-range evolutionary patterns of TKGs using an attention-based recurrent network. In addition, we integrate local structures and evolutionary patterns to strengthen the representation learning of facts, which in turn improves the performance of temporal link prediction.

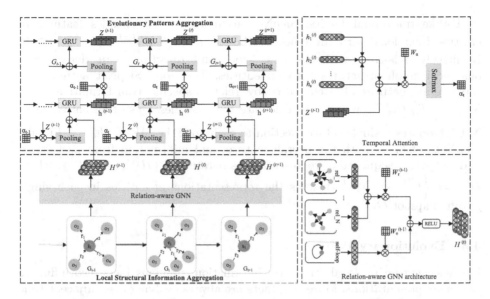

Fig. 2. The architecture of the SiepNet temporal link prediction model.

4.2   Local Structural Information

To aggregate local structural information from multi-relational and multi-hop neighbors in each graph snapshot G_t, SiepNet seeks to make two linked nodes share similar representations. To achieve this, we let each node representation $h_o^{(t)}$ in G_t aggregates neighbors and past messages, and then calculate its new representation. Initially, $h_o^{(0)}$ is set to trainable embedding vector for each node. SiepNet calculates the forward-pass update of an entity denoted by v_o in a multi-relational graph, based on the following message-passing neural network:

$$h_o^{(t)} = \sigma\Big(\sum_{s \in N_{o,r}^t} \mathcal{F}_{str}(h_s^{(t-1)}, r^{(t-1)}) + W_o^{(t-1)} h_o^{(t-1)} \Big) \tag{1}$$

where $h_o^{(t)}$ is the intermediate representation of node v_o at time slice t, combining local structural messages $h_s^{(t-1)}$ from all neighbors $N_{o,r}^t$ under relation $r \in E_t$ and its past messages $h_o^{(t-1)}$. $W_o^{(t-1)}$ is a learnable parameter, indicating the past weight. To comprehensively aggregate the local structural messages of node v_o, we implement the message function $\mathcal{F}_{str}(.,.)$ by

$$\mathcal{F}_{str}(h_s^{(t-1)}, r^{(t-1)}) = \frac{1}{c_{o,r}^t} W_r^{(t-1)}[h_s^{(t-1)} \times r^{(t-1)}] + b_{str} \tag{2}$$

where $h_s^{(t-1)} \times r^{(t-1)}$ is the local structural messages, while $W_r^{(t-1)}$ and b_{str} are the learnable parameters, indicating the local weight and bias. $c_{s,r}$ is a normalizing factor that can either be learned or chosen in advance (e.g., $c_{o,r}^t = |N_{o,r}^t|$).

Unlike traditional GCNs, SiepNet accumulates and encodes features of entities from local structural neighborhoods, i.e., $\frac{1}{c_{s,r}}W_r^{(t-1)}[h_s^{(t-1)} \times r^{(t-1)}]$. Intuitively, relations with different types and directions can derive various local graph structures between entities. Therefore, SiepNet accumulates the overall features of each entity by relation-specific transformations, i.e., $\sum_{s \in N_{o,r}^t} \mathcal{F}_{str}(h_s^{(t-1)}, r^{(t-1)})$. To calculate the past messages of an entity, Siep-Net introduces a single self-connection to each node, i.e., $W_o^{(t-1)}h_o^{(t-1)}$. Finally, SiepNet combines both the overall features and information from past steps, and outputs a sequence of representations notated as $\{H^{(1)}, \cdots, H^{(t)}\}$, where $H^{(t)} = \{h_1^{(t)}, \cdots, h_n^{(t)}\}$ denotes the representations of entities in each single graph snapshot G_t.

4.3   Evolutionary Patterns

Besides aggregating local structural information, previous facts also influence current representations. Moreover, facts are always evolving over adjacent time windows, further changing the local structural information of the current graph snapshot. Intuitively, we should capture these two evolutionary patterns, i.e., long-range historical dependence and short-range structural dependence. To achieve this, we design an attention-based recurrent block in SiepNet to capture evolutionary patterns in TKGs. Formally, SiepNet combines the local structural representation $h_o^{(t)}$ and the historical representation $(h_o^{(t-1)}, Z^{(t-1)})$:

$$h_o^{(t)}, Z^{(t)} := \mathcal{F}_{evo}(h_o^{(t)}, h_o^{(t-1)}, Z^{(t-1)}) \tag{3}$$

where $\mathcal{F}_{evo}$ is a recurrent operator, which allows SiepNet to learn long-range dependencies of sequence data and explore the evolving patterns of temporal knowledge graphs to update current representations. When there are few structural dependencies from neighbor nodes (i.e., $h_o^{(t)} \longrightarrow 0$), current representations $(Z^{(t)}, h_o^{(t)})$ will be greatly influenced by long- and short-range historical dependencies $(Z^{(t-1)}, h_o^{(t)})$. Otherwise, local structural dependences $h_o^{(t)}$ will have a greater impact on current representations.

Most existing works use simple recurrent neural networks to implement $\mathcal{F}_{evo}$ in message propagation, e.g., RE-NET [14] uses GRU [2], EvoNet [11] uses LSTM [10], etc. For historical snapshot propagation, these methods only summarize the current representations of nodes, i.e., $Z^{(t)} = \sum_{o \in V_t} h_o^{(t)}$, ignoring dynamic interactions of nodes across time windows. However, both long-range historical dependence and short-range dynamic dependence present different temporal information, influencing the evolution of facts. To improve the ability of temporal link prediction, $\mathcal{F}_{evo}$ should consider historically long-range and short-range dependence of previous facts $G_{1:t}$ when modelling snapshot propagation, and thus influence current representations through local dynamic dependence of node interactions. Specifically, $\mathcal{F}_{evo}$ can be implemented by

$$\mathcal{F}_{evo}(h_o^{(t)}, h_o^{(t-1)}, Z^{(t-1)}) = \begin{cases} Z^{(t)} = \text{RNN}\left(Z^{(t-1)}, G_t \oplus g(\alpha_t \sum_{o \in V_t} h_o^{(t)})\right) \\ \\ h_o^{(t)} = \text{RNN}\left((1 - \alpha_t)h_o^{(t-1)}, h_o^{(t)} \oplus g(\alpha_t Z^{(t-1)})\right) \end{cases}$$

$$(4)$$

where $\oplus$ denotes the concatenation operator and $g(*)$ is an element-wise max-pooling operator. We use a recurrent model RNN to update current representations $h_o^{(t)}$ based on historical representation $(h_o^{(t-1)}, Z^{(t-1)})$ and current local structural representation $h_o^{(t)}$, and capture evolutionary patterns $Z^{(t)}$ based on long-range and short-range dependencies $(Z^{(t-1)}, h_o^{(t)})$ as well as current facts G_t.

Typically, the impact of long-range historical dependence and short-range structural dependence on current representations varies over time. Accordingly, we design the following temporal attention mechanism as follows to capture temporal information in node interactions, which in turn helps to model the long-range and short-range evolutionary patterns of facts.

$$\alpha_t = \text{softmax}(W_\alpha(Z^{(t-1)} \oplus \sum_{s \in N_{o,r}^t} h_s^{(t)})) \tag{5}$$

where W_α is a independent parameter matrix, updated automatically by back-propagation. The attention score α_t re-weights the two evolutionary patterns, which is calculated based on long-range evolutionary dependencies and short-range structural dependencies.

The recurrent model RNN aims at smoothing two input vectors at each time step, which can be implemented using many existing methods. Here, we utilize GRU to update $h_o^{(t)}$ as an example.

$$h_o^{(t)} : \begin{cases} a^{(t)} = h_o^{(t)} \oplus g(\alpha_t Z^{(t-1)}) \\ i^{(t)} = \sigma(W_i a^{(t)} + U_z(1 - \alpha_t)h_o^{(t-1)}) \\ r^{(t)} = \sigma(W_r a^{(t)} + U_r(1 - \alpha_t)h_o^{(t-1)}) \\ h_o^{(t)} = (1 - i^{(t)}) \circ (1 - \alpha_t)h_o^{(t-1)} + i^{(t)} \circ \tanh(W_h a^{(t)} + U_h(r^{(t)} \circ h_o^{(t-1)})) \end{cases}$$

$$(6)$$

where $i^{(t)}$ and $r^{(t)}$ are update gate and reset gate respectively, while $\circ$ is a Hadamard operator. The current node representations are updated by receiving their currently local structure dependencies and historical evolution dependencies, with a temporal attention score regulating the weight of long-range and short-range dependencies.

Consequently, both the representations $h_o^{(t)}$ and $Z^{(t)}$ capture the evolutionary patterns and local structural dependencies up to the t-th time step, which in turn can be used to predict the facts G_{t+1} at the next time step. Then, we encode the current graph snapshot G_t as representation $\mathbf{H}_G^{(t)}$ with a fully connected layer, which can be formulated as

$$\mathbf{H}_G^{(t)} = \text{FCL}_n(Z^{(t)} \oplus \sum_{o \in V_t} h_o^{(t)}; \theta_n) \tag{7}$$

where the input are the concatenated features of all $h_o^{(t)}$ and $Z^{(t)}$, while θ_n denotes the parameters of FCL_n. Then we use a classifier to estimate the probability of the next graph snapshot $\mathbf{P}(G_{t+1} \mid \mathbf{H}_G^{(t)})$.

4.4 Model Optimization

As the topology of TKGs changes over time, SiepNet model should continuously update its parameters to accommodate the evolutionary patterns of TKGs. Furthermore, note that the snapshots closer to the next time slice $(t + 1)$ have more similar characteristics than those farther from the ground truth. Hence, we introduce the first l graph snapshots $G_{t-l+1}^{t+1} = \{G_{t-l+1}, G_{t-l+2}, \cdots, G_{t+1}\}$ as the input, which is close to the next time slice $(t + 1)$, based on minimizing the cross-entropy loss $\mathcal{L}$ for training.

$$\mathcal{L} = - \sum_{\tau=(t-l)}^{t} \hat{G}_{\tau+1} \log \mathbf{P}(G_{\tau+1} \mid \mathbf{H}_G^{(\tau)}) + (1 - \hat{G}_{\tau+1}) \log(1 - \mathbf{P}(G_{\tau+1} \mid \mathbf{H}_G^{(\tau)})) \quad (8)$$

where $\hat{G}_{\tau+1} \in \mathbb{R}^{|G_{\tau+1}|}$ is the label set of ground truths with elements of 1 if the fact occurs and 0 otherwise. SiepNet can fully aggregate the latest temporal information of the dynamic network, according to the sequence of previous snapshots G_{t-l+1}^{t+1}, which is considered as the most similar characteristics to the actual snapshots of G_{t+1}.

As in previous work on regularization, we employ dropout [9] to alleviate overfitting while capturing local structural information and evolutionary patterns.

5 Experiments

5.1 Experimental Setup

Datasets. In our experiments, we used five widely use TKG datasets, including three event-based TKGs (i.e., GDELT [18], ICEWS14 [27], and ICEWS18 [29]) and two public TKGs (i.e., WIKI [17] and YAGO [24]) specifically.

Evaluation Setting and Metrics. Following the prior work [34], we split each dataset except ICEWS14 into a training set, a validation set, and a test set at a ratio of 80%/10%/10%, respectively. For dataset ICEWS14, we directly utilize the splitting provided in [27]. We report a widely used filtered settings [8,14,34] of Mean Reciprocal Rank (MRR) and Hits at K (Hits@K), which are standard evaluation metrics for link prediction.

Baselines. We compare our proposed model SiepNet with a variety of static KG models and TKG models. Static KG models include DistMult [32], R-GCN [23], ConvE [4] and RotatE [26]. TKG models include TTransE [13], TA-DistMult [5], TA-TransE [5], HyTE [3], RE-NET [14], TeMP [30], RE-GCN [20], xERTE [6], TANGO-TuckER [7], TANGO-Distmult [7], CyGNet [34], EvoKG [22] and TLogic [21].

Model Configurations. Initially, we set the length of the history l to 10, which means that SiepNet saves the sequence of 10 previous snapshots. The dropout rate is set to 0.5, and the embedding size is set to 200 to match the baseline methods set in [34]. The model parameters are optimized using Adam optimizer [15] with a learning rate of 0.001. The training epoch is set to 20, which is sufficient for convergence in most cases. All experiments are conducted on GeForce GTX 3080 Ti. The baseline results are also adopted from [33].

5.2   Performance Evaluation

Overall Performance. Table 1 and Table 2 show the temporal link prediction performance of SiepNet and baselines on five real-world TKGs, where the best results are shown in **bold**. We use "–" instead of experimental results that are not run out within a day. Remarkably, SiepNet consistently outperforms the baselines in most cases, which convincingly validates its effectiveness.

Table 1. Performance (in percentage) for temporal link prediction on YAGO and WIKI datasets under the filtered settings

Method	YAGO			WIKI		
	MRR	Hits@1	Hits@3	MRR	Hits@1	Hits@3
DisMult [2015]	59.47	52.97	60.91	46.12	37.24	49.81
R-GCN [2018]	41.30	32.56	44.44	37.57	28.15	39.66
ConvE [2018]	62.32	56.19	63.97	47.57	38.76	50.10
RotatE [2018]	65.09	62.21	65.67	50.67	48.17	50.71
TTransE [2016]	32.57	27.94	43.39	31.74	22.57	36.25
TA-DisMult [2018]	61.72	50.57	65.32	48.09	45.97	49.51
TA-TransE [2018]	56.61	46.76	65.95	24.24	1.74	47.18
HyTE [2018]	23.16	10.78	45.74	43.02	28.81	45.74
RE-NET [2020]	65.16	63.29	65.63	51.97	48.01	52.07
TeMP [2020]	62.25	55.39	64.63	49.61	46.96	50.24
RE-GCN [2021]	65.29	59.98	68.70	44.86	39.82	46.75
xERTE [2021]	58.75	58.46	58.85	–	–	–
TANGO-TuckER [2021]	67.21	65.56	67.59	53.28	52.21	53.61
TANGO-Distmult [2021]	68.34	67.05	68.39	54.05	51.52	53.84
CyGNet [2021]	63.47	64.26	65.71	45.50	50.48	50.79
EvoKG [2022]	55.11	54.37	**81.38**	50.66	12.21	**63.84**
TLogic [2022]	1.29	0.49	0.85	51.07	50.13	51.18
SiepNet (ours)	**73.77**	**71.65**	74.65	**54.46**	**52.35**	62.73

Specifically, static KG methods usually show promising results, but lag behind the best-performing TKG method SiepNet to a large extent, as they

Table 2. Performance (in percentage) for temporal link prediction on ICEWS14, ICEWS18 and GDELT datasets under the filtered settings

Method	ICEWS14			ICEWS18			GDELT		
	MRR	Hits@1	Hits@3	MRR	Hits@1	Hits@3	MRR	Hits@1	Hits@3
DisMult	19.06	10.09	22.00	22.16	12.13	26.00	18.71	11.59	20.05
R-GCN	26.31	18.23	30.43	23.19	16.36	25.34	23.31	17.24	24.96
ConvE	40.73	33.20	43.92	36.67	28.51	39.80	35.99	27.05	39.32
RotatE	29.56	22.14	32.92	23.10	14.33	27.61	22.33	16.68	23.89
TTransE	6.35	1.23	5.80	8.36	1.94	8.71	5.52	0.47	5.01
TA-DistMult	20.78	13.43	22.80	28.53	20.30	34.57	29.35	22.11	34.56
TA-TransE	15.99	0.00	26.39	17.69	0.01	30.14	19.18	0.00	33.20
HyTE	11.48	5.64	13.04	7.31	3.10	7.50	6.37	0.00	6.72
RE-NET	45.71	38.42	49.06	42.93	36.19	45.47	40.12	32.43	43.40
TeMP	43.13	35.67	45.79	40.48	33.97	42.63	37.56	29.82	40.15
RE-GCN	32.37	24.43	35.05	32.78	24.99	35.54	29.46	21.74	32.01
xERTE	32.92	26.44	36.58	36.95	30.71	40.38	–	–	–
TANGO-TuckER	46.42	38.94	50.25	44.56	37.87	47.46	38.00	28.02	43.91
TANGO-Distmult	46.68	41.20	48.64	44.00	38.64	45.78	41.16	35.11	43.02
CyGNet	<u>48.63</u>	<u>41.77</u>	<u>52.50</u>	<u>46.69</u>	<u>40.58</u>	<u>49.82</u>	<u>50.29</u>	<u>44.53</u>	**54.69**
EvoKG	18.30	6.30	19.43	29.67	12.92	33.08	11.29	2.93	10.84
TLogic	38.19	32.23	41.05	37.52	30.09	40.87	22.73	17.65	24.66
SicpNet	**49.97**	**42.65**	**53.28**	**47.93**	**43.41**	**52.36**	**50.79**	**45.10**	53.11

cannot capture the sequential patterns across time-stamps. Surprisingly, almost static KG methods normally perform better than two TKG methods (i.e., TTransE and HyTE) on five TKG datasets. It owes to the fact that TTransE and HyTE learn representations for each snapshot independently, instead of capturing long-range historical dependencies. Besides, the experimental results of TA-DistMult and DistMult validate the effectiveness of incorporating temporal information for temporal link prediction, where TA-DistMult is a temporal-aware version of static KG method DistMult.

In addition, SiepNet drastically outperforms other TKG methods, although they all consider dynamic features of facts. Especially on YAGO dataset with the most facts, SiepNet leads to improvements of 2.70% in MRR, 6.97% in Hits@1, and 5.10% in Hits@3 compared with the best baseline. We believe this is due to that SiepNet considers dynamic long-range and short-range historical dependencies using temporal attention, while other TKG models ignore the evolutionary patterns. The excellent performance of SiepNet and RE-NET validate the importance of long-range dependencies for link prediction. Although our performance in Hits@3 of YAGO, WIKI, and GDELT dataset are not the best, the remarkable performances in Hits@1 and MRR prove that our algorithm SiepNet is able to predict future facts more accurately. The main reason is that there is a large number of repetitive facts in these datasets. Thus, CyGNet and EvoKG perform well on Hits@3, but they cannot predict more accurate facts, resulting in Hits@1 much lower than ours. TeMP is designed to handle knowledge graph complementation tasks (graph interpolation) rather than predicting future events, so it does not perform as well as extrapolation models. Although xERTE supports a certain degree of predictive interpretation capability, it cannot efficiently handle large-scale datasets, such as GDELT and WIKI.

Note that static KG model and TKG model perform similarly well on YAGO and WIKI, but poorly on ICEWS14, ICEWS18 and GDELT. As discussed in [22], the time intervals of YAGO and WIKI datasets are much larger than other datasets. Therefore, each time-stamp in YAGO and WIKI has more local structural information than the other three datasets. Besides, ICEWS14 and ICEWS18 are extracted from the Integrated Crisis Early Warning System (ICEWS), which records many recurring political events with time stamps. Accordingly, only modelling repetitive patterns or 1-hop neighbors will lose a significant amount of evolutionary patterns and structural information. The experimental results show that SiepNet is able to better model these datasets, which contain complex dynamic dependencies over concurrent facts.

Performance over Time. To further evaluate the performance of SiepNet over time, we compared the performance in percentage of different timestamps, using filtered Hits@3 on YAGO, WIKI, and ICEWS18. As shown in Fig. 3, SiepNet consistently outperforms baselines over different timestamps. The performance of each method varies with the entities in the test set at each timestamp. In addition, the difference between our TKG model SiepNet and static KG model ConvE evolves slowly as time goes by, as shown in Fig. 3. We believe that further facts in the future are even harder to predict.

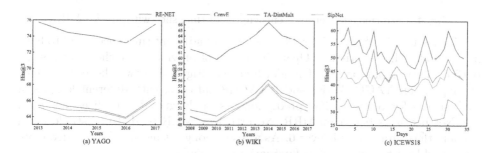

Fig. 3. Performance over specific timestamps with filtered Hits@3.

Specifically, each method shows a significant performance improvement at a particular timestamp in the future. We believe this is because facts from the past tend to reappear at the future timestamps. As shown in Fig. 3(a), all methods perform poorly in 2016, but in 2017 surpass their performance in 2013.

5.3   Ablation Study

To eliminate the effect of different model components of SiepNet, we create variants of SiepNet by adapting the use of model components and report the performances (in percentage) on YAGO dataset.

Table 3. Ablation study for temporal link prediction

Method	YAGO			
	MRR	Hits@1	Hits@3	Hits@10
SiepNet w. R	66.25	64.92	66.53	68.54
SiepNet w. B	73.53	71.66	74.66	76.92
SiepNet w/o TA	64.30	62.41	64.86	67.52
SiepNet	73.77	71.65	74.65	77.24

Evolutionary Patterns. To demonstrate how evolutionary patterns affect the final results of SiepNet, we conduct experiments using l random past graph snapshots rather than l snapshots closest to the current graph snapshot. The results denoted as SiepNet w. R are presented in Table 3. Obviously, SiepNet w. R hurts model quality, suggesting that modelling the snapshots closer to the current time slice can improve performance.

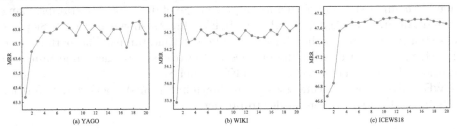

Fig. 4. Performance over different lengths of time slice with filtered MRR.

As described in Sect. 4.5, graph snapshots of adjacent time slices tend to have more similar characteristics. Thus, the length of previous time slice l affects the performance of our proposed model SiepNet. Figure 4 shows the performance of SiepNet on YAGO, WIKI and ICEWS18 datasets, with different lengths of time slices l for temporal link prediction. As the length of time slices increases, SiepNet performs better on MRR. Nevertheless, MRR tends to be stable when the length of time slices is over 6. As a result, longer time slices introduce more noise and lead to performance fluctuations of SiepNet.

Evolutionary Directions. SiepNet w. B in Table 3 indicates the variant of SiepNet using Bi-GRU instead of GRU to explore evolving patterns of TKGs. The experimental results of SiepNet w. B and SiepNet are similarly well on YAGO, as compared with other variants of SiepNet. Therefore, combining forward and backward snapshot information has less significant impacts on the performance of SiepNet and more computational overhead.

Temporal Attention. The results denoted as SiepNet w/o TA in Table 3 demonstrate the performance of SiepNet without temporal attention component. It can be seen that SiepNet w/o TA performs noticeably worse than SiepNet on YAGO datasets, which justifies the necessity of temporal attention component to model long-range and short-range dependencies.

6   Conclusion

In this paper, we propose a novel temporal link prediction model SiepNet, which adapts to the evolutionary process of dynamic facts by modelling temporal adjacency facts with associated semantic and informational patterns. Specifically, SiepNet explores the local structural information based on a relation-aware GNN architecture. In addition, SiepNet incorporates temporal attention to help with modelling long-range and short-range historical dependencies hidden in TKGs. The experimental results on seventeen baselines demonstrate the significant advantages and promising performance of SiepNet in temporal link prediction. In future work, we will explore the persistence modelling of facts, rather than just predicting missing facts at a certain time slice t.

Acknowledgment. This work was supported by the National Natural Science Foundation of China (No. U2003208 and No. 62172451), the Scientific and Technological Innovation 2030-Major software of New Generation Artificial Intelligence (No. 2020AAA0109601), and the Open Research software of Zhejiang Lab (No. 2022KG0AB01).

References

1. Bordes, A., Usunier, N., Garcia-Duran, A., Weston, J., Yakhnenko, O.: Translating embeddings for modeling multi-relational data. In: Neural Information Processing Systems (NIPS), Lake Tahoe, America, pp. 1–9. ACM (2013)
2. Chung, J., Gulcehre, C., Cho, K., Bengio, Y.: Gated feedback recurrent neural networks. In: International Conference on Machine Learning, pp. 2067–2075. PMLR (2015)
3. Dasgupta, S.S., Ray, S.N., Talukdar, P.: HyTE: hyperplane-based temporally aware knowledge graph embedding. In: Proceedings of the 2018 Conference on Empirical Methods in Natural Language Processing, Brussels, Belgium, pp. 2001–2011. ACL (2018)
4. Dettmers, T., Minervini, P., Stenetorp, P., Riedel, S.: Convolutional 2d knowledge graph embeddings. In: Proceedings of the Thirty-Second AAAI Conference on Artificial Intelligence and Thirtieth Innovative Applications of Artificial Intelligence Conference and Eighth AAAI Symposium on Educational Advances in Artificial Intelligence, New Orleans, USA, pp. 1811–1818. AAAI (2018)
5. García-Durán, A., Dumancic, S., Niepert, M.: Learning sequence encoders for temporal knowledge graph completion. In: Proceedings of the 2018 Conference on Empirical Methods in Natural Language Processing, Brussels, Belgium, pp. 4816–4821. ACL (2018)

6. Han, Z., Chen, P., Ma, Y., Tresp, V.: Explainable subgraph reasoning for forecasting on temporal knowledge graphs. In: International Conference on Learning Representations (ICLR) (2021)
7. Han, Z., Ding, Z., Ma, Y., Gu, Y., Tresp, V.: Learning neural ordinary equations for forecasting future links on temporal knowledge graphs. In: Proceedings of the 2021 Conference on Empirical Methods in Natural Language Processing, pp. 8352–8364 (2021)
8. He, Y., Zhang, P., Liu, L., Liang, Q., Zhang, W., Zhang, C.: HIP network: historical information passing network for extrapolation reasoning on temporal knowledge graph. In: Proceedings of the Thirtieth International Joint Conference on Artificial Intelligence, (IJCAI), pp. 1915–1921 (2021)
9. Hinton, G., NitishSrivastava, A., Salakhutdinov, I.R.R.: Improving neural networks by preventing co-adaptation of feature detectors. Comput. Sci. **3**(4), 212–223 (2012)
10. Hochreiter, S., Schmidhuber, J.: Long short-term memory. Neural Comput. **9**(8), 1735–1780 (1997)
11. Hu, W., Yang, Y., Cheng, Z., Yang, C., Ren, X.: Time-series event prediction with evolutionary state graph. In: Proceedings of the 14th ACM International Conference on Web Search and Data Mining, pp. 580–588 (2021)
12. Hui, B., Zhang, L., Zhou, X., Wen, X., Nian, Y.: Personalized recommendation system based on knowledge embedding and historical behavior. Appl. Intell. **52**(1), 954–966 (2022)
13. Jiang, T., et al.: Towards time-aware knowledge graph completion. In: Proceedings of COLING 2016, the 26th International Conference on Computational Linguistics: Technical Papers, Osaka, Japan, pp. 1715–1724. ACL (2016)
14. Jin, W., Qu, M., Jin, X., Ren, X.: Recurrent event network: autoregressive structure inferenceover temporal knowledge graphs. In: Proceedings of the 2020 Conference on Empirical Methods in Natural Language Processing (EMNLP), pp. 6669–6683. ACL, Virtual (2020)
15. Kingma, D.P., Ba, J.: Adam: a method for stochastic optimization. In: International Conference on Learning Representations (Poster), San Diego, California, USA. Openview (2015)
16. Lautenschlager, J., Shellman, S., Ward, M.: ICEWS event aggregations. Harvard Dataverse 3 (2015)
17. Leblay, J., Chekol, M.W.: Deriving validity time in knowledge graph. In: Companion Proceedings of the the Web Conference 2018, Lyon, France, pp. 1771–1776. ACM (2018)
18. Leetaru, K., Schrodt, P.A.: GDELT: global data on events, location, and tone, 1979–2012. In: International Studies Association, San Francisco, California, USA, pp. 1–49 (2013)
19. Li, Z., et al.: Search from history and reason for future: two-stage reasoning on temporal knowledge graphs. In: Proceedings of the 59th Annual Meeting of the Association for Computational Linguistics and the 11th International Joint Conference on Natural Language Processing (Volume 1: Long Papers), pp. 4732–4743, Berkeley Hotel, Bangkok, Thailand. ACL (2021)
20. Li, Z., et al.: Temporal knowledge graph reasoning based on evolutional representation learning. In: Proceedings of the 44th International ACM SIGIR Conference on Research and Development in Information Retrieval, pp. 408–417. ACM, Virtual (2021)

21. Liu, Y., Ma, Y., Hildebrandt, M., Joblin, M., Tresp, V.: TLogic: temporal logical rules for explainable link forecasting on temporal knowledge graphs. In: Proceedings of the AAAI Conference on Artificial Intelligence, vol. 36, pp. 4120–4127 (2022)

22. Park, N., Liu, F., Mehta, P., Cristofor, D., Faloutsos, C., Dong, Y.: EvoKG: jointly modeling event time and network structure for reasoning over temporal knowledge graphs. In: Proceedings of the Fifteenth ACM International Conference on Web Search and Data Mining, pp. 794–803 (2022)

23. Schlichtkrull, M., Kipf, T.N., Bloem, P., van den Berg, R., Titov, I., Welling, M.: Modeling relational data with graph convolutional networks. In: Gangemi, A., et al. (eds.) ESWC 2018. LNCS, vol. 10843, pp. 593–607. Springer, Cham (2018). https://doi.org/10.1007/978-3-319-93417-4_38

24. Suchanek, F.M., Kasneci, G., Weikum, G.: Yago: a core of semantic knowledge. In: Proceedings of the 16th International Conference on World Wide Web, Banff, Alberta, Canada, pp. 697–706. ACM (2007)

25. Sun, H., Zhong, J., Ma, Y., Han, Z., He, K.: Timetraveler: reinforcement learning for temporal knowledge graph forecasting. In: Proceedings of the 2021 Conference on Empirical Methods in Natural Language Processing, Barceló Bávaro Convention Centre, Punta Cana, Dominican Republic, pp. 8306–8319. ACL (2021)

26. Sun, Z., Deng, Z.H., Nie, J.Y., Tang, J.: RotatE: knowledge graph embedding by relational rotation in complex space. In: International Conference on Learning Representations, Vancouver, Canada. OpenReview (2018)

27. Trivedi, R., Dai, H., Wang, Y., Song, L.: Know-evolve: deep temporal reasoning for dynamic knowledge graphs. In: International Conference on Machine Learning, Sydney, Australia, pp. 3462–3471. PMLR (2017)

28. Trivedi, R., Farajtabar, M., Biswal, P., Zha, H.: DyRep: learning representations over dynamic graphs. In: International Conference on Learning Representations, New Orleans, Louisiana, USA. OpenReview (2019)

29. Ward, M.D., Beger, A., Cutler, J., Dickenson, M., Dorff, C., Radford, B.: Comparing GDELT and ICEWS event data. Analysis 21(1), 267–297 (2013)

30. Wu, J., Cao, M., Cheung, J.C.K., Hamilton, W.L.: TeMP: temporal message passing for temporal knowledge graph completion. In: Proceedings of the 2020 Conference on Empirical Methods in Natural Language Processing (EMNLP), pp. 5730–5746 (2020)

31. Xiao, H., Chen, Y., Shi, X.: Knowledge graph embedding based on multi-view clustering framework. IEEE Trans. Knowl. Data Eng. 33(2), 585–596 (2019)

32. Yang, B., Yih, S.W.t., He, X., Gao, J., Deng, L.: Embedding entities and relations for learning and inference in knowledge bases. In: Proceedings of the International Conference on Learning Representations (ICLR) 2015, San Diego, California, USA. OpenReview (2015)

33. Xu, Y., Ou, J., Xu, H., Fu, L.: Temporal knowledge graph reasoning with historical contrastive learning. CoRR (2022)

34. Zhu, C., Chen, M., Fan, C., Cheng, G., Zhang, Y.: Learning from history: modeling temporal knowledge graphs with sequential copy-generation networks. In: Proceedings of the AAAI Conference on Artificial Intelligence, vol. 35, pp. 4732–4740. AAAI, Virtual (2021)

Multimedia

DQN-Based Stitching Algorithm for Unmanned Aerial Vehicle Images

Ji Ma⬤, Wenci Liu(✉) ⬤, and Tingwei Chen⬤

Liaoning University, Shenyang 110036, Liaoning, China
liuwenci1225@163.com

Abstract. Based on Global Positioning System (GPS), Inertial Measurement Unit (IMU) and altitude calculation to obtain the attitude and position information of Unmanned Aerial Vehicle (UAV) for image alignment, this method significantly improves the stitching speed and stitching quality compared with the traditional image matching. However, as UAVs can take thousands of images, the images can be blurred and distorted, making manual screening time consuming and difficult to ensure that the right reference image is selected. In order to be able to select the right reference image for the final stitching quality. This paper proposed a Deep Q-Network (DQN)-based image stitching algorithm for UAVs (IMDQN). The UAV captured images were preprocessed and fed into DQN, and the globally optimal reference image was decided by the action selection model, Root Mean Square Error (RMSE) metrics were also used as an incentive mechanism. The action selection model was used to select the optimal reference image to integrate the input to Position and Orientation System (POS) data for image alignment operation, while the final stitching was accomplished using the weighted average fusion algorithm for picture fusion. The study's findings demonstrate the algorithm's clear autonomy and usefulness above the conventional approach, particularly given the enhanced image stitching quality. The study's findings aid in the creation of later picture stitching methods.

Keywords: UAV · Image stitching · DQN · POS data

1 Introduction

Unmanned aerial vehicles (UAVs) have seen increased application recently across many industries. Its flexible takeoff and landing, low operating costs, and convenient operation have made it highly popular. By carrying high-definition cameras to observe the low-altitude ground, UAVs can obtain a large amount of observation data and achieve functions that ordinary cameras cannot achieve [1]. However, the coverage of a single image was constrained by the camera's focal length restrictions, high resolution, and flight height. A wide-field scene image must be created by stitching together several photos with overlapping ranges in order to monitor broad target regions and obtain a thorough picture of the overall situation. Therefore, image stitching techniques are critical, but those currently available are still challenging.

© The Author(s), under exclusive license to Springer Nature Switzerland AG 2023
X. Yang et al. (Eds.): ADMA 2023, LNAI 14177, pp. 125–138, 2023.
https://doi.org/10.1007/978-3-031-46664-9_9

Traditional image stitching focused on the extraction of feature points, the correctness of matching pairs, and the alignment of images [2–5]. As an example, Lan [6] proposed a Grid-based Motion Statistics (GMS)-Random Sample Consensus (RANSAC) based aerial image mosaic algorithm for UAVs, which improved the accuracy of image alignment by obtaining a high-quality set of correct interior points. To improve the SIFT algorithm, they used a down-sampling method and gradient normalization-based feature descriptors to achieve feature point matching, thereby reducing computational effort and improving matching accuracy [7]. The improved Shape-Preserving Half-Projective (SPHP) algorithm can reduce ghosting in the overlapping areas of the image, making the stitched image more natural [8]. When faced with narrow image overlap, severe image shadow or containing large areas with texture defects such as water, forest, desert, etc., failure to extract a large number of feature points would result in time-consuming and prone to matching errors or even inability to perform image stitching. Therefore, these methods used local affine distortion techniques and smooth transformation of overlapping regions to mitigate blurring artifact effects and optimize the transformation matrix to achieve large scale image stitching [9].

With respect to the above mentioned, Kim et al. [10] proposed an image Mosaic method using resampling grids to solve the Mosaic problem of narrow overlapping UAV images. Wu et al. [11] proposed texture adaptive extraction method successfully overcame the issue of scarce feature points and challenging matching. The approach could not only extract more feature points but also guarantee their uniform distribution.

Even though the aforementioned study findings have made some instances of image stitching easier, they were unable to resolve the challenging issue of feature point extraction and matching. In order to incorporate graph theory and Position and Orientation System (POS) data into the image stitching process, Yu et al. [12] proposed the idea. The experimental findings demonstrated the method's superiority in lowering projection distortion and cutting alignment time, as well as its ability to tackle the mosaic problem of small area overlap and texture defect images. When compared, Ren et al. [13] described a novel offline calibration method that quickly calibrates a system by regressing the global transformation matrix to the optimal transformation matrix calculated by a feature-based technique employing a Multilayer Perceptron (MLP) neural network. However, because the method did not take into account the reference image selection principle, the error accumulation phenomena was not effectively eliminated.

All of the above problems were not well eliminated by the accumulation of errors that led to missing scenes, distortions, artifacts and other problems in the image. So, to better solve these problems. This paper proposed a Deep Q-Network based stitching algorithm for UAV images (IMDQN). The globally optimal reference picture was located using the IMDQN technique, and the well-aligned image was found using POS data. Which well be stitched into a complete panoramic image by a weighted average fusion algorithm. The IMDQN algorithm effectively solves the problems of image stitching distortion caused by the error of reference image selection.

To sum up, the contributions of this work are as follows:

- Reference image selection. The images captured by the UAV are pre-processed and input to DQN, and the global optimal reference image is determined by the action selection model.

- Incentive mechanism. The root mean square error (RMSE) metric is used as an incentive mechanism, and the smaller the RMSE value, the larger the reward value.
- Global alignment method of images. The imaging model is recovered using high-precision POS data, and the homology matrix between adjacent images is derived to perform alignment of multiple images.

The remainder of this essay is structured as follows. In Sect. 2 of this essay, some work is presented that is closely related to this topic. The overall architecture of picture stitching, the alignment procedure based on POS data, and the IMDQN training procedure are all described in Sect. 3. Discussion of the experimental findings can be found in Sect. 4. In Sect. 5 described conclusion.

2 Related Work

2.1 DQN

DQN is Deep Q Network, and DON is a kind of deep reinforcement learning [14–16]. It combines Q-learning and deep learning [17]. The high-dimensional input data is used as the state in reinforcement learning, as the input of the neural network (Agent), and then the neural network model outputs the value (Q value) corresponding to each action to cause the action to be performed. This is done by utilizing the robust characterization capabilities of neural networks. The goal of intensive learning is to get the most out of learning [18]. The core of its algorithm is:

- Objective function: Construction of deep learning learnable functions based on Q-learning algorithms;
- Target network: Creating a target Q value using a convolutional neural network and comparing it to the Q value of the subsequent state;
- Experience return mechanism: handles the non-stationary data distribution and correlation issues.

In fact, the primary challenge with DQN is how to update the weight parameters in the value network. DQN is still essentially an extension of the Q-learning concept. According to the Bellman equation, the state information and reward information R into the value network and output Q value to obtain the loss function L(θ) are needed to solve this problem.

$$L(\theta) = E[(T \arg etQ - Q(s, a, \theta))^2] \tag{1}$$

where the target Q is: $R(s, a) + \sigma \max(s_-, a_-, \theta_-)$. θ and θ_- denote the weights of the evaluated network and the target network, respectively. The gradient descent approach can be used to directly solve the weight parameters of the convolutional neural network once the loss function has been determined.

2.2 POS Data

POS data is a positioning and fixing system, which is a high-precision position and attitude measurement system of Inertial Measurement Unit (IMU)/Differential Global Positioning System (DGPS) combination. Global Positioning System (GPS) satellite signals are observed continuously using GPS receivers mounted on the aircraft and synchronized with GPS receivers located on one or more base stations on the ground. Precision positioning mainly uses differential GPS positioning (DGPS) technology, while attitude measurement mainly uses inertial measurement units (IMUs) to sense the acceleration of aircraft or other carriers, and obtain information such as the carrier's speed and attitude after integration operations. Processing of UAV images is impacted by the precision of POS data. While differential UAVs are capable of obtaining precise station coordinates, pose acquisition is not as efficient. In some places, we can employ POS to resolve the image mosaic problem when the image feature points cannot be retrieved and utilized.

3   Image Stitching Framework

Because of its small weight, the UAV is more likely to experience unstable flight conditions. Large tilt angles can cause faults in the image that cannot be corrected by perspective transformation and may even cause perspective distortion over a significant portion of the image. The reference image that is used will have a direct impact on the stitching's aesthetic effect, particularly for low altitude flights. Therefore, it is essential to choose a good reference image for stitching throughout the image stitching procedure. The following Fig. 1 is the general framework diagram of image stitching in this paper.

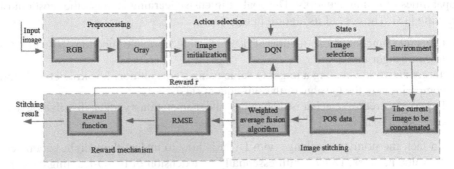

Fig. 1. Image stitching general framework diagram

For a given set of images, a randomly selected image is preprocessed and initialized to input into the DQN, and the action model will decide on the next reference image. The image alignment and fusion model is used to output the stitched image; a reward value is also generated. Then the action model selects another image to generate a new stitched image and repeats the process several times to be able to find all the reference images that are good for stitching [19]. Integrating all the reference images to make the best stitching effect. The framework of this paper consists of the following four sections:

- Preprocessing. Each RGB image is converted to grayscale and reduced to a size that reduces the computation and improves the performance of the model.
- Action selection. The action selection model uses the overlap pattern to predict the stitching results of different actions and decides the reference image based on the optimal prediction.

- Image alignment and fusion. POS data alignment technology can be used to locate and correct these image data, thus improving the stitching accuracy of the images. The original image's texture features and detail information can both be preserved using the weighted average fusion process, resulting in a fused image that as closely as possible replicates the original image's information.
- Reward mechanism. Action selection is modeled as a sequential decision process, and action selection is modeled as the selection of new reference images by the rewards of existing images.

3.1 Action Selection and Reward Mechanism

Action Selection

In this paper, we propose the IMDQN model, in which the action selection model uses a deep reinforcement learning method for finding reference images that help image stitching. The learning process is similar to that of Q-learning. For the input image, the agent observes the environment and selects an action from the action space and obtains a reward for that action. The agent's objective is to maximize the reward by selecting the most effective method for discovering the following helpful reference image. The action selection model is trained using overlapping patterns to find the best strategy. The overlap pattern includes four factors that determine whether the position of the image to be stitched is top or bottom, the overlap region, as well as the image's rotation angle. The overlap functions are as follows:

$$C = w_1 P_s + w_2 P_x + w_3 A + w_4 I \tag{2}$$

where C represents the degree of overlap, P_s indicates the top side of the image, and P_x indicates a downward image bias. Location information can be obtained based on the GPS that comes with the UAV. A is the area size of the image overlap, and I indicate the rotation angle of the image; this parameter can be obtained according to IMU. w_1, w_2, w_3, w_4 denotes the weight of each of the four parameters and $w_1 + w_2 + w_3 + w_4 = 1$。

The prerequisite for extracting the amount of image overlap of aerial UAV images is to know the UAV's separation from the ground. Therefore, we first calculate the UAV flight altitude, which is calculated as follows:

$$a = \frac{GSD * f}{c} \tag{3}$$

where GSD is the ground resolution, which refers to the minimum distance between two targets on the captured image. f is the camera's focal length, and c is the pixel size of the camera. When the camera and its pixels are determined on board the UAV, f and c are also determined. From the above formula, it can be seen that the higher the altitude of the UAV, the lower the ground resolution, i.e., the lower the clarity and accuracy of the image. When the UAV flies at a low altitude, let the height and width of the image ground coverage area H and W, the camera angle of view β; from the above formula, we can know the flight altitude of the UAV a; then, the image resolution, where h and w are expressed as the vertical and horizontal pixel values of the image, respectively, can be derived according to the trigonometric function.

$$W = \frac{2a \tan(\beta/2)}{h^2 + w^2} \tag{4}$$

$$H = \frac{Wh}{w} \qquad (5)$$

According to (4) (5), we can know what is the overlapping area A of the image.

$$A = \frac{2aWh\tan(\beta/2)}{(h^2 + w^2)w} \qquad (6)$$

Figure 2 shows the three overlapping cases of common aerial UAV areas. We can locate the position of the UAV by GPS coordinates and calculate the size of the image's overlapping area appropriately, which may be done by applying the formula above.

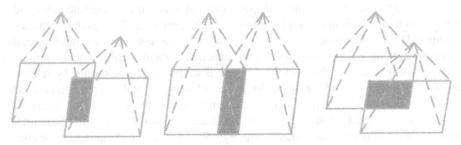

Fig. 2. Overlapping areas of different locations of the UAVs

By adjusting the weights each time, the IMDQN algorithm can find an optimal decision strategy according to the size of the reward value. The final weight adjustments after several training sessions are 0.2, 0.2, 0.4, and 04, and its splicing is the best.

Reward Mechanism

Each action space is accompanied by a reward value, where the reward function in this paper sets the Root Mean Square Error (RMSE). The award value increases as the RMSE value decreases. The formula for RMSE is as follows.

$$RMSE = \sqrt{\frac{1}{MN} \sum_{i=1}^{MN} [f(i,j) - f'(i,j)]^2} \qquad (7)$$

where M and N provide the image's height and width., respectively. $f'(i,j)$ and $f(i,j)$ denote the coordinates of the image to be evaluated and the coordinates of the original image, respectively.

The next state reward value of s_- is less than the reward value of state s. Stating that the newly selected reference image is valid for the task, the agent receives a positive reward and the newly selected reference image is accepted, expressed as:

$$r(s, s_-) = \begin{cases} 0 \ r(s) \leq r(s_-) \\ 1 \ r(s) > r(s_-) \end{cases} \qquad (8)$$

where $r(s)$ is the reward's value at the s state, and $r(s_-)$ is the reward's worth at the s_- state. The reward is 1 when the transfer of states yields correct output results and higher stitching quality (RMSE reduction). Otherwise, when the stitching quality is not improved, the reward is 0. For positive reward, the currently selected image is stored and the reference image collection is updated.

After each training, since an image is selected for the first training, the second training should avoid repeated selection to avoid wasting time. The IMDQN algorithm in this paper sets up a set G specifically for placing the selected images, and each selected image is compared with this set G. If there is a duplicate image, it is skipped. Otherwise, it is continued.

3.2 Image Alignment

For image alignment methods performed by feature matching, prior to stitching, tuning optimization can be used to uniformly modify the transformation matrix. Using POS data, image stitching is implemented in this paper [20]. We made an effort to cut down on the number of photos used in the stitching due to the high degree of image overlap, which cut down on alignment time and helped us advance our goal of lowering error accumulation. Based on the IMU and GPS, we can derive the flight attitude and geographical coordinates of the UAV, from which the homology matrix between images can be calculated. Two images I_1 and I_2 are taken by the UAV at L_1 and L_2 positions, for a location $P(x_w, y_w, z_w)$ on the ground in the world coordinate system. The coordinates on I_1 and I_2 are denoted as $l_1(u_1, v_1, 1)$ and $l_2(u_2, v_2, 1)$. The homology matrix H from I_1 to I_2 can be expressed as:

$$\begin{bmatrix} u_1 \\ v_1 \\ 1 \end{bmatrix} = \lambda H \begin{bmatrix} u_2 \\ v_2 \\ 1 \end{bmatrix} \tag{9}$$

where λ is an unreliable constant. We set I_1 to the metric system of the earth. The following is discovered when the world coordinate system is converted into the pixel coordinate system:

$$Z_{c1} \begin{bmatrix} u_1 \\ v_1 \\ 1 \end{bmatrix} = M \begin{bmatrix} x_w \\ y_w \\ 1 \end{bmatrix} \tag{10}$$

The camera's internal parameter matrix, M, is broken down into the internal and external parameter matrices. The internal parameter matrix is shown in the following equation:

$$M = NW = \begin{bmatrix} f_x & 0 & c_x & 0 \\ 0 & f_y & c_y & 0 \\ 0 & 0 & 1 & 0 \end{bmatrix} \begin{bmatrix} R & T \\ 0 & 1 \end{bmatrix} \tag{11}$$

N is the internal reference matrix for the camera, and W is the exterior reference matrix. The flight attitude and geographical coordinates of the UAV are known from

IMU and GPS. From this, we get the position change s of l_1 moved to l_2, where the flight angle of the UAV also changes, taking into account that the rotation matrix R with respect to the x, y and z axes of the world coordinates, which is a 3 × 3 matrix with 3 degrees of freedom.

$$R = R_X R_Y R_Z = \begin{bmatrix} 1 & 0 & 0 \\ 0 & \cos\theta_x & -\sin\theta_x \\ 0 & \sin\theta_x & \cos\theta_x \end{bmatrix} \begin{bmatrix} \cos\theta_y & 0 & \sin\theta_y \\ 0 & 1 & 0 \\ -\sin\theta_y & 0 & \cos\theta_y \end{bmatrix} \begin{bmatrix} \cos\theta_z & -\sin\theta_z & 0 \\ \sin\theta_z & \cos\theta_z & 0 \\ 0 & 0 & 1 \end{bmatrix} \quad (12)$$

From the above equation, the photographic equation at l_2.

$$Z_{c2} \begin{bmatrix} u_2 \\ v_2 \\ 1 \end{bmatrix} = MR \begin{bmatrix} x_w \\ y_w \\ 1 \end{bmatrix} - KRs \quad (13)$$

And the photographic ground at this point is represented as:

$$\frac{n^T}{d} \begin{bmatrix} x_w \\ y_w \\ z_w \end{bmatrix} = -1 \quad (14)$$

where n is the ground's normal vector, $\begin{bmatrix} x_w \\ y_w \\ z_w \end{bmatrix}$ is the target point's world coordinate system, and d represents the distance from the world coordinate system's origin to the surface. Integrating the above equations yields the complete equation at l_2.

$$Z_{c2} \begin{bmatrix} u_2 \\ v_2 \\ 1 \end{bmatrix} = KR \begin{bmatrix} x_w \\ y_w \\ z_w \end{bmatrix} + \frac{KRSn^T}{d} \begin{bmatrix} x_w \\ y_w \\ z_w \end{bmatrix} \quad (15)$$

The relationship between l_1 and l_2 can be deduced from the above equation.

$$\begin{bmatrix} u_2 \\ v_2 \\ 1 \end{bmatrix} = \frac{Z_{c1}}{Z_{c2}} KR(1 + \frac{s}{d}n^T)K^{-1} \begin{bmatrix} u_1 \\ v_1 \\ 1 \end{bmatrix} \quad (16)$$

$\frac{Z_{c1}}{Z_{c2}}$ is a number that is not zero, so the homology matrix yields.

$$H = \frac{\frac{Z_{c1}}{Z_{c2}} KR(1 + \frac{s}{d}n^T)K^{-1}}{\lambda} \quad (17)$$

The resulting homology matrix between the two images is obtained, and the images are thus aligned.

3.3 IMDQN Training Process

The purpose of training the IMDQN is to allow the intelligence to learn from experience so that it can automatically find valuable reference images based on the state. In this algorithm, images are first selected randomly from the training sample for image initialization, forming the initialized image as the initial state s, which is then fed into the IMDQN initializes the parameter output action A_i, selects the image to generate the next state $s\_$, and obtains the reward R by image alignment and fusion and reward function. $s\_$ and R are again used as input update parameters for IMDQN, which is trained iteratively. In this algorithm, the intelligence acquires rewards by exploring the environment (state spaces) until it arrives at a correct prediction with high confidence. In each round of iteration, each image in the image set is trained once with the aim of improving the intelligence level of the intelligence being expressed as a state action value function Q. Figure 3 is a diagram of the training process.

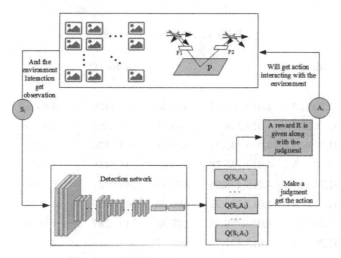

Fig. 3. IMDQN training process diagram

4 Experiment

On a computer running Windows 10 with an Intel Core i5 processor clocked at 3.6 GHz, 64-bit, 8 GB of RAM, all experiments were carried out. The programming language used in this research is Python 3.6, while PyCharm 2020 serves as the development environment. All of the methods are based on OpenCV 3.4.2. This paper will compare 2 aspects:

- The results of the IMDQN comparing the algorithm used in this study to the final stitched images of the Minimum Spanning Tree (MST) algorithm and the Multilayer Perceptron (MLP) algorithm;
- RASE and Mean Absolute Error (MAE) are used as the evaluation metric for splicing accuracy. The approach performs better the lower the RASE and MAE values are.

4.1 Reference Image Alignment

In this paper, the processing steps for image alignment using reference images of good quality selected by IMDQN are as follows:

- The set of image relations was constructed using POS data with reference to the selection of images and the determination of data.
- Image alignment. We can determine the homologous transformation matrix between nearby images using Eq. (14), and we can further realize the projection transformation from each image to the data using Eq. (15). Table 2 displays the global alignment outcomes from the aforementioned data.
- Fusion of images. We apply the fusion approach to achieve a seamless transition between images after projecting all of the images to remove artifacts and blurring in the overlapped regions and create a pleasing visual impact. To do this, this research specifically uses the weighted average fusion method [21] (Table 1).

Table 1. POS data of images

Images	Location(m)			Gesture (°)		
	X	Y	Z	α	β	θ
1.jpg	432,704.27	4,013,502.33	887.56	345.1426	3.6458	0.1256
2.jpg	432,702.21	4,013,348.43	886.32	347.2589	− 1.1354	− 2.1429
3.jpg	432,703.47	4,013,445.21	885.31	343.6526	1.4589	1.3579
4.jpg	432,709.35	4,013,456.78	888.45	341.0456	2.3478	2.6584
5.jpg	432,710.25	4,013,478.26	887.89	343.9421	5.2264	0.3214
6.jpg	432,712.05	4,013,504.22	886.96	345.8739	− 0.2654	− 2.6598
7.jpg	432,708.46	4,013,389.45	887.38	349.4287	− 1.9546	0.5266
8.jpg	432,709.41	4,013,421.76	886.97	344.7749	2.3588	1.9945

After the above steps, we finished the stitching of the reference image based on POS and got the global stitching image, as shown in Figs. 4, 5 and 6 below for the three images. Figure 3 shows that the algorithm of this paper makes the stitching effect smooth, with no ghosting, no distortion, and excellent stitching quality, while Fig. 4, as framed in red, has obvious stitching traces and serious distortion. Figure 5 has blurred edges, serious distortion, and serious error accumulation. Compared with the final results achieved by the other two algorithms. It is obvious that the algorithm in this work is superior.

The above stitched together three groups of images are stitched together based on the three groups of images selected below. From Figs. 7, 8 and 9, we can see that different algorithms select different reference images, so the stitched effect image is also different.

4.2 Evaluation Indicators

The algorithm evaluation metrics in this paper include two aspects: RMSE and MAE.

Table 2. Global alignment of images

Global Matrix	Results			Global Matrix	Results		
H_1	0.9546	− 0.0788	− 7.8841	H_3	1.0542	− 0.0215	− 17.8936
	0.0421	0.9415	750.2554		0.0702	1.0921	88.1245
	0.0000	− 0.0002	0.8945		0.0000	0.0001	1.0212
H_2	0.9512	− 0.1533	54.4482	H_4	0.0000	0.0000	1.0000
	0.0842	0.8422	390.2215		0.0000	1.0000	0.0000
	0.0000	− 0.0001	0.8654		1.0000	0.0000	0.0000

Fig. 4. IMDQN algorithm

Fig. 5. MST algorithm

- MAE: The absolute errors between the predicted and observed values are averaged out to form the MAE, or mean absolute error.

$$MAE = \frac{1}{N} \sum_{i=1}^{N} |f(i,j) - f\prime(i,j)| \qquad (18)$$

Fig. 6. MLP algorithm

Fig. 7. Reference image selected by IMDQN

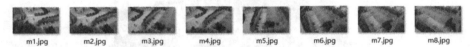

Fig. 8. Reference image selected by MST

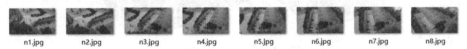

Fig. 9. Reference image selected by MLP

From Table 3, we can see that the error values of the bolded RSME and MAE are small compared with those of the other two algorithms, which means that the IMDQN algorithm in this paper has small error and high splicing quality.

Table 3. Comparison of evaluation indicators0

method	RMSE	MAE
IMDQN	**0.3988**	**0.3244**
MST	0.4033	0.4279
MLP	0.6254	0.5239

5 Conclusion

In order to fix the issue of blurred and distorted images of aerial photography, which leads to poor stitching effect. In this paper, we proposed IMDQN algorithm. Firstly, the images were pre-processed and inputted to DQN for training. The action selection model was to select new reference images by the reward of existing images, and finally, the global best reference image was decided. Next, the POS data was used to align the reference images to be stitched together, and because POS data is accurate, a relatively accurate homologous transformation matrix between neighbouring images was obtained, and further calculations were made to determine the projection transformation relationship between the global image and the reference. Finally, a scene with a larger field of view was obtained by image fusion. This study combines experiment and theory, and finally, it not only improves the quality of image stitching, but also the algorithm has obvious autonomy and adaptability. It provides some references for the subsequent research of image stitching.

References

1. Liu, C., Zhang, S., Akbar, A.: Ground feature oriented path planning for unmanned aerial vehicle mapping. IEEE J. Sel. Top. Appl. Earth Obser. Remote Sens. 12(4), 1175–1187 (2019)
2. Cui, H., Li, Y., Zhang, K.: A fast UAV aerial image mosaic method based on improved KAZE. In: 2019 Chinese Automation Congress, pp. 2427–2432 (2019)
3. Yuanting, X., Yi, L., Kun, Y., Chun Xue, S.: Research on image mosaic of low altitude UAV based on Harris corner detection. In: 2019 14th IEEE International Conference on Electronic Measurement & Instruments, pp. 639–645, IEEE, Changsha (2019)
4. Fangjie, C., Jun, H., Zuwu, W., Guoqiang, Z., Jianlian, C.: Image registration algorithm based on improved GMS and weighted projection transform. Laser & Optoelectronics Progress 55(11), 180–186 (2018)
5. Zhenyu, L., Yuan, T., Fang-jie, C., Jun, H.: Uav aerial image Mosaic algorithm based on improved ORB and PROSAC. Adv. Laser Optoelectronics 56(23), 91–99 (2019)
6. Lan, X., Guo, B., Huang, Z., Zhang, S.: An improved UAV aerial image mosaic algorithm based on GMS-PROSAC. In: 2020 IEEE 5th International Conference on Signal and Image Processing, pp.148–152. IEEE, Nanjing (2020)
7. Liu, Y.M., Wang, H.Y., Sun, Y., Gao, X.: Farmland Aerial Images Fast-Stitching Method and Application Based on Improved SIFT Algorithm. IEEE Access 10, 95411–95424 (2022)
8. Leng, J., Wang, S.: UAV remote sensing image mosaic technology combined with improved SPHP algorithm. In: 2020 IEEE International Conference on Mechatronics and Automation, pp. 1155–1160. IEEE, Beijing (2020)
9. Xiangdong, F., Chunying, Wei.: UAV Aerial image stitching method based on adaptive optimization of coverage area. In: Radio Engineering, pp.1–8 (2022)
10. Kim, J., Kim, T. Shin, D.: Fast and robust geometric correction for mosaicking UAV images with narrow overlaps. Remote Sens, pp. 2557–2576 (2017)
11. Wu, X.: Research on UAV image positioning optimization technology. PLA Information Engineering University (2017)
12. Yu, G., Ha, C., Shi, C., Gong, L., Yu, L.: A fast and robust UAV images mosaic method. In: Wang, L., Wu, Y., Gong, J. (eds.) Proceedings of the 7th China High Resolution Earth Observation Conference (CHREOC 2020). CHREOC 2020. LNEE, vol. 757, pp. 229–245. Springer, Singapore (2022). https://doi.org/10.1007/978-981-16-5735-1_17
13. Ren, M., Li, J., Song, L., Li, H., Xu, T.: MLP-Based Efficient Stitching Method for UAV Images. IEEE Geosci. Remote Sens. Lett. 19, 1–5 (2022)
14. Quan, L., et al.: A survey on deep reinforcement learning. J. Comput. Sci. 1–28 (2017)
15. Morales, E.F., Murrieta-Cid, R., Becerra, I., et al.: A survey on deep learning and deep reinforcement learning in robotics with a tutorial on deep reinforcement learning. Intel. Serv. Robotics. 14, 773–805 (2021)

16. Le, N., Rathour, V.S., Yamazaki, K., et al.: Deep reinforcement learning in computer vision: a comprehensive survey. Artif. Intell. Rev. **55**, 2733–2819 (2022)
17. Chen, S.A., Tangkaratt, V., Lin, H.T., et al.: Active deep Q-learning with demonstration. Mach. Learn. **109**, 1699–1725 (2020)
18. Mishra, S., Arora, A.: A Huber reward function-driven deep reinforcement learning solution for cart-pole balancing problem. Neural Comput. Appl. **35**, 16705–16722 (2022)
19. Bellver, M., Giroinieto, X., Marques, F.et al.: Hierarchical object detection with deep reinforcement learning. ZarXivpreprintarXiv:1611, p. 03718 (2016)
20. Hongya, T., Yun-cai, L.: A two-step rectification algorithm for airborne linear images with POS data. Zheijang Univ. **6**, 492–496 (2005)
21. Pei, H-X., Liu, J-D., Ge, J-L., et al.: A review of image stitching techniques. J. Zhengzhou Univ. (Science Edition) **51**(4), 1–10 (2019)

HM-QCNN: Hybrid Multi-branches Quantum-Classical Neural Network for Image Classification

Haowen Liu[1], Yufei Gao[1(✉)], Lei Shi[1], Lin Wei[1], Zheng Shan[2], and Bo Zhao[2]

[1] School of Cyber Science and Engineering, Songshan Laboratory, Zhengzhou University, Zhengzhou, China
{yfgao,shilei}@zzu.edu.cn
[2] State Key Laboratory of Mathematical Engineering and Advanced Computing, Zhengzhou, China

Abstract. Quantum machine learning has been developing in recent years, demonstrating great potential in various research domains and promising applications for pattern recognition. However, due to the constraints of quantum hardware, the input qubits are restricted caused by small circuit size, and the fuzziness in all dimensions caused by the features that are difficult to be effectively mined. Besides, previous studies focus on binary classification, but multi-classification received little attention. To address the difficulty in multi-classification, this paper proposed a hybrid multi-branches quantum-classical neural network (HM-QCNN) that utilizes a multi-branch strategy to construct the convolutional part. The part consists of three branches to extract the features of different scales and morphologies. Two quantum convolutional layers apply quantum CRZ gates and rotational gates to design a random quantum circuit (RQC) with 4 qubits and full qubits measurements. The experiments on three public datasets (MNIST, Fashion MNIST, and MedMNIST) demonstrate that HM-QCNN outperforms other prevalent methods with accuracy, precision, and convergence speed. Compared with the classical CNN and the hybrid neural network without multi-branches, HM-QCNN reached 97.40% and improved the accuracy of classification by 6.45% and 1.36% on the MNIST dataset, respectively.

Keywords: Quantum machine learning · Multi-classification · Hybrid quantum neural network · Medical images

1 Introduction

As quantum computing improves by leaps and bounds, the development of quantum algorithms that uses noisy intermediate-scale quantum (NISQ) to perform useful computational tasks is entering a boom period [1]. In this stage, quantum machine learning (QML) is a promising applications of quantum computing in the era of NISQ, which attempts to use quantum hardware to achieve computational acceleration or better performance for tasks in machine learning, while random quantum circuits (RQC) provide

X. Yang et al. (Eds.): ADMA 2023, LNAI 14177, pp. 139–151, 2023.
https://doi.org/10.1007/978-3-031-46664-9_10

a prospective path [2–4]. Compared with classical machine learning, QML algorithms based on RQC have two potential advantages, i.e., greater expressiveness [5] and more computational power [6, 7], which originate from the superposition principle of quantum mechanics.

Recently, inspired by CNNs, quantum convolutional neural networks (QCNNs) have been proposed. These networks employed both classical and quantum hardware, and encapsulated parts of complex neural networks in quantum devices to exploit the super-position and entanglement of quantum systems, thus speeding up computation [8]. The central idea is to implement a quantum convolutional layer by applying shallow RQC, and the corresponding feature mapping is implemented by measuring the output quantum state of the RQC. The output of the quantum convolutional layer is classical data and thus can be directly adapted to the structure in CNNs, while also exploiting the capabilities of hardware of the current NISQ.

A proliferation of studies using QCNNs for binary classification, and an increasing number of research scholars devote themselves to studying the task of pattern recognition on images. The research on multi-classification is further complex because the distinction between multiple categories needs to be considered, and the classifier needs to make additional decisions. Therefore, for the multi-classification task, a framework combining classical computer and quantum hardware is introduced, which has been widely used in recent QML studies [9, 10], and the classifier needs to make more decisions, studies on multi-classification are more complex and fewer than binary classification. For the multi-classification task, a framework combined with classical computers and quantum hardware is introduced, which has been widely used in recent QML studies, and helps to explore the potential computational power of the NISQ computer. As shown in Fig. 1, it can be divided into two parts: the encoding model and the HNN model. The former is responsible for processing the input data, and the latter is the module for training.

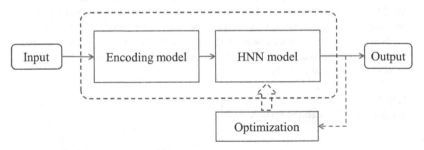

Fig. 1. Framework of the quantum convolutional neural network

The contributions of the current study are summarized in the following four folds:

1. The proposed HM-QCNN introduces multiple branches to construct networks, which implemented by RQCs, and two different scales of convolution kernels, in order to learn the syncretic features.
2. To verify the applicability of the model to the multi-classification, the experiments are conducted both on natural image datasets (MNIST, Fashion MNIST) and medical image dataset (MedMNIST).

3. Compared with previous approaches, HM-QCNN achieves better performance of accuracy, precision, and convergence speed.
4. To the best of our knowledge, this study is the first to explore the effectiveness of QNNs on medical images.

The remainder of the paper is organized as follows. Recent works related to QML and QCNNs are reviewed and summarized in Sect. 2. And Sect. 3 describes the encoding model and proposed HM-QCNN architecture in detail. The experiments of this work are presented in Sect. 4, comparing and demonstrating the various performances of HNN for image classification, and discussing the results. In Sect. 5 conclusions are drawn and directions for future work are suggested.

2 Related Work

The current volume of data is growing at an overwhelming rate, and the computational power required by machine learning algorithms increases with the data, which is gradually becoming limited for classical machine learning. And with the computational potential of quantum computers exceeding that of any classical computer, QML as a research frontier in AI has emerged as a prospective solution to the challenge of increasing data volumes [11]. QML has received a lot of attention in recent years, including quantum autoencoders [12, 13], quantum Boltzmann machines [14], quantum generative adversarial learning [7, 15, 16], and quantum kernel methods [17, 18].

Among them, lots of studies focused on the applications of QML in classification tasks, such as Edward Grant et al. [19] concluded that more expressive circuits have better accuracy and established hierarchical quantum circuits for binary classification of classical datasets IRIS and MNIST. Moreover, Yang et al. [20] organized SRA images into a data tensor and proposed a deep sparse tensor filter network for image classification.

In addition, motivated by the learning capability of CNNs and the potential power of QML, the hybrid quantum-classical neural network framework has emerged as a promising approach for classification tasks. Liu et al. [21] designed a hybrid quantum-classical convolutional neural network (QCCNN) that is friendly to current NISQ computers in terms of quantum bits and circuit depth, adapting to quantum computing to enhance the process of feature mapping while retaining the nonlinearity and scalability of classical CNN. Wei et al. [22] presented a quantum convolutional neural network (QCNN), which greatly reduces the computational complexity compared to classical. And applied it for image processing with numerical simulations for spatial filtering and edge detection. Finally, the model was verified on MNIST to have some robustness in image recognition. Cong et al. [23] analyzed the performance of the QCNN beyond existing methods and demonstrated that it could accurately identify quantum states associated with a one-dimensional topological phases. Francesco et al. [24] proposed a network model based on a variational circuit that reduces the circuit depth required for data encoding, using quantum neural networks for classification methods on recent quantum hardware. MacCormack et al. [9] offered the branching quantum convolutional neural network bQCNN inspired by QCNN with higher expressiveness.

Most of the existing studies are focused on the tasks of pattern recognition and image binary classification, the solution to multi-classification problems through quantum neural networks is still being explored, and the research on the recognition and classification of traditional natural images is also deficient. This work explored and designed a network model based on a quantum convolution filter fabricated by RQC combining a quantum convolution layer with a traditional network model structure for the multi-classification problems of handwritten digits and some natural images.

3   Method

3.1   Encoding Model

Quantum encoding is a process of converting classical information into quantum states, which is a very important step in the process of solving classical problems using quantum algorithms. Most encoding methods could be seen as parameterized circuits acting on and the parameters are determined by the classical information. The task of the encoding model in the framework is to map classical morphological data to quantum states in Hilbert space, and here three different encoding methods will be presented to achieve this transformation.

The first and most efficient in spatial terms method is to encode classical data in superimposed amplitudes by associating the normalized input data with the probability amplitudes of the quantum states, called the amplitude encoding method (AE) [25]. This approach encodes an N-dimensional classical vector x to a quantum state with n quantum bits, where $n = log_2(N)$ and $|x = \sum_i^N x_i|i$. Here $|i$ is a set of computational bases in Hilbert space and needs to satisfy $|x|^2 = 1$. However, depending on the quantum classifier used, the computational cost of preparing the data to quantum form will offset the speedup obtained in the classification process in general.

Another simpler approach is basic encoding, where the data is encoded onto the substrate of a quantum state. Each classical data vector will be encoded in each quantum bit, with the two fundamental states 0 and 1 will be considered as $|0\rangle$ and $|1\rangle$ of the quantum bit. This type of encoding method transforms a binary string of length n into a quantum state $|x\rangle = |i_x\rangle$ with n quantum bits, and is therefore inefficient in terms of space, yet efficient in terms of time [25].

The third encoding method is angle encoding, which employs quantum rotation gates to encode classical information x. The angle of these quantum gates is determined by the classical information. $|x\rangle = \overset{n}{\otimes} R(x_i)|0^n\rangle$, here any one of Rx, Ry, and Rz can be used as R. Usually the number of quantum bits is equal to the classical information dimension.

As the experimental framework shown in Fig. 1, this paper tried each of the above three methods in the encoding module to compare and analyze their performance in the multi-classification task. Among them, the basic encoding applied X gate and angle encoding used Ry gate rotating around the y-axis. All of them are constructed by RQC, whose circuits are shown in the Fig. 2.

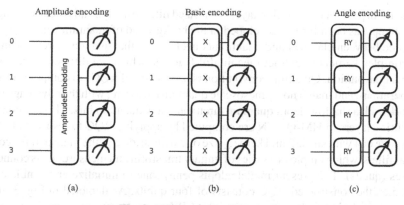

Fig. 2. Different encoding circuits

3.2 Hm-Qcnn

After encoding the classical data into quantum states, different gate operations are employed to each qubit corresponding to these data to form a quantum convolutional layer. In most previous works, the network is a quantum convolutional replacement of one traditional convolutional layer in the traditional network structure so that the whole structure contains at least one quantum convolution. The hybrid network structure applied in this work is based on hybrid computation, which consists of two parts, quantum and classical networks. The quantum part is responsible for the quantum convolution and the classical network part uses the convolutional and fully connected layers with the classical CNN structure. Here, unlike previous works, three branches are constructed in the HM-QCNN model, as shown in Fig. 3, two of which are quantum convolutional layers composed of quantum circuit and the other is a conventional convolutional layer.

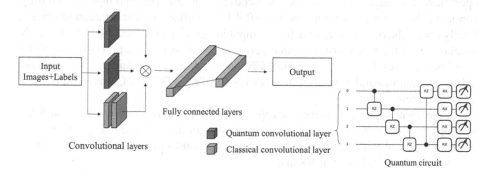

Fig. 3. HM-QCNN architecture

The main point of convolution layers is utilized filter to analyze all patches of images. This concept has been further developed in the background of quantum computing. The difference between classical and quantum convolution is that quantum circuits can produce complex kernels to extract meaningful features, which are difficult to handle by classical convolution. Quantum convolution is used as small RQCs to compute convolution operations, and match noisy mesoscale quantum hardware, with the advantage that it can work with a shallow depth quantum circuit and few quantum bits. The two quantum convolution layers in HM-QCNN are computed by applying RQC to respectively build kernel_size of 4 with stride 2 and kernel_size of 2 with stride 2 as the main part for convolutional filters, which employs a series of unitary transform and measurements connected by wires (qubits). The present model adopts pennyLane to initialize and simulate four qubits, i.e., the constructed RQC consists of four qubits. As depicted in Fig. 3, in the quantum convolutional filter, first a two-qubit CRZ quantum gate operation is employed, in other words, CRZ quantum gates are operated on each pair of adjacent qubits, which enables to capture of the relevant information on the same layer of the network. Then the RX quantum rotational gates are applied to operate on each qubit, embedding valid information into the quantum system. The final measurement phase, also known as the decoding phase, refers to the conversion of the quantum data into classical form [26]. Pauli matrix can be used as a measurement method, unlike other works with single qubit measurements, all-qubit are measured in this work, taking expectations by using Pauli-Z measurements for each qubit to obtain enough hidden information from the quantum system. The results of measurement are not yet direct representations of the predicted labels and therefore need to be further input to the classical network for processing.

The classical convolution layer is the key and important layer to extract features in the part of CNNs, which performs the convolution operation on the input features with kernels. Features are extracted from the images and map them to the next layer as complex features. The traditional convolutional layer branches in this model consist of two convolutional kernels of sizes 1 and 4 with strides 1 and 2, respectively. After these operations, the outputs of these three branches are concatenated and input to two fully-connected layers for classifying, and leakyReLU are utilized as the activation function to finally obtain the predicted results for the input images. The fully-connected layer is the second part of the CNN structure, that performs the classification process by applying weights to predict the classes. Classical CNN network with the equivalent structure and hybrid quantum neural network (QUANV1 − CONV1 − FC1 − FC2) are compared in this experiment.

In the learning phase, the cross-entropy loss is utilized as the loss function, and Adam is adopted as the optimizer for parameter optimization. During training, the network model is updated with parameters by backpropagation to minimize the error between the output results and the real results.

4 Experiments

4.1 Experiments Setting

Three independent models are compared in this paper, the proposed HM-QCNN, classical CNN with the equivalent structure, and HNN without multiple branches, i.e. HNN (w/o multi). Accuracy, precision, recall, and F1 score are used as evaluation metrics to assess classification performance of the model.

Setup. The experimental environment used in this work is Python 3.8, PyTorch 1.12.0, CUDA 11.6, batch size set to 32, the learning rate of 0.5, with a total of 50 epochs trained. Numerical simulations of the experiments are performed with PennyLane [27].

Datasets. Experiments are conducted on three public datasets MNIST, Fashion MNIST and MedMNIST. Different triple-classification tasks are performed on different datasets in this work. For example, the MNIST dataset is randomly generated in three experiments, the first experiment contains numbers {1,7,9}, the second task kept numbers {3,5,8}, and the third performed classification experiments on numbers {2,4,6}, which are described as E1, E2, and E3, respectively. Similarly, three tasks were generated on the Fashion MNIST dataset: the first task retained "T-shirt/top" "Trouser" and "Pullover"; the second task classified "Dress" "Coat" and "Sandal"; the third task kept the data of "Shirt", "Sneaker" and "Bag", which are denoted as E4, E5, and E6, accordingly.

4.2 Results and Discussion

This section discusses the performance of HM-QCNN on image multi-classification tasks. The experimental results demonstrate that the proposed model can be used to solve many types of image classification problems, and good results can be obtained not only on handwritten digital images, but also on natural images.

Three independent models are tested in the experiment with accuracy, precision, recall and F1-score as evaluation indicators shown in Table 1. The number of optimal performances is bolded. On MNIST dataset, HM-QCNN achieves 95.73%, 93.75%, 97.40% accuracy, which are 2.08%, 9.27% and 6.45% higher than the classical CNN model, and better than the HNN (w/o multi) model by 2.6%, 2.6%, and 1.36%, respectively. The optimal result is presented in E3 with accuracy, precision, recall and f1 scores of 97.40%, 97.40%, 97.39% and 97.40%. Moreover, on Fashion MNIST dataset, HM-QCNN performs slightly poor than HNN (w/o multi) in E4 and E5, but achieves 98.44%, 98.47%, 98.46% and 98.46% for accuracy, precision, recall and f1-score in E6, which is the best result among these methods.

Similarly, as shown in Fig. 4, the experimental data after model training are visualized with the t-SNE technique. Combined with the results in Table 1 to analyze, the data of E2 is not aggregated and also corresponds to the lower accuracy in the table. Compared with other groups of experiments, considering the reason of which, the original data of E2 is more disorganized, the results after classification are relatively poor.

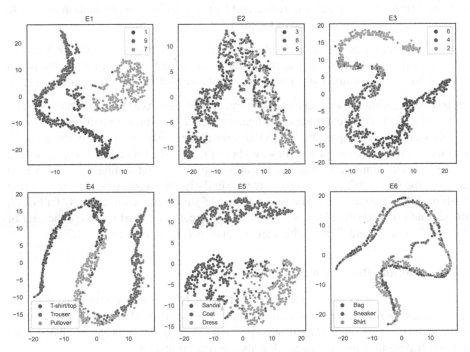

Fig. 4. Visualization with t-SNE of experimental datasets after training

Meanwhile, as shown in Fig. 5, it can be clearly seen that the performance of HM-QCNN is superior to classical CNN and the HNN without multiple branches, and all the classification accuracy can reach more than 93%. It indicates that the proposed HM-QCNN can effectively improve network performance and better solve classification problems in images. In addition, from the comparison of running time in the Fig. 6, it can be seen that the proposed model can significantly reduce training time and speed up convergence, which will help to classify images faster in practical applications. However, the gap between the execution time of HNN and the classical CNN is large, and the reason for this is the experiments are conducted with quantum numerical simulation, which speed cannot reach the real quantum computing hardware. Moreover, the speed is also affected by the limitation on the input qubits. However, in the future, with the development of quantum hardware, more qubits can be used to process images, thus improving the performance of the HNN.

Table 1. Performance evaluation of experiments

Experiment		Model	Acc. (%)	Pre. (%)	Re. (%)	F1-score (%)
MNIST		CNN	93.65	93.66	93.48	93.55
	E1	HNN (w/o multi)	93.13	92.95	92.03	92.97
		HM-QCNN	**95.73**	**95.66**	**95.64**	**95.65**
		CNN	84.48	84.54	84.51	84.52
	E2	HNN (w/o multi)	91.15	91.13	91.24	91.26
		HM-QCNN	**93.75**	**93.75**	**93.73**	**93.74**
		CNN	90.94	90.95	90.96	90.94
	E3	HNN (w/o multi)	96.04	96.10	96.03	96.05
		HM-QCNN	**97.40**	**97.40**	**97.39**	**97.40**
Fashion MNIST		CNN	94.06	94.08	94.01	94.02
	E4	HNN (w/o multi)	**94.17**	**94.30**	**94.19**	**94.24**
		HM-QCNN	93.95	93.65	93.65	93.65
		CNN	94.48	94.48	94.47	94.47
	E5	HNN (w/o multi)	**96.98**	**96.99**	**96.97**	**96.97**
		HM-QCNN	95.83	95.84	95.83	95.83
		CNN	97.40	97.43	97.50	97.46
	E6	HNN (w/o multi)	96.25	96.35	96.39	96.36
		HM-QCNN	**98.44**	**98.47**	**98.46**	**98.46**

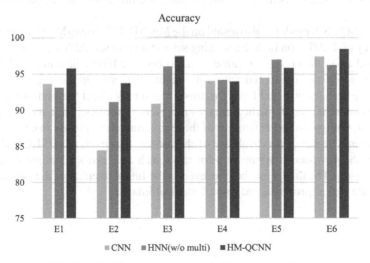

Fig. 5. Classification accuracy of different experiments

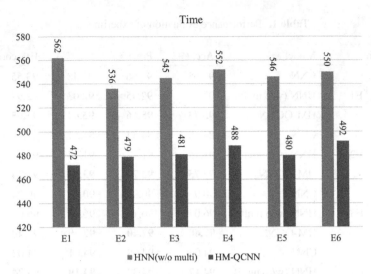

Fig. 6. Comparison of the execution time

In addition, experiments were also conducted on different encoding methods for triple-classification on MNIST and visualized the training results on the {1,7,9} sub-dataset. The training accuracy of the three different encoding methods is depicted in Fig. 7(a), and the training loss curves of the three different encoding methods with continuous reduction are shown in Fig. 7(b). The figure demonstrates that the angle encoding converges faster and achieves higher accuracy with smaller loss values. While amplitude encoding converges slower, but the training accuracy exceeds basic encoding to reach 99.98% at 25 epochs. Therefore, in this experiment, the angle encoding method works better.

The HM-QCNN model is also tested on the MedMNIST- breastMNIST dataset, with an accuracy of 73.08% on both the training set and testing set. Although the accuracy is not as good as on the other two datasets, this is because biomedical images have more special characteristics compared with other natural images. On the one hand, medical images have higher noise and lower contrast, which may affect the performance of the model. On the other hand, medical images represent structures inside the human body, and the morphology and other features of these structures vary greatly from case to case, which requires higher generalizability of the model. However, the HM-QCNN model still offers the prospect of application for tasks such as classification and diagnosis in medical images. The model can be improved in the future to enhance the generalization performance and improve the analysis of medical images.

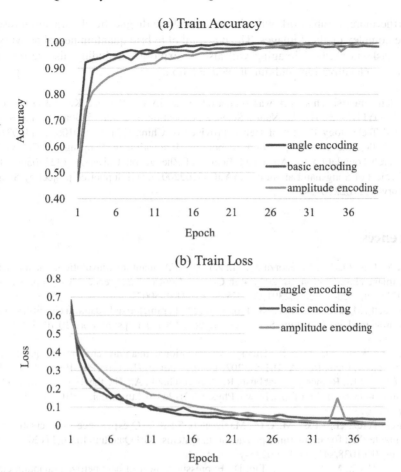

Fig. 7. Visualization of the learning curve for different encodings

5   Conclusion

To effectively improve the efficiency of classical CNNs while ensuring accuracy and precision, this paper develops the structure of hybrid quantum neural networks with multibranch by constructing parameterized quantum circuits. And conducts some experiments for multi-classification tasks. The results indicate that the HM-QCNN model achieves better accuracy in both MNIST and Fashion MNIST and outperforms the HNN without branches in terms of execution time.

In the NISQ era, due to the limitations of the quantum hardware for the input qubits, the size of natural images is too large for existing devices, so relevant operations like dimensionality reduction are required before inputting to the model, which may adversely affect the model performance. However, in the near future, as the algorithms continue to be explored, lower qubit algorithms suitable for quantum hardware will be studied and designed.

Furthermore, future work will aim to expand the diagnostic classification research to more complex medical images. The potential of hybrid quantum neural networks for various tasks in medical imaging will also be explored, including disease diagnosis, lesion region localization, and tumor segmentation.

Acknowledgements. This work was supported in part by the National Key R&D Program of China (2020YFB1712401), the Nature Science Foundation of China (62006210), the Key Scientific and Technology Project of Henan Province of China (221100210100, 221100211200, 221100210600), the Key Project of Collaborative Innovation in Nanyang (22XTCX12001), the Research Foundation for Advanced Talents of Zhengzhou University (32340306), Preresearch Project of Songshan Laboratory (YYJC022022001), and Supported project by Songshan Laboratory (232102210154).

References

1. Lü, Y., Gao, Q., Lü, J., Ogorzałek, M., Zheng, J.: A quantum convolutional neural network for image classification. In: 2021 40th Chinese Control Conference (CCC), pp. 6329–6334 (2021). https://doi.org/10.23919/CCC52363.2021.9550027
2. Benedetti, M., Lloyd, E., Sack, S., Fiorentini, M.: Parameterized quantum circuits as machine learning models. Quantum Sci. Technol. **5**(1) (2019) https://doi.org/10.1088/2058-9565/ab5944
3. Liu, Y., Wang, D., Xue, S., Huang, A.: Variational quantum circuits for quantum state tomography. Phys. Rev. A. **101**(5) (2020). https://doi.org/10.1103/PhysRevA.101.052316
4. McClean, J.R., Romero, J., Babbush, R., Aspuru-Guzik, A.: The theory of variational hybrid quantum-classical algorithms. New J. Phys. **18**(2) (2016). https://doi.org/10.1088/1367-2630/18/2/023023
5. García-Pérez, G., Rossi, M.A. C., Maniscalco, S.: IBM Q experience as a versatile experimental testbed for simulating open quantum systems. NPJ Quantum Inf. **6**(1) (2020). https://doi.org/10.1038/s41534-019-0235-y
6. Du, Y., Hsieh, M.-H., Liu, T., Tao, D.: Expressive power of parametrized quantum circuits. Phys. Rev. Res. **2**(3) (2020). https://doi.org/10.1103/PhysRevResearch.2.033125
7. Lloyd, S., Weedbrook, C.: Quantum generative adversarial learning. Phys. Rev. Lett. **121**(4) (2018). https://doi.org/10.1103/PhysRevLett.121.040502
8. Trochun, Y., Stirenko, S., Rokovyi, O., Alienin, O.: Hybrid classic-quantum neural networks for image classification. In: 2021 11th IEEE International Conference on Intelligent Data Acquisition and Advanced Computing Systems: Technology and Applications (IDAACS), pp. 968–972 (2021). https://doi.org/10.1109/idaacs53288.2021.9661011
9. MacCormack, I., Delaney, C., Galda, A., Aggarwal, N., Narang, P.: Branching quantum convolutional neural networks. Phys. Rev. Res. **4**(1) (2022). https://doi.org/10.1103/PhysRevResearch.4.013117
10. Henderson, M., Shakya, S., Pradhan, S., Cook, T.: Quanvolutional neural networks: powering image recognition with quantum circuits. Quantum Mach. Intell. **2**(2) (2020). https://doi.org/10.1007/s42484-020-00012-y
11. Hur, T., Kim, L., Park, D.K.: Quantum convolutional neural network for classical data classification. Quantum Mach. Intell. **4**(1) (2022). https://doi.org/10.1007/s42484-021-00061-x
12. Romero, J., Olson, J.P., Aspuru-Guzik, A.: Quantum autoencoders for efficient compression of quantum data. Quantum Sci. Technol. **2**(4) (2017). https://doi.org/10.1088/2058-9565/aa8072

13. Ding, Y., Lamata, L., Sanz, M., Chen, X., Solano, E.: Experimental implementation of a quantum autoencoder via quantum adders. Adv. Quantum Technol. **2**(7–8) (2019). https://doi.org/10.1002/qute.201800065

14. Jain, S., Ziauddin, J., Leonchyk, P., Yenkanchi, S., Geraci, J.: Quantum and classical machine learning for the classification of non-small-cell lung cancer patients. SN Appl. Sci. **2**(6) (2020). https://doi.org/10.1007/s42452-020-2847-4

15. Pandian, A., Kanchanadevi, K., Mohan, V.C., Krishna, P.H. and Govardhan, E.:Quantum generative adversarial network and quantum neural network for image classification. In: 2022 International Conference on Sustainable Computing and Data Communication Systems (ICSCDS), pp.473–478 (2022). https://doi.org/10.1109/icscds53736.2022.9760943

16. Jonathan Romero, A.A.-G.: Variational quantum generators: generative adversarial quantum machine learning for continuous distributions. Adv. Quantum Technol. **4**(1) (2020). https://doi.org/10.1002/qute.202000003

17. Patrick Rebentrost, M.M., Lloyd, S.: Quantum support vector machine for big data classification. Phys Rev Lett. **113**(13) (2014). https://doi.org/10.1103/10.1103/PhysRevLett.113.130 503113.130503

18. Havlíček, V., et al.: Supervised learning with quantum-enhanced feature spaces. Nature **567**(209–212) (2019). https://doi.org/10.1038/s41586-019-0980-2

19. Grant, E., Benedetti, M., Cao, S., Hallam, A.: Hierarchical quantum classifiers. NPJ Quantum Inf. **4**(1) (2018). https://doi.org/10.1038/s41534-018-0116-9

20. Yang, S., Wang, M., Feng, Z., Liu, Z., Rundong, L.: Deep sparse tensor filtering network for synthetic aperture radar images classification. IEEE Trans. Neural Netw. Learn. Syst. **29**, 3919–3924 (2018). https://doi.org/10.1109/TNNLS.2017.2688466

21. Liu, J., Lim, K. H., Wood, K. L., Huang, W.: Hybrid quantum-classical convolutional neural networks. Sci. China Phys. Mech. Astron. **64**(9) (2021). https://doi.org/10.1007/s11433-021-1734-3

22. Wei, S., Chen, Y., Zhou, Z., Long, G.: A quantum convolutional neural network on NISQ devices. AAPPS Bull. **32**(1) (2022). https://doi.org/10.1007/s43673-021-00030-3

23. Cong, I., Choi, S., Lukin, M.D.: Quantum convolutional neural networks. Nat. Phys. **15**(1273–1278) (2019). https://doi.org/10.1038/s41567-019-0648-8

24. Tacchino, F., Barkoutsos, P. K., Macchiavello, C., Gerace, D.: Variational learning for quantum artificial neural networks. In: 2020 IEEE International Conference on Quantum Computing and Engineering (QCE), pp.130–136 (2020). https://doi.org/10.1109/qce49297.2020.00026

25. Jian, Z., Zhao-Yun, C., Xi-Ning, Z., Cheng, X.: Quantum state preparation and its prospects in quantum machine learning. Acta Phys. Sin. **70**(14) (2021). https://doi.org/10.7498/aps.70.20210958

26. Yuki Takeuchi, T.M.: Quantum computational universality of hypergraph states with Pauli-X and Z basis measurements. Sci. Rep. **13585**(9) (2019). https://doi.org/10.1038/s41598-019-49968-3

27. Bergholm, V., Izaac, J., Schuld, M., Gogolin, C., Ahmed, S.: PennyLane: automatic differentiation of hybrid quantum-classical computations. ArXiv. (2022). https://doi.org/10.48550/arXiv.1811.04968

DetOH: An Anchor-Free Object Detector with Only Heatmaps

Ruohao Wu[1], Xi Xiao[1(✉)], Guangwu Hu[3], Hanqing Zhao[2], Han Zhang[2], and Yongqing Peng[2]

[1] Shenzhen International Graduate School, Tsinghua University, Shenzhen 518055, China
xiaox@sz.tsinghua.edu.cn
[2] Beijing Research Institute of Telemetry, Beijing 100076, China
[3] Shenzhen Institute of Information Technology, Shenzhen 518172, China

Abstract. In object detection, the anchor-based method relies on too much manual design, and the training and prediction process is too inefficient. In recent years, one-stage anchor-free methods such as Fully Convolutional One-stage Object Detector (FCOS) and CenterNet have made a splash in object detection. They not only have a simple structural design, but also demonstrate competitive performance. They have exceeded many two-stage or anchor-based approaches. However, in industrial applications, the design of its multiple output heads hinders the installation of the model. At the same time, different output heads mean a combination of multiple loss functions. This introduces problems in training. Here, we propose an anchor-free object Detector with Only Heatmaps (DetOH) to solve object detection. Bounding box parameters are calculated by post-processing. The design of the single output head allows object detection to use a semantic segmentation network, realizing the unification of the two frameworks. In addition, compared with CenterNet, we have greatly improved the speed of object detection (6 vs. 32 Frames Per Second) with 3.2% Average Precision boost. The proposed DetOH framework can be applied to multi-target tracking, key point detection and other tasks.

Keywords: Heatmaps · Anchor-free · Object Detector

1 Introduction

Object detection is an algorithm that predicts the bounding box position and category label for each instance of interest in an image. The classical algorithms mainly rely on sliding windows [1, 2], which classify every possible position and therefore require high speed. This also established the position of the anchor in object detection. After the advent of deep learning, detection has shifted to the use of FCN (Fully Convolutional Networks) since Faster R-CNN [3]. Many current anchor-based detectors such as Faster R-CNN, SSD [4], YOLOv2 [5], and v3 [6] rely on a predefined set of anchor boxes.

Many anchor-free detectors have also appeared in recent years, and their performance has gradually surpassed that of anchor-based detectors. For example, CenterNet [7, 8]

X. Yang et al. (Eds.): ADMA 2023, LNAI 14177, pp. 152–167, 2023.
https://doi.org/10.1007/978-3-031-46664-9_11

predicts promising performance with hourglass for predicting center, offsets, and object size. Fully Convolutional One-stage Object Detector (FCOS) [9, 10] uses a FCN to demonstrate the commonality between semantic segmentation tasks and object detection tasks. However, compared with the simple output head and loss function of semantic segmentation, an anchor-free detector requires multiple loss function combinations to assist training due to its multiple outputs. It is natural to ask the question: can we really do object detection like semantic segmentation? The answer is yes.

We found that the reason for the multiple outputs of the anchor-free detector is the deviation of the center positioning and the parameters of the object box. The former is caused by the output feature map being smaller than the input. The latter can be solved by mining information in classification heatmaps. Therefore, we designed the polynomial heatmaps so that the boundary of the object box can be obtained by post-processing. At the same time, their size is equal to the input, thus avoiding the offset of the center positioning. To highlight the portability and speed/precision balance of our model, we applied it to a smart security system. The detailed contributions are as follows.

1) We proposed an anchor-free object detector with only heatmaps. This not only simplifies the process of object detection but also allows for greater flexibility in detecting objects of varying sizes and shapes. We also introduced the concept of polynomial heatmaps for object detection which helps in post-processing and precise predictions of object boundaries.

2) We designed the Center Point (CP) loss function to effectively improve the detector performance. It draws on the positioning of the CP in the inference process and effectively alleviates the problem of CP offset. This greatly improves model accuracy.

3) We conducted experiments on the Microsoft COCO (Common Objects in COntext) dataset and applied the model to embedded devices to demonstrate the effectiveness of our work. Compared to CenterNet that use the same backbone network, we improved model performance by 3.2% Average Presicion (AP), increasing speeds from 6 to 32 FPS (Frames Per Second).

This paper is organized as follows. Related work and methods are analyzed in Sect. 2. In Sect. 3, the specific methods and core innovations are introduced. The experiments are shown in Sect. 4 followed by the ablation study in Sect. 5.

2   Related Work

In this section, we introduce some work related to our method. They are mainly divided into two parts: anchor-based detectors and anchor-free detectors.

2.1   Anchor-Based Detectors

Many anchor-based detectors achieve a good balance of speed and accuracy, such as Fast Region-based Convolutional Network (R-CNN) [13] and Deconvolutional Single Shot Detector (DSSD) [15]. Anchor boxes can be considered suggested regions, and they are classified as correct or negative patches. The anchor utilizes and avoids duplicate feature calculations, greatly speeding up the detection process. But it is worth noting that anchor-based detectors have some drawbacks:

As shown by Faster R-CNN [3] and RetinaNet [11], the performance of the detection is sensitive to the size, proportion, and number of anchor boxes. These hyperparameters can greatly affect the AP performance on the COCO dataset [12]. Therefore, these hyperparameters need to be fine-tuned in anchor-based detectors.

Detectors difficulty handling object candidates with large shape variations, especially small objects. Therefore, they need to set different sizes or aspect ratios for the same kind of object. To achieve high accuracy, anchor boxes need to be densely placed on the input image. Meanwhile, a lot of negative samples are generated, which brings imbalance problems to training. Anchor boxes also introduce complex computations into the training process and loss functions, such as calculating IoU (Intersection over Union).

2.2　Anchor-Free Detectors

The earliest anchor-free detector was probably YOLOv1 [6], which predicts points near the center of the object's bounding box because they are believed to be able to produce higher quality detections. But using only points close to the center resulted in low recall. CornerNet [16] uses a pair of corner points to detect boundaries and groups them to form the final detected box. CornerNet learns an additional distance metric, the purpose of which is to find pairs of corner points belonging to the same instance. This requires more complex post-processing. Another detector, Unitbox [17], is based on DenseBox [18]. Unitbox is considered unsuitable for general object detection because of the difficulty of handling overlapping bounding boxes and the relatively low recall. Feature Selective Anchor-Free (FSAF) [19] proposes to add an anchor-based detection branch to anchor-free detectors. As they consider that completely anchor-free detectors do not achieve good performance, they also utilize feature selection modules to improve the performance of anchor-free branching. So anchor-free detectors have comparable performance to anchor-based detectors. RepPoints [20] indicates that the box consists of a set of points and uses a conversion function to obtain the object box. Corner Proposal Network (CPN) [21] and HoughNet [22] require grouping or post-voting processing, so they are quite complex and slow. CenterNet [7] is a concurrent anchor-free detector. It uses a clean network structure to demonstrate the performance to be expected. A similar model, FCOS [9] adds center-ness branching, enabling a better accuracy/speed trade-off. There are many subsequent work based on FCOS [25, 26]. These models enhance the detection characteristics, loss function or allocation strategy of FCOS to further improve the performance of anchor-free detectors.

However, these detectors often have complex output heads which bring trouble to model training and loss function design. Besides, its multiple output heads are usually not supported by embedded chips, making it difficult for industrial applications.

3　Our Approach

In this section, our approach is divided into four parts. The model overview is first introduced, followed by our heatmaps and the calculation of the object size, and finally the loss function in our work.

3.1 Model Overview

In this paper, we propose an anchor-free object detection method with only heatmaps. Similar to other object detectors that use heatmaps, our model is divided into two parts, the feature extraction backbone network and the output head. The backbone network is mainly used to extract the image features of various scales, mostly using mature convolutional networks, such as ResNet [28], Deep Layer Aggregation (DLA) [29] and Hourglass [30]. FPN (Feature Pyramid Network) [23] or BiFPN [24] are sometimes added. The output head converts the feature map extracted by the backbone into a feature map that can obtain object box information. This is the main difference between these detectors. CenterNet and FCOS are the most representative models that use heatmaps for object detection.

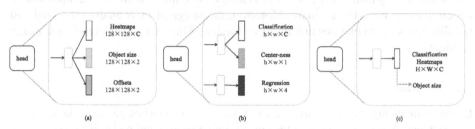

Fig. 1. Output head structure in different models. (a) shows CenterNet's output head, the prediction module that contains Heatmaps, Object size, and Offsets. (b) shows the output head of FCOS that contains Classification, Center-ness and Regression at different scales. (c) shows the output head of our method, the prediction module only contains Classification Heatmaps.

As shown in the Fig. 1, CenterNet's output head, i.e., the prediction module, contains three branches: Heatmaps, Object size, and Offsets. The center of the object is obtained through the heatmaps, and then the offset is corrected moderately. Finally, the size of the object is added to get a complete object box. For an input image of $512 \times 512 \times 3$ pixels, the output heatmaps size is $128 \times 128 \times C$ pixels, where C is the number of categories.

The output head of FCOS also contains three parts, i.e., Classification, Center-ness and Regression. The center of the object is obtained through the product feature maps of classification and center-ness, and then the offset and size information are obtained by regression. Then a complete object box is obtained, which utilizes a feature pyramid structure and thus contains several output heads of multiple scales. The size of the feature map varies from 1/8 to 1/128 of the input image.

However, the output head of our model only contains Classification Heatmaps, from which the object size can be calculated by post-processing. It is worth noting that the classification heatmaps and input images are of equal size.

We design the heatmaps to be as large as the input image, so that we can directly get to the object CP without offsets according to the heatmaps. Meanwhile, unlike the Gaussian circles generated by CenterNet, we plan to generate polynomial heatmaps. This means that a heat spot representing an object is no longer isotropic. The rate at which it decays in different directions is related to its dimensions in the corresponding direction, so that the width and height can be calculated from the parameters of the polynomial. Thus, the output of object size feature map is not required. After removing the object size feature map and the offset feature map, the model only needs one branch, i.e., the classification heatmaps. This greatly simplifies the model.

This simplification is meaningful in many aspects. First, it can support more embedded devices. Some embedded chips do not support models with multiple output heads, which causes them to encounter some obstacles in industrial applications. Secondly, its training mode is end-to-end, which can directly calculate the loss function between the output value and the predicted value. Many models design different loss functions for different output heads. And different positive and negative samples need to be generated to meet their training requirements. This brings a lot of inconvenience to fine tuning in industrial applications. Finally, since the input and output are equally large, the model can be realized by using the FCN. There is little need to limit the size of the input image, which does not need to be square. Moreover, the design of FCN enables the model to use different sizes of pictures in training and testing. This enables high resolution image applications.

3.2   Heatmaps

Before introducing our heatmaps, we can review how CenterNet generates heatmaps.

As CenterNet first filled the input image into a square and resize it to 512×512. Its output heatmaps was 1/4 the size of the input after downsampling. Therefore, there was a certain offset error between the position of the center from the heatmaps and the real object center. The offset feature map was set to compensate for the offset error. At the same time, the radius of the Gaussian heat circle was determined by the height and width of the object. Thus, the object size feature map was needed to generate the height and width (Fig. 2).

Fig. 2. Heatmap between CenterNet and our method. CenterNet limited the size of the input image to 512×512 by padding and resizing. The heatmap size was 128×128, and the heat spots of the two objects were circular. Our method does not need to limit the input image size, and the heat map is of the same size. The heat spot shape of an object is related to its horizontal and vertical sizes.

In our method, there is no need to fill the input image into a square. The size of classification heatmaps is set to be equal to the input image, and the position of the center obtained from the heatmaps is the real object center without offset error. So there is no need to output the offset feature map. Meanwhile, the way we generate the heatmaps is different from CenterNet. It used Gaussian distributions in heatmaps, but we use polynomial heatmaps. The radius of the Gaussian circle is jointly determined by the height and width of the object. The polynomial heatmaps in our method can be correlated with the length and width in the corresponding direction. Then the object size feature map is not needed, because the length and width of the object can be calculated according to the polynomial heatmaps.

To calculate the size parameters in both horizontal and vertical directions from the heatmaps, we define a simple polynomial which is related to the width and height of the object. For the object box whose CP is located at (x_c, y_c) and whose width is w and height is h, we define the probability heat value generation mode as polynomial. It can be expressed as,

$$v = 1 - \alpha(\frac{|x - x_c|}{w})^r - \alpha(\frac{|y - y_c|}{h})^r \tag{1}$$

where v represents the probability heat value at the position (x, y), r represents the degree of the polynomial, α represents the size attenuation coefficient, and its value should be greater than 0.5. When its value is 0.5, the probability heat value attenuates to 0 at the midpoint of the edge of the object box. All values greater than 0 form an inscribed graph of the object box. To avoid having a heat value less than zero, we set the value less than 0 to 0, to keep the probability between 0 and 1.

By adjusting r (the degree of the polynomial) and α (the attenuation coefficient) in Eq. (1), we can get different types of heat spots. A few examples of the heatmaps that can be generated with different hyperparameters are shown in Fig. 3.

(a)   (b)   (c)   (d)   (e)   (f)

Fig. 3. Examples of the heatmap with different r and α. The luminance in the figure reflects the magnitude of the probability heat value. (a) is the image slice of an object box; (b) and (c) are the heatmaps when r is 1, but their α are different with values of 0.5 and 1, respectively; (d) is the heatmap when r is 0.5; (e) and (f) are the heatmaps when r is 2, but their α are different with values of 0.5 and 1, respectively.

When two objects are too close, the problem of overlapping heat regions occurs. Our approach is to take the larger value at the overlapping position. This tries to avoid the influence of one on the other. As long as the centers do not coincide exactly, we can locate the centers of both. But when it comes to calculating the size of an object, it might oversize.

3.3 Object Size

The whole process is divided into three parts: Locating the CP of the object; Calculating its width and height according to the heatmaps; Finally, fine-tuning the boundaries of the object.

CP Position. The process for locating the CP of an object is slightly different from CenterNet. Before the maxpooling step, we first perform a large kernel mean filtering on the heatmaps. This is because there are multiple identical maximum values near the CP of the large object, causing duplicate object boxes. After that, it is maxpooled. By comparing the maxpooling result with the mean filtering result, the position with the same value is the local maximum point, which is the center of the object.

Size Parameters. After locating the center of the object box, we need to get the width and height of the object. On the heatmaps, the probability heat value decreases in the corresponding direction from the center of the object. When traversing from the center of the object along the horizontal or vertical direction, the probability heat value change at this time is independent of the coordinates in the other direction. This is easy to get from Eq. (1). According to the mapping relationship between the value and the distance from the CP, the corresponding direction size parameter is determined. The width can be calculated by the following formula:

$$w = \frac{\sqrt[r]{\alpha}}{2} \left(\frac{(x_c - x_l)}{\sqrt[r]{1 - v_l}} + \frac{(x_r - x_c)}{\sqrt[r]{1 - v_r}} \right) \tag{2}$$

where v_l represents the probability heat value at the position (x_l, y_c) on the left of the center, v_r represents the probability heat value at the position (x_r, y_c) on the right of the center, and α is the attenuation coefficient. When the probability heat value decays to a more reliable value, i.e., the probability threshold such as 0.5, the width of the object frame can be calculated by Eq. (2). Similarly, the height can be calculated by the following formula:

$$h = \frac{\sqrt[r]{\alpha}}{2} \left(\frac{(y_c - y_t)}{\sqrt[r]{1 - v_t}} + \frac{(y_b - y_c)}{\sqrt[r]{1 - v_b}} \right) \tag{3}$$

where v_t represents the probability heat value at the position (x_c, y_t) on the top of the center, v_b represents the probability heat value at the position (x_c, y_b) below the center, and α is the attenuation coefficient.

Bounding Box. With the CP position and width and height, it is easy to calculate the position of the object's bounding box. However, in the experiment, we found that fine-tuning the bounding box can alleviate the problem of CP positioning bias. For example, when the CP is unbiased, the values of the two terms in parentheses in Eq. (2) should be equal. But when offset, the two are unequal. Taking the left border as an example, its value is considered to be related to the width calculated by the left position (x_l, y_c). The formula is expressed as:

$$b_l = x_c - \frac{\sqrt[r]{\alpha}}{2} \frac{(x_c - x_l)}{\sqrt[r]{1 - v_l}} \tag{4}$$

where b_l represents the left boundary of the object.

This way, when the CP is left off, the left border is skewed to the right accordingly, and vice visa. The other three boundaries can be calculated in a similar way. With all the boundary parameters, we get a complete object box.

3.4 Loss Function

The loss function adopted in this paper is a combination of two loss functions. One is the MSE (Mean Squared Error) loss, and the other is the CP Loss designed for the model proposed in this paper. The linear combination of the two functions is the loss function adopted in the training of our model, which can be expressed by the formula:

$$L(y, p) = a \cdot L_{MSE}(y, p) + L_{CP}(y, p) \tag{5}$$

where y is the ground truth, p is the predicted value, and a is a hyperparameter used to coordinate the difference of orders of magnitude between the two loss functions. In this paper, the value is set to 100.

The reason why the MSE loss is selected is that the probability heat value is continuous. While the classical CE (Cross Entropy) loss function is suitable for the calculation of discrete variables, i.e., the case of only 0 or 1. Similarly, the Dice loss [31] widely used in image segmentation is less applicable to continuous variables. To confirm this, we conducted ablation experiments on the loss function to prove its effectiveness, as detailed in Subsect. 5.2.

During the experiment, we found that if the CP position of a object deviated, it would lead to problems in the calculation of width and height, and thus greatly reducing the accuracy. To alleviate this problem, we propose a CP loss function, which draws on the center position in model inference and the Dice loss design.

The specific method is as follows: Firstly, the maxpooling layer with step size of one is used to process the maximum value of the neighborhood of the output heatmaps. Secondly, the mask matrix is defined as the position of the maximum point. In other words, the position of the pooled heatmaps that is equal to the original heatmaps is assigned the value of 1, and the rest is 0. It can be expressed by the following formula:

$$mask(x, y) = \begin{cases} 0, & mp(hm(x, y)) \neq hm(x, y) \\ 1, & mp(hm(x, y)) = hm(x, y) \end{cases} \tag{6}$$

where, $mask(x, y)$ represents the mask matrix value at the position (x, y), $hm(x, y)$ represents the output heatmaps value at the position (x, y), and $mp()$ represents the maxpooling operation, whose kernel size is set to the same as in the locating CP.

After obtaining the mask matrix, we can calculate the cross entropy of the predicted value and the true value after the mask. Specifically, multiplying the predicted value and the true value by the mask matrix and dividing the calculation result of the cross-entropy loss function by the modulus of the mask matrix, we can get the CP loss function in Eq. (7).

$$L_{CP}(y, p) = \frac{L_{CE}(y \cdot mask, p \cdot mask)}{mask} \tag{7}$$

In other words, the CP loss function only considers the error at the maximum point of the classification heatmaps. When the maximum point is closer to the object center, the loss function value is smaller, and vice versa. Meanwhile, it also limits the number of maximum points. Too many maximum values will lead to the increase of 1s of the mask matrix, thus increasing the value of the loss function. The loss function of CP greatly improves the CP positioning, which is also confirmed by the ablation experiments.

4 Evaluation

This section shows the experimental result: firstly, data and experimental environment we adopt and its settings, and then the accuracy and speed of each model.

4.1 Data and Settings

The experimental data set uses the COCO dataset [12]. We use the COCO train2017 split for training and val2017 split as validation for our evaluation study. We report our main results on the test-dev split by uploading our detection results to the evaluation server. All models are trained on the PyTorch 1.9.1 framework and an NVIDIA 2080Ti GPU with 11 GB memory.

4.2 Accuracy and Speed

The accuracy index we use is mAP (mean Average Precision). The speed uses the common metric FPS. AP_{50} refers to the AP when the value of the IoU is 50%, and the same is true for AP_{75}. AP_S, AP_M and AP_L are the AP values of three different scale objects of small, medium and large. The speed/accuracy comparison shows between DetOH and some of the most recent detection methods in Fig. 4.

Using the backbone DLA-34 [29], DetOH can achieve 40.3% of AP at 68 FPS on a single 2080Ti GPU graphics card. We further replaced DLA-34 with a deeper network DLA-60, resulting in a better speed/accuracy trade-off (43.3% AP at 32 FPS). Compared to CenterNet [32], we improved network performance by 3.2% AP, increasing speeds from 6 to 32 FPS. This means that with improved accuracy, DetOH is 433% faster than CenterNet when using the same backbone network. DetOH also outperforms other methods in terms of speed and accuracy, including anchor-based methods.

To achieve higher accuracy, we use deeper backbone networks and a more efficient feature extraction structure. The specific results are shown in Table 1. As it can be seen,, our model AP is higher than all classic models. Specifically, it is 1.2% AP higher than CenterNet and 2.9% AP higher than FCOS. It is worth noting that our approach is significantly ahead in the performance of small objects. This may be due to the beneficial gain brought about by increasing the output resolution. Some of the latest methods [27] with a particularly high AP use a lot of tricks, such as data augmentation during the testing phase, increasing deformable convolution.

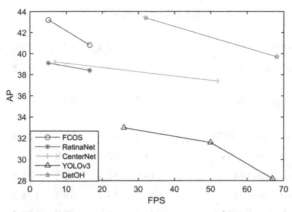

Fig. 4. Speed/accuracy trade-off between DetOH and several recent methods: FCOS [9], CenterNet [7], YOLOv3 [6] and RetinaNet [11]. Speed is measured on a NVIDIA 2080Ti GPU. For fair comparison, we only measure the network latency for all detectors. DetOH achieves competitive performance compared with recent methods including anchor-based ones.

Table 1. DetOH vs. Other State-of-the-art Two-stage or One-stage Detectors. DetOH outperforms a few recent anchor-based and anchor-free detectors.

Method	Backbone	AP	AP_{50}	AP_{75}	AP_S	AP_M	AP_L
Two-stage methods:							
Faster R-CNN by G-RMI [33]	Inception-ResNet-v2 [28]	34.7	55.5	36.7	13.5	38.1	52.0
Faster R-CNN + + + [28]	ResNet-101	34.9	55.7	37.4	15.6	38.7	50.9
Faster R-CNN w/ FPN [23]	ResNet-101-FPN	36.2	59.1	39.0	18.2	39.0	48.2
Faster R-CNN w/ TDM [14]	Inception-ResNet-v2-TDM	36.8	57.7	39.2	16.2	39.8	52.1
One-stage methods:							
YOLOv2 [5]	DarkNet-19	21.6	44.0	19.2	5.0	22.4	35.5
SSD [4]	ResNet-101-SSD	31.2	50.4	33.3	10.2	34.5	49.8
YOLOv3 [6]	Darknet-53	33.0	57.9	34.4	18.3	35.4	41.9
DSSD [15]	ResNet-101-DSSD	33.2	53.3	35.2	13.0	35.4	51.1

(continued)

Table 1. (*continued*)

Method	Backbone	AP	AP$_{50}$	AP$_{75}$	AP$_S$	AP$_M$	AP$_L$
RetinaNet [11]	ResNet-101-FPN	39.1	59.1	42.3	21.8	42.7	50.2
CornerNet [16]	Hourglass-104	40.5	56.5	43.1	19.4	42.7	53.9
FSAF [19]	ResNeXt-101-FPN	42.9	63.8	46.3	26.6	46.2	52.7
FCOS [9]	ResNet-101-FPN	43.2	62.4	46.8	26.1	46.2	52.8
CenterNet [8]	Hourglass-104	44.9	62.4	48.1	25.6	47.4	**57.4**
DetOH	Hourglass-104	44.7	63.1	48.1	28.3	48.4	54.7
DetOH	ResNet-101-BiFPN	**46.1**	**65.3**	**49.8**	**29.2**	**49.6**	56.5

4.3 Application

In the multi-source sensor intelligent security system, we apply the DetOH model to detect objects in the image stream. We apply the trained model to the digital processing chip Hi3519AV100 and successfully achieve object detection, proving a good balance of speed and accuracy. Although Hi3519AV100 has 1.7 TOPS neural network compu-ting performance, it supports the Caffe framework only and is based on Caffe-1.0. It does not support the attention mechanism and channel shuffling, which makes most of the recent mobile models difficult to apply. Considering the huge data processing tasks with only 1 GB memory, it is unfriendly to large models.

Our model runs on the chip's AI-accelerated unit, while post-processing can run on CPU. This is another advantage of DetOH, which can handle different detection processes with different arithmetic units. We used pipeline acceleration for model inference and post-processing. For a single frame size of 1920 × 1080 pixels, the processing time can be reduced to 380 ms.

5 Ablation Study

In this section, we conduct an ablation study on the three main ideas in this work. The first is to compare different heatmaps. The effects of different loss functions on the results are then compared. Finally, we look at the difference that boundary fine-tuning brings.

We adjusted the dataset for faster ablation studies. The COCO dataset was too large and the training time was too long, so we replaced it with a small private dataset. The training set of this dataset contains 3,518 images, and the test set contains 704 images, all of which are of 640 × 360 pixels. And the data set covers only three categories: person, vehicle, and aircraft. At the same time, we used a smaller backbone network, MobileUnet [32]. It has fewer parameters and is easier to converge. This means that the training time per ablation experiment can be reduced.

5.1 Heatmaps

This section will introduce the influence of heatmap hyperparameters on the object detection, i.e., the degree of polynomial and the attenuation coefficient in Eq. (1). To see the change of its probability heat value more conveniently, the change curve of the probability heat value with coordinates is plotted. Figure 5 shows the probability heat value curves with different hyperparameters.

From the derivative at the CP, it is not derivable when the polynomial degree is 1 or 0.5, and its value is 0 when the polynomial degree is 2. Our experiments show that neural networks fit a derivable function more easily. To evaluate the impact of heatmaps generated by different hyperparameter combinations on object detection accuracy, an ablation study was conducted. The statistical results of the influence of different hyperparameter combinations on accuracy are shown in Table 2.

From the degree of polynomials, the best performer is the quadratic elliptic heatmaps, followed by the linear diamond heatmaps, and the worst is the square star heatmaps. It can be seen that CNNs are better at fitting derivable convex function graphs. When the probability heat value does not fall smoothly at the CP, it is difficult to simulate the effect of this mutation.

From the attenuation coefficient, when it is large, the heat spot area will be more concentrated near the center of the object, and the accuracy rate is higher. When it is small, the heat spots are more dispersed and the accuracy rate decreases. This is because there is overlap between multiple objects of the same kind. When the two object areas partially overlap, the scattered hot spots will also be more likely to intersect. Although the probability heat value of the intersection area to the maximum value can reduce the influence of the CP position offset, it still brings trouble to the calculation of size parameters. If the two targets are too close together, the size calculation will be on the larger side.

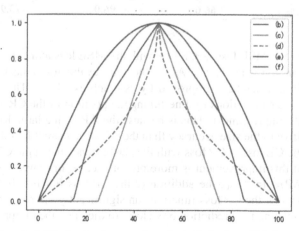

Fig. 5. Curve of probability heat value as a function of abscissa. This is a target with a width of 101 and the ordinate is on the central axis. (b) (c) (d) (e) (f) corresponds to that in Fig. 3, respectively.

Table 2. Quantitative results from Different Hyperparameter Combinations.

	Degree of polynomial	Attenuation coefficient	AP	AP_{50}	AP_{75}
(b)	$r = 1$	$\alpha = 0.5$	66.6	96.0	73.9
(c)	$r = 1$	$\alpha = 1$	72.5	**98.3**	84.3
(d)	$r = 0.5$	$\alpha = 0.5$	52.2	94.1	49.2
(e)	$r = 2$	$\alpha = 0.5$	72.3	93.7	86.8
(f)	$r = 2$	$\alpha = 1$	**74.4**	97.5	**92.7**

5.2 Loss Function

The loss function adopted in this work is a combined loss function, which is replaced by a different loss function to study its effect on the result. The statistical results of the influence of different loss functions on accuracy are shown in Table 3.

Table 3. Quantitative Results from Different Loss Functions.

Loss Function	AP	AP_{50}	AP_{75}
CE loss	28.0	71.1	22.3
Dice loss	0.483	2.45	0.121
Focal loss	23.7	61.5	24.4
MSE loss	56.0	94.7	45.0
Focal loss + CP loss	44.3	82.3	39.6
CE loss + CP loss	55.3	91.8	58.6
MSE loss + CP loss	**66.6**	**96.0**	**73.9**

By comparing the MSE loss + CP loss with the MSE loss alone, the improvement of the CP loss function is huge. The AP as the main evaluation indicator is improved by 18.9%. In AP_{75}, it has been improved by as much as 64.2%. This shows a huge improvement in size calculations by fine-tuning the position of the CP. But in AP_{50}, it brought only 1.4% improvement. This is because the AP with a large IoU threshold is much more sensitive to the size accuracy than the AP with a small IoU threshold.

Comparing the CE loss + CP loss with the performance using the CE loss alone, it can be found that this improvement is more pronounced. It improved by 97.5% in AP and 162.8% in AP_{75}. In AP_{50}, the addition of the CP loss function brought a 29.1% boost. This shows that the CP loss function can significantly improve the accuracy of size calculation, making the prediction box closer to the bbox then improving the IoU.

5.3 Boundary Fine-Tune

To evaluate the impact of boundary fine-tuning on object detection accuracy, an ablation study was conducted.

Table 4. Quantitative Results from Boundary Fine-tuning.

Boundary Fine-tune	AP	AP_{50}	AP_{75}
	74.4	97.5	92.7
✓	**76.5**	**97.6**	**94.5**

As can be seen from Table 4, boundary fine-tuning brings improvement to AP at high IoU threshold, which illustrates its effectiveness in bounding box fitting. These findings suggest that boundary fine-tuning can play a critical role in improving the accuracy of object detection algorithms.

6 Conclusion

In this work, we propose an anchor-free object detector with only heatmaps (DetOH). Our experiments demonstrate that DetOH is superior to widely used anchor-free object detectors, including CenterNet and FCOS, but with much less model complexity. The single-head design makes it easier to be applied to embedded devices. Now it has been applied to the field of security, making human life more secure. Its network architecture is also suitable for other intensive prediction tasks, such as semantic segmentation. Given its effectiveness and efficiency, we hope it soon be applied to high-level tasks such as multi-object tracking, motion recognition, and behavior understanding.

Acknowledgments. This work was supported inpart by the National Natural Science Foundation of China (61972219), the Research and Development Program of Shenzhen (JCYJ20190813174403598), the Overseas Research Cooperation Fund of Tsinghua Shenzhen International Graduate School (HW2021013), the Guangdong Basic and Applied Basic Research Foundation (2022A1515010417), the Key Project of Shenzhen Municipality (JSGG20211029095545002), the Science and Technology Research Project of Henan Province(222102210096).

References

1. Viola, P., Jones, M.J.: Robust real-time face detection. Int. J. Comput. Vision **57**, 137–154 (2004)
2. Doll´ar, P., Appel, R., Belongie, S., Perona, P.: Fast feature pyramids for object detection. IEEE Trans. Pattern Anal. Mach. Intell. **36**(8), 1532–1545 (2014)
3. Ren, S., He, K., Girshick, R., Sun, J.: Faster R-CNN: towards real-time object detection with region proposal networks. In: Advances in Neural Information Processing Systems, vol. 28 (2015)
4. Liu, W., Anguelov, D., Erhan, D., Szegedy, C., Reed, S., Fu, C.-Y., Berg, A.C.: Ssd: Single shot multibox detector. In: Leibe, B., Matas, J., Sebe, N., Welling, M. (eds.) Computer Vision – ECCV 2016. Lecture Notes in Computer Science, vol. 9905, pp. 21–37. Springer, Cham (2016). https://doi.org/10.1007/978-3-319-46448-0_2

5. Redmon, J., Farhadi, A.: Yolo9000: better, faster, stronger. In: Proceedings of the IEEE Conference on Computer Vision and Pattern Recognition, pp. 7263–7271 (2017)
6. Redmon, J., Divvala, S., Girshick, R., Farhadi, A.: You only look once: unified, real-time object detection. In: Proceedings of the IEEE Conference on Computer Vision and Pattern Recognition, pp. 779–788 (2016)
7. Zhou, X., Wang, D., Krähenbühl, P.: Objects as points. arXiv preprint: arXiv:1904.07850 (2019)
8. Duan, K., Bai, S., Xie, L., Qi, H., Huang, Q., Tian, Q.: CenterNet: keypoint triplets for object detection. In: Proceedings of the IEEE/CVF International Conference on Computer Vision, pp. 6569–6578 (2019)
9. Tian, Z., Shen, C., Chen, H., He, T.: FCOS: fully convolutional one-stage object detection. In: Proceedings of the IEEE/CVF International Conference on Computer Vision, pp. 9627–9636 (2019)
10. Tian, Z., Shen, C., Chen, H., He, T.: FCOS: a simple and strong anchorfree object detector. IEEE Trans. Pattern Anal. Mach. Intell. **44**(4), 1922–1933 (2020)
11. Lin, T.Y., Goyal, P., Girshick, R., He, K., Doll'ar, P.: Focal loss for dense object detection. In: Proceedings of the IEEE International Conference on Computer Vision, pp. 2980–2988 (2017)
12. Lin, T.-Y., Maire, M., Belongie, S., Hays, J., Perona, P., Ramanan, D., Dollár, P., Zitnick, C.L.: Microsoft coco: Common objects in context. In: Fleet, D., Pajdla, T., Schiele, B., Tuytelaars, T. (eds.) ECCV 2014. Lecture Notes in Computer Science, vol. 8693, pp. 740–755. Springer, Cham (2014). https://doi.org/10.1007/978-3-319-10602-1_48
13. Girshick, R.: Fast R-CNN. In: Proceedings of the IEEE International Conference on Computer Vision, pp. 1440–1448 (2015)
14. Shrivastava, A., Sukthankar, R., Malik, J., Gupta, A.: Beyond skip connections: top-down modulation for object detection. arXiv preprint: arXiv:1612.06851 (2016)
15. Fu, C.Y., Liu, W., Ranga, A., Tyagi, A., Berg, A.C.: DSSD: deconvolutional single shot detector. arXiv preprint: arXiv:1701.06659 (2017)
16. Law, H., Deng, J.: CornerNet: detecting objects as paired keypoints. In: Ferrari, V., Hebert, M., Sminchisescu, C., Weiss, Y. (eds.) Computer Vision –. Lecture Notes in Computer Science, vol. 11218, pp. 765–781. Springer, Cham (2018). https://doi.org/10.1007/978-3-030-01264-9_45
17. Yu, J., Jiang, Y., Wang, Z., Cao, Z., Huang, T.: Unitbox: an advanced object detection network. In: Proceedings of the 24th ACM International Conference on Multimedia, pp. 516–520 (2016)
18. Huang, L., Yang, Y., Deng, Y., Yu, Y.: DenseBox: unifying landmark localization with end to end object detection. arXiv preprint: arXiv:1509.04874 (2015)
19. Zhu, C., He, Y., Savvides, M.: Feature selective anchor-free module for single-shot object detection. In: Proceedings of the IEEE/CVF Conference on Computer Vision and Pattern Recognition, pp. 840–849 (2019)
20. Yang, Z., Liu, S., Hu, H., Wang, L., Lin, S.: RepPoints: point set representation for object detection. In: Proceedings of the IEEE/CVF International Conference on Computer Vision, pp. 9657–9666 (2019)
21. Duan, K., Xie, L., Qi, H., Bai, S., Huang, Q., Tian, Q.: Corner proposal network for anchor-free, two-stage object detection. In: Vedaldi, A., Bischof, H., Brox, T., Frahm, J.-M. (eds.) Computer Vision – ECCV 2020. Lecture Notes in Computer Science, vol. 12348, pp. 399–416. Springer, Cham (2020). https://doi.org/10.1007/978-3-030-58580-8_24
22. Samet, N., Hicsonmez, S., Akbas, E.: Houghnet: Integrating near and long-range evidence for bottom-up object detection. In: Vedaldi, A., Bischof, H., Brox, T., Frahm, J.-M. (eds.) Computer Vision – ECCV 2020. Lecture Notes in Computer Science, vol. 12370, pp. 406–423. Springer, Cham (2020). https://doi.org/10.1007/978-3-030-58595-2_25

23. Lin, T.Y., Doll'ar, P., Girshick, R., He, K., Hariharan, B., Belongie, S.: Feature pyramid networks for object detection. In: Proceedings of the IEEE Conference on Computer Vision and Pattern Recognition, pp. 2117–2125 (2017)
24. Tan, M., Pang, R., Le, Q.V.: EfficientDet: scalable and efficient object detection. In: Proceedings of the IEEE/CVF Conference on Computer Vision and Pattern Recognition, pp. 10781–10790 (2020)
25. Qiu, H., Ma, Y., Li, Z., Liu, S., Sun, J.: Borderdet: Border feature for dense object detection. In: Vedaldi, A., Bischof, H., Brox, T., Frahm, J.-M. (eds.) Computer Vision – ECCV 2020. Lecture Notes in Computer Science, vol. 12346, pp. 549–564. Springer, Cham (2020). https://doi.org/10.1007/978-3-030-58452-8_32
26. Li, X., et al.: Generalized focal loss: Learning qualified and distributed bounding boxes for dense object detection. In: Advances in Neural Information Processing Systems, vol. 33, pp. 21002–21012 (2020)
27. Dai, X., et al.: Dynamic head: Unifying object detection heads with attentions. In: Proceedings of the IEEE/CVF Conference on Computer Vision and Pattern Recognition, pp. 7373–7382 (2021)
28. Szegedy, C., Ioffe, S., Vanhoucke, V., Alemi, A.: Inception-v4, inceptionresnet and the impact of residual connections on learning. In: Proceedings of the AAAI Conference on Artificial Intelligence, vol. 31 (2017)
29. He, K., Zhang, X., Ren, S., Sun, J.: Deep residual learning for image recognition. In: Proceedings of the IEEE Conference on Computer Vision and Pattern Recognition, pp. 770–778 (2016)
30. Yu, F., Wang, D., Shelhamer, E., Darrell, T.: Deep layer aggregation. In: Proceedings of the IEEE Conference on Computer Vision and Pattern Recognition, pp. 2403–2412 (2018)
31. Newell, A., Yang, K., Deng, J.: Stacked hourglass networks for human pose estimation. In: Leibe, B., Matas, J., Sebe, N., Welling, M. (eds.) Computer Vision – ECCV 2016. Lecture Notes in Computer Science, vol. 9912, pp. 483–499. Springer, Cham (2016). https://doi.org/10.1007/978-3-319-46484-8_29
32. Abdollahi, A., Pradhan, B., Alamri, A.: VNet: an end-to-end fully convolutional neural network for road extraction from high-resolution remote sensing data. IEEE Access **8**, 179424–179436 (2020)
33. Jing, J., Wang, Z., R"atsch, M., Zhang, H.: Mobile-Unet: an efficient convolutional neural network for fabric defect detection. Text. Res. J. **92**(1–2), 30–42 (2022)
34. Huang, J., et al.: Speed/accuracy trade-offs for modern convolutional object detectors. In: Proceedings of the IEEE Conference on Computer Vision and Pattern Recognition, pp. 7310–7311 (2017)

LAANet: An Efficient Automatic Modulation Recognition Model Based on LSTM-Autoencoder and Attention Mechanism

Qing Li⑩ and Xin Zhou(✉)

Institute of Software Chinese Academy of Sciences, Beijing, China
2066736985@qq.com

Abstract. This paper focuses on the task of automatic modulation recognition. Existing studies have shown low recognition accuracy at low signal-to-noise ratios, and models with a large number of parameters usually demand substantial computational resources, resulting in slower reasoning processes. In this paper, we propose a novel architecture for efficient recognition based on LSTM-Autoencoder and attention mechanism to address these challenges. Experimental results on benchmark datasets show that the proposed method achieves an average recognition accuracy of 62.43% and 64.49% on the RadioML2016.10a and RadioML2016.10b datasets, respectively. On the RadioML2016.10a dataset, the proposed model outperforms other SOTA models with a 2 ∼ 6% points improvement in recognition accuracy. The model also demonstrates superior recognition accuracy for both QAM64 and QAM16 modulation schemes and effectively increases average recognition accuracy by 1–2 percentage points in the -8dB to 2(±2) dB lower SNR range, indicating its noise robustness. On the RadioML2016.10b dataset, the proposed method's recognition accuracy is slightly higher than the SOTA model, demonstrating good performance.

Keywords: Automatic modulation recognition · Autoencoder · Attention

1 Introduction

Automatic modulation recognition (AMR) is essential in wireless communication as it identifies the pattern and characteristics of received signals for more efficient transmission and reception of messages. It can be generally grouped into two categories: likelihood-based (LB) methods and feature-based (FB) methods. LB methods [1–3] are optimal in the sense of Bayesian estimation but rely heavily on prior knowledge and parameter estimation. In contrast, FB methods are more computationally efficient and require less prior knowledge. FB methods use feature extraction techniques such as principal component analysis (PCA), recursive feature elimination, and wavelet decomposition [4] to identify the most

X. Yang et al. (Eds.): ADMA 2023, LNAI 14177, pp. 168–182, 2023.
https://doi.org/10.1007/978-3-031-46664-9_12

significant features and then use classification algorithms, such as support vector machines [5], neural networks and decision trees, to classify the data.

In the past few years, deep learning has been rapidly developed and widely used in computer vision, natural language processing, speech processing, etc., researchers have also attempted to apply deep learning-based methods to AMR. The relevant methods include CNN (convolutional neural network)-based [6,7], RNN (recurrent neural network)-based [8], and hybrid frameworks [9]. The research on modulation recognition based on deep learning was pioneered by the O'Shea group. In 2016, they constructed a modulation signal recognition dataset, RadioML2016.10a [10], using open-source software GNU-Radio and then utilized a 2-layer CNN to perform end-to-end processing of the waveform of the modulation signal, achieving good recognition results [6]. Next, a deep residual network framework was proposed [11] with a significant improvement in recognition performance over [6]. Generally, the CNN-based models directly take the original I/Q (in-phase/quadrature symbols of the modulated signal [12]) data as input. However, if the size of the convolution kernel can cover the two dimensions (I/Q) of the input, the addition operation in the convolution process will weigh the different dimensions of the feature but ignore the inherent characteristics of the two different dimensions of the I/Q input. To address this problem, MCNet [13] adopts 1D convolutional layers and 2D convolutional layers to extract features from single and joint dimensions, respectively. In addition, recent studies have carefully designed a series of CNN-based structures for modulation recognition tasks [7,14,15]. While CNN-based models are capable of capturing spatial features, they have been found to be inadequate at capturing temporal information. RNNs, on the other hand, are capable of extracting temporal features of time series data, particularly Long Short-Term Memory (LSTM) networks [16]. Paper [17] utilizes a 2-layer LSTM on amplitude and phase data for modulation classification, outperforming the proposed model [18]. Furthermore, to exploit both temporal and spatial features, CLDNN and CLDNN2 [15,18], which are combinations of CNN, LSTM, and deep neural networks (DNN), have been proposed as novel frameworks for modulation recognition. As the complexity of models increases, the speed of training and inference decreases. Therefore, designing lightweight and efficient recognition models has become a research focus. The LSTMDAE [19] proposes a multi-task learning framework based on the LSTM denoising autoencoder, which uses the features learned by the autoencoder for modulation classification. Subsequently, in order to further enhance the performance of AMR, ConvLSTMAE [20] proposed a parallel autoencoder architecture that leverages the temporal and spatial features extracted from two different encoders — an LSTM autoencoder and convolutional autoencoder. Despite these efforts, the recognition accuracy remains low under the low signal-to-noise ratio (SNR). Improving recognition accuracy under low SNR conditions remains a pressing problem in AMR tasks. To solve the above two problems, this paper presents a novel framework for effective modulation recognition that is based on the combination of the autoencoder [21] and attention mechanism. The main contributions of this research are as follows:

(1) The design of a novel architecture that combines an autoencoder and attention mechanism for automatic modulation recognition. Experiment results demonstrate the proposed framework achieves better recognition performance than SoTA frameworks.

(2) Investigation of the fine-tuning process for integrating the learned hidden features from the autoencoder with subsequent neural networks to improve performance.

(3) Exploration of the potential for attention mechanisms to further enhance the performance of automatic modulation recognition systems.

The remainder of the paper is organized as follows: in Sect. 2, we have a quick review of signal theory and the definition of the modulation recognition task. Section 3 presents our model LAANet in detail. In Sect. 4, we describe details about the setup of the experimental evaluation and the experimental results. Finally, the conclusion is given in Sect. 5.

2   Problem Definition

2.1   Definition of the AMR Problem

Automatic modulation recognition can be viewed as a multi-class classification problem. The received signal can be represented as (1).

$$r(t) = s(t) + n(t), \tag{1}$$

where $s(t)$ is the noise-free complex base-band envelope of the received signal, also known as the valid signal, and $n(t)$ is the noise, including the receiver's internal noise and noise from the antenna and external environment. The goal of modulation recognition is to determine which class among N the received signal belongs to. It can be represented by calculating

$$P(y = C_k | r(t)), \tag{2}$$

where C_k denotes the kth class and y is the true class of the signal.

3   Model

In this section, we explicate the architecture of the proposed model, LAANet, in detail. As depicted in Fig. 1, LAANet is comprised of three primary components: the LSTM autoencoder, the intermediate part, and the classifier. The LSTM autoencoder is formed by the first two LSTM layers and a shared dense layer, minimizing the error between the input and its reconstructed output to learn a compact data representation. The intermediate part comprises a Gaussian dropout layer, an Lstm layer, and an attention layer, performing model regularization and fine-tuning the information learned by the autoencoder. The attention layer then weighs the output of the last LSTM layer to obtain the final features for classification. Finally, the classifier consists of two fully connected

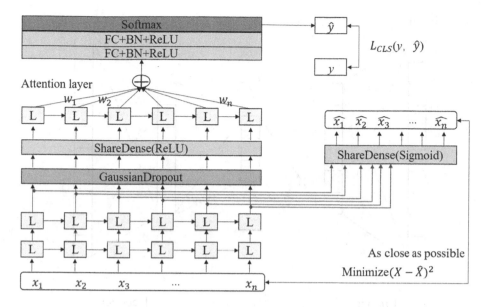

Fig. 1. An illustration of LAANet.

layers, followed by a softmax activation function to produce category probabilities. The overarching framework is a multi-task framework comprising two tasks: the task of reconstructing inputs in the LSTM autoencoder and the task of modulation classification. The total loss is calculated as the weighted sum of the classification loss and the reconstruction loss of the LSTM autoencoder.

3.1 An LSTM AutoEncoder-LSTMAE

The architecture of the autoencoder consists of three components: an encoder, hidden features, and a decoder, as illustrated in Fig. 2. The encoder and decoder are typically composed of stacked neural network layers, which can be selected from a range of options, such as CNN, LSTM, and DNN, depending on the task. The autoencoder can be easily trained using gradient back-propagation. During the training process, the encoder converts input data into a compact hidden representation, which the decoder then employs to reconstruct the original input as accurately as possible. Generally, we can define the encoding and decoding processes as (3) and (4).

$$h_i = g(x_i) \tag{3}$$

$$\hat{x}_i = f(h_i) = f(g(x_i)) \tag{4}$$

where x_i denotes the input of the autoencoder. The encoder and decoder are modeled as the functions g and f respectively. The output of the encoder, i.e., the hidden feature vector, is represented by h_i. The output of the decoder is represented by $\hat{x}_i$. The learning process of the autoencoder can be viewed as finding two functions, g and f, that minimize the following objective function:

Input   Output

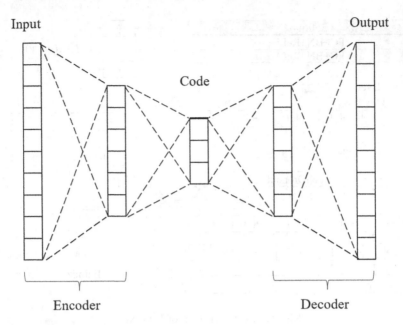

Fig. 2. Architecture of the autoencoder.

$$argmin \Sigma \Delta(x_i, \hat{x}_i), \tag{5}$$

where Δ is a metric used to calculate the difference between the input and reconstructed output of the autoencoder, with mean squared error (MSE) being a commonly used metric.

Based on the characteristics of the autoencoder and LSTM network, this study designs an LSTM autoencoder to effectively capture the temporal dependencies of wireless radio signals, which are considered time sequences. In LSTMAE, the first two LSTM layers act as the encoder, while a shared dense layer serves as the decoder. The encoder learns intrinsic features of inputs by attempting to reconstruct them as accurately as possible. This information is hugely consequential in the final classification.

3.2 Attention

LSTM with attention is a powerful deep learning architecture used for various natural language processing tasks. Traditionally, the output vector of the last step of the LSTM layer is directly fed into a classifier for classification. In contrast, the attention module first calculates the weight for each time step and then computes the weighted sum of vectors from all time steps to obtain the attention vector, which is subsequently input to the classifier. The essence of the attention mechanism lies in its ability to assign flexible weights to different portions of the input sequence, thereby enabling the network to concentrate on the utmostly relevant information to the current task. In our implementation, the attention layer is incorporated after the last LSTM layer. The structure of the attention layer is depicted in Fig. 3 and the calculation process can be succinctly described as follows: Let H be a matrix consisting of output vectors $[h_1, h_2, h_3, ..., h_t]$ that the LSTM layer produced, where t denotes the length of the input signal. The attention vector r is formed by a weighted sum of these output vectors:

$$M = tanh(H) \tag{6}$$

$$\alpha = softmax(w^T M) \tag{7}$$

$$H' = H\alpha \tag{8}$$

$$r = sum(H') \tag{9}$$

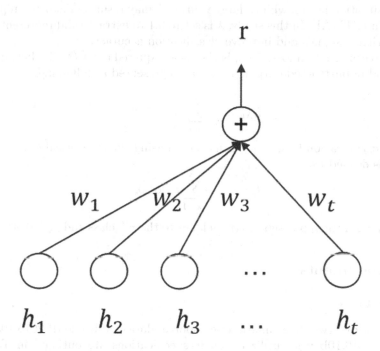

Fig. 3. An illustration of the attention mechanism.

where $H \in R^{d^w \times t}$, d^w is the dimension of the out vectors of the LSTM, w is a trained parameter vector, and w^T is a transpose. The dimension of w, α, r is d^w, t, d^w separately.

3.3   Loss

As previously mentioned, the proposed model is in a multi-task framework that includes signal modulation classification and input signal reconstruction in the

LSTMAE. In this structure, the primary task is classification, while the LST-MAE aims to learn the essential characteristics of the input signals throughout the reconstruction process. The objective of LSTMAE is to minimize the error between its input and output, with the reconstruction loss using mean squared error. On the other hand, the classification task uses cross-entropy loss to gauge the accuracy of identifying the modulation type of the input signal.

Overall, the total loss is the weighted sum of the reconstruction loss L_{AE} and the classification loss L_{CLS}, represented as follows:

$$L = (1 - \lambda)L_{AE} + \lambda L_{CLS}, \tag{10}$$

where λ is a weighting factor that balances the relative importance of the reconstruction and classification tasks. A small value of λ minimizes the influence of the classification layers, while a large value of λ may result in distorted representations in LSTMAE. In this study, λ is set to 0.1 to extract valid representations of the original signals and improve classification accuracy.

The reconstruction loss, L_{AE}, is the mean-squared error (MSE) between the input and reconstructed output. It can be represented as follows:

$$L_{AE} = \frac{1}{n} \sum_{i=1}^{n} (x_i - \hat{x}_i)^2 \tag{11}$$

The classification loss, L_{CLS}, is measured using the categorical cross-entropy loss. It is defined as:

$$L_{CLS} = - \sum_{k=1}^{K} p_k \log \hat{p}_k, \tag{12}$$

where $p_k = 1$ if the input signal $x(t)$ belongs to the kth class and $p_k = 0$ otherwise.

4   Experiments

4.1   DataSet

In this study, two benchmark open-source datasets RadioML2016.10a and RadioML2016.10b were utilized. Their specifications are outlined in Table.1. The RadioML2016.10a dataset contains 220,000 modulation signals with SNRs ranging from -20dB to 18dB, including 11 commonly used modulation types: WBFM, AM-DSB, AM-SSB, BPSK, CPFSK, GFSK, 4-PAM, 16-QAM, 64-QAM, QPSK, and 8PSK. The RadioML2016.10b dataset is an extension of the RadioML2016.10a dataset, including 1.2 million signals and 10 modulation types, excluding AM-SSB. Each signal in both datasets has 128 complex float time I/Q samples and was generated under adverse simulated propagation environments, affected by AWGN, multipath fading, sample rate offset, and center frequency offset, similar to real-world scenarios. The datasets were split into a 6:2:2 ratio, for training, validation, and testing purposes, respectively. Specifically, for each

SNR and modulation type, 600 (3600) signals were randomly selected as training data, 200 (1200) signals as validation data, and 200 (1200) signals as test data.

Table 1. Information of RadioML2016.10a and RadioML2016.10b.

dataset	modulation	sample	size	SNR(dB)
RadioML 2016.10a	11 classes (8PSK, BPSK, CPFSK, GFSK, PAM4, AM-DSB, AM-SSB, 16QAM, 64QAM, QPSK, WBFM)	2×128	220000	$-20{:}2{:}18$
RadioML 2016.10b	10 classes (8PSK, BPSK, CPFSK, GFSK, PAM4, AM-DSB, 16QAM, 64QAM, QPSK, WBFM)	2×128	1200000	$-20{:}2{:}18$

4.2   Details

In the proposed model, the hidden dimensions of the three LSTM layers are set to 64, 32, and 32 respectively. In addition, the number of nodes of the shared dense layer used for the decoder layer is 2, while the number of nodes in the shared dense layer after the Gaussian dropout layer is 64. The number of nodes in the fully connected layers of the classifier is set to 32, 16, and K(number of categories in the dataset), respectively.

The input data is in the form of normalized amplitude and phase features to assist in learning time-dependent relationships [17], instead of using the original I/Q components.

The optimization algorithm used in all experiments is the Adam optimizer. The initial learning rate is set to 0.001, and a batch size of 400 is used throughout the experiments. The rate of the Gaussian dropout is set to 0.2. In case the validation loss does not decrease within 5 epochs, the learning rate is halved. The training process is terminated if the validation loss remains stable for 50 epochs. The experiments were implemented using the Tesla P100 GPU and Keras with Tensorflow as the backend.

4.3   Main Results and Discussion

We initially trained our model using the RadioML2016.10a dataset with the SNRs ranging from -20dB to 18dB. To prove the effectiveness of the proposed model, a series of experiments were conducted on the RadioML2016.10b dataset with exactly the same experimental configuration.

Table 2. Highest and Average accuracy comparison on the two datasets.

Model	RadioML2016.10a		RadioML2016.10b	
	Highest accuracy	Average accuracy	Highest accuracy	Average accuracy
CGDNet	84.05%	56.55%	89.64%	61.12%
PET-CGDNN	90.68%	60.22%	93.18%	63.84%
IC-AMCNET	84.55%	56.58%	92.83%	62.29%
MCLDNN	91.64%	60.87%	93.74%	64.61%
MCNET	83.27%	55.84%	88.89%	61.04%
LSTM	90.68%	60.30%	93.48%	64.30%
LSTMDAE	85.41%	57.10%	93.43%	64.11%
LAANet(Ours)	**92.73%**	**62.43%**	**93.88%**	**64.69%**

Table 3. TOP-1 CLASSIFICATION ACCURACY COMPARISON OF OUR PRO-POSED MODEL (LAANet) VS.EXISTING MODELS ON RADIOML2016.10A DATASET. THE HIGHEST AVERAGE TOP-1 CLASSIFICATION ACCURACY FOR EACH SNR IS MARKED IN BOLD.

Model	-20	-18	-16	-14	-12	-10	-8	-6	-4	-2
CGDNet	9.64	9.41	9.27	11.41	15.41	22.05	34.68	50.27	65.05	74.77
PET-CGDNN	9.36	9.23	8.95	11.00	14.18	23.91	36.32	52.55	65.68	77.05
IC-AMCNET	9.18	9.09	9.68	10.95	14.50	23.64	35.05	49.27	63.32	72.45
MCLDNN	9.45	9.32	8.82	10.45	13.95	23.55	**37.55**	53.82	66.59	81.14
MCNET	9.50	9.45	9.50	11.00	15.05	**25.23**	36.95	51.32	62.55	73.09
LSTM	9.05	**9.82**	10.41	12.41	16.14	24.36	35.05	51.36	63.55	78.05
LSTMDAE	**9.86**	9.45	9.68	11.73	16.00	24.73	36.27	51.45	61.18	73.82
LAANet(our model)	9.73	9.36	**10.45**	**13.18**	**16.82**	23.86	36.36	**55.05**	**69.27**	**82.77**

Model	0	2	4	6	8	10	12	14	16	18
CGDNet	79.95	82.18	83.00	83.09	82.73	83.77	83.55	83.32	83.36	84.05
PET-CGDNN	85.64	88.32	89.95	89.95	90.45	90.68	90.23	90.59	90.14	90.18
IC-AMCNET	80.77	82.45	83.00	83.73	83.86	84.05	84.55	84.27	83.68	84.09
MCLDNN	87.50	88.36	90.50	90.36	91.27	90.77	90.95	91.64	90.82	90.64
MCNET	77.14	80.27	81.64	80.95	82.68	83.27	82.68	82.68	82.09	81.82
LSTM	85.73	88.91	89.73	90.68	90.59	89.64	90.00	90.45	89.14	90.00
LSTMDAE	79.41	81.68	83.50	84.14	84.77	84.23	85.73	84.91	84.14	85.41
LAANet(our model)	**89.82**	**91.23**	**92.73**	**92.59**	**92.36**	**92.64**	**93.14**	**92.55**	**92.32**	**92.55**

Recognition Accuracy. The results of the proposed model were compared with several state-of-the-art models, including CGDNet [22], PET-CGDNN [23], IC-AMCNET [7], MCLDNN [9], MCNET [13], LSTM [17], LSTMDAE [19], as summarized in Table.2. The best results for each metric on both datasets are emphasized. The proposed model achieved the highest recognition performance on both datasets, with an average classification accuracy of 62.43% and 64.69%, respectively. It is worth noting that on the RadioML2016.10a dataset, the proposed model achieved an average recognition accuracy that is 2–6 percentage points higher than other models. On the RadioML2016.10b dataset, the proposed model performed comparably to MCLDNN, LSTM, and LSTMDAE, which could be due to the larger data scale of the dataset allowing the model to learn adequately. This insight also leads us to consider the importance of data in deep learning: the larger and more comprehensive the dataset, the more beneficial it is for the training and generalization performance of the model. Figure 4 illustrates the model's recognition performance on these two datasets, consistent with Table 2. Table.3 presents a detailed report of the classification performance of our proposed model at varying SNRs. The results indicate that our model achieved an average accuracy rate of 92.19% within the SNR span of 0dB to 18dB. Notably, the proposed model outperforms other models across the range of -6dB to 18dB SNRs. However, it is important to recognize that achieving higher accuracy rates in extremely low SNR environments (-20dB to -10dB) poses a significant challenge. At present, the accuracy rate from -8dB to $2(\pm2)$dB represents the main indicator of our model's recognition performance at low SNRs. It can be observed that our model effectively enhances the average recognition accuracy rate by 1–2 percentage points from -8dB to $2(\pm2)$dB SNRs, demonstrating its robustness against noise.

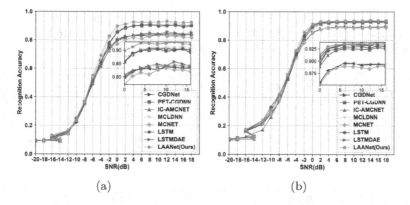

(a)   (b)

Fig. 4. Recognition accuracy on both datasets(-20dB to 18dB) (a)RadioML2016.10a (b)RadioML2016.10b.

Confusion Matrix. Figure 5 presents a group of confusion matrices on the RadioML201610.10a and RadioML201610.10b datasets at -2dB and 18dB. For each confusion matrix, each row represents the true modulation type, while each column represents the predicted modulation type. This provides insight into the classification performance of the proposed model for each modulation scheme. For the SNR of 18dB, the majority of values are located along the diagonal line in the confusion matrix, indicating that nearly all categories are accurately recognized. Despite this, some confusion in the recognition of WBFM signals still exists, which is primarily attributed to the silent periods in the audio signal. Besides, there is some level of confusion between QAM16 and QAM64 since QAM16 is a subset of QAM64. As indicated in Fig. 6, the proposed model demonstrates improved performance in differentiating between these two signal types from a general perspective.

Model Parameters and Complexity Analysis. Figure 7 presents a comparison of the number of learned parameters and training time among various models. The proposed model requires more training time compared to existing models, such as IC-AMCNET [7], MCNET [13], CGDNN [22], and PET-CGDNN [23], which constitute convolutional neural networks with lower computational complexity. Nevertheless, this increase in time cost is counterbalanced by the

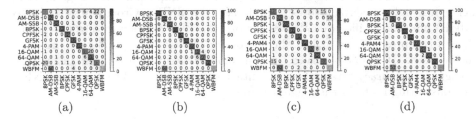

Fig. 5. Confusion matrixes of the proposed model for different SNRs on different datasets (a) RadioML2016.10a & −2dB (b) RadioML2016.10a & 18dB (c)RadioML2016.10b & −2dB (d)RadioML2016.10b & 18dB

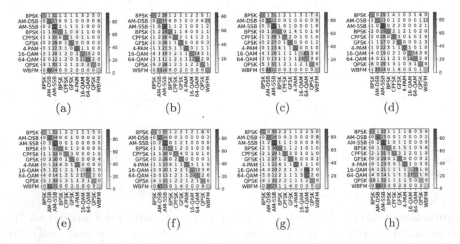

Fig. 6. Confusion matrixes of different models for all SNRs on RadioML2016.10a dataset (a) LAANet(Ours) (b) LSTM (c) LSTMDAE (d) MCLDNN (e) CGDNN (f) PET-CGDNN (g)ICAMC (h)MCNET.

observed higher recognition accuracy across all tested cases, coupled with the virtue of a relatively small parameter scale.

4.4   Ablation Experiment

In terms of model design, this study employs various technical means, including data preprocessing, autoencoders, regularization, and attention mechanisms. In order to verify the effectiveness of these components, this section conducts ablation experiments on different components in the model. Data preprocessing was previously explained, where the normalization amplitude of the signal and phase features were found to be more suitable for the LSTM structure to learn.

The proposed model in this paper is based on LSTMDAE [19]. Unlike the original model, the number of units in the first two LSTM layers is set to 64 and 32, respectively. Furthermore, the input is not subjected to any added noise preprocessing, serving as the baseline (Base). To enhance the performance of the base structure, several functional components were incorporated into the model, such as extra LSTM layers, attention mechanisms, and regularization layers. The research process is described below:

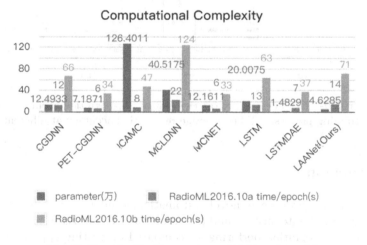

Fig. 7. Complexity Comparison of different models

– Base+A: Based on the base structure, a shared dense layer and an LSTM layer were added.
– Base+A+B: Gaussian dropout regularization layer was then added.
– Base+A+B+C (LAANet): The attention mechanism was introduced.

Figure 8 illustrates the performance comparison of models at different stages throughout the research process. As shown in Fig. 8a, the recognition accuracy gradually improves during the training process, and its performance is significantly enhanced after incorporating the LSTM layer. This improvement can be attributed to the LSTM layer's ability to refine the features that the autoencoder has learned through signal reconstruction tasks and extract additional features. Moreover, the introduction of attention mechanisms is found to effectively improve the model's overall performance. Meanwhile, the incorporation of regularization layers allows the model to learn meaningful features more rapidly and accelerates the model's convergence speed. Figure 8b presents the average recognition accuracy at different SNRs, from which we can draw similar conclusions to those drawn from Fig. 8a. Overall, the ablation experiments demonstrate that the integration of LSTM layers, attention mechanisms, and regularization can significantly improve recognition accuracy. The synergistic effects of these techniques boost the model's feature extraction and learning capabilities, leading to better adaptation to signal recognition tasks.

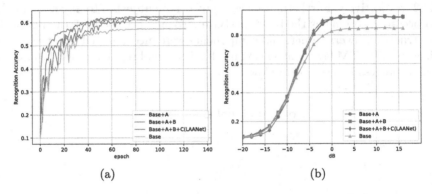

(a)   (b)

Fig. 8. The results of ablation study (a) the change in recognition accuracy over time during the training process. (b) The average recognition accuracy at different SNRs.

5   Conclusion

In this paper, we propose a novel automatic modulation recognition model based on an LSTM autoencoder and attention mechanism. Experimental results show improved recognition performance compared to existing approaches at relatively few parameters. However, its computation and inference process is time-consuming due to the LSTM-based framework. Future research can explore various avenues to address the limitations and improve the performance of the AMR task.

References

1. Abdi, A., Dobre, O.A., Choudhry, R., Bar-Ness, Y., Su, W.: Modulation classification in fading channels using antenna arrays. In: IEEE MILCOM 2004. Military Communications Conference, 2004. vol. 1, pp. 211–217. IEEE (2004)

2. Beidas, B.F., Weber, C.L.: Higher-order correlation-based classification of asynchronous MFSK signals. In: Proceedings of MILCOM'96 IEEE Military Communications Conference, vol. 3, pp. 1003–1009. IEEE (1996)
3. Lay, N.E., Polydoros, A.: Modulation classification of signals in unknown ISI environments. In: Proceedings of MILCOM'95, vol. 1, pp. 170–174. IEEE (1995)
4. Hassanpour, S., Pezeshk, A.M., Behnia, F.: Automatic digital modulation recognition based on novel features and support vector machine. In: 2016 12th International Conference on Signal-Image Technology & Internet-Based Systems (SITIS), pp. 172–177. IEEE (2016)
5. Moser, E., Moran, M.K., Hillen, E., Li, D., Wu, Z.: Automatic modulation classification via instantaneous features. In: 2015 National Aerospace and Electronics Conference (NAECON), pp. 218–223. IEEE (2015)
6. O'Shea, T.J., Corgan, J., Clancy, T.C.: Convolutional radio modulation recognition networks. In: Engineering Applications of Neural Networks: 17th International Conference, EANN 2016, Aberdeen, UK, September 2–5, 2016, Proceedings 17. pp. 213–226. Springer (2016)
7. Hermawan, A.P., Ginanjar, R.R., Kim, D.S., Lee, J.M.: CNN-based automatic modulation classification for beyond 5G communications. IEEE Commun. Lett. 24(5), 1038–1041 (2020)
8. Hong, D., Zhang, Z., Xu, X.: Automatic modulation classification using recurrent neural networks. In: 2017 3rd IEEE International Conference on Computer and Communications (ICCC), pp. 695–700. IEEE (2017)
9. Xu, J., Luo, C., Parr, G., Luo, Y.: A spatiotemporal multi-channel learning framework for automatic modulation recognition. IEEE Wireless Commun. Lett. 9(10), 1629–1632 (2020)
10. O'shea, T.J., West, N.: Radio machine learning dataset generation with gnu radio. In: Proceedings of the GNU Radio Conference, vol. 1 (2016)
11. O'Shea, T.J., Roy, T., Clancy, T.C.: Over-the-air deep learning based radio signal classification. IEEE J. Sel. Topics Signal Process. 12(1), 168–179 (2018)
12. Viterbi, A.J., Omura, J.K.: Principles of digital communication and coding. Courier Corporation (2013)
13. Huynh-The, T., Hua, C.H., Pham, Q.V., Kim, D.S.: MCNET: an efficient CNN architecture for robust automatic modulation classification. IEEE Commun. Lett. 24(4), 811–815 (2020)
14. Tekbıyık, K., Ekti, A.R., Görçin, A., Kurt, G.K., Keçeci, C.: Robust and fast automatic modulation classification with CNN under multipath fading channels. In: 2020 IEEE 91st Vehicular Technology Conference (VTC2020-Spring), pp. 1–6. IEEE (2020)
15. Liu, X., Yang, D., El Gamal, A.: Deep neural network architectures for modulation classification. In: 2017 51st Asilomar Conference on Signals, Systems, and Computers, pp. 915–919. IEEE (2017)
16. Hochreiter, S., Schmidhuber, J.: Long short-term memory. Neural Comput. 9(8), 1735–1780 (1997)
17. Rajendran, S., Meert, W., Giustiniano, D., Lenders, V., Pollin, S.: Deep learning models for wireless signal classification with distributed low-cost spectrum sensors. IEEE Trans. Cognitive Commun. Netw. 4(3), 433–445 (2018)
18. West, N.E., O'shea, T.: Deep architectures for modulation recognition. In: 2017 IEEE International Symposium on Dynamic Spectrum Access Networks (DySPAN), pp. 1–6. IEEE (2017)
19. Ke, Z., Vikalo, H.: Real-time radio technology and modulation classification via an LSTM auto-encoder. IEEE Trans. Wireless Commun. 21(1), 370–382 (2021)

20. Yunhao, S., Hua, X., Lei, J., Zisen, Q.: ConvLSTMAE: a spatiotemporal parallel autoencoders for automatic modulation classification. IEEE Commun. Lett. **26**(8), 1804–1808 (2022)
21. Goodfellow, I., Bengio, Y., Courville, A.: Deep learning. MIT press (2016)
22. Njoku, J.N., Morocho-Cayamcela, M.E., Lim, W.: CGDNET: efficient hybrid deep learning model for robust automatic modulation recognition. IEEE Netw. Lett. **3**(2), 47–51 (2021)
23. Zhang, F., Luo, C., Xu, J., Luo, Y.: An efficient deep learning model for automatic modulation recognition based on parameter estimation and transformation. IEEE Commun. Lett. **25**(10), 3287–3290 (2021)

A Compact Phoneme-To-Audio Aligner
for Singing Voice

Meizhen Zheng, Peng Bai, and Xiaodong Shi[✉]

Department of Artificial Intelligence, School of Informatics, Xiamen University,
Xiamen 361005, China
{midon,baipeng}@stu.xmu.edu.cn, mandel@xmu.edu.cn

Abstract. The phoneme-to-audio alignment task aims to align every
phoneme to a corresponding speech or singing audio segment. It has
many applications in the research and commercial field. Here we propose
an easy-to-train and compact phoneme-to-audio alignment model that
is especially effective for singing audio alignment tasks. Specifically, we
design a compact model with simple encoder-decoder architecture without a popular but redundant attention component. The model can be well
trained in relatively few epochs for different datasets with a combination
of CTC loss and mel-spectrogram reconstruction loss. We apply a dedicated dynamic programming algorithm to the output likelihood matrix
from the model to acquire alignment results. We conduct extensive experiments to verify the effectiveness of our method. Experiments show that
our method outperforms the baseline models on different datasets. Our
codes are available on github.

Keywords: phoneme-to-audio · alignment · compact · dynamic
programming · singing · multi-scale information fusion

1 Introduction

Phoneme-to-audio alignment, also known as *Phoneme Segmentation* or *Phoneme
Boundary Detection* in some literature, is designed to learn a function that accurately maps the phoneme sequence to the target audio segment boundaries. It
is an essential preliminary task for many applications, such as training singing
voice synthesis (SVS) models [14,20], subtitle calibration, and lyrics-singing synchronization[1].

Several tools exist for phoneme-level forced alignment [6,11,16,22]. Most of
them rely on Hidden Markov Models - Gaussian Mixture Model (HMM-GMM)
to infer hidden states from likelihood scores derived from features computed
on raw audio or spectral representations. However, we found that these tools
perform well in alignment tasks for speaking audio but degrade performance
when applied to singing audio. We argue that this is because of the inherent

---

[1] https://github.com/zhengmidon/singaligner.

X. Yang et al. (Eds.): ADMA 2023, LNAI 14177, pp. 183–197, 2023.
https://doi.org/10.1007/978-3-031-46664-9_13

acoustic characteristic discrepancy between the singing voice and the speaking voice. Specifically, 1) pitch change in the singing audio of a phoneme is more complicated than that in the speech audio of the same phoneme; 2) the pitch range of singing voice is wider, i.e., the mel-spectrogram of singing voice has relatively affluent information in the high-frequency region; 3) phonemes voiced in singing audio have a longer duration on average. HMM-GMMs may not be capable of handling these complex features.

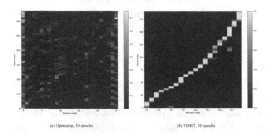

(a) Opencpop, 10 epochs   (b) TIMIT, 10 epochs

Fig. 1. Attention matrix from our re-implementation of model in [26]. The model learns clear alignment for speech data but fails to capture alignment for singing data. (a) model trained on singing dataset Opencpop [28] for 10 epochs; (b) model trained on speech dataset TIMIT [31] for 10 epochs.

Recent research has focused on neural aligners which employ neural networks(e.g., LSTM [9]) to handle acoustic features(e.g., mel-spectrograms) and text information jointly to obtain alignment results [23,26,30]. Attention mechanisms are exploited in [23,26] to realize acoustic-textual interaction, which is expected to get frame-wise and phoneme-wise alignment information. However, when we reproduced the model in [26] and tested it with singing data and speech data, we found that for speech phoneme-audio pairs, the attention matrix emerges with a clear monotonic alignment trend after several epochs. But it fails to acquire alignment when trained with singing data. Figure 1 illustrates the details. We argue that it can be attributed to the intrinsic nature discrepancy between singing and speech audios mentioned before.

To deal with this problem, in this work, we discard the attention mechanism and the text encoder mentioned in the previous research and design an acoustic encoder architecture with **multi-scale information fusion** to effectively capture the complicated acoustic features. As a result, we reduce the model size, making it so compact that it has only 2.57m parameters (which is 37% less than that model in [26] has) in our experiments and, therefore, easy to train. Our model can be well-trained in dozens of epochs, while this number is several hundred in previous work.

Inspired by [26], we employ the mel-spectrogram reconstruction loss to assist CTC loss [7] for model training. Thanks to the compact structure of the model, it takes less time to train it, which is relatively efficient. We exert a dynamic programming algorithm directly(i.e., no need for additional processing) on the likelihood matrix, which is used to compute CTC loss to achieve phoneme-to-audio alignment. Experiment results on different singing datasets in different languages imply the effectiveness and generality of our method.

In summary, the contributions of this work are:

- a compact and general phoneme-to-audio aligner for the singing voice.
- an acoustic encoder with **multi-scale information fusion** which is effective in processing singing voice features.
- a dynamic programming algorithm that directly utilizes the likelihood matrix for alignment search.
- experiments on different datasets in different languages to validate the effectiveness of our method.

The paper is structured as follows. In Sect. 2, we review several statistic and neural phoneme-to-audio aligners; detailed model architecture and the alignment algorithm are represented in Sect. 3; we conduct experiments to validate the performance of our approach in Sect. 4; The work is concluded in Sect. 5

2   Related Works

2.1   Statistical Phoneme-to-audio Aligners

To the best of our knowledge, representative statistical phoneme-to-audio aligners are proposed in [6,11,16,22]. The majority of them follow an HMM-GMM workflow. Take the Montreal Forced Aligner (MFA) [16] for example. A monophonic GMM-HMM is first trained while iteratively re-estimating the alignment. Then, provided by the alignment from the monophonic model, triphone models are trained iteratively. Speaker adaptation is performed as a last step if the speaker identities are known. It is built upon Kaldi toolkit [19], which is efficient and easy to use. We experimentally observe that this tool is highly effective in phoneme-to-audio alignment for speech voice. But when we apply it to singing data, its performance is unsatisfactory.

2.2   Neural Phoneme-to-audio Aligners

Neural phoneme-to-audio aligners can be divided into two categories, namely, the *phoneme-uninformed* aligners and *phoneme-informed* aligners. Models of the former type are designed to predict alignment results provided by only the ground truth phoneme duration, but without the phonemes themselves, we refer readers to [1,10,12,17]. While in the latter setup, models are fed with phoneme duration information coupled with presumed phonemes [3,13,23,24,26,30]. This work only discusses the latter category to which our model belongs.

[3] reduces the phoneme segmentation task to a binary classification at each time step with the aid of deep bidirectional LSTM. [13] proposes an RNN-based neural architecture coupled with a parameterized structured loss function to learn segmental representations for phoneme boundary detection. Directed

against the alignment of wild corrupted speech data, [23] proposes to perform phoneme-informed speech-music separation and phoneme alignment jointly using recurrent neural networks and the attention mechanism. These studies carefully research alignment scenarios of speech data but have not conducted experiments on singing audio. Like [23,24] unites phoneme level lyrics alignment with text-informed singing voice separation with the help of a new DTW-attention mechanism. It performs well in singing voice alignment. In the same structure of text-audio cross attention, [26] suggests adding a mel-spectrogram reconstruction loss to CTC loss to improve the alignment performance. It seems that the attention mechanism is popular in the alignment model. But we find that it may help few in the context of singing voice alignment, as illustrated in Fig. 1. So we abandon this design and shift attention to the audio encoder structure layout search.

3   Method

In this section, we first present the mathematical definition of the phoneme-to-audio task. Then, we introduce our overall model structure as well as the audio encoder. After that, loss functions are described, and a dynamic programming algorithm is represented at the end.

3.1   Task Definition

Let $M \in \mathbb{R}^{\mathcal{N} \times \mathcal{S}}$ be the mel-spectrogram where $\mathcal{N}$ is number of singing audio frames and $\mathcal{S}$ represents number of mel bins. We rewrite it as vectors so that $M = \{m_0, m_1, m_2, \ldots, m_{\mathcal{N}-1}\}$. Let $Y = \{y_0, y_1, y_2, \ldots, y_{\mathcal{T}-1}\}$ be the transcription phoneme sequence where $\mathcal{T}$ is the sequence length and $y_i \in A$ where A is the phoneme alphabet. The phoneme-to-audio alignment task aims to learn a function f so that $f(m_i) = \hat{y}_i, \hat{y}_i \in \{0, 1, 2, \ldots, \mathcal{T}-1\}, \hat{y}_i \leqslant \hat{y}_{i+1}$. Here we constrain $\hat{y}_i \leqslant \hat{y}_{i+1}$ as the alignment is monotonic.

3.2   Overall Model

Figure 2 depicts the overall model structure. As explained in Sect. 1, we do not adopt the popular attention mechanism mentioned in previous works in our model design. Removing the text encoder and the attention component makes our model more compact. There are three main modules in our model: 1) the audio encoder which will be detailedly described in the following subsection; 2) the CTC decoder that incorporates hidden states from the audio encoder and outputs a log-likelihood matrix for CTC loss computation; 3) the mel decoder which is expected to reconstruct the mel-spectrogram as precisely as possible. The CTC decoder and the mel decoder are composed of a two-layer bidirectional LSTM and a linear layer.

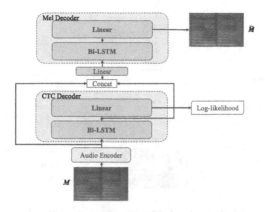

Fig. 2. The overall model architecture. *Concat* means *concatenate*.

3.3   Audio Encoder

The acoustic features of the singing voice are more complex than those of the speech voice. For example, a singing voice has a wider pitch range, more prosodic change, and longer phoneme duration. So we argue that it is necessary to design an encoder that can efficiently capture and encode acoustic information in mel-spectrograms.

We design an audio encoder illustrated in Fig. 3. It mainly comprises convolution layers, max-pooling layers, and bidirectional LSTMs. The key idea of the design is **multi-scale information fusion**. We arrange n groups of convolution layer and max pooling layer **in parallel**. Each of them is expected to obtain acoustic features of a specific scale from the mel-spectrograms. We *unsqueeze* the mel-spectrogram to extend an extra *channel* dimension. So the parameter *input channels* of the convolution layer is set to 1. Different channels can acquire information from different modes on a specific scale. The kernel of the 2d convolution layer is two-dimensional. One dimension convolves along the time scale, capturing prosodic information, and another convolves along the pitch scale, capturing information from different pitch regions.

We attach great importance to different settings of the kernel size k_c^i in different groups. We split the mel-spectrogram into several pitch regions along the pitch dimension. Convolution stride s_i is set to a large value in pitch dimension so that the kernel can obtain acoustic features from a specific pitch region at each stride. The following max-pooling layer is designed to reduce information from all pitch regions. After max-pooling, information from all groups is added up. The pitch dimension is reduced to 1 after convolving and pooling. So we *squeeze* it to maintain the dimensions of the feature map. Following a non-linear transformation, the feature map is fed to LSTM layers to fuse information along the time scale.

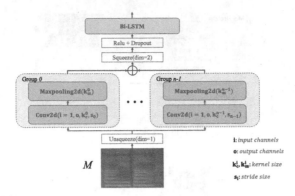

Fig. 3. The audio encoder structure. Operator *unsqueeze* extends a target extra dimension on the data; operator *squeeze* removes a target dimension from the data.

3.4 Loss Functions

The CTC loss [7] has been successfully employed in lyrics-to-audio alignment tasks [25,27]. We continue to use this loss function in our model training because of the similarity between lyrics-to-audio alignment and phoneme-to-audio alignment. Formally, when using CTC loss, the model is required to predict an extra *blank* label ε, which is absent in the alphabet A. As a result, the predicted sequence of model becomes $\bar{Y} = \{\bar{y}_0, \bar{y}_1, \ldots, \bar{y}_{T-1}\}, \bar{y}_i \in \{\varepsilon\} \bigcup A$. For example, it can be $\bar{Y} = \{\varepsilon, \varepsilon, a, a, a, \varepsilon, \varepsilon, b, b, \varepsilon, \varepsilon\}$ that corresponds to the target sequence $Y = \{a, b\}$. Define a reduce function $\mathcal{R}$ so that $\mathcal{R}(\bar{Y}) = Y$. Given M, the posterior probability can be expressed as:

$$\mathbb{P}(Y|M) = \sum_{\bar{Y}, \mathcal{R}(\bar{Y})=Y} \prod_{i=0}^{T-1} \mathbb{P}(\bar{y}_i|M) \tag{1}$$

Then CTC loss is computed as:

$$\mathcal{L}_{CTC} = -log\mathbb{P}(Y|M) \tag{2}$$

CTC loss only ensures that the sequence decoded from probabilities $\bar{Y}$ is close to the correct one Y. As a result, the time locations in the phonetic posteriogram do not contribute to the loss value. Whereas they directly have an impact on the alignment quality [26]. It is important to incorporate this information into loss functions. Inspired by [26], we also employ a mel-spectrogram reconstruction loss to help the model predict phonemes at their accurate position. Let $\bar{M}$ be the mel-spectrogram predicted by the model, the reconstruction loss is a simple $L2$ distance between $\bar{M}$ and M:

$$\mathcal{L}_{REC} = \|\bar{M} - M\|_2^2 \tag{3}$$

Eventually, the loss function for the model is written as:

$$\mathcal{L} = \mathcal{L}_{CTC} + \lambda * \mathcal{L}_{REC} \tag{4}$$

where parameter λ balances the effect of two losses.

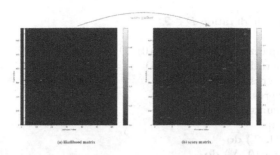

Fig. 4. The likelihood matrix L and the score matrix $\mathcal{C}$ acquired by the alignment algorithm. The bright yellow line in Figure (a) indicates the index of *blank* label. Notice that *Alphabet Index* in figure (a) represents the index of the alphabet of the model. We add *blank* and a few other symbols to the dataset alphabet to build the alphabet of the model.

3.5   Alignment Algorithm

The model is trained with CTC loss which is usually employed in Automatic Speech Recognition(ASR) tasks. There is a gap between this and phoneme-to-audio alignment. [26] calculates a cost matrix from the likelihood matrix according to the definition of CTC loss, then removes *blank* labels from the cost matrix. Finally, beam search decoding [4] is employed to retrieve phoneme durations.

We observe that though CTC loss tends to guide the model to predict *blank* label, the model will learn to assign a larger probability to the correct label for every audio frame. Hence we propose a dynamic programming algorithm that directly deals with the likelihood matrix. The algorithm is mainly motivated by [21]. We add some matrix processing steps to the original algorithm and represent it in Algorithm 1. It is written in PyTorch [18] style. The input likelihood matrix $L \in \mathbb{R}^{\mathcal{N} \times |A|}$, where $|A|$ represents the size of phoneme alphabet, is the output of CTC decoder. Input vector $\mathcal{I}$ is composed of indices of phonemes in the alphabet. The key step of the algorithm is **line 5** in Algorithm 1, which acquires a score matrix $\mathcal{C}$. Every entry of $\mathcal{C}$ records the score that every frame gains from every phoneme. We visualize this transformation in Fig. 4. As the figure illustrates, the score matrix is roughly monotonic hence suitable for dynamic programming. The rest of the algorithm is dynamic programming. For better understanding, we interpret the matrix symbols: $\mathcal{R}$ is a reward matrix that records the reward of every path; the boundary matrix is marked as $\mathcal{B}$, which keeps the optimal path; $\mathcal{P}$ represents accumulative score matrix recording the accumulative sum of entries in score matrix $\mathcal{C}$.

Algorithm 1 DP for Phoneme Duration Extraction

1: **Input**: Likelihood matrix $L \in \mathbb{R}^{\mathcal{N} \times |A|}$, alphabet indices of phoneme sequence $\mathcal{I} \in \mathbb{R}^{\mathcal{T}}$
2: **Output**: Phoneme duration $D \in \mathbb{R}^{\mathcal{T}}$
3: **Initialize**:
4:      $\mathcal{I}_t = \mathcal{I}$.unsqueeze(0).repeat($A$.size(0), 1)
5:      $\mathcal{C} = A$.gather(dim = 1, index = $\mathcal{I}_t$)
6:      $\mathcal{R} = $ torch.zeros_like($\mathcal{C}$)
7:      $\mathcal{B} = $ torch.zeros_like($\mathcal{C}$)
8:      $\mathcal{P} = $ torch.cumsum($\mathcal{C}$, dim=1)
9: $\mathcal{R}[0, :] = \mathcal{P}[0, :]$
10: **for** i in range(1, $\mathcal{T}$) **do**
11:     **for** j in range(0, $\mathcal{N}$) **do**
12:         **for** k in range(0, j + 1) **do**
13:             $r = \mathcal{R}[i\text{-}1, k] + \mathcal{P}[i, j] - \mathcal{P}[i, k]$
14:             **if** $r > \mathcal{R}[i, j]$ **then**
15:                 $\mathcal{R}[i, j] = r$
16:                 $\mathcal{B}[i, j] = k$
17:             **end if**
18:         **end for**
19:     **end for**
20: **end for**
21: $P = \mathcal{N} - 1$
22: **for** i in range($\mathcal{T}$, 0, -1) **do**
23:     $D = P - \mathcal{B}[i - 1, P]$
24:     $P = \mathcal{B}[i - 1, P]$
25: **end for**
26: D.reverse()
27: **return** D

---

4   Experiments

4.1   Datasets

To verify the generality of our method, we select four singing alignment datasets in Chinese, English, Hokkien, and Japanese, respectively.

Opencpop [28] is a singing corpus that consists of 100 popular **Mandarin** songs performed by a female professional singer. Audio files are recorded with studio quality at a sampling rate of 44,100 Hz. All singing recordings have been cut into segments shorter than 15 s, and phonetically annotated with phoneme and syllable (note) boundaries. It adopts the *initials* and *finals* in Chinese Pinyin for phonetic annotation. The total duration of the recording is around 5.2 h, and the number of segments is 3756.

The English singing dataset NUS48E [2] is a 169-minute collection of audio recordings of the sung and spoken lyrics of 48 (20 unique) **English** songs recorded at 44,100 Hz by 12 subjects and a complete set of transcriptions and

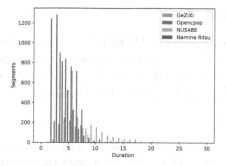

Fig. 5. Segment duration distribution of the four datasets. Duration is expressed in second(s).

duration annotations at the phone level for all recordings of sung lyrics, comprising 25,474 phoneme instances. It adopts the 39-phoneme set used by the CMU Dictionary for phonetic annotation [29]. We use the singing part in our experiments which consists of 115-minute audio. We split the songs into 1227 5 s-10 s fragments for training convenience.

We collect a dataset named GeZiXi which contains audio recordings of 4.54 h. The audio content is the a cappella Gezi Opera, a traditional Chinese opera performed in **Hokkien**. The 1938 audio segments are recorded at 44,100 Hz by five subjects and aligned with text manually. Audio duration spans between 1 s and 29 s.

NamineRitsu [32] is a Japanese singing dataset composed of 107 **Japanese** songs. Recordings are performed by a single female singer without accompaniment, constituting a 4.29-hour dataset. The songs are recorded at 44,100 Hz. We split the origin songs into 4571 segments longer than 1 s but shorter than 30 s according to the silence parts.

The segment duration distribution of the four datasets is illustrated in Fig. 5. The datasets have different duration distributions. Thus, they are qualified for performance evaluation.

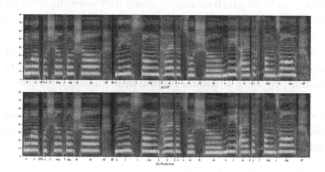

Fig. 6. Alignment result comparison between the ground truth and prediction from our model on Opencpop data. (a) is the ground truth phoneme durations; (b) is the predicted phoneme durations from our model.

4.2 Experiment Setups

Model Architecture Setups. All Bi-LSTMs are designed with two layers whose hidden sizes are 256. The number of groups in the audio encoder is set to 2, and output channels o is set to 256. The network parameters of the audio encoder are set as follows: $k_c^0 = (20, 3), s_0 = (10, 1), k_m^0 = (7, 3); k_c^1 = (60, 1), s_1 = (10, 1), k_m^1 = (3, 3)$.

Other Setups. Raw audios are converted into mel-spectrograms with Short Term Fourier Transform (STFT). We re-sample the audio at a sampling rate of 24k. We use Hanning window with size 512. FFT size is set to 512, and hop length is set to 128. We adopt the PyTorch implementation of CTC loss in our experiments and set the hyperparameter λ to 1. Training is done with the AdamW [15] optimizer in initial learning rate 1e-3, which decays with an exponential rate of 0.93. Training is done with the help of the early stopping strategy according to alignment performance on the validation set. All experiments are carried out with batch size 16 on a single GeForce RTX 2080 Ti GPU.

4.3 Results

Experiment results are shown in Table 1. The performances of the models are evaluated with six main assessment metrics for the alignment task, namely, Mean Average Error (MAE), which is the average time imprecision in predictions; MEDian average error (MED), which is the median time imprecision in predictions; Percentage of Correctly Aligned Segments (PCAS) [5], which measures the percentage of overlap of ground truth and estimated segments. MAE and MED are expressed in milliseconds (ms), and PCAS is a percentage indicator. We compute the mean of the three metrics on the four datasets and put it in the table. Notice that the results of Teytaut et al. [26] are from our re-implementation model. We find that metrics on the GeZiXi dataset significantly differ from those on other datasets. Because samples from the GeZiXi dataset have a longer duration(refer to Fig. 5), the phoneme intervals are further segmented into smaller intervals according to their pitch changes, which increases the difficulty in boundary prediction. Our model outperforms the baseline models by a large margin on the four datasets in the metrics on average.

Notably, when we add the attention mechanism to our model, its performance does not improve(see the bottom of the table). It indicates that the attention mechanism can not provide additional helpful information but increase the model size.

We visualize an example of the alignment result on Opencpop data in Fig. 6. It can be found that our model tends to predict phoneme boundaries according to spectral similarity, as illustrated in the last part of the figure.

Table 1. Test results on the four datasets. MAE and MED are expressed in milliseconds(ms), and PCAS is expressed in percentages(%). The *Mean* row records the average values of the corresponding metrics on four datasets. w/ attn means with an attention mechanism.

Method	Dataset	MAE↓	MED↓	PCAS↑
MFA	*Opencpop*	84.5	40.4	74.3
	GeZiXi	183.8	81.0	64.8
	NUS48E	90.0	36.1	72.4
	NamineRitsu	43.0	22.0	86.5
	Mean	100.3	44.9	74.5
Teytaut et al. [26]	*Opencpop*	154.1	71.9	47.7
	GeZiXi	442.5	236.0	26.9
	NUS48E	157.1	72.7	64.6
	NamineRitsu	49.9	23.7	82.3
	Mean	200.9	101.1	55.4
Ours	*Opencpop*	56.1	36.2	79.4
	GeZiXi	167.0	65.3	71.1
	NUS48E	38.4	23.0	88.6
	NamineRitsu	30.9	14.6	89.0
	Mean	**73.1**	34.8	**82.0**
Ours-w/ attn	*Opencpop*	60.5	34.4	77.7
	GeZiXi	165.6	63.0	71.0
	NUS48E	38.0	21.7	89.2
	NamineRitsu	30.8	17.7	89.5
	Mean	73.7	**34.2**	81.9

4.4 Ablation Studies

Effectiveness of the Proposed Audio Encoder. To evaluate how much improvement the model gains from the proposed audio encoder, we simply replace the audio encoder with other components such as Linear + LSTM and Conformer [8]. We test model performance on the datasets introduced in Subsect. 4.1. Experiment setups are the same as that in Subsect. 4.2. The number of layers and hidden size are set to the same value. As shown in Table 2, our proposed audio encoder outperforms other encoders on average thanks to the well-designed convolution structure.

Effectiveness of the Alignment Algorithm. We apply our alignment algorithm to the output of the model in [26] to validate the effectiveness of the algorithm. We conduct the experiment on a speech dataset TIMIT [31], which contains 5-hour multi-speaker speech audios and the corresponding phonetic

Table 2. Effectiveness of different encoders on different datasets. MAE and MED are expressed in milliseconds(ms), and PCAS is expressed in percentages(%). The *Mean* row records the average values of the corresponding metrics on four datasets.

Encoder	Dataset	MAE↓	MED↓	PCAS↑
Linear+LSTM	*Opencpop*	56.3	35.4	80.3
	GeZiXi	174.6	76.0	69.1
	NUS48E	42.4	25.1	88.2
	NamineRitsu	32.7	15.7	89.4
	Mean	76.5	38.1	81.7
Conv1d+LSTM	*Opencpop*	62.5	33.5	78.7
	GeZiXi	165.8	65.7	70.5
	NUS48E	41.1	24.7	89.2
	NamineRitsu	35.0	19.0	88.6
	Mean	76.1	35.7	81.8
LSTM	*Opencpop*	55.5	32.1	81.0
	GeZiXi	174.0	89.0	68.6
	NUS48E	39.2	21.1	87.1
	NamineRitsu	34.0	18.0	89.2
	Mean	75.7	40.1	81.5
Conformer [8]	*Opencpop*	62.8	36.1	74.7
	GeZiXi	184.1	78.7	67.8
	NUS48E	47.9	21.5	87.3
	NamineRitsu	28.6	13.0	89.9
	Mean	80.9	37.3	80.6
Ours	*Opencpop*	56.1	36.2	79.4
	GeZiXi	167.0	65.3	71.1
	NUS48E	38.4	23.0	88.6
	NamineRitsu	30.9	14.6	89.0
	Mean	**73.1**	**34.8**	**82.0**

Table 3. Effectiveness of the alignment algorithm on TIMIT dataset. The column *Training Time* represents the training time consumption expressed in hours(h). The units of MAE and MED are milliseconds(ms), and the unit of PCAS is percentages(%).

Method	MAE↓	MED↓	PCAS↑	Training Time(h)
MFA	16.3	15.7	93.7	1.6
Teytaut et al. [26]	16.3	11.8	94.1	4
Ours	16.4	12.1	93.9	0.6

transcriptions. Experiment setups are the same as in Subsect. 4.2 except that the sampling rate is changed to 16k. Table 3 records the results. The alignment algorithm of our method is competitive with that of the baseline method. And the training time can be reduced a lot with our alignment algorithm using the early stopping strategy.

Table 4. Performance of model trained on different sizes of dataset. The units of MAE and MED are milliseconds(ms), and the unit of PCAS is percentages(%).

Dataset	Size(h)	MAE↓	MED↓	PCAS↑
NamineRitsu	4.29	30.9	14.6	89.0
NamineRitsu-M	2.16	35.7	19.3	87.6
NamineRitsu-S	1.14	39.0	19.7	87.0

The Impact of Dataset Size. We remove part of the data from the original NamineRitsu dataset to build two smaller datasets, namely, NamineRitsu-M and NamineRitsu-S, whose size is 2.16 h and 1.14 h, respectively. We conduct experiments on them to research the impact of different dataset sizes on our method. Experiment setups are the same as that in Subsect. 4.2. The results are illustrated in Table 4. The model performance decreases as the dataset becomes smaller. But our model still outperforms the baseline even if trained on a much smaller dataset, which validates the robustness of our method to different data sizes.

5   Conclusion

In this paper, we present an easy-to-train and compact model specially designed for phoneme-to-audio alignment tasks for the singing voice. We analyze the attention mechanism employed by previous works and find that it is not indispensable for singing alignment as it is not able to learn valid alignment information between phonemes and audio. We discard the attention components and highlight on the design of an efficient audio encoder. We introduce an alignment algorithm that takes the likelihood matrix as input and outputs the alignment result. We test our method on datasets of different languages. Experiment results indicate the effectiveness of our method.

Acknowledgment. This work is supported by the National key R&D Program of China (Grant No.2020AAA0107904), the Key Support Project of NSFC-Liaoning Joint Foundation (Grant No. U1908216), and the Major Scientific Research Project of the State Language Commission in the 13th Five-Year Plan (Grant No. WT135-38).

References

1. Bhati, S., Villalba, J., Żelasko, P., Moro-Velazquez, L., Dehak, N.: Segmental contrastive predictive coding for unsupervised word segmentation. arXiv preprint arXiv:2106.02170 (2021)
2. Duan, Z., Fang, H., Li, B., Sim, K.C., Wang, Y.: The nus sung and spoken lyrics corpus: a quantitative comparison of singing and speech. In: 2013 Asia-Pacific Signal and Information Processing Association Annual Summit and Conference, pp. 1–9. IEEE (2013)
3. Franke, J., Mueller, M., Hamlaoui, F., Stueker, S., Waibel, A.: Phoneme boundary detection using deep bidirectional LSTMs. In: Speech Communication; 12. ITG Symposium, pp. 1–5. VDE (2016)
4. Freitag, M., Al-Onaizan, Y.: Beam search strategies for neural machine translation. arXiv preprint arXiv:1702.01806 (2017)
5. Fujihara, H., Goto, M., Ogata, J., Okuno, H.G.: LyricSynchronizer: automatic synchronization system between musical audio signals and lyrics. IEEE J. Sel. Topics Signal Process. **5**(6), 1252–1261 (2011)
6. Gorman, K., Howell, J., Wagner, M.: Prosodylab-aligner: a tool for forced alignment of laboratory speech. Can. Acoust. **39**(3), 192–193 (2011)
7. Graves, A., Fernández, S., Gomez, F., Schmidhuber, J.: Connectionist temporal classification: labelling unsegmented sequence data with recurrent neural networks. In: Proceedings of the 23rd International Conference on Machine Learning, pp. 369–376 (2006)
8. Gulati, et al.: Conformer: Convolution-augmented transformer for speech recognition. arXiv preprint arXiv:2005.08100 (2020)
9. Hochreiter, S., Schmidhuber, J.: Long short-term memory. Neural Comput. **9**(8), 1735–1780 (1997)
10. Kamper, H., van Niekerk, B.: Towards unsupervised phone and word segmentation using self-supervised vector-quantized neural networks. arXiv preprint arXiv:2012.07551 (2020)
11. Kisler, T., Schiel, F., Sloetjes, H.: Signal processing via web services: the use case webMAUS. In: Digital Humanities Conference (2012)
12. Kreuk, F., Keshet, J., Adi, Y.: Self-supervised contrastive learning for unsupervised phoneme segmentation. arXiv preprint arXiv:2007.13465 (2020)
13. Kreuk, F., Sheena, Y., Keshet, J., Adi, Y.: Phoneme boundary detection using learnable segmental features. In: ICASSP 2020–2020 IEEE International Conference on Acoustics, Speech and Signal Processing (ICASSP), pp. 8089–8093. IEEE (2020)
14. Liu, J., Li, C., Ren, Y., Chen, F., Zhao, Z.: DiffSinger: Singing voice synthesis via shallow diffusion mechanism. In: Proceedings of the AAAI Conference on Artificial Intelligence, vol. 36, pp. 11020–11028 (2022)
15. Loshchilov, I., Hutter, F.: Decoupled weight decay regularization. arXiv preprint arXiv:1711.05101 (2017)
16. McAuliffe, M., Socolof, M., Mihuc, S., Wagner, M., Sonderegger, M.: Montreal forced aligner: trainable text-speech alignment using kaldi. In: Interspeech, vol. 2017, pp. 498–502 (2017)
17. Michel, P., Räsänen, O., Thiollière, R., Dupoux, E.: Blind phoneme segmentation with temporal prediction errors. In: Proceedings of ACL 2017, Student Research Workshop, pp. 62–68 (2017)

18. Paszke, A., et al.: PyTorch: an imperative style, high-performance deep learning library. In: Advances in Neural Information Processing Systems 32 (2019)
19. Povey, D., et al.: The kaldi speech recognition toolkit. In: IEEE 2011 Workshop on Automatic Speech Recognition and Understanding. IEEE Signal Processing Society (2011)
20. Ren, Y., et al.: FastSpeech 2: Fast and high-quality end-to-end text to speech. arXiv preprint arXiv:2006.04558 (2020)
21. Ren, Y., Tan, X., Qin, T., Luan, J., Zhao, Z., Liu, T.Y.: DeepSinger: Singing voice synthesis with data mined from the web. In: Proceedings of the 26th ACM SIGKDD International Conference on Knowledge Discovery & Data Mining, pp. 1979–1989 (2020)
22. Rosenfelder, I., Fruehwald, J., Evanini, K., Yuan, J.: FAVE (forced alignment and vowel extraction) program suite. URL http://fave.ling.upenn.edu (2011)
23. Schulze-Forster, K., Doire, C.S., Richard, G., Badeau, R.: Joint phoneme alignment and text-informed speech separation on highly corrupted speech. In: ICASSP 2020–2020 IEEE International Conference on Acoustics, Speech and Signal Processing (ICASSP), pp. 7274–7278. IEEE (2020)
24. Schulze-Forster, K., Doire, C.S., Richard, G., Badeau, R.: Phoneme level lyrics alignment and text-informed singing voice separation. IEEE/ACM Trans. Audio Speech Lang. Process. **29**, 2382–2395 (2021)
25. Stoller, D., Durand, S., Ewert, S.: End-to-end lyrics alignment for polyphonic music using an audio-to-character recognition model. In: ICASSP 2019–2019 IEEE International Conference on Acoustics, Speech and Signal Processing (ICASSP), pp. 181–185. IEEE (2019)
26. Teytaut, Y., Roebel, A.: Phoneme-to-audio alignment with recurrent neural networks for speaking and singing voice. In: Proceedings of Interspeech 2021, pp. 61–65. International Speech Communication Association; ISCA (2021)
27. Vaglio, A., Hennequin, R., Moussallam, M., Richard, G., d'Alché Buc, F.: Multilingual lyrics-to-audio alignment. In: International Society for Music Information Retrieval Conference (ISMIR) (2020)
28. Wang, Y., et al.: Opencpop: a high-quality open source chinese popular song corpus for singing voice synthesis. arXiv preprint arXiv:2201.07429 (2022)
29. Weide, R., et al.: The carnegie mellon pronouncing dictionary. release 0.6, https://www.cs.cmu.edu/ (1998)
30. Zhu, J., Zhang, C., Jurgens, D.: Phone-to-audio alignment without text: a semi-supervised approach. In: ICASSP 2022–2022 IEEE International Conference on Acoustics, Speech and Signal Processing (ICASSP), pp. 8167–8171. IEEE (2022)
31. Zue, V., Seneff, S., Glass, J.: Speech database development at MIT: timit and beyond. Speech Commun. **9**(4), 351–356 (1990)
32. Canon, [NamineRitsu] Blue (YOASOBI) [ENUNU model Ver. 2, Singing DBVer.2 release], https://www.youtube.com/watch?v=pKeo9IE_L1I, Accessed: 2022.10.06

Song-to-Video Translation: Writing a Video from Song Lyrics Based on Multimodal Pre-training

Feifei Fu[iD], Zelong Sun[iD], Guoxing Yang[iD], Xiaolong He[iD], and Zhiwu Lu[(✉)][iD]

Gaoling School of Artificial Intelligence, Renmin University of China, Beijing, China
{fufeifei,luzhiwu}@ruc.edu.cn

Abstract. Writing a video from text script (i.e., video editing) is an important but challenging multimedia-related task. Although a number of recent works have started to develop deep learning models for video editing, they mainly focus on writing a video from generic text script, not suitable for some specific domains (e.g., song lyrics). In this paper, we thus introduce a novel video editing task called song-to-video translation (S2VT), which aims to write a video from song lyrics based on multimodal pre-training. Similar to generic video editing, this S2VT task also has three main steps: lyric-to-shot retrieval, shot selection, and shot stitching. However, it has a large difference from generic video editing in that: the song lyrics are often more abstract to understand than the common text script, and thus a large-scale multimodal pre-training model is needed for lyric-to-shot retrieval. To facilitate the research on S2VT, we construct a benchmark dataset with human annotations according to three evaluation metrics (i.e., semantic-consistence, content-coherence, and rhythm-matching). Further, a baseline method for S2VT is proposed by training three classifiers (each for a metric) and developing a beam shot-selection algorithm based on the trained classifiers. Extensive experiments are conducted to show the effectiveness of the proposed baseline method in the S2VT task.

Keywords: Video editing · Song-to-video translation · Multimodal pre-training

1 Introduction

As a major communication medium, video has a wide range of applications in our daily life. Particularly, short videos are getting more and more popular in social media, which leads to a large demanding for efficient video editing. However, generating a well-edited video often requires the user to have professional skills in manual video shot selection/cutting and shot stitching, which are rather boring and time-consuming. To meet the need of non-professional users and reduce the boring editing time, we are in need of automatic/semi-automatic video editing system. With such video editing system, users can select video shots w.r.t. the

X. Yang et al. (Eds.): ADMA 2023, LNAI 14177, pp. 198–213, 2023.
https://doi.org/10.1007/978-3-031-46664-9_14

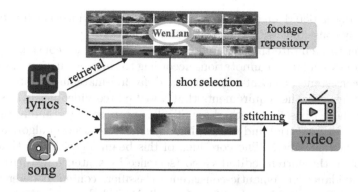

Fig. 1. The schematic illustration of our song-to-video translation (S2VT) framework. It aims to write/generate a target video from a song and its paired lyrics.

text script, drag them into one existing template, and then synthesize a target video with favorite background music selected.

Many video editing methods have been proposed to write videos from text scripts [8,11,13,21,22]. From traditional machine learning methods to recent deep learning ones, the workflow of video editing has been largely improved. For example, earlier work [9] resorts to a semi-automatic method for home video editing by analyzing a set of candidate videos and computing their 'unsuitability' scores. With the development of deep learning, more advancing video editing methods [3,8,11,21,22] have been devised for automatic video generation from text scripts (e.g., text instruction or single semantic label map). Note that these cutting-edge works mainly focus on writing/generating a video from generic text script, not suitable for some specific domains (e.g., song lyrics). In this paper, we thus introduce a novel video editing task termed song-to-video translation (S2VT), which aims to automatically generate a video from song lyrics based on multimodal pre-training (e.g., WenLan [7] and CLIP [18]).

Similar to generic video editing, the S2VT task also has three main steps: cross-modal shot retrieval (i.e., lyric-to-shot retrieval), shot selection, and shot stitching, as shown in Fig. 1. Among them, cross-modal shot retrieval is a critical step, which is very difficult even for experts. For each sentence of the text script, the semantic meaning of the retrieved video shots should be consistent with that of this sentence. To overcome the challenge in cross-modal shot retrieval, Write-A-Video [21] resorts to a vision-semantic embedding approach (VSE++) [6] and Transcript-to-Video [22] devises a vision-language embedding module for generating videos from generic text scripts. However, compared to generic text scripts, the song lyrics are often more abstract to understand, and thus lyric-to-shot retrieval is still difficult for these vision-language embedding modules (with small model parameters or training data) used in [21,22]. Therefore, in this work, we choose to deploy a large-scale multimodal pre-training model WenLan [7] for lyric-to-shot retrieval. With this very-large vision-language embedding model pre-trained over 650M image-text pairs, the generated video is expected to be semantically consistent with the song lyrics.

In order to facilitate the research on the S2VT task, we construct a benchmark dataset with human annotations according to three evaluation metrics (i.e., semantic-consistence, content-coherence, and rhythm-matching). Note that these metrics are exactly the requirements that a well-edited video needs to satisfy in S2VT. Further, a baseline method for S2VT is proposed by training three classifiers (each for a metric) and developing a beam shot-selection algorithm based on the trained classifiers. The core idea of this beam algorithm is that adding a new shot to the current edited video (sentence by sentence in turn) is under the joint guidance of semantic-consistence classifier, content-coherence classifier, and rhythm-matching classifier. Overall, the whole generation process of S2VT is illustrated in Fig. 1. Specifically, for each sentence of the song lyrics, we retrieve the top-6 shots that are semantically consistent with this sentence from the footage repository by deploying the multimodal pre-training model WenLan. Further, from the retrieved top-6 shots, the most suitable shot is selected with the beam shot-selection algorithm. Finally, all the selected shots are stitched together to synthesize a complete and artistic video.

Our main contributions are three-fold: **(1)** We introduce a novel video editing task called song-to-video translation (S2VT). For this challenging task, we propose a baseline method to automatically generate videos from song lyrics based on multimodal pre-training, which is a good start of the study of S2VT. **(2)** We construct a song-to-video benchmark dataset with human annotations according to three evaluation metrics (i.e., semantic-consistence, content-coherence, and rhythm-matching), which greatly facilitates the research on S2VT. **(3)** We design an automatic evaluation system to evaluate the performance of S2VT. It consists of three evaluation metrics including semantic-consistence, content-coherence, and rhythm-matching. Extensive experiments are conducted to show the effectiveness of the proposed baseline method.

2   Related Work

2.1   Music-to-Video Generation

Over the past few years, many studies have chosen to explore the association patterns between audio data and video data to perform a variety of tasks [1,14,17]. Among them, music video generation [2,5,10,16] has been studied most extensively. Specifically, [14] devises an automatic MTV (Music Television) generation system by mining the association patterns between music and video clips in professional MTV. [15] proposes a music-driven method to generate video montage. That is, with a set of video clips and a background music as input, video montage is synthesized by analyzing the music requirements and video content. It can be seen that these generation methods mainly utilize the rhythm and melody features of music to match the video content features (i.e., rhythm-matching). In contrast, our current work focuses on song-to-video translation from song lyrics: the generated video content is not only matched to the musical rhythm, but also semantically consistent with the song lyrics (i.e., semantic-consistence).

2.2 Music-Video Dataset

To facilitate the research on music retrieval and recommendation, Music-Video Dataset (MVD) has been constructed in [19,20]. The music videos of MVD are manually collected from the western music artists, which cover multiple languages (English, Thai, French, German, etc.), and the average duration of music videos is 4 min. In this work, we construct a song-to-video dataset, which can also be regarded as a music-video dataset. However, our dataset is quite different from MVD in both collection and usage: it is automatically generated by our system with Chinese lyrics as input, and is constructed to facilitate the development of song-to-video translation. More importantly, the generated videos are annotated according to three metrics (i.e., semantic-consistence, content-coherence, and rhythm-matching). The high-quality videos in our dataset (with high scores on three metrics) are thus very close to the human-level. With our proposed dataset (the average duration of videos is 30 s), any model can be trained by simply adopting videos with high scores as ground-truth. Therefore, our proposed dataset is indeed realistic for future work on S2VT.

3 Song-to-Video Translation

Song-to-Video translation (i.e., S2VT) is a novel video editing task, which aims to create an artistic video from song lyrics (Chinese song lyrics in this paper). Due to abstract nature of song lyrics, a multimodal pre-trained model Wen-Lan [7] is deployed to match lyrics with video shots. Note that WenLan is a very-large Chinese multimodal (vision-language) embedding model pre-trained over 650M image-text pairs collected, thus it has a good cross-modal semantic understanding ability (more details can refer to [7]). The pipeline of our S2VT framework is illustrated in Fig. 2, which consists of three main steps: lyric-to-shot retrieval, shot selection, and shot stitching.

Lyric-to-Shot Retrieval. In the cross-modal retrieval phase, the large-scale multimodal pre-trained model WenLan is utilized to ensure the semantic consistency between video content and song lyrics: the image and text encoders of WenLan are first used as feature extractors to extract lyric features and shot features, respectively; the similarity scores between lyrics and shots are then computed by the cosine distance; the top-6 shots are retrieved from the footage repository for each sentence of the song lyrics. More specifically, for shot feature extraction, we divide each shot (in the footage repository) into 8 segments equally, and extract the middle frame of each segment, resulting in 8 frames per shot in total. We then feed them into the image encoder of WenLan to obtain each frame feature, and take the average feature as the shot feature. Since the footage repository is very large, we extract the shot features in advance and save them as npy files, which is convenient for cross-modal retrieval. With WenLan, we can retrieve the top-6 shots that most semantically match each sentence of song lyrics from the footage repository (which will be introduced in Subsect. 4.1). To avoid repeated retrieval for similar sentences in song lyrics, we have processed the similar sentences. If the similarity of two adjacent sentences is greater than

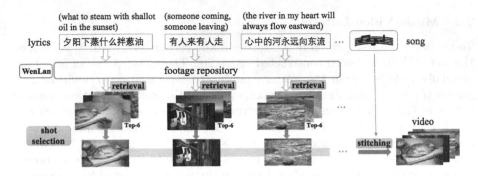

Fig. 2. Illustration of the pipeline of our song-to-video translation (S2VT) framework. For each sentence of the song lyrics, the top-6 shots are retrieved through WenLan [7] from the footage repository. Further, with certain shot selection strategy, 1–2 shots are often selected from the top-6 shots for each sentence (with duration) of the song lyrics. Finally, all selected shots are stitched together to composite a complete and artistic video (with the corresponding song being fused).

certain threshold, they are spiced together (simultaneously feed them into Wen-Lan to retrieve shots). Here, we also use the cosine distance to calculate the similarity between sentences, and set the threshold to 0.85. Further, to avoid the repeated use of some shots in an edited video during retrieval, we put the used shots into a history list for record. If the current shot falls in the history list, we then pass it and select the next best shot.

Shot Selection. In the shot selection phase, there are different selection strategies for selecting shots from the retrieved top-6 shots per sentence (of the song lyrics). Two simple strategies are Top-1 Selection and Random Selection. Concretely, Top-1 Selection indicates that we always select the top-1 shot that best matches the given sentence from the top-6 shots. Random Selection indicates that we randomly select a shot for the given sentence from the top-6 shots. In addition to these two simple selection strategies, we propose another more sound selection strategy in Sect. 5, i.e., beam shot-selection under the joint guidance of certain classifiers for improving the quality of generated videos.

Shot Stitching. In the shot stitching phase, the required shot frames are first obtained according to the duration of each sentence of the song lyrics. Generally, each sentence requires only one shot, but sometimes two shots are needed to cover the duration of this sentence. Such two-shot-per-sentence stitching may cause bad results (e.g., the second shot may pass by in a flash). To overcome this drawback, we choose to keep a balance between the required frames of the first shot and those of the second shot, and simply delete the redundant frames from the two shots. Finally, we stitch all the obtained shot frames in turn, write them to a video file, and add the corresponding song clip. In this way, a complete and artistic video can be composited. Here, the video frame rate is set to 30 fps, and the frame image size is resized to $(1920, 1080)$.

As we have mentioned, a well-edited video must meet three requirements: (a) **semantic-consistence** – the video content is consistent with the semantics and context of song lyrics; (b) **content-coherence** – the video content is coherent on its own; (c) **rhythm-matching** – the video content matches the music rhythm. However, the video generation methods developed based on Random Selection and Top-1 Selection have distinct drawbacks, because WenLan [7] only pays attention to requirement (a) (i.e., semantic-consistence) and ignores the other two important requirements (b) and (c) (i.e., content-coherence and rhythm-matching). Therefore, we choose to improve these methods and propose a beam shot-selection algorithm for solving the S2VT task, so that all three requirements are considered during video generation. By inputting song and paired lyrics, our system can generate an composite and artistic video. In addition, we construct a song-to-video benchmark dataset to facilitate the research on S2VT and design an automatic evaluation system to evaluate the performance of S2VT task.

4 Song-to-Video Dataset

In this section, we construct a song-to-video benchmark dataset with human annotations according to the three evaluation metrics. The whole dataset construction is divided into three parts: song and footage collection, video generation, and video annotation. For video generation, we adopt Random Selection for shot selection, and keep the other steps the same as Sect. 3. Below we only describe the other two parts of dataset construction in detail.

4.1 Song and Footage Collection

Song Collection. The collection of song clips is very important for video generation. A qualified song clip can make the composited video more attractive to users. The collection of song clips is divided into two steps: the collection of the entire song files and corresponding lyrics files, and the cutting of songs. We mainly describe the second step – the cutting of songs. Concretely, we first cut the entire lyric file into several parts (each part has several sentences), and then calculate the duration of each lyric clip. If the duration of a lyric clip exceeds 40 s, we cut it again to ensure that the duration of final song clip is kept between 20 s and 40 s. For those clips shorter than 20 s, we directly discard them. After this cutting step, a song (with its lyrics) is cut into several different song clips (with their clip lyrics). Overall, we collect a total of about 400 Chinese songs.

Footage Collection. The collection of footage is also vital for video generation. Rich and diverse footage is beneficial to improving the quality of generated videos, making them more attractive. We have crawled about 20,000 video shots from several websites such as pexels.com. These shots contain rich contents, and a set of keywords are used for shot crawling: nature, flower, drink, club, highway, sunset, sing, climb, etc. Note that we pre-process all video shots before retrieval. That is, we filter out some bad shots (e.g., pornography and violence) from the huge footage repository. We present a number of shot examples in Fig. 3(a), and also visualize the shot duration distribution of the entire footage repository in

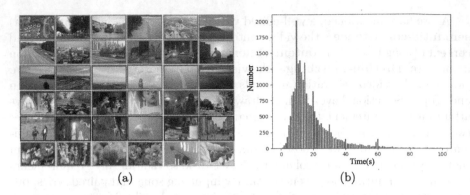

(a)   (b)

Fig. 3. (a) The shot examples from the footage repository. (b) The distribution of shot duration for the footage repository.

Table 1. The statistics of positive and negative samples under each metric in the training set and test set.

Metrics	semantic-consistence	content-coherence	rhythm-matching
# Positive (Training)	1002	852	1278
# Negative (Training)	771	921	495
# Positive (Test)	124	95	156
# Negative (Test)	76	105	44

Fig. 3(b). From the shot duration distribution, we can see that the duration of most video shots is between 8 s and 20 s.

4.2   Video Annotation

With our song and footage collection, we further generate about 2,000 videos (each for a song clip). The song-to-video translation method (with Random Selection) described in Sect. 3 is used for automatic video generation.

To evaluate the quality of generated videos and also obtain the training data for subsequent classifier training, we manually annotate the generated videos according to three evaluation metrics. Here, we use the three requirements mentioned in Sect. 3 as the evaluation metrics of the video quality, including semantic-consistence, content-coherence, and rhythm-matching. We score each video according to these three metrics by three professional annotation persons (with payment). These professional persons are independent of each other. The video scoring process is as follows: if a metric out of the three ones is met, we mark it as 1, otherwise mark it as −1. We count the total score of each video under each metric. If the total score is less than 0, we annotate its label to 0 (negative sample), otherwise annotate its label to 1 (positive sample). Note that we select those videos with more than 5 shots to better judge the content-coherence metric. Further, we arrange 1,773 videos as the training set and 200 videos as

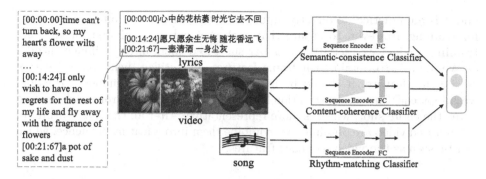

Fig. 4. The schematic illustration of the three classifiers used for the beam algorithm. The semantic-consistence classifier takes the lyric features and video sequence features as input, the content-coherence classifier takes the video sequence features as input, and the rhythm-matching classifier takes the video sequence features and musical rhythm features as input.

the test set. To avoid the overfitting to the songs, all song clips from the same song are forced to fall into either the training set or the test set. The statistics of positive and negative samples under each metric are shown in Table 1. Our constructed benchmark dataset is available at an anonymous GitHub link[1].

5   Baseline Method

As we have mentioned in Sect. 3, we can propose a baseline method for song-to-video translation by instantiating the shot selection strategy with a beam shot-selection algorithm. This is now made possible by training three classifiers according to the three metrics (i.e., semantic consistence, content-coherence, and rhythm-matching) over our constructed dataset. Note that the trained classifiers can be used as the scoring functions of the beam algorithm for measuring the quality of the (partial) generated videos. In the following, we give the details of classifier training (see Fig. 4) and beam shot-selection algorithm.

5.1   Classifier Training

As illustrated in Fig. 4, we train three auxiliary classifiers over the annotated video dataset according to the three metrics, which are called as semantic-consistence classifier, content-coherence classifier, and rhythm-matching classifier respectively. These classifiers are used as the scoring functions for selecting shots in the beam algorithm to improve the quality of generated videos.

During training all three classifiers, video features are always needed. In this regard, first of all, we describe the extraction of video features. Specifically, a generated video usually contains a large number of frames from multiple shots,

---

[1] https://github.com/S2VTouser/Video-Dataset.

and it is not practical to directly input all shot frame features into the classifiers for training. However, to ensure that each shot can be exploited for classifier training, we first extract 10 frames per shot, feed them into the image encoder of WenLan [7] to obtain their frame features, and utilize their average feature as the shot feature. It should be noted that, instead of continuous extraction, we adopt equal-interval extraction for extracting 10 frames from each shot to cover the entire shot, which is more representative. After all the shot features of a generated video are obtained, we combine them into a feature sequence, which can be seen as the video sequence features.

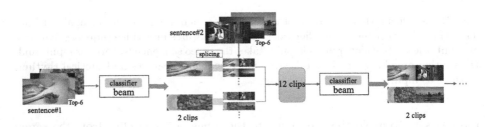

Fig. 5. The beam shot-selection algorithm used in the proposed baseline method for song-to-video translation.

The composition of the three classifiers is similar. Due to the limited amount of annotated data, the overall structure of each classifier is kept as simple as possible to prevent over-fitting, which mainly includes a sequence encoder layer for extracting sequence information and a fully connected layer with two outputs (i.e., positive or negative) for classification of the input video. Here, the gated recurrent unit (GRU) [4] module is used as the sequence encoder. As shown in Fig. 4, when training the semantic-consistence classifier, we take the obtained video shot sequence features and lyric features (extracted with the text encoder of WenLan) as input. When training the content-coherence classifier, we only take the obtained video sequence features as input. When training the rhythm-matching classifier, we take the obtained video sequence features and musical rhythm features (extracted with the pre-trained PANNs [12]) as input. Overall, after a cross-entropy loss is defined over the output layer for each classifier, we train it over the annotated video data (their performance is reported in Table 3).

5.2   Beam Shot-Selection Algorithm

The beam shot-selection algorithm is the core of our proposed baseline method for song-to-video translation. By using the trained three classifiers as the scoring functions, the beam shot-selection algorithm is able to select the 'best' shots combinations for generating a video with better quality, as shown in Fig. 5. We set the beam size to 2. Specifically, for each song clip, we retrieve the top-6 shots with the first sentence of its song lyrics, feed them into a set of trained classifiers to obtain their scores (averaged over the set of trained classifiers), and select the 'best' Top-2 shots with the highest scores. For the retrieved top-6 shots with the second sentence, we splice the selected shots of the first sentence respectively with these 6 shots, feed them into the set of trained classifier, and select the 'best' Top-2 shot combination with the highest scores from 12 video clips. This step can be repeated in turn. Finally, we obtain the 'best' spliced whole video. Note that the set of trained classifiers used in the beam shot-selection algorithm may include only one arbitrary classifier, two arbitrary classifiers, or even all three classifiers. However, we find that the joint guidance of all three classifiers leads to the best average performance of video generation (see Table 2).

Table 2. The ablation study results among Random Selection, Top-1 Selection, and Beam Selection within the song-to-video translation (S2VT) framework.

Algorithm	semantic-consistence	content-coherence	rhythm-matching	average
Random Selection	0.6711	0.4607	0.7906	0.6408
Top-1 Selection	0.6090	0.5195	0.7852	0.6379
Beam-sem	**0.9554**	0.6086	0.8210	0.7950
Beam-coh	0.7046	**0.9943**	0.8358	0.8449
Beam-rhy	0.7282	0.5890	**0.9979**	0.7717
Beam-sem-coh	0.8850	0.9792	0.8370	0.9004
Beam-sem-rhy	0.9112	0.5993	0.9877	0.8327
Beam-coh-rhy	0.7199	0.9852	0.9733	0.8928
Beam-sem-coh-rhy	0.8888	0.9394	0.9742	**0.9341**

6 Experimental Results

We establish an evaluation system to measure the performance of video generation under the aforementioned three metrics. These evaluation metrics can be defined with the trained semantic-consistence classifier, content-coherence classifier and rhythm-matching classifier (see Subsect. 5.1), respectively. Specifically, the evaluation process is as follows: we feed each generated video with a song lyrics (random select 80 different song files and corresponding lyrics files) into a classifier (corresponding to a metric), obtain the outputted score that this video belongs to the positive class, and use this score to evaluate the quality of the generated video under the corresponding metric.

Ablation Study. We conduct ablation study to compare the quality of generated videos by Random Selection, Top-1 Selection, and Beam Selection within the same song-to-video translation framework (see Fig. 2). Among them, Random Selection means that the 'best' shot is randomly selected from the retrieved top-6 shots for each sentence of song lyrics. Top-1 Selection means that the 'best' shot is set to top-1 shot from the retrieved top-6 shots for each sentence of song lyrics. Beam Selection means that under the guidance of a set of

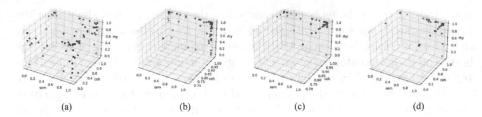

Fig. 6. The scatter plots of different score distributions (of the generated videos with 80 lyrics files) according to the three evaluation metrics. (a), (b), (c), and (d) correspond to Random Selection, Beam-coh, Beam-coh-rhy, and Beam-sem-coh-rhy (these notations are exactly the same as Table 2), respectively. The three dimensions sem, coh and rhy in each sub-graph denote the semantic-consistence, content-coherence, and rhythm-matching metrics, respectively.

diverse classifiers, the 'best' shot with the highest average score is selected at each beam step. The ablation study results are shown in Table 2. Firstly, Beam-sem indicates the guidance of semantic-consistence classifier, Beam-coh indicates the guidance of content-coherence classifier, and Beam-rhy indicates the guidance of rhythm-matching classifier. Secondly, Beam-sem-coh indicates the guidance of semantic-consistence classifier and content-coherence classifier, and Beam-sem-rhy/Beam-coh-rhy is defined similarly. Finally, Beam-sem-coh-thy indicates the joint guidance of semantic-consistence classifier, content-coherence classifier, and rhythm-matching classifier.

From Table 2, we have the following observations: (1) The average scores obtained by Random Selection and Top-1 Selection are very similar. Among the three evaluation metrics, Random Selection outperforms Top-1 Selection in terms of semantic-consistence and rhythm-matching. This means that Random Selection achieves improvement on semantic-consistence and rhythm-matching, but leads to degradation on content-coherence, as compared to Top-1 Selection. (2) Under the guidance of a single classifier, Beam Selection yields significant improvements over both Random Selection and Top-1 Selection. Particularly, under the guidance of content-coherence classifier, the Beam-coh method even doubles the score of Random Selection in terms of content-coherence, indicating the strong coherence enhancement of video content generated with the Beam-coh method. Moreover, under the guidance of semantic-consistence classifier, the performance of the Beam-sem method is nearly twenty-eight percent higher than that of Random Selection in terms of semantic-consistence, showing that the video content generated by the Beam-sem algorithm is more consistent with the song lyrics (and also its context). (3) Under the joint guidance of any two classifiers, the Beam Selection method generally outperforms the method under a single classifier in terms of the average performance, indicating the scalability of Beam Selection for song-to-video translation (S2VT). This observation is further verified by the superior overall performance of the Beam-sem-coh-rhy method under the joint guidance of all three classifiers. In this paper, we thus take the

S2VT method with Beam-sem-coh-rhy for shot selection as our baseline method. Along with our constructed benchmark dataset, the proposed baseline method is expected to facilitate the research on the S2VT task.

Statistic Analysis. From the compared methods in Table 2, we select several representative ones for further statistical analysis of the videos generated by each method. As shown in Fig. 6, we present the 3D scatter plots based on the obtained score of each video under each metric. Figure 6(a) indicates that the score distribution is relatively scattered for Random Selection. After using the Beam-coh method, the 3D data points are concentrated at the higher part along the coh axis, as illustrated in Fig. 6(b). After using the Beam-coh-rhy method, the 3D data points are further clustered at the higher part along the rhy axis, as illustrated in Fig. 6(c). Finally, on the basis of Fig. 6(c), Fig. 6(d) shows that the 3D data points are further clustered to the higher part along the sem axis, after using Beam-sem-coh-rhy. This directly demonstrates the effectiveness of our Beam-sem-coh-rhy method. That is, our proposed method has a better overall performance over all three metrics, as compared with Random Selection.

Fig. 7. The video content coherence comparison between Random Selection and the Beam-coh method. We present an example of two videos generated with the same lyrics in the first two rows (or in the last two rows).

Visualization Results. Through the ablation study, we can see that our proposed method can achieve a large improvement on three metrics over Random Selection (see Table 2). To compare Random and Beam methods more intuitively, we present some examples on the content-coherence and semantic-consistence in Fig. 7 and Fig. 8, respectively. Firstly, we can observe from Fig. 7 that the scene change in the pictures are more natural, according to Beam-coh vs. Random Selection. That is, the video content generated by the Beam-coh method has better coherence than Random method. Secondly, we can observe from Fig. 8 that, compared with Random Selection, the Beam-sem method makes the video content better match the semantics of the song lyrics. For example, in the first

Fig. 8. The semantic consistency comparison between Random Selection and the Beam-sem method. We present an example of two videos generated with the same lyrics in the first two rows (or in the last two rows).

two rows, for the lyrics like 'the rain turns the whole city upside down', and 'write down my grief', the semantic expression of Beam-sem is more accurate than Random Selection. In last two rows, for the lyrics like 'look at the sea and the sky' and 'I loosen the ropes of time', the last row is clearly more contextually appropriate. In addition, several example videos generated by our proposed baseline method can be found at the anonymous GitHub link[2]. From these results, we can intuitively see that WenLan has great ability of generalization to the Chinese song lyrics. More importantly, tens of generated videos with our baseline have 2M plays and 250K likes on TikTok, which demonstrates adopting WenLan without fine-tuning is sufficient.

Table 3. The training and test accuracy of three classifiers.

Classifier	semantic-consistence	content-coherence	rhythm-matching
Training accuracy	96.95	97.86	96.00
Test accuracy	80.50	82.00	83.00

Classification Accuracy Analysis. We separately train three classifiers (i.e., semantic-consistence, content-coherence and rhythm-matching classifiers) for the beam algorithm based on the human-labeled song-to-video dataset. As shown in Table 1, the ratio of positive and negative examples for the training set under each metric is almost the same as that for the test set. Moreover, as shown in Fig. 4, each classifier only has two layers: one is the sequence encoder layer for extracting sequence information, and the other is the fully connected layer for classification. The input dimension of the classifier is 2,560, and the output dimension is 2. The training and test accuracies of the three classifiers are shown in Table 3. We can see that all three classifiers achieve a test accuracy over 80%.

7 Conclusions

In this paper, we introduce a novel video editing task called song-to-video translation (S2VT), which aims to write a video from song lyrics based on multimodal pre-training. To facilitate the research on S2VT, we construct a song-to-video benchmark dataset with human annotations according to three evaluation metrics (i.e., semantic-consistence, content-coherence, and rhythm-matching). With this dataset, we propose a baseline method to solve the S2VT task by training three classifiers (each for a metric) and developing a beam shot-selection algorithm based on the trained classifiers. We design an automatic evaluation system to evaluate the performance of S2VT. Extensive experiments are conducted to show the effectiveness of the proposed baseline method in the S2VT task.

8 Limitations

In this paper, we propose a novel video editing task called S2VT. Although the proposed method achieve great results in the S2VT task, it still exists the limitations. Concretely, the song lyrics we adopted in this paper is Chinese song lyrics (lyrics themselves are often more abstract), which makes them difficult for existing video editing methods using generic sentences as inputs to understand and apply them in the S2VT task (i.e., they completely failed in the S2VT task). Meanwhile, this is the reason why we do not compare with more video editing methods in the baseline results. In future work, we will improve this issue. For example, we could manually translate Chinese song lyrics to English ones, and feed them into the other video editing methods for generation. Finally, the quality of generated videos by the two ways (other video editing methods and ours) will be compared based on the corresponding metrics.

Moreover, the evaluation metrics (i.e., semantic-consistence, content-coherence, and rhythm-matching) are not comprehensive enough. In future work, we will explore more metrics to evaluate the quality of generated videos more comprehensively. On the other hand, we will continue to optimize the shot selection algorithm to make the generated videos better and more artistic.

Acknowledgements. This work was supported by National Natural Science Foundation of China (61976220).

References

1. Aytar, Y., Vondrick, C., Torralba, A.: Soundnet: learning sound representations from unlabeled video. In: Advances in Neural Information Processing Systems 29 (2016)
2. Bellini, R., Kleiman, Y., Cohen-Or, D.: Dance to the beat: Synchronizing motion to audio. Comput. Visual Media 4(3), 197–208 (2018)
3. Chen, Q., Wu, Q., Chen, J., Wu, Q., van den Hengel, A., Tan, M.: Scripted video generation with a bottom-up generative adversarial network. IEEE Trans. Image Process. 29, 7454–7467 (2020)
4. Chung, J., Gulcehre, C., Cho, K., Bengio, Y.: Empirical evaluation of gated recurrent neural networks on sequence modeling. arXiv preprint arXiv:1412.3555 (2014)
5. Davis, A., Agrawala, M.: Visual rhythm and beat. In: Proceedings of the IEEE Conference on Computer Vision and Pattern Recognition Workshops, pp. 2532–2535 (2018)
6. Faghri, F., Fleet, D.J., Kiros, J.R., Fidler, S.: Vse++: improving visual-semantic embeddings with hard negatives. arXiv preprint arXiv:1707.05612 (2017)
7. Fei, N., et al.: Wenlan 2.0: Make ai imagine via a multimodal foundation model. arXiv preprint arXiv:2110.14378 (2021)
8. Fu, T.J., Wang, X.E., Grafton, S.T., Eckstein, M.P., Wang, W.Y.: Language-based video editing via multi-modal multi-level transformer. arXiv preprint arXiv:2104.01122 (2021)
9. Girgensohn, A., et al.: A semi-automatic approach to home video editing. In: Proceedings of the 13th Annual ACM Symposium on User Interface Software and Technology, pp. 81–89 (2000)
10. Hua, X.S., Lu, L., Zhang, H.J.: Automatic music video generation based on temporal pattern analysis. In: Proceedings of the 12th Annual ACM International Conference on Multimedia, pp. 472–475 (2004)
11. Kim, D., Joo, D., Kim, J.: Tivgan: text to image to video generation with step-by-step evolutionary generator. IEEE Access 8, 153113–153122 (2020)
12. Kong, Q., Cao, Y., Iqbal, T., Wang, Y., Wang, W., Plumbley, M.D.: Panns: large-scale pretrained audio neural networks for audio pattern recognition. IEEE/ACM Trans. Audio Speech Lang. Process. 28, 2880–2894 (2020)
13. Leake, M., Davis, A., Truong, A., Agrawala, M.: Computational video editing for dialogue-driven scenes. ACM Trans. Graph. 36(4), 130–1 (2017)
14. Liao, C., Wang, P.P., Zhang, Y.: Mining association patterns between music and video clips in professional MTV. In: Huet, B., Smeaton, A., Mayer-Patel, K., Avrithis, Y. (eds.) MMM 2009. LNCS, vol. 5371, pp. 401–412. Springer, Heidelberg (2009). https://doi.org/10.1007/978-3-540-92892-8_41
15. Liao, Z., Yu, Y., Gong, B., Cheng, L.: Audeosynth: music-driven video montage. ACM Trans. Graph. (TOG) 34(4), 1–10 (2015)
16. Lin, J.C., Wei, W.L., Wang, H.M.: Automatic music video generation based on emotion-oriented pseudo song prediction and matching. In: Proceedings of the 24th ACM International Conference on Multimedia, pp. 372–376 (2016)
17. Parekh, S., Essid, S., Ozerov, A., Duong, N.Q., Pérez, P., Richard, G.: Weakly supervised representation learning for audio-visual scene analysis. IEEE/ACM Trans. Audio Speech Lang. Process. 28, 416–428 (2019)
18. Radford, A., et al.: Learning transferable visual models from natural language supervision. In: ICML, pp. 8748–8763 (2021)

19. Schindler, A.: Multi-modal music information retrieval: augmenting audio-analysis with visual computing for improved music video analysis. arXiv preprint arXiv:2002.00251 (2020)
20. Schindler, A., Rauber, A.: Harnessing music-related visual stereotypes for music information retrieval. ACM Trans. Intell. Syst. Technol. (TIST) **8**(2), 1–21 (2016)
21. Wang, M., Yang, G.W., Hu, S.M., Yau, S.T., Shamir, A.: Write-a-video: computational video montage from themed text. ACM Trans. Graph. **38**(6), 177–1 (2019)
22. Xiong, Y., Heilbron, F.C., Lin, D.: Transcript to video: efficient clip sequencing from texts. arXiv preprint arXiv:2107.11851 (2021)

Medical Image Analysis

Breast Cancer Histopathology Image Classification Using Frequency Attention Convolution Network

Ruidong Lu[1], Qiule Sun[2], Xueyan Ding[1], and Jianxin Zhang[1,3(✉)] 🆔

[1] School of Computer Science and Engineering, Dalian Minzu University,
Dalian, China
jxzhang0411@163.com
[2] School of Information and Communication Engineering,
Dalian University of Technology, Dalian, China
[3] SEAC Key Laboratory of Big Data Applied Technology, Dalian Minzu University,
Dalian, China

Abstract. The existing deep learning works mainly capture breast cancer histopathology image features in the spatial domain, and they rarely consider the frequency domain feature representation of histopathology images. According to the classical digital signal processing theory, frequency domain features may outperform spatial domain features in analyzing texture images. Motivated by this, we attempt to mine frequency domain features for the breast cancer histopathology image classification application, and further propose a novel frequency-attention convolutional network called SeFFT-Net by combining the Fourier transform with the channel attention mechanism. The core of SeFFT-Net consists of a newly constructed frequency-based squeeze and excitation (SeFFT) module, which first performs Fourier transform with residual construction to capture deep features in the frequency domain of histopathology images, followed by a squeeze-and-excitation attention operator to further enhance important frequency features. We extensively evaluate the proposed SeFFT-Net model on the public BreakHis breast cancer histopathology dataset, and it achieves the optimal image-level and patient-level classification accuracy of 98.67% and 98.16%, respectively. Meanwhile, ablation studies also well demonstrate the effectiveness of introducing frequency transforms for this medical image application.

Keywords: Breast cancer · Histopathology image classification · Convolutional neural network · Frequency domain · Channel attention

1 Introduction

Globally, breast cancer is the most common malignancy in women and the cancer with the highest mortality rate [1]. Early diagnosis and treatment of breast cancer is essential in augmenting the survival rate of patients, while pathological

X. Yang et al. (Eds.): ADMA 2023, LNAI 14177, pp. 217–229, 2023.
https://doi.org/10.1007/978-3-031-46664-9_15

diagnosis is still seen as the definitive method for breast cancer diagnosis [2]. Since the traditional pathological diagnosis mainly relies on the experience of pathologists, which is time-consuming and laborious, with the rapid growth of the demand for pathological diagnosis, computer-aided diagnosis of breast cancer histopathology images is becoming more and more important.

Recently, breast cancer histopathology image classification methods related to deep learning have achieved great success and gradually become the mainstream. Among them, some works employ typical convolutional neural networks (CNN), such as AlexNet, VGGNet, and ResNet, to pre-train on large-scale natural image datasets as feature extractors, and then use machine learning classifiers to distinguish the extracted deep feature image [3,8]. Deniz et al. utilize pre-trained AlexNet and VGG16 models to capture deep features of breast cancer histopathology images, and then employ support vector machine to distinguish the deep features [8]. As a counterpart, Gupta and Bhavsar employ residuals and dense networks to capture deep features of histopathology images, followed by XGBoost as a feature classifier, and they achieve the best patient-level classification result of 96.76% [3]. In order to narrow the gap between the extracted image features and the classifiers, researchers further leverage learnable CNNs for breast cancer histopathology image classification. Considering that the histopathological images used for model training are limited, transfer learning is usually used to improve performance. For example, Shalu and Mehra explored the effect of transfer learning on breast cancer histopathology images compared with fully trained networks using VGG16, VGG19 and ResNet50 models [5]. Subsequently, Chukwu et al. utilize pre-trained DenseNet and transfer learning technology to obtain the best accuracy rate of 97.42% on the public breast cancer histopathology image dataset [6]. Meanwhile, considering the characteristics of breast cancer histopathology images, some works attempt to build novel CNN models for this medical task. Spanhol et al. [7] construct a simple plain CNN model with five trainable layers, and experimental results demonstrate that it outperforms conventional methods. Likewise, Budak et al. [8] propose a learnable model combining fully convolutional networks and bidirectional long-short-term memory, and they achieve average results of 94.98% on the breast cancer histopathology image database. Moreover, to focus on important discriminative deep features, attention mechanisms are also widely introduced to classify histopathology images with excellent performance [9]. In general, deep learning-related models have recently greatly promoted the development of computer-aided histopathology diagnosis of breast cancer, showing obvious advantages in classification accuracy compared with traditional work [10–12].

However, current deep learning-related breast cancer histopathology image classification methods are mainly implemented in spatial domain, while rarely consider the frequency domain features of histopathological images. According to the theory of digital signal processing, frequency domain is more suitable for analyzing texture images than spatial domain. Actually, some researchers have recently attempted to explore frequency-domain deep learning methods for computer vision applications [13–15]. Gueguen et al. learn CNNs directly on the

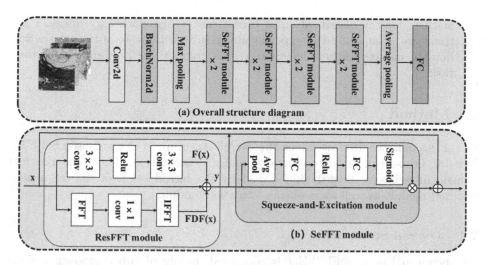

Fig. 1. Overall architecture of SeFFT-Net for breast cancer histopathology image classification. SeFFT-Net leverages ResNet18 as the backbone and embeds SeFFT modules at multiple layers that integrate residual Fourier transform (ResFFT) with channel attention to compute important frequency-domain features.

discrete cosine transform (DCT) of deep features for effective image classification [13], while Ehrlich et al. [14] propose a new method using frequency-domain compressed the image is used as input to the residual network. Additionally, Zhong et al. [15] explore frequency domain features as additional cues to better solve camouflaged object detection task.

Inspired by these works, we try to study breast cancer histopathology image classification task by introducing frequency domain deep features. In this work, we propose a novel frequency-domain attention convolutional network, namely SeFFT-Net, which firstly utilizes Fourier transform to capture the frequency-domain deep features of histopathology images, followed by an attention module [16] is used to further enhance important frequency features. The overall architecture of the given SeFFT-Net model is shown in Fig. 1. The main contributions of this paper can be encapsulated in three facets.

(1) This work attempts to explore frequency-domain deep features for breast cancer histopathology image classification applications, and further utilizes Fourier transform and channel attention mechanism to propose a novel frequency-domain attention convolution network called SeFFT-Net.
(2) SeFFT-Net first performs a Fast Fourier transform operator combined with residual construction (ResFFT) to compute deep features in the frequency domain of histopathology images, and then further enhances high-valued frequency feature impact with a squeeze and excite attention module to obtain more promising classification results.
(3) We extensively evaluate SeFFT-Net on the public BreakHis dataset. Ablation studies demonstrate the effectiveness of introducing frequency-domain

features for the classification of breast cancer histopathology image. Furthermore, comparing the experimental results with state-of-the-art spatial domain models further demonstrates its competitive performance in this task.

2   Method

In this section, we first introduce the overall structure of the proposed SeFFT-Net for breast cancer histopathology image classification. Then, the Fourier transform is briefly described, followed by the introduction of the frequency residual module as well as frequency attention module.

2.1   Overall Structure

As shown in Fig. 1, SeFFT-Net is composed mainly of two components, i.e., a backbone model and a SeFFT module. ResNet18 [17] is used as the backbone model owing to its superiority in the breast cancer histopathology image classification task. Actually, breast cancer pathological images have complex frequency distributions, and the essential information of such images is mainly concentrated in the low-frequency area. Therefore, capturing frequency domain features becomes critical, how to interact with convolutional features should be deeply considered.

We achieve that by proposed SeFFT module. We endeavor to replace the residual module of the backbone model with the newly constructed SeFFT module, which well integrates Fourier transform and channel attention mechanism, thereby capturing the frequency domain depth features of histopathological images and interacting with convolutional features. Specifically, the frequency domain is applied by the ResFFT module in SeFFT module, which can simultaneously process the images in the space domain and the frequency domain. Frequency domain features are captured by Fast Fourier Transform (FFT) and Inverse Fast Fourier Transform (IFFT). Initial interaction $\mathbf{Y}$ between frequency domain features and convolutional features is obtained here by element-wise addition. In addition, in order to enhance high-value information consequences and neglect low-value information consequences, we present a channel attention module, i.e., Squeeze-and-Excitation module, on the result of the ResFFT module. This will further interaction between the two types of features. It is noteworthy that our SeFFT-Net network structure is very flexible and can insert any convolutional neural networks applied to other medical image classification tasks.

2.2   Fourier Transform

The Fourier transform is the most basic and widely used frequency transform operator in image processing and analysis. During image processing and analysis, the Fourier transform decomposes the image into sine component and cosine component. The frequency feature of the digital image after Fourier transform is a complex numbers. The frequency domain network [13–15,18] can perform various operations on real and imaginary images on the basis of the original network architecture, such as complex number operations, so as to learn more robust frequency domain features. In addition, we can transform the image from the spatial domain to the frequency domain using the Fourier transform, and processing the image in the frequency domain. Afterwards, the frequency domain image is restored to the space domain image by inverse Fourier transform. Given an input image patch m, of size $M \times N$, denoted as $f_m(\text{x}, \text{y})$, the following is the calculation expression of the discrete Fourier transform [18]:

$$f_m(u, v) = \sum_{x=0}^{M-1} \sum_{y=0}^{N-1} f_m(x, y) e^{-j2\pi(\frac{ux}{M} + \frac{vy}{N})} \tag{1}$$

The Fast Fourier Transform (FFT) is a fast algorithm for the discrete Fourier transform. It is obtained by improving the original algorithm according to the characteristics of discrete Fourier transform, which greatly reduces the calculation amount of the computer. Efficient fast Fourier transforms can model interactions between spatial locations with log-linear complexity. By using Fast Fourier Transform (FFT), the image is divided into real image and imaginary image, so that a series of feature extraction operations such as convolution, batch normalization, and activation can be performed on the image in the frequency domain. This enables the network to extract richer frequency feature information. Afterwards, the real and imaginary images are mixed. Finally, the Inverse Fast Fourier Transform (IFFT) can effectively aggregate local information and improve the learning ability of non-local information.

2.3   SeFFT Module

In this section, we mainly present the specific structure of SeFFT module. The SeFFT contains a ResFFT module and a channel attention module. The main purpose of ResFFT module is to capture frequency domain features. In ResFFT module, the Fourier transform is responsible for this purpose. The channel attention module aims to highlight important features from the ResFFT module or submitting for a classifier.

In Fig. 1, the SeFFT module integrates Fast Fourier Transform with the channel attention mechanism, which outputs meaningful frequency domain information by processing images in both spatial and frequency domains simultaneously. As shown in the blue box in the Fig. 1, to obtain more detailed frequency domain features, we add the Fourier Transform operation to the residual block of ResNet18 backbone [17], which can assist the network to concentrate on critical local features and enhance recognition accuracy. We first give the formulation of the classical residual block in the ResNet architecture. The conventional residual block is expressed as:

$$\mathbf{Y} = \mathbf{X} + F(\mathbf{X}). \tag{2}$$

Here, $\mathbf{X} \in R^{C \times H \times W}$ and $\mathbf{Y} \in R^{C \times H \times W}$ are input and output tensors, where C, H, W are channel number, height and width, respectively. Besides, F is a residual learning block. Then, ResFFT module improves the above residual block by adding a frequency domain branch that captures representative frequency domain features of breast cancer histopathological images. It is formulated as:

$$\mathbf{Y} = \mathbf{X} + F(\mathbf{X}) + FDF(\mathbf{X}) \tag{3}$$

$$FDF(\mathbf{X}) = IFFT(conv(FFT(\mathbf{X}))) \tag{4}$$

From above equation, ResFFT module fuses the input $\mathbf{X}$, convolution learning block F and frequency domain features learning block FDF, also providing interaction between convolution features and frequency domain features. For capturing frequency domain features, ResFFT module first performs fast Fourier transform FFT to convert images from the space domain to the frequency domain, then applies efficient 1×1 convolution (conv.) to compress the number of channels and add network nonlinearity, and finally utilises inverse fast Fourier transform $IFFT$ to convert the information in frequency domain to space domain. By doing so, the operation done in the frequency domain is presented on the image after inverse fast Fourier transform.

Next, after the ResFFT module, we introduce a classic Squeeze-and-Excitation module [16] to further enhance deep features. Specifically, the squeeze operation compresses image features from outputs of the ResFFT module by average pooling, followed by two FC layers and Relu layers for interaction between channel responses and increasing nonlinearity respectively. The excitation operation generates weights via sigmoid function for each feature channel, which fully captures the dependency between channels and outputs the same number of weights as the input characteristics. Thus, it automatically obtains the importance spatial and frequency information of each channel through learning, and suppresses the characteristics that are useless for the current task according to this. And then, in the weighting operation before output, we establish the connection of input, ResFFT module and SeFFT module, which is helpful for the reverse propagation of gradient in the training process and the realization of feature reuse through the connection of features.

3   Experimental Results and Discussion

First, we describe the public breast cancer histopathology image dataset used to evaluate the SeFFT-Net model. Then, the parameter settings and evaluation metrics are briefly introduced. Finally, we report and analyze the experimental results in detail, including ablation experiment results, Comparison with advanced spatial domain methods, as well as visualization results.

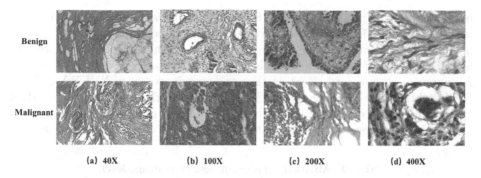

Benign

Malignant

(a) 40X   (b) 100X   (c) 200X   (d) 400X

Fig. 2. Typical breast cancer histopathology images at four magnification factors in the BreakHis dataset.

3.1 Dataset

A commonly used breast cancer histopathological image dataset, namely BreakHis, is adopted to evaluate SeFFT-Net in this work. The BreakKHis dataset is a publicly available large-scale non-global breast cancer histopathology image dataset (http://web.inf.ufpr.br/vri/databases/breast-cancer-histopatho logical-database-breakhis), which provides a good benchmark for this medical application. The BreakHis dataset contains 7909 histopathological images from 82 patients, each of which is labeled with benign tumors (fibroadenoma, adenoma, tubular adenoma and trichoma) or malignant tumors (lobular carcinoma, ductal carcinoma, papillary carcinoma and mucinous carcinoma). In addition, 2480 samples belong to benign images, and the remaining 5429 samples are malignant images. Each sample image has an RGB channel mode with the size of 700×460 pixels in size, and the color depth of each channel is 8 bits. According to the different magnification, the samples of each patient can be divided into four groups of 40 times ($40\times$), 100 times ($100\times$), 200 times ($200\times$) and 400 times ($400\times$). Figure 2 shows some typical breast cancer histopathology images at different magnification factors in the BreakHis dataset.

3.2 Experimental Settings

The original data set of BreakHis is randomly divided into a training set and a test set at each magnification factor. The training set consists of 70% images, and the rest 30% images constitute the test set. In addition, 25% of the training set images are retained for cross validation to select model parameters. All experiments utilize the same training data set and test data set. In the image preprocessing stage, to reduce the impact of possible over fitting problems, we perform simple crop and flip operations to increase the sample size of the training set. For network training, the initial learning rate is set to LR = 0.001, and the learning rate decays to half of the current learning rate after every five iterations. The data set is randomly scrambled to avoid any negative impact on learning by using orderly training data. Besides, the loss function is optimized using a stochastic gradient descent (SGD) algorithm with a batch size of 8. The momentum factor is set to 0.9 to prevent the loss function from falling into a local optimal solution, and control the loss function to reach the global minimum. All models are trained for cosine annealing learning rate attenuation in 100 cycles. All experiments are carried out on the server configured with NVIDIA GeForce RTX 2080Ti using the Python deep learning framework. Additionally, we adopt two commonly used classification accuracy indicators of image-level recognition rate and patient-level recognition rate to evaluate the model performance.

Table 1. Ablation experiment results at image level.

Method	40× (%)	100× (%)	200× (%)	400× (%)
ResNet18	95.99	95.68	97.35	93.77
ResFFT-Net	96.49	96.80	98.01	94.87
SENet	96.49	96.96	98.01	94.87
SeFFT-Net	96.99	98.08	98.67	95.24

Table 2. Ablation experiment results at the patient level.

Method	40× (%)	100× (%)	200× (%)	400× (%)
ResNet18	95.62	96.06	97.57	94.52
ResFFT-Net	96.77	96.94	97.10	95.53
SENet	96.04	96.61	98.01	95.47
SeFFT-Net	96.44	98.16	98.14	95.57

3.3   Experimental Results

Ablation Experiment Results. To prove the effectiveness of SeFFT-Net as well as the frequency domain features for this medical task, we first conduct image-level and patient-level ablation experiments on the BreakHis dataset, whose results are reported in Table 1 and Table 2, respectively. In the two tables, we first employ the typical ResNet18 model as the baseline. Then, we embed the Fourier transform module into the model to construct ResFFT-Net, and further integrate the ResFFT module with squeeze-and-excitation channel attention to construct SeFFT-Net. In addition, we also introduce SE-Net as a counterpart to better show the effectiveness of frequency domain features.

As shown in Table 1, the baseline of ResNet18 achieves the image-level recognition rates of 95.99%, 95.68%, 97.35% and 93.77% on 40X, 100X, 200X and 400X data sets, respectively. After introducing the frequency domain features, the ResFFT model gains the corresponding accuracy results of 96.49%, 96.80%, 98.01% and 94.87%, which outperforms ResNet18 on the four data sets, thus showing the effectiveness of introducing frequency domain features. By simultaneously integrating frequency transform and attention mechanism, SeFFT-Net

Table 3. Comparisons with advanced spatial domain methods at both image-level and patient-level.

Reference	Year	Image-Level (%)				Patient-Level (%)			
		40×	100×	200×	400×	40×	100×	200×	400×
Spanhol et al. [7]	2017	84.60	84.80	84.20	81.60	84.00	83.90	86.30	82.10
Han et al. [23]	2017	95.80	96.90	96.70	94.90	97.10	95.70	96.50	95.70
Gupta et al. [3]	2018	–	–	–	–	94.71	95.90	96.76	89.11
Lichtblau et al. [20]	2019	85.60	87.40	89.80	87.00	83.90	86.00	89.10	86.60
Alom et al. [24]	2019	97.95	97.57	97.32	**97.36**	**97.60**	97.65	97.56	**97.62**
Zhang et al. [19]	2020	95.03	90.41	88.48	85.00	95.50	91.57	89.20	89.20
Hou [21]	2020	90.89	90.99	91.00	90.97	91.00	91.00	91.00	91.00
Man et al. [28]	2020	**99.13**	96.39	86.38	85.20	96.32	95.89	86.91	85.16
Li et al. [27]	2021	87.85	86.68	87.75	85.30	87.93	87.41	88.76	85.55
Chukwu et al. [6]	2021	93.64	97.42	95.87	94.67	94.23	97.86	96.35	95.24
Sharma and Kumar [29]	2021	96.25	96.25	95.74	94.11	–	–	–	–
Boumaraf et al. [26]	2021	98.13	97.39	96.63	94.05	–	–	–	–
Saxena et al. [22]	2021	88.36	87.14	90.02	84.16	92.88	83.61	89.98	81.63
Xu et al. [30]	2022	94.94	94.18	95.38	92.64	–	–	–	–
Hao et al. [25]	2022	96.75	95.21	96.57	93.15	96.33	95.26	96.09	92.99
Chhipa et al. [4]	2022	93.00	93.26	92.28	88.74	93.26	93.45	92.45	89.57
SeFFT-Net (Ours)	–	96.99	**98.08**	**98.67**	95.24	96.44	**98.16**	**98.14**	95.57

further improves the classification performance to 96.99%, 98.08%, 98.67% and 95.24%, which is also better than the Squeeze-and-Excitation network (SENet) [16]. Compared with the baseline, SeFFT-Net can gain classification accuracy improvement of 1.00%, 2.40%, 1.32% and 1.47%, respectively. Thereby, the image-level ablation experimental results well demonstrate the effects of the frequency domain features as well as the proposed SeFFT-Net for breast cancer histopathology image classification.

When it comes to the patient-level ablation results listed in Table 2, SeFFT-Net respectively gains the accuracy values of 96.44%, 98.16%, 98.14% and 95.57% on the 40X, 100X, 200X and 400X datasets, which also shows the best performance among the four models. SeFFT-Net outperforms the baseline by 0.82%, 2.10%, 0.57% and 1.05% gains on the four data sets, respectively. Meanwhile, it is superior to SENet with average accuracy improvement of 0.55%. Additionally, ResFFT-Net averagely outperforms the baseline ResNet18 model by 0.64% classification accuracy. The above results again well prove the effectiveness of SeFFT-Net and the frequency domain features.

Comparison with Typical Spatial Domain Methods. To further show the performance of SeFFT-Net on breast cancer pathological image classification task, we compare it with a variety of advanced spatial domain methods proposed in recent years. The detailed results at both image-level and patient-level are demonstrated in Table 3.

As shown in Table 3, SeFFT-Net has a notable competitive performance compared to the previously representative spatial domain methods. Specifically, it is worth noting that the given model achieves the image-level classification accuracy of 96.99%, 98.08%, 98.67% and 95.24% on 40×, 100×, 200× and 400× data sets, which are significantly better than results in literature [7,8,19–22]. Besides, SeFFT-Net gains the best accuracy values on both 100× and 200× data sets among all the works. Despite not achieving optimal results on the 40× and 400× data bases, it ranks fourth and second on the two data sets at the image level, respectively. Moreover, when it comes to the patient-level evaluation results, SeFFT-Net achieves recognition rates of 96.44%, 98.16%, 98.14% and 95.57% on four multiples, and it is also superior to other methods on 100× and 200× data sets. Meanwhile, SeFFT-Net ranks the third place on both 40× and 400× data bases. Among these spatial CNN-based methods, CSDCNN, IRRCNN+Aug., VGG 19 and DenseNet in literature [6,23,24,26] obtain the most promising results with the average recognition rate around 96%. However, SeFFT-Net overall shows very competitive or better performance over the three works. According to the above results, we can see that the SeFFT-Net model is effective for breast cancer pathological image classification task, which can be attributed to the frequency domain feature to some extent.

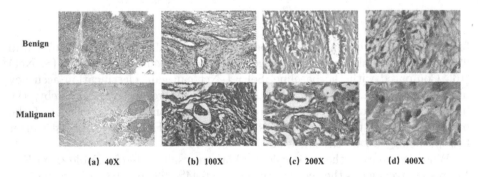

Fig. 3. Histopathological images that are incorrectly classified by ResNet18 but can be correctly classified by SeFFT-Net.

Visualization Results. In this section, we manage these breast cancer pathology tissue slices that are misclassified by the baseline but can be rightly distinguished by SeFFT-Net, and show eight typical images at four magnifications in Fig. 3. In the figure, images in the first row are labeled as benign tumors, while images in the second row belong to malignant tumors. Due to the complexity and irregularity of breast cancer histopathology images, the baseline model can not well distinguish some histopathology images, specially for those containing blank areas. After introducing frequency transform and channel attention modules, the SeFFT-Net model can well classify these breast cancer pathology tissue

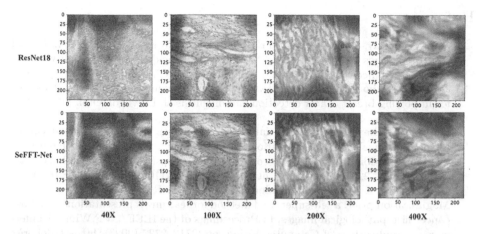

Fig. 4. Visualized heatmap results of ResNet18 and SeFFT-Net deep feature activations at four magnifications.

slices, and the classification accuracy is significantly improved compared to the baseline. Then, we also visualize heatmaps of deep features in Fig. 4, aiming to further display the regions of interest of different networks and hope to provide a valuable reference for classification results.

4   Conclusion

This paper attempts to explore the application of frequency domain related deep learning methods in breast cancer histopathology image classification tasks, and further propose a novel frequency attention network called SeFFT-Net by combining the advantages of frequency transformation and channel attention mechanism. SeFFT-Net adds Fourier transform on the spatial residual structure to extract the frequency-domain features of histopathology images, and then enhances the feature representation with an attention operator to obtain more promising classification performance. Experimental results on the public dataset BreakHis demonstrate the effectiveness of SeFFT-Net in this medical image application, while ablation studies on two landmark spatial counterparts provide a good demonstration of the effect of introducing frequency-domain features. In the future, we will attempt to capture more discriminant frequency features for breast cancer histopathology image classification. Besides, it is also interesting to explore the combination of frequency features with transformer models.

Acknowledgements. This work was supported in part by the National Natural Science Foundation of China under Grant 61972062, the Applied Basic Research Project of Liaoning Province under Grant 2023JH2/101300191 and 2023JH2/101300193.

References

1. Bray, F., Ferlay, J., Soerjomataram, I., Siegel, R.L., Torre, L.A., Jemal, A.: Global cancer statistics 2018: GLOBOCAN estimates of incidence and mortality worldwide for 36 cancers in 185 countries. CA: Cancer J. Clin. **68**(6), 394–424 (2018)
2. Joy, J.E., Penhoet, E.E., Petitti, D.B., Ebrary, I.: Saving women's lives: strategies for improving breast cancer detection and diagnosis. J. Laryngol. Otol. **86**(2), 105–19 (2005)
3. Gupta, V., Bhavsar, A.: Sequential modeling of deep features for breast cancer histopathological image classification. In Proceedings of the IEEE Conference on Computer Vision and Pattern Recognition Workshops(CVPRW), Salt Lake City, UT, USA, pp. 2254–2261 (2018). https://doi.org/10.1109/CVPRW.2018.00302
4. Chhipa, P.C., Upadhyay, R., Pihlgren, G.G., Saini, R., Uchida, S., Liwicki, M.: Magnification prior: a self-supervised method for learning representations on breast cancer histopathological images. In Proceedings of the IEEE/CVF Winter Conference on Applications of Computer Vision, pp. 2717–2727 (2023). https://doi.org/10.48550/arXiv.2203.07707
5. Shallu, M.R.: Breast cancer histology images classification: training from scratch or transfer learning? ICT Exp. **4**(4), 247–254 (2018)
6. Chukwu, J.K., Sani, F.B., Nuhu, A.S.: Breast cancer classification using deep convolutional neural networks. FUOYE J. Eng. Technol. **6**(2), 35–38 (2021)
7. Spanhol, F.A., Oliveira, L.S., Cavalin, P.R., Petitjean, C., Heutte, L.: Deep features for breast cancer histopathological image classification. In 2017 IEEE International Conference on Systems, Man, and Cybernetics (SMC), Banff, AB, Canada, pp. 1868–1873 (2017). https://doi.org/10.1109/SMC.2017.8122889
8. Deniz, E., Şengür, A., Kadiroğlu, Z., Guo, Y., Bajaj, V., Budak, Ü.: Transfer learning based histopathologic image classification for breast cancer detection. Health Inf. Sci. Syst. **6**(1), 1–7 (2018)
9. Jiang, Y., Chen, L., Zhang, H., Xiao, X.: Breast cancer histopathological image classification using convolutional neural networks with small SE-ResNet module. PloS One **14**(3), e0214587 (2019)
10. Sohail, A., Khan, A., Wahab, N., Zameer, A., Khan, S.: A multi-phase deep CNN based mitosis detection framework for breast cancer histopathological images. Sci. Rep. **11**(1), 1–18 (2021)
11. Juppet, Q., De Martino, F., Marcandalli, E., Weigert, M., Burri, O., Unser, M.: Deep learning enables individual xenograft cell classification in histological images by analysis of contextual features. J. Mammary Gland Biol. Neoplasia **26**(2), 101–112 (2021)
12. Hirra, I., Ahmad, M., Hussain, A., Ashraf, M.U., Saeed, I.A., Qadri, S.F.: Breast cancer classification from histopathological images using patch-based deep learning modeling. IEEE Access **9**, 24273–24287 (2021)
13. Gueguen, L., Sergeev, A., Kadlec, B., Liu, R., Yosinski, J.: Faster neural networks straight from JPEG. In: 32nd Conference on Neural Information Processing Systems, pp. 1–13 (2018)
14. Ehrlich, M., Davis, L. S.: Deep residual learning in the JPEG transform domain. In: Proceedings of the IEEE/CVF International Conference on Computer Vision, Seoul, Korea, pp. 3484–3493 (2019). https://doi.org/10.1109/ICCV.2019.00358
15. Zhong, Y., Li, B., Tang, L., Kuang, S., Wu, S., Ding, S.: Detecting camouflaged object in frequency domain. In Proceedings of the IEEE Conference on Computer Vision and Pattern Recognition, New Orleans, LA, USA, pp. 4504–4513 (2022). https://doi.org/10.1109/CVPR52688.2022.00446

16. Hu, J., Shen, L., Sun, G.J., Albanie, S., Wu, E.H., Sun, G.: Squeeze-and-excitation networks. In: Proceedings of the IEEE Conference on Computer Vision and Pattern Recognition, pp. 7132–7141. (2018). https://doi.org/10.48550/arXiv.1709.01507

17. He, K., Zhang, X., Ren, S., Sun, J.: Deep residual learning for image recognition. In: Proceedings of the IEEE Conference on Computer Vision and Pattern Recognition, pp. 770–778 (2015). 10.48550/arXiv.1512.03385

18. Wang, K.N., He, Y., Zhuang, S., Miao, J., He, X., Zhou, P.: FFCNet: fourier transform-based frequency learning and complex convolutional network for colon disease classification. In: Proceedings of the 25th International Conference of Medical Image Computing and Computer Assisted Intervention-MICCAI 2022: 25th International Conference, Singapore, 18–22 September 2022, Proceedings, Part III, pp. 78–87 (2022). https://doi.org/10.1007/978-3-031-16437-8_8

19. Zhang, J., Wei, X., Dong, J., Liu, B.: Aggregated deep global feature representation for breast cancer histopathology image classification. J. Med. Imaging Health Inf. **10**(11), 2778–2783 (2020)

20. Lichtblau, D., Stoean, C., Magalhaes, M.: Cancer diagnosis through a tandem of classifiers for digitized histopathological slides. PLoS ONE **14**(1), e0209274 (2019)

21. Hou, Y.: Breast cancer pathological image classification based on deep learning. J. Xray Sci. Technol. **28**(4), 727–738 (2020)

22. Saxena, S., Shukla, S., Gyanchandani, M.: Breast cancer histopathology image classification using kernelized weighted extreme learning machine. Int. J. Imaging Syst. Technol. **31**(1), 168–179 (2021)

23. Han, Z., Wei, B., Zheng, Y., Yin, Y., Li, K., Li, S.: Breast cancer multi-classification from histopathological images with structured deep learning model. Sci. Rep. **7**(1), 4172 (2017)

24. Alom, M.Z., Yakopcic, C., Nasrin, M.S., Taha, T.M., Asari, V.K.: Breast cancer classification from histopathological images with inception recurrent residual convolutional neural network. J. Digit. Imaging **32**(5), 605–617 (2019)

25. Hao, Y., Zhang, L., Qiao, S., Bai, Y., Cheng, R., Xue, H.: Breast cancer histopathological images classification based on deep semantic features and gray level co-occurrence matrix. Plos One **17**(5), e0267955 (2022)

26. Boumaraf, S., Liu, X., Wan, Y., Zheng, Z., Ferkous, C., Ma, X.: Conventional machine learning versus deep learning for magnification dependent histopathological breast cancer image classification: a comparative study with visual explanation. Diagnostics **11**(3), 528 (2021)

27. Li, X., Li, H., Cui, W., Cai, Z., Jia, M.: Classification on digital pathological images of breast cancer based on deep features of different levels. Math. Prob. Eng. **2021**, 1–13 (2021)

28. Man, R., Yang, P., Xu, B.: Classification of breast cancer histopathological images using discriminative patches screened by generative adversarial networks. IEEE Access **8**, 155362–155377 (2020)

29. Sharma, S., Kumar, S.: The Xception model: a potential feature extractor in breast cancer histology images classification. ICT Exp. **8**(1), 101–108 (2022)

30. Xu, Y., dos Santos, M.A., Souza, L.F.F., Marques, A.G., Zhang, L., da Costa Nascimento, J.J.: New fully automatic approach for tissue identification in histopathological examinations using transfer learning. IET Image Process. **16**(11), 2875–2889 (2022)

Wavelet-SVDD: Anomaly Detection and Segmentation with Frequency Domain Attention

Linhui Zhou[1], Weiyu Guo[1(✉)], Jing Cao[2], Xinyue Zhang[1], and Yue Wang[1]

[1] School of Information, Central University of Finance and Economics, Beijing 102206, People's Republic of China
zhoulinhui@email.cufe.edu.cn, weiyu.guo@cufe.edu.cn
[2] China United Network Communications Group Co., Ltd., Beijing, China
caoj33@chinaunicom.cn

Abstract. Anomaly detection is a formidable challenge that entails the formulation of a model capable of detecting anomalous patterns in datasets, even when anomalous data points are absent. Traditional algorithms focused on learning knowledge regarding the typical features that arise in images, such as texture, shape, and color, to distinguish between normal and anomalous examples. However, there is untapped potential in frequency domain features for differentiating anomalous patterns, and current methodologies have not exhaustively exploited this avenue. In this work, we present an extension of the deep learning version of support vector data description (SVDD), a prevalent algorithm used for anomaly detection, through the introduction of Wavelet transformation and frequency domain attentions in the feature learning network. This extension allows for the consideration of frequency domain patterns in defect detection, and improves detection performance significantly. We performed extensive experiments on the MVTecAD dataset, and the results revealed that our approach attained advanced performance in both anomaly detection and segmentation localization, thereby confirming the efficacy of our proposed innovative designs.

Keywords: Anomaly detection · Wavelet transformation · Frequency domain attention

1 Introduction

Anomaly detection constitutes a pivotal binary classification issue that aims to detect the abnormalities in the data. This challenge persists across various industries such as finance, manufacturing, and video surveillance. Notably, a significant number of abnormal instances are either unattainable or inadequate for distribution modeling during training, anomaly detection is typically formulated as a semi-supervised or one-class classification task [8]. The identification of anomalies is particularly challenging in image data, as the difference between

© The Author(s), under exclusive license to Springer Nature Switzerland AG 2023
X. Yang et al. (Eds.): ADMA 2023, LNAI 14177, pp. 230–243, 2023.
https://doi.org/10.1007/978-3-031-46664-9_16

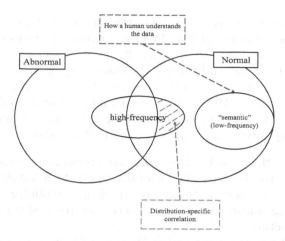

Fig. 1. Distribution of different frequencies

normal and anomalous patterns is often subtle, and defects can be nuanced, particularly in high-resolution images. Consequently, anomaly detection represents a distinctive binary classification problem that requires careful consideration, particularly in image data analysis.

Considering the diversity and scarcity of anomaly samples, a common strategy in such cases is to model the distribution of normal data and detect anomalies by identifying outliers. The pivotal aspect of this approach is to learn a concise boundary for normal data. In this regard, the support vector data description (SVDD) [12] and its extensions [10,15] have been employed as classical algorithms for one-class classification. These methods construct a data-enclosing hypersphere in the kernel space, enveloping most of the normal samples, for the purpose of anomaly detection. Nonetheless, existing works primarily focus on detecting semantic outliers, such as visual objects from distinct classes, in object-centric natural images, with little regard for the finer details, such as changes in texture, within an image. However, recent study [13] has illuminated that the features which can afford insight into the rate of transitions between pixels in an image are also useful to distinguish the abnormality from normal data. The potential for frequency domain features to effectively distinguish between normal and abnormal images deserves consideration.

As shown in Fig. 1, the low-frequency portion of an image is the primary source of semantic information perceived by the human visual system. This implies that, in anomaly detection tasks, the low-frequency features of normal images, which are consistent with the human visual system, share the same distribution, whereas its high-frequency features may not. However, tranditional Convolutional Neural Networks (CNNs) are probably capable of learning features that contain mixed high-frequency information [13], which may interfere with the construction of distribution of normal samples with the ability to distinguish abnormal samples by the deep SVDD model. In this work, we aim to tackle the challenges related to the detection of image abnormalities and segmentation by means of integrating frequency domain features into CNNs. We present an innovative Wavelet attention that enables a more sophisticated distinction between normal and abnormal instances by incorporating frequency domain features. In this regard, a Wavelet Transform based network is proposed that extends the deep SVDD model to learn a precise boundary for normal data by considering both visual objects and frequency domain features. In a nutshell, our contributions in this work can be summarized as follows:

- We investigate the impact of frequency domain characteristics on the efficacy of anomaly detection, and put forward a multi-stage wavelet network that employs Wavelet attention to acquire knowledge pertaining to both the frequency domain features and visual objects, with the goal of improving image anomaly detection.
- We extend the classical method of Deep SVDD [10] for anomaly detection to frequency domain learning, and propose our Wavelet SVDD, which makes a good distinction between normal and anomalous in the feature space containing frequency domain features.
- A series of experiments are conducted to validate the effectiveness of the proposed method and the key designs, which demonstrate that our approach attained advanced performance in both anomaly detection and segmentation localization.

2   Related Works

This work aims to enhance the precision of anomaly detection through the incorporation of frequency domain feature learning into the framework of deep neural network-based Support Vector Data Description (SVDD). Its related work can be classified into three distinct categories: distance metric, frequency domain analysis, and frequency domain learning methods.

2.1   Distance Metric Based Methods

Distance-based methods focus on the training of a feature extractor that learns compact distribution of feature vectors derived from normal images by minimizing intra-class distances between samples. During the testing phase, the majority of methods employ the distance between the features of the sample undergoing evaluation and the normal features as a metric for detecting anomalies.

Deep support vector data description (Deep SVDD) [10] is a widely used technique in this domain. The authors of this approach artificially assign a point in the feature space as the feature center, and reduce the distance between the normal sample features and the center by mapping them to the center. Jihun et al. [15] expanded Deep SVDD to operate at the patch level by learning of the relative position semantics of patches through a self-supervised approach, thus avoiding the use of artificial centers by minimizing the distance between semantically similar patches. However, notwithstanding the efficacy of the frequency domain feature analysis in detecting anomalies, existing distance-based methodologies have demonstrated a proclivity towards neglecting this avenue.

2.2 Frequency Domain Analysis Based Methods

The focus of frequency domain analysis based methods is on identifying irregularities in areas with regular textures. Previous methods [6,14] primarily rely on the manipulation of frequency spectrum information of the image, with the aim of removing periodic background textures and enhancing the visibility of anomalous regions. For example, Tianxiao et al. [14] involves the removal of frequency spectrum information of the background to highlight abnormal regions, while Chenlei et al. [6] employ only the phase spectrum to eliminate repetitive backgrounds in the inverse Fourier transform. However, these techniques have certain limitations in the case of image backgrounds and often require manual intervention for constructing periodic images. In contrast, our method learn the frequency domain features of the image, rather than relying solely on the spectrogram of the image.

2.3 Frequency Domain Learning Based Methods

The discrete Wavelet transform (DWT) [1] and Fourier transform (FT) [11] are widely employed image processing technique utilized for frequency domain analysis, which can transform an image from spatial domain to the frequency domain. Since the DWT can easily realize with the multi-level downsampling style, which is harmonious with deep convolutional neural networks (CNNs), it has been frequently combined with convolutional networks to deal with the tasks of computing vision.

For example, in order to enhance performance in the tasks of texture classification and image annotation, Shin et al. [5] proposed a wavelet-CNN architecture which incorporates a multiscale wavelet transform applied to the input image. This design has demonstrated superior performance compared to non-wavelet CNNs in these areas. Li et al. [9] presented an innovative solution to counteract the problem of feature loss encountered in wavelet-CNN [5]. They proposed to replace the downsampling features of CNNs with the low-frequency component of the discrete wavelet transform (DWT) and combining it with regular convolution, as opposed to spanwise convolution, resulting in improved feature retention. For a better fusion of spatial features and frequency domain features of the image, Zhao et al. [16] proposed an attention based network structure , i.e., Wavelet Attention (WA) block. The WA block first effectuates a decomposition of the feature map into low and high-frequency components through DWT's down-sampling operation. Subsequently, the high-frequency details of the feature map in the high-frequency component are selectively captured, while the essential information of the feature map residing in the low-frequency component remains undisturbed.

Previous research has demonstrated that high frequency information significantly impacts image classification, whereas we proposed a Wavalet Attention based SVDD approach to utilize an attention mechanism on the frequency domain to identify the relevant part of the high-frequency information for anomaly detection.

3   Methodology

Problem Formulation. The task of detecting anomalies is akin to binary classification in that it involves accurately distinguishing between normal and anomalous data. In the case of image anomaly detection, images that exhibit minor defects or those that fall outside the semantic distribution are typically deemed anomalous. To this end, various techniques have been proposed to learn a score function $\mathcal{A}_\theta$ to assess the level of anomaly in an image. Specifically, a high value of $\mathcal{A}_\theta(x)$ indicates that the image is anomalous during testing. Presently, the area under the receiver operating characteristic curve (AUROC) [3] is the standard metric employed to evaluate the efficacy of the $\mathcal{A}_\theta$ function in detecting anomalies, which is defined as:

$$AUROC(\mathcal{A}_\theta) = P(\mathcal{A}_\theta(X_{normal}) < \mathcal{A}_\theta(X_{abnormal})) \tag{1}$$

Ideally, an effective score function should be capable of assigning low and high scores to normal and anomalous input images, respectively. Moreover, for anomaly localization, the corresponding anomaly score is determined for each pixel.

Model Overview. As shown in Fig. 2, the proposed Wavelet SVDD involves two primary components: feature learning and anomaly calculation. Initially, the model employs a novel wavelet attention network to learn the feature distribution of normal images from both frequency domain features and visual objects. During the testing phase, we follow the pradigm of patch SVDD [15] to divide trained images into several patches and acquire the feature vectors of these patches by the Wavelet SVDD network, thereby enabling the separation of normal images into distinct patch distributions. Next, we extract the features of a testing image by the Wavelet SVDD network in manner of sliding windows. Finally, the distance between these extracted features from the testing image and the distribution of normal patches is treated as the abnormality score. The segmentation of abnormal pixels and the abnormality score of the entire image can be realized in manner of differentiation between testing image and the trained normal images.

3.1   Wavelet Transformation Network

The previous work [13] has demonstrated that the low-frequency portion of an image is the primary source of semantic information perceived by the human

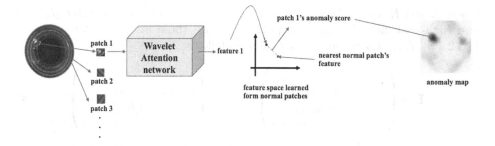

Fig. 2. Overall flow of the proposed model

visual system, and highlighted that high-frequency features are not extraneous noise; rather, a substantial number of them are correlated with the data distribution. Thus, we aim to enhance discrimination between normal and abnormal feature distributions by learning to filter out high-frequency information from the features obtained by CNNs, and select the effective high-frequency information from the filtered information.

The use of Discrete Wavelet Transform (DWT) in image processing has proven to be effective in obtaining high-quality down-sampling information while minimizing information loss in Convolutional Neural Networks (CNN). In this study, we aim to integrate the Discrete Wavelet Transform (DWT) into convolutional neural network (CNN) for frequency domain learning to enable the CNN to autonomously learn the components proficient in distinguishing anomalies among the frequency features generated by DWT. Specifically, we propose a Wavelet block which incorporates DWT operations into the feature extraction layers of CNN to enhance its performance. As illustrated in Fig. 3, we applies the DWT technique [1] with CNN to extract relevant features in the frequency domain. The Wavelet block first decomposes feature maps of CNN into low-frequency and high-frequency components by the DWT. The low-frequency component (X_{ll}) retains the primary information structure of feature maps, while the high-frequency components (X_{lh}, X_{hl}, and X_{hh}) store detailed information along with noise. Following the DWT, a 1×1 convolution layer and an Inverse Wavelet Transform (IWT) operation are stacked to select frequency features and convert them back into the spatial domain, respectively.

In line with previous work [9], the input of 2D-DWT $X \in R^{n \times n}$ can be obtained as follows:

$$\mathbf{X}_{ll} = \mathbf{LXL}^{\mathrm{T}}, \ \mathbf{X}_{lh} = \mathbf{HXL}^{\mathrm{T}}$$
$$\mathbf{X}_{hl} = \mathbf{LXH}^{\mathrm{T}}, \ \mathbf{X}_{hh} = \mathbf{HXH}^{\mathrm{T}} \tag{2}$$

As a result of the biorthogonal property inherent in the Discrete Wavelet Transform (DWT), it is possible to reconstruct the original feature $\mathbf{X}$ with high accuracy and without any loss of information using the Inverse Wavelet Transform (IWT). The 2D-IWT is applied in accordance with the following procedure:

$$\mathbf{X} = \mathbf{L}^{\mathrm{T}}\mathbf{X}_{ll}\mathbf{L} + \mathbf{H}^{\mathrm{T}}\mathbf{X}_{lh}\mathbf{L} + \mathbf{L}^{\mathrm{T}}\mathbf{X}_{hl}\mathbf{H} + \mathbf{H}^{\mathrm{T}}\mathbf{X}_{hh}\mathbf{H} \tag{3}$$

where L and H are cyclic matrices composed of wavelet low-pass filter $\{l_k\}_{k\in Z}$ and high-pass filter $\{h_k\}_{k\in Z}$, respectively. Both these matrices have a size of $\lfloor N/2 \rfloor \times N$. L and H can be expanded as follows:

$$\mathbf{L} = \begin{pmatrix} \cdots\cdots\cdots \\ \cdots\; l_0 \;\; l_1 \;\cdots \\ \cdots\; l_0 \;\; l_1 \;\cdots \\ \cdots\cdots\cdots \end{pmatrix}, \quad \mathbf{H} = \begin{pmatrix} \cdots\cdots\cdots \\ \cdots\; h_0 \;\; h_1 \;\cdots \\ \cdots\; h_0 \;\; h_1 \;\cdots \\ \cdots\cdots\cdots \end{pmatrix} \quad (4)$$

The Discrete Wavelet Transform (DWT) and Inverse Wavelet Transform (IWT) can be implemented as DWT and IWT layers in deep learning frameworks such as PyTorch, respectively. These layers operate on multichannel data on a per-channel basis. It should be noted that the wavelets chosen for use must possess finite filters to ensure that the size of the generated matrices is $\lfloor N/2 \rfloor \times N$. An example of a simple wavelet family is the Haar wavelet, which is characterized by a low-pass filter of $\{l_k\}_{k\in Z} = \{1/\sqrt{2}, 1/\sqrt{2}\}$ and a high-pass filter of $\{h_k\}_{k\in Z} = \{1/\sqrt{2}, -1/\sqrt{2}\}$.

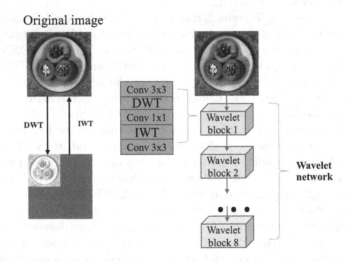

Fig. 3. Wavelet block

3.2   Wavelet Attention

In the presented network architecture, discerning and extracting valuable features from the frequency domain that are pertinent to anomaly detection has been successfully achieved. Nevertheless, Wang et al. [13] have observed that convolutional neural networks (CNNs) tend to prioritize learning low-frequency features in images. However, for anomaly detection, identifying high-frequency features, such as minute defects, is crucial in discriminating between anomalous and normal instances. As a remedy, we further introduce an attention mechanism into the proposed Wavelet block to enable its CNN to concentrate more attention on the high-frequency elements.

Inspired by the Wavelet Attention mechanism proposed by Zhao et al. [16], we propose an enhanced Wavelet block based Wavelet Attention, which captures the detailed information of feature maps in the high-frequency component as the attention information, and the main information of feature maps in the low-frequency component $\mathbf{X}_{ll} = \{\mathbf{x}^{ll}\}_{i=1}^{N_p}$, is not affected. As Fig. 4 shown, the high-frequency components, i.e., $\mathbf{X}_{lh} = \{\mathbf{x}^{lh}\}_{i=1}^{N_p}$, $\mathbf{X}_{hl} = \{\mathbf{x}^{hl}\}_{i=1}^{N_p}$ and $\mathbf{X}_{hh} = \{\mathbf{x}^{hh}\}_{i=1}^{N_p}$, are selected and integrated into the low-frequency feature maps by an attention structure, which can be defined as:

$$\mathbf{z}_i - \mathbf{x}_i^{ll} + \frac{\exp(\mathbf{x}_i^{hl} + \mathbf{x}_i^{lh} + \mathbf{x}_i^{hh})}{\Sigma_{m=1}^{N_p} \exp(\mathbf{x}_m^{hl} + \mathbf{x}_m^{lh} + \mathbf{x}_m^{hh})}\mathbf{x}_i^{ll} \qquad (5)$$

where $Np = H \times W$ is the number of elements on frequency feature maps.

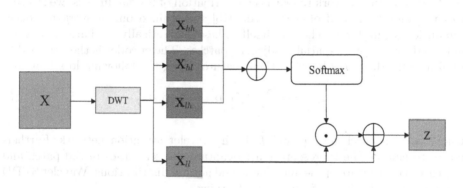

Fig. 4. Wavelet attention, $\oplus$ denotes broadcast element-wise addition, and $\odot$ denotes broadcast element-wise multiplication.

3.3   Wavelet SVDD

After analyzing the above information, we replaced the DWT decomposition part of our wavelet network with our wavelet attention to form our basic wavelet attention block. This basic wavelet attention block is then superimposed with the above wavelet network block to form a deep wavelet attention network as a feature learning network. We experimented with the depth and size of the network, as well as the number of wavelet attention network additions. Ultimately, we determined our feature learning network to be a network consisting of 4 layers of wavelet attention blocks and 4 layers of wavelet network blocks. Wavelet attention is placed in layers 2, 3, 6, and 7 after our experiments. The final network structure is shown in Fig. 5.

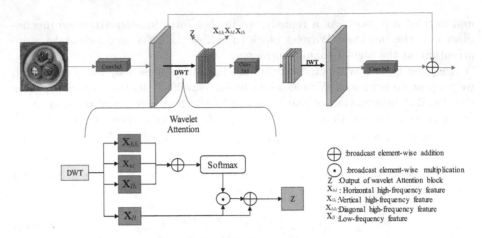

Fig. 5. Wavelet attention block

To enable our network to learn the distribution of normal images, we referred to the training method of patch svdd [15] and trained our network to collect semantically similar patches by itself. These semantically similar patches are obtained by sampling spatially adjacent patches. The encoder is then trained to minimize the distances between their features using the following loss function:

$$\mathcal{L}_{\text{SVDD}} = \sum_{i,i'} \| f_\theta(\mathbf{p}_i) - f_\theta(\mathbf{p}_{i'}) \|_2 \qquad (6)$$

where $\mathbf{p}_{i'}$ is a patch near $\mathbf{p}$ and f_θ is the wavelet attention network. Furthermore, to enforce the representation to capture the semantics of the patch and improve the structure of the anomalous and normal distributions, Wavelet SVDD appends the following self-supervised learning.

We followed the practice in patch SVDD based on Doersch et al. [4] and trained an encoder and classifier pair to predict the relative positions of two patches. A well-performing encoder pair means that the trained encoder can extract useful features for location prediction. For a randomly sampled patch $\mathbf{p}_1$, Doersch et al. [4] drew another patch $\mathbf{p}_2$ from a 3×3 grid in one of its 8 neighborhoods. If we let the true relative position be $y \in \{0, ..., 7\}$, the classifier C_ϕ is trained to correctly predict $y = C_\phi(f_\theta(\mathbf{p}_1), f_\theta(\mathbf{p}_2))$. We added a self-supervised learning signal by adding the following loss term:

$$\mathcal{L}_{\text{SSL}} = \text{Cross-entropy}\left(y, C_\phi\left(f_\theta(\mathbf{p}_1), f_\theta(\mathbf{p}_2)\right)\right) \qquad (7)$$

As a result, the encoder is trained using a combination of two losses with the scaling hyperparameter λ, as presented in Eq. 8. This optimization is performed using stochastic gradient descent and the Adam optimizer [7].

$$\mathcal{L}_{\text{Wavelet\ Psvdd}} = \lambda \mathcal{L}_{\text{SVDD}} + \mathcal{L}_{\text{SSL}} \qquad (8)$$

3.4 Calculate Anomaly Score

After training the feature learning network, the representations from the network are used to detect anomalies. First, the representation of every normal train patch [15], $f_\theta(\mathbf{p}_{\text{normal}})|\mathbf{p}_{\text{normal}})$, is calculated and stored. Given a query image x, for every patch p with a stride s within x, the L_2 distance to the nearest normal patch in the feature space is defined as its anomaly score using Eq. 9.

$$\mathcal{A}_\theta^{\text{patch}}(\mathbf{p}) \doteq \min_{\mathbf{p}_{\text{normal}}} \|f_\theta(\mathbf{p}) - f_\theta(\mathbf{p}_{\text{normal}})\|_2 \tag{9}$$

At the same time, to improve the stability of our method and avoid the appearance of query patches being affected by noise in the normal distribution, we also set another anomaly score calculation function $\mathcal{A}2_\theta^{\text{patch}}$. The difference between this and the above anomaly score calculation function is that $\mathcal{A}2_\theta^{\text{patch}}$ considers the next closest patches in addition to the closest patches to the query patches. This reduces the influence of noise in the training data to a certain extent. Therefore, $\mathcal{A}2_\theta^{\text{patch}}$ is defined as:

$$\mathcal{A}2_\theta^{\text{patch}}(\mathbf{p}) \doteq \frac{1}{2} \times \min_{\mathbf{p}_{\text{normal1}}\mathbf{p}_{\text{normal2}}} \|f_\theta(\mathbf{p}) - f_\theta(\mathbf{p}_{\text{normal1}}) - f_\theta(\mathbf{p}_{\text{normal2}})\|_2 \tag{10}$$

Patch-wise calculated anomaly scores are then distributed to the pixels. As a result, pixels receive the average anomaly scores of every patch to which they belong. We use $\mathcal{M}$ and $\mathcal{M}2$, calculated from the two scoring methods $\mathcal{A}$ and $\mathcal{A}2$, respectively, to represent the resulting anomaly maps.

We divided the size of 32 and 64 patches input into the network, respectively, to obtain different sizes of anomaly maps. We aggregate multiple maps using element-wise multiplication. The resulting anomaly map, M_{multi}, provides the answer to the problem of anomaly segmentation:

$$\begin{aligned}
\mathcal{M}1_{\text{multi}} &\doteq \mathcal{M}1_{\text{small}} \odot \mathcal{M}1_{\text{hig}} \\
\mathcal{M}2_{\text{multi}} &\doteq \mathcal{M}2_{\text{small}} \odot \mathcal{M}2_{\text{big}} \\
\mathcal{M}\text{blend}_{\text{multi}} &\doteq \mathcal{M}1_{\text{multi}} \odot \mathcal{M}2_{\text{multi}}
\end{aligned} \tag{11}$$

where M_{small} and M_{big} are the generated anomaly maps with different scales of patches, respectively. The pixels with high anomaly scores in the map $M_{multi} = \{\mathcal{M}1_{\text{multi}}, \mathcal{M}2_{\text{multi}}, \mathcal{M}\text{blend}_{\text{multi}}\}$ are deemed to contain defects.

It is straightforward to address the problem of anomaly detection. The maximum anomaly score of the pixels in an image is its anomaly score, which can be expressed as:

$$\mathcal{A}_\theta^{\text{image}}(\mathbf{x}) \doteq \max_{i,j} \mathcal{M}_{\text{multi}}(\mathbf{x})_{ij} \tag{12}$$

4 Experiments

We selected the MVTecAD dataset [2] to test the effect of our improvements. This dataset consists of 15 classes of industrial images, each class categorized as either an object or texture. Ten object classes contain regularly positioned objects, while the texture classes contain repetitive patterns.

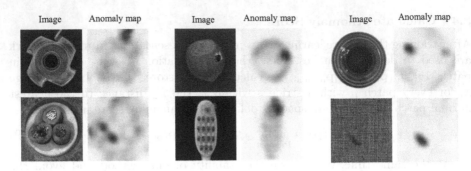

Fig. 6. Anomaly maps

Table 1. Detection and segmentation performance on MVTec AD

Classes	Det.	Seg.
bottle	0.996	0.987
cable	0.953	0.969
capsule	0.935	0.962
carpet	0.946	0.964
grid	0.949	0.965
hazelnut	0.964	0.978
leather	0.975	0.976
metal_nut	0.963	0.986
pill	0.946	0.965
screw	0.934	0.959
tile	0.984	0.941
toothbrush	1.000	0.983
transistor	0.943	0.969
wood	0.974	0.962
zipper	0.983	0.958
Average	0.963	0.968

Table 2. Detection and segmentation performance compared with baselines

Method	Det.	Seg.
InTra	0.950	0.966
PyramFlow	0.960	0.945
RegAD	0.927	0.966
CutPaste	0.961	0.883
Patch SVDD	0.921	0.951
Wavelet SVDD (Ours)	0.963	0.968

4.1 Anomaly Detection and Segmentation Results

Table 1 shows the detection performance of our method in each type of MVTecAD dataset in terms of AUROC. As shown in Fig. 6, the anomaly maps generated using the proposed method indicate that defects are properly localized, regardless of their size. Table 2 shows the detection and segmentation performances for the MVTecAD dataset compared with baselines in terms of AUROC.

4.2 Effect of Wavelet Attention

To explore the effect of our Wavelet attention block, we compared the performance of the network without the Wavelet attention block to that of the network with the Wavelet attention block added at different positions. Our network has mainly 8 wavelet layers. We compared the network without the Wavelet attention block to the network with the Wavelet attention block in layers 2, 3, 6, and 7, as well as in layers 3 and 6, in layers 2 and 7, and in all layers, respectively. The results on MVTec are shown in Fig. 7.

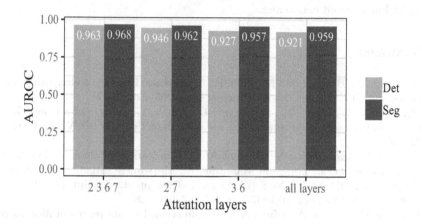

Fig. 7. Effects of different attention layers

The experimental results show that the accuracy of anomaly detection and segmentation can be improved by adding a Wavelet attention block in layers 2, 3, 6, and 7, respectively. The improvement of WA blocks in layers 2, 3, 6, and 7 is better than that in layers 2 and 7. In contrast, adding Wavelet attention blocks in all layers decreases accuracy. One possible explanation is that the influence of multiple Wavelet attentions creates a shortcut path dependence, which weakens the learning effect. Additionally, the frequency domain information in shallow layers may not be as useful for distinguishing anomalies as in deep layers. In conclusion, this experiment verifies the usability of Wavelet attention.

5 Conclusion

In this work, we present a novel technique for image anomaly detection and segmentation called Wavalet Attention SVDD. Instead of only relying on the conventional features extracted by convolutional network, we improve the patch SVDD [15] by involving the frequency domain characteristics of images to differentiate anomalies. We extensively evaluated our method on the MVTecAD dataset and observed that our approach outperformed existing techniques in both anomaly detection and segmentation localization. These results validate the effectiveness of our innovative designs. However, the present approach inherits the inference architecture of patch SVDD, which necessitates anomaly detection inference based on feature database retrieval, thus resulting in time consumption. In future work, we plan to enhance detection inference by incorporating an Auto-Encoder structure into our detection model and accomplishing end-to-end learning and inference.

Acknowledgements. This work is jointly supported by National Natural Science Foundation of China(62106290) and Program for Innovation Research in Central University of Finance and Economics.

References

1. Antonini, M., Barlaud, M., Mathieu, P., Daubechies, I.: Image coding using wavelet transform. IEEE Trans. Image Process. **1**(2), 205–220 (1992)
2. Bergmann, P., Fauser, M., Sattlegger, D., Steger, C.: Mvtec ad-a comprehensive real-world dataset for unsupervised anomaly detection. In: ICCV, pp. 9592–9600 (2019)
3. Calders, T., Jaroszewicz, S.: Efficient AUC optimization for classification. In: Kok, J.N., Koronacki, J., Lopez de Mantaras, R., Matwin, S., Mladenič, D., Skowron, A. (eds.) PKDD 2007. LNCS (LNAI), vol. 4702, pp. 42–53. Springer, Heidelberg (2007). https://doi.org/10.1007/978-3-540-74976-9_8
4. Doersch, C., Gupta, A., Efros, A.A.: Unsupervised visual representation learning by context prediction. In: ICCV, pp. 1422–1430 (2015)
5. Fujieda, S., Takayama, K., Hachisuka, T.: Wavelet convolutional neural networks. arXiv preprint arXiv:1805.08620 (2018)
6. Guo, C., Ma, Q., Zhang, L.: Spatio-temporal saliency detection using phase spectrum of quaternion Fourier transform. In: CVPR, pp. 1–8 (2008)
7. Kingma, D.P., Ba, J.: Adam: a method for stochastic optimization. arXiv preprint arXiv:1412.6980 (2014)
8. Li, C.-L., Sohn, K., Yoon, J., Pfister, T.: Cutpaste: self-supervised learning for anomaly detection and localization. In: CVPR, pp. 9664–9674 (2021)
9. Li, Q., Shen, L., Guo, S., Lai, Z.: Wavelet integrated CNNs for noise-robust image classification. In: CVPR, pp. 7245–7254 (2020)
10. Ruff, L., Vandermeulen, R., Goernitz, N., et al.: Deep one-class classification. In: ICML, pp. 4393–4402 (2018)
11. Tao, R., Zhao, X., Li, W., et al.: Hyperspectral anomaly detection by fractional Fourier entropy. IEEE J. Sel. Topics Appl. Earth Obs. Remote Sens. **12**(12), 4920–4929 (2019)
12. Tax, D.M., Duin, R.P.: Support vector data description. Mach. Learn. **54**, 45–66 (2004)
13. Wang, H., Wu, X., Huang, Z., Xing, E.P.: High-frequency component helps explain the generalization of convolutional neural networks. In: CVPR, pp. 8684–8694 (2020)

14. Wu, T., Wen, M., Wang, Y., et al.: Spectra-difference based anomaly-detection for infrared hyperspectral dim-moving-point-target detection. Infrared Phys. Technol. **128**, 104489 (2023)

15. Yi, J., Yoon, S.: Patch SVDD: patch-level SVDD for anomaly detection and segmentation. In: ACCV (2020)

16. Zhao, X., Huang, P., Shu, X.: Wavelet-attention CNN for image classification. Multimedia Syst. **28**(3), 915–924 (2022)

Anatomical-Functional Fusion Network for Lesion Segmentation Using Dual-View CEUS

Peng Wan[1], Chunrui Liu[2], and Daoqiang Zhang[1(✉)]

[1] Key Laboratory of Brain-Machine Intelligence Technology, Ministry of Education, College of Computer Science and Technology, Nanjing University of Aeronautics and Astronautics, Nanjing, China
dqzhang@nuaa.edu.cn
[2] Department of Ultrasound, Affiliated Drum Tower Hospital, Medical School of Nanjing University, Nanjing, China

Abstract. Dual-view contrast-enhanced ultrasound (CEUS) has been widely applied in lesion detection and characterization due to the provided anatomical and functional information of lesions. Accurate delineation of lesion contour is important to assess lesion morphology and perfusion dynamics. Although the last decade has witnessed the unprecedented progress of deep learning methods in 2D ultrasound imaging segmentation, there are few attempts to discriminate tissue perfusion discrepancy using dynamic CEUS imaging. Combined with the side-by-side gray-scale US view, we propose a novel anatomical-functional fusion network (AFF-Net) to fuse complementary imaging characteristics from dual-view dynamic CEUS imaging. Towards a comprehensive characterization of lesions, our method mainly tackles with two challenges: 1) how to effectively represent and aggregate enhancement features of the dynamic CEUS view; 2) how to efficiently fuse them with the morphology features of the US view. Correspondingly, we design the channel-wise perfusion (PE) gate and anatomical-functional fusion (AFF) module with the goal to exploit dynamic blood flow characteristics and perform layer-level fusion of the two modalities, respectively. The effectiveness of the AFF-Net method on lesion segmentation is validated on our collected thyroid nodule dataset with superior performance compared with existing methods.

Keywords: Multi-modality Fusion · Nodule Segmentation · Contrast-enhanced ultrasound · Co-attention

1 Introduction

Ultrasound (US), as the first-line diagnostic tool in early screening and diagnosis, has become increasingly important in clinical assessment due to the advantages of cost-effectiveness, portability, non-ionizing radiation, and real-time assessment. Thyroid nodule are a common finding in the general population with a

---

P. Wan and C. Liu—Contributed equally to this work.

© The Author(s), under exclusive license to Springer Nature Switzerland AG 2023
X. Yang et al. (Eds.): ADMA 2023, LNAI 14177, pp. 244–256, 2023.
https://doi.org/10.1007/978-3-031-46664-9_17

detection rate of 50% to 60% [1,2]. Ultrasonic features like nodule size, location, shape regularity, margin smoothness, and extra-thyroidal extension are important imaging findings for malignancy risk prediction [3], postoperative assessment [5], and fine-needle aspiration biopsy planning [4]. Thus, accurate nodule segmentation is an indispensable step in clinical practice. In addition to traditional anatomical imaging (B-mode ultrasound, BUS), the emerging functional imaging (contrast-enhanced ultrasound, CEUS) allows for a real-time observation of microvascular perfusion within thyroid gland by enhancing blood flow signals from small vessels [6,7]. Generally, radiologists perform a comprehensive analysis of morphology features in gray-scale US and perfusion features in contrast-enhanced US, but this step requires a high level of expertise and is susceptible to subjective errors.

Although several machine learning or deep learning techniques have been proposed for segmenting thyroid nodules using US imaging, including active contours [10], fuzzy clustering [9], and fully convolution network [8,11–13] etc., segmentation performances of these methods are still limited. One major limitation is that these methods have not fully exploited ultrasonic characteristics complementarity in the segmentation task. Taking cystic nodules as example, gray-scale US is more sensitive to internal hypoechoic regions. Nevertheless, due to the infiltrative growth pattern, we might observe a vague or incomplete boundary since marginal echoic intensity differences become much smaller. In case of that, contrast-enhanced US could complement this by highlighting the varying hemodynamic changes around marginal regions, assisting nodule localization and boundary delineation. Another limitation is that existing CEUS based segmentation methods depend on a preselected a reference frame with relatively distinguished contours, ignoring dynamic blood perfusion information. Actually, perfusion discrepancy might consists in initial enhancement, progression to ultimate wash out. Therefore, it is necessary to reason over the whole perfusion process to sufficiently mine enhancement discrepancy between nodule and thyroid gland.

From the perspective of multi-modality imaging segmentation, it is of great importance to exploit the complementarity of different sort of imaging. Towards this goal, Dolz et al. [39] extend the definition of dense connectivity to multimodal streams, such that dense connectivity within each stream and across different streams could enhances the modality information flow while facilitates the network training. On the other hand, attention mechanism also arouse considerable interest in exploiting inter-dependencies of different modalities, instead of simple summation or concatenation operation. Chen et al. [40] proposes a 3D convolutional block to produce the spatial map highlighting relevant image regions from multiple sources. According to imaging prior knowledge, one MR modality is picked as the master modality and the other is treated as an assistant modality. Information fusion is conducted by transferring the attention map learned from the master stream (teacher network) to supervise the training of the assistant stream (student network). Intuitive, sufficient inter-modality interaction at different-level feature abstraction could ensure enough freedom to capture

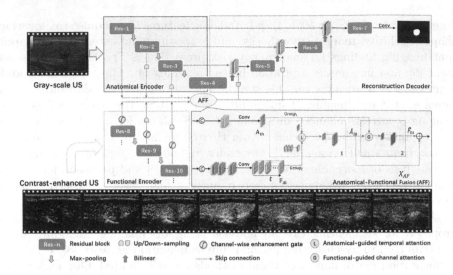

Fig. 1. Illustration of the proposed Anatomical-Functional Fusion Network (AFF-Net).

complex dependencies between modalities. Nonetheless, the optimal layer-level fusion method would vary with specific modalities, leaving an open question for our dual-view CEUS segmentation problem.

In this paper, we propose an anatomical-functional fusion network (AFF-Net) for thyroid nodule segmentation using dual-screen CEUS imaging. For simplicity, we term the morphological and echoic characteristics in gray-scale US view as anatomical features, and dynamic enhancement patterns depicting the real-time blood supply in contrast-enhanced US view as functional features. Figure 1 shows a schematic diagram of our AFF-Net model, which consists of modality-specific encoders and reconstruction decoder, as well as the specifically designed anatomical-functional fusion (AFF) module. By sequentially attending to feature representations of dynamic enhancement patterns and static morphological features, the introduced co-attention mechanism in AFF module integrates multiple US modalities in a layer-level fusion manner. To fully exploit enhancement characteristics, we also introduce a channel-wise enhancement (CE) gate to jointly model enhancement appearances at single point and intensity variations among adjacent frames. We validate the model performance on our collected dual-view thyroid dataset.

2   Related Work

2.1   Medical Imaging Segmentation

For almost a decade, deep learning methods represented by fully convolutional networks (FCN) have pushed medical imaging segmentation into a considerable maturity level both in accuracy and robustness [21,22]. Characterized by a U-shape encoder-decoder architecture, FCN has becomes the basic architecture in various medical segmentation tasks, including cardiac MRI [23], thyroid US [24] and abdominal CT [25]. The former encoder is responsible for representation learning by enhancing pixel-wise discrimination ability, while the latter decoder is the founding part to fuse features from multiple encoding hierarchies. To be aware of different-scale objects, a series of multi-scale representation learning strategies have been proposed, including Gaussian (Laplacian) image pyramid [26,27], atrous spatial pyramid pooling [28], dilated convolution [29], and pyramidal convolution [30]. As for the global context modeling, global context network (GCNet) combines a simplified self-attention mechanism and squeeze-excitation mechanism. As for the basic convolution operation, SEgmentation TRansformer (SETR) [32] replaces it with a pure transformer structure, which also achieves competitive performance.

2.2 Multi-modality Imaging Segmentation

Multi-modal medical imaging (e.g., CT, PET, MRI and US, et al.) has achieved extensive application in comprehensive characterization of morphological, patho-physiological and molecular features of tumors. To exploit the complementarity of different sort of imaging, an increasing number of deep multi-modal methods have emerged recently [33–35]. As mentioned above, feature fusion can be realized at three stages. Among which, early-fusion refers to stacking raw imaging or low-level features channel-wisely by assuming a linear inter-modality relationship [36]. Actually, imaging characteristics from distinct modalities are heterogeneous more than complementary since the imaging acquisition processes differ greatly from each other. To model inter-modality correlation in a higher level feature space, the rest two fusion strategies adopt a multi-path network structure so as to extract a hierarchical representation separately using the state-of-the-art design of each modality. For late fusion, high-level feature maps from different paths are fused only at the stage of model prediction. To facilitate knowledge transfer among different streams (modalities), information fusion is performed in a hierarchical way in the layer-level fusion. As suggested in studies [37] [38], layer-level fusion has the potential to be the optimal fusion way.

3   Materials and Method

3.1   Dataset

In this study, we totally collected 114 dual-screen CEUS videos from patients who attended xx Hospital for thyroid ultrasound examination. All examinations were performed on a Philips iU22 scanner (Philips Medical Systems, Best, the Netherlands) at a low mechanical index ≤ 0.12 using the second-generation contrast agents SonoVue (Bracco SpA, Milan, Italy). Dual-screen CEUS videos were exported as AVI video files with the spatial resolution 600×800. Each video has a duration at least 3 min with a framerate of 15 fps, recording the complete thyroid perfusion process. Each examination was performed by an expert with over 10-year clinical experience, and annotated by at least two senior radiologists to reduce inter-observer variabilities. Each radiologist first reviewed the whole CEUS video, and then selected an optimal frame to contour the boundary. Approval was obtained by the ethics review board of local hospital and the informed consent was obtained from patients before this study.

3.2   Anatomical-Functional Fusion Network (AFF-Net)

Architecture. As illustrated in Fig. 1, we adopt a two-stream U-shape structure to construct our AFF-Net model. In the *encoding* phase, the backbone of Anatomical Encoder consists of four residual blocks separated by 2×2 max-pooling layer. Each block has two 3×3 Conv layers (all with unit stride and zero-padding), followed by the batch normalization and ReLU activation. Each layer is connected to the input of the previous layer. The number of channels is $[16, 16; 32, 32; 64, 64; 64, 64]$. As for Functional Encoder, we adopt three stacked residual blocks with the channel number $[16, 16; 32, 32; 64, 64]$. Besides, we introduce the channel-wise enhancement (CE) gate to explicitly represent inter-frame intensity variations. In the *decoding* phase, anatomical-functional fusion (AFF) module is used to fuse dual-modal feature maps from multiple encoding scales. Along the up-sampling path, to-be-fused ultrasonic representations comprise three components, 1) up-sampled anatomical map generated by the deconvolution layer; 2) high-resolution anatomical map passed by the skip connection; and 3) down-scaled multi-modal map output by the AFF module. The up-sampling path is composed of three sequential residual blocks (channels: $[64, 64, 32, 32, 16, 16]$) separated by the deconvolution layer. Finally, pixel-wise category map P is reconstructed on the fused multi-modal features using a 1-channel 1×1 Conv layer, normalized by a *sigmoid* layer.

Channel-Wise Enhancement Gate. Given sequential enhancement appearance feature maps $\mathbf{M} \in R^{T \times C \times H \times W}$, where T, C, H, W denote the temporal, channel and two spatial dimensions respectively, we first apply a 1×1 2D convolution to reduce feature channels $\mathbf{M}^r(t) = Conv_r * \mathbf{M}(t)$, r is the reduction factor set to 4. Based on that, feature-level enhancement dynamics $E(t)$ is approximately represented as inter-frame feature difference between time step t and $t + 1$,

$$\mathbf{E}(t) = Conv_c * \mathbf{M}^r(t + 1) - \mathbf{M}^r(t), t \in [1, T - 1] \tag{1}$$

where $\mathbf{E}(t) \in R^{C/r \times H \times W}$ is the enhancement map at time step t, $Conv_c$ is a 3×3 channel-wise convolution. In this way, we could obtain $T - 1$ enhancement variations representations. To keep temporal consistency, we append an all-zero enhancement map $\mathbf{E}(T)$ at time step T. Then, sequential variations representation maps are convolved by a 1×1 Conv layer to restore channel dimension to C. Finally, we obtain the combined perfusion representation $\mathbf{F} = \mathbf{M} + \mathbf{E}$ via an element-wise summation between the input enhancement appearance M and the sequential enhancement variation $\mathbf{E} = [\mathbf{E}(1), \mathbf{E}(2), \ldots \mathbf{E}(T)] \in R^{T \times C \times H \times W}$. The behind intuition is that significant intensity variations of contrast-enhanced US view correlate with the real-time changes of spatial distribution of contrast agents, which is expected to trace enhancement discrepancy between lesion and normal tissues.

3.3 Anatomical-Functional Fusion Module

It is worth noting that conventional gray-scale US and contrast-enhanced US actually reflect the thyroid nodule status by complementarily different views. That is, morphological features in gray-scale US are intrinsically correlated with blood flow features in contrast-enhanced US. Therefore, leveraging the semantic consistency between modalities, alternating co-attention mechanism is adopted in our anatomical-functional fusion (AFF) module, which co-attends to both modalities sequentially to distinguish important components for nodule boundary recognition.

Multi-scale Fusion and Grid Split. Given anatomical (functional) features A^s (F_t^s) from different scales s, we rescale them into a common resolution (equaling to the output of the first residual block) by bilinear interpolation, and merge them along channels, $\mathbf{A}_m = Conv_r [A^1; A^2; A^3]$ ($\mathbf{F}_{m,t} = Conv_r [F_t^1; F_t^2; F_t^3]$), where r is channel reduction factor set to 16. Considering the spatial correspondences, our AFF module restricts inter-modal interactions within the same region, which is greatly different from co-attention mechanism in Visual Question Answering [14,15,19,20] that builds associations between all pairs of image-question locations. Thus, we split the multi-scale anatomical (functional) map $\mathbf{A}(\mathbf{F})$ into N regular grids to co-attend both modalities.

Anatomical-Guided Temporal Attention. To evaluate which contrast frames should be attended or overlooked, the first step is to generate temporal attention under the anatomical guidance. For each i-th grid, we summarize the anatomical-guide attention operation as $\hat{\mathbf{F}}^i = L(\mathbf{F}^i, p^i)$, where $i = \{1, 2, \ldots N\}$, $\mathbf{F}^i$ and p^i denote the combined enhancement representation and anatomical feature, respectively. Specifically, global average pooling (GAP) is used to summarize the spatial information of $\mathbf{A}^i$, which is then transformed by a fully-connected layer W_A to generate the anatomical guidance p^i,

$$p^i = W_A * GAP\left(\mathbf{A}^i\right) \tag{2}$$

Based on that, temporal attention score $\mathbf{s}$ is calculated by the dot-product between p^i and the respective enhancement descriptor $f_t^i = GAP\left(\mathbf{F}^i\right)$, aiming at highlighting temporal points with significant appearance or intensity variations.

$$s_t = \sigma\left(\langle p^i, f_t^i \rangle\right) \tag{3}$$

where $\sigma(\cdot)$ denotes the *sigmoid* function for normalization. And thus, attentive enhancement representation $\hat{\mathbf{F}}^i$ is calculated by the weighted sum $\hat{\mathbf{F}}^i = \sum_{t=1}^{T} s_t \cdot \mathbf{F}_t^i$.

Functional-Guided Channel Attention. Apart from identifying salient contrast frames to focus on, we also need to emphasize important channels of gray-scale US map, which are closely associated with essential attributes, such as some kind of edges, low echoes, and boundaries. Similar to [17], we propose the functional-guide channel attention operator in each grid, described as $\hat{\mathbf{A}}^i = G\left(\mathbf{A}^i, q^i\right)$, where q^i is the functional guidance from the global average pooling of $\hat{\mathbf{F}}^i$, $\hat{\mathbf{A}}^i$ is the recalibrated anatomical map.

To generate recalibration signal, we first squeeze the spatial information of A^i into the deep anatomical descriptors a^i using GAP, and then predict a joint representation based on anatomical descriptor a^i and functional guidance q^i as follows,

$$K = W_F\left[a^i; q^i\right] \tag{4}$$

where $K \in R^{C^B}$, $W_F \in R^{C^B \times C^{F+B}}$. Finally, K is normalized by sigmoid layer to recalibrate the anatomical map, producing the recalibration representation $\hat{\mathbf{A}}_i = \mathbf{A}_i \odot \sigma\left(K\right)$, where $\odot$ denotes the channel-wise product operation.

As described above, the alternating anatomical-functional attention mechanism is independently performed in each spatial grouping. Finally, the AFF module outputs the fused representation $\mathbf{X}_{AF}$ by combining attended anatomical and functional features $\hat{\mathbf{A}}_i$ and $\hat{\mathbf{F}}^i$ via element-wise summation. In the decoding stage, $\mathbf{X}_{AF}$ is rescaled to the match the resolution before each residual block.

3.4 Implementation and Loss Function

The proposed AFF-Net was implemented using deep learning framework Pytorch and run on a single GPU (NVIDIA TITAN RTX, 24 GB). Considering temporal redundancy of raw CEUS videos, we adopted a temporal pruning strategy [18] to screen out informative contrast subsequences with the length of $T = 7$. Accordingly, one single gray-scale US image and the accompanying contrast-enhanced US subsequence were fed as two modalities into our AFF-Net, as was common in the baseline and competing methods. Model parameters are updated using the Adam optimizer with the default parameters. The learning rate was initialized to 0.001 and adjusted using cosine annealing schedule every 30 epochs. We used a small batch size of 2 and terminated the learning process when validation performance begins to convergence. Our AFF-Net was trained using the Dice loss,

$$L_{Dice} = 1 - \frac{2\sum_{i=1}^{N} p_i y_i + \varepsilon}{\sum_{i=1}^{N} p_i + \sum_{i=1}^{N} y_i + \varepsilon} \tag{5}$$

where N is the number of pixels in the image, $p_i \in [0,1]$ is the predicted probability of i_{th} pixel belonging to the lesion area, $y_i \in \{0,1\}$.

4 Experiments and Results

Experimental Setup. In our experiments, we adopted the standard setup of 5-fold cross-validation for performance evaluation and comparison of our method

Table 1. Comparison with State-of-the-art Methods and Baselines on the task of thyroid nodule segmentation.

Methods	Fusion	Thyroid nodule			
		DSC(%)	IoU(%)	HD	$p \leq 0.05$
MC-CNN	$Conv_0$	75.36 ± 2.38	61.17 ± 3.10	9.39 ± 0.51	*
	$Conv_1$	76.38 ± 2.43	62.53 ± 3.13	8.95 ± 0.41	*
MB-CNN	Average	76.99 ± 2.42	63.09 ± 3.09	8.75 ± 0.28	*
	Majority	76.10 ± 2.08	61.84 ± 2.60	9.151 ± 0.45	*
HyperDenseNet	–	79.76 ± 1.99	$66.54 \pm 2.55^*$	8.22 ± 0.37	–
Co-learning	–	77.34 ± 2.19	63.76 ± 2.83	9.39 ± 0.47	*
MMTM	–	78.04 ± 2.81	64.26 ± 3.61	9.33 ± 0.41	*
AFF-Net	–	$\mathbf{81.74 \pm 1.73}$	$\mathbf{69.40 \pm 2.18}$	$\mathbf{8.50 \pm 0.36}$	–

* denotes a significant difference compared with our method, the last column denotes significant comparisons for all three metrics.

and competing methods, as well as all baselines. In this paper, segmentation performance was evaluated by three metrics, including Dice Similarity Coefficie (DSC), Intersection over Union (IoU) and Hausdorff distance (HD) [16]. The first three metrics measures the degree of overlap between segmentation result S and ground truth Y, and HD measures boundary distances. For all experimental comparisons, we computed the p-value with the two-sample t-test.

We first compared our AFF-Net method with several fusion baselines. 1) Multi-channel (MC) CNN, implementing multi-modal US fusion via channel-wise concatenation at the network input ($Conv_0$) or after first convolution block ($Conv_1$); 2) Multi-channel (MB) CNN, implementing a late fusion of segmentation results by average or majority voting, where each modality was processed separately. Then, we compare it with more complex layer-wise fusion structures, including 1) HyperDenseNet that extends the dense connectivity to a multi-branch structure; 2) Co-learning Network that derives a spatially varying fusion map at each decoding scale; 3) Multimodal transfer module (MMTM) that recalibrates multi-modal tensors along the channel dimension. In our implementation, we replace the original 2D convolution with 3D ones, aiming at learn spatial-temporal features from dynamic contrast-enhanced US view.

Baselines and Competing Methods: Quantitative segmentation results are summarized in Table 1. We observe that the layer-level fusion of deep features from different modalities achieves a superior performance over the manner of early-level and late-level fusion. And our proposed AFF-Net achieves the largest overall improvements, these improvements are statistically significant compared to all baselines, verifying the effectiveness of cross-modality imaging fusion and enhancement dynamics representation in the task of thyroid nodule segmentation. By allowing dense connectivity between encoding streams, Hyper-DenseNet achieves the smallest mean boundary distance of 8.22, and comparable

Table 2. Comparative results of ablation analysis.

Methods	Thyroid nodule			
	DSC(%)	IoU(%)	HD	$p \leq 0.05$
A-Net	74.22 ± 3.13	59.72 ± 4.03	9.65 ± 0.36	*
F-Net	73.88 ± 2.77	59.40 ± 3.54	9.43 ± 0.44	*
AFF-Net-C	79.08 ± 1.79	$65.70 \pm 2.26^*$	$8.74 \pm 0.42^*$	–
AFF-Net	$\mathbf{81.74 \pm 1.73}$	$\mathbf{69.40 \pm 2.18}$	$\mathbf{8.50 \pm 0.36}$	–

* denotes a significant difference compared with our method, the last column denotes significant comparisons for all three metrics.

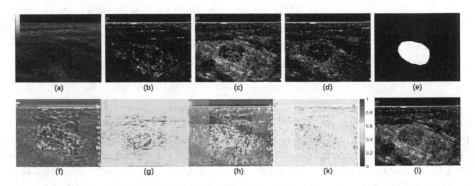

Fig. 2. (a) Gray-scale US; (b–d) Dynamic contrast-enhanced US; (e) Ground-truth; (f) Highlighted anatomical channel $\mathbf{A}_c$; (g)Significant enhancement point $\mathbf{F}_t$; (h) Attend anatomical map $\hat{\mathbf{A}}$; (k) Temporally aggregated functional map $\hat{\mathbf{F}}$; (L)Segmentation result $\mathbf{P}$. For illustration, we normalize the feature values into the range of $[0-1]$.

performances in terms of mean DSC 79.76% vs. 81.74% and IoU 66.54% vs. 69.40%, respectively. Another interesting finding is that channel-wise attention in MMTM outperforms spatial-channel-wise attention in Co-learning method. It indicates that deriving a more complex weighting tensor might not be well suitable for feature fusion in the task of nodule segmentation using dual-screen CEUS imaging. For an more intuitive understanding, we provide an visualization of our model in Fig. 2, including the model prediction and the intermediate feature maps generated by the AFF module.

Ablation Analysis: To evaluate the usefulness of multi-modal US fusion and two major components of our method (i.e., CE gate and AFF module), we compare AFF-Net with its three variants, i.e., 1) A-Net, which removes the branch of enhancement features learning from contrast-enhanced US view; 2) F-Net, which removes the branch of morphological features representation from gray-scale US view; 3) AFF-Net-C, which removes CE gate for enhancement variations modeling.

From Table 2, we observe that fusing deeper-layer features of gray-scale US and contrast-enhanced US provides a clear improvement over the single-path version, with an increase on performance of nearly 7%. Even compared with MC-CNN with early fusion in Table 1, depending on single US modality (A-Net or F-Net) still show inferior performance, further validating the advantage of fusion of morphological features and microvascular perfusion features in our task of thyroid nodule segmentation. When adding channel-wise enhancement gate for explicit perfusion differences representation learning, we could see a significantly higher IoU score (p ¡ 0.05) 69.4% than that of the baseline AFF-Net-C that removes CE gate directly, demonstrating its effectiveness to capture enhancement discrepancy between thyroid nodules and normal gland.

5   Conclusion

In this paper, we have proposed an anatomical-functional fusion network to automatically segment thyroid nodules using dual-screen contrast-enhanced US imaging. Experimental results on our collected datasets have demonstrated the effectiveness of our method in both dynamic enhancement modeling and complementary feature fusion (morphology and perfusion). As the future work, we will extend our current model to a multi-task architecture that jointly detects lesion regions and predicts clinical status for thyroid nodule treatment.

Acknowledgement. This work was supported by the National Natural Science Foundation of China (Nos. 62136004, 62276130, 61732006, 61876082), and also by the Key Research and Development Plan of Jiangsu Province (No. BE2022842).

References

1. Haugen, B.R., Alexander, E.K., Bible, K.C., et al.: 2015 American thyroid association management guidelines for adult patients with thyroid nodules and differentiated thyroid cancer: the American thyroid association guidelines task force on thyroid nodules and differentiated thyroid cancer. Thyroid **26**(1), 1–133 (2016)
2. Liang, X.W., Cai, Y.Y., Yu, J.S., Liao, J.Y., Chen, Z.Y.: Update on thyroid ultrasound: a narrative review from diagnostic criteria to artificial intelligence techniques. Chin. Med. J. **132**(16), 1974–1982 (2019)
3. Wang, M., Sun, P., Zhao, X., Sun, Y.: Ultrasound parameters of thyroid nodules and the risk of malignancy: a retrospective analysis. Cancer Control **27**(1), 1073274820945976 (2020)
4. Ha, E.J., Na, D.G., Baek, J.H., Sung, J.Y., Kim, J., et al.: US fine-needle aspiration biopsy for thyroid malignancy: diagnostic performance of seven society guidelines applied to 2000 thyroid nodules. Radiology **287**(3), 893–900 (2018)
5. Kant, R., Davis, A., Verma, V.: Thyroid nodules: advances in evaluation and management. Am. Fam. Physician **102**(5), 298–304 (2020)
6. Sorrenti, S., Dolcetti, V., Fresilli, D., et al.: The role of CEUS in the evaluation of thyroid cancer: from diagnosis to local staging. J. Clin. Med. **10**(19), 4559 (2021)

7. Radzina, M., Ratniece, M., Putrins, D.S., Saule, L., Cantisani, V.: Performance of contrast-enhanced ultrasound in thyroid nodules: review of current state and future perspectives. Cancers **13**(21), 5469 (2021)

8. Ma, J., Wu, F., Jiang, T., et al.: Ultrasound image-based thyroid nodule automatic segmentation using convolutional neural networks. Int. J. Comput. Assist. Radiol. Surg. **12**, 1895–1910 (2017)

9. Koundal, D., Sharma, B., Guo, Y.: Intuitionistic based segmentation of thyroid nodules in ultrasound images. Comput. Biol. Med. **121**, 103776 (2020)

10. Mahmood, N.H., Rusli, A.H.: Segmentation and area measurement for thyroid ultrasound image. Int. J. Sci. Eng. Res. **2**(12), 1–8 (2011)

11. Mi, S., Bao, Q., Wei, Z., Xu, F., Yang, W.: MBFF-Net: multi-branch feature fusion network for carotid plaque segmentation in ultrasound. In: de Bruijne, M., et al. (eds.) MICCAI 2021. LNCS, vol. 12905, pp. 313–322. Springer, Cham (2021). https://doi.org/10.1007/978-3-030-87240-3_30

12. Li, H., et al.: Contrastive rendering for ultrasound image segmentation. In: Martel, A.L., et al. (eds.) MICCAI 2020. LNCS, vol. 12263, pp. 563–572. Springer, Cham (2020). https://doi.org/10.1007/978-3-030-59716-0_54

13. Lu, J., Ouyang, X., Liu, T., Shen, D.: Identifying thyroid nodules in ultrasound images through segmentation-guided discriminative localization. In: Shusharina, N., Heinrich, M.P., Huang, R. (eds.) MICCAI 2020. LNCS, vol. 12587, pp. 135–144. Springer, Cham (2021). https://doi.org/10.1007/978-3-030-71827-5_18

14. Lu, J., Yang, J., Batra, D., et al.: Hierarchical question-image co-attention for visual question answering. In: 29th International Proceedings on Advances in Neural Information Processing Systems, Barcelona, Spain. Curran Associates Inc. (2016)

15. Liu, Y., Zhang, X., Zhang, Q., et al.: Dual self-attention with co-attention networks for visual question answering. Pattern Recogn. **117**, 107956 (2021)

16. Aspert, N., Santa-Cruz, D., Ebrahimi, T.: Mesh: measuring errors between surfaces using the hausdorff distance. In: 29th IEEE International Conference on Multimedia and Expo, Lausanne, Switzerland, pp. 705–708. IEEE (2002)

17. Joze, H.R.V., Shaban, A., Iuzzolino, M.L., et al.: MMTM: multimodal transfer module for CNN fusion. In: 33th IEEE/CVF Conference on Computer Vision and Pattern Recognition, pp. 13289–13299. IEEE (2020)

18. Liang, X., Lin, L., Cao, Q., Huang, R., Wang, Y.: Recognizing focal liver lesions in CEUS with dynamically trained latent structured models. IEEE Trans. Med. Imaging **35**(3), 713–27 (2016)

19. Nguyen, D.K., Okatani, T.: Improved fusion of visual and language representations by dense symmetric co-attention for visual question answering. In: 32th Proceedings of the IEEE Conference on Computer Vision and Pattern Recognition, Salt Lake City, UT, pp. 6087–6096. IEEE (2018)

20. Yu, Z., Yu, J., Cui, Y., Tao, D., Tian, Q.: Deep modular co-attention networks for visual question answering. In: 32th Proceedings of the IEEE Conference on Computer Vision and Pattern Recognition, Long Beach, CA, pp. 6274–6283. IEEE (2019)

21. Long, J., Shelhamer, E., Darrell, T.: Fully convolutional networks for semantic segmentation. In: Proceedings of the IEEE Conference on Computer Vision and Pattern Recognition, pp. 3431–3440 (2015)

22. Ronneberger, O., Fischer, P., Brox, T.: U-Net: convolutional networks for biomedical image segmentation. In: Navab, N., Hornegger, J., Wells, W.M., Frangi, A.F. (eds.) MICCAI 2015. LNCS, vol. 9351, pp. 234–241. Springer, Cham (2015). https://doi.org/10.1007/978-3-319-24574-4_28

23. Zheng, Q., Delingette, H., Duchateau, N., et al.: 3-D consistent and robust segmentation of cardiac images by deep learning with spatial propagation. IEEE Trans. Med. Imaging **37**(9), 2137–2148 (2018)
24. Zhou, S., Wu, H., Gong, J., et al.: Mark-guided segmentation of ultrasonic thyroid nodules using deep learning. In: Proceedings of the 2nd International Symposium on Image Computing and Digital Medicine, pp. 21–26 (2018)
25. Oktay, O., Schlemper, J., Folgoc, L.L., et al.: Attention u-net: learning where to look for the pancreas. arXiv preprint arXiv:1804.03999 (2018)
26. Lin, G., Shen, C., Van Den Hengel, A., et al.: Efficient piecewise training of deep structured models for semantic segmentation. In: Proceedings of the IEEE Conference on Computer Vision and Pattern Recognition, pp. 3194–3203 (2016)
27. Chen, L.C., Yang, Y., Wang, J., et al.: Attention to scale: scale-aware semantic image segmentation. In: Proceedings of the IEEE Conference on Computer Vision and Pattern Recognition, pp. 3640–3649 (2016)
28. Zhao, H., Shi, J., Qi, X., et al.: Pyramid scene parsing network. In: Proceedings of the IEEE Conference on Computer Vision and Pattern Recognition, pp. 2881–2890 (2017)
29. Qin, Y., et al.: Autofocus layer for semantic segmentation. In: Frangi, A.F., Schnabel, J.A., Davatzikos, C., Alberola-López, C., Fichtinger, G. (eds.) MICCAI 2018. LNCS, vol. 11072, pp. 603–611. Springer, Cham (2018). https://doi.org/10.1007/978-3-030-00931-1_69
30. Duta, I.C., Liu, L., Zhu, F., et al.: Pyramidal convolution: Rethinking convolutional neural networks for visual recognition. arXiv preprint arXiv:2006.11538 (2020)
31. Ni, J., Wu, J., Tong, J., et al.: GC-Net: global context network for medical image segmentation. Comput. Methods Programs Biomed. **190**, 105121 (2020)
32. Zheng, S., Lu, J., Zhao, H., et al.: Rethinking semantic segmentation from a sequence-to-sequence perspective with transformers. In: Proceedings of the IEEE/CVF Conference on Computer Vision and Pattern Recognition, pp. 6881–6890 (2021)
33. Kumar, A., Fulham, M., Feng, D., et al.: Co-learning feature fusion maps from PET-CT images of lung cancer. IEEE Trans. Med. Imaging **39**(1), 204–217 (2019)
34. Zhong, Z., Kim, Y., Zhou, L., et al.: 3D fully convolutional networks for co-segmentation of tumors on PET-CT images. In: Proceeding of the 2018 IEEE 15th International Symposium on Biomedical Imaging (ISBI 2018), pp. 228–231. IEEE (2018)
35. Zhao, X., Li, L., Lu, W., et al.: Tumor co-segmentation in PET/CT using multi-modality fully convolutional neural network. Phys. Med. Biol. **64**(1), 015011 (2018)
36. Zhang, W., Li, R., Deng, H., et al.: Deep convolutional neural networks for multi-modality isointense infant brain image segmentation. Neuroimage **108**, 214–224 (2015)
37. Yang, X., Molchanov, P., Kautz, J.: Multilayer and multimodal fusion of deep neural networks for video classification. In: Proceedings of the 24th ACM International Conference on Multimedia, pp. 978–987 (2016)
38. Joze, H.R.V., Shaban, A., Iuzzolino, M.L., et al.: MMTM: multimodal transfer module for CNN fusion. In: Proceedings of the IEEE/CVF Conference on Computer Vision and Pattern Recognition, pp. 13289–13299 (2020)

39. Dolz, J., Gopinath, K., Yuan, J., et al.: HyperDense-Net: a hyper-densely connected CNN for multi-modal image segmentation. IEEE Trans. Med. Imaging **38**(5), 1116–1126 (2018)
40. Li, C., Sun, H., Liu, Z., Wang, M., Zheng, H., Wang, S.: Learning cross-modal deep representations for multi-modal MR image segmentation. In: Shen, D., et al. (eds.) MICCAI 2019. LNCS, vol. 11765, pp. 57–65. Springer, Cham (2019). https://doi.org/10.1007/978-3-030-32245-8_7

SpMVNet: Spatial Multi-view Network for Head and Neck Organs at Risk Segmentation

Hongzhi Liu[1], Shiyu Zhu[2], Qianjin Feng[3(✉)], and Yang Chen[1,4]

[1] School of Computer Science and Engineering, Southeast University, Nanjing, China
chenyang.list@seu.edu.cn
[2] National Key Laboratory of Transient of Physics, Nanjing University of Science and Technology, Nanjing, China
[3] School of Biomedical Engineering, Southern Medical University, Guangzhou, China
fengqj99@fimmu.com
[4] Key Laboratory of Computer Network and Information Integration (Ministry of Education), Southeast University, Nanjing, China

Abstract. Head and neck (HaN) cancers are often treated with radiotherapy. Since radiation inevitably causes damage to human organs, it is necessary to control the dose of radiation in different areas during radiation therapy to protect organs at risk (OARs). To solve these incompatible problems, we proposed an end-to-end spatial multi-view network for head and neck organs at risk segmentation, named SpMVNet, to take advantage of both spatial continuous context and multi-view relevance in whole volume CT images. The proposed method includes a symmetric segmentation network (SymNet) and a continuous context network (CCNet), making full use of organs' structural symmetry in CT slices and spatial contextual information of volume data. Our proposed method is validated on the MICCAI 2015 Head and Neck Automatic Segmentation Challenge datasets. Extensive experiments show that it achieves lower error range for most organ segmentation with better evaluation metrics than state-of-the-art methods. This proposed method is helpful to improve the precision of organ segmentation in radiotherapy.

Keywords: Automated segmentation · Organs at risk · Head and neck CT images

1 Introduction

Cancer is a common disease in the world, with a high fatality rate threatening human life and health. More than millions of people die of cancer every year, among which head and neck (HaN) cancer is one of the most difficult cancers to treat because of its complex anatomical structure [15]. And for clinical treatment, the high precision radiotherapy is often the preferred treatment for head and neck cancer, but it is necessary to limit the radiation dose to avoid damage

X. Yang et al. (Eds.): ADMA 2023, LNAI 14177, pp. 257–271, 2023.
https://doi.org/10.1007/978-3-031-46664-9_18

to the organs at risk (OARs), as well as reduce sequelae and complications. It can be seen that accurately delineating the areas of organs at risk is particularly important for the design of radiotherapy schedules. Organs at risk are highly sensitive to radiation, such as the optic nerve and optic chiasm, which cannot tolerate excessive radiation. And the key step in radiation therapy planning is the identification of the boundaries of high-risk organs. Therefore, the automatic segmentation of high-risk organs helps reduce the workload of doctors in radiation therapy planning, resulting in a reduction in the overall cost of radiation therapy from both a time and economic perspective.

CT imaging overcomes the problem of human anatomical structure information overlapping in X-ray imaging, and has the characteristics of high acquisition speed, high spatial accuracy and resolution. Its three-dimensional (3D) data can clearly display the spatial density and accurate position information of human organs, and two-dimensional (2D) plain scans can be used to detect suspicious lesions. Therefore, computed tomography (CT)-based treatment planning remains to be the mainstream in current clinical treatment.

For the multi-target segmentation task in this paper, how to extract the representation of human organs from CT images is a thought-provoking problem due to the large sizes and shape differences of human organs and the complex spatial structure positions. For 2D neural network, it processes slice images layer by layer, which cannot learn the correlation between successive slices, resulting in the loss of spatial information. However, for the 3D framework of voxel-by-voxel image processing, patch training is usually used to counter the large increase in parameters caused by the network, and the maximum receiving range of the network will be limited by computing resources, thus it is easy to lose the global information of large organs.

In actual clinical practice, radiologists usually manually segment the OARs on the each layer of CT images, which is time-consuming and lies on rich experience. Even so, this process of segmentation could also lead to incorrect and misdiagnosis problems. Our proposed method can accurately delineate the organs at risk for radiotherapy schedules according to the prior knowledge of doctors, which can save time and labor cost while explaining the objectivity and interpretability of the method. The research in this paper is based on a publicly available dataset. The aim is to perform the aforementioned blade segmentation on head and neck computed tomography (CT) images. The example of CT images and labels are shown in Fig. 1.

Deep learning methods represented by convolutional neural networks (CNN) in recent years, have made great achievements in the field of medical segmentation [1,9,13,14,22], and CNN has also been applied for OARs segmentation in head and neck CT images [10,18,20,23]. At present, some researchers have completed the related works on this task. The first [10] using deep learning methods proposed a 2D CNN for OARs segmentation from HaN CT images, but it only got a slight improvement in right submandibular gland and right optic nerve, and the performance for the other OARs was similar to that of the

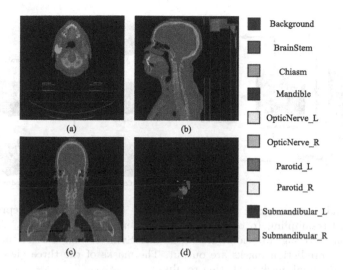

Fig. 1. Segmentation labels of nine organs at risk in head and neck CT images. (a)–(d) are axial, coronal and sagittal views as well as 3D masks, respectively. Different colors on the right represent relevant organs at risk.

traditional methods. Zhu *et at.* [23] proposed the end-to-end method Anato-myNet, a three dimensional squeeze-and-excitation U-Net (3D SE U-Net) based on the SE attention mechanism, combining dice loss and focal loss as optimization constraints. Tong *et at.* [20] designed a fully convolutional neural networks framework with stacked auto-encoder as a shape latent representation model for HaN radiotherapy. However, these existing deep learning-based methods usually produce accurate segmentation maps for large organs and ignore the characteristics of different views of CT data, which have influence on accuracy of small organs and may not be helpful for segmentation of symmetrical OARs.

In this paper, we proposed an end-to-end spatial multi-view network for OARs segmentation, named SpMVNet. The challenging head and neck organs segmentation problem is divided into three views as branches of processing. We first design a symmetric segmentation network (SymNet) to take advantage of the symmetric anatomical structure features of the axial and coronal views, and divide the input network into two parts to make it easier for the network to learn similar features of the symmetric structure. We raise a continuous context network (CCNet) to make full use of the spatially continuous structural information of CT images to make the segmentation masks to be continuous. And the proposed method shows great performance on MICCAI 2015 challenge datasets.

2   Method

In this section, we describe the method of OARs segmentation for head and neck CT images. Our strategy is to simulate the way experienced doctors observe, that is, to predict and locate OARs in different views of volume CT and then output

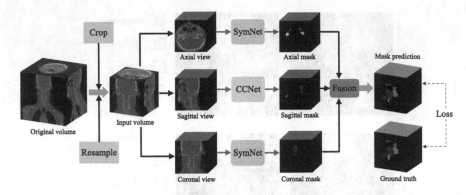

Fig. 2. Overall framework of our SpMVNet. The origin volume is preprocessed by cropping and resampling to obtain the input volume of the network. The data of axial, coronal and sagittal views are input into SymNet and CCNet module, then corresponding prediction masks are output. The masks of the three views are fused and used as the final mask prediction results.

the most probable segmentation results by fusing the masks of three branches at the same spatial position. The overall framework of the proposed SpMVNet has two main components, symmetric segmentation network (SymNet) and continuous context segmentation network (CCNet).

2.1 SpMVNet

We propose a novel end-to-end spatial multi-view network (SpMVNet) for HaN OARs segmentation and its structure is illustrated in Fig. 2. The input volumes are obtained from the origin volumes through image preprocessing, preserving the information of the key parts in HaN CT volumes. After our observation and consultation with hospital experts, we explore the segmentation network using the features of different views and divide the segmentation task into two main sub-networks, namely symmetric segmentation network (SymNet) and continuous context segmentation network (CCNet). We notice that OARs such as the parotid, optic nerve and submandibular have left-right symmetrical physiological structures, so that the CT volumes divided into left and right slices along the midline of the brain for feature learning in axial and coronal views.

SpMVNet for segmentation of HaN OARs can be interpreted as a mathematical theoretical model: a CT medical image I as input and a group of representation constraints C_i $(i = 1, 2, \cdots)$, and the segmentation of I is to acquire a delineation of it, which can be expressed by the following Eq. 1:

$$\bigcup_{x=1}^{N} R_x = I, \quad R_x \bigcap R_y = \varnothing,$$

$$\forall x \neq y, \quad x, y \in [1, N]. \tag{1}$$

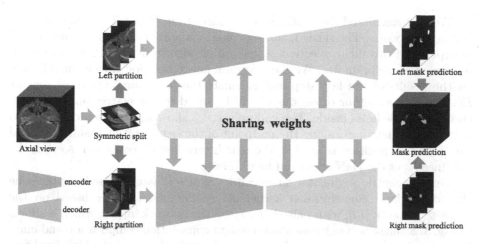

Fig. 3. The structure of our SymNet. The input volume is split into left and right sectors by the brain midline and fed into the siamese network to get the predicted masks of the left and right partitions respectively, and finally merged into the labels.

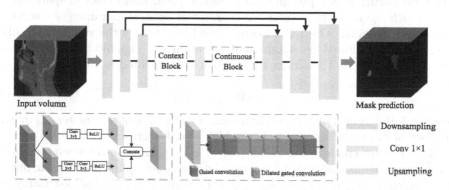

Fig. 4. Illustration of the proposed CCNet. It is on the basis of the segmentation network 3D U-Net with a context block and a continuous block. Input volumes are fed into a feature encoder module, where the ResNet-34 block pretrained from ImageNet [5] is used to replace the original U-Net encoder block.

Herein, R_x satisfies both sets of pixels of the HaN CT images I in the constraint C_i and so does R_y. There is no intersection between R_x and R_y. And x, y are used to distinguish the different regions. N indicates the number of classification including background and nine OARs.

2.2 SymNet

The head and neck CT images have structural symmetry in the axial and coronal views, so the images can be segmented along the midline of brains to obtain the left or right OARs structures, which inspires us to design a symmetrical network for organ feature extraction.

We first calculated the midlines of 2D slices from HaN CT volumes in axial and coronal views. The slices I_s are processed via automatic nonparametric and unsupervised threshold selection segmentation algorithm OTSU [16] to obtain regions of whole brain M_h. We then perform image inflation with a small kennel on the results of the last step and calculate the maximum connected regions R_c. The outer contour of regions C will be saved to the matrix m_c and filled, eliminating the holes inside. The end points P_{up} and P_{down} are searched up and down in the matrix m_c along the midpoint of the segmentation results M_h, and the boundary position is used as the search termination condition, so that the midlines l_m of the HaN slices can be obtained.

Our symmetric segmentation network (SymNet) is composed of the same shared weighted convolutional kernel of encoder and decoder based on the Siamese Network [12], with paired (I_1, I_2) as the network inputs, which is shown in Fig. 3. Siamese network uses shared weight convolution computation and maximum pooling procedures to calculate the similarity between the high-level features (F_1, F_2) of the input images.

We then divide the two-dimensional slices along the dissection line into left and right partition as input and joint the two prediction masks in spatial position, which significantly reduces the amount of network parameters compared with other methods. The L1 distance is used to estimate the similarity of high level features, followed by weight multiplication and sigmoid function to map the value into [0, 1]. The similarity function is formulated as Eq. 2:

$$p = \sigma(W \cdot |f_1 - f_2|), \tag{2}$$

where σ is the sigmoid activation function, W is the weight parameters, is matrix product of two matrices, and f_i is the high level feature F.

The SymNet employs the U-Net [19] with long skip-connections as the baseline network. U-net consists of downsampling and upsampling processes to obtain the predicted segmentation masks. The skip-connection from the downsampling part to the upsampling part has several advantages in fusing local and global features for accurate segmentation with details and resolving the gradient vanishing problem in deep learning models. In our approach, In our approach, the network learns the similar anatomical shape of the left and right portions of the HaN organs, reducing the difference in segmentation results while transferring the convolutional features of the downsampling to the upsampling phase.

2.3   CCNet

The proposed CCNet consists of three major parts: Downsampling module, context block, continuous block and upsampling module. And its detailed illustration are shown in Fig. 4.

A challenge in OARs segmentation is the large variation of object sizes in HaN CT image. For example, a tumor in middle or late stage can be much larger than that in early stage. Motivated by the feature pyramids and multi-scale feature concatenation, we propose novel context block to encode the high-level semantic feature maps. For segmentation task, the receptive field of the high-level network is relatively large, and the semantic information representation ability is strong, but the representation ability of geometric information is weak while low-level network is relatively small, and the geometric detail information representation ability is strong. In order to enable the network to fully learn features of different scales and improve the effectiveness of the features, our method extracts the feature maps output by stage3, and stage5 based on the Resnet network. For the input of 512×512 size, the output feature map size They are $64 \times 64 \times 512$ and $16 \times 16 \times 2048$, which correspond to the shallow texture features, intermediate transition features and deep semantic features of the image, and are input to the subsequent self-attention module for each layer features for further channel filtering. By combining the convolution of different rates, the context block is able to extract features for objects with various sizes.

Vanilla convolutions in a U-Net [19] have no significant effect for multi-organ segmentation. The inputs of skip connections are almost zeros thus cannot propagate detailed color or texture information to the decoder of that region. Therefore, We customize a continuous block with gated convolution and dilated gated convolution [21]. Gated convolution learns a dynamic feature selection mechanism for each channel and each spatial location and the mask feature output O_M can be formulated as Eq. 3.

$$
\begin{aligned}
Gating_M &= \sum \sum W_g \cdot V_{HaN}, \\
Feature_M &= \sum \sum W_f \cdot V_{HaN}, \\
O_M &= \phi(Feature_M) \odot \sigma(Gating_M),
\end{aligned}
\tag{3}
$$

where $Gating_M$ and $Feature_M$ represent two type of features extracted from corresponding convolution filter W_g and W_f for the same input volume V_{HaN}. Besides, ϕ and σ mean sigmoid function and activation function.

For delineation boundaries of HaN OARs, our encoder-decoder architecture equipped with context block and continuous block is sufficient to obtain reasonably continuous segmentation results.

2.4 Loss Functions

As illustrated in Fig. 2, our approach needs to train the proposed network to predict each pixel in the CT images to be background or nine OARs, which is a pixel-wise classification problem. And a widely used loss function is cross entropy loss. However, the objects in this task such as chiasm and optic nerve often take up small regions in the CT images. In this paper, we use the dice coefficient loss function [2,4] to optimize network parameters, which helps to constrain the multi-organ masks from the ground truth. The comparison experiments and discussions are also conducted in the following section. The dice coefficient is a measure of overlap widely used to assess segmentation performance when ground truth is available, as in Eq. 4:

$$L_{dice} = 1 - \sum_k^K \frac{2\omega_k \sum_i^N p(k,i)g(k,i)}{\sum_i^N p^2_{(k,i)} + \sum_i Ng^2_{(k,i)}}, \tag{4}$$

where N is the pixel number, $p(k,i) \in [0,1]$ and $g(k,i) \in {0,1}$ denote predicted probability and ground truth label for class k, respectively. K is the class number, and $\sum_k \omega_k = 1$ are the class weights. In our paper, we set $\omega_k = \frac{1}{K}$ empirically.

We use shape-aware loss [7] to take shape of organs into account. In general, all the loss function values are calculated from the pixels in the image, but shape-aware loss calculates the average point to curve Euclidean distance D among points around curve of predicted segmentation $\hat{C}$ to the ground truth C_{GT} and use it as coefficient to cross-entropy loss function. It is defined as follows:

$$E_i = D(\hat{C}, C_{GT}),$$
$$L_{shape} = -\sum_i^N [CE(\hat{y}, y) - iE_i CE(\hat{y}, y)]. \tag{5}$$

Using E_i the network learns to produce a prediction masks similar to the training shapes.

The final loss function is defined as:

$$L_{loss} = L_{dice} + L_{shape} + L_{reg}. \tag{6}$$

Herein, L_{reg} represents the regularization loss (also called to weight decay) [8] used to avoid overfitting.

3   Experiments

In this section, we conduct evaluation experiments to evaluate the performance of the different methods on MICCAI 2015 Head and Neck Auto-Segmentation Challenge dataset [17]. Nine anatomical segmentation structures in the dataset are highly relevant OARs for radiation therapy treatment in the head and neck, including brainstem, mandible, chiasm, left and right optic nerves, left and right parotid glands, as well as left and right submandibular. And manual contouring data used are segmented by three different medical imaging experts. For fair comparison, all methods are trained and validated using the same data and condition settings. The predicted segmentation results are quantitatively evaluated by two widely used metrics. Furthermore, we demonstrates the outperformance of our proposed approach through segmentation visualization and ablation study.

3.1   Dataset Preprocessing

The dataset consists of 48 CT scan sequences, of which 38 cases are used as training set and 10 cases as testing set following [6]. In this work, nine anatomical structures are considered as segmentation targets, including brainstem, mandible, chiasm, bilateral optic nerves, bilateral parotid glands, and bilateral

Table 1. Dice score coefficient (%) ↑ of results by different compared methods on MICCAI 2015 dataset. The larger the value, the more accurate the segmentation.

Organ	3D U-Net	AnatomyNet	FocusNetv2	Ours
Brain Stem	0.814	0.863	0.879	**0.895**
chiasm	0.508	0.541	0.708	**0.719**
Mandible	0.801	0.921	0.940	**0.952**
Optic nerve left	0.613	0.721	0.788	**0.793**
Optic nerve right	0.608	0.691	**0.809**	0.805
Parotid glands left	0.836	0.878	0.887	**0.895**
Parotid glands right	0.802	0.872	0.892	**0.908**
Submandibular glands left	0.759	0.808	0.836	**0.842**
Submandibular glands right	0.771	0.807	0.829	**0.833**

Table 2. 95% HD score (mm) ↓ of results by different compared methods on MICCAI'15 dataset. The smaller the value, the more accurate the segmentation.

Organ	3D U-Net	AnatomyNet	FocusNetv2	Ours
Brain Stem	11.122	8.396	1.839	**0.574**
chiasm	4.418	1.741	1.144	**0.996**
Optic nerve left	3.539	2.549	2.980	**2.080**
Optic nerve right	1.157	2.827	1.909	**0.855**
Mandible	1.074	0.578	**0.511**	0.531
Parotid glands left	4.716	6.447	4.106	**3.715**
Parotid glands right	8.045	4.177	5.732	**4.108**
Submandibular glands left	5.479	2.938	1.819	**1.406**
Submandibular glands right	3.322	1.534	1.321	**0.908**

submandibular glands. We first convert the original imaging data to NIfTI format, keeping the same size 512×512 pixels with $110 - 190$ slices. And in-plane pixel spacing varied between 0.76×0.76 mm and 1.27×1.27 mm. We then normalized the data to satisfy a standard normal distribution with a mean of 0 and variance of 1. In addition, we normalized the grayscale values of the images.

3.2   Implementation Details

We implemented our model with PyTorch framework. Batch size was set to be 1 because of different sizes of whole-volume CT images. We first used SGD optimizer with momentum 0.9, learning rate 0.001 and the number of epochs being 50. Then, Adam optimizer [11] was used for training, with $\beta_1 = 0.5$ and $\beta_2 = 0.999$, and the number of epochs 600. During training process, we apply the following image augmentations to enhance the training set: random resize with

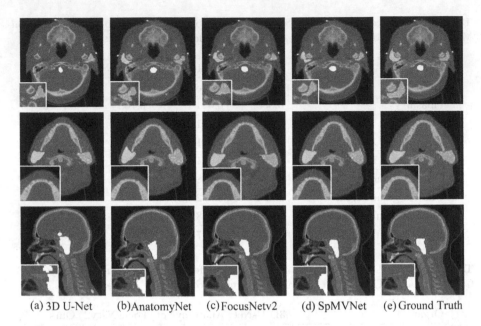

(a) 3D U-Net   (b)AnatomyNet   (c)FocusNetv2   (d) SpMVNet   (e) Ground Truth

Fig. 5. Comparison of different methods for visualization on miccai dataset. (a)–(e) are the mask predicition of 3D U-Net, AnatomyNet, FocusNetv2, the proposed SpMVNet and ground truth, respectively.

Table 3. Dice score coefficient (%) ↑ of results by baseline and improved methods on MICCAI 2015 dataset. The larger the value, the more accurate the segmentation.

Organ	baseline	baseline w CCNet	baseline w SymNet	Ours
Brain Stem	0.751	0.830	0.864	**0.895**
chiasm	0.424	0.575	0.617	**0.719**
Mandible	0.742	0.847	0.881	**0.952**
Optic nerve left	0.569	0.694	0.769	**0.793**
Optic nerve right	0.608	0.627	0.696	**0.805**
Parotid glands left	0.771	0.776	0.842	**0.895**
Parotid glands right	0.726	0.799	0.852	**0.908**
Submandibular glands left	0.637	0.706	0.788	**0.842**
Submandibular glands right	0.688	0.690	0.775	**0.833**

scale range [0.5, 2.0], crop, and horizontal flipping with probability 0.5. The label images should do the same transformation as CT images. All the experiments were performed on a standard desktop with Ubuntu 16.04, using one NVIDIA GeForce RTX 3090 GPU with 24 GB memory.

Table 4. 95% HD score (mm) ↓ of results by baseline and improved methods on MICCAI 2015 dataset. The smaller the value, the more accurate the segmentation.

Organ	baseline	baseline w SymNet	baseline w CCNet	Ours
Brain Stem	14.227	5.733	3.357	**0.574**
chiasm	1.349	1.239	1.048	**0.996**
Optic nerve left	3.386	3.539	2.973	**2.080**
Optic nerve right	1.225	0.886	0.891	**0.855**
Mandible	1.234	0.974	0.612	**0.531**
Parotid glands left	6.852	5.496	3.909	**3.715**
Parotid glands right	9.645	6.594	5.394	**4.108**
Submandibular glands left	8.190	7.218	3.521	**1.406**
Submandibular glands right	1.839	1.241	1.091	**0.908**

3.3 Evaluation Metrics

In order to accurately evaluate the segmentation results, this article uses two evaluation indexes, Dice Similarity Coefficient and 95% Hausdorff Distance, to evaluate the segmentation results. They are the most common used metrics for evaluating 3D medical image segmentations and include volumeand overlap-based metric types. Multiple metrics are used because different metrics reflect different types of errors. For example, when segmentations are small, distance-based metrics such as HD are recommended over overlap-based metrics such as Dice coefficient. Overlap-based metrics are recommended if volume-based statistics are important. In the following, the metrics used are described in more detail:

The Dice coefficient measures the volumetric overlap between the automatic and manual segmentation. It is defined as:

$$Dice = \frac{2|A \cap B|}{|A| + |B|}, \tag{7}$$

where A and B are the labeled regions that are compared and $|.|$ is the volume of a region. The Dice coefficient can have values between 0 (no overlap) and 1 (complete overlap).

The maximum HD measures the maximum distance of a point in a set A to the nearest point in a second set B. Commonly it is defined as:

$$H(A, B) = \max(h(A, B), h(B, A)),$$
$$h(A, B) = \max_{a \in A} \min_{b \in B} \| a - b \|, \tag{8}$$

where $\|.\|$ is the Euclidean distance, a and b are points on the boundary of A and B, and h (A, B) is often called the directed HD. It should be mentioned that maximum HD is sensitive to outliers but appropriate for nonsolid segmentations.

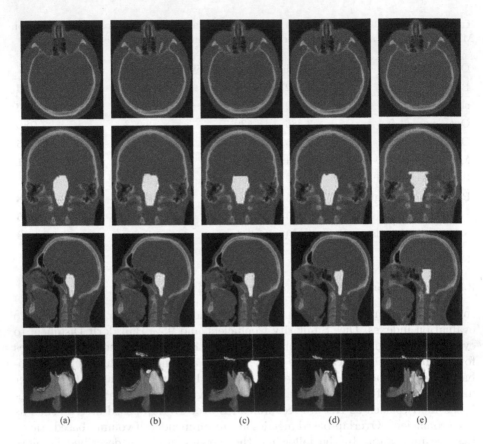

Fig. 6. Results of ablation experiments of our method. From top to bottom are the axial, coronal and sagittal views as well as 3D mask results. (a)–(e) are the mask predicition of baseline, baseline with CCNet, baseline with SymNet, our proposed method and ground truth, respectively.

The 95% HD is similar to maximum HD. However, in contrast to maximum HD, 95% HD is based on the calculation of the 95th percentile of the distances between boundary points in A and B. The purpose for using this metric is to eliminate the impact of a very small subset of inaccurate segmentations on the evaluation of the overall segmentation quality.

3.4   Quantitative Comparison

We compared our framework with three head and neck relevant segmentation methods, including 3D U-Net [3], AnatomyNet [23] and FocusNetv2 [6]. Note that we used the official code and results of 3D U-Net [3], AnatomyNet [23] as well as FocusNetv2 [6]. Compared with current state-of-the-art methods, our approach achieves effective improvements in the quantitative metrics of most OARs.

Table 1 and Table 2 shows the quantitative comparison of these methods. Most of the compared algorithms achieved above 0.7 on the Dice score coefficient of organs at risk except for chiasm. 3D U-Net [3] treats small and large organs equally, which will affect the segmentation results on small objects, and even organs with symmetrical structures are far inferior to other methods. Compared with FocusNetv2 [6], our framework achieves better performance on most organs at risk without using a complex multiple network architecture, corroborating that our strategy has the full capability to draw out the rich information from the CT data.

3.5   Qualitative Comparison

As shown in Fig. 5, our method shows the best visualized on parotid gland and optic nerve. As can be seen from the axial views in the first two rows, 3D U-Net [3] cannot identify this OARs, thus losing the information of OARs segmentation, and the segmentation on the mandible is discontinuous, so the OARs cannot be segmented completely. The problem of discontinuous segmentation also exists in AnatomyNet [23] and FocusNetv2 [6], and the segmentation information is incorrectly labeled at the position of the crania, which affects the segmentation results. Compared other methods, our method exploits the feature information of symmetrical OARs in the head and neck to help train network better to approximate the reference labels on left and right parotid. Our method extracts spatial context structure information and obtains convincing continuous OARs segmentation masks, which can achieve better segmentation results. Therefore, Our method is able to produce higher-quality OARs segmentation masks compared with other methods.

3.6   Ablation Study

We design ablation experiments to verify the effectiveness of each part of the proposed method, as shown in Fig. 6. Firstly, the baseline method is to replace SymNet and CCNet with U-Net [19] and 3D U-Net [3] for segmentation of three views. However, the visualization results show that the baseline network does not have sufficient ability to recognize some obvious OARs structures leading to poor segmentation results.

Then we add our proposed SymNet and CCNet to the baseline model to analyze the continuity and symmetry of the segmentation results. Figure 6(b) and (c) show that CCNet and SymNet can make up for the lack of spatial context information and the inability to identify symmetric organs in the baseline method. It can be seen from the Fig. 6(d) that our proposed method is closer to the ground truth and the brainstem segmentation is more complete, but there is still the problem that small OARs cannot be identified.

Similar to the comparison method, we also calculated dice score coefficient and 95% HD scores in the ablation experiments, as shown in Table 3 and Table 4. Our method can take advantage of SymNet and CCNet to achieve promising results.

4   Conclusion

In this paper, we propose an end-to-end spatial multi-view network segmentation framework SpMVNet. Focusing on head and neck CT images, we explore a Sym-Net to combine multi-view probabilistic symmetry maps for mask predicition of specific organ volumes symmetrically distributed along the midline of the brain. The method innovatively improves the siamese network for OARs segmentation and takes the 2D slices on the left and right sides as input, and then synthesizes the 3D segmentation prediction results. We also solve the problem of lack of continuity in the segmentation of some OARs and achieve higher segmentation metrics through CCNet. We also reduce the segmentation errors of existing methods for OARs, and achieve a certain improvement in the accuracy of symmetric OARs segmentation. The evaluation results demonstrate the effectiveness of the proposed method in our paper. In this paper, an effective method is proposed to solve the difficulties of organ endangerment in radiotherapy, which will be helpful to the analysis and processing of biological information.

Acknowledgment. The work was supported in part by the State Key Project of Research and Development Plan under Grants 2022YFC2401600 and 2022YFC2408500, in part by the National Natural Science Foundation of China under Grant T2225025, in part by the Key Research and Development Programs in Jiangsu Province of China under Grant BE2021703 and BE2022768.

References

1. Cai, J., Lu, L., Zhang, Z., Xing, F., Yang, L., Yin, Q.: Pancreas segmentation in MRI using graph-based decision fusion on convolutional neural networks. In: Ourselin, S., Joskowicz, L., Sabuncu, M.R., Unal, G., Wells, W. (eds.) MICCAI 2016. LNCS, vol. 9901, pp. 442–450. Springer, Cham (2016). https://doi.org/10.1007/978-3-319-46723-8_51

2. Chen, L.C., Papandreou, G., Kokkinos, I., Murphy, K., Yuille, A.L.: Deeplab: semantic image segmentation with deep convolutional nets, atrous convolution, and fully connected CRFs. IEEE Trans. Pattern Anal. Mach. Intell. **40**(4), 834–848 (2017)

3. Çiçek, Ö., Abdulkadir, A., Lienkamp, S.S., Brox, T., Ronneberger, O.: 3D U-Net: learning dense volumetric segmentation from sparse annotation. In: Ourselin, S., Joskowicz, L., Sabuncu, M.R., Unal, G., Wells, W. (eds.) MICCAI 2016. LNCS, vol. 9901, pp. 424–432. Springer, Cham (2016). https://doi.org/10.1007/978-3-319-46723-8_49

4. Crum, W.R., Camara, O., Hill, D.L.: Generalized overlap measures for evaluation and validation in medical image analysis. IEEE Trans. Med. Imaging **25**(11), 1451–1461 (2006)

5. Deng, J., Dong, W., Socher, R., Li, L.J., Li, K., Fei-Fei, L.: Imagenet: a large-scale hierarchical image database. In: 2009 IEEE Conference on Computer Vision and Pattern Recognition, pp. 248–255. IEEE (2009)

6. Gao, Y., et al.: Focusnetv 2: imbalanced large and small organ segmentation with adversarial shape constraint for head and neck CT images. Med. Image Anal. **67**, 101831 (2021)

7. Hayder, Z., He, X., Salzmann, M.: Shape-aware instance segmentation. arXiv preprint arXiv:1412.6980 (2016)
8. Hoerl, A.E., Kennard, R.W.: Ridge regression: biased estimation for nonorthogonal problems. Technometrics **12**(1), 55–67 (1970)
9. Hu, P., Wu, F., Peng, J., Bao, Y., Chen, F., Kong, D.: Automatic abdominal multi-organ segmentation using deep convolutional neural network and time-implicit level sets. Int. J. Comput. Assist. Radiol. Surg. **12**(3), 399–411 (2017)
10. Ibragimov, B., Xing, L.: Segmentation of organs-at-risks in head and neck CT images using convolutional neural networks. Med. Phys. **44**(2), 547–557 (2017)
11. Kingma, D.P., Ba, J.: Adam: a method for stochastic optimization. arXiv preprint arXiv:1412.6980 (2014)
12. Koch, G., Zemel, R., Salakhutdinov, R., et al.: Siamese neural networks for one-shot image recognition. In: ICML Deep Learning Workshop, Lille, vol. 2 (2015)
13. Milletari, F., et al.: Hough-CNN: deep learning for segmentation of deep brain regions in MRI and ultrasound. Comput. Vis. Image Underst. **164**, 92–102 (2017)
14. Milletari, F., Navab, N., Ahmadi, S.A.: V-net: fully convolutional neural networks for volumetric medical image segmentation. In: 2016 Fourth International Conference on 3D Vision (3DV), pp. 565–571. IEEE (2016)
15. World Health Organization: World health statistics 2019: monitoring health for the SDGs, sustainable development goals. World Health Organization (2019)
16. Otsu, N.: A threshold selection method from gray-level histograms. IEEE Trans. Syst. Man Cybern. **9**(1), 62–66 (1979)
17. Raudaschl, P.F., et al.: Evaluation of segmentation methods on head and neck CT: auto-segmentation challenge 2015. Med. Phys. **44**(5), 2020–2036 (2017)
18. Ren, X., et al.: Interleaved 3D-CNNs for joint segmentation of small-volume structures in head and neck CT images. Med. Phys. **45**(5), 2063–2075 (2018)
19. Ronneberger, O., Fischer, P., Brox, T.: U-Net: convolutional networks for biomedical image segmentation. In: Navab, N., Hornegger, J., Wells, W.M., Frangi, A.F. (eds.) MICCAI 2015. LNCS, vol. 9351, pp. 234–241. Springer, Cham (2015). https://doi.org/10.1007/978-3-319-24574-4_28
20. Tong, N., Gou, S., Yang, S., Ruan, D., Sheng, K.: Fully automatic multi-organ segmentation for head and neck cancer radiotherapy using shape representation model constrained fully convolutional neural networks. Med. Phys. **45**(10), 4558–4567 (2018)
21. Yu, J., Lin, Z., Yang, J., Shen, X., Lu, X., Huang, T.S.: Free-form image inpainting with gated convolution. In: Proceedings of the IEEE/CVF International Conference on Computer Vision, pp. 4471–4480 (2019)
22. Zhu, Q., Du, B., Turkbey, B., Choyke, P.L., Yan, P.: Deeply-supervised CNN for prostate segmentation. In: 2017 International Joint Conference on Neural Networks (IJCNN), pp. 178–184. IEEE (2017)
23. Zhu, W., et al.: Anatomynet: deep learning for fast and fully automated whole-volume segmentation of head and neck anatomy. Med. Phys. **46**(2), 576–589 (2019)

SNN-BS: A Clinical Terminology Standardization Method Using Siamese Networks with Batch Sampling Strategy

Xiao Wei, Xiaoxin Wang, and Nengjun Zhu[✉]

School of Computer Engineering and Science, Shanghai University, 333 Nanchen
Road, Baoshan District, 200444 Shanghai, China
{xwei,wangxiaoxin,zhu_nj}@shu.edu.cn

Abstract. Clinical terminology standardization is important for effective integration and sharing of medical information. It aims to convert clinical colloquial descriptions into standard clinical terminologies. However, the accuracy and efficiency of this task are challenged by the gap between colloquial descriptions and standard terminologies, the slight discrepancy across standard terminologies, and the low efficiency of terminology retrieval. To address these challenges, we propose a novel method called SNN-BS for standardizing clinical terminology based on a Siamese network with a batch sampling strategy. SNN-BS enhances its discrimination ability by sampling a set of terminologies to form a retrieval set with the target terminology. By combing two kinds of similarities, we amplify the differences in features between colloquial descriptions and clinical terminologies while considering deeper semantic relationships. Moreover, we use the lighter Bert-tiny model to encode the terminologies and improve the efficiency of terminology retrieval by reducing comparison numbers through regarding it as a question-and-answer selection task. Finally, we conducted experiments on two datasets to evaluate the performance of our model. The experimental results demonstrate that our method achieves a high level of accuracy, reaching 91.30% and 90.24%, respectively, which outperforms the baselines.

Keywords: Clinical terminology standardization · Siamese network · Batch sampling strategy

1 Introduction

Efficient processing of clinical medical texts using intelligent technologies has become a hot topic in recent years, with standardization of clinical terminology serving as its cornerstone. Its target is to transform the spoken description in clinical medicine (i.e., the origin word) into a standardized description (i.e., the target terminology) in the ICD (International Classification of Diseases) standard terminologies coding set. The standard terminology set classifies diseases according to certain rules based on certain characteristics of clinical diseases

X. Yang et al. (Eds.): ADMA 2023, LNAI 14177, pp. 272–287, 2023.
https://doi.org/10.1007/978-3-031-46664-9_19

and uses coding methods to represent them. It plays a key role in integrating, exchanging, sharing, and statistics of medical information [14]. Table 1 lists several examples of clinical terminology standardization tasks.

Table 1. Standardized examples of clinical terms

Origin word	Target terminology
HIFU	
(HIFU for primary liver cancer)	(Ultrasonic scalpel therapy for liver damage)
(Left Ventricular Drainage)	(Ventricular extracranial shunt)
DJ	D-J
(Pull out DJ tube under cystoscope)	(Cystoscopy D-J tube extraction)

Although clinical terminology standardization has achieved some progress in recent years, existing methods still encounters some challenges, which significantly limit the accuracy and efficiency of this task: 1) The oral expression of the same target terminology is various, and some of them may vary significantly in grammar. It is challenging to associate the origin word solely by analyzing identical tokens without related clinical knowledge. 2) The standard terminologies within the same clinical category often appear very similar, resulting in confusion and difficulty in distinguishing them accurately. The standard terminologies in the same category are mistakenly linked to the same origin word if the origin word is not clearly described. 3) The vast amount of terminologies challenges the efficiency of terminologies standardization. When performing terminology standardization tasks, matching and retrieving target terminologies consume much time, making it challenging to meet the efficiency requirements of clinical medicine.

Existing methods has mainly used the traditional text similarity or semantic similarity evaluation to solve these problems recently [8,13]. However, the degree of specialization in clinical medicine is high, and standard terminologies within the same major category have close similarities. Thus, these methods cannot distinguish similar standard terminologies. Moreover, these methods usually use the form of "$[CLS]o[SEP]d_i[SEP]$" as a standard pair, where the standard terminology $d_i \in D$. Once terminology task requires encoding and comparative learning of $|D|$ standard pairs. While taking the ICD-9 coding set as an example, $|D| \approx 10000$, such methods cannot meet the efficiency requirements of clinical terminology standardization, much less the ICD standard terminology set is constantly expanding.

Our research falls into the category of semantic similarity modeling. To address the mentioned issues, we propose a method named SNN-BS for standardizing clinical terminology based on Siamese networks with batch sampling strategy. The Siamese network has two Siamese subnets with the same structure and shares parameters on the left and right. It inputs two pieces of similar text through the left and right sub-net, and maps them into a new space for feature representation and comparison [7]. It has natural advantages for determining the semantic similarity of the same type of text. However, the traditional Siamese network's approach to this task is not different from other semantic similarity methods. Also, it lacks good recognition ability for relatively similar terminology standard terminologies, which still cannot complete the challenges.

Therefore, SNN-BS includes three customized modules, i.e., Data sampling unit, Bert encoder unit, and Simple feature fusion module. The Data sampling unit is responsible for the sampling generation of the training set. It divides the standard terminology set $|D|$ into k candidate terminology sets W_i and fuses the target terminology t into each W_i. This design reduces the number of encoding and comparison learning from $|D|$ times to k times ($k \ll |D|$). Also, the ability of our model to identify similar standard terminologies can be enhanced. The Bert encoder unit is responsible for tokenizing and encoding words and terminologies. We choose Bert-tiny as the encoder. Its advantage is that the smaller the parameter number of this model is, the faster the inference speed will be. Compared with other methods, it can greatly optimize the efficiency problem and minimize the loss of model accuracy due to its excellent migration effect on the Bert model. The Similar feature fusion module is responsible for calculating the similarity between the origin word o and D_i by combing two kinds of similarities. We use a two-layer similarity calculation method to describe the similarity between clinical texts from different dimensions and through the feed-forward neural network fusion. It can effectively alleviate the problem of low accuracy caused by the large semantic difference between origin word and target terminology. By integrating the three parts of the above design, we can address the above-mentioned challenges and achieve a good balance between the accuracy and efficiency of terminology standardization.

In this paper, our work makes the following contributions: 1) The design of data sampling and Bert encoder unit in SNN-BS also care about the efficiency issue when optimizing the accuracy of clinical terminology standardization. Our method reduces the number of coding and comparison learning by batch sampling the standard terminologies. We adopted a lighter Bert-tiny model with better volume and transfer effect for encoding, which considers the accuracy and efficiency of terminology standardization. 2) We propose a method of randomly mixing the target terminology into each sampling candidate terminology set to enhance the model's ability to select answers for confusing standard terminologies. Besides, we highlight the differences between the origin word and target terminology through two-layer similarity fusion, which can alleviate the problem of low accuracy caused by the highly specialized clinical terminology standardization task and improve the ability to capture long-distance semantic standard terminologies. 3) To better evaluate the method's effect, we tested our model on the Yidu-N7K and the self-built ICD9-INT dataset. The experimental results show that the accuracy of our method in this paper has reached 91.30% and 90.24%, both exceeding the SOTA model in accuracy rate.

2 Related Work

Aiming at the standardization task of clinical terminology standardization, academia mainly adopts two methods based on text similarity and semantic similarity in many academic papers.

Yan et al. [11] introduced a deep generative model to generate the core semantics of the description text and obtained a standard terminology candidate set, and then used the BERT-based semantic similarity algorithm to reorder the candidate set to obtain the final standard terminology. Huang et al. [3] proposed a method for standardizing origin words based on combined semantic similarity technology. It is mainly based on domain knowledge base combined with word segmentation, entity recognition and word vector representation technology to calculate the similarity between origin word and standard terminology. Devlin [1] proposed a pre-trained model BERT, which can predict the similarity between sentence pairs through text classification. Sun et al. [8] proposed to select candidate terminologies based on the Jaccard similarity algorithm, and obtain standard terminology-matching results based on the Bert model. Liu et al. [4] used the method of N-gram and Bert to optimize the solution of many-to-many matching between origin words and target terminologies in ICD-9 encoding.

According to the current results, searching for target terminology of clinical terms by mining semantic information focuses on text structure, but there are knowledge errors in the over-spoken description that are difficult to mine; The similarity comparison method lacks support from clinical expertise, making it challenging to distinguish between easily confused standard terminologies. It can be seen that terminology standardization has been regarded as a natural language processing task, but there is still much room for improvement in terms of semantics or similarity calculation.

Therefore, we propose using the Siamese network that can take into account both semantic information and distinguish the confusing words in the professional field. We will combine this with the Bert-tiny model to complete the encoding of clinical terminology.

3 SNN-BS Method

Our SNN-BS method incorporates two sub-nets (i.e., left-subnet and right-subnet) based on the paradigm of Siamese networks, in which some learnable parameters are shared to encode origin words and standard terminology set, respectively. It solves the problem by adopting a batch sampling strategy and combining Euclidean similarity and dot product similarity to calculate the absolute distance and relative distance between them, which alleviates the problem of significant differences in origin word expression of the same target terminology and high similarity of standard terminologies in the medical field. Besides, we use the Binary Cross entropy loss function to abstract the semantic similarity problem of the clinic terminology standardization task into a question-and-answer selection, i.e., the origin word as a question, and each candidate terminology set as an answer set. This approach allows the model to focus on the detailed differences between each answer while making it easier for parallel operations to improve recognition efficiency.

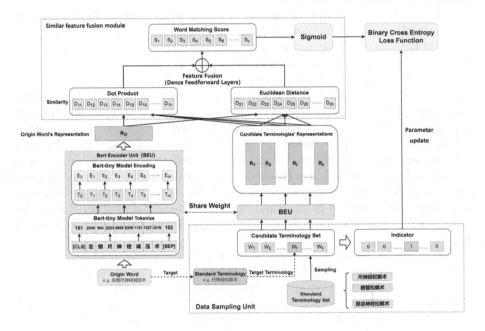

Fig. 1. Overall framework diagram of SNN-BS

Our model includes four parts shown in Fig. 1: the data sampling unit, the Bert encoder unit (BEU), the similar feature fusion module, and the parameter update unit.

3.1 Data Sampling Unit

The data diversity of the ICD standard terminology set is very limited [10]. Therefore, in the Data Sampling Unit, we sample the standard terminology set D according to a specific size and randomly mix the target terminology into each candidate terminology set W_i. This sampling strategy can enhance the training data and meet the model's generalization ability to the expansion requirements of the future ICD standard terminology set and other similar tasks [12]. Figure 2 shows the strategy for data sampling.

We sample the ICD standard terminology set D according to the size of batch (i.e., S_b, default is 512) parameter set in advance. After that, we divide the standard terminology set into k blocks, where: $k = \lceil |D|/(S_b - 1) \rceil$.

After sampling, we insert the target terminology into each block. Then we get k candidate terminology sets. After that, we generate the corresponding

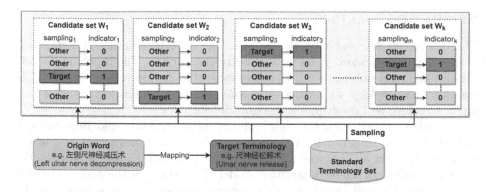

Fig. 2. Sampling strategy for training data: Divide the standard terminology set into multiple candidate terminology set, and randomly insert the target terminology into each set

indicator list as the calculation basis for the subsequent binary cross entropy loss function, where:

$$Indicator_i = \begin{cases} 1 & Candidate_i = TargetTerminology \\ 0 & Candidate_i \neq TargetTerminology \end{cases} \qquad (1)$$

We randomly shuffle all the words in each candidate terminology set to ensure the position of the target terminology t in each candidate terminology set has a certain degree of randomness and avoid wrong fitting of the model during loss calculation. Finally we get k candidates word set W_i, where $W_i = D_i \cup \{t\}$. For example, the origin word o input by the user, i.e., "" ("Left ulnar nerve decompression"), we insert its corresponding target terminology t, i.e., "" ("Ulnar nerve release") at a random position in each candidate terminology set W_i. Then we set the value of t in the corresponding position of the indicator list to 1 and others to 0.

In Bert Encoder Unit (BEU), we tokenize and encode the origin word o and k candidate terminology sets. For example, the origin word "" ("Ulnar nerve release") will be tokenized through the Bert-tiny model in the form of "$[CLS]$ $[SEP]$". After obtaining the id of each Chinese character, it is sent to the encoding layer for vectorization processing and input into the Siamese network. During the vectorization process, we set a maximum length limit of 40 characters, and truncation processing will be performed on terminologies longer than the maximum limit.

In the later model training process, the model takes an origin word o and a candidate terminology set W_i as input for a training session. This measure aims to prevent excessive parameter amounts from preventing training when the standard terminology set is too large. Also, it can increase the frequency of occurrence and the comparison of the target terminology, which can effectively strengthen the prediction effect of the model under the limited training set.

3.2 Bert Encoder Unit

The traditional Siamese network structure uses an origin word and a standard terminology as standard pairs for similarity comparison. The input layer's left sub-net and right sub-net are of the same type of text. However, due to the ICD standard terminology set having numerous standard terminologies, using this input structure will result in low recognition efficiency.

Therefore, we redesign the right sub-net input structure and Bert encoder unit of the Siamese network. We take the candidate terminology set as a whole input by the right sub-net and form a standard pair with the origin word o. We use the smaller parameter quantities model "Bert-tiny" as the encoder, which reduces the times of encoding and comparative learning times from the $|D|$ times required by the traditional Siamese network to k times ($k \ll |D|$). This measure can significantly improve the model's efficiency for this task without losing accuracy.

As shown in Fig. 1, we input the origin word o that needs to be standardized into the left sub-net and input the candidate terminology set W_i into the right sub-net. Our method transforms the task of terminology standardization into a question-and-answer selection task by embedding the target terminology t into the input matrix. To encode the input words, we use the Bert-tiny model with a smaller number of parameters to ensure the efficiency of standardization. After encoding the origin word with the Bert Encoder Unit, we vectorize the candidate terminology set W_i input by the right sub-net through the shared parameter weight. In order to avoid the loss of relevance between text and text context, our method adopts the method of Position Embedding to record the position of key information and enhance the dependence between texts.

We use the same weight value and parameters for each set of origin word o and candidate terminology set W_i to tokenize each word according to the Bert-tiny vector representation specification. After this step, each origin word and standard terminology is represented by a S_h-dimensional vector. Where S_h represents the hidden layer size of the pre-training model. We make the following definitions: x represents the vector representation of the origin word, Y_i represents the vector representation of the i-th candidate terminology set, and y_j represents the vector representation of the j-th candidate word in W_i, so we get:$Shape(x) = shape(y_j) = [1, S_h]$, $Shape(Y) = [S_b, S_h]$. For example, the Bert-tiny model's hidden size is 128 dimensions. S_b represents the size of each candidate terminology set.

3.3 Similar Feature Fusion Module

We integrate Euclidean similarity and dot product similarity methods in the Similar Feature Fusion module. When calculating similarity features for vector representations of o and W_i, we also consider the absolute and relative distances between them to further determine the similarity coefficient of the target terminology corresponding to the origin word. Then, we use a fully connected layer to integrate the origin word and target terminology, which highlights the corresponding features between them and improves the capture ability of long-distance semantic target terminologies. Finally, we complete the loss calculation in the training phase.

As shown in Fig. 3, We calculate the absolute and relative distance features between each standard pair. We extract the key information features of them through the dot product similarity and Euclidean similarity so that the model can focus on the key information in the description. Then we calculate the similarity score after the fusion of the two through the fully connected layer.

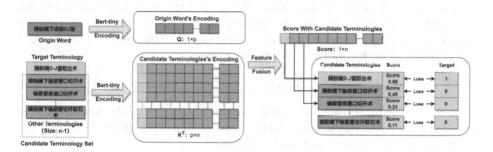

Fig. 3. Feature fusion and loss calculation: The closer the Score value is to 1, the closer the fusion feature of the standard terminology is to the fusion feature of the origin word

Euclidean Distance The Euclidean similarity feature refers to the domain features in clinical terminology descriptors, which can effectively reflect the absolute difference in semantic features between o and each W_i to narrow the selection range of standard terminologies. We calculate the Euclidean distance according to the following formula:

$$Score_{ED}(x, y_j) = \sqrt{\sum_{j+1}^{n}(x - y_j)^2} \tag{2}$$

After that, we get a vector of $[1, S_b]$ dimension with Euclidean similarity feature, where each column in the vector represents the Euclidean distance between the origin word o and the S_b standard terminologies in W_i.

Dot Product The dot product similarity feature measures the individual difference between o and W_i. It can correct the prediction results, which has an enormous description difference between o and W_i. We use its insensitive characteristics to absolute values to filter out the standard terminologies with large differences in distance features with origin word. Let $Q = \{x\}$, $K = Y_i = \{y_1, y_2, ..., y_{512}\}$, then we get: $Shape(Q) = [1, S_h]$, $Shape(K) = [S_b, S_h]$. We calculate the dot product similarity score according to the following formula:

$$Score_{DP}(x, y_i) = softmax(Q \cdot K^T) \tag{3}$$

Similarly, we also get a vector of $[1, S_b]$ dimension with dot product similar feature.

Feature Fusion. To better measure the similarity distance between origin word o and W_i, we set up two layers of full connection layer, align the Euclidean similarity score and the point-product similarity score calculated above, and get the feature vector of $[2, S_b]$ dimension. We use the two-layer feed-forward neural network to map further, express the feature, and output it to the one-dimensional sample space. Finally, we get the similarity score after the vector of origin word x and vector of each candidate word y_i are fused:

$$Matching_i = Contact(Score_{ED}(x, y_i), Score_{DP}(x, y_i)) \tag{4}$$

After obtaining the similar characteristics of the $[1, S_b]$ dimension, we map the matching score of each candidate word to the $[0,1]$ interval through the sigmoid function and record it as $Indicator_j$:

$$Indicator_j = Sigmoid(Matching_i) = \frac{1}{1 + e^{-Matching_j}} \tag{5}$$

For the score of $Indicator_j$, the closer it is to 1, the closer its corresponding candidate word is to the origin word, and vice versa. In Fig. 3, the similar confidence of origin word "" ("Pull out DJ tube under cystoscope") and candidate word "" ("Cystoscopy D-J tube extraction") is the highest. Finally, we calculate the loss function and update the weight of the indicator score and indicator target, which is regarded as completing a round of loss calculation.

3.4   Parameter Update Unit

The Parameter update unit is responsible for calculating losses and updating parameters. In the pre-processing stage, we obtain k candidate terminology sets W_i. Each W_i contains target terminology and other interference standard terminologies. Therefore, we need to calculate the indicator score value for each W_i according to the steps in Sect. 3.2. After that, we calculate the loss with the corresponding indicator target value to complete the update of the training parameters.

When selecting a loss function, scholars often use Contrastive loss for loss calculation in the Siamese network. However, in our method, we regard the similarity comparison task as a multi-category task with more categories and use Binary Cross Entropy Loss for optimization learning, which enhances the ability to distinguish similar standard terminologies [6]:

$$BCELoss = -[y \cdot logp(y) + (1 - y) \cdot log(1 - p(y))] \tag{6}$$

where y is the ground truth value corresponding to the origin word, and $p(y)$ is the predicted value of the model output. When $y_i = 1$, the word is the target terminology corresponding to the origin word, then $BCELoss = -logp(1)$. if $p(y_i)$ approaching 1, then the value of $BCELoss$ Approaching 0; if $p(y_i)$ approaching 0, the value of $BCELoss$ Approaching 1, and vice versa. It can be seen that compared with the contrastive loss function commonly used in a Siamese network, binary cross entropy loss has the characteristic of approaching the real label, which plays a key role in the prediction of standard terminologies.

By repeating the above process, one loss calculation for all candidate terminology sets of a single origin word o is regarded as a training batch, and one loss calculation for all origin words is regarded as an epoch.

3.5 Reasoning

The inference structure of our model is similar to the training structure. The standard terminology set is divided into k candidate terminology set W_i. The difference is that our method uses batch sampling and regression prediction to enhance the ability to distinguish confusing words. We take the Top p standard terminologies with the highest confidence for the prediction results of each candidate terminology set W_i to form a new candidate terminology set W_{k+1}, and use it as the input for the next round. Which defines: $p = \lceil S_b/50 \rceil$

For the $k*p$ candidate standard terminologies of k candidate terminology sets, we input them again into the model structure for secondary reasoning to obtain a new matching score. Then the model will output p standard terminologies with the highest confidence. We merge these outputs according to the output number requirements to form the final prediction result.

4   Experimental Results and Analysis

4.1   Dataset

Our test dataset consists of two parts: the YiduN7K dataset from CHIP2019 and the self-built dataset based on the ICD-9 international dictionary set, which is named ICD9-INT.

YiduN7K Dataset. The YiduN7K dataset is one of the clinical medical information processing evaluation tasks from CHIP2019 (China Conference on Health Information Processing). The origin words are all from the real medical data of Grade III A hospitals, including 4000 training word pairs, 1000 validation word pairs, and 2000 testing word pairs. The data structure is also presented in the form of <origin word, target terminology>.

Since there are many-to-many standard terminology prediction entries in the CHIP2019 dataset, the accuracy rate is defined as the total number of origin words to be predicted divided by the combination of origin words and target terminologies:

$$A = \frac{1}{N} \sum_{i=1}^{N} \frac{|P_i \cap G_i|}{max(|P_i|, |G_i|)} \tag{7}$$

where P_i is the standard terminology set predicted by the origin word i, and G_i is the real target terminology set of the origin word i.

ICD9-INT Dataset. ICD9-INT dataset is built according to the ICD-9-CM-3 international version coding standard [2]. The ICD-9-CM-3 international version contains 4875 standard terminologies. For each origin word, we use the NLPCDA data enhancement tool to generate two corresponding confusion words, build a 9,750-size dataset, and generate 7,800 training word pairs and 1950 testing word pairs through random segmentation. The data structure is also presented in the form of <origin word, target terminology> pair as shown in Table 1. We still use the formula 7 to calculate the accuracy.

4.2   Main Result

We compare our model with the baseline model of existing methods for ICD terminology standardization tasks:

HCAN [5] completes semantic matching between origin word and target terminology through multi-granularity importance weight measurement and models short text similarity to select target terminology. The ABTSBM model [4] utilizes neural networks to train the original term combination splitting method based on named entity recognition and part of speech tagging for the ICD dataset of many-to-many. The Bert-target method [1] pre-trains bidirectional representations through left and right contexts and selects target terminology through question answering and inference after fine-tuning. The Bert with Longest common sub-sequence method [11] analogizes the clinical terminology standardization task to a translation task. It introduces a deep generative model to generate the core semantics of the description text and reorder the candidate set by using the Bert-based semantic similarity algorithm to obtain the final target terminologies. The Bert with Jaccard algorithm method [8] calculates the Jaccard similarity coefficient between the origin word and target terminology to be standardized. It generates a set of candidate standard terminologies, and uses the Bert model for matching and classification.

To optimize the training process, we set the S_b to 512, the learning rate to 1e-5, and the epochs to 50. The model parameters are saved using the early stop method. For other methods, all parameters are tuned to achieve their best performance. Then, the comparison results are shown in Table 2.

We can have the following main observations:

First, our method outperforms all the compared methods on both datasets. On CHIP2019, it achieves very high accuracy, i.e., 91.30%, which increases that of HCAN by 17.8%. The reason behind this is that compared with simple semantic modeling, using pretrained models can better explore the correlation between the

Table 2. Encoding Efficiency Comparison of Different Pre-training Models

Paper	Method	$Acc_{YiduN7K}$	$Acc_{ICD9-INT}$
Rao et al., 2018 [9]	HCAN	73.50%	74.51%
Devlin et al., 2018 [1]	Bert-target	88.00%	86.10%
Yijia Liu, 2021 [4]	ABTSBM	87.50%	86.61%
YAN Jinghui, 2021 [11]	Longest common sub-sequence	89.00%	87.18%
SUN Yuejun, 2021 [8]	Jaccard algorithm	90.04%	88.82%
Ours	**SNN-BS**	**91.30%**	**90.24%**

origin word and the target terminology. Second, Devlin et al.'s Bert-target model achieved good results at the time, i.e., 88.00%. It utilizes the Bert pre-training model to transform terminology standardization tasks into text classification tasks, and introduces the pre-training model into terminology annotation tasks.

Third, the Bert with Jaccard algorithm has also achieved good results, combining Jaccard algorithm and Bert encoder, which can better utilize the good feature extraction characteristics of Bert encoder in terminology standardization tasks.

Although our model uses the Bert-Tiny model with fewer parameters and lighter weight, it still has improved prediction accuracy compared with the SOTA model. This is mainly due to our redesign of the Siamese network structure for the clinical terminology standardization task, so that by generating multiple loss calculations under the data sample unit, the model is able to fit better to the data. After encoding the origin word and candidate terminology set, we fused the absolute and relative distance features. It makes our model not only consider the explicit similarity of the text, but also capture the deeper semantic features in the clinical terminology vocabulary, which has better generalization ability for the test samples with obvious features but difficult similarity matching.

4.3 Contrast and Ablation Experiment

In order to verify the reasoning ability of our model, we compared Bert-tiny with the common Bert-base and Robert-small model reasoning speed under the same framework. We simulated the efficiency of one-to-one similarity comparison in the SOTA method. Table 3 shows the experimental results.

Table 3. Comparison of Encoding efficiency of different pre-training models

Model	S_b	$Size_{code}$	Time Cost
SNN-BS(+Bert-Tiny)	512	2000	144.42 s
SNN-BS(+Bert-Tiny)	1	2000	5324.52 s
SNN-BS(+Robert-Small)	512	2000	722.79 s
SNN-BS(+Bert-Base)	512	2000	1278.70 s

According to the results in Table 3, it can be seen that using the Bert-tiny model can significantly reduce the efficiency of model reasoning compared with Robert-small and Bert-base. With the same Bert-tiny model, using candidate terminology set as the right-subnet input (set the S_b to 512) has a significant improvement in efficiency compared with using single standard terminology as the right-subnet input (set the S_b to 1).

We conducted ablation experiments for the data sampling unit and similar feature fusion module we adopted. Table 4 shows the experimental results.

Table 4. Comparison of Encoding efficiency of different pre-training models

Model	$Acc_{YiduN7K}$	$Acc_{ICD9-INT}$
SNN-BS(+Dot Product)	84.55%	84.10%
SNN-BS(+Euclidean Distance)	86.25%	85.84%
SNN-BS(+Dot Product & Euclidean Distance)	89.50%	89.02%
SNN-BS(+Data Sampling & Dot Product)	86.65%	86.21%
SNN-BS(+Data Sampling & Euclidean Distance)	88.05%	87.48%
SNN-BS(+All)	**91.30%**	**90.24%**

According to the results in Table 4, it can be seen that the accuracy of the method using the data sampling unit has increased by about 1.78% compared with the method without this strategy; The accuracy rate of only considering dot product similarity is about 3.11% higher than that of considering fusion similarity; The accuracy rate of only considering the Euclidean distance similarity is increased by about 4.64% compared with that of considering the fusion similarity. Furthermore, the model's prediction accuracy can reach the best effect of 91.30% when using the data sampling unit and similar feature fusion module.

4.4   Case Study

For the target terminology results predicted by our method, we selected some origin word samples with the incorrect prediction of SOTA model. We compared the prediction results of Bert-target [1] method, Siamese network without batch sampling strategy, and Siamese network with batch sampling strategy. Table 5 shows the experimental result.

From the result, our method can effectively match those standard pairs with similar meanings between origin words and target terminologies. After the data sampling unit and similar feature fusion module, the ability to distinguish "all" or "local" operations in target terminology can be effectively improved. For origin word with vague body parts, it can also match the corresponding target terminology. For example, the word "" ("Right lower leg amputation") should focus on the characteristics of "" ("thigh") rather than "" ("lower body"). However, in some cases where there are ambiguous and mixed descriptions in the

Table 5. Sampling comparison of prediction results

Origin word/Target terminology	Method	Forecast Word	Result
(O)[1]	Bert-target	7	False
(T)[2]	SNN-BS(initial)	2	True
	SNN-BS(tuning)	2	True
(O)[3]	Bert-target	8	False
(T)[4]	SNN-BS(initial)	8	False
	SNN-BS(tuning)	4	True
(O)[5]	Bert-target	9	False
(T)[6]	SNN-BS(initial)	9	False
	SNN-BS(tuning)	9	False

[1]: Endoscopic Assisted Thyroglossal Duct Cyst Resection
[2]: Thyroglossal duct lesion resection
[3]: Right lower leg amputation
[4]: Thigh amputation
[5]: D-J tube implantation under cystoscope
[6]: Transurethral ureteral stenting
[7]: Thyroglossal duct resection
[8]: Lower extremity amputation
[9]: Cystoscopy D-J tube extraction

origin word, its terminology standardization ability still needs to be improved due to lacking external domain knowledge. For example, the origin word, i.e., "("D-J tube placement under cystoscopy") is too concerned about the semantic characteristics of ""("D-J tube"), which leads to the neglect of global semantics in the prediction process, resulting in the wrong result.

5   Conclusion

Aiming to address the accuracy and efficiency problems of clinical terminology standardization task, we reduced the number of encoding and comparison learning by sampling the standard terminology set. We used the Bert-tiny model, which is lighter and has a better migration effect for encoding. We randomly mixed the target terminology into each sampling candidate terminology set to strengthen the model's ability to select answers for confusing standard terminologies. Through two-level similarity fusion, it highlight the corresponding characteristics between the origin word and the target terminology, alleviating the low accuracy problem caused by the strong professionalism of clinical terminology standardization tasks and improving the capture ability of remote semantic standard terminologies.

We take into account the accuracy and efficiency of terminology standardization. The experimental results of this method on the Yidu-N7K dataset and the ICD9-INT dataset show that our method is superior to the SOTA model in accuracy. Also, it can effectively improve the standardization accuracy of the origin word with unclear descriptions of surgical sites. However, the accuracy needs to be improved for test cases with clinical abbreviations in the origin word. In the next step. We can consider combining abbreviations in the clinical medical field and external knowledge bases of abbreviations to further improve the terminology standardization ability of the model.

Acknowledgemment. This work is partially supported by National Natural Science Foundation of China under Grant No. 62202282 and Shanghai Youth Science and Technology Talents Sailing Program under Grant No. 22YF1413700.

References

1. Devlin, J., Chang, M.W., Lee, K., Toutanova, K.: BERT: pre-training of deep bidirectional transformers for language understanding. arXiv:1810.04805 (2018)
2. Gao, Y., Fu, X., Liu, X., Wu, J.: Multi-features-based automatic clinical coding for Chinese ICD-9-CM-3. In: Farkaš, I., Masulli, P., Otte, S., Wermter, S. (eds.) ICANN 2021. LNCS, vol. 12895, pp. 473–486. Springer, Cham (2021). https://doi.org/10.1007/978-3-030-86383-8_38
3. Huang, J.: Automatic encoding of disease terminology based on combined semantic similarity calculation. Micro Comput. Appl. **36**(08), 157–160 (2020)
4. Liu, Y., Li, S., Yu, J., Tan, Y., Ma, J., Wu, Q.: Many-to-many Chinese ICD-9 terminology standardization based on neural networks. In: Huang, D.-S., Jo, K.-H., Li, J., Gribova, V., Hussain, A. (eds.) ICIC 2021. LNCS, vol. 12837, pp. 430–441. Springer, Cham (2021). https://doi.org/10.1007/978-3-030-84529-2_36
5. Rao, J., Liu, L., Tay, Y., Yang, W., Shi, P., Lin, J.: Bridging the gap between relevance matching and semantic matching for short text similarity modeling. In: Proceedings of the 2019 Conference on Empirical Methods in Natural Language Processing and the 9th International Joint Conference on Natural Language Processing (EMNLP-IJCNLP), pp. 5370–5381 (2019)
6. Ruby, U., Yendapalli, V.: Binary cross entropy with deep learning technique for image classification. Int. J. Adv. Trends. Comput. Sci. Eng. **9**(10), 1–8 (2020)
7. Sinha, R., Desai, U., Tamilselvam, S., Mani, S.: Evaluation of Siamese networks for semantic code search. arXiv preprint arXiv:2011.01043 (2020)
8. Sun, Y., Liu, Z., Yang, Z., Lin, H.: Standardization of clinical terminology based on BERT. Chinese J. Inf. Technol. **35**(4), 75–82 (2021)
9. Vaswani, A., et al.: Attention is all you need. In: Advances in Neural Information Processing Systems, vol. 30 (2017)
10. Wullschleger, P., Lionetti, S., Daly, D., Volpe, F., Caro, G.: Auto-regressive self-attention models for diagnosis prediction on electronic health records. In: 2022 IEEE International Conference on Big Data, pp. 1950–1956. IEEE (2022)
11. Yan, J., Xiang, L., Zhou, Y., Sun, J., Chen, S., Xue, C.: Application of deep generative model in clinical terminology standardization. Chinese J. Inf. Technol. **35**(5), 77–85 (2021)
12. Zhang, Z., Liu, J., Razavian, N.: Bert-xml: Large scale automated ICD coding using BERT pretraining. arXiv preprint arXiv:2006.03685 (2020)

13. Zhou, L., Qu, W., Wei, T., Zhou, J., Gu, Y., Li, B.: A review on named entity recognition in Chinese medical text. In: Sun, X., Zhang, X., Xia, Z., Bertino, E. (eds.) ICAIS 2021. CCIS, vol. 1422, pp. 39–51. Springer, Cham (2021). https://doi.org/10.1007/978-3-030-78615-1_4

14. Zhu, N., Cao, J., Shen, K., Chen, X., Zhu, S.: A decision support system with intelligent recommendation for multi-disciplinary medical treatment. ACM Trans. Multim. Comput. Commun. Appl. **16**(1s), 33:1-33:23 (2020)

Natural Language

Read Then Respond: Multi-granularity Grounding Prediction for Knowledge-Grounded Dialogue Generation

Yiyang Du[1(✉)], Shiwei Zhang[2], Xianjie Wu[1], Zhao Yan[3], Yunbo Cao[3], and Zhoujun Li[1]

[1] State Key Lab of Software Development Environment, Beihang University, Beijing, China
duyiyang@buaa.edu.cn
[2] Baidu Inc., Beijing, China
[3] Tencent Cloud Xiaowei, Beijing, China

Abstract. Retrieval-augmented generative models have shown promising results in knowledge-grounding dialogue systems. However, identifying and utilizing exact knowledge from multiple passages based on dialogue context remains challenging due to the semantic dependency of the dialogue context. Existing research has observed that increasing the number of retrieved passages promotes the recall of relevant knowledge, but the performance of response generation improvement becomes marginal or even worse when the number reaches a certain threshold. In this paper, we present a multi-grained knowledge grounding identification method, in which the coarse-grained selects the most relevant knowledge from each retrieval passage separately, and the fine-grained refines the coarse-grained and identifies final knowledge as grounding in generation stage. To further guide the response generation with predicted grounding, we introduce a grounding-augmented copy mechanism in the decoding stage of dialogue generation. Empirical results on MultiDoc2Dial and WoW benchmarks show that our method outperforms state-of-the-art methods.

Keywords: Knowledge-grounded dialogue · Retrieval-augmented · Grounding prediction

1 Introduction

Dialogue generation task faces the problem of producing non-informative or hallucinatory response [10,18]. Inspired by the retrieval-then-generation framework in open-domain QA [11,15,28], recent efforts have been made to address these concerns by knowledge-based dialogue generation. Those approaches typically

---

Y. Du and S. Zhang—Equal contribution.

© The Author(s), under exclusive license to Springer Nature Switzerland AG 2023
X. Yang et al. (Eds.): ADMA 2023, LNAI 14177, pp. 291–306, 2023.
https://doi.org/10.1007/978-3-031-46664-9_20

involve knowledge searching, finding relevant knowledge according to the dialogue context, and producing final contextual responses [24, 27].

Unexpectedly, as the number of retrieved documents increases, the performance of existing models either saturates or even degrades. As reported in [25, 27], adding more knowledge documents to a vanilla response generation model leads to a more severe problem of hallucinations, i.e. plausible statements with factual errors. This might be because incorrectly retrieved passages with high lexical overlap with the input dialogue context can mislead the response generator, rather than providing reasonable knowledge. So, how to identify relevant knowledge from the numerous retrieved results to guide the response generation becomes a critical problem.

To identify relevant knowledge and improve response performance, paragraph-level methods are proposed to filter passages which contain knowledge related to the dialogue. EviGui-G [2] exclude noisy documents from retrieval results by predicting whether a retrieved document provides relevant evidence to response as an auxiliary task. Re^{2}G [9] purposes a retriever-ranker-generator framework to filter the retrieved knowledge fed to the generator and applies knowledge distillation to train the ranker and retriever jointly. DIALKI [29] extracts knowledge by first selecting the most relevant passage to the dialogue context and then selecting the final knowledge string within the selected passage to guide response generation. By selecting a exclusive span from multiple passages as grounding, this token-level method further locks the scope of relevant knowledge and achieves good results, especially in long documents. However, as a result of this method of selecting only one grounding, there is the risk of error propagation, which will contaminate the response once irrelevant knowledge is chosen.

In this paper, we propose a novel **Multi-granularity Grounding Guided Generation** (MG^4) model which introduces two types of token-level knowledge, namely coarse-grained and fine-grained groundings, and fuse them with weighted attention to encourage the generator to consider the importance of knowledge in different dialogue contexts. Our method has the ability to extract critical grounding information from a vast array of knowledge documents in a coarse-to-fine manner, thereby assisting in the generation of final responses. Furthermore, our experiments have shown that the coarse-to-fine approach outperforms any of its individual components. The framework imitates the process of human search for answers using a browser. Initially, it reads each relevant document retrieved and identifies the most relevant knowledge in each document as coarse-grained groundings for the query. Next, it assesses the importance of each piece of knowledge and combines the understanding from each document with a fine-grained grounding to generate a response.

Concretely, we introduce two distinct granularities of grounding: coarse-grained and fine-grained grounding. The former aims to extract different spans of evidence from every retrieved passage through a question-and-answer system. To further identify the most relevant evidence from retrieved passages, we introduce a fine-grained grounding predicting task during the encoding phase of generation, which can locate exclusive grounding as knowledge from all retrieved passages. Additionally, to enhance the guidance of responses by grounding, we devise a grounding-augmented copy mechanism during the decoding phase of generation to encourage the generator to utilize the predicted grounding when producing responses explicitly. Our experimental results demonstrate that different granularities of grounding can effectively direct the generator to improve response performance. Our best model achieves the state of the art on both MultiDoc2Dial [8] and WoW [20] at the time of writing. Our contributions are summed up as follows:

1. We propose a grounding-guided dialogue generation model based on two different granularities knowledge(coarse-grained and fine-grained).
2. We further incorporate grounding to guide generation by introducing a grounding-augmented copy mechanism, which give additional attention to two granularities grounding and the retrieved original paragraph text.
3. We achieve a new state-of-the-art on MultiDoc2Dial and WoW in automated metrics. Our method generates more accurate dialogue responses and alleviates hallucination problems in human evaluation and verification.

2   Related Work

Retrieval-augmented generation. The retrieval-augmented generator is a two-stage pipeline framework: (i) first to retrieve relevant passages from the knowledge source (the retriever) [3,12,13,26,30]; and (ii) second to generate an answer based on retrieved passages with the original query (the generator) [14,22]. RAG [15] retrieves relevant passages from external sources [13] and then generate the final response in a sequence-to-sequence style with marginalizing generation probabilities from different retrieved documents. FiD [11] retrieves a larger number of passages, encodes them independently, and then fuses the encoder results of multiple passages in the decoder phase. EMDR^2 [28] purpose an end-to-end training method and updates the retriever and reader parameters using an expectation-maximization algorithm. Recent work improves the retrieval component [19] or introduces passage re-ranking modules [7] for further improvements.

Knowledge-grounded Dialogues. Knowledge-grounded dialogue systems aim to generate knowledgeable and engaging responses based on context, and external knowledge [4,5,21,31–33]. EviGui-G [2] introduces a joint task whether a candidate passage provides relevant evidence to enhance the ability to identify gold passages. K2R [1] proposes a knowledge to response modular model to generate a knowledge sequence, then attends to its own generated knowledge sequence to produce a final response. To address knowledge identification in conversational systems with long grounding documents, DIALKI [29] extends multi-passage reader models in open question answering to obtain dense encodings of different spans in multiple passages in the grounding document, and it contextualizes them with the dialogue history. Re^{2}G [9] introduces a reranking mechanism between the retriever and the reader, which permits merging retrieval results from sources with incomparable scores (Fig. 1).

3   Method

3.1   Overview

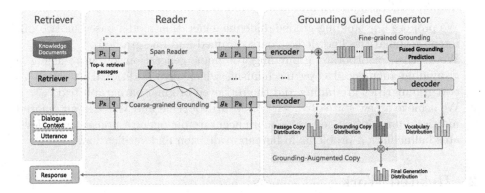

Fig. 1. Overview architecture of MG^4 framework

Problem Description. In knowledge-grounded response generation task, given a set of knowledge documents D, dialogue context U consisting of dialogue history $\{u_1, \ldots, u_{T-1}\}$ and user's utterance of current turn u_T, the goal is to generate response u_{T+1}. The probability of the generated responses can be written as:

$$p(y_t \mid P, U) = \prod_{t=1}^{n} p(y_t \mid P, U, y_1, \ldots, y_{t-1}) \tag{1}$$

where y_t is the t-th token in the agent response u_{T+1}, P is the split results of documents D. In order to distinguish the above D and P, we use "document" and "passage" respectively to denote the text of before and after segmentation. As the dialogue is knowledge-guided, the response is entailed by the grounding evidence in gold document among the provided multiple documents. In the gold passage related to the question, the grounding G_g is a span evidence to guide response generation. In this paper, we predict and exploit the grounding evidence in a multi-stage style to enhance the final response generation.

Method Overview. We propose a grounding-guided framework which extends the retrieval-augmented generation paradigm by adding a reader module to predict grounding - the token level evidence from retrieved passages to guide response generation. **Firstly**, the retriever retrieves the top-K knowledge passages (segments of the document) related to the last turn utterance and the dialogue history. **Secondly**, the reader module (Sect. 3.2) is used to predict coarse-grained grounding evidence from every retrieved passage independently. It is worth noticing that coarse-grained grounding may be inaccurate considering that there is an error prediction of the reader, or the retrieved passage may not include the grounding evidence. **Thirdly**, the response generator finds the most relevant evidence from multiple retrieved passages. We propose using the grounding-guided encoder (Sect. 3.3) and copy-augmented decoder (Sect. 3.4) in the generator to produce the final response. The grounding-guided encoder uses the encoder representation of the generator to predict fine-grained grounding, and the copy-augmented decoder encourages the generator to borrow words from the predicted grounding explicitly.

3.2 Coarse-Grained Grounding Prediction in Reader

Firstly, by taking current utterance u_T with dialogue history $\{u_1, \ldots, u_{T-1}\}$ and a retrieved passage p_i as input, the grounding reader aims to infer important grounding evidence span from each retrieved passage p_i. We train our reader to use all the three tuples of dialogue context, gold passage, and grounding evidence span of gold passage in the training set. The grounding evidence span can be obtained in most cases since the response is written by human based on its provenance.

We use span-based reading comprehension model to predict coarse-grained grounding. The start and end probability are calculated by a linear projection from the last hidden states of reader's encoder:

$$\hat{p}^{\ start} = \sigma(\varphi(H)) \quad \hat{p}^{\ end} = \sigma(\varphi(H)) \tag{2}$$

where $\hat{p}^{\ start}$ and $\hat{p}^{\ end}$ is start and end probability distribution, H is the representation of reader's encoder, σ is softmax function and $\varphi(\circ)$ is MLP. The cost function is defined as :

$$J(\boldsymbol{\theta}) = -\frac{1}{T} \sum_{t=1}^{T} \log\left(\hat{p}_{y_t^s}^{start}\right) + \log\left(\hat{p}_{y_t^e}^{end}\right) \tag{3}$$

where T is the number of training samples, y_t^s and y_t^e are the true start and end position of the t-th sample.

Then we use the well-trained reader to infer grounding evidence G for every retrieved passage in the training and evaluation set. The usage of coarse-grained grounding evidence G will be introduced in Sect. 3.3.

3.3 Fine-Grained Grounding Prediction in Generator Encoder

The generator is an encoder-decoder structure where the encoder part encodes every retrieved passage independently with the dialogue context and the coarse-grained grounding predicted in Sect. 3.2. The representation of j-th passage $h_{enc}^j \in R^{d \times l_j}$ can be calculated by encoder:

$$h_{enc}^j = \text{Encoder}\,(C; p_j; g_j) \tag{4}$$

where C is the dialogue context, p_j is the j-th passage and g_j is the predicted coarse-grained grounding in j-th passage from reader. The input form of j-th passage feed to the generator encoder can be described in detail as follows:

$$[\bar{S}_u, u_1, ... u_T; \bar{S}_p, p_j^0, p_j^1, ... \bar{S}_g, g_j^s, ... g_j^e, \bar{E}_g, ... p_j^{l_j}] \tag{5}$$

where u_T is the dialogue utterance of T-th turn, $p_j = \{p_j^0, ..., p_j^{l_j}\}$ is the context tokens of j-th retrieved passage with length l_j. $g_j = \{g_j^s, ..., g_j^e\}$ is coarse-grained grounding predicted by reader in the passage with the start and end position. $\bar{S}_u, \bar{S}_p, \bar{S}_g, \bar{E}_g$ are special tokens to indicate the start position of dialogue and passage context, the start and end position of coarse-grained grounding.

Fine-Grained Grounding Prediction. The encoder part of the generator incorporates fine-grained grounding prediction to identify the most relevant grounding evidence from all the retrieved passages to generate a response. Fine-grained grounding prediction can also fuse and denoise the coarse-grained groundings as it can be jointly trained with the response generation part. The error in coarse-grained groundings can arise from two sources: (1) errors in the prediction from the reader module, and (2) the possibility that the retrieved passage may not inherently contain any grounding evidence.

In the training phase, we consider that some retrieved passages may not contain the exact gold grounding evidence but rather similar useful information. Therefore, we use a token-level matching method to identify tokens present in the gold grounding and use them as fine-grained grounding labels. In the validation phase, the predicted grounding evidence is leveraged in the generator decoder part described in Sect. 3.4. The fine-grained grounding prediction is composed of a linear layer and a sigmoid function, which acts on the representation from the generator encoder. Since tokens included in the gold grounding accounts for a small proportion of the tokens in all retrieved passages, we sample negative tokens and apply focal loss [17] to train the grounding evidence prediction. The loss function can be defined as follows:

$$\boldsymbol{p}_{\mathrm{g}}(i) = \sigma(W_g h_i + b_g) \tag{6}$$

$$J(\boldsymbol{\theta}) = \sum_{y_i=1}^{M} \alpha \left(1 - \boldsymbol{p}_{\mathrm{g}}(i)\right)^\gamma \log \boldsymbol{p}_{\mathrm{g}}(i) + \sum_{y_i=0}^{N} (1-\alpha) \boldsymbol{p}_{\mathrm{g}}(i)^\gamma \log \left(1 - \boldsymbol{p}_{\mathrm{g}}(i)\right) \tag{7}$$

where h_i is the i-th position's representation from generator encoder, W_g and b_g are trainable parameters, σ is sigmoid function, $J(\boldsymbol{\theta})$ is the loss objective contributed by M positive grounding tokens and N negative tokens. α and γ is the hyperparameters in focal loss.

3.4   Copy Grounding Evidence in Generator Decoder

The decoder part of generator jointly decodes all encoded features of retrieved knowledge to generate response. We fuse the encoder inputs in a Fusion-in-Decoder style [11] to empower the decoder to attend all input passages and get cross-attention result within a linear time complexity. As described in Sect. 3.3, the representation of j-th passage $h_{enc}^j \in R^{d \times l_j}$ can be calculated as follows :

$$h_{enc}^j = \text{Encoder} \left(C; p_j; g_j \right) \tag{8}$$

Then concatenate the h_{enc}^j to produce h_{enc} for decoder:

$$h_{enc} = h_{enc}^1 \circ h_{enc}^2 \circ h_{enc}^3 \dots h_{enc}^K \tag{9}$$

Our decoder is based on transformer style, so the cross-attention result can be calculated in the transformer layer itself:

$$e_{t,i} = \frac{(W_s s_t)^T W_h h_i}{\sqrt{d_k}} \tag{10}$$

$$\alpha_{t,i} = \text{softmax} \left(e_{t,i} \right) \tag{11}$$

where the h_i is the i-th position's representation of h_{enc}, s_t is the t-th step representation of h_{dec} calculated by self-attention and layer-normalization. W_s and W_h are learnable weights. d_k is the hidden size of k-th head, where we take out the last layer of transformer and the average of heads as the cross-attention output including cross-attention weights $e_{t,i}$ and cross-attention probs $\alpha_{t,i}$.

Grounding Augmented Copy Mechanism. We propose a grounding-augmented copy mechanism to encourage generator to explicitly borrow words from the predicted grounding. Let $L = \sum_{i=0}^{k} l_i$ denote the total encoder length after concatenation, g be the fine-grained grounding introduced by Sect. 3.3 to identify whether a token is present in gold grounding. The attention score from the response to predicted grounding can obtained by re-normalizing the cross attention weights in grounding token positions.

$$m_{t,i} = \begin{cases} 1, & g(i) = 1 \\ -\infty, & g(i) = 0 \end{cases} \tag{12}$$

$$n_{t,i} = \begin{cases} 1, & g(i) = 0 \\ -\infty, & g(i) = 1 \end{cases} \tag{13}$$

$$\beta_{t,i} = \text{softmax} \left(e_{t,i} \cdot m_{t,i} \right) \tag{14}$$

$$\gamma_{t,i} = \text{softmax} \left(e_{t,i} \cdot n_{t,i} \right) \tag{15}$$

The cross-attention probability from decoder time step t to token i in the fine-grained grounding evidence is denoted as $\beta_{t,i}$, while $\gamma_{t,i}$ represents the same probability for tokens in the other part of the passage except for the grounding evidence. This cross-attention probability can be used as a copied probability to contribute to the final probability distribution.

$$\boldsymbol{p}_{grounding}(w) = \sum_{i:x_i=w} \beta_{t,i}, \quad \boldsymbol{p}_{passage}(w) = \sum_{i:x_i=w} \gamma_{t,i} \tag{16}$$

where $P_{grounding}(w)$ is the vocabulary probability distribution by copying grounding evidence and $P_{passage}(w)$ is the vocabulary probability distribution by copying the other part of encoder input including passages and dialogue context. We reserve the distribution from passages and dialogue context because not all response words come from grounding and they may come from dialogue or other parts in passages. Then we add the copy vocabulary probability distribution to the generator vocabulary probability distribution with a learnable 3-way gate.

$$p_1, p_2, p_3 = \text{softmax}\left(W_{\text{gate}}^3 \cdot h_t^{\text{dec}} + b_{\text{gate}}^3\right) \tag{17}$$

$$\boldsymbol{p}_{generate}(w) = \text{lm}_{head}\left(h_t^{dec}\right) \tag{18}$$

$$\boldsymbol{p}(w) = p_1 \cdot \boldsymbol{p}_{generate}(w) + p_2 \cdot \boldsymbol{p}_{grounding}(w) + p_3 \cdot \boldsymbol{p}_{passage}(w) \tag{19}$$

where $W_{\text{gate}}^3 \in \mathbb{R}^{d \times 3}$, $b_{\text{gate}}^3 \in \mathbb{R}^{d \times 3}$ are learnable parameters, lm_{head} is the output layer in transformer to calculate target vocab distribution. p_1, p_2, p_3 are 3-way gate probability. $\boldsymbol{p}(w)$ is the final target vocab distribution considering the contribution of generation, the predicted grounding and retrieved passages. According to $\boldsymbol{p}(w)$ to decode a word w step by step, the final response is generated.

4   Experiment

4.1   Datasets

MultiDoc2Dial. [8] is a new goal-oriented dialogue dataset based on multiple documents, containing 29,748 queries in 4800 dialogues with an average of 14 turns based on 488 documents from different domains. Each dialogue turn annotates the dialog data with the roles, dialogue behavior, human speech, and the grounding span with document information.

WoW. [6] is a large conversational dataset based on knowledge retrieved from Wikipedia. It covers a wide range of topics (a total of 1365), comprising 22311 dialogues and 201999 rounds. We verify the performance of our model on the WoW KILT version [20]. The KILT version requires model to find and fuse knowledge from all of Wikipedia pages rather than the provided knowledge candidates for each turn in original dataset, which is more suitable for our setting.

4.2   Baselines

RAG. [15] retrieves relevant passages from external sources and then generate the final response in a sequence-to-sequence style with marginalizing generation probabilities from different retrieved documents. **FiD** [11] Fusion-in-Decoder encodes all retrieved passages independently and then fuses the encoder result of multiple passages in the decoder phase. **EMDR^2** [28] provides an end-to-end approach to optimize retriever and generator parameters using model feedback itself as "pseudo-labels" for latent variables. **DIALKI** [29] identifies the most relevant passage and grounding span in the passage from multiple documents and then only use the single passage and span to generate response. **EviGui-G** [2] incorporate evidentiality of passages and introduces a leave-one-out method to create pseudo evidentiality labels for model training.

Re²G. [9] applies a retriever-ranker-generator framework to filter the retrieved knowledge fed to the generator and applies knowledge distillation to jointly train the ranker and retriever.

Table 1. Main Results. Results of automatic metrics on test set of MultiDoc2Dial and WoW. [†] denotes the model is based on T5-base while [‡] denotes T5-large and [§] denotes BART-large;

	MultiDoc2Dial		WoW	
	F1	R-L	F1	R-L
RAG [§]	34.25	31.85	13.11	11.57
FiD[†]	41.74	40.37	16.52	15.16
DIALKI [§]	38.95	37.64	17.04	15.65
EviGui-G[†]	43.14	41.33	17.30	15.93
EMDR² [‡]	43.76	41.86	-	-
Re²G [§]	44.26	42.40	18.90	16.76
MG⁴ base[†]	45.30	43.38	18.69	16.80
MG⁴ [‡]	**45.72**	**43.94**	**19.28**	**17.26**

4.3   Experiment Setting

For the evaluation of our knowledge-based dialogue system, we evaluate the generated responses against the reference responses with automatic metrics, including token-level F1 score(F1) [23], Rouge-L(R-L) [16].

We report the results of RAG, FiD, and EviGui-G from [2], as well as Re2G from [9]. We reproduced the evaluation results using the same hyper-parameters, averaging over five runs with different seeds, and conducting a t-test with a p-value less than 0.05. The result of our model on the WoW dataset is from the KILT [20] version, which provides an online submission board[1].

To train the MG⁴ model, we use the Adam optimizer with a learning rate of 5e-5. The number of top-k passages is set to 50. The input length of dialogue context and a single passage is set to 512, while the grounding span max length is set to 128, and the maximum response length is set to 50. α and γ in focal loss are set to 0.25 and 2.

---

[1] https://eval.ai/web/challenges/challenge-page/689/leaderboard/1909.

4.4 Quantitative Results

According to the Table 1, The end-to-end EMDR^2 model has a slight advantage over FiD by 1.12 F1 and 1.19 Rouge-L in MultiDoc2Dial, indicating that the end-to-end model may somewhat mitigate the problem of accumulating pipeline framework errors. Our MG^4 method outperforms nearly all benchmark results on both MultiDoc2dial and WoW. MG^4 outperforms DIALKI model on both Multidoc2Dial and WoW, indicating that selecting only one grounding from the most relevant passage has the risk of error propagation and multi-granularity grounding with weighted attention grounding copy mechanism can effectively identify multiple related information and improve the quality of generation. In particular, MG^4 outperforms the end-to-end model EMDR^2 by 2.46 F1 and 2.08 Rouge-L on MultiDoc2Dial, illustrating that grounding knowledge in retrieved passages can bring more performance gains than just training retriever and generator in an end-to-end way. Re^{2}G gets good performance on both MultiDoc2Dial and WoW by adding a reranker module to filter the retrieved knowledge, while MG^4 can also highlight some token-level evidence in retrieved knowledge and outperform Re^{2}G by 1.46 F1, 1.54 Rouge-L and 0.38 F1 and 0.50 Rouge-L.

Table 2. Ablation results. G^{coarse} denotes introducing the coarse-grained to the generator predicted by reader; *Copy* denotes introducing copy mechanism in the generator to copy words from fine-grained grounding. G^{fine} denotes fine-grained grounding prediction. **MG^4-CG** doesn't remove any module but replaces copying mechanism from fine-grained grounding with coarse-grained grounding.

	MultiDoc2Dial		WoW	
	F1	R-L	F1	R-L
MG^4	45.72	43.94	19.28	17.26
w/o G^{coarse}	43.63	41.81	17.72	16.38
w/o *Copy*	45.27	43.51	18.74	17.19
w/o G^{fine}	44.22	42.28	18.02	16.65
MG^4-CG	44.97	43.11	18.37	16.82

Table 2 presents the results of the ablation experiments. There is a clear drop when removing coarse-grounding, i.e. w/o G^{coarse}, illustrating the effectiveness of reader module and the influence of reader to generator. Removing grounding augmented copy mechanism, i.e. w/o *Copy*, drops the performance on both datasets, proving that copy mechanism can enhance the guidance from grounding to response generation. In w/o *Copy* setting, furtherly removing the fine-grained grounding prediction task, w/o G^{fine}, will continue to bring performance drop, indicating that training via joint tasks to predict evidentiality labels can bring help to the generation task. Finally, we conduct experiments on copying from coarse-grained grounding, i.e. MG^4-CG. It's performance is lower than MG^4, which can be understood as the superiority of fine-grained grounding compared to coarse-grained grounding in guiding response generation.

4.5   Human Evaluation

Human annotators are asked to evaluate our model by quantifying the three aspects of generated responses, as described below: (i) **Fluency**, a measure of whether the response is consistent and less repetitive. (ii) **Relevance**, which measures the relevance of the response to the dialogue context. (iii) **Factuality** measures the correctness and faithfulness of all facts involved in the generated response.

Table 3. Absolute human valuation results for MG^4 versus $EMDR^2$ on MultiDoc2Dial. The table presents each metric average value for all annotators and samples out of 3 points. The Fleiss' kappa between annotators is 0.58.

Model	Fluency	Relevance	Factuality
$EMDR^2$	2.54	2.33	2.13
MG^4	**2.67**	**2.73**	**2.51**

Table 4. Comparative evaluation results between MG^4 and $EMDR^2$, where the percentage indicates the proportion of preference by all evaluators.

Aspect	Win	Lose	Tie
Fluency	**32%**	20%	48%
Relevance	**57%**	13%	30%
Factuality	**62%**	22%	16%

We choose $EMDR^2$, which is the most important reference in terms of automatic measurements, for comparative purposes. We sample the evaluation dialogue turns from the MultiDoc2Dial, which is factually supported by knowledgeable customer service documents. Table 3 shows the absolute evaluation results of human annotation. To reduce the evaluation inconsistency caused by different evaluators, we also conduct a comparative evaluation with results shown in Table 4. We found that MG^4 outperformed $EMDR^2$ in both evaluation dimensions, indicating that MG^4 can improve knowledge utilization through our coarse-to-fine grounding prediction method and grounding-augmented copy mechanism. It is noteworthy that our model has a significant improvement on the factuality metric, demonstrating its ability to alleviate the dialogue hallucination problem to some extent.

5   Further Analysis

5.1   Can Grounding Guide the Response Generation?

Table 5. Oracle experiments to explore the upper bound impact of gold grounding on MultiDoc2dial and WoW dev dataset.

Model	Dataset	F1	R-L
FiD	MultiDoc	42.14	40.67
FiD with Gold-G	MultiDoc	51.39	49.79
FiD	WoW	16.15	15.86
FiD with Gold-G	WoW	29.53	28.46

We conduct two experiments to illustrate the influence of grounding information to final response generation.

Firstly, we introduce gold grounding to the generator in the way as Sect. 3.2 and conduct oracle experiments to explore the influence of grounding. As shown in Table 5, the grounding-guided model significantly improved in performance by 9.25 F1, 9.12 Rouge-L on MultiDoc2Dial and 13.38 F1, 12.6 Rouge-L on WoW. According to the experimental results, the introduction of gold grounding can significantly improve the generation performance in the knowledge-grounded dialogue generation task.

Secondly, we leverage a span-based model called reader in Sect. 3.2 to predict grounding as a replacement of gold grounding. According to the ablation experiment results in Table 2, we can find that the performance gain from the reader module is most notable, proving that knowledge-grounded dialogue can benefit from the coarse-grained grounding from the extractive reader module.

5.2   Why We Need a Multi-granularity Grounding Prediction?

In our paper, multi-granularity grounding includes coarse-grained and fine-grained grounding. We add a fine-grained grounding prediction introduced in Sect. 3.3 in the generator encoder to find most relevant evidence from all retrieved passages. Figure 2 shows an actual case of the predicted coarse-grained grounding and fine-grained grounding in WoW, and the bottom right shows their contribution to the final response by taking out the cross-attention weight (average in the output sequence dimension) in the generator. In Table 2, we compare the performance between MG^4-CG and MG^4 in which MG^4-CG means copying from coarse-grained grounding and MG^4 means copying from fine-grained grounding. The introduction of fine-grained grounding can improve 0.75 F1 and 0.83 Rouge-L in MultiDoc2dial as well as 0.91 F1 and 0.44 Rouge-L in WoW compared to coarse-grained grounding which illustrates the validity of our coarse-to-fine method to predict token level evidence from multiple retrieved passages.

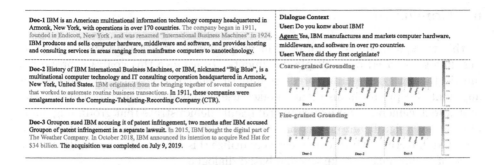

Fig. 2. Coarse-grained vs. Fine-grained. Retrieved passages are located on the left side. The blue portion is the coarse-grained grounding predicted from multiple passages while the light green portion is the fine-grained grounding predicted from grounding evidence denoiser. (Color figure online)

5.3   MG^4 Performs Better with More Passages

Figure 3 shows the Rouge-L score of our MG^4 model and the FiD model in different passage number settings. From the figure, we can see that the performance gains influenced by retrieved passage numbers is marginal as the number increases. It's worth noticing that our MG^4 model can get even higher improvement compared to FiD with larger retrieved passage numbers. The improvement is 1.92 Rouge-L in 10 passages setting and 3.27 Rouge-L in 50 passages setting. It can be interpreted as that larger number of retrieved passages means larger amount of relevant knowledge information as well as noise, which will bring more burden to the generator module. While our MG^4 can alleviate this problem by providing token-level multi-granularity grounding from retrieved passages to the generator.

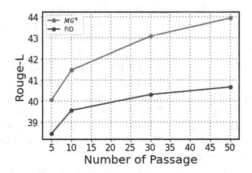

Fig. 3. Impact of the input passage number to response performance on MultiDoc2Dial.

6 Conclusion

In this work, our aim is to address the grounding identification issue in generating dialogues based on multiple documents. To achieve this goal, we propose a multi-granularity grounding prediction method in conjunction with a grounding-augmented copy mechanism that makes use of predicted key information from multiple documents. Our experimental results demonstrate that grounding information has a significant impact on guiding dialogue generation and that our proposed architecture, MG^4, can effectively utilize this information and mitigate the issue of hallucination in knowledge-based dialogue.

References

1. Adolphs, L., Shuster, K., Urbanek, J., Szlam, A., Weston, J.: Reason first, then respond: Modular generation for knowledge-infused dialogue. arXiv preprint arXiv:2111.05204 (2021)
2. Asai, A., Gardner, M., Hajishirzi, H.: Evidentiality-guided generation for knowledge-intensive NLP tasks. arXiv preprint arXiv:2112.08688 (2021)
3. Brown, T., et al.: Language models are few-shot learners. Adv. Neural. Inf. Process. Syst. **33**, 1877–1901 (2020)
4. Chen, X., et al.: Bridging the gap between prior and posterior knowledge selection for knowledge-grounded dialogue generation. In: Proceedings of the 2020 Conference on Empirical Methods in Natural Language Processing (EMNLP), pp. 3426–3437 (2020)
5. Davison, J., Feldman, J., Rush, A.M.: Commonsense knowledge mining from pretrained models. In: Proceedings of the 2019 Conference on Empirical Methods in Natural Language Processing and the 9th International Joint Conference on Natural Language Processing (EMNLP-IJCNLP), pp. 1173–1178 (2019)
6. Dinan, E., Roller, S., Shuster, K., Fan, A., Auli, M., Weston, J.: Wizard of Wikipedia: Knowledge-powered conversational agents. arXiv preprint arXiv:1811.01241 (2018)
7. Fajcik, M., Docekal, M., Ondrej, K., Smrz, P.: R2-d2: a modular baseline for open-domain question answering. arXiv preprint arXiv:2109.03502 (2021)
8. Feng, S., Patel, S.S., Wan, H., Joshi, S.: MultiDoc2Dial: modeling dialogues grounded in multiple documents. In: EMNLP (2021)
9. Glass, M., Rossiello, G., Chowdhury, M.F.M., Naik, A., Cai, P., Gliozzo, A.: Re2G: retrieve, rerank, generate. In: Proceedings of the 2022 Conference of the North American Chapter of the Association for Computational Linguistics: Human Language Technologies, pp. 2701–2715. Association for Computational Linguistics, Seattle, United States, July 2022. https://aclanthology.org/2022.naacl-main.194
10. Holtzman, A., Buys, J., Du, L., Forbes, M., Choi, Y.: The curious case of neural text degeneration. arXiv preprint arXiv:1904.09751 (2019)
11. Izacard, G., Grave, E.: Leveraging passage retrieval with generative models for open domain question answering. arXiv preprint arXiv:2007.01282 (2020)
12. Jones, K.S.: A statistical interpretation of term specificity and its application in retrieval. J. Doc. **28**, 1–11 (1972)
13. Karpukhin, V., et al.: Dense passage retrieval for open-domain question answering. arXiv preprint arXiv:2004.04906 (2020)

14. Lewis, M., et al.: Bart: Denoising sequence-to-sequence pre-training for natural language generation, translation, and comprehension. arXiv preprint arXiv:1910.13461 (2019)
15. Lewis, P., et al.: Retrieval-augmented generation for knowledge-intensive NLP tasks. Adv. Neural. Inf. Process. Syst. **33**, 9459–9474 (2020)
16. Lin, C.Y.: Rouge: a package for automatic evaluation of summaries. In: Text Summarization Branches Out, pp. 74–81 (2004)
17. Lin, T.Y., Goyal, P., Girshick, R., He, K., Dollár, P.: Focal loss for dense object detection. In: Proceedings of the IEEE International Conference on Computer Vision, pp. 2980–2988 (2017)
18. Ma, Y., Nguyen, K.L., Xing, F.Z., Cambria, E.: A survey on empathetic dialogue systems. Inf. Fusion **64**, 50–70 (2020)
19. Paranjape, A., Khattab, O., Potts, C., Zaharia, M., Manning, C.D.: Hindsight: posterior-guided training of retrievers for improved open-ended generation. arXiv preprint arXiv:2110.07752 (2021)
20. Petroni, F., et al.: Kilt: a benchmark for knowledge intensive language tasks. In: Proceedings of the 2021 Conference of the North American Chapter of the Association for Computational Linguistics: Human Language Technologies, pp. 2523–2544 (2021)
21. Prabhumoye, S., Hashimoto, K., Zhou, Y., Black, A.W., Salakhutdinov, R.: Focused attention improves document-grounded generation. arXiv preprint arXiv:2104.12714 (2021)
22. Raffel, C., et al.: Exploring the limits of transfer learning with a unified text-to-text transformer. arXiv preprint arXiv:1910.10683 (2019)
23. Rajpurkar, P., Zhang, J., Lopyrev, K., Liang, P.: Squad: 100,000+ questions for machine comprehension of text. arXiv preprint arXiv:1606.05250 (2016)
24. Rashkin, H., Reitter, D., Tomar, G.S., Das, D.: Increasing faithfulness in knowledge-grounded dialogue with controllable features. arXiv preprint arXiv:2107.06963 (2021)
25. Reimers, N., Gurevych, I.: The curse of dense low-dimensional information retrieval for large index sizes. arXiv preprint arXiv:2012.14210 (2020)
26. Robertson, S.E., Walker, S., Jones, S., Hancock-Beaulieu, M.M., Gatford, M., et al.: Okapi at TREC-3. NIST Spec. Publ. SP **109**, 109 (1995)
27. Shuster, K., Poff, S., Chen, M., Kiela, D., Weston, J.: Retrieval augmentation reduces hallucination in conversation. arXiv preprint arXiv:2104.07567 (2021)
28. Singh, D., Reddy, S., Hamilton, W., Dyer, C., Yogatama, D.: End-to-end training of multi-document reader and retriever for open-domain question answering. In: Advances in Neural Information Processing Systems, vol. 34 (2021)
29. Wu, Z., Lu, B.R., Hajishirzi, H., Ostendorf, M.: DIALKI: Knowledge identification in conversational systems through dialogue-document contextualization. In: Proceedings of the 2021 Conference on Empirical Methods in Natural Language Processing, pp. 1852–1863. Association for Computational Linguistics, Online and Punta Cana, Dominican Republic, November 2021. https://doi.org/10.18653/v1/2021.emnlp-main.140, https://aclanthology.org/2021.emnlp-main.140
30. Xiong, L., et al.: Approximate nearest neighbor negative contrastive learning for dense text retrieval. arXiv preprint arXiv:2007.00808 (2020)
31. Zhan, H., Zhang, H., Chen, H., Ding, Z., Bao, Y., Lan, Y.: Augmenting knowledge-grounded conversations with sequential knowledge transition. In: Proceedings of the 2021 Conference of the North American Chapter of the Association for Computational Linguistics: Human Language Technologies, pp. 5621–5630 (2021)

32. Zhang, S., Du, Y., Liu, G., Yan, Z., Cao, Y.: G4: Grounding-guided goal-oriented dialogues generation with multiple documents. In: Proceedings of the Second DialDoc Workshop on Document-grounded Dialogue and Conversational Question Answering, pp. 108–114. Association for Computational Linguistics, Dublin, Ireland, May 2022. https://doi.org/10.18653/v1/2022.dialdoc-1.11, https://aclanthology.org/2022.dialdoc-1.11

33. Zhu, W., Mo, K., Zhang, Y., Zhu, Z., Peng, X., Yang, Q.: Flexible end-to-end dialogue system for knowledge grounded conversation. arXiv preprint arXiv:1709.04264 (2017)

SUMOPE: Enhanced Hierarchical Summarization Model for Long Texts

Chao Chang[1,2], Junming Zhou[1], Xiangwei Zeng[1], and Yong Tang[1,2]([✉])

[1] South China Normal University, Guangzhou 510631, Guangdong, China
{changchao,ytang}@m.scnu.edu.cn
[2] Pazhou Lab, Guangzhou 510330, Guangdong, China

Abstract. With the progress of the times, the ever-advancing and improving Internet technology and the ever-generating social media network platforms have made the amount of information in the network explode, which contains a massive scale of redundant content. Then how to quickly extract the key information from the huge amount of data becomes crucial. In this paper, we propose a novel enhanced hierarchical summarization model SUMOPE for long texts, which combines both extractive and abstractive methods to deal with long texts. Our model first uses an extractive method called SUMO to select key sentences from the long text and form a bridging document. Then, our model uses an abstractive method based on PEGASUS with a copy mechanism to generate the final summary from the bridging document. Our model can effectively capture the important information and relations in the long text and produce coherent and concise summaries. We evaluate our model on two datasets and show that it outperforms the state-of-the-art methods in terms of ROUGE scores and human evaluation.

Keywords: Text summarization · Transformer · Natural language processing

1 Introduction

Text summarization has been a well-established focus of research in the field of natural language processing, involving the creation of concise and coherent summaries for lengthy texts while preserving essential information. With the growing volume of online information, there is an increasing demand for efficient and accurate methods to summarize large amounts of textual data.

Currently, text summarization is approached through two primary methodologies: extractive and generative. Extractive summarization is based on statistical methods. It involves calculating the relevance of each sentence in the text based on certain extraction rules, such as keywords, position, and similarity to the overall text. It also involves selecting the top-ranked sentences as a summary. This method is relatively simple and has strong interpretability since the

X. Yang et al. (Eds.): ADMA 2023, LNAI 14177, pp. 307–319, 2023.
https://doi.org/10.1007/978-3-031-46664-9_21

extracted summary is faithful to the original text. However, extractive summarization depends largely on the quality of the source text. It may suffer from incoherent semantics and repetition in poorly structured text. Generative summarization methods can overcome these issues by not simply using the words and phrases from the source text to create the summary, but rather by extracting the meaning from the text and generating the summary one word at a time. Generative summarization is typically achieved through sequence-to-sequence models, but it may encounter problems such as out-of-vocabulary (OOV) and long-distance dependency issues. Furthermore, the encoding phase of summary generation can lead to a notable information loss due to the challenge of long-distance dependencies.

To address these challenges, we propose a enhanced hierarchical summarization model for long text, called SUMOPE. The first stage uses a hierarchical encoder-decoder architecture to extract salient sentences from the input text, and the second stage refines the selected sentences to produce a high-quality summary. Our model incorporates attention mechanisms and reinforcement learning to improve sentence selection and refinement. We evaluate SUMOPE on benchmark datasets and compare it with state-of-the-art models. Our experiments show that SUMOPE outperforms existing methods in automatic metrics and human evaluations.

The contributions of this paper can be summarized as follows:

- The paper proposes a novel enhanced hierarchical summarization model SUMOPE for long text, which addresses the challenge of generating high-quality summaries for lengthy texts.
- The enhanced hierarchical summarization model integrates both extractive and abstractive methods, leveraging the advantages of both to improve the quality of the generated summaries.
- The proposed model achieves state-of-the-art performance on two datasets, demonstrating its effectiveness and practicality for real-world applications.

2   Related Work

In the initial stages, extractive summarization methods were mostly unsupervised and based on statistics. These methods mainly relied on calculating the word frequency and the position of sentences to determine the score of each sentence in the text. Subsequently, they amalgamated the sentences with the most elevated scores to formulate a summary. Luhn, Jones et al. [1,2] completed the task of text summarization by identifying keywords with significant information content in the text.

As research in machine learning and deep learning advances, supervised extractive summarization has become the mainstream research approach. In 2015, Can et al. [3] proposed a ranking framework for multi-document summarization. It uses Recursive Neural Networks to perform hierarchical regression and measure the salience of sentences and phrases in the parsing tree. The model learns ranking features automatically and concatenates them with hand-crafted features of words to conduct hierarchical regressions. In 2017, Nallapati et al. [4] proposed an extractive summarization model based on Recurrent Neural Networks, which enables visualization of its predictions based on abstract features such as information content, salience, and novelty. Additionally, the model can be trained abstractively using human-generated reference summaries, eliminating the need for sentence-level extractive labels. In 2019, Liu et al. [5] presented a comprehensive framework for applying BERT, a pre-trained language model, to text summarization, covering both extractive and abstractive models. A document-level encoder based on BERT is introduced to capture the semantics of a document and obtain sentence embedding vector. For extractive summarization, inter-sentence transformer layers are stacked on top of the encoder. For abstractive summarization, a new fine-tuning schedule is proposed to handle the mismatch between the pre-trained encoder and the decoder. In 2021, Huang et al. [6] proposed an approach for extractive summarization that integrates discourse and coreference relationships by modeling the relations between text spans in a document using a heterogeneous graph. The graph contains three types of nodes, each corresponding to text spans of different granularity.

With further research into extractive summarization, researchers have discovered problems such as repetitive generation and lack of semantic coherence. In contrast, abstractive summarization, which generates new words and expressions based on the understanding of the text, is closer to human summarization thinking and emphasizes consistency and coherence [7]. In 2019, Dong et al. [8] proposed a comprehensive pre-trained language model capable of fine-tuned for both natural language comprehension and generation tasks. It utilizes a shared Transformer network along with specific self-attention masks to manage contextual information. In 2020, Zhang et al. [9] proposed a large Transformer-based encoder-decoder model that is pre-trained on massive text corpora with a new self-supervised objective tailored for abstractive text summarization. It generates summaries by removing/masking important sentences from the input document and generating them together as one output sequence from the remaining sentences. In 2020, Liu et al. [10] proposed a training paradigm for abstractive summarization models, which assumes a non-deterministic distribution to assign probability mass to different candidate summaries based on their quality.

While generative summarization models are capable of generating more accurate and readable summaries, they are limited by deep-learning techniques in obtaining text representations for long documents. In 2018, chen et al. [11] proposed a summarization model that follows a two-stage approach where salient sentences are selected and then rephrased to generate a concise summary. A sentence-level policy gradient method bridges the computation between the two neural networks while maintaining fluency. In 2021, Li et al. [12] proposed an extractive-abstractive approach to address the interpretability issue in abstractive summarization while avoiding the redundancy and lack of coherence in extractive summarization. The framework uses the Information Bottleneck principle to jointly train extraction and abstraction in an end-to-end fashion. It first extracts a pre-defined amount of evidence spans and then generates a summary using only the evidence. In 2022, Xiong et al. [13] proposed a summarization model that uses elementary discourse units (EDUs) as the textual unit of content selection to generate high-quality summaries. The model first uses an EDU selector to choose salient content and then a generator model to rewrite the selected EDUs into the final summary. The group tag embedding is applied to determine the relevancy of each EDU in the entire document, allowing the generator to ingest the entire original document.

3   Proposed Technique

Inspired by SUMO [14] and PEGASUS [9], in this paper, we design a novel framework named SUMOPE to implement long text summarization, depicted in Fig. 1. Specifically, the extraction model based on SUMO extracts key sentences from long texts. These extracted sentences are then used as inputs to the generation model to produce the final summary. The transition document represents the set of extracted sentences from the extraction model. Its length falls between that of the original text and the summary, encompassing a significant portion of the crucial information found in the input document.

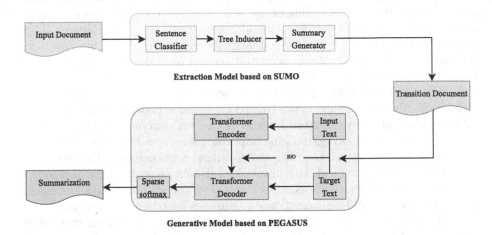

Fig. 1. The overview architecture of our proposed SUMOPE framework is with two modules: extraction model based on SUMO and generative model based on PEGASUS.

3.1   Extraction Model Based on SUMO

The extraction model based on SUMO is an approach to single-document summarization that uses tree induction to generate multi-root dependency trees that capture the connections between summary sentences and related content. This technique hinges on the concept of framing extractive summarization as a tree induction challenge, where each root node within the tree symbolizes a summary sentence and the attached subtrees to it represent sentences whose content is related to and covered by the summary sentence.

The module comprises three main components: a sentence classifier, a tree inducer, and a summary generator. The sentence classifier uses a Transformer model with multi-head attention to classify each sentence in the input document as summary-worthy or not. The tree inducer then induces a multi-root dependency tree that captures the relationships between summary sentences and related content through an iterative refinement process that builds latent structures while using information learned in previous iterations. Finally, the summary generator selects the highest-scoring summary-worthy sentences from the induced tree and ensures that the selected sentences are coherent and cover all relevant aspects of the input document.

The SUMO algorithm generates these subtrees through iterative refinement and builds latent structures using information learned in previous iterations.

First, we decompose the input document D into individual sentences s_i. We then compute a score s_i^* for each sentence s_i, which reflects its importance for generating the summary. Precisely, we use the subsequent formula to calculate the score:

$$s_i^* = \sum_{j=1}^{n} w_j f_j(s_i) \tag{1}$$

where n is the number of features, w_j is the weight of feature j, and $f_j(s_i)$ is the value of feature j on sentence s_i.

Next, we select the highest-scoring sentence as a new root node and add all sentences dependent on it to form a new subtree. We then remove all sentences in this subtree from the document and add them to the summary set S. This process is repeated until all documents have been processed and all relevant subtrees have been added to S.

Finally, we use gradient descent to optimize feature weights and latent structures for generating more accurate, coherent, and diverse summaries. Specifically, we use a loss function L that balances coherence and diversity across documents and subtrees:

$$L = \sum_{i=1}^{m} \alpha_i L_i + \beta L_{div} \tag{2}$$

where α_i is the weight of document i, L_i is its loss function, β is a balancing factor, and L_{div} is the diversity loss function across all subtrees in S. We use the following formula to calculate L_{div}:

$$L_{div} = \sum_{i=1}^{k} \sum_{j=i+1}^{k} \frac{1}{d(T_i, T_j)} \tag{3}$$

where k is the number of subtrees in S, T_i and T_j are the i-th and j-th subtrees, and $d(T_i, T_j)$ is the distance between them.

One of the key advantages of this module is its ability to capture complex relationships between sentences in an input document. By inducing multi-root dependency trees, this approach can identify not only which sentences are most important for summarization but also how they relate to each other. This allows for more informative summaries that capture all important aspects of an input document while still being concise and easy to read.

3.2 Generative Model Based on PEGASUS

The generative model built upon the PEGASUS is a sequence-to-sequence architecture with gap-sentences generation as a pre-training objective tailored for abstractive text summarization. This technique involves the pre-training of expansive Transformer-based encoder-decoder models using extensive text datasets, all guided by a novel self-supervised objective.

The module architecture is based on a standard Transformer encoder-decoder. The encoder processes the input text, producing a sequence of hidden states that the decoder subsequently utilizes to produce the summary. The pre-training objective of PEGASUS involves generating gap-sentences, which are sentences that have been removed from the original text and replaced with special tokens. The module is trained to predict these gap-sentences given the surrounding context.

Formally, let $X = \{x_1, x_2, ..., x_n\}$ be an input document consisting of n sentences, and let $Y = \{y_1, y_2, ..., y_m\}$ be its corresponding summary consisting of m sentences. The goal of the PEGASUS algorithm is to learn a conditional probability distribution $p(Y|X)$ that generates a summary given an input document. This distribution can be factorized as follows:

$$p(Y|X) = \prod_{i=1}^{m} p(y_i|y_{<i}, X) \tag{4}$$

where $y_{<i}$ denotes the previously generated summary sentences.

To pre-train the model using gap-sentences generation, we first randomly select some sentences from the input document and replace them with special tokens. We then use this modified document as input and train the model to generate the missing sentences given the remaining context. The objective function used for pre-training is the negative log-likelihood of the ground-truth missing sentences:

$$\mathcal{L}_{pre} = -\sum_{i=1}^{k} \log p(y_i^*|y_{<i}^*, X) \tag{5}$$

where y_i^* denotes the ground-truth missing sentence and k is the number of missing sentences.

After pre-training, the model is fine-tuned on a specific summarization task using supervised learning. During fine-tuning, we use a similar objective function as in pre-training, but with the ground-truth summary sentences as targets:

$$\mathcal{L}_{fine} = -\sum_{i=1}^{m} \log p(y_i|y_{<i}, X) \tag{6}$$

where y_i denotes the ground-truth summary sentence.

The copy mechanism allows for direct copying of certain segments from the original text to the generated summary, thereby avoiding simple summarization. By preserving more original text information and avoiding information loss, especially for rare or non-existent words, the use of copy mechanism enhances the completeness and accuracy of the generated summary. In the decoder, a new label distribution is added for each token, as shown below:

$$p\left(y_t, z_t \mid y_{\rhd t}, x\right) = p\left(y_t \mid y_{\langle t}, x\right) \cdot p\left(z_t \mid y_{\langle t}, x\right) \tag{7}$$

where B represents the token copied from the source text, I represents the token copied from the source text and forming a continuous segment with the previous tokens, and O represents the token not copied from the source text.

During the training phase, the model adds a sequence prediction task, and by calculating the longest common subsequence between the original text and the summary, corresponding BIO tags are obtained. During the prediction phase, for each step, the label Zt is predicted first. If Zt is O, no further processing is needed. If Zt is B, it means that words that have never appeared in the original text need to be masked. If Zt is I, it means that all corresponding n-grams unrelated to the original text need to be masked.

4   Experimental Evaluation

4.1   Experimental Setting

Data. In order to verify the effectiveness of our proposed approach, we conduct extensive experiments on two datasets, which are TTNews [15] and Scholat-News[1]. As our model is designed for medium to long text, we controlled the length and clarity of the TTnews and SchoaltNews datasets by filtering out all articles with a length of less than 800 words. The datasets used in the experimentation are listed in Table 1.

Baselines

- LEAD is a classic method for text summarization that relies on the assumption that the first few sentences of a document contain the most important information. It involves selecting the first N sentences of a document as the summary, where N is a pre-defined number.

---

[1] https://www.scholat.com/.

Table 1. Data statistics.

Datasets		docs	avgArt	maxArt	minArt	avgSum	maxSum	minSum
TTNews	train	21359	1916	22312	800	42	78	21
	test	862	1909	17204	800	35	65	21
SchoaltNews	train	5988	3024	202089	800	41	122	2
	test	1216	1612	80831	800	45	98	2

- BertSum [5] is a text summarization approach that makes use of the Bidirectional Encoder Representations from Transformers (BERT) model. It introduces a document-level encoder rooted in BERT, capable of encoding an entire document and obtain deriving sentence representations. The extractive model is constructed atop this encoder by layering multiple inter-sentence Transformer layers, effectively capturing document-level attributes for sentence extraction.
- LongformerSum [16] is a text summarization method that leverages the Longformer model, which is designed to handle long sequences, for generating summaries of text documents.
- PGN [17] is a sequence-to-sequence framework that employs a soft attention distribution to generate an output sequence comprising elements sourced from the input document. PGN combines extractive and abstractive summarization methods by allowing the model to copy words directly from the source document while also generating new words to form a coherent summary.
- UniLM [8] is a pre-trained language model adaptable for tasks involving both comprehension and generation of natural language. The model is pre-trained via three distinct forms of language modeling tasks, making use of a common Transformer network alongside targeted self-attention masks, all strategically employed to regulate contextual understanding.
- BART [18] is a pre-training technique tailored for sequence-to-sequence models, seamlessly merging bidirectional and auto-regressive transformers. It uses a denoising autoencoder architecture, where text is corrupted with an arbitrary noising function and a sequence-to-sequence model is learned to reconstruct the original text.
- SUMO [14] is an extractive text summarization method that generates a summary by identifying key sentences in a document and organizing them into multiple subtrees. Each subtree consists of one or more root nodes, which are sentences relevant to the summary.
- PEGASUS [9] is a pre-training algorithm for abstractive text summarization that uses gap-sentences generation as a self-supervised objective. The model is trained to generate missing sentences given the remaining context, which allows it to capture the salient information in the input document. During fine-tuning, the model is optimized to generate a summary that captures the most important information in the input document.

Evaluations. ROUGE-N and ROUGE-L are two commonly used evaluation metrics for measuring the quality of text generation, such as machine translation, automatic summarization, question answering, and so on. They both compare the model-generated output with reference answers and calculate corresponding scores. In this paper, we use ROUGE-1, ROUGE-2, and ROUGE-L as evaluation metrics, where a higher score indicates higher quality of the generated text.

Environment and Parameter. The computer used in this study is equipped with an Intel(R) Xeon(R) Gold 5218 CPU @ 2.30 GHz and 256 GB of memory, with a Tesla V100 GPU with 32 GB of memory. Pycharm 2022 was used as the compiler, and Pytorch was adopted as the deep learning framework, with Python version 3.7. Experimental analysis and comparison were conducted using third-party libraries, including jieba2, bert4torch3, and Fengshenbang-LM4.

In the extraction model based on SUMO, the vocabulary size is 30,000, the maximum number of sentences is 200, the batch size is 256, the learning rate is 0.1, and the size of both EMD and hidden layers is 128. The transformer layers used to obtain sentence representations are set to 3, and the model is trained for 5 iterations.

In the generative model based on PEGASUS, we use the Chinese version of PEGASUS-BASE as the pre-trained model and adopt the Adam optimizer with a learning rate of 2e−5. The batch size for training is 32, the epoch is 10, and the beam search width is 3. The maximum length of the generated summary is set to 90.

4.2   Result Analysis

As shown in Table 2 and Table 3, our model outperforms other models in terms of ROUGE-1, ROUGE-2, and ROUGE-L evaluation metrics on the TTnews and ScholatNews datasets.

Compared to the LEAD algorithm, BertSum demonstrates better performance, indicating that using Bert for summarization can greatly improve extraction accuracy. When comparing LongformerSum and BertSum, results show that replacing Bert with the Longformer model leads to improvement in evaluation metrics on two different long-text datasets. This is mainly due to the fact that Bert only retains the first 512 tokens, while Longformer can allow input with up to 4096 tokens, allowing the model to obtain more information. The results of LongformerSum and SUMO models demonstrate that treating extractive text summarization as a tree induction problem can produce results comparable to methods that use large-scale pre-trained models. Furthermore, the SUMO model outperforms LongformerSum in terms of training time and parameter size.

The performance of generative summarization models is excellent on text summarization tasks. Compared to BART and UniLM, the PEGASUS model has better performance, which demonstrates that specialized pre-training models may be more effective for specific tasks than general pre-training models. Therefore, the generation phase of our proposed model is optimized based on PEGA-SUS. Comparison of PEGASUS and our proposed model shows that SUMOPE

Table 2. Performance evaluation based on TTNews dataset.

Method	ROUGE-1	ROUGE-2	ROUGE-L
LEAD	21.63	7.2	16.97
BertSum	32.53	18.01	30.7
LongformerSum	37.45	21.76	31.91
PGN	30.91	16.88	28.47
UniLM	46.52	33.12	42.34
BART	54.8	39.14	51.33
SUMO	37.23	22.14	30.96
PEGASUS	55.54	41.34	52.81
SUMOPE	**56.19**	**43.32**	**54.59**

Table 3. Performance evaluation based on ScholatNews dataset.

Method	ROUGE-1	ROUGE-2	ROUGE-L
LEAD	37.27	26.44	32.67
BertSum	35.63	24.6	33.72
LongformerSum	38.33	27.4	32.73
PGN	36.36	28.75	32.89
UniLM	51.68	42.12	46.91
BART	60.59	54.05	57.92
SUMO	38.1	27.61	33.22
PEGASUS	63.23	58.37	62.06
SUMOPE	**65.39**	**59.49**	**63.23**

has improved to some extent in the three evaluation dimensions of Rouge-1, Rouge-2, and Rouge-L. Although most of the key information in the article is concentrated in the first 512 words, some critical information still exists in the second half of the article (contrasts and conclusions). Since our proposed model first preserves the key information of the article through extraction, the score of the generated summary will be higher.

Considering the limitations of solely using ROUGE metrics to evaluate the quality of generated summaries, as it cannot comprehensively assess whether the summary corresponds to the main theme of the article, this chapter adopts a subjective approach to evaluate the fidelity and fluency of the generated summaries. The study invites 20 students to subjectively evaluate the generated summaries based on fidelity and fluency, among other criteria. The participants are required to rate the summaries generated by different models for 9 randomly selected articles from the TTNEWS and ScholatNews test sets.

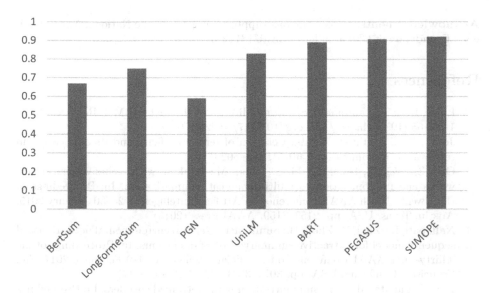

Fig. 2. The human evaluation results of the model are presented.

Based on the data presented in Fig. 2 it can be concluded that the summaries generated by the proposed model in the NLPCC and ScholatNews datasets are more in line with the main content of the articles, with a higher degree of matching with the standard summaries, more complete retention of key information, smoother semantic flow, and lower redundancy. Therefore, to some extent, it validates the effectiveness of the proposed model in generating high-quality summaries.

5 Conclusion

In this paper, we have proposed a enhanced hierarchical summarization model for long texts, SUMOPE, which combines extractive and abstractive methods to deal with long texts. The first stage of our model, called SUMO, selects key sentences from the input text to form a bridging document, and the second stage uses an abstractive method based on PEGASUS with a copy mechanism to generate the final summary. Our model has been evaluated on two datasets and has been shown to outperform the state-of-the-art methods in terms of ROUGE scores and human evaluation.

In future work, we plan to investigate the effectiveness of our model in other languages and domains. Additionally, we aim to explore the possibility of improving the performance of the model by incorporating other advanced techniques, such as reinforcement learning, to further enhance the selection and refinement of sentences. Overall, we believe that the proposed model has significant potential for improving the efficiency and accuracy of text summarization in various applications.

318   C. Chang et al.

Acknowledgements. This work was supported in part by the National Natural Science Foundation of China under Grant U1811263.

References

1. Luhn, H.P.: The automatic creation of literature abstracts. IBM J. Res. Dev. **2**(2), 159–165 (1958). https://doi.org/10.1147/rd.22.0159
2. Jones, K.S.: A statistical interpretation of term specificity and its application in retrieval. J. Documentation **60**(5), 493–502 (2004)
3. Cao, Z., Wei, F., Dong, L., Li, S., Zhou, M.: Ranking with recursive neural networks and its application to multi-document summarization. In: Proceedings of the Twenty-Ninth AAAI Conference on Artificial Intelligence, 25–30 January 2015, Austin, Texas, USA, pp. 2153–2159. AAAI Press (2015)
4. Nallapati, R., Zhai, F., Zhou, B.: SummaRuNNer: a recurrent neural network based sequence model for extractive summarization of documents. In: Proceedings of the Thirty-First AAAI Conference on Artificial Intelligence, 4–9 February 2017, San Francisco, California, USA, pp. 3075–3081. AAAI Press (2017)
5. Liu, Y., Lapata, M.: Text summarization with pretrained encoders. In: Proceedings of the 2019 Conference on Empirical Methods in Natural Language Processing and the 9th International Joint Conference on Natural Language Processing, EMNLP-IJCNLP 2019, Hong Kong, China, 3–7 November 2019, pp. 3728–3738. Association for Computational Linguistics (2019)
6. Huang, Y.J., Kurohashi, S.: Extractive summarization considering discourse and coreference relations based on heterogeneous graph. In: Proceedings of the 16th Conference of the European Chapter of the Association for Computational Linguistics: Main Volume, EACL 2021, Online, 19–23 April 2021, pp. 3046–3052. Association for Computational Linguistics (2021)
7. Ma, C., Zhang, W.E., Guo, M., Wang, H., Sheng, Q.Z.: Multi-document summarization via deep learning techniques: a survey. ACM Comput. Surv. **55**(5), 102:1–102:37 (2023). https://doi.org/10.1145/3529754
8. Dong, L., et al.: Unified language model pre-training for natural language understanding and generation. In: Advances in Neural Information Processing Systems 32: Annual Conference on Neural Information Processing Systems 2019, NeurIPS 2019, Vancouver, BC, Canada, 8–14 December 2019, pp. 13042–13054 (2019)
9. Zhang, J., Zhao, Y., Saleh, M., Liu, P.J.: PEGASUS: pre-training with extracted gap-sentences for abstractive summarization. In: Proceedings of the 37th International Conference on Machine Learning, ICML 2020, 13–18 July 2020, Virtual Event. Proceedings of Machine Learning Research, vol. 119, pp. 11328–11339. PMLR (2020)
10. Liu, Y., Liu, P., Radev, D.R., Neubig, G.: BRIO: bringing order to abstractive summarization. In: Proceedings of the 60th Annual Meeting of the Association for Computational Linguistics (Volume 1: Long Papers), ACL 2022, Dublin, Ireland, 22–27 May 2022, pp. 2890–2903. Association for Computational Linguistics (2022)
11. Chen, Y., Bansal, M.: Fast abstractive summarization with reinforce-selected sentence rewriting. In: Proceedings of the 56th Annual Meeting of the Association for Computational Linguistics, ACL 2018, Melbourne, Australia, 15–20 July 2018, Volume 1: Long Papers, pp. 675–686. Association for Computational Linguistics (2018)

12. Li, H., et al.: EASE: extractive-abstractive summarization end-to-end using the information bottleneck principle. In: Proceedings of the Third Workshop on New Frontiers in Summarization, pp. 85–95 (2021)
13. Xiong, Y., Racharak, T., Nguyen, M.L.: Extractive elementary discourse units for improving abstractive summarization. In: SIGIR 2022: The 45th International ACM SIGIR Conference on Research and Development in Information Retrieval, Madrid, Spain, 11–15 July 2022, pp. 2675–2679. ACM (2022)
14. Liu, Y., Titov, I., Lapata, M.: Single document summarization as tree induction. In: Proceedings of the 2019 Conference of the North American Chapter of the Association for Computational Linguistics: Human Language Technologies, NAACL-HLT 2019, Minneapolis, MN, USA, 2–7 June 2019, Volume 1 (Long and Short Papers), pp. 1745–1755. Association for Computational Linguistics (2019)
15. Hua, L., Wan, X., Li, L.: Overview of the NLPCC 2017 shared task: single document summarization. In: Huang, X., Jiang, J., Zhao, D., Feng, Y., Hong, Yu. (eds.) NLPCC 2017. LNCS (LNAI), vol. 10619, pp. 942–947. Springer, Cham (2018). https://doi.org/10.1007/978-3-319-73618-1_84
16. Wei, F., Yang, J., Mao, Q., Qin, H., Dabrowski, A.: An empirical comparison of distilBERT, longformer and logistic regression for predictive coding. In: IEEE International Conference on Big Data, Big Data 2022, Osaka, Japan, 17–20 December 2022, pp. 3336–3340. IEEE (2022)
17. See, A., Liu, P.J., Manning, C.D.: Get to the point: summarization with pointer-generator networks. In: Proceedings of the 55th Annual Meeting of the Association for Computational Linguistics, ACL 2017, Vancouver, Canada, 30 July–4 August, Volume 1: Long Papers, pp. 1073–1083. Association for Computational Linguistics (2017)
18. Lewis, M., et al.: BART: denoising sequence-to-sequence pre-training for natural language generation, translation, and comprehension. In: Proceedings of the 58th Annual Meeting of the Association for Computational Linguistics, pp. 7871–7880 (2020)

An Extractive Automatic Summarization Method for Chinese Long Text

Jizhao Zhu[1,2], Wenyu Duan[1(✉)], Naitong Yu[2], Xinlong Pan[3], and Chunlong Fan[1]

[1] Shenyang Aerospace University, Shenyang, China
qddwy15@126.com
[2] Key Laboratory of Network Data Science and Technology, Institute of Computing Technology, Beijing, China
[3] Naval Aviation University, Yantai, China

Abstract. The extractive automatic summarization method is capable of quickly and efficiently generating summaries through the steps of scoring, extracting and eliminating redundant sentences. Currently, most extractive methods utilize deep learning technology to treat automatic summarization as a binary classification task. However, the effectiveness of automatic summarization for Chinese long text is limited by the maximum input length of the model, and it requires a large amount of training data. This paper proposes an unsupervised extractive automatic summarization method which solves the long text encoding problem by incorporating contextual semantics into sentence-level encoding. Firstly, we obtain the semantic representation of sentences by using the RoBERTa model. Secondly, we propose an improved k-Means algorithm to cluster sentence representations. By defining sparse and dense clusters, we improve the accuracy of summary sentence selection while preserving maximum semantic information from the original text. Experimental results on the CAIL2020 dataset show that our method outperforms baselines by 6.64/7.68/7.14% respectively on ROUGE-1/2/L. Moreover, we further enhance the automatic summarization results by 4.5/5.36/3.24% by adding domain rules tailored to the dataset's characteristics.

Keywords: automatic summarization · extractive · Chinese long text · RoBERTa

1 Introduction

With the rapid development of Internet technology, the explosion of all kinds of content-rich information has caused problems related to information overload. Text information, as an important source of resources for people, contains rich content, but how to efficiently obtain and use effective information from the huge amount of text has become a challenge that has long plagued us. Automatic text summarization technology can extract the key information from a large amount of text and obtain a summary that expresses the main content of the original text, which can help people to quickly filter and capture important information.

Automatic text summarization methods can be categorized as extractive and abstractive summarization according to the method of extraction. Extractive summarization

X. Yang et al. (Eds.): ADMA 2023, LNAI 14177, pp. 320–333, 2023.
https://doi.org/10.1007/978-3-031-46664-9_22

takes a limited number of sentences from the original text without any modification and arrange them in the order in which they appear in the original text. The current mainstream extractive summarization method scores the importance of each sentence in the original text, then ranks the sentences according to their scores, and finally selects the top sentences with higher scores as the summary. Another approach treats extractive summarization as a binary classification task, using deep learning techniques to predict sentence labels, with label 1 indicating the sentence should be included in the summary and label 0 indicating the opposite, finally, all sentences with label 1 are summarized in the order in which they appear in the original text. Abstractive summarization uses linguistic generation models to obtain summaries based on semantic understanding of the original text, the content of which does not consist directly of sentences from the original text. With the development of deep learning techniques, in recent years abstractive summarization tasks have been mainly based on sequence-to-sequence neural architectures, generating summaries by training encoder-decoder models. Abstractive summarization is based on the understanding of the global semantics of the original text, which is concise and similar to the process of human summary writing, but limited by the processing power of language generation models and the reliance of deep learning on large-scale tagging data, abstractive summarization still suffers from semantic errors and incorrect sequencing, and is difficult to expand in the application field. The content of the extractive summarization is derived from the original text, without the aforementioned issues, and is more reliable and cost-effective compared to the abstractive summarization. Therefore, this paper investigates extractive automatic text summarization.

At present, extractive summarization research work mainly uses supervised learning. Nallapati et al. [1] constructed an extractive summarization model based on gated recurrent neural network, and proposed a mechanism for extractive summarization using a abstractive summarization training pattern for the problem of insufficient training data. However, calculating the summary sentence probability based on the model, problems such as domain adaptation occur when testing on different datasets. Yang et al. [2] designed an extractive summarization model with hierarchical representation using Transformer and Longformer as encoding layers, which can support input text length up to 4096 characters, making it possible for the model to encode long text directly. But if the length of the input text exceeds the model's maximum supported length, it still needs to be truncated or discarded. Gu et al. [3] reduced redundancy by modeling extractive summarization as a multi-step iterative process of scoring and selecting sentences using reinforcement learning, while considering historical information. Although the above approaches achieve relatively promising results on extractive summarization, they require a large amount of labelled contexts for training the models, thus the transferability of the models tends to be weak.

In this paper, we propose an unsupervised approach to extractive automatic text summarization. Based on the pre-training model RoBERTa, we firstly obtain the sentence encoding representation incorporating contextual semantic information, and use sentence-level encoding to make the model capable of encoding long texts; then, we use the improved k-Means clustering algorithm to cluster the sentence representation, improve the accuracy of summary sentence selection and reduce summary redundancy by defining sparse and dense clusters, while preserving the original text information as much as possible. The model in this paper enhances the transferability by combining pre-training techniques with unsupervised learning. Experimental results on publicly available datasets show that the method in this paper improved 6.64/7.68/7.14% on ROUGE-1/2/L compared to the baseline. In addition, the performance of summary extraction is further improved by 4.5/5.36/3.24% according to the domain characteristics of the dataset by adding domain rule knowledge to enhance the text summary generation capability of the model. The approach can be extended to other verticals as a general solution.

2 Related Work

Early research on automatic text summarization started with the classification of important sentences in the text, primarily by using simple statistical methods to make simple statistics on word frequency, topic words, word position and other information, thus judging the importance of the sentences. With the development of automatic text summarization technology, simple statistical methods no longer satisfied the practical needs, researchers began to use external resources to assist in text summarization, such as TFIDF, lexical chains, etc., but these methods are still at a relatively light level for the analysis of text semantics and structure. In recent years, machine learning has provided new ideas for text summarization tasks, especially the application of supervised learning approaches in the field of natural language processing. With the proposal of architectures such as RNN, LSTM, CRF, and Transformer [4, 5]. Cho et al. [6] proposed a sequence-to-sequence model with encoder and decoder structures that utilize local and global information of input sequences to obtain the output [7]. Rush et al. [8] constructed a fully data-driven, extractive summary generation system combining neurolinguistic models and other encoders in this context, and applied the above models to the text summarization task for the first time. The research on text summarization is primarily based on the relationship between textual contexts to learn document representation. Cheng et al. [9] proposed a data-driven approach based on neural networks and continuous sentence features to compose a summary extraction framework by building hierarchical document encoders and extractors based on attention mechanisms. This framework can develop different classes of summary models and can propose sentences and words from a large number of documents, which still achieve relatively good results without any linguistic annotations. Effective text encoding is also the key focus of the text summarization task since the quality of text encoding is crucial to the summary extraction results. Mikolov et al. [10] used simple neural networks without hidden layers to train high-quality word vectors to maximize the accuracy of vector operations in order to learn high-quality vector representations from a large dataset that keeps a linear pattern between words. Pennington et al. [11] proposed a new word vector learning method using global statistical and local contextual information to construct word co-occurrence matrices based on corpus and then obtain word vectors. Following the achievement of deep neural networks with pre-trained models in natural language processing tasks, the approaches of deep learning are now played an important role in automatic text summarization tasks. Deep learning models not only obtain contextual and semantic information of the learned text corpus, but also embed text structure, word level, focus word, and topic information into the representation vector of the text. Zhang et al. [12] presented hierarchical BERT for document encoding, which was trained employing unlabeled data, abstracted the model to a sentence prediction task, and then classified the sentences. Liu et al. [13] proposed BERT-based document-level encoder, which overlays several inter-sentence Transformer layers to capture document-level features of sentences on top of extractive and abstractive encoders. A new fine tuning model is proposed for abstractive summarization, and finally extractive and abstractive methods are combined to further improve the summarization quality through two-stage strategy. Reinforcement learning based approaches are gradually proposed, since most of the work on extractive automatic text summarization in recent years has been built by combining different neural networks to build automatic text summarization models. Narayan et al. [14] proposed a novel training algorithm by conceptualizing the extractive summarization task as sentence ranking task. The model consists of sentence encoder, document encoder and sentence extractor, which uses reinforcement learning combined with maximum likelihood cross-entropy loss to directly optimize the evaluation metrics. As current extractive summarization models usually only extract sentences that meet the requirements, but ignore sentence-to-sentence coherence. Wu et al. [15] proposed neural coherence model using reinforcement learning algorithm for the above problem, to improve the coherence and readability of sentences with model output and evaluation index ROUGE as reward for training. Chen et al. [16] proposed a sentence-level reinforcement learning model for obtaining summaries in response to the problems of redundancy as well as inaccuracy of longer texts as summaries, which enables combining extractive and abstractive summarization without the need to annotate the key sentence corpus.

Extractive summarization currently suffers from problems such as redundancy and semantic incoherence [17]. Since the extractive summarization is composed of sentences of higher importance, it is obvious that these sentences of higher importance are closer to the topic and indicate content similar to the central theme of the original text, thus a higher probability of repeated expressions, which makes the summary redundant in terms of content. The common way is to rank the sentences after scoring or binary classification, according to the order in which the summary sentences appear in the original text, but the adjacent sentences in the summary may be relatively far away from each other in the original text, which results in semantic incoherence. Therefore, the extractive automatic text summarization method still needs further research and improvement.

3   Methodology

This paper proposes an extractive automatic text summarization method based on the pre-training model RoBERTa [18] and the improved K-Means clustering algorithm for Chinese long text. Encoding module obtains sentence encoding representation set incorporating contextual semantic information by utilizing RoBERTa model. Improved Clustering module treats sentence-level vector set obtained in the previous step as input, clusters sentence encoding representation set, improves the accuracy of summary extraction by defining sparse and dense clusters, and finally outputs a set of summary sentences that meet the requirements and are ordered. The architecture of our model is illustrated in Fig. 1.

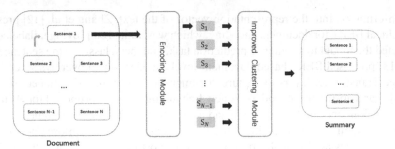

Fig. 1. Overall architecture of the model

3.1 Encoding Module

This paper utilizes the pre-training model RoBERTa for sentence-level encoding of documents, which is a deep bi-directional language representation model based on Transformer and belongs to the modified version of the pre-training model BERT [19]. RoBERTa has made main optimizations to BERT as follows: (a) Dynamic Masking. BERT generated the mask matrix during data preprocessing, thus only one random mask was performed for each sample, but the improved masking dynamically generates a new mask method each time when a sequence is fed into the model. (b) Cancel NSP task. Previous work argued that NSP loss was an important factor in training the original BERT model in which removing the NSP task would impair the model performance. However, RoBERTa challenged this conclusion and demonstrated after a series of comparative experiments that the overall model performance was improved after the NSP task was removed. (c) Expand batch size. Several previous works have shown that boosting the model batch size can improve the performance of the optimization as well as the final task when the learning rate is appropriately increased, and BERT is also applicable to large batch training, so it has been experimentally demonstrated that scaling up the batch size significantly improves the performance of the model. (d) More datasets. Further scaling the training data to 160GB, which is 10 times the size of the BERT. Therefore, we adopted the pre-trained model RoBERTa in the encoding module.

For the extractive summarization task, if the document consists of N sentences, denoted as $\{sent_1, sent_2, ..., sent_N\}$, where $sent_i$ is labeled 1 denotes that this sentence should be included in the summary and the label 0 denotes the opposite. For as shown in Fig. 1, N sentences are input to the encoding layer sequentially, and to ensure the reliability of the embedding, we make character substitution for special characters as well as numbers in the sentences to prevent noise signals from affecting the encoding results. Taking a sentence encoding process as an example, assume that $D = \{s_1, s_2, ..., s_N\}$ is represented as a document representation containing N sentences, where $s_i = \{w_1, w_2, ..., w_L\}$ is the semantic representation of the *i-th* sentence in the document and is also the input to the RoBERTa encoding layer, where L denotes the character length of the sentence. The input of the RoBERTa encoding layer is the numerical sum of the word embedding $\{E_{w_1}, E_{w_2}, ..., E_{w_L}\}$ and the position embedding $\{E_{p1}, E_{p2}, ..., E_{pL}\}$, and a feature vector representation $S_i = \{T_{w_1}, T_{w_2}, ..., T_{w_L}\}$ is obtained by the encoding layer incorporating the contextual semantic information, as shown in Fig. 2. The N sentences in the document are input to the RoBERTa encoding layer in turn, and the semantic representation $T = \{S_1, S_2, ..., S_N\}$ of all sentences is obtained, and this output will be used as the input for the next stage of the model.

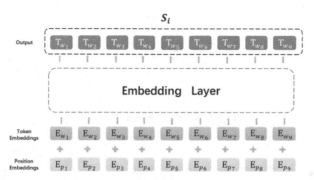

Fig. 2. RoBERTa Embedding Layer.

3.2 Improved Clustering Module

Text clustering algorithms usually calculate the similarity between nodes. The main idea is to divide the coded representation into several clusters so that the text within the cluster has as much similarity as possible and the text outside the cluster has as little similarity as possible. Text clustering algorithms include partitioning methods, distribution-based methods, hierarchical methods, density-based methods, graph-based methods, etc. The partitioning methods are used to classify the sample data into specific categories by continuously iteratively updating the data center. The distribution-based methods determine data classes based on different distributions of data. The hierarchical methods cluster data by constructing a hierarchy of data. The density-based clustering methods divide data in high-density areas into the same class. The graph-based methods achieve clustering by using a similarity matrix to feature decomposition of the data. Traditional clustering algorithms all have their own advantages and shortcomings, so it is important to choose the right clustering algorithm according to the requirements in different situations.

Traditional sentence representations employ word co-occurrence, which causes the loss of similarity of many other semantically related sentences, especially those related to a given topic, because of their lack of word co-occurrence. This paper adopts RoBERTa for the encoding procedure, which obtains the semantic representation of the sentence containing both its own semantic information and contextual information, thus the representation is richer. The sentence-level representation vector of the document is considered as the input to the text clustering module after setting the extraction percentage *Ratio* $\in (0, 1)$, denoted as $T = \{S_1, S_2, ..., S_N\}$. The k-Means clustering algorithm separates the given N samples into r classes, which is shown in Fig. 3. We also applied other algorithms for sentence encoding clustering, such as Gaussian Mixture Model, and the

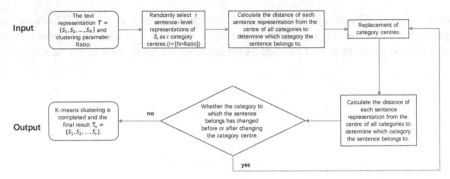

Fig. 3. Clustering flow diagram.

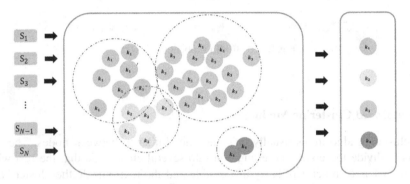

Fig. 4. Clustering process diagram.

experimental results show that their results are no better than k-Means clustering algorithm in this paper's approach. The clustering algorithm pursues the principle of less variance within clusters and more difference outside clusters, where sentence representation vectors in the same cluster are similar. To effectively reduce the extractive summarization redundancy, the closest sentence representation to the center of mass in each cluster is selected as the summary sentence for consideration, denoted as $T_s = \{S_1, S_2, \ldots, S_r\}$, where $r = [N \times Ratio]$ means the largest integer no than $N \times Ratio$. As shown in the example of Fig. 4, there are r sentences selected as summary sentence, here $r = 4$. The Manhattan distance is used to measure the distance from the sample point to the center of mass of the cluster, which is calculated as shown in Eq. 1, where x is sample point, μ is the centroid of the cluster, n is the number of features in the sample point, and i is each of the features that make up the sample point.

$$d(x, \mu) = \sum_{i=1}^{n} (|x_i - \mu_i|) \tag{1}$$

As explained in the previous section, extractive text summarization methods generally suffer from the deficiency of redundancy in summary content. Miller et al. proposed method selects the closest sentence representation in each cluster to the center of mass as the summary sentence to be selected in the clustering module to solve the problem of sentence redundancy in extractive summary selection, but this approach still has short-comings [22]. It is clearly unbalanced to select only the sentence closest to the center of mass in each cluster as the content representation of that cluster, irrespective of the number of nodes in the cluster. In addition, one sentence is selected in the cluster even if there are few nodes in the cluster and it deviates from the original clustering center, which leads to irrelevant content being selected as the summary sentence. Therefore, we define sparse and dense clusters for the above two situations. We define clusters with less than c_θ nodes in the cluster as sparse clusters and clusters with more than 50% of the total number of nodes in the original text as dense clusters. The documents represent $T = \{S_1, S_2, ..., S_N\}$ are clustered into k groups as shown in Fig. 5, where $c_j (j = 1, 2, \ldots, k)$ denotes the k clustering centers and c^* indicates the semantic center of the document.

For sparse clusters, we implement the following strategy:

$$l_{c_j} = \begin{cases} 0, & dis(c_j, c^*) > d_\theta \text{ and } count(c_j) < c_\theta \\ 1, & otherwise \end{cases} \tag{2}$$

$$dis(x, \mu) = \sqrt{\sum_{i=1}^{n} (x_i - \mu_i)^2} \tag{3}$$

$$d_\theta = \frac{\alpha}{j} \sum_{i=1}^{j} dis(c_i, c^*) \tag{4}$$

As shown in Eq. 2, $l_{c_j} = 1$ indicates that the sentence is selected from that cluster as the summary sentence after completion of clustering, and $l_{c_j} = 0$ indicates that the sentence is excluded from that cluster. The distance from the center of mass c_j of the j-th cluster to the semantic center of the text c^* is represented by $dis(c_j, c^*)$, and the Euclidean Metric is used to represent this distance, which is calculated as shown in Eq. 3 and 4. $count(c_j)$ indicates the number of nodes in the j-th cluster, c_θ and α indicate the hyperparameter, and $c_\theta = 3$, $\alpha = 3/2$ are set in this paper. The sentence will not be extracted in the cluster when the distance between the center of prime of this sparse cluster and the semantic center of the original text is more than d_θ. The d_θ calculation formula is shown in Eq. 4.

For dense clusters, we adopt the following strategy:

$$Add_{c_j} = argMin(sim(S, S^*)), S \in C_\theta \tag{5}$$

$$sim(S, S^*) = \frac{\sum_{i=1}^{n} (S_i \times S_i^*)}{\sqrt{\sum_{i=1}^{n} S_i^2} \times \sqrt{\sum_{i=1}^{n} (S_i^*)^2}} \tag{6}$$

We consider that only one sentence extracted from a dense cluster is not enough to represent the information of the whole cluster. As shown in Eq. 5, Add_{c_j} indicates the semantic representation vector supplementally extracted in the cluster, which is selected from the set of candidate vectors C_θ, which consists of the three sample points closest to the Euclidean distance from the center of this dense cluster. We select the sentence with the least similarity to the semantic representation of the extracted sentences in the set of candidate vectors as our supplementary extraction, which can reduce the summary redundancy in part. As shown in Eq. 6, S^* denotes the semantic representation of the sentences that have been extracted and $sim(S, S^*)$ denotes the cosine similarity. The improved clustering process is shown in Fig. 5.

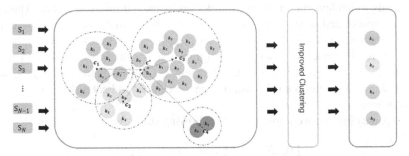

Fig. 5. Improved clustering process diagram.

4   Experiment

4.1   Dataset

This paper uses the CAIL2020 text summarization dataset, which is composed of 4047 labeled civil adjudication documents produced, mainly consisting of body text, expert summary, individual sentence, and labels for whether each sentence should be included in the summary. The dataset is a long text dataset, and its specific statistical information is shown in Table 1, where Avg. doc. and Avg. summ. are statistics at the word level and sentence level for both the original document and the extractive summarization.

Table 1.   An overview of the data set used in this article.

Dataset	Avg. doc. length		Avg. summ. length	
	# of words	# of sent.	# of words	# of sent.
CAIL2020	2629	59	767	12

4.2   Evaluation Standard

This paper applies the *ROUGE* metric as a reference standard for evaluating the quality of abstracts, which is calculated by the primary concept of comparing the extracted summaries of the model with the given reference summaries in the dataset [20]. We take the sentences with label 1 in the dataset and concatenate them sequentially as a reference summary, which judges the summary model performance by counting the number of base units that overlap between it and the model output. The *ROUGE* evaluation criteria consists of a series of evaluation methods, such as *ROUGE-N*, *ROUGE-L*, *ROUGE-S*, *ROUGE-W*, and *ROUGE-SU*. This paper utilizes *ROUGE-1*, *ROUGE-2* and *ROUGE-L* to evaluate the summary quality. The detailed calculation process of *ROUGE-N* is shown in Eq. 7, and *ROUGE-L* is shown in Eq. 8–10.

$$ROUGE - N = \frac{\sum_{S \in \{Ref\}} \sum_{gram_n \in S} Count_{match}(gram_n)}{\sum_{S \in \{Ref\}} \sum_{gram_n \in S} Count(gram_n)} \tag{7}$$

Ref in Eq. 7 is our standard summary, n is the length of the n-gram, $Count_{match}(gram_n)$ is the number of n-gram that appear in both the standard summary and the model-extracted summary, and $Count(gram_n)$ is the number of n-gram that appear in the standard summary. The *ROUGE* also includes three metrics: accuracy P (precision), recall R (recall) and F-value.

$$R_{lcs} = \frac{LCS(X, Y)}{m} \tag{8}$$

$$P_{lcs} = \frac{LCS(X, Y)}{n} \tag{9}$$

$$F_{lcs} = \frac{(1 + \beta^2)R_{lcs}P_{lcs}}{R_{lcs} + \beta^2 P_{lcs}} \tag{10}$$

Equation 8–10 is the detailed formula of *ROUGE-L*. *lcs* refers to the common longest subsequence, where *LCS (X, Y)* refers to the length of the longest common subsequence in the reference summary and the model-extracted summary, m and n denote the character lengths of the reference summary and the model-extracted summary, and β is set to a large number. R_{lcs} and P_{lcs} are the recall and accuracy, and F_{lcs} is the *ROUGE-L*.

4.3 Experimental Design

We implemented experiments on the CAIL2020 text summary dataset to prove the effectiveness of the proposed model in this paper, and the following models were selected for comparison. **Lead-3**: Simply extracting the first three sentences from the original text as a summary has achieved better results in the field of journalism, which even surpasses some deep learning models and usually serves as a baseline model for extractive automatic summarization. **TextRank**: An unsupervised graph-based ranking algorithm, that treats each sentence in the text as a node in the graph, calculates the score of each node by computing the similarity matrix between them, and finally selects a number of sentences with the highest score to form summary [21]. **Baseline**: Miller et al. [22] proposed an extractive lecture-oriented summarization model based on pre-training model BERT with clustering algorithm. **Our Model_BERT**: The encoding module of the proposed model in this paper is replaced with the pre-trained model BERT. **Our Model + Domain Regular**: Based on the model in this paper for extractive automatic summarization, domain rules for datasets are added as an additional extraction method in the extraction process. In order to have an objective comparison, the experiments all extract about 15% of the original text to constitute the target summary, that is, set Ratio = 0.15, to ensure the readability of the summary as much as possible with concise and precise content. We split the dataset into three groups and randomly selected 1000 documents as input each time, and finally took the average Rouge-1/2/L scores of the three extractions as our evaluation metric, as shown in Table 2 for the experimental results.

Table 2. Models performance based on dataset.

Model	Rouge-1	Rouge-2	Rouge-L
Lead-3	29.26	19.04	20.02
TextRank	38.43	23.76	24.05
BaseLine	45.38	28.16	31.01
Our Model_BERT	49.88	33.29	35.79
Our Model	**52.02**	**35.84**	**38.15**
Our Model + Domain Regular	56.52	41.20	42.39

4.4 Result Analysis

As shown in Table 2, the model proposed in this paper achieves the best results on the dataset without considering the addition of domain rule extraction results, and its performance improves by 6.64/7.68/7.14% in the Rouge-1/2/L scores, compared to the baseline model, which shows that the method proposed in this paper is effective. The comparison shows that the proposed model improves the Rouge-1/2/L scores by 2.14/2.55/2.36% compared with Our Model_BERT model, which means that the summarization results of the method proposed in this paper depend on the semantic encoding model, therefore, more suitable semantic encoding representation can obtain higher quality text summarization. The worst result of Lead-3 indicates that this method only works well for data in the news domain. Since the majority of information in the news corpus is concentrated in the head of the text, this method does not have better robustness for data in other domains and is hardly applicable as a general summarization method. The result of TextRank is better than Lead-3 and can be applied to different domain data, however, it is susceptible to factors such as word separation and high-frequency words, so that the summarization result is not satisfactory. The model proposed in this paper is better than TextRank method and can be used for extractive automatic summarization in general domain text. In the application field, we supplemented a domain rule-based approach, which added the rules of that data domain (including but not limited to information on the structure of the data type, common knowledge, etc.) to the extraction process, which further improved the extraction results, improving the Rouge-1/2/L scores by 4.5/5.36/3.24%, thus confirming that this approach can be extended to the process of concrete domain applications.

We visualize the sentence-level embeddings obtained from the encoding module using the t-Distributed Stochastic Neighbor Embedding (t-SNE), which is a machine learning algorithm used for dimensionality reduction and is highly suitable for transform-ing high-dimensional data into two or three dimensions [23]. The document's sentence representation set is clustered with improved k-Means, and then the 768 dimensional sentence-level representation vector is reduced to two dimensions and visualized using the t-SNE algorithm.

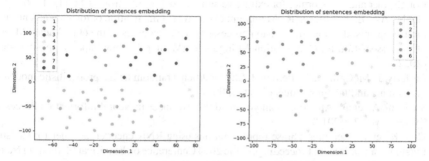

Fig. 6. t-SNE projection of sentences embedding

We find that the clustering distribution of each sentence-level representation vector after dimensionality reduction has obvious boundaries, and a few nodes are distributed in the edge positions. As shown in Fig. 6, we picked two documents during the experiment and visualized their sentence-level embedding distribution. As shown in the left plot of Fig. 6, the visualization results of the fourth class marked in yellow and the eighth class marked in dark gray are similar to the case of the presence of sparse clusters defined in this paper, while the first class marked in green exceeds 50% of the total number of summary points, which is consistent with the supplementary extraction of dense clusters defined in this paper. As shown in the right plot of Fig. 6, the third class of nodes marked as purple is also similar to the case of sparse cluster definition, and all need to be handled according to the corresponding strategies, which all prove the theoretical basis of the proposed method in this paper sideways.

5   Conclusion

In this paper, we propose an unsupervised automatic text summarization method for Chinese long text, utilizing pre-trained models combined with improved clustering algo-rithms for extractive summarization. The experimental results show that the extractive summarization method based on RoBERTa model and improved k-Means clustering algorithm proposed in this paper can obtain good results for extractive summarization of Chinese long text. However, summary extraction results depend on the encoding qual-ity of the text data, the process of feature encoding mapping for long texts still causes certain degree of information loss that affects the final results. In the future, we will consider adding global information to the encoding representation process to improve the extractive summarization performance.

Acknowledgement. This research work is supported by the National Natural Science Foundation of China (Grant No. 61972266).

References

1. Nallapati, R., Zhai, F., Zhou, B.: SummaRuNNer: a recurrent neural network based sequence model for extractive summarization of documents. In: Proceedings of the AAAI Conference on Artificial Intelligence, vol. 31 (2017)
2. Yang, S., Zhang, S., Fang, M., Yang, F., Liu, S.: A hierarchical representation model based on Longformer and transformer for extractive summarization. Electronics 11(11), 1706 (2022)
3. Gu, N., Ash, E., Hahnloser, R.: MemSum: extractive summarization of long documents using multi-step episodic Markov decision processes. In: Proceedings of the 60th Annual Meeting of the Association for Computational Linguistics (Volume 1: Long Papers), pp. 6507–6522 (2022)
4. Lafferty, J., McCallum, A., Pereira, F.C.: Conditional random fields: probabilistic models for segmenting and labeling sequence data (2001)
5. Vaswani, A., et al.: Attention is all you need. In: Advances in Neural Information Processing Systems, vol. 30 (2017)
6. Cho, K., et al.: Learning phrase representations using RNN encoder–decoder for statistical machine translation. In: Proceedings of the 2014 Conference on Empirical Methods in Natural Language Processing (EMNLP), pp. 1724–1734 (2014)
7. Sutskever, I., Vinyals, O., Le, Q.V.: Sequence to sequence learning with neural networks. In: Advances in Neural Information Processing Systems, vol. 27 (2014)
8. Rush, A.M., Chopra, S., Weston, J.: A neural attention model for abstractive sentence summarization. In: Proceedings of the 2015 Conference on Empirical Methods in Natural Language Processing, pp. 379–389 (2015)
9. Cheng, J., Lapata, M.: Neural summarization by extracting sentences and words. In: Proceedings of the 54th Annual Meeting of the Association for Computational Linguistics (Volume 1: Long Papers), pp. 484–494 (2016)
10. Mikolov, T., Chen, K., Corrado, G., Dean, J.: Efficient estimation of word representations in vector space. arXiv preprint arXiv:1301.3781 (2013)
11. Pennington, J., Socher, R., Manning, C.D.: Glove: global vectors for word representation. In: Proceedings of the 2014 Conference on Empirical Methods in Natural Language Processing (EMNLP), pp. 1532–1543 (2014)
12. Zhang, X., Wei, F., Zhou, M.: HiBERT: document level pre-training of hierarchical bidirectional transformers for document summarization. In: Proceedings of the 57th Annual Meeting of the Association for Computational Linguistics, pp. 5059–5069 (2019)
13. Liu, Y., Lapata, M.: Text summarization with pretrained encoders. In: Proceedings of the 2019 Conference on Empirical Methods in Natural Language Processing and the 9th International Joint Conference on Natural Language Processing (EMNLPIJCNLP), pp. 3730–3740 (2019)
14. Narayan, S., Cohen, S.B., Lapata, M.: Ranking sentences for extractive summarization with reinforcement learning. In: Proceedings of the 2018 Conference of the North American Chapter of the Association for Computational Linguistics: Human Language Technologies, Volume 1 (Long Papers), pp. 1747–1759 (2018)
15. Wu, Y., Hu, B.: Learning to extract coherent summary via deep reinforcement learning. In: Proceedings of the AAAI Conference on Artificial Intelligence, vol. 32 (2018)
16. Chen, Y.C., Bansal, M.: Fast abstractive summarization with reinforce-selected sentence rewriting. In: Proceedings of the 56th Annual Meeting of the Association for Computational Linguistics (Volume 1: Long Papers), pp. 675–686 (2018)
17. Liang, X., Li, J., Wu, S., Li, M., Li, Z.: Improving unsupervised extractive summarization by jointly modeling facet and redundancy. IEEE/ACM Trans. Audio Speech Lang. Process. 30, 1546–1557 (2021)

18. Liu, Y., et al.: RoBERTa: a robustly optimized BERT pretraining approach. arXiv preprint arXiv:1907.11692 (2019)
19. Kenton, J.D.M.W.C., Toutanova, L.K.: BERT: pre-training of deep bidirectional transformers for language understanding. In: Proceedings of NAACL-HLT, pp. 4171–4186 (2019)
20. Rouge, L.C.: A package for automatic evaluation of summaries. In: Proceedings of Workshop on Text Summarization of ACL, Spain (2004)
21. Mihalcea, R., Tarau, P.: TextRank: bringing order into text. In: Proceedings of the 2004 Conference on Empirical Methods in Natural Language Processing, pp. 404–411 (2004)
22. Miller, D.: Leveraging bert for extractive text summarization on lectures. arXiv preprint arXiv: 1906.04165 (2019)
23. Van der Maaten, L., Hinton, G.: Visualizing data using t-SNE. J. Mach. Learn. Res. 9(11) (2008)

A Likelihood Probability-Based Online Summarization Ranking Model

Shuhao Yue, Dunhui Yu[✉], and Di Xie

Hubei University, Hubei Wuhan 430062, China
yumhy@163.com

Abstract. Abstractive summarization models frequently utilize decoding strategies, like beam search, which cause issues, such as a vast search space and exposure bias during decoding. To address this problem, we propose an online ranking model for summarization based on the likelihood probability. Our approach aims to establish a correlation between the output probability of a candidate summary and its quality, assigning higher output probabilities to summaries of better quality. Consequently, the ranking model can assess and select the best summary from several candidate summaries by contrasting their output probabilities during the ranking stage, thereby enhancing the performance of the summary model across various metrics. Simultaneously, our model adopts online sampling at each training step and incorporates information from the inference stage into the training process, which effectively mitigates the exposure bias that arises from the inconsistency between the model training and inference processes. Empirical results show that the proposed model performs impressively on the CNNDM and LCSTS public datasets. Compared with the baseline model, our online summarization ranking model yielded a 3.97, 2.55, and 4.12 increase in ROUGE-1, ROUGE-2, and ROUGE-L, respectively. Overall, experiments reaffirm the relevance of the ranking stage and the impact of the model in optimizing other metrics.

Keywords: Beam search · Exposure bias · Ranking model · Contrasting · Online sampling

1 Introduction

Abstractive summarization is a conditional generation task that involves rewriting the original text into a shorter form while retaining its main information. It is typically modeled as a sequence-to-sequence(Seq2Seq) learning problem [1], that is, after compressing the original text information into a dense vector [2], the encoder streamlines its output to guide the autoregressive method utilized by the decoder to gradually summarize the text. Despite large-scale pre-trained language models such as BART [3] and PEGASUS [4] are moving toward human-like performance, models trained under Seq2Seq frameworks and

© The Author(s), under exclusive license to Springer Nature Switzerland AG 2023
X. Yang et al. (Eds.): ADMA 2023, LNAI 14177, pp. 334–346, 2023.
https://doi.org/10.1007/978-3-031-46664-9_23

maximum likelihood estimation(MLE) methods still encounter three primary issues: 1) The model training phase involves a reference summary to guide the decoder, while the inference phase relies on the model's past predictions, leading to exposure bias [5]. 2) Model optimization utilizes a word-level loss function during training, whereas sequence-level evaluation methods, such as ROUGE [6], are employed during testing. 3) Common inference strategies such as Greedy or Beam Search(BS) do not guarantee the summaries with the highest score under these strategies will achieve the highest evaluation score.

To address the aforementioned concerns, we introduces LonRanker (**Likelihood-based Online Ranker**), a likelihood-based online ranking model that includes a ranking stage in the summarization process following the generation phase. Unlike previous encoder-only ranking models, our proposed approach utilizes the encoder-decoder structured BART as the foundational model framework, which offers two key benefits. Firstly, adopting BART as a generative model for scoring candidate summaries aligns more closely with the first-stage summary generation process and enables more optimal utilization of its parameters. Furthermore, BART has the capability of generating candidate summaries during training stage. Rather than traditional ranking models, BART can learn from the latest candidate summary distributions, making it suitable for use by online setting. Moreover, reusing the summaries produced via the model's inference during training helps mitigate the issue of exposure bias. After LonRanker generates a new batch of candidate summaries, we evaluate these summaries using various metrics such as ROUGE [6] or BERTScore [7]. Then, the original text should be given as input to the encoder in LonRanker, while the candidate summary should be used as input to the decoder. Afterward, compute the probability of the candidate summary, which will be used to score the model for that summary with respect to its original text. Subsequently, employing contrastive learning [8], the summary with the highest ROUGE score is chosen as the positive sample, while the remaining candidate summaries form the negative sample. The model maximizes the LonRanker score of the positive sample candidate summary to significantly enhance the output probability of higher quality summaries.

2   Related Work

Natural Language Generation(NLG) has undergone significant development over a long history. In recent years, the widespread use of attention mechanisms has made the Transformer [9] model based on attention the mainstream method in this field. As compared to the traditional RNN model, it possesses the advantages of being able to capture long-distance dependencies and conduct parallel training. Transformer model has always encountered exposure bias issues due to the use of the *teacher-forcing*, and the utilization of the word-level MLE loss to optimize the model is incongruous with its sequence-level evaluation. Bengio et al. [10] proposed the scheduled sampling to resolve the issues discussed earlier, which entails utilizing the model's own predictions to substitute the reference sequence in the later stages of training. Several reinforcement learning approaches [11] take ROUGE or BLUE scores as rewards to guide models, which addresses the non-differentiability challenge of these evaluation metrics. In addition, Wiseman et al. [12] introduced a technique for optimizing beam search throughout the training phase, leading to enhanced performance of inference stage.

Ranking candidate summaries is one of the mainstream methods in recent research in abstarctive summarization. According to Cohen and colleagues' observations [13], the use of the beam search to obtain the candidate summaries with the highest score does not guarantee the selection of the summary with the best ROUGE score. This finding highlights the significance of ranking the candidate summaries to ensure a potentially better quality summary is ultimately produced. If the ranking model is consistently able to choose the optimal, or superior, summary among a plethora of potential summaries, it will have a direct impact on the summarization model's performance regarding ROUGE metrics. SimCLS [14] is a two-stage model that utilizes beam search strategy in the generation phase to generate a set of candidate summaries. In the ranking phase, it utilizes the RoBERTa [15] model to encode the source text and candidate summaries, and uses a loss function based on contrastive learning to rank the candidate summaries. SummaReranker [16] employs diverse decoding techniques to extract features of candidate summaries with advantageous evaluation metrics. A multi-task learning framework, based on a hybrid expert system architecture, is developed for summary reranking, which can jointly optimize several evaluation metrics. JGR [17] utilizes a joint training method which first fixes the generator and optimizes the ranker with ranking loss, and then fixes the ranker and leverages reinforcement learning to update the generator using the ranker's feedback as guidance for optimization. Moreover, aside from ranking-based summarization models, there are additional works that approach summarization from alternative perspectives. GSum [18] optimizes the training of its summarization model by incorporating supplementary guidance in the form of critical sentences, which are predicted by the model itself. ConSum [19] and SeqCo [20] are both sequence-level contrastive learning [8] models for abstarctive summarization, which aim to minimize the distance between the original document and its summary, as well as the distance between the generated summary and the original document during training.

Contrastive learning [8] has been integrated into natural language processing tasks in recent years, yielding remarkable results. In abstarctive summarization, previous works [19–21] typically formulated the reference summary as a positive example, while treating candidate summaries generated by the model as negative examples. This approach involves reducing the distance to positive examples in vector space. Alternatively, it is possible to combine the List-wise ranking loss to align the beam search score and ROUGE score of candidate summaries. Both of these methods provide sequence-level supervision to traditional summarization models and alleviate the problem of exposure bias. A range of decoding techniques can be employed for natural language generation tasks. Greedy strategy, nucleus sampling [22], beam search(BS), and the diverse beam search [23] (DBS) are some examples. In comparison, while the greedy strategy can produce only one summary, nucleus sampling generates candidate summaries of significantly lower quality. The issue with BS is its tendency to generate overly repetitive summaries, resulting in computation and resource wastage during subsequent ranking. Consequently, we adopt DBS for online sampling, generating a fresh round of candidate summaries dynamically.

3   Model

In this section, we will introduce the proposed likelihood-based online ranking model, LonRanker. The initial section provides an overview of fundamental methods utilized for abstarctive summarization, followed by a detailed description of the training process of the proposed model.

The model training process is shown in Fig. 1, which includes a generator and a ranker, both of which share a single Seq2Seq model. LonRanker's training process is as follows: 1) The generator decodes a single original text to generate multiple candidates; 2) The candidates were evaluated, their ROUGE scores were calculated and sorted; 3) Input the original text wit candidates into the ranker together, and get the output probability of the candidate abstract as the evaluation score; 4) By employing contrastive learning, candidates with the highest ROUGE scores are assigned the highest output probabilities.

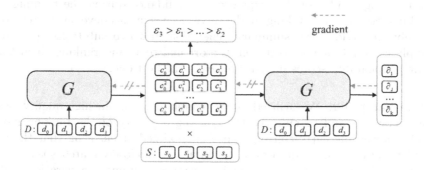

Fig. 1. Model training process

3.1   Abstractive Summarization

Seq2Seq architecture is commonly used in existing abstarctive summarization models, which accepts a lengthy text sequence and outputs a shorter, related sequence following a auto-regression methodology. During training stage, assuming that the original document is given as $D = (d_1, d_2, \ldots, d_{|D|})$ and its reference summary $S = (s_1, s_2, \ldots, S_{|S|})$, then the conditional probability $P(S \mid D; G)$ is estimated as follows:

$$P(S \mid D; G) = \prod_{t=1}^{|S|} p(s_t \mid s_{\leq t}, D; G) \tag{1}$$

where G represents the model parameters, and $S_{\leq t}$ represents all outputs before time step t. Then, the model can be optimized by minimizing the negative log-likelihood loss (NLL):

$$\mathcal{L}^{\mathrm{NLL}}(G) = -\frac{1}{|S|} \sum_{t=1}^{|S|} \log p(S \mid D; G) \tag{2}$$

During inference, the model need start from scratch and output the summary $\hat{S}$ in a autoregressive manner step by step, solely depending on previous prediction $\hat{S}_{<i}$ while forecasting, disregarding $S_{<i}$ which was used during the training process. This discrepancy causes exposure bias, a problem that will be analyzed in-depth further on.

3.2   Training Process of LonRanker

In order to address the aforementioned challenges in abstarctive summarization, we developed a summary ranking model that incorporates information from the inference stage and provides sequence-level guidance during the training process. Furthermore, the ranking model serves as a sequence-level evaluator. More precisely, we divided the summary generation into two sub-tasks: generating multiple candidate summaries from the original text, and ranking them based on their probability scores in order to select the best one.

Online Sampling. We utilized online sampling to obtain high-quality candidate summaries and to enhance the training difficulty of the ranking model, thereby preventing early convergence. It was aimed at ensuring that the difficulty level of the sample data for each training step closely matches the current ability of the ranking model. To achieve a strong summary generation ability, the model G is trained for a specific period of time before the actual training begins, focusing on optimizing the $\mathcal{L}^{\mathrm{NLL}}(G)$ objective.

Before each training step in the subsequent phase of the ranking model training, we fix the parameters of model G and use the DBS decoding strategy to produce several candidate summaries $\mathbb{C} = \{C_1, C_2 \ldots, C_k\}$ based on the source document D where k is number of candidate summaries. For each C_i in the set of candidate summaries $\mathbb{C}$, we calculate its ROUGE score relative to its source document D, and refer to it as $\mathcal{E}(D, C_i)$, which is later abbreviated as $\mathcal{E}_i$. To rank the summaries in the candidate set $\mathbb{C}$ according to their score $\mathcal{E}_i$, we assign the summary with the highest score as the positive sample C^+. Similarly, we designate the lowest j candidates as negative samples C^-. This ranking step is crucial in order to align the probability and quality in the subsequent stages.

Ranking Candidate Summaries. Numerous works have been done on learning to rank, with many methods such as Point-wise, Pair-wise, and List-wise being prominent. Earlier ranking models for abstarctive summarization often employed List-wise or multiple binary classifiers. Considering that the model only needs to output single summary and to preserve training resources, we take contrastive learning to maximize the output probability of the positive sample C^+. Further, optimization solely occurs for the optimal summary as opposed to other summaries, without regarding their relationships with each other.

Given the original document D and candidate summary C_i, they are fed into the encoder and decoder of the model G, respectively. The output probability, which is also the evaluation score $\partial(D, C_i)$ are then calculated as:

$$\partial_i \sim P(D, C_i) = \frac{1}{|C|^\alpha} \sum_{t=1}^{|C|} p(c_t \mid c_{\leq t}, D; G) \tag{3}$$

where α is a penalty term that adjusts the summary length based on the model preference towards short or long summaries. The score of the C^+ model evaluation is recorded as ∂^+, and the score of the C^- model is recorded as ∂^-. By contrastive learning to maximize ∂^+ while minimizing ∂^-, the optimization goal of the model can be expressed as:

$$G = \arg\max_G \log \frac{\exp(\partial^+/\tau)}{\exp(\partial^+/\tau) + \sum_{i=1}^j \exp(\partial_i^-/\tau)} \tag{4}$$

where exp() represents the logarithmic function and τ representing the temperature coefficient. Therefore, the ranking loss of model G can be denoted as:

$$\mathcal{L}^{rank}(G) = -\log \frac{\exp(\partial^+/\tau)}{\exp(\partial^+/\tau) + \sum_{i=1}^j \exp(\partial_i^-/\tau)} \tag{5}$$

Our rank method achieves a time complexity of $\mathcal{O}(n)$, which is a significant improvement over the $\mathcal{O}(n^2)$ time complexity of the List-wise loss function where n denote the number of candidates. Finally, to reach the goal of generating candidate summaries continuously online, it is essential to preserve the model's ability to create summaries to some extent while training the ranking model. Therefore, we combine Eq. 2 and Eq. 5 to obtain the final total loss function $\mathcal{L}$:

$$\mathcal{L} = \mathcal{L}^{rank} + \lambda \mathcal{L}^{NLL} \tag{6}$$

where λ denote the weight for summarization generation learning.

4   Experiment

4.1   Dataset

CNNDM. [24] is currently one of the largest datasets for single-document English summaries available. It includes news articles from the CNN and Daily

Mail websites, along with their corresponding artificially generated summaries. The dataset includes a total of 287,226 training data points, 13,368 validation data points, and 11,490 test data points. The non-anonymous version of the dataset was utilized in outr study.

XSum. [25] is a large-scale dataset comprising around 220,000 English news articles and their corresponding single-sentence summaries spanning diverse topics and domains. In contrast to the CNNDM dataset, the XSum dataset features shorter article length and prioritizes content richness and summary concision.

LCSTS. [26] is a comprehensive dataset of Chinese short text summaries. It comprises genuine Chinese short texts retrieved from Sina Weibo, along with their summaries authored by those who posted them. With over 2.4 million pairs of (short text, summary), the dataset contains over 10,000 pairs with manually annotated relevance scores.

4.2   Implementation

We trained and evaluated PEGASUS [4] and BART [3] as the skeleton models, respectively, on the CNNDM [24], XSum [25] and LCSTS [26] datasets, and initialized them with pre-trained parameters. Prior to commencing formal experiments, the dataset was trained for 0.5 epochs using $\mathcal{L}^{NLL}(G)$ as the optimization objective. In all subsequent experiments, a learning rate of 10^{-4}, a batch size of 128, and the Adam optimizer were seted, with warm-up strategy applied in the first 10% of training steps. To conduct online sampling, we employ the DBS [23] decoding technique whereby k is set to 16 for candidate summary data volume and j is equal to 12 for negative sample count. For quantitative measurement of summary similarity to the target, we utilize ROUGE-1, ROUGE-2, and ROUGE-L [6], as well as their arithmetic mean.

4.3   Result

To assess the effectiveness of the model proposed for abstractive summarization tasks, we will compare LoneRanker with other well-known models. The experimental results for applying LonRanker to different abstarctive summarization models are shown in Table 1. LonRanker-R serves as a model for ranking sentences, and a model that generates summaries using either BART [3] or PEGASUS [4]. LonRanker-G is directly applied to summary generation, serving as a standard Seq2Seq model.

The results suggest that the proposed model in our study outperformed other models in ROUGE-N metrics. Within the CNNDM and XSum dataset, SimCLS [14], SummaReranker [16], and LonRanker-R all act as models for the ranking stage. They employ the same generator, PEGASUS [4], to obtain candidate summaries, and select the final output based on the ranking model's evaluation score. LonRanker-R is unique compared to prior encoder-only models, as it employs the

PEGASUS model with an encoder-decoder structure that closely resembles the generation process of the first stage. Additionally, it is capable of online sampling and immediately acquiring more recent, higher-quality candidate summaries as training data, which enhances model training. GSum [18], SeqCo [20], BRIO [27] and LonRanker-G are all single-stage models that do not need to generate additional candidate summaries, resulting in resource savings and other benefits. GSum has an extra supervisor, compared to the latter models, that is responsible for using guide information to control the content and structure of the summary. SeqCo treats a document, its gold summary, and its model-generated summary as different views of the same meaning representation, and during training, we maximize the similarity between them. During training, both BRIO and LonRanker-G align the probability and quality of generated summaries while retaining the ability to create summaries. However, BRIO utilizes pre-generated candidate summaries and List-wise ranking loss, whereas LonRanker-G applies online sampling and contrastive learning to maximize the probability output of the optimal candidate summary compared to other candidates. On the LCSTS dataset, RNN-Context [26] employs simple GRU architecture, in addition to possessing an extra network for attention mechanism and LCCN [28] uses a network that is able to replicate single or multiple characters at the same time to improve the method of creating summaries of generated text. Reinforced-Topic [29] utilizes convolutional neural networks to incorporate topic models into the summarization model, and employs reinforcement learning methods to optimize the model for the ROUGE metrics. LonRanker also exhibits significant performance advantages over similar models, without the need for supplementary information or extra parameters.

5   Analysis

5.1   Why Is Ranking Needed?

To demonstrate the effectiveness of the ranking phase, we used PEGASUS [4] to generate 16 candidate summaries for the entire corpus of the CNNDM [24] dataset and analyzed the results and significant disparities in evaluation scores of previously generated summaries. The experimental results are presented in Table 2.

As in Table 2, Min and Max denote the least and most acceptable summaries selected from the set of candidate summaries, while AVG represents the average score of all candidates. Random assumes a random selection from the candidate summaries, and PEGAUSU corresponds to choosing the summary with the highest score in DBS. The results show that the highest-scoring summary in DBS does not always have the highest ROUGE score, but it is still more reliable than AVG and Random. An ideal ranking model that consistently selects the best summary could improve the current baseline by over 11% points, highlighting the significant research potential of ranking work.

Table 1. Performance of Different Models on CNNDM, Xsum and LCSTS Datasets

Model	ROUGE-1	ROUGE-2	ROUGE-L
CNNDM			
SimCLS	46.67	22.15	43.54
SummaReranker	47.16	22.55	43.87
GSum	45.94	22.32	42.58
BRIO	47.28	22.3	44.15
PEGAUSU	44.17	21.47	41.11
LonRanker-G	47.37	23.45	44.78
LonRanker-R	**48.14**	**24.02**	**45.23**
XSum			
GSum	45.40	21.89	36.67
SeqCo	45.65	22.41	37.04
SimClS	47.61	24.57	39.44
PEGAUSU	47.21	24.56	39.25
LonRanker-G	47.89	25.04	39.97
LonRanker-R	**48.13**	**25.78**	**40.32**
LCSTS			
RNN	46.67	22.15	43.54
LCCN	47.16	22.55	43.87
Reinforced-Topic	45.12	33.08	42.68
BART	44.62	29.76	40.76
LonRanker-G	46.73	32.98	43.36
LonRanker-R	**47.04**	**33.57**	**43.89**

Table 2. Statistical analysis of summary generation scores

Method	ROUGE-1	ROUGE-2	ROUGE-L
Min	35.16	12.65	32.16
Max	55.67	28.73	51.03
AVG	44.05	21.37	40.87
Random	43.87	20.45	41.76
PEGAUSU	44.17	21.47	41.11
LonRanker	**48.14**	**24.02**	**45.23**

5.2   How Many Candidats Do We Need?

In the previous experiment, we treated the number of candidate summaries, which was set to $k = 16$, as a hyperparameter that significantly affects the final model performance. In our study, we selected 6 values $\{1, 4, 8, 12, 16, 20\}$ and performed experiments with each of them. Figure 2 illustrates the results.

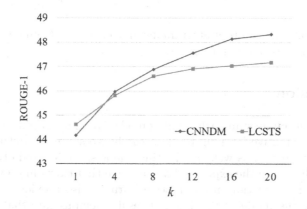

Fig. 2. Performance of the model on the two datasets for different values

Figure 2 shows that the model's performance consistently improves as the number of candidate summaries increases, without any evident decrease even at a maximum value of 20. Nevertheless, in the comparative experiments, we ultimately set this value to 16. On the one hand, this was necessary to ensure fairness with other ranking models (SimCLS, BRIO), which also adopted the same value. On the other hand, as this value increases, the associated cost of training and inference increases, while an excessively large value of k is not pragmatically meaningful.

5.3   What About the Other Metrics?

All of the previous experiments utilized ROUGE scores as the standard for ranking models. However, abstractive summarization tasks are evaluated using various other metrics, including BERTScore [7] and FactCC [30]. FactCC is a metric designed to evaluate the consistency of factual information in generated summaries. To evaluate the effectiveness of our method in optimizing other indicators, we replaced the ROUGE scores that were utilized during training with the outputs produced by FactCC in this section. Table 3 shows the experimental results, LR-ROUGE and LR-FactCC respectively indicate that the LonRanker uses ROUGE or FactCC as the ranking standard.

The results in Table. 3 indicate that, despite a slight decrease in the ROUGE scores, the FactCC scores significantly improved when the ranking standard was changed from ROUGE to FactCC. This suggests that the proposed method can

Table 3. Performance of the model on FactCC

Method	ROUGE-1	ROUGE-2	ROUGE-L	FactCC
PEGAUSU	44.17	21.47	41.11	50.98
LR-ROUGE	**48.14**	**24.02**	**45.23**	51.74
LR-FactCC	47.28	22.75	43.76	**53.12**

tailor and adjust the appropriate ranking model during training based on the priority of fluency or factual accuracy.

6   Conclusion

We proposed an online summary ranking model based on likelihood probability. By utilizing contrastive learning to align candidate summary output probabilities and their qualities, as well as adopting online sampling and other methods, it effectively alleviated the exposure bias problem that exists in previous models. Moreover, the introduction of contrastive learning also provides sequence-level supervision for the model. Experimental results demonstrate that this method effectively improves model performance on CNNDM and LCSTS datasets. The approach of summarization can be implemented on any Seq2Seq framework without the need for any other dataset processing, and it has high portability. In the second phase, LonRanker as a ranking model and can improve other evaluation metrics by modifying the training objective. Nevertheless, generate-ranking's two-stage inference mode has a high cost. Subsequent studies will focus on exploring ways to improve the efficiency and speed of inference performance.

References

1. Sutskever, I., Vinyals, O., Le, Q.V.: Sequence to sequence learning with neural networks. In: Advances in Neural Information Processing Systems, pp. 27–35 (2014)
2. Ma, S., Sun, X., Li, W.: Query and output: generating words by querying distributed word representations for paraphrase generation. In: Proceedings of the 2018 Conference of the North American Chapter of the Association for Computational Linguistics: Human Language Technologies, Volume 1 (Long Papers), pp. 196–206 (2018)
3. Lewis, M., Liu, Y., Goyal, N.: BART: denoising sequence-to-sequence pre-training for natural language generation, translation, and comprehension. In: Proceedings of the 58th Annual Meeting of the Association for Computational Linguistics, pp. 7871–7880 (2020)
4. Zhang, J., Zhao, Y., Saleh, M., Liu, P.: Pegasus: pre-training with extracted gap-sentences for abstractive summarization. In: International Conference on Machine Learning, pp. 11328–11339 (2020)
5. Ranzato, M., Chopra, S., Auli, M., Zaremba, W.: Sequence level training with recurrent neural networks. arXiv preprint arXiv:1511.06732 (2015)

6. Lin, C.Y., Hovy, E.: Automatic evaluation of summaries using n-gram co-occurrence statistics. In: Proceedings of the 2003 Human Language Technology Conference of the North American Chapter of the Association for Computational Linguistics, pp. 150–157 (2003)
7. Zhang, T., Kishore, V., Wu, F., Weinberger, K.Q., Artzi, Y.: BERTscore: evaluating text generation with BERT. In: International Conference on Learning Representations, pp. 4781–4788 (2020)
8. Hadsell, R., Chopra, S., LeCun, Y.: Dimensionality reduction by learning an invariant mapping. In: 2006 IEEE Computer Society Conference on Computer Vision and Pattern Recognition (CVPR 2006), vol. 2, pp. 1735–1742. IEEE (2006)
9. Vaswani, A., et al.: Attention is all you need. Adv. Neural. Inf. Process. Syst. **30**, 30–40 (2017)
10. Bengio, S., Vinyals, O., Jaitly, N., Shazeer, N.: Scheduled sampling for sequence prediction with recurrent neural networks. Adv. Neural. Inf. Process. Syst. **28**, 28–36 (2015)
11. Li, S., Lei, D., Qin, P., Wang, W.Y.: Deep reinforcement learning with distributional semantic rewards for abstractive summarization. In: Proceedings of the 2019 Conference on Empirical Methods in Natural Language Processing and the 9th International Joint Conference on Natural Language Processing (EMNLP-IJCNLP), pp. 6038–6044 (2019)
12. Wiseman, S., Rush, A.M.: Sequence-to-sequence learning as beam-search optimization. arXiv preprint arXiv:1606.02960 (2016)
13. Cohen, E., Beck, J.C.: Empirical analysis of beam search performance degradation in neural sequence models. In: International Conference on Machine Learning, pp. 1290–1299 (2019)
14. Liu, Y., Liu, P.: SimCLS: a simple framework for contrastive learning of abstractive summarization. In: Proceedings of the 59th Annual Meeting of the Association for Computational Linguistics and the 11th International Joint Conference on Natural Language Processing (Volume 2: Short Papers), pp. 1065–1072 (2021)
15. Liu, Y., et al.: Roberta: a robustly optimized BERT pretraining approach. arXiv preprint arXiv:1907.11692 (2019)
16. Ravaut, M., Joty, S., Chen, N.: SummaReranker: a multi-task mixture-of-experts re-ranking framework for abstractive summarization. In: Proceedings of the 60th Annual Meeting of the Association for Computational Linguistics (Volume 1: Long Papers), pp. 4504–4524 (2022)
17. Shen, W., et al.: Joint generator-ranker learning for natural language generation. arXiv preprint arXiv:2206.13974 (2022)
18. Dou, Z.Y., Liu, P., Hayashi, H., Jiang, Z., Neubig, G.: GSum: a general framework for guided neural abstractive summarization. In: Proceedings of the 2021 Conference of the North American Chapter of the Association for Computational Linguistics: Human Language Technologies, pp. 4830–4842 (2021)
19. Sun, S., Li, W.: Alleviating exposure bias via contrastive learning for abstractive text summarization. arXiv preprint arXiv:2108.11846 (2021)
20. Xu, S., Zhang, X., Wu, Y., Wei, F.: Sequence level contrastive learning for text summarization. In: Proceedings of the AAAI Conference on Artificial Intelligence, pp. 11556–11565 (2022)
21. Cao, S., Wang, L.: CLIFF: Contrastive learning for improving faithfulness and factuality in abstractive summarization. In: Proceedings of the 2021 Conference on Empirical Methods in Natural Language Processing, pp. 6633–6649 (2021)
22. Holtzman, A., Buys, J., Du, L., Forbes, M., Choi, Y.: The curious case of neural text degeneration. arXiv preprint arXiv:1904.09751 (2019)

23. Vijayakumar, A.K., Cogswell, M., Selvaraju, R.R., Sun, Q., Lee, S., Crandall, D., Batra, D.: Diverse beam search: Decoding diverse solutions from neural sequence models. arXiv preprint arXiv:1610.02424 (2016)

24. See, A., Liu, P.J., Manning, C.D.: Get to the point: summarization with pointer-generator networks. In: Proceedings of the 55th Annual Meeting of the Association for Computational Linguistics (Volume 1: Long Papers), pp. 1073–1083 (2017)

25. Narayan, S., Cohen, S.B., Lapata, M.: Don't give me the details, just the summary! topic-aware convolutional neural networks for extreme summarization. In: Proceedings of the 2018 Conference on Empirical Methods in Natural Language Processing, pp. 1797–1807 (2018)

26. Hu, B., Chen, Q., Zhu, F.: LCSTS: a large scale Chinese short text summarization dataset. In: Proceedings of the 2015 Conference on Empirical Methods in Natural Language Processing, pp. 1967–1972 (2015)

27. Liu, Y., Liu, P., Radev, D., Neubig, G.: BRIO: bringing order to abstractive summarization. In: Proceedings of the 60th Annual Meeting of the Association for Computational Linguistics (Volume 1: Long Papers), pp. 2890–2903 (2022)

28. Wan, B., Tang, Z., Yang, L.: Lexicon-constrained copying network for Chinese abstractive summarization. arXiv preprint arXiv:2010.08197 (2020)

29. Wang, L., Yao, J., Tao, Y., Zhong, L., Liu, W., Du, Q.: A reinforced topic-aware convolutional sequence-to-sequence model for abstractive text summarization. In: Proceedings of the Twenty-Seventh International Joint Conference on Artificial Intelligence (IJCAI-2018), pp. 4453–4460 (2018)

30. Kryscinski, W., McCann, B., Xiong, C., Socher, R.: Evaluating the factual consistency of abstractive text summarization. In: Proceedings of the 2020 Conference on Empirical Methods in Natural Language Processing (EMNLP), pp. 9332–9346 (2020)

Spatial Commonsense Reasoning
for Machine Reading Comprehension

Miaopei Lin[1], Meng-xiang Wang[3], Jianxing Yu[2,4]([✉]), Shiqi Wang[2],
Hanjiang Lai[1,4], Wei Liu[4], and Jian Yin[2,4]

[1] School of Computer Science and Engineering, Sun Yat-sen University,
Guangzhou, China
linmp3@mail2.sysu.edu.cn, laihanj3@mail.sysu.edu.cn
[2] School of Artificial Intelligence, Sun Yat-sen University, Zhuhai, China
yujx26@mail.sysu.edu.cn, wangshq25@mail2.sysu.edu.cn
[3] China National Institute of Standardization, Beijing 100088, China
wangmx@cnis.ac.cn
[4] Guangdong Key Laboratory of Big Data Analysis and Processing,
Guangzhou, China
{liuw259,issjyin}@mail.sysu.edu.cn

Abstract. This paper studies the problem of spatial commonsense reasoning for the machine reading comprehension task. Spatial commonsense is the human-shared but latent knowledge of object shape, size, distance, and position. Reasoning this abstract knowledge can facilitate machines better perceive their surroundings, which is crucial for general intelligence. However, this valuable topic is challenging and has been less studied. To bridge this research gap, we focus on this topic and propose a new method to realize spatial reasoning. Given a text, we first build a potential reasoning graph based on its parsing tree. To better support spatial reasoning, we retrieve the related commonsense entities and relations from external knowledge sources, including the pre-trained language model (LM) and knowledge graph (KG). LM covers all kinds of factual knowledge and KG has abundant commonsense relations. We then propose a new fusion method called LEGRN (LM Edge-GNN Reasoner Networks) to fuse the text and graph. LEGRN adopts layer-based attention to integrate the LM text encoder and KG graph encoder, which can capture correlations between LM text context and KG graph structure. Considering that spatial relations involve a variety of attributes, we propose an attribute-aware inferential network to deduce the correct answers. To evaluate our approach, we construct a new large-scale dataset named CRCSpatial, consisting of 40k spatial reasoning questions. Experiment results illustrated the effectiveness of our approach.

Keywords: spatial commonsense reasoning · question answering

1 Introduction

With the rapid development of the Web, a large number of text data are produced rapidly. Acquisition of the knowledge embodied in the text has become an extensive requirement. A feasible direction for this requirement is machine

X. Yang et al. (Eds.): ADMA 2023, LNAI 14177, pp. 347–361, 2023.
https://doi.org/10.1007/978-3-031-46664-9_24

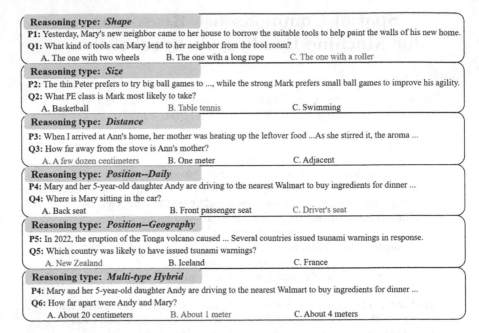

Reasoning type: *Shape*
P1: Yesterday, Mary's new neighbor came to her house to borrow the suitable tools to help paint the walls of his new home.
Q1: What kind of tools can Mary lend to her neighbor from the tool room?
    A. The one with two wheels   B. The one with a long rope   C. The one with a roller

Reasoning type: *Size*
P2: The thin Peter prefers to try big ball games to ..., while the strong Mark prefers small ball games to improve his agility.
Q2: What PE class is Mark most likely to take?
    A. Basketball   B. Table tennis   C. Swimming

Reasoning type: *Distance*
P3: When I arrived at Ann's home, her mother was heating up the leftover food ...As she stirred it, the aroma ...
Q3: How far away from the stove is Ann's mother?
    A. A few dozen centimeters   B. One meter   C. Adjacent

Reasoning type: *Position--Daily*
P4: Mary and her 5-year-old daughter Andy are driving to the nearest Walmart to buy ingredients for dinner ...
Q4: Where is Mary sitting in the car?
    A. Back seat   B. Front passenger seat   C. Driver's seat

Reasoning type: *Position--Geography*
P5: In 2022, the eruption of the Tonga volcano caused ... Several countries issued tsunami warnings in response.
Q5: Which country was likely to have issued tsunami warnings?
    A. New Zealand   B. Iceland   C. France

Reasoning type: *Multi-type Hybrid*
P4: Mary and her 5-year-old daughter Andy are driving to the nearest Walmart to buy ingredients for dinner ...
Q6: How far apart were Andy and Mary?
    A. About 20 centimeters   B. About 1 meter   C. About 4 meters

Fig. 1. Examples with various spatial reasoning types in CRCSpatial

reading comprehension (MRC-QA). In this research topic, current work has well-studied simple questions, but not complex questions, such as spatial reasoning ones. This kind of question asks about the object's shape, size, distance, position, etc. Its answer has to reason over complex text contexts and external spatial commonsense knowledge. Humans use this knowledge to perceive space and understand the surrounding environment in their daily life. It is the prerequisite to build the general intelligent agent. For example, when someone is stirring the heated food, we can infer that the appropriate distance from the stove is a few dozen centimeters, not one meter away or adjacent to it (see Q3 in Fig. 1). Existing work has studied background knowledge on causality [5], hyponymy [7], social psychology [14,20], and temporal [12,28]. Little work studies spatial commonsense questions. Thus, this topic has great research value.

For this task, it is crucial to acquire sufficient commonsense knowledge. A straightforward source is the knowledge graphs (KGs), such as Concept-Net [15] that contains abundant parent-child relations. Since most KGs are hand-made, their scale is limited and the coverage is insufficient to answer various kinds of questions. We can resort to the large pre-trained language models (PLMs) [7,10,16], which are good at capturing various kinds of knowledge inherent in the large-scale corpus. This knowledge is implicit and full of noise. Traditional PLMs-based methods can well tackle simple questions whose answers can be found by matching the text and PLMs retrieved contents. However, our spatial commonsense questions multi-hop reasoning [26] over complex contexts, where some contexts do not appear in the given text but can be inferred from their meaning. Moreover, the noises in PLMs would harm the performance. One simple way is to integrate PLMs and KGs. That can simultaneously take advantage of the high coverage in PLMs and high-quality knowledge in KGs. Traditional methods often first represent contextual features using a text encoder (e.g. PLMs). They then extract subgraphs from KG and represent them via a graph network (e.g. GNN-based model) [8,27]. They next concatenate these two representations to make predictions. However, these representations cover various kinds of spatial commonsense knowledge in different structures. The edges in the spatial commonsense KG contain multiple instead of single features. For example, the edges that represent position relationship between *"work"* and *"house"* include *"in"*, *"outside"*, *"around"*, and with different weights. Without considering the differences in edge types, it is hard to support complex reasoning over spatial relations. That would harm the performance on the spatial QA task.

To address these challenges, we introduce a new framework named LEGRN (LM Edge-GNN Reasoner Networks). Our model enables layer-level deep interaction between the text encoder and the graph encoder. It consists of n-layer LM and GNN. For each layer, we design layer-based attention to bi-directionally fuse text and graph representations. In this way, the LM layer can interactively encode the graph-aware semantics of the given context by referring to the graph, and the GNN layer can enrich the node representation with the text representation. Considering that spatial relations contain various features, the graph edges have plentiful semantics. We thus encode both node and edge at the GNN layers and augment the node representation with edges encoded correlations. That can help us learn the structural features of the graph and gain better spatial reasoning ability. Considering the lack of benchmark in this new topic, we build a large-scale new dataset called CRCSpatial (Spatial Commonsense Reading Comprehension). There are 40k multiple-choice questions. As shown in Fig. 1, the questions require reasoning about the four types of spatial commonsense, including *shape, size, distance,* and *position.* Such spatial commonsense knowledge is crucial to find the correct answer. We conduct evaluations on this dataset to verify the effectiveness of our approach.

In summary, the main contributions of this paper include:

- To the best of our knowledge, we are the first to propose the topic of spatial commonsense reasoning, which has great value for the task of MRC-QA.
- We propose an effective method for this new topic, which uses layer-based attention to integrate text context and graph spatial knowledge. That can facilitate reasoning of hidden commonsense and complex semantics.
- We build a new large-scale dataset named CRCSpatial for our new task. Extensive experiments are conducted on it to evaluate our method fully.

2   Dataset Construction

We first create a dataset named CRCSpatial with 40,713 multi-choice questions that require spatial reasoning ability.

2.1 Creation Process

We adopt four steps to yield the spatial reasoning samples.

Paragraph Collection: Considering the popular MCScript2.0 [12] contains rich spatial context, we reused its 3,487 paragraphs, which were high-quality stories about daily life. The text content is less noisy and is convenient to better measure the performance on spatial reasoning tasks. In addition, we collected 511 paragraphs from travel blogs on three websites[1] which cover adequate geographic commonsense knowledge. In total, we obtained 3,998 paragraphs with spatial-related context.

Question and Answer Generation: First, we tasked student assistants with proposing questions based on the given paragraph. We instructed them to generate effective questions that (1) require at least one of four spatial commonsense types we defined to answer correctly, and (2) cannot be solved using just the words or phrases from the paragraph. Furthermore, for each question, we required the assistants to create one correct answer and two incorrect answers. In this stage, we constructed a set of 4,000 samples in total. We used this sample set to fine-tune a pre-trained commonsense reasoning-based question generation model [25]. This model can generate a question given a paragraph and a manually annotated correct answer. It first retrieves relevant commonsense contents from the external knowledge bases (*ConceptNet and WebChild*). Based on them, it then utilizes a Transformer-based model to yield questions. We asked assistants to provide 20 possible answer options for each paragraph, using words or phrases related to the described scene that did not appear in the paragraph. To ensure diversity in the question types, we required the constructed correct answers to cover as many types as possible, such as person names, place names, location phrases, and so on. Finally, we obtained a foundational QA set automatically generated by the model.

Distractor Generation: To collect distractor candidate options for the questions, we fine-tuned the T5 transformer model on the aforementioned set of 4,000 samples. The model took the answer, question, and context paragraph as input: *answer + [SEP] + question + paragraph*, and generated two candidate options as output: *distractor1 + [SEP] + distractor2*.

Validation and Refinement: We assigned 10 student assistants to inspect, filter, and modify the entire raw dataset. Specifically, this included: (1) We filtered out invalid questions, including those that are meaningless, unrelated to spatial commonsense, or unrelated to the paragraph. (2) We also revised simple or erroneous questions, including those that can be answered solely based on

---

[1] https://expertvagabond.com, https://www.wtcf.org.cn, https://www.tripadvisor.com.

the text, solely through spatial commonsense knowledge, or through only one-step reasoning. (3) Additionally, we checked and modified candidate answers, including verifying the correctness of the correct candidates and ensuring that the distractor options are incorrect and of the same type as the correct answer. Finally, all paragraphs, questions, and answers were spell-checked and corrected by running aSpell2 and spell-checker.

2.2 Data Statistics

There are 40,713 questions for the given 3,998 paragraphs. That is, each paragraph has about 10 questions. We split the dataset into training, development, and test sets based on an 8:1:1 ratio, with each paragraph appearing in only one of the three subsets. Table 1 presents the basic statistics.

Table 1. Statistics of CRCSpatial dataset

	Total	Train	Dev	Test
# of text	3,998	3,201	402	395
# of questions	40,713	32,571	4,070	4,072
Avg. paragraph length	155.1	155.2	152.9	156.3
Avg. question length	11.5	11.5	11.6	11.6
Avg. answer length	2.7	2.7	2.7	2.7

We determined the question types based on the question words using simple heuristics. In particular, yes/no questions start with an auxiliary verb (do, does, did) or a modal verb (is, are, can, was, were, could). We found that the question types in the CRCSpatial dataset are diverse, including What-38%, Where-15%, Why-8%, Who-8%, How far/close-8%, When-6%, Which-6%, How-5%, Yes/no-3%, Other-3%.

We then randomly sampled 500 instances to manually label the spatial commonsense types required to answer the questions. Figure 1 shows typical examples. The distribution of types of spatial commonsense reasoning is as follows: 12% for *Shape*, 10% for *Size*, 14% for *Distance*, 42% for *Position* (including 37% for daily scenes and 5% for geographic scenes), and 22% for *hybrid reasoning*. Position reasoning and hybrid reasoning account for a relatively large proportion, which is more in line with the complex questions people face in actual situations that require multi-types of spatial commonsense reasoning.

3   Approach

As shown in Fig. 2, to answer the spatial questions, we first derive the related commonsense reasoning graph. We then encode the input texts and the created graph. Their correlations are captured by co-attention to better decode the answer. Next, we define some notations and elaborate details on each part.

3.1   Problem Formulation

Given a paragraph p, a question q, and a set of answer options A, the goal of our task is to find the correct answer $a \in A$, where a can answer q by reasoning over the context p. Considering that some contexts involve the spatial commonsense, we construct a reasoning graph $G = (V, E)$ to capture them via the external knowledge sources. Here V is the set of entity nodes in the graph; $E \subseteq V \times R \times V$ is the set of edges that connect nodes in V, where R represents a set of relation types. Formally, we can compute the plausibility scores between p, q, and each option, where the score on the answer $a \in A$ is the largest.

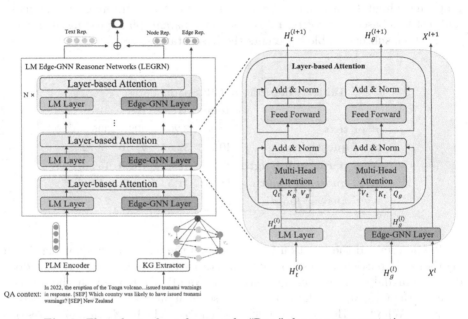

Fig. 2. Flow chart of our framework. "Rep." denotes representation.

To calculate the score, we first apply PLM (*e.g., RoBERTa*) to encode the inputs $[p; q; a]$. We then retrieve the related spatial reasoning graph G from the external sources [23]. Afterward, we use N layers LM Edge-GNN Reasoner Networks (LEGRN) to jointly capture the correlations among the textual inputs and G. Each layer consists of an LM layer (Sect. 3.3), an Edge-GNN layer (Sect. 3.4), and a Layer-based Attention layer (Sect. 3.5). We design a layer-based attention mechanism to bi-directionally connect the LM and Edge-GNN layers. That can fuse learned features between the two. Finally, we use the representations from the last LEGRN layer to predict the answer.

3.2   Acquisition of Spatial Reasoning Knowledge Graph

To answer the spatial question, we derive a potential reasoning graph covering plausible evidential clues. Following previous works [23], we first finely identify the entities mentioned in the paragraph, question, and candidate options, as V_{pqa}. To find the necessary but hidden commonsense clues, we then retrieve a multi-hop subgraph from the external KG related to these entities, as $G_{pqa} = (V_{pqa}, E_{pqa})$. It comprises all nodes on the k-hop paths between nodes in V_{pqa}.

Afterward, we connect the entities in the inputs to the retrieved graph. That can facilitate us to deduce the answers by reasoning from the text to the hidden commonsense knowledge. Given the input text (represented as c), we link it to each entity in V_{pqa} by three relation types r_{cp}, r_{cq} and r_{ca}, so as to form a reasoning graph $G_{Rea} = (V_{Rea}, E_{Rea})$, where $V_{Rea} = V \cup c$ and $E_{Rea} = E \cup (c, r_{cp}, v)|v \in V_p \cup (c, r_{cq}, v)|v \in V_q \cup (c, r_{ca}, v)|v \in V_a$. We use the entity embedding to initialize the node and edge, where c is initialized as zero.

3.3   Representation of Text Context

We use PLM (i.e., RoBERTa) to embed the input texts as $H_t^{(0)} \in \mathbb{R}^{m \times d_p}$, m is the hidden layer size of PLM, d_p is the hidden size. To unify the hidden size, we feed it into an MLP network to get the new $H_t^{(0)} \in \mathbb{R}^{m \times d}$. We then utilize a Transformer [19] with co-attention to capture the context with regard to the graph. For the l-th layer, we compute graph-aware representation $H_t^{(l)}$ by the layer-based attention based on $H_t^{(l-1)}$ and the node representation $H_g^{(l-1)}$, as $H_t^{(l)} = Transformer(H_t^{(l-1)}, H_g^{(l-1)})$.

3.4   Representation of Graph Structure

The spatial relations involve a variety of attributes and the answering process may involve reasoning over them. Thus we propose an attribute-aware network to capture this structural reasoning context.

Given the reasoning graph G_{Rea}, we first initialize the node embedding $H_g^{(0)} \in \mathbb{R}^{n \times d_{in}}$ and edge embedding $X^0 \in \mathbb{R}^{n \times n \times S}$, where n is the size of nodes, d_{in} is the vector size, and S is the dimension of edges vector. Let $X_{ij} \in \mathbb{R}^S, i = 1, 2, \ldots, n, j = 1, 2, \ldots, n$ denote the edge vector with S-dimension, linking the i-th and j-th nodes. X_{ijs} represents the s-th channel of the edge vector in $X_{ij.}$. When there is no edge connecting the i-th and j-th nodes, $X_{ij.} = \mathbf{0}$, where "." is an operator used to select the range *slice* of a vector dimension.

We then jointly feed the node $H_g^{(0)}$ and edge X^0 into the proposed N layer LM Edge-GNN network. In l-th layer, we capture the edge and node correlations by layer-based attention, so as to obtain edge-aware representation of the node and node-aware representation of the edge. We employ an edge-enhanced graph encoder, EGNN(A) [4], to incorporate multi-dimensional edge features in the graph with multiple reasoning relations. In detail, let $H_g^{(l)} = \left[h_1^{(l)}; \cdots ; h_n^{(l)} \right]$, the node representation is updated via aggregation operation defined as follows:

$$H_g^{(l+1)} = \sigma \left[\|_{s=1}^S \left(\alpha_{..s}^l \left(H_g^{(l)}, X_{..s}^{(l)} \right) H_g^{(l)} W^l \right) \right], \quad (1)$$

354   M. Lin et al.

where σ is a non-linear activation function, $\|$ denotes the concatenation opera-
tor, W^l is a learnable weight matrix. α is used to yield the $n \times n \times S$ tensor. In
Eq. (1), α^l denotes the attention coefficients, where each element α^l_{ijs} is a function of the original $h_i^{(l)}$, $h_j^{(l)}$ and X_{ijs}. Considering that spatial commonsense has
many attributes, there are multi-dimensional features on the edge. To capture
this structural context, we regard each feature as a multi-channel signal and
use multi-dimensional attention to calculate their correlations. The attention for
each feature is computed as Eq. (2), where the W^l is the shared weight.

$$\alpha^l_{ijs} = \exp\left\{\text{LeakyReLU}\left(a^T\left[h_{i\cdot}^{(l)}W^l\|h_{j\cdot}^{(l)}W^l\right]\right)\right\}X^l_{ijs}, \tag{2}$$

By passing through the Edge-GNN layer, the node can be updated as $H_g^{(l)} = H_g^{(l+1)}$, and the new edge is updated by the attention coefficients as $X^l = \alpha^l$.

3.5   Layer-Based Attention to Capture Structural Correlations

Considering that the spatial answer may need to deduce over the text and
graph, independent representation of them may not capture this hidden reasoning knowledge. We thus propose layer-based attention to elicit this sharing
information by fusing the LM text and Edge-GNN graph features. We compute
the multi-layer interactions on the text and graph by propagation. That allows
us to find out the spatial commonsense knowledge related to the text from the
graph. Inversely, we can utilize the context in the text to improve reasoning on
the graph.

In detail, given the text representation $H_t^{(l)}$ from the l-th LM layer and node
representation $H_g^{(l)}$ from the l-th Edge-GNN layer, we first leverage a transformer
to capture their context as Eq. (3), where Q^t, Q^g, key K^t, K^g, V^t, V^g are the
intermediate matrices, $W_i^Q, W_i^K, W_i^V \in \mathbb{R}^{d \times d_k}$ are trainable matrices.

$$Q_i = H^{(l)}W_i^Q, \quad K_i = H^{(l)}W_i^K, \quad V_i = H^{(l)}W_i^V, \tag{3}$$

Then, we utilize two transformers with multi-head attention to compute the
correlations of the vectors. In detail, the input is first divided into multiple
segments. We then map the segmented input into a latent space by a linear
transformation, and calculate the attention score of each segment by as Eq. (4),
where Q^t, K^g, V^g are the metrics for the text transformer, Q^g, K^t, V^t are those
for the graph transformer.

$$T_i(Q_i^t, K_i^g, v_i^g) = softmax(Q_i^t K_i^{g^T}/\sqrt{d_k})v_i^g$$
$$G_i(Q_i^g, K_i^t, v_i^t) = softmax(Q_i^g K_i^{t^T}/\sqrt{d_k})v_i^t \tag{4}$$

Afterward, we collect the outputs of each attention head and take the
weighted sum of them as $O_t^{(l)} = \|_{i=1}^h(T_i)W_t^O$ and $O_g^{(l)} = \|_{i=1}^h(G_i)W_g^O$, where $\|$
is the concatenation operator, $W_t^O, W_g^O \in \mathbb{R}^{hd_k \times d}$ are the weight matrices. To
avoid gradient divergence, we add normalization to encourage the data falling

outside the saturation zone of the activation function. Besides, we utilize the residual block to prevent degeneration in network training. We employ two residual add functions to get the fused representation of the text and graph, as $H_t^{(l)} = LN(H_t^{(l)} + O_t^{(l)})$ and $H_g^{(l)} = LN(H_g^{(l)} + O_g^{(l)})$, respectively, where LN denotes the layer normalization. To jointly grasp all the information, we feed them into two feed-forward networks (i.e., MLP), and output as Eq. (5), where $H_t^{(l+1)} \in \mathbb{R}^{m \times d}$ and $H_g^{(l+1)} \in \mathbb{R}^{n \times d}$ are the input of $(l + 1)$-th layer of LM and Edge-GNN, respectively. In this way, we can fuse different levels of semantic features from the text and graph to obtain a more comprehensive representation.

$$H_t^{(l+1)} = LN(H_t^{(l)} + MLP(H_t^{(l)})), \quad H_g^{(l+1)} = LN(H_g^{(l)} + MLP(H_g^{(l)})), \quad (5)$$

3.6 Training and Prediction

Based on the above representations, we concatenate them as the input to a classifier and get the output probability for each option of the multi-choice question, as $p(a|p, q) = Classifier(Pool(H_t^{(L)}) \oplus Pool(H_g^{(L)}))$, where the classifier consists of a two-layer fully connected network with ReLU activation function, Pool is the mean pooling operation which can well capture the key classified information. The highest score is predicted as the answer. In the training phase, we update the model parameters according to cross-entropy loss. The objective function is to minimize the cross-entropy loss to correctly predict the answer, as Eq. (6), where p_i denotes the prediction, a_i is the ground truth answer for the i-th sample.

$$\mathcal{L}(\theta) = -\frac{1}{N}\sum_i a_i \log p(\hat{a}_i|p_i, q_i), \quad (6)$$

4 Experiments

Next, we fully evaluated the effectiveness of our method in various aspects.

4.1 Dataset and Experimental Settings

Since this is the first work on the spatial reasoning MRC-QA task, no public dataset is available to evaluate the spatial reasoning ability. We then conducted evaluations on our created CRCSpatial dataset (Sect. 2). The dataset consisted of 40k examples, each requiring spatial commonsense reasoning ability. Statistic details of the dataset were displayed in Table 1. To collect the spatial commonsense, we resorted to the external KG of WebChild2.0. It used a network of over 18 million assertions to describe 2 million unambiguous concepts. Most of them are spatial related, including the shape of objects (e.g. *hasShape*), size of objects (e.g. *hasSize*), relationships between objects (e.g. *largerThan, partOf, nextTo, above*). We utilized the method of QA-GNN [23] to initialize node and edge embeddings. We applied PLM to embed all triples in WebChild2.0 and then obtained a pooled representation for each entity. When constructing the subgraph (Sect. 3.2), we set the hop size $k = 2$. We employed accuracy as the evaluation metric.

4.2   Implementation Details

Model configurations were demonstrated as follows. Our model was trained on 24 GB Nvidia RTX 3090 GPU. Based on the *HuggingFace* PyTorch API, we implemented the *transformer*. We used pre-trained vectors in RoBERTa-large [10] to embed the words. We set the dimension ($D = 100$) and the number of layers ($L = 4$) of our LM Edge-GNN networks. The dropout rate was 0.2. We empirically took the Adam as the optimizer. We set the batch size to 64 and fixed the embedding dimension to 64. The learning rate for the LM module and GNN module were set to 10^{-5} and 10^{-3}, respectively.

4.3   Evaluation Baselines

We adopted two kinds of methods for comparison: fine-tuned PLMs which do not use the KG, and KG-based models that use external knowledge for reasoning.

Group1: Fine-Tuned PLMs. We verified that as an external source of knowledge, whether KG is more effective than the pre-trained model (PLMs). We compared our model against the vanilla fine-tuned LMs, including GPT-2 (2019) [13], BERT-large (2019) [6], RoBERTa-large (2019) [10], ERNIE2.0-large (2020) [16]. Among them, ERNIE2.0 was a semantic knowledge-augmented pre-trained model based on BERT, which supported continuous learning.

Group2: KG-Based Models. We evaluated whether the models with a combination of LM and KG had stronger representation power than our LM Edge-GNN method. These models include: (1) GconAttn (2019) [21] used for knowledge concept matching based on the Match-LSTM framework; (2) KagNet (2019) [8] extracted QA-related subgraphs from the KG and derived the relational paths by GCN and LSTM; (3) MHGRN (2020) [3] utilized a reasoning path to facilitate multi-hop deduction; (4) QA-GNN (2021) [23] constructed a reasoning graph and introduced a graph network to joint encode the text and graph for multi-hop prediction; (5) GreaseLM (2022) [26] fused embeddings of LM and GNN by a two-layer MLP.

Human Evaluation. To measure the difficulty of the CRCSpatial dataset, we conducted the human evaluation. A sampling protocol was enacted wherein 200 sets of questions are drawn at random from the test set. Subsequently, we recruited three annotators to identify the most plausible and answer for each question. We used Randolph's free-marginal kappa metric to measure the agreements of various annotators. By majority voting, we reported the human performance.

4.4   Comparison Results Against Various Baselines

Table 2 showed the comparison results. We observed that we obtained consistent improvements over fine-tuned PLMs and KG-based models. We had outperformance in terms of ~10.56% dev accuracy, ~11.37% test accuracy on RoBERTa, and ~7.30% dev accuracy, ~7.28% test accuracy against the previous best KG-based system, GreaseLM. The improvement over PLMs suggested that incorporating external knowledge can facilitate understanding spatial commonsense. In addition, the improvement over the GreaseLM model was significant. That indicated our model can perform spatial reasoning better by using a KG graph.

Table 2. Dev and Test accuracy on CRCSpatial.

Group	Model	Dev Acc. (%)	Test Acc. (%)
Fine-tuned PLMs	GPT-2 (2019) [13]	54.51(±0.39)	54.13(±0.21)
	BERT-large (2019) [6]	61.18(±0.76)	58.74(±0.10)
	RoBERTa-large (2019) [10]	66.76(±0.71)	65.20(±0.45)
	ERNIE2.0-large (2020) [16]	66.85(±1.12)	63.27(±1.09)
KG-based models	GconAttn (2019) [21]	64.13(+0.93)	63.85(±0.21)
	KagNet (2019) [8]	65.47(±0.81)	65.23(±0.56)
	MHGRN (2020) [3]	67.75(±0.86)	66.08(±0.10)
	QA-GNN (2021) [23]	69.00(±0.24)	66.15(±0.74)
	GreaseLM (2022) [26]	70.02(±0.52)	69.29(±1.16)
	LEGRN (Ours)	**77.32(±0.73)**	**76.57(±0.39)**
	Human	-	93.00(±2.00)

4.5   Ablation Study

We further conducted an ablation study to better analyze the gain of each part of our model, including the connection of the LM to the GNN, the layer-based attention. As presented in Table 3, we found that removing the attention would result in more than ~2.5% performance drop. When we dropped the text module, the performance would degrade performance more. That indicated the usefulness of our module design. We also tested replacing multi-layer layer-based attention by one layer, resulting in a 4.41% drop in performance. We can infer that the multi-level design possessed the capability to effectively engage with varying levels of spatial attributes. Furthermore, if we removed all the layer-based attention layers, the performance dropped significantly by 5.29%. That showed the importance of connecting LM and GNN modules to collaborate on reasoning. We also evaluated the effectiveness of edge embeddings in the GNN module for enhancing the context representation of graph nodes. Removing the edge embedding of GNN would result in 2.86% performance degradation. Finally, when we

Fig. 3. Effect of # of LEGRN Layers.

Table 3. Ablation study.

Model	Dev Acc.(%)
LEGRN (final)	**76.57**
w/o Layer-based Attention (Text)	73.89
w/o Layer-based Attention (Graph)	74.02
w/o Layer-based Attention (Multi-level)	72.16
w/o Layer-based Attention	71.28
w/o Edge embedding of GNN	73.71
w/o Layer-based Attention and Edge embedding of GNN	68.96

removed both the layer-based attention and edge embedding of GNN modules, the performance significantly decreased from 76.57% to 68.96%. In addition, we fine-tuned the number of LEGRN layers and found that L = 4 worked best on the dev set (see Fig. 3).

4.6 Case Study

Moreover, we conducted a case study to examine the detail of the performance gain. As shown in Fig. 4, by performing multiple layer attention to connect LM and GNN modules, our model performed well on multi-hop spatial commonsense reasoning. For example, the Q1.2 in Fig. 4, to answer the question, our model went through multiple reasoning steps: *his mon → heating up the food → in the kitchen → Microwave, so the microwave is close to his mom*. That indicated the usefulness of external knowledge sources from WebChild2.0.

correct case	**P1:** This past weekend, my family made so much food that there was still plenty of it going into the week... On Monday after getting home from work, rather than making a new meal, my mom took some of the food that we had saved from the weekend and put it into a pan, and put that pan on the stove... **Q1.1:** Where was his mom most likely standing when heating food? A. Behind the gas stove. B. In front of the refrigerator. ✔C. In front of the gas stove. **Q1.2:** What might be close to his mom when heating up the food? A. Bed ✔ B. Microwave C. Computer
error case	**P2:** When I arrived at Ann's home, her mother was heating up the leftover food ...As she stirred it, the aroma ... **Q2:** How far away from the stove is Ann's mother? ✔ A. A few dozen centimeters B. One meter ✘C. Adjacent **P3:** Mary and her 5-year-old daughter Andy are driving to the nearest Walmart to buy ingredients for dinner ... **Q3:** How far apart were Andy and Mary? ✘ A. About 20 centimeters ✔B. About 1 meter C. About 4 meters

Fig. 4. Examples predictions of our model. ✔ denotes the correct answers and ✘ marks the prediction errors.

4.7 Error Analysis and Discussion

To analyze the pros and cons of the model, we collected the predicted results and examine the errors. Subsequently, four error cases are identified as part of the analysis: (1) Lack of evidence (36%). For example, the Q2 in Fig. 4 requires commonsense knowledge about distance. However, our model was unable to extract relevant knowledge from the external KG we used. Mitigating this issue could be achieved through the implementation of more sophisticated extraction strategies and the incorporation of a broader array of knowledge sources. (2) Inconsistent with Human Commonsense (27%). For questions where there are two correct candidate answers, humans choose the plausible one based on commonsense, while our model mistakenly selected the choice which is incongruous with human commonsense. (3) Relatively weak in multi-type hybrid reasoning (24%). For example, Q3 in Fig. 4 requires the model to determine the position of Mary and Andy and then infer the distance between them. (4) Unanswerable Questions (13%). The model encounters difficulty in effectively processing the phrase "None of the above." It is challenging to directly infer this response option from either the provided passage or the posed question. We will investigate them in future work.

5 Related Work

Previous works have investigated some aspects of spatial commonsense, including size comparison [1], lengths distributional of objects [2], position relationships from object co-occurrences [24], and position relationships between people and objects in different events [9]. These works focus on the extraction of spatial commonsense knowledge from visual or textual sources, while this paper focuses on spatial commonsense reasoning and application. In addition, a commonsense MRC dataset, Commonsense QA [17], considers the position relationship of "location at". Instead, we consider more comprehensive position relationships, as well as three other types of spatial knowledge including shape, size, and distance.

The reasoning ability of PLMs is weak. To address this issue, many works attempt to integrate structural KG which naturally has relations to facilitate multi-hop reasoning. Some works explore pre-train PLMs to learn structured knowledge implicitly [7,16]. There are also many works that explicitly encode KG information for structured reasoning. The simple way is to directly concatenate the representation of KG and the QA context from two independent modules [11,27]. A further way to fuse KG and LM is to augment one modality with the other, such as utilizing embeddings of subgraphs retrieved from KG to enhance the input text representation [8,22], or utilizing text representations to enhance the reasoning ability of graphs module [3,18,23]. Recent works explore the bidirectional interaction of the two modalities. GreaseLM [26] proposes to fuse the LM and GNN representations via message passing. However, its interaction mode is still relatively shallow since it fuses embeddings of LM and GNN with a simple MLP. Different from previous methods, we adopt Layer-based

Attention to bidirectionally connect each layer of LM and GNN, achieving a deep fusion of text and graph representation.

6   Conclusion

We have proposed a novel topic on spatial commonsense reasoning for the task of MRC-QA. For this topic, we proposed a practical model called LEGRN. It can capture correlations between LM text context and KG graph structure by using a graph network with fused layers. Our model also adopted an attribute-aware network to deduce the correct answers. In the evaluation, we constructed a new large-scale dataset, named CRCSpatial, consisting of 40k spatial reasoning questions. Experimental results illustrated the effectiveness of our approach.

Acknowlendgement. This work is supported by the National Natural Science Foundation of China (62276279, 62002396), the Key-Area Research and Development Program of Guangdong Province (2020B0101100001), the Tencent WeChat Rhino-Bird Focused Research Program (WXG-FR-2023-06), and Zhuhai Industry-University-Research Cooperation Project (2220004002549).

References

1. Bagherinezhad, H., Hajishirzi, H., Choi, Y., Farhadi, A.: Are elephants bigger than butterflies? reasoning about sizes of objects. In: Proceedings of the AAAI, vol. 30 (2016)
2. Elazar, Y., Mahabal, A., Ramachandran, D., Bedrax-Weiss, T., Roth, D.: How large are lions? inducing distributions over quantitative attributes. In: Proceedings of the 57th ACL, pp. 3973–3983 (2019)
3. Feng, Y., Chen, X., Lin, B.Y., Wang, P., Yan, J., Ren, X.: Scalable multi-hop relational reasoning for knowledge-aware question answering. In: Proceedings of the EMNLP, pp. 1295–1309 (2020)
4. Gong, L., Cheng, Q.: Exploiting edge features for graph neural networks. In: Proceedings of the IEEE/CVF, pp. 9211–9219 (2019)
5. Huang, L., Le Bras, R., Bhagavatula, C., Choi, Y.: Cosmos QA: machine reading comprehension with contextual commonsense reasoning. In: Proceedings of the 9th EMNLP-IJCNLP, pp. 2391–2401 (2019)
6. Kenton, J.D.M.W.C., Toutanova, L.K.: Bert: pre-training of deep bidirectional transformers for language understanding. In: Proceedings of NAACL-HLT, pp. 4171–4186 (2019)
7. Levine, Y., et al.: SenseBert: driving some sense into Bert. In: Proceedings of the 58th ACL, pp. 4656–4667 (2020)
8. Lin, B.Y., Chen, X., Chen, J., Ren, X.: KagNet: knowledge-aware graph networks for commonsense reasoning. In: Proceedings of the 9th EMNLP-IJCNLP, pp. 2829–2839 (2019)
9. Liu, X., Yin, D., Feng, Y., Zhao, D.: Things not written in text: exploring spatial commonsense from visual signals. In: Proceedings of the 60th ACL, pp. 2365–2376 (2022)
10. Liu, Y., et al.: Roberta: a robustly optimized Bert pretraining approach. arXiv preprint arXiv:1907.11692 (2019)

11. Mihaylov, T., Frank, A.: Knowledgeable reader: enhancing cloze-style reading comprehension with external commonsense knowledge. In: Proceedings of the 56th ACL, pp. 821–832 (2018)
12. Ostermann, S., Roth, M., Pinkal, M.: Mcscript2. 0: a machine comprehension corpus focused on script events and participants. In: Proceedings of * SEM 2019, pp. 103–117 (2019)
13. Radford, A., Wu, J., Child, R., Luan, D., Amodei, D., Sutskever, I., et al.: Language models are unsupervised multitask learners. OpenAI Blog **1**(8), 9 (2019)
14. Sap, M., Rashkin, H., Chen, D., Le Bras, R., Choi, Y.: Social IQA: commonsense reasoning about social interactions. In: Proceedings of the 9th EMNLP-IJCNLP, pp. 4463–4473 (2019)
15. Speer, R., Chin, J., Havasi, C.: ConceptNet 5.5: an open multilingual graph of general knowledge. In: Proceedings of the AAAI, vol. 31 (2017)
16. Sun, Y., et al.: Ernie 2.0: a continual pre-training framework for language understanding. In: Proceedings of the AAAI, vol. 34, pp. 8968–8975 (2020)
17. Talmor, A., Herzig, J., Lourie, N., Berant, J.: CommonsenseQA: a question answering challenge targeting commonsense knowledge. In: Proceedings of NAACL-HLT, pp. 4149–4158 (2019)
18. Taunk, D., Khanna, L., Kandru, S.V.P.K., Varma, V., Sharma, C., Tapaswi, M.: GrapeQA: graph augmentation and pruning to enhance question-answering. In: Companion Proceedings of the ACM Web Conference 2023, pp. 1138–1144 (2023)
19. Vaswani, A., et al.: Attention is all you need. In: Advances in Neural Information Processing Systems, vol. 30 (2017)
20. Wang, G., Hou, X., Yang, D., Mckeown, K., Huang, J.: Semantic categorization of social knowledge for commonsense question answering. In: Proceedings of the 2nd Workshop on Simple and Efficient Natural Language Processing, pp. 79–85 (2021)
21. Wang, X., et al.: Improving natural language inference using external knowledge in the science questions domain. In: Proceedings of the AAAI, vol. 33, pp. 7208–7215 (2019)
22. Yang, A., et al.: Enhancing pre-trained language representations with rich knowledge for machine reading comprehension. In: Proceedings of the 57th ACL, pp. 2346–2357 (2019)
23. Yasunaga, M., Ren, H., Bosselut, A., Liang, P., Leskovec, J.: QA-GNN: reasoning with language models and knowledge graphs for question answering. In: Proceedings of the NAACL 2021, pp. 535–546 (2021)
24. Yatskar, M., Ordonez, V., Farhadi, A.: Stating the obvious: extracting visual common sense knowledge. In: Proceedings of the NAACL 2016, pp. 193–198 (2016)
25. Yu, J., Liu, W., Zheng, L., Su, Q., Zhao, B., Yin, J.: Generating deep questions with commonsense reasoning ability from the text by disentangled adversarial inference. arXiv preprint (2023)
26. Zhang, X., et al.: GreaseLM: graph reasoning enhanced language models for question answering. In: Proceedings of ICLR (2022)
27. Zhong, W., Tang, D., Duan, N., Zhou, M., Wang, J., Yin, J.: Improving question answering by commonsense-based pre-training. In: Tang, J., Kan, M.-Y., Zhao, D., Li, S., Zan, H. (eds.) NLPCC 2019. LNCS (LNAI), vol. 11838, pp. 16–28. Springer, Cham (2019). https://doi.org/10.1007/978-3-030-32233-5_2
28. Zhou, B., Khashabi, D., Ning, Q., Roth, D.: "going on a vacation" takes longer than "going for a walk": a study of temporal commonsense understanding. In: Proceedings of the 9th EMNLP-IJCNLP, pp. 3363–3369 (2019)

Multimodal Learning for Automatic Summarization: A Survey

Zhicheng Zhang(✉) ⓘ, Yibo Sun ⓘ, and Shiyan Su ⓘ

The University of Queensland, Brisbane, QLD 4072, Australia
zhicheng.zhang3@uqconnect.edu.au

Abstract. With the widespread availability of multiple data sources, such as image, audio-video, and text data, automatic summarization of multimodal data is becoming an important technology in decision support. This paper presents a comprehensive survey and summary of the main articles in the field of multimodal summarization techniques in recent years. Firstly, we define multimodal summarization and briefly describe the development process. Then, we survey existing techniques and their applicability in different domains. Additionally, we provide an analysis of their results and discuss the insights of those approaches, along with the challenges and future research directions. Based on our study, we found that the encoder-decoder approach is currently the best approach for automated summarization. In the future, we believe that the applications of multimodal summarization could develop rapidly in many different fields, particularly in medicine. In our case studies, we demonstrate that multimodal learning is a promising research direction for providing timely and accurate summarizations compared to unimodal approaches.

Keywords: Multimodal Summarization · Feature Engineering · Foundation Models · Attention Mechanism

1 Introduction

In past years, text-based unimodal automatic summarization has been developed and extensively researched [20]. Then, multimodal summarization has begun to receive increasing attention [3,13]. Multimodal automatic summarization can process and correlate information from multiple modalities, such as text, images, audio and video, to produce more coherent and accurate summaries with a high level of information. This approach has shown promising results in improving the quality and effectiveness of automated summaries. It includes a few steps: multimodal input, feature engineering, main model, fine-tuning models, and multimodal summary.

The aim of this article is to provide a comprehensive review of recent approaches in multimodal automatic summarization. We present a comprehensive overview of the main model and application areas of existing methods, categorizing the techniques into different types: methods based on neural networks,

© The Author(s), under exclusive license to Springer Nature Switzerland AG 2023
X. Yang et al. (Eds.): ADMA 2023, LNAI 14177, pp. 362–376, 2023.
https://doi.org/10.1007/978-3-031-46664-9_25

method based on integer linear programming (ILP), method based on submodule optimization, graph- based approaches, method based on LDA Topic Models, and some domain-specific techniques. The application scenarios are considered as universal, news, meetings, movies, sports, medical, and others.

The paper also discusses challenges and future directions of multi-modal automatic summarization, identifies some important datasets, and provides possible directions for improvements of performance and quality with respect to the newly developed technologies.

2   Process of Multimodal Summarization

2.1   Multimodal Summarization

In 2009 Kay L. O'Halloran stated in his article that multimodality generally refers to different properties of the same medium and is a more precise and subdivided concept for representing something through multiple dimensions [40]. It can be expressed as different information properties, data, or representations that describe the same matter or object.

In Mani's book [34], automatic summarization is defined as the process of condensing a group or large amount of information and presenting its most important parts to the user in a short form. Examples include condensing a long report or collection of books into a concise text or presenting the statistics of a season of NBA games in a condensed form as a single image. Therefore, the output of automatic summarization is not limited to text, as numerous studies have shown that even better results can be achieved using images, videos, or multimodality as output. Zhu [55] claims that graphical summaries can increase user satisfaction by an average of 12.4%compared to text-only summaries.

Multimodal summarization can be defined as a computerized method of presenting a large amount of information in many different forms to the user in a streamlined manner. The input to the method must contain multiple forms, and the output can be in any form such as text, images, video or a combination of forms.

2.2   Development of Multimodal Summarization

The concept of multimodality can be traced back as far as the speeches of ancient Greece in BC and is used to express the diversity of behavior [3]. However, with the invention of the computer and the explosion of information flow in the information age, multimodal information has replaced traditional monotypic information in all aspects of life [4].

From the 1980s, the audio-visual speech recognition (AVSR) approach became the beginning of multimodal research [54]. Researchers found that when the demonstrator's lip movements did not match the articulation, the results received by the observed subjects would be affected. When the demonstrator mouthed [ba] and the dubbing was [ga], most subjects would mishear [da], which certainly suggests that multimodality can have a large impact on the results [36].

With the development of neural convolutional networks, multimodality was applied to automatic summarization in conference proceedings in 2003 [35]. By this time the authors had begun to model interactions using Hidden Markov Models and reduced the action error rate on the test set to 5.7%, providing ample evidence of the feasibility and promising future of the project.

It was not until 2006 that the concept of deep learning was introduced [24]. Since then, CNNs [23] and RNNs [19] have started to develop rapidly, encoder-decoder models, weighted attention mechanisms, and transformers [2] have been proposed, and Multimodal Summarization has started to evolve faster.

3   Methods

3.1   Method Based on Neural Networks

Neural network-based approaches are often preferred by researchers in the generation of multimodal summaries.

Neural network frameworks generally consist of an encoder-decoder, with the addition of a multimodal fusion module to form a complete architecture.

In 2003, McCowan [35] proposed the use of the Hidden Markov Models to model meeting behavior in the meeting domain. Early integration approaches combined the features of all participants in a single HMM and trained them. In 2009, Evangelopoulos [15] applied the spatio-temporal attention mechanism to film summaries, which improved their precision and avoided skimming caused by unimodal or visual-auditory-only modalities. In 2013, Evangelopoulos [14] further improved the method in the same area.

In the general domain, Nallapati [39] started using RNNs to summarize text in 2017, and a year later, Chen [11] used bidirectional RNNs to encode text and sentences, using a convolutional neural network VGGNet [46] to process images. This approach allows for the summarization of documents containing images and outperforms the SummaRuNNer method [39] in ROUGE scoring. Li [26] used VGG19 to extract image features, and Tsai [52] used a Transformer-based model for summarization. Additionally, Khullar [22] proposed a MAST method, which can summarize three modalities of "text-audio-video".

In the field of news summaries, good progress has also been made in multimodal research. Chen [10] used an attentional hierarchical encoder-decoder model to process text-centered information complemented by images, resulting in multimodal summaries. Zhu's MSMO method [55] uses a visual overlay mechanism to select suitable images from the output to supplement the summary. Palaskar [41] used a ResNeXt-101 3D convolutional neural network for video encoding. Another approach used by Chen [12] was to input text and images and use the then state-of-the-art Oxford VGGNet for image vector representation, which greatly improved the processing speed. Zhu [57] improved his MSMO [55] method proposed in 2018 in 2021 (Fig. 1).

In the medical field, Fan [16] proposed the FW-Net method to fuse CT images and MRI images to produce summaries with minimal loss of information, which

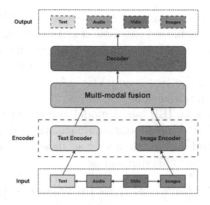

Fig. 1. Framework of Multimodality Automatic Summarization

resulted in good performance. The core of their algorithm is a two-layer U-Net algorithm that follows an encoder-decoder architecture.

Furthermore, Liu [32] also used CNN for the fusion summarization of multimodal medical images, including CT and MRI images. Their algorithm uses CNN to process two images and a weight map through a Siamese network, which uses Gaussian pyramid decomposition. They then perform Laplacian pyramid decomposition on each of the two images and finally perform another Laplacian pyramid decomposition on the resulting fused summary image. Torres [51] also used the DECU framework based on the CNN algorithm to generate an automatic summary of patient activity and determine the patient's health status by collecting other physiological parameters from video acquired by cameras and multiple sensors.

In other domains, Libovický [29] uses the seq2seq method to process instructional videos to generate tutorial summaries. Li [25] employs techniques such as R-CNN, ResNet, encoder-decoder and attention to produce product summaries in the e-commerce domain using a unique dataset. Song [47] utilizes the Swin Transformer and a Generative Pre-trained Language Model (GPLM) to generate product summaries in the e-commerce domain. Gao [17] employs the Sim Net network approach to implement code summaries in programming. Additionally, Ma [33] Gao [18] uses the Transformer architecture for code summarization.

3.2 Method Based on Integer Linear Programming (ILP)

Integer linear programming (ILP) is a method belonging to operations research that requires the decision variables to be integers. Unlike seq2seq, this approach directly intercepts the textual content, avoiding the problem of incoherent statements. This method was first used only for text summaries. Until Boudin [8] proposed an approximation algorithm that solved the NP-hard problem, and showed that it was not limited to text, but could also be applied to multimodal problems.

It wasn't until 2020 that Jangra [21] proposed a JILP-Multimodal summarization framework that achieved the task of summarizing multimodalities using ILP, and named the task TIVS, i.e. summarizing text-image-video. The method also simply uses a neural network approach to the output in a pre-processing phase, such as encoding the text using VGG. At its core, it uses the Joint-ILP Framework for core summarization.

Allawadi [1] also uses an ILP based model and has the same inputs and outputs as the previous JILP-Multimodal summarization framework [40]. However, his model is more refined and yields better accuracy and recall on ROGUE.

The decision variable of this method is:

$$M_{txt} = \left[m_{i,j}^{txt}\right] ; M_{img} = \left[m_{i,j}^{img}\right] ; M_c = \left[m_{ij}^c\right] \tag{1}$$

M(txt,img) is a binary square matrix of x*x. Whether txt or img is exemplar. c represents the cross-model, representing the correlation threshold between the image and the sentence. Its core function is:

$$
\begin{aligned}
f(x) = \text{Arg}_{\max} \Big\{ & \lambda_1 * m * k_{txt}^2 * \Big(\left[\textstyle\sum_{i=1}^n Mtxt_i * SIM_{\text{cosine}}\left(s_i, \quad O_{txt}\right)\right]^{(\alpha)} \\
& + \left[\textstyle\sum_{i=1}^n Mimg_{i,i} * SIM_{\text{cosine}}\left(s_i, O_{img}\right)\right]^{(\beta)} \Big) + \\
& \lambda_2 * (k_{txtt} + k_{img}) * k_{txt}^2 * \Big(\left[\textstyle\sum_{i=1}^n \sum_{j=1}^p M_{i,i}^c * SIM_{\text{cosine}}\left(s_i, \quad I_j\right)\right]^{(\gamma))} \\
& -\lambda_3 * (k_{txt} + k_{img}) * m * \Big(\left[\textstyle\sum_{i=1}^n \sum_{j=1}^n Mtxt_i * Mtxt_j * SIM_{\text{costne}}\left(s_i, I_j\right)\right]^{(\delta)} \Big) \Big\}
\end{aligned}
\tag{2}
$$

In the formula, α and β represent the salience score of the text-set and image-set, respectively. To avoid the problem of modal deviation, the coefficients m and ktxt+kimg are introduced. γ represents the cross-modal correlation score, δ represents the redundant part of the summary.

3.3   Method Based on Submodule Optimization

The submodule function is an aggregation function that provides a more tangible representation of diminishing marginal utility in the economic domain. Similarly to ILP, in 2010, Lin [31] first proposed applying this modified greedy algorithm to text summarization.

Until 2016, in the field of journalism, Modani [38] proposed an approach that uses a five-part submodule function to generate a summary of both text and image modalities. The method innovatively defines the image coverage term and an image diversity reward term for images. Allowing for the generic generation of a bimodal summary of text-image composition. Subsequently, the new method proposed by Li [27] reached new heights by being able to process four modalities: text, image, audio, and video. Chali [9] uses three measures of importance, coverage, and non-redundancy as submodule functions to detect sentence summaries. Tiwari [49] proposed a method for generating final summaries using three measures of coverage, novelty, and significance as submodule functions. The formula is:

$$f_{\text{coverage}}\left(S\right) = \frac{\left|\left\{w \in S | w \in \left(V^{txt} \cup V^{vis}\right)\right\}\right|}{\left|V^{txt} \cup V^{vis}\right|}. \tag{3}$$

$$f_{\text{novelty}}\left(S\right) = \sum_w \max_{d \in S}\left\{0, \min_{d' \in S - \{d\}}\left\{\phi(d, w) - \phi\left(d', w\right)\right\}\right\}. \tag{4}$$

$$f_{\text{sig}}\left(S\right) = \log\left(c_{sig}\right) + \cos\left(\overrightarrow{d^{txt}}, \overrightarrow{v^{txt}},\right) + \cos\left(\overrightarrow{d^{vis}}, \overrightarrow{v^{vis}}\right). \tag{5}$$

They given a summary S. Coverage is as the fraction of $stxt$ and $svis$ of the vocabulary covered by the summary. Novelty means that the model should give preference to sentences with new information. w is a textual or visual word that appears in the document d of the summary.

They model the vectors $dtxt$ and $dvis$ and calculate their weighted cosine similarity. Capturing the importance of the document to the topic.

The authors use a Markov Random Fields-based similarity measure to compare different descriptions of the same or similar content across different platforms and track events over time to reconstruct the full event. Finally, the final content is selected using a submodule function-based approach. The core function is computed as follows:

$$f(A \cup \{s\}) - f(A) \geq f(B \cup \{s\}) - f(B) \tag{6}$$

In this formula, $A, B \subseteq s$, $A \subseteq B$, $s \in \frac{S}{B}$. S is a set.

3.4   Method Based on Graphs

Graph-based approaches have been used in the field of automatic summarization for a longer period. In 2004, the Graph based approach was applied to the field of journalism, and the Textrank method proposed by Mihalcea [37] has been able to extract important sentences from large news articles.

Until 2016, the Graph-based approaches was heavily used for multimodal automatic summarization. It was in the above-mentioned work by Modani [38] that a modified graph-based approach and a modification to the submodular approach were used to summarize both text and image modalities. Moreover, the proposed graph-based approach could handle not only images but also documents. The approach sets up images and documents as nodes into the graph, uses the connections between the nodes as weights based on similarity, and sets a reward score, as well as attach a cost. Finally, a greedy algorithm is used to select the most appropriate summary. Schinas [45] also proposed an MGraph framework for the textual, visual, temporal, and social multimodal content in social networking sites for visual summary summarization.

Subsequently, another paper by Li [27], mentioned above, also used the Graph based approach and summarized the four modalities of text-image-audio-video. In this case, the GBA (Graph based approach) is used to calculate the salience score of a text set. The text set here includes text documents, but also a large amount of text that may be incorrectly transcribed from speech. These sentences are treated as nodes to form a graph. The formula for calculating the salience score is computed as follows:

$$\begin{cases} Sa\,(t_i) = \mu \sum_j Sa\,(t_j) \cdot M_{ji} + \frac{1-\mu}{N} \\ M_{ji} = \mathrm{sim}\,(t_j, t_i) \end{cases} \tag{7}$$

In this formula, $\mu = 0.85$, N is total number of the text units; Mji is the relationship between text unit ti and ti; Ti is averaging the embedding of the words in ti. And $Sim(,)$ means cosine similarity between two texts.

Zhu [56] proposed an unsupervised graph-based multimodal summarization model, which does not require the dataset to contain annotations in order to perform summarization. The method classifies modal summarization into modal-mixed and modal non-mixed according to the form of output and can perform either unimodal or multimodal output to suit different application scenarios. Additionally, the method can also measure the similarity between text and images through the model.

Recently, Sun [48] applied the Graph based approach to the field of remote sensing images with Multimodal change detection for remote sensing for Earth observation. The approach performs a regression summary of different modal satellite images for regression summarization.

3.5   Method Based on LDA Topic Model

The Latent Dirichlet allocation (LDA) Topic Model [7] can be utilized for extracting visual words from images through feature extraction and clustering algorithms, thereby facilitating multimodal summarization.

Their approach towards Multimodal summarization has mainly been applied in the field of journalism. Bian [5] proposed the multimodal-LDA method for summarizing social media data in microblogs. The article first detects events, then quickly summarizes the most representative sub-topics, and generates a fluent summary text based on it to restore the entire process of the event quickly. On this basis, different summary focuses are selected based on the type of news to provide a more realistic picture of the event. However, the method may face difficultly in distinguishing the focus for mixed events or events in borderlands. Additionally, the summarization performance is significantly reduced for news with inconsistent text and images. The model Inference formula is as follows:

$$\varphi_k^{TS}(w) = \frac{N^w(Z = k, R = S) + \lambda^{TS}}{\sum_{t \in V^t}(N^t(Z = k, R = S) + \lambda^{TS})} \tag{8}$$

$$\varphi_k^{VS}(u) = \frac{N^u(Z = k, Q = S) + \lambda^{VS}}{\sum_{u \in V^v}(N^u(Z = k, Q = S) + \lambda^{VS})} \tag{9}$$

In the formula, $\phi_k^{TS}(w)$ represents the probability of w occurring in the kth specific text distribution, while $\phi_k^{VS}(u)$ represents the probability of the visual distribution. Where Vt and Vv denote text words and visual words, respectively.

$N^w(Z = k, R = S)$, $N^u(Z = k, Q = S)$ denote the number of text words after the sampling process.

Bian [6] proposed a method for removing latent noise images using a spectral filtering model as the core method, which allowed the algorithm to address the above problem well. In another work, Li [28] proposed the hierarchical latent Dirichlet allocation (HLDA) model to analyze the subject structure of news and then used subsequent methods such as crawlers and MST algorithms to process the subject matter. Wadagave [53] proposed the multimodal-LDA (MMLDA) summarization using the TWITTER API, which can also generate visual summaries.

3.6 Domain Specific Techniques

We can see that the above techniques are the dominant approaches to Multimodal summarization, but there are specific times when researchers use their own unique techniques suited to situations and particular data sets. These techniques are often related to relevant characteristics within the domain.

In sports, key sporting moments are often replayed in slow motion, and spectators will remain silent before a serve and then cheer loudly after a goal is scored. These specific phenomena can help the model to better identify key highlights of sports. Tjondronegoro [50] used this idea, together with the Video/Text Alignment Module, Social Media Classification Module and Text Analysis Module to complete automatic summaries of sports matches. Sanabria [44] also uses similar ideas and completes multimodal summaries with methods such as multi-instance learning neural networks. There are also specific features that can indicate the presence of key content in a session to avoid watching meaningless video content from start to finish. Erol [13] suggests that this can be done by analyzing sound direction and audio amplitude, local luminance variations, and term frequency-inverse document frequency measure, or even the video's movements of the characters to identify key sections for summary output.

In the field of e-commerce, Li [25] not only used a method based on neural networks but also adopted an aspect-based reward augmented maximum likelihood (RAML) training method [50], which effectively summarizes the aspects such as "Capacity", "Control", "Motor" and so on.

In the field of film summarization, Evangelopoulos [14] also used the concept of saliency to analyze three perspectives - auditory, visual, and textual - to obtain key frames of the film and summarize them. A multimodal fusion technique was eventually used to generate a comprehensive attention model. Finally, a summary is generated by extracting the most important scenes and episodes from the film based on the attention weights.

4   Taxonomy of the Methods

Although there are currently some articles that review similar topic, they are generally published too early or do not describe some areas. In this survey, I browsed through over 200 articles from 2003 to 2022 and selected the most valuable nearly 50 of them for classification statistics. They were classified by the main method into: Method Based on ILP, Method Based on Submodule Optimization, Method Based on Graphs, Method Based on LDA Topic Model, Domain Specific Techniques, and on this basis they are divided chronologically by application area. In the selection of papers, for the early years, we chose articles with high citation numbers. For the less cited articles of the last two years, we chose to use articles from higher quality publications. However, the datasets used for multimodal summarization tasks are not uniform.

Table 1. A list of methods, datasets, input and output patterns and their applications, with T(Text), I(Image), A(Audio), V(video) data.

Paper	Method	Input	Output	Datasets	Application
[35]	neural	A,V	T	60 meeting recordings (30 recordings × 2 participant sets)	Meetings
[15]	neural	T, A, V	A, V	3 movie segments	Movies
[14]	neural	T, A, V	A, V	7 half hour segments of movies	Movies
[26]	neural	T, I	T	66,000 triplets (sentence, image and summary)	News
[22]	neural	T, I, V	T	300 h of short instructional videos spanning different domains	General
[10]	neural	T, I	T, I	219k documents	News
[51]	neural	T, V	T	ICU patient Data set (author created)	Medical
[47]	neural	T, I	T	1.4 million products covering three coarse categories	Other
[17]	neural	T, C	C	10 open source Java projects, 40932 Ethereum Smart Contracts (ESC) code	Other
[21]	ILP	T, I, A, V	T, I, A, V	25 themes (500 documents, 151images, 139 videos)	News
[27]	sub/graph	T, I, A, V	T	66,000 triplets (sentence, image and summary)	News
[56]	graph	T, I	T, I	293,965 document,1,928,356 image	General
[6]	LDA	T, I	T, I	20 topics	News
[50]	specific	T, A, V	T	313k document, 2.0m image, news document, image title pair, sentence summary	Sport

Table 1 shows the basic information on commonly used datasets, the input and output modes of the paper, and the fields and sources of the paper. Method based on neural networks are still the dominant methods in the current methodology and are the focus of research in this survey. Figure 2 shows the current percent of each method.

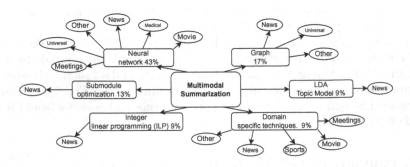

Fig. 2. Framework of Multimodality Automatic Summarization

5   Challenges and Issues

5.1   Challenges

Evaluation Criteria: There is no single correct answer to a multimodal summary. Even for manual assessments, there is no absolute perfect answer, limited by the personal preferences of the reader. And for machine assessment, there are different measures such as Rouge [30] scores. Furthermore, the output of multimodal summaries is also often multimodal, and it is difficult to measure the strengths and weaknesses between different modalities; in many cases, evaluation criteria do not allow for a comparison between methods that output text and images.

Currently, more advanced evaluation criteria [38] use vector function and reward mechanisms and they avoid to use Ground Truth. The methodological equation is:

$$\mu_M = \mu_T + \mu_I + \sigma_{T,I} \tag{10}$$

$$\mu_{(T,I)} = \sum_{w \in T,I} \hat{R}_{v,w} * \max_{x \in S,I}\{\text{Sim}(x,y)\} \tag{11}$$

$$\sigma_{T,I} = \sum_{v \in S} \sum_{w \in I} \left\{ \text{Sim}(v,w) * R_v * \hat{R}_w \right\} \tag{12}$$

where μT and μI are diversity-aware information coverage measures for the text and image parts of the summary, respectively. σT, I denotes the sum of the similarity between sentences and images in the summary across all pairs. $R *$ $\max_{x \in S,I}\{\text{Sim}(x,y)\}$ denotes the maximum similarity between sentence in the document text and any sentence in the summary. R is the reward value. However, this kind of methods lack normalization, and the results are heavily influenced by the length of the content.

No New Image Generated: Many methods in multimodal summarization use multimodal output when outputting, and their output often contains images. However, these methods generally output images by selecting the relevant image in the input video or image, and in the network, for output. The problem is that when there is no suitable image in the input video, the output becomes difficult and the quality of the output becomes lower, even if there is a mismatch between the text and the image.

Poor Quality Data Set: For machine learning, having datasets that perform well across various domains is crucial. However, currently, there are many datasets that are too small [21,27] or lack specific domains, with few datasets available.

Modal Alignment: Most methods have difficulty dealing with asynchronous modal data that is not aligned. Cross-modal alignment requires the resolution of timing asynchronies and scale inconsistencies between modalities.

Multimodal Semantic Understanding: The process of generating summaries requires semantic understanding and analysis of multimodal data. This includes the recognition and understanding of objects, scenes, etc. in images and videos, and the modelling of semantics in text.

5.2   Issues

Application Technologies: New technologies such as chat GPT [42] and DALL-E 2 [43] can solve problems such as no new images being generated, poor output text and difficulties with human-computer interaction. Neural network, with encoder-decoder as the core or use the transformer method are likely to become the mainstream approach to summarisation.

Deeper Applications in Medicine: Current approaches in the medical mainly use cnn for fusion abstraction of images from different modalities [16], while the critical text is neglected. In the future, summarising and outputting text modalities and images acquired by multiple sensors as a reference for doctors' decision making in routine examinations and ICUs will reduce doctors' decision making time.

Multimodal Alignment: Data heterogeneity, modal imbalance and semantic splitting make it difficult for multimodal approaches to achieve alignment. The performance and stability of multimodal alignment can be improved through data pre-processing, feature fusion and migration learning.

Real Time Summarization: Facing the sports domain, multi-modal real-time summarisation can be performed through specific scenarios, broadcast in different languages for different groups of people. Using machine learning algorithms, natural language processing (NLP) techniques, attention mechanisms, combined with text generation models, concise and accurate summaries of the competition can be generated. The generated text summaries are translated into the target language using machine translation models and natural and fluent speech announcements are generated using speech synthesis techniques.

Post-Joint Representation Approach: This can be addressed using joint representation learning or stepwise fusion strategies. The model considers the relationship between multiple modes at the same time in the training process, rather than just fusion information in the later stage.

Better Evaluation Criteria: Evaluation criteria will become more comprehensive and accurate. We could continue to use the vector function and reward mechanism from the reinforcement Learning. In addition representation learning can be introduced to extract Low-dimensional representations. Convergence modelling using multiple methods

Datasets Expanded: Datasets will cover more areas and a variety of data forms, and many new datasets will be constructed.

6 Conclusions

Multimodal summarization tasks allow people to navigate information from text, images, audio, and video more effectively. In this paper, we defined the problem and analysed the extent to which existing mainstream methods are used in different domains with the datasets provided. We identified a number of papers using new approaches and application areas that have not been summarized before. Through reflection and analysis, we enumerated the challenges currently faced by existing technologies, predicted possible future trends, and described some research issues and directions for future development.

References

1. Allawadi, S., Rana, V., Jain, M., et al.: Multimedia data summarization using joint integer linear programming. In: 2021 5th International Conference on Computing Methodologies and Communication (ICCMC), pp. 1462–1466. IEEE (2021)
2. Bahdanau, D., Cho, K., Bengio, Y.: Neural machine translation by jointly learning to align and translate. arXiv preprint arXiv:1409.0473 (2014)
3. Baltrušaitis, T., Ahuja, C., Morency, L.P.: Multimodal machine learning: a survey and taxonomy. IEEE Trans. Pattern Anal. Mach. Intell. **41**(2), 423–443 (2018)
4. Bateman, J.A.: Multimodality and Genre. Palgrave Macmillan UK, London (2008). https://doi.org/10.1057/9780230582323
5. Bian, J., Yang, Y., Chua, T.S.: Multimedia summarization for trending topics in microblogs. In: Proceedings of the 22nd ACM International Conference on Information and Knowledge Management, pp. 1807–1812 (2013)
6. Bian, J., Yang, Y., Zhang, H., Chua, T.S.: Multimedia summarization for social events in microblog stream. IEEE Trans. Multimedia **17**(2), 216–228 (2014)
7. Blei, D.M., Ng, A.Y., Jordan, M.I.: Latent dirichlet allocation. J. Mach. Learn. Res. **3**(Jan), 993–1022 (2003)
8. Boudin, F., Mougard, H., Favre, B.: Concept-based summarization using integer linear programming: from concept pruning to multiple optimal solutions. In: Conference on Empirical Methods in Natural Language Processing (EMNLP) 2015 (2015)
9. Chali, Y., Tanvee, M., Nayeem, M.T.: Towards abstractive multi-document summarization using submodular function-based framework, sentence compression and merging. In: Proceedings of the Eighth International Joint Conference on Natural Language Processing (Volume 2: Short Papers), pp. 418–424 (2017)

10. Chen, J., Zhuge, H.: Abstractive text-image summarization using multi-modal attentional hierarchical RNN. In: Proceedings of the 2018 conference on empirical methods in natural language processing, pp. 4046–4056 (2018)

11. Chen, J., Zhuge, H.: Extractive text-image summarization using multi-modal RNN. In: 2018 14th International Conference on Semantics, Knowledge and Grids (SKG), pp. 245–248. IEEE (2018)

12. Chen, J., Zhuge, H.: News image captioning based on text summarization using image as query. In: 2019 15th International Conference on Semantics, Knowledge and Grids (SKG), pp. 123–126. IEEE (2019)

13. Erol, B., Lee, D.S., Hull, J.: Multimodal summarization of meeting recordings. In: 2003 International Conference on Multimedia and Expo. ICME 2003. Proceedings (Cat. No. 03TH8698), vol. 3, pp. III-25. IEEE (2003)

14. Evangelopoulos, G., et al.: Multimodal saliency and fusion for movie summarization based on aural, visual, and textual attention. IEEE Trans. Multimedia **15**(7), 1553–1568 (2013)

15. Evangelopoulos, G., et al.: Video event detection and summarization using audio, visual and text saliency. In: 2009 IEEE International Conference on Acoustics, Speech and Signal Processing, pp. 3553–3556. IEEE (2009)

16. Fan, F., et al.: A semantic-based medical image fusion approach. arXiv preprint arXiv:1906.00225 (2019)

17. Gao, X., Jiang, X., Wu, Q., Wang, X., Lyu, C., Lyu, L.: Gt-SimNet: improving code automatic summarization via multi-modal similarity networks. J. Syst. Softw. **194**, 111495 (2022)

18. Gao, Y., Lyu, C.: M2TS: multi-scale multi-modal approach based on transformer for source code summarization. In: Proceedings of the 30th IEEE/ACM International Conference on Program Comprehension, pp. 24–35 (2022)

19. Greff, K., Srivastava, R.K., Koutník, J., Steunebrink, B.R., Schmidhuber, J.: LSTM: a search space odyssey. IEEE Trans. Neural Networks Learn. Syst. **28**(10), 2222–2232 (2016)

20. Hahn, U., Mani, I.: The challenges of automatic summarization. Computer **33**(11), 29–36 (2000)

21. Jangra, A., Jatowt, A., Hasanuzzaman, M., Saha, S.: Text-image-video summary generation using joint integer linear programming. In: Jose, J.M., et al. (eds.) ECIR 2020. LNCS, vol. 12036, pp. 190–198. Springer, Cham (2020). https://doi.org/10.1007/978-3-030-45442-5_24

22. Khullar, A., Arora, U.: Mast: multimodal abstractive summarization with trimodal hierarchical attention. arXiv preprint arXiv:2010.08021 (2020)

23. Krizhevsky, A., Sutskever, I., Hinton, G.E.: ImageNet classification with deep convolutional neural networks. Commun. ACM **60**(6), 84–90 (2017)

24. LeCun, Y., Bengio, Y., Hinton, G.: Deep learning. Nature **521**(7553), 436–444 (2015)

25. Li, H., Yuan, P., Xu, S., Wu, Y., He, X., Zhou, B.: Aspect-aware multimodal summarization for Chinese e-commerce products. In: Proceedings of the AAAI Conference on Artificial Intelligence, vol. 34, pp. 8188–8195 (2020)

26. Li, H., Zhu, J., Liu, T., Zhang, J., Zong, C., et al.: Multi-modal sentence summarization with modality attention and image filtering. In: IJCAI, pp. 4152–4158 (2018)

27. Li, H., Zhu, J., Ma, C., Zhang, J., Zong, C.: Multi-modal summarization for asynchronous collection of text, image, audio and video. In: Proceedings of the 2017 Conference on Empirical Methods in Natural Language Processing, pp. 1092–1102 (2017)

28. Li, Z., Tang, J., Wang, X., Liu, J., Lu, H.: Multimedia news summarization in search. ACM Trans. Intell. Syst. Technol. **7**(3), 1–20 (2016)
29. Libovický, J., Palaskar, S., Gella, S., Metze, F.: Multimodal abstractive summarization for open-domain videos. In: Proceedings of the Workshop on Visually Grounded Interaction and Language (ViGIL). NIPS (2018)
30. Lin, C.Y.: Rouge: a package for automatic evaluation of summaries. In: Text Summarization Branches Out, pp. 74–81 (2004)
31. Lin, H., Bilmes, J.: Multi-document summarization via budgeted maximization of submodular functions. In: Human Language Technologies: The 2010 Annual Conference of the North American Chapter of the Association for Computational Linguistics, pp. 912–920 (2010)
32. Liu, Y., Chen, X., Cheng, J., Peng, H.: A medical image fusion method based on convolutional neural networks. In: 2017 20th International Conference on Information Fusion (Fusion), pp. 1–7. IEEE (2017)
33. Ma, Z., Gao, Y., Lyu, L., Lyu, C.: MMF3: neural code summarization based on multi-modal fine-grained feature fusion. In: Proceedings of the 16th ACM/IEEE International Symposium on Empirical Software Engineering and Measurement, pp. 171–182 (2022)
34. Mani, I.: Automatic Summarization, vol. 3. John Benjamins Publishing (2001)
35. McCowan, I., et al.: Modeling human interaction in meetings. In: 2003 IEEE International Conference on Acoustics, Speech, and Signal Processing, 2003. Proceedings. (ICASSP 2003), vol. 4, pp. IV-748. IEEE (2003)
36. McGurk, H., MacDonald, J.: Hearing lips and seeing voices. Nature **264**(5588), 746–748 (1976)
37. Mihalcea, R., Tarau, P.: TextRank: bringing order into text. In: Proceedings of the 2004 Conference on Empirical Methods in Natural Language Processing, pp. 404–411 (2004)
38. Modani, N., et al.: Summarizing multimedia content. In: Cellary, W., Mokbel, M., Wang, J., Wang, H., Zhou, R., Zhang, Y. (eds.) Web Information Systems Engineering-WISE 2016: 17th International Conference, Shanghai, China, November 8–10, 2016, Proceedings, LNCS, Part II 17, pp. 340–348. Springer, Cham (2016). https://doi.org/10.1007/978-3-319-48743-4_27
39. Nallapati, R., Zhai, F., Zhou, B.: SummaRuNNer: a recurrent neural network based sequence model for extractive summarization of documents. In: Proceedings of the AAAI Conference on Artificial Intelligence, vol. 31 (2017)
40. O'Halloran, K.L.: Interdependence, interaction and metaphor in multisemiotic texts. Soc. Semiot. **9**(3), 317–354 (1999)
41. Palaskar, S., Libovický, J., Gella, S., Metze, F.: Multimodal abstractive summarization for how2 videos. arXiv preprint arXiv:1906.07901 (2019)
42. Radford, A., Narasimhan, K., Salimans, T., Sutskever, I.: Improving language understanding by generative pre-training (2018)
43. Ramesh, A., Dhariwal, P., Nichol, A., Chu, C., Chen, M.: Hierarchical text-conditional image generation with clip latents. arXiv preprint arXiv:2204.06125 (2022)
44. Sanabria, M., Sherly, Precioso, F., Menguy, T.: A deep architecture for multimodal summarization of soccer games. In: Proceedings Proceedings of the 2nd International Workshop on Multimedia Content Analysis in Sports, pp. 16–24 (2019)
45. Schinas, M., Papadopoulos, S., Kompatsiaris, Y., Mitkas, P.A.: MGraph: multimodal event summarization in social media using topic models and graph-based ranking. Int. J. Multimedia Inf. Retrieval **5**, 51–69 (2016)

46. Simonyan, K., Zisserman, A.: Very deep convolutional networks for large-scale image recognition. arXiv preprint arXiv:1409.1556 (2014)

47. Song, X., Jing, L., Lin, D., Zhao, Z., Chen, H., Nie, L.: V2P: vision-to-prompt based multi-modal product summary generation. In: Proceedings of the 45th International ACM SIGIR Conference on Research and Development in Information Retrieval, pp. 992–1001 (2022)

48. Sun, Y., Lei, L., Tan, X., Guan, D., Wu, J., Kuang, G.: Structured graph based image regression for unsupervised multimodal change detection. ISPRS J. Photogramm. Remote. Sens. 185, 16–31 (2022)

49. Tiwari, A., Weth, C.V.D., Kankanhalli, M.S.: Multimodal multiplatform social media event summarization. ACM Trans. Multimedia Comput. Commun. Appl. 14(2s), 1–23 (2018)

50. Tjondronegoro, D., Tao, X., Sasongko, J., Lau, C.H.: Multi-modal summarization of key events and top players in sports tournament videos. In: 2011 IEEE Workshop on Applications of Computer Vision (WACV), pp. 471–478. IEEE (2011)

51. Torres, C., Rose, K., Fried, J.C., Manjunath, B.: Summarization of ICU patient motion from multimodal multiview videos. arXiv preprint arXiv:1706.09430 (2017)

52. Tsai, Y.H.H., Bai, S., Liang, P.P., Kolter, J.Z., Morency, L.P., Salakhutdinov, R.: Multimodal transformer for unaligned multimodal language sequences. In: Proceedings of the Conference. Association for Computational Linguistics. Meeting, vol. 2019, p. 6558. NIH Public Access (2019)

53. Wadagave, P., Garg, B.: A heterogeneous data summarization system to generate automatic summary of data using twitter API using tweets, images etc. Harbin Gongye Daxue Xuebao J. Harbin Inst. Technol. 54(6), 167–172 (2022)

54. Yuhas, B.P., Goldstein, M.H., Sejnowski, T.J.: Integration of acoustic and visual speech signals using neural networks. IEEE Commun. Mag. 27(11), 65–71 (1989)

55. Zhu, J., Li, H., Liu, T., Zhou, Y., Zhang, J., Zong, C.: MSMO: multimodal summarization with multimodal output. In: Proceedings of the 2018 conference on empirical methods in natural language processing, pp. 4154–4164 (2018)

56. Zhu, J., Xiang, L., Zhou, Y., Zhang, J., Zong, C.: Graph-based multimodal ranking models for multimodal summarization. Trans. Asian Low Resour. Lang. Inf. Process. 20(4), 1–21 (2021)

57. Zhu, J., Zhou, Y., Zhang, J., Li, H., Zong, C., Li, C.: Multimodal summarization with guidance of multimodal reference. In: Proceedings of the AAAI Conference on Artificial Intelligence, vol. 34, pp. 9749–9756 (2020)

Deep Knowledge Tracing with Concept Trees

Yupei Zhang[1,2,3], Rui An[1,2,3], Wenxin Zhang[1,2,3], Shuhui Liu[1,2,3], and Xuequn Shang[1,2,3(✉)]

[1] School of Computer Science, Northwestern Polytechnical University, Xi'an, China
{ypzhaang,shang,zwx}@nwpu.edu.cn, anrui@mail.nwpu.edu.cn
[2] MIIT Big Data Storage and Management Lab, Xi'an, China
[3] Department of Automation, Tsinghua University, Beijing, China
shliu620@tsinghua.edu.cn

Abstract. Knowledge tracing aims to diagnose the student's knowledge status and predict the responses to the next questions, which is a critical task in personalized learning. The existing studies consider more academic features, while this paper introduces DKCT, a deep knowledge tracing model with concept trees, to integrate the hierarchical concept tree that describes the structure of concepts in a question. DKCT casts the knowledge concept tree (KCT) in a question from the views of feature, breadth, and difficulty into a KCT representation at first. Then, DKCT is composed of an encoder network with multi-head attention on the question representations and a decoder network with multi-head attention on the interaction embeddings. Finally, DKCT integrates the student embeddings by using fully connected networks to predict the responses to the next questions. Extensive experiments conducted on two real-world educational datasets show that DKCT has a higher prediction accuracy than the currently popular KT models. This work paves the way to consider KCT for knowledge tracing.

Keywords: Data Mining · Intelligent Education · Concept Tree · Knowledge Tracing · Personalized Exercises Recommendation

1 Introduction

Student knowledge tracing (KT) aims to acquire the current knowledge state from question-answering records to predict the responses to the next questions [8,22]. With KT models, an automatic online-learning system guides a student with suitable exercises to smooth the learning progress [3,11]. Inspired by the powerful deep learning strategy, Piech et al. proposed the well-known model of deep knowledge tracking (DKT) by using the recurrent neural network (RNN) to learn student knowledge representation [14]. DKT paves the way to learning deep features from the sequential exercising records [19], resulting in user-friendly performance on exercise recommendations. Its kernel position in personalized learning caused a large number of studies in recent years, such as the regularized DKT (DKT+) [18], exercise-aware KT (EKT) [7], and self-attentional neural knowledge tracking (SAINT) [2].

© The Author(s), under exclusive license to Springer Nature Switzerland AG 2023
X. Yang et al. (Eds.): ADMA 2023, LNAI 14177, pp. 377–390, 2023.
https://doi.org/10.1007/978-3-031-46664-9_26

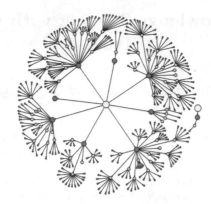

Fig. 1. The KC trees in Eedi [17]. The color points, i.e. Nattierblue-blue-Yellow-Green, represent the knowledge concepts at the different levels of the KC trees.

The current studies are mainly concentrated on enhancing the RNN structure, integrating the side information, and fusing the data relationship [8]. DKT introduces the deep learning model into the field of knowledge tracing [14]. To boost the DKT model, DKT+ introduced two regularization terms of reconstruction and waviness to remove the inconsistency of the knowledge state along time series [18]. To integrate more useful information, EKT explores students' exercising records and the question's text contents to improve the perdition performance [7]. While interaction features are more important to the raw KT, Pandey et al. [13] proposed a self-attention KT model (SAKT) to capture the long-term dependency among interactions; Choi et al. proposed a separated self-attentional neural KT (SAINT) model to embed exercises and interactions by encoder-decoder attention networks [2]; Hamid et al. proposed to leverage the graph convolutional network (GCN) to embed high-order relations into the question and student representations for KT [6]. Besides, dynamic key-value memory network (DKVMN) introduces a key matrix of knowledge concepts (KCs) and a value matrix of mastery states into a memory-augmented neural network [15] to improve the interpretability [20].

However, a few investigations take into account the structured features of KCs, which is indeed the important target in a KT task. In the real world, a KC can usually be organized into a hierarchical tree [10], called a knowledge concept tree (KCT), as shown in Fig. 1. KCT provides all sub-KCs for a specific KC from coarse to fine grain. Without consideration of the KCT, traditional KT methods often fail to capture fine-grained concepts, leading to inaccurate and hard-understanding KT results. As shown in Fig. 2, traditional KT models mark the student as having mastery status of "inequality" when getting a correct response to Question 4, but the student lefts most sub-KCs of "inequality" with no mastery. Such a problem harms the KT models in real-world applications.

In this paper, we proposed deep knowledge tracing with concept trees, dubbed DKCT, to integrate the KCTs into a KT model. DKCT first encodes

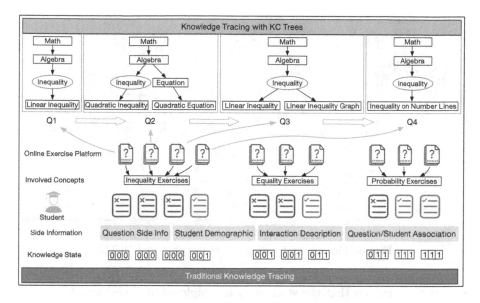

Fig. 2. The concerned KC trees in knowledge tracing.

KC trees by constructing tree-structure tensors, performing upward cumulative average operations, and doing branch-level aggregation, and then integrates the features from questions and interactions by extending the transformer architecture [16], followed by training a linear classifier on both features of decoder and student. Our contributions lie in two-fold as follows:

1. This paper introduces a new KT route by considering the knowledge-concept tree (KCT) embedded in a question. KCT could help improve the performance and interpretation of knowledge tracing.
2. We propose an implementation method, DKCT, by using the popular Transformer model to integrate the KCT embedding. Wherein we encode the features from multiple views, including features, breadths, and depths.

In addition, DKCT achieves higher prediction accuracy on two public educational datasets than the popular KT models. The rest of this paper is organized as follows. Section 2 reviews the DKT problem and presents the proposed method. Then we present the experiments and the results in Sect. 3. Finally, we conclude this study and discuss the results in Sect. 4.

2   The Proposed Method

2.1   Problem Definition

DKCT can be formulated into the following supervised learning problem. Denote the j-th student's past $t - 1$ interactions by $\mathcal{I}_{t-1}^{j} = \{I_1, I_2, \cdots, I_{t-1}\}$, where $I_t = (q_t, kct_t, r_t, d_t, s_j)$. Note that here

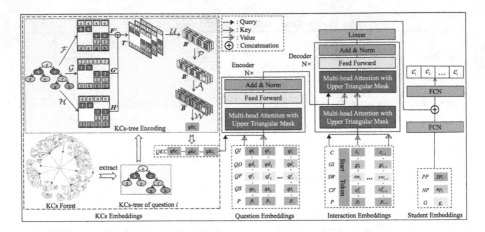

Fig. 3. The schematic diagram of the proposed DKCT.

- q_t is the question that the student s_j is completed at timestamp t;
- kct_t is a subtree of the KC trees embedded in question q_t, as in Fig. 1;
- r_t is the correctness of the student j's response, i.e., being 1 for a correct response and 0 for an incorrect response;
- d_t is the side information of interactions, e.g., response time;
- s_j is the side information of the student j.

The goal of DKCT is then to predict the probability P that the student j gives a correct response to the next question t, i.e., $P(r_{t+1} = 1 | q_{t+1}, kct_t, \mathcal{I}_{t-1}^j)$.

2.2   KC-Tree Encoded from Multiple Views

To encode the KC tree kct_t of the question t, we traversed the tree hierarchically and defined three sets of tree nodes, i.e., a set of leaf nodes denoted by $\mathcal{L}^t = \{l_1, l_2, \cdots, l_m\} \in \mathbb{R}^m$, a set of non-leaf nodes denoted by $\mathcal{N}^t = \{n_1, n_2, \cdots, n_k\} \in \mathbb{R}^k$, and a set of descendant nodes of all non-leaf nodes $\mathcal{C}^t = \{C_1, C_2, \cdots, C_k\}$ where C_i is the set of descendant nodes of the non-leaf node i.

In this study, we encoded the KC tree from multiple views, i.e., KC features, KC breadths, and KC levels, by introducing three encoding tensors into DKCT, respectively. The detail on encoding KCT has three steps as follows.

(1) The *KC-feature view* is corresponding to the KCs contained in a question and the structural relationships between KCs. We constructed the tensor $F^t \in \mathbb{R}^{(k+1) \times m \times d}$ with a map $\mathcal{F}$ by

$$f_{ij} := \mathcal{F}(\mathcal{L}^t, \mathcal{N}^t, \mathcal{C}^t)_{ij} = \begin{cases} x_j^L, & if \ i = k+1 \\ x_i^N, & elseif \ l_j \in c_i \\ \mathbf{0}, & otherwise \end{cases} \tag{1}$$

where $f_{ij}, (i = \{1, 2, \cdots, k+1\}, j = \{1, 2, \cdots, m\})$ is the element located at the i-th row and the j-th column of the tensor F^t; $x_j^L \in \mathbb{R}^d$ and $x_i^N \in \mathbb{R}^d$ are the embedding of the leaf nodes l_j and the non-leaf nodes n_i; $\mathbf{0}$ denotes $d \in \mathbb{R}^d$ with all entries being zero.

(2) The *KC-breadth view* is to mention the involved scope of a knowledge concept, i.e., the number of fine concepts in a KC, which is a common prior in the real world. Meanwhile considering the tree branch, we constructed the tensor $G^t \in \mathbb{R}^{(k+1) \times m \times d}$ with a map $\mathcal{G}$ by

$$g_{ij} := \mathcal{G}(\mathcal{L}^t, \mathcal{N}^t, \mathcal{C}^t) = \begin{cases} x_{ij}^A, & \text{if } i < k+1 \text{ and } l_j \in c_i \\ \mathbf{0}, & \text{otherwise} \end{cases} \quad (2)$$

where $g_{ij}, (i \in \{1, 2, \cdots, k+1\}, j \in \{1, 2, \cdots, m\})$ is the element of the tensor G^t; $x_{ij}^A \in \mathbb{R}^d$ is the embeddings of $|v_{ij}| = |\{n_p | n_p \in C_i \text{ and } l_j \in C_p\}|$, v_{ij} is the set of l_j's ancestors up to n_i, and $|\bullet|$ denotes the number of elements in the set.

(3) The *KC-level view* aims to encode the intrinsic difficulty of a knowledge concept, i.e., the depth of a concept from its root in the KC tree that is shown in Fig. 1. We constructed $H^t \in \mathbb{R}^{(k+1) \times m \times d}$ with a map $\mathcal{H}$ by

$$h_{ij} := \mathcal{H}(\mathcal{L}^t, \mathcal{N}^t, \mathcal{C}^t)_{ij} = \begin{cases} x_j^{L-level}, & \text{if } i = k+1 \\ x_i^{N-level}, & \text{else if } l_j \in C_i \\ \mathbf{0}, & \text{otherwise} \end{cases} \quad (3)$$

where $h_{ij}, (i \in \{1, 2, \cdots, k+1\}, j \in \{1, 2, \cdots, m\})$ is the element of the tensor H^t; $x_j^{L-level} \in \mathbb{R}^d$ and $x_i^{N-level} \in \mathbb{R}^d$ are the embeddings of $level(l_j)$ and $level(n_i)$, where $level(\cdot)$ is the hierarchy of the input node in the KC tree.

Finally, the KC embeddings from the above three views are combined into T^t by

$$T^t = F^t \oplus G^t \oplus H^t \quad (4)$$

where $\oplus$ denotes the concatenation operation in the feature dimension. This study sets the embedding dimensions to $2/d$, $4/d$, and $4/d$, respectively.

2.3   DKCT: Deep Knowledge Tracing with Concept Trees

In our study, we assume the KC trees are given, which organizes all knowledge concepts involved in all questions into hierarchical structures, as shown in Fig. 3 (left below). Then, our proposed DKCT method is composed of the following main steps.

In the first step, we calculated the multi-view KC embeddings by using Eq. (4) based on the given KC trees, referring to Subsect. 2.2.

In the second step, we performed the upward cumulative-average operation $\mathcal{U}$ on T^t to knit the representation vector for the KC tree t in a bottom-up way [12]. The resulting vector $B^t = (b_1, b_2, \cdots, b_m) \in \mathbb{R}^{m \times d}$ indicates branch features, where each element in B^t is obtained by averaging all its descendants in a particular branch, i.e.,

$$b_j := \mathcal{U}(T^t)_{ij} = \frac{\sum_{i=1}^{k+1} t_{ij}}{\#\{\mathbf{I}(t_{ij}) | i = \{1, 2, \cdots, k+1\}\}} \quad (5)$$

where b_j is the j-th element of the vector B^t; $\mathcal{U}$ is the map; $\mathbf{I}(\cdot)$ is the Indicator function, being 1 for $t_{ij} > 0$ otherwise 0; $\#\{A\}$ counts the number of non-zero elements in the set A.

In the third step, we reshape all branch-level vectors B^t's from the multiple questions into a uniform length. Towards this end, we defined a zero padding function $\mathcal{P}$ to take B^t as input and return $\hat{B}^t = (b_1, b_2, \cdots, b_m, p_{m+1}, \cdots, p_z) \in \mathbb{R}^{z \times d}$, where z is the maximum m in all questions and p_i is padding vector composed of the predefined value.

In the final step, we combine all branch-level representations. We defined an attention function $\mathcal{A}^t$ on $\hat{B}^t$ to achieve $\overline{B}^t = (\overline{b}_1, \overline{b}_2, \cdots, \overline{b}_z) \in \mathbb{R}^{z \times d}$ as

$$\overline{B}^t = \mathcal{A}^t(\hat{B}^t) = Softmax(mask(\frac{Q_t^T K_t}{\sqrt{d}})) V_t \qquad (6)$$

where $Q_t = [q_1^t, q_2^t, \cdots, q_z^t] = \hat{B}^t W_t^Q$, $K_t = [k_1^t, k_2^t, \cdots, k_z^t] = \hat{B}^t W_t^K$, and $V_t = [v_1^t, v_2^t, \cdots, v_z^t] = \hat{B}^t V_t^V$, where W_t^Q, W_t^K, $W_t^V \in \mathbb{R}^{d \times d}$ are the query, key, and value projection matrices, respectively [16]. The mask function is defined by

$$mask(a_{ij}) = \begin{cases} -\infty, & if\, q_i^t = p\ or\ k_j^t = p \\ a_{ij}, & otherwise \end{cases} \qquad (7)$$

where a_{ij} is the affinity value between query q_i^t and key k_j^t, and p is the padding value. With Eqs. (6) and (7), the question knowledge concept qkc_t for the question t is obtained by aggregating elements in $\overline{B}^t$ by the mapping function $\mathcal{W}$

$$qkc_t := \mathcal{W}(\overline{B}^t) = \sum_{j=1}^{k} \overline{b}_j \qquad (8)$$

In summary, the hierarchical accumulation process of encoding the KC tree in a question can be simply expressed as

$$qkc_t = \mathcal{W}(\mathcal{A}^t(\mathcal{P}(\mathcal{U}(T^t)))) \qquad (9)$$

2.4   Model Implementation

The proposed DKCT, shown in Fig. 3, is implemented based on the Transformer architecture [16]. Since the model works with inputs of a fixed length, for a sequence whose length n_0 is less than t, we repetitively added padding of question-answer pair to the left of the sequence; for a sequence whose length n_0 is greater than t, we partitioned the sequence into sub-sequences of length t. All obtained sub-sequences are served as inputs to the DKCT model.

With the pre-processed inputs, the encoder of DKCT takes the sequence $QKC = [qkc_1, qkc_2, \cdots, qkc_t] \in \mathbb{R}^{t \times d}$ of KC embedding as attention query and sequence, and question embedding $QE = [qe_1, qe_2, \cdots, qe_t] \in \mathbb{R}^{t \times d}$ as attention key and value, followed by feeding the encoder output $O^{enc} =$

$[o_1^{enc}, o_2^{enc}, \cdots, o_t^{enc}] \in \mathbb{R}^{t \times d}$ to the decoder. The decoder takes O^{enc} and interaction embedding $IE \in \mathbb{R}^{t \times d}$ with the start token and finally combines the decoder output $O^{dec} \in \mathbb{R}^{t \times d}$ with student embedding $SE \in \mathbb{R}^{t \times d}$ into a fully connected layer classifier to predict the response correctness $C' = [c_1', c_2', \cdots, c_t']$. This process could be simply written into

$$O^{enc} = Encoder(QKC, \; QE),$$
$$O^{dec} = Decoder(O^{enc}, \; IE), \qquad (10)$$
$$C' = FCN(O^{dec}, \; SE).$$

The main components of DKCT in Fig. 3 are described as follows.

Multi-head Attention with Upper Triangular Mask. The multi-head attention networks are simply the attention networks applied h times to the same input sequence with different projection matrices. For each $1 \le i \le h$, the query and key-value pairs using the following equations:

$$Q_i = QW_i^Q, \quad K_i = KW_i^K, \quad V_i = VW_i^V \qquad (11)$$

where W_i^Q, W_i^K, and $W_i^V \in \mathbb{R}^{d \times d}$ are weight matrices of query, key, and value, respectively. Then, the attention head i can be

$$head_i = Softmax(mask(\frac{Q_i K_i}{\sqrt{d}})) V_i \qquad (12)$$

where the masking mechanism replaces the upper triangular part of matrix $Q_i K_i$ from the dot-product with $-\infty$, i.e., masking inputs information from the future for all multi-head attention networks to prevent invalid attending. The attention $head_i$ is the values V_i multiplied by the masked attention weights, and the multi-head attention O^{att} is a linear combination of all head matrices,

$$O^{att} = Concat(head_i, \; head_2, \; \cdots, head_h) W^O \qquad (13)$$

where d is the dimension of the query vectors and the key vectors and $W^O \in \mathbb{R}^{hd \times d}$ is a weight matrix.

Feed-Forward Networks. Feed-forward network (FFN) is defined by:

$$F = FFN(O^{att}) = ReLU(O^{att} W_1 + b_1) W_2 + b_2, \qquad (14)$$

where $W_1, W_2 \in \mathbb{R}^{d \times d}$, $b_1, b_2 \in \mathbb{R}^d$ are parameters learned during training.

Encoder. The encoder is a stack of N identical layers. A single encoder layer can be expressed into

$$O^{att-enc} = SkipConct(Multihead(Q^{enc}, \; K^{enc}, \; V^{enc})),$$
$$O^{enc} = SkipConct(FFN(LayerNorm(O^{att-enc}))), \qquad (15)$$

where $Q^{enc} = QKC$, $K^{enc} = V^{enc} = QE$. Skip connection [5] and layer normalization [1] are applied to each sub-layer.

Table 1. Details of Eedi and Junyi dataset.

Statistics	Eedi	Junyi
#Questions	27,613	837
#Students	118,971	247,606
#Responses	15,867,850	25,925,993
# Knowledge Concepts	1,125	837
# KCs Level	4	4
Experiment Sequence Length	40	40
# Train Students	95,176	198,084
# Test Students	23,795	49,522
# Train Interaction	274,519	460,036
# Test Interaction	69,019	116,407

Decoder. Similar to the encoder, the decoder is a sequence of N identical decoder layers, skip connection, and layer normalization are applied to each sub-layer. A single decoder layer is shown as follows,

$$O_1^{att-dec} = SkipConct(Multihead(Q^{dec},\ K^{dec},\ V^{dec})),$$
$$O_2^{att-dec} = SkipConct(Multihead(O_1^{att-dec},\ O^{enc},\ O^{enc})), \qquad (16)$$
$$O^{dec} = SkipConct(FFN(LayerNorm(O_2^{att-dec}))),$$

where $Q^{dec} = K^{dec} = V^{dec} = IE$. Finally, the output of the last decoder layer is connected to the student embedding, which is passed to a fully connected layer to produce the final output $C' = [c'_1, c'_2, \cdots, c'_t]$.

3   Experiment

3.1   Dataset

We conducted experiments on two big-size educational datasets. The datasets were divided by the ratio 8:2 from the student wise. The statistical details of these datasets are provided in Table 1. The datasets are described as follows.

Eedi. This dataset was provided by the online learning platform Eedi [17], an online education provider currently used in tens of thousands of schools. This dataset is composed of 15,867,850 responses from 118,971 students on 27,613 questions about mathematics. Each question in the dataset is associated with a list of knowledge concepts. Experts arrange these knowledge concepts in a tree structure based on the knowledge hierarchy relationship. As illustrated in Fig. 1, the KCT has 1,125 knowledge concepts, and the 4 knowledge hierarchies are marked with different colors. The dataset also includes student features and interaction features.

JunyiAcademy (Junyi). This dataset was provided by the e-learning platform JunyiAcademy. This dataset is composed of 25,925,993 responses from 247,606 students on 837 questions about mathematics. Each question in the dataset is associated with a list of prerequisite knowledge concepts. We construct the prerequisite knowledge concept tree using the 3rd-order prior knowledge concepts. The dataset also includes interactive features.

3.2 Model Training and Evaluation

We implemented the proposed DKCT with Pytorch. For the network architecture, the stack of N for encoder and decoder is 8, and the number of heads for multi-head attention is 8. Without a specific claim, the hyper-parameters in experiments are set as follows: the learning rate is 0.001, the decaying rate is 0.99 every 100 steps, the embedding dimension is 256, and the batch size is 64. While the interaction sequences in the dataset are of different lengths, we partitioned or padded all sequences into a unified length of 40.

To compare with the state-of-the-art models [8], we conducted the same experiments by using DKT [14], DKT+ [18], DKVMN [20], SAKT [13], SAINT [2], and DOPE [6]. For these approaches, we used the same hyper-parameters as that reported in their respective papers. We ran all experiments ten times and then calculated the average accuracy (ACC) and the average area under the curve (AUC) [4]. Besides, the t-test is used to check the statistical significance.

3.3 Prediction Accuracy

Table 2 shows evaluation results for all mentioned methods. As is shown, there are improvements in both ACC and AUC by considering more side information. Both SAINT and DOPE achieve higher ACC and AUC than the other traditional methods. While DKCT delivers the highest performance in terms of ACC and AUC among all methods (p-values < 0.05). On the Eedi dataset, our model gains 1.83% in ACC and 1.86% in AUC compared with the GNN-based DOPE, while gains 2.11% in ACC and 2.97% in AUC compared with the transformer-based SAINT. In comparison with DKT and DKT+, our model gains more than 4% in ACC and more than 6% in AUC. On the Junyi dataset, our model gains 1.13% in ACC and 3.13% in AUC compared with DOPE, while gains 0.91% in ACC and 3.96% in AUC compared with SAINT. In comparison with DKT and DKT+, our model gains more than 1.74% in ACC and more than 4.8% in AUC. From these results, DKCT could benefit from KCTs and achieves a better performance than the other methods.

3.4 Parameter Observation

This subsection provides surveys on the two parameters, i.e., the number of stack layers N and the input dimension d in the proposed model. We varied $N \in \{4, 6, 8\}$ and $d \in \{6, 128, 256\}$ by a grid search. At each pair of (N, d), we

Table 2. Performance comparison with all mentioned models.

method	Eedi		Junyi	
	ACC	AUC	ACC	AUC
DKT	0.7002	0.7351	0.8375	0.7461
DKT+	0.7100	0.7501	0.8396	0.7516
SAKT	0.7136	0.7581	0.8407	0.7587
DKVMN	0.7294	0.7831	0.8446	0.7677
SAINT	0.7331	0.7857	0.8479	0.7600
DOPE	0.7359	0.7968	0.8457	0.7683
DKCT	**0.7542**	**0.8154**	**0.8570**	**0.7996**

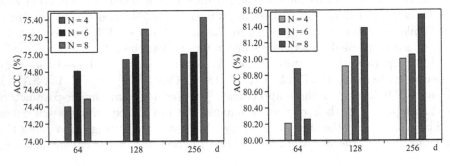

Fig. 4. Parameter discussion on the number of stack layers N and the input dimension d in the proposed model.

trained TSAKT on the training set and calculated the average ACC and the average AUC. The results are plotted in Fig. 4. It is observed from the results: 1) there are small improvements when fixing N or d; 2) TSAKT benefits from a bigger N and d on both ACC and AUC; 3) the best performance is achieved at $N = 8$ and $d = 256$. Here, we did not yet investigate the parameters on bigger values due to the computation limitation.

3.5   Ablation Study

Ablation study on the KC tree encoding aims to check the effectiveness of encoding the KC tree from three views on the Eedi dataset. The results are shown in Table 3, where KC-feature encodes KCT from the feature view, KC-feature-breadth encodes KCT from both the views of feature and breadth, KC-feature-level encodes from both the views of feature and level, and KC-feature-breadth-level encodes all the three views. As is shown, DKCT with encoding KC-feature delivers a better result than DKT; DKCT with a combined view obtains a boosted performance, while DKCT with encoding the three views gains more improvements. These observations manifest that every view of the KC tree contributes to the prediction in KT.

Table 3. Ablation study on KC tree encoding.

KC Tree view	ACC	AUC
KC-feature	0.7447	0.8025
KC-feature-breadth	0.7500	0.8100
KC-feature-level	0.7523	0.8134
KC-feature-breadth-level	**0.7542**	**0.8154**

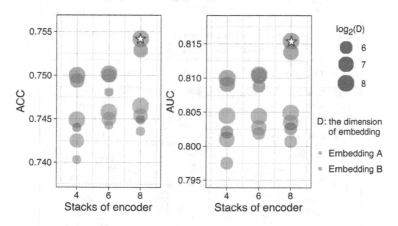

Fig. 5. Ablation study on integrated embeddings.

Ablation study of side information is to evaluate DKCT performance with side information or without side information on the Eedi dataset. Two embeddings are set as follows:

- Embedding A: In this setting, we only consider the questions with question id (QI) and positional information (P) and the interactions with answer correctly (C) and positional information (P).
- Embedding B: In addition to Embedding A, we also consider the side information of questions (e.g. question difficulty (QD), question subject (QS), and question frequency (QF)), and the side information of interactions (e.g. group id (GI), scheme of work id (SW), and confidence score given for the answer (CF)).

Figure 5 shows the evaluation results of DKCT with Embedding A or Embedding B against the encoder's stacks and the input embedding dimension. As is shown, Embedding B delivers an improvement on both ACC and AUC comprised with Embedding A. Therefore, DKCT benefits from the side information integration and the parameter rising.

3.6 Attention Analysis

Understanding the reasons behind prediction results could provide helpful suggestions, and evidence supports for educational interventions and learning plans

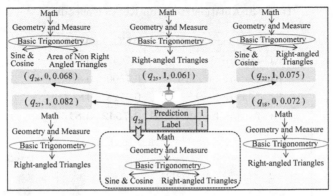

(a) Explanation of why the student answered q_{28} correctly

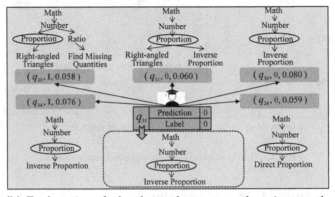

(b) Explanation of why the student answered q_{35} incorrectly

Fig. 6. The relationship between questions for prediction explanation. $(q_k, 1/0, \text{weight})$ are the question, the correct/incorrect answer, and the attention weight.

[21]. The encoder multi-head attention layer learns the contributions degrees to the target question for the past questions. The attention weights potentially answer the question, "Why is a student predicted to be correct or incorrect on the response to the target question?" Figure 6 shows the top-5 attention weights from the encoder multi-head attention layer for two cases on the Eedi dataset.

Figure 6a shows the reason why the student answered q_{28} correctly. The top-5 attention weights for q_{28} together with the corresponding KC tree and their answers are shown in Fig 6a. The KC tree of the top-5 question and q_{28} are all related to the knowledge concept "Basic Trigonometry". It shows that the multi-head attention layer of the encoder could capture the same knowledge concepts related to the target question. The student answered q_{22} correctly, and the q_{28} KC tree has the same structure as q_{22}, so it is reasonable to believe that the student has mastered the "Sine and Cosine Rules", and "Right-angled Triangles". In addition, although the students answered q_{26} incorrectly, we found that

q_{26} focus on another knowledge concept, "Area of Non Right-angled Triangles". Therefore, there is no direct connection between q_{28} and q_{26}.

Figure 6b shows the reason why the student answered q_{35} incorrectly. The KC tree of the top-5 question and q_{35} are all related to the knowledge concept "Proportion." The student answered incorrectly at q_{24}, q_{30}, q_{31} and the KC tree of q_{35} has the same structure as q_{30}, so the student may not have mastered "Inverse Proportion". Although the student answered correctly in subsequent q_{33} and q_{34} through practice, the student may have mastered the "direct Proportion", which is not directly related to the "Inverse Proportion". Thus, the model predicts that the student will answer incorrectly.

From the two cases, it is observed that The multi-head attention layer in the encoder could capture the same knowledge concepts related to the target question and identify important degrees for the question related to the target question, leading to an explanation for the predicted result. Incorrect answers may be due to a lack of mastery of relevant knowledge concepts. Our findings are the potential for developing a personalized learning platform.

4   Discussion and Conclusion

This paper introduces a novel KT model with concept trees, DKCT, to integrate KC trees. DKCT trains an encoder with KCTs and question features, a decoder with interaction features, and a fully connected network with student features. Experiment results show that DKCT benefits from the KCT encoding and the side information, resulting in the highest accuracy compared to other KT models. The embedded attention is helpful in explaining the reason why a student is predicted to have a correct response or an incorrect response to a question.

However, there are two limitations. One is that DKCT fails to consider the multi-mode features. In the Eedi dataset, a few questions are expressed by pictures where the image features could provide the visual features for typical geometrical questions. Another problem is that DKCT fails to consider the similarity between questions and students, which has been proven to be useful for various pattern recognition tasks [9].

Overall, the proposed DKCT model integrates the KC trees and achieves a higher KT performance than the popular method. In the future, we are to solve the two limitations above to enhance the KT performance on more datasets. The interpretability of the KT model based on the KCs and the knowledge structure of a student are also our future studies.

Acknowledgment. This study was funded in part by the National Natural Science Foundation of China (62272392, U1811262, 61802313), the Key Research and Development Program of China (2020AAA0108500), the Reformation Research on Education and Teaching at Northwestern Polytechnical University (2022JGY62), the Higher Research Funding on International Talent cultivation at Northwestern Polytechnical University (GJGZZD202202).

References

1. Ba, J.L., Kiros, J.R., Hinton, G.E.: Layer normalization. arXiv preprint arXiv:1607.06450 (2016)
2. Choi, Y., et al.: Towards an appropriate query, key, and value computation for knowledge tracing. In: Proceedings of the Seventh ACM Conference on L@S, pp. 341–344 (2020)
3. Emanuel, E.J.: MOOCs taken by educated few. Nature **503**(7476), 342–342 (2013)
4. Hastie, T., Tibshirani, R., Friedman, J.: The Elements of Statistical Learning. SSS, Springer, New York (2009). https://doi.org/10.1007/978-0-387-84858-7
5. He, K., Zhang, X., Ren, S., Sun, J.: Deep residual learning for image recognition. In: Proceedings of the IEEE conference on CVPR, pp. 770–778 (2016)
6. Karimi, H., Derr, T., Huang, J., Tang, J.: Online academic course performance prediction using relational graph convolutional neural network. In: EDM (2020)
7. Liu, Q., et al.: EKT: exercise-aware knowledge tracing for student performance prediction. IEEE TKDE **33**(1), 100–115 (2021)
8. Liu, Q., Shen, S., Huang, Z., Chen, E., Zheng, Y.: A survey of knowledge tracing. arXiv preprint arXiv:2105.15106 (2021)
9. McInnes, L., Healy, J., Melville, J.: UMAP: uniform manifold approximation and projection for dimension reduction. arXiv preprint arXiv:1802.03426 (2018)
10. Nayak, G., Dutta, S., Ajwani, D., Nicholson, P., Sala, A.: Automated assessment of knowledge hierarchy evolution: comparing directed acyclic graphs. Inf. Retrieval J. **22**(3), 256–284 (2019)
11. Nguyen, T.: The effectiveness of online learning: beyond no significant difference and future horizons. MERLOT J. Online Learn. Teach. **11**, 309–319 (2015)
12. Nguyen, X.P., Joty, S., Hoi, S.C., Socher, R.: Tree-structured attention with hierarchical accumulation. arXiv preprint arXiv:2002.08046 (2020)
13. Pandey, S., Karypis, G.: A self-attentive model for knowledge tracing. arXiv preprint arXiv:1907.06837 (2019)
14. Piech, C., et al.: Deep knowledge tracing. arXiv preprint arXiv:1506.05908 (2015)
15. Santoro, A., Bartunov, S., Botvinick, M., Wierstra, D., Lillicrap, T.: One-shot learning with memory-augmented neural networks. arXiv preprint arXiv:1605.06065 (2016)
16. Vaswani, A., et al.: Attention is all you need. In: Advances in Neural Information Processing Systems, pp. 5998–6008 (2017)
17. Wang, Z., et al.: Instructions and guide for diagnostic questions: the NeurIPS 2020 education challenge. arXiv preprint arXiv:2007.12061 (2020)
18. Yeung, C.K., Yeung, D.Y.: Addressing two problems in deep knowledge tracing via prediction-consistent regularization. In: Proceedings of the Fifth Annual ACM Conference on Learning at Scale, pp. 1–10 (2018)
19. Zaremba, W., Sutskever, I., Vinyals, O.: Recurrent neural network regularization. arXiv preprint arXiv:1409.2329 (2014)
20. Zhang, J., Shi, X., King, I., Yeung, D.Y.: Dynamic key-value memory networks for knowledge tracing. In: Proceedings of the 26th International Conference on World Wide Web, pp. 765–774 (2017)
21. Zhang, Y., An, R., Cui, J., Shang, X.: Undergraduate grade prediction in Chinese higher education using convolutional neural networks. In: LAK21, pp. 462–468 (2021)
22. Zhang, Y., Dai, H., Yun, Y., Liu, S., Lan, A., Shang, X.: Meta-knowledge dictionary learning on 1-bit response data for student knowledge diagnosis. Knowl. Based Syst. **205**, 106290 (2020)

Privacy and Security

Cryptography-Inspired Federated Learning for Generative Adversarial Networks and Meta Learning

Yu Zheng[1], Wei Song[2], Minxin Du[1], Sherman S. M. Chow[1(✉)], Qian Lou[3], Yongjun Zhao[4], and Xiuhua Wang[5]

[1] Department of Information Engineering, Chinese University of Hong Kong, N.T., Hong Kong
yuzheng404@link.cuhk.edu.hk
[2] School of Computer Science and Engineering, Northeastern University, Shenyang, China
songwei@mail.neu.edu.cn
[3] Department of Computer Science, University of Central Florida, Orlando, USA
[4] Shanghai, China
[5] School of Cyber Science and Engineering, Huazhong University of Science and Technology,Wuhan, China

Abstract. Federated learning (FL) aims to derive a "better" global model without direct access to individuals' training data. It is traditionally done by aggregation over individual gradients with differentially private (DP) noises. We study an FL variant as a new point in the privacy-performance space. Namely, cryptographic aggregation is over local models instead of gradients; each contributor then locally trains their model using a DP version of Adam upon the "feedback" (e.g., fake samples from GAN – generative adversarial networks) derived from the securely-aggregated global model. Intuitively, this achieves the best of both worlds – more "expressive" models are processed in the encrypted domain instead of just gradients, without DP's shortcoming, while heavy-weight cryptography is minimized (at only the first step instead of the entire process). Practically, we showcase this new FL variant over GAN and meta-learning, for securing new data and new tasks.

Keywords: Federated learning · Cryptography · Differential privacy

1 Introduction

Training an accurate model often needs massive data, which may be sensitive (*e.g.*, medical records [36,46]) and hard to collect. Federated learning (FL) [29] allows a central server to train a global model by aggregating local model parameters/gradients derived by data owners. The server does not directly access the

S. S. M. Chow—Supported in part by the General Research Funds (CUHK 14210621 and 14209918), University Grants Committee, Hong Kong. Wei Song is supported by the Fundamental Research Funds for the Central Universities (N2316010).

X. Yang et al. (Eds.): ADMA 2023, LNAI 14177, pp. 393–407, 2023.
https://doi.org/10.1007/978-3-031-46664-9_27

training data, which can still be inferred from leakages of shared gradients [32] or parameters [18]. Differential privacy (DP) [14], notably DP stochastic gradient descent (DP-SGD) [3], adds calibrated noise to gradients, ensuring that an adversary cannot determine whether an example is in the training data. Yet, when applying DP-SGD in FL, the noisy model parameters/gradients shared for each aggregation can still reveal sensitive information (*e.g.*, demographic features [32]) of the training data.

Cryptographic techniques such as multiparty computation (MPC) [40] or homomorphic encryption (HE) [20] allow parties to jointly compute the output of a function without revealing their private inputs. They can protect input data/models by offering indistinguishability [6] instead of just neighbor ones in DP. However, the overhead of processing every step in the encrypted domain can be prohibitive, either through heavy communication or computation. While the training process is protected, the established model is released and still vulnerable to inference attack [35] or dataset-property attack [45] since it could be considered purely trained without protection.

Only a few recent works [39,43] consider strong protection for FL by incorporating DP and cryptography to avoid potential leakages from noisy updates. They use dedicated yet still costly cryptography to process *each* user's noisy update in multiple rounds. One could use a tailored secure aggregation protocol [7]. Its "mutual-cancellation" mechanism is prone to failures either by network errors or malicious attacks. A general remedy is to use secret sharing [7], which may introduce large communication overheads to recover "missing pieces" from redundant shares.

We jump out of "the folklore FL approach" – working over gradients with DP noises and start on a new route of stronger protection. We aim to employ cryptography for secure aggregation while DP protects model outputs concerning a small set of data points. Intuitively, local models capture more information than just gradients, which are now cryptography-protected. Our next insight is to minimize the use of cryptography for communication-efficient FL. We are intrigued to explore specific FL applications, expecting "simple-enough" subsequent contributions from individuals. Specifically, we consider generative adversarial networks (GANs) [21] and meta-learning [34]). Both expect only a probability distribution function (PDF) after the initial local model aggregation from training-data contributors. For further model tuning, we use DP (without cryptography) for its much less overhead. Our contributions are summarized below:

1. We put forth an intuitive yet rarely explored FL protection.
2. This work details DP-Adam, replacing the *de facto* DP-SGD for training with example-level privacy and faster convergence.
3. We showcase our protection with instantiations of GAN and meta-learning.

2   Related Works

FL allows clients to learn a shared model collaboratively while addressing the data heterogeneity issue and being communication-efficient [29]. Typically, the global model is transmitted among all parties for latter aggregation over local models. Various applications have emerged, such as GANs, which generate more representative fake data by mitigating "domain shifts" across data contributors [17]. For GANs [21], the generator is trained to generate new data with the same statistics as the training data and the discriminator is trained to tell how "realistic" the input seems via probabilities.

For privacy, many prior works consider perturbing gradients by Gaussian noise [3]. McMahan *et al.* [30] adapt DP-SGD to FL, later improved with the adaptive clipping of gradients [4]. Both add a small amount of noise to each individual gradient, "insufficient" for meaningful privacy guarantees for users [37]. Stronger protections [24,41] thus inject enough noise to gradients, yet with the much worse performance of trained models. Some works [5,8] started to apply DP to GANs. Other approaches consider adding DP noise to ADMM [22] or the model parameters [38]. However, these DP-only designs may still suffer from risks [19] beyond membership inference.

Another line of research resorts to cryptographic primitives [13,33]. Bonawitz et al. [7] build a protocol on secret sharing for aggregating users' models (without learning each individual one). PrivFL [28] and BatchCrypt [44] respectively encrypt models and gradients in *every* aggregation by HE since they can be summed directly over encrypted data. However, they are not practical for gigantic model sizes or numbers of users. Moreover, aggregated updates (decrypted at a certain iteration) and final outputs can still reveal information (*e.g.*, membership) about private inputs [27,35].

For defense, two concurrent works [39,43] incorporate DP and cryptography: they respectively use (multi-input) functional encryption and (threshold) additive HE to encrypt local models trained by DP-SGD. However, as in those cryptographic designs, costly encryption is run for every user's upload (in each aggregation), limiting the practicality. We are thus motivated by devising a stronger defense while being more efficient.

3   Preliminary

3.1   Differential Privacy and DP-SGD

DP [14] has become the *de facto* standard to protect individual privacy – the contribution of a single data point/example only incurs a limited impact on (statistical) analytics performed on the entire dataset of all individuals.

Definition 1. *A randomized mechanism $\mathcal{M}$ ensures (ϵ, δ)-DP if for any two neighboring datasets $D \simeq D'$ differing in a single example, $\forall O \subseteq Range(\mathcal{M})$, the output space of $\mathcal{M}$ satisfying $\Pr(\mathcal{M}(D) \in O) \leq e^\epsilon \cdot \Pr(\mathcal{M}(D') \in O) + \delta$, where $\epsilon \geq 0$ and $0 \leq \delta \leq 1$ are two privacy parameters.*

For (ϵ, δ)-DP, Gaussian mechanism (GM) [15] is to add i.i.d. noise drawn from a Gaussian distribution $\mathcal{N}(0, \sigma^2 \cdot S_2^2(f))$ to the output of a (vector-valued) function $f(\cdot)$, where $\sigma^2 = 2\ln(1.25/\delta)/\epsilon^2$ and $S_2(f)$ is L_2-sensitivity $\sup_{D \simeq D'} \|f(D)$

$- f(D')\|_2$. DP has two useful properties [15]: 1) the free post-processing post-processing allows computations on the output of a DP mechanism without extra privacy loss; 2) sequential composition helps to build more complex DP mechanisms (say, DP-SGD [3]) atop basic ones (*e.g.*, GM) operating on same data.

DP-SGD can be used to train models with *example-level* privacy: an adversary cannot determine whether any specific example is in the training dataset. It only slightly modifies the raw minibatch optimization process by using GM to perturb gradients. Specifically, at each training step, we first clip per-example gradients to a fixed maximum norm C (a tunable parameter) such that $S_2(f) = C$. Then, DP-SGD adds Gaussian noise to the gradient aggregated over a batch of randomly-sampled examples. To analyze a tight privacy bound [16], DP-SGD devises the moments accountant, later generalized to Rényi DP, to prove $(O(q\epsilon\sqrt{T}), \delta)$-DP [31] with a sampling probability q (for building a batch) in T-step training.

3.2   Homomorphic Encryption (HE)

HE [20] allows performing arithmetics (such as '+' and '×') on encrypted data, yielding results as those derived on plaintexts when decrypted. We adopt an efficient instance called CKKS [9], an easy-to-use HE library offered by Microsoft SEAL [2], for summing up encrypted real numbers in $\mathcal{R}$ and training/evaluating models on encrypted data. An HE scheme is defined by the below syntax.

Definition 2. *(Somewhat) Homomorphic encryption (HE), supporting a number of multiplications and additions, is defined by the probabilistic polynomial time algorithms below.*

KeyGen(1^λ): *Given a security parameter λ, it outputs a secret key* sk, *a public key* pk, *and an evaluation key* evk[1].
Enc$_{\mathsf{pk}}(m)$: *For a message $m \in \mathcal{R}$, it encrypts m to a ciphertext $c \in \mathcal{R}_{q_L}^k$ which is also denoted by* $[\![m]\!]$.
Dec$_{\mathsf{sk}}([\![m]\!])$: *With* sk, *it decrypts a ciphertext* $[\![m]\!]$ *to m.*
Add$([\![m_1]\!], [\![m_2]\!])$: *It outputs a ciphertext* $[\![m_1 + m_2]\!]$.

4   Differentially Private Adam over GAN

Our approach first lets each user train GANs on its own data using DP-Adam locally. It then asks each user to upload the HE-encrypted generator, which is aggregated to a global one at the server homomorphically. In the later interactive update (or tuning) phase, our insight is that we do not need to protect the local discriminators' outputs, deemed non-sensitive, for updating the global generator. This can significantly boost efficiency compared to the works [39,43] encrypting each user's update. Each discriminator is further updated with DP-Adam and the generated data.

---

[1] It can be used to evaluate ciphertext multiplications.

Algorithm 1: Differentially Private Adam (Outline)

Input: Examples $\{x_1, \ldots, x_N\}$, loss function $\mathcal{L}(\Theta)$, step size α, decay rates β_1, β_2, a constant γ, batch size L, clip norm C, noise scale σ

Output: Model parameters Θ_T at step T

1   Initialize Θ_0 randomly;
2   **for** $t \in [T]$ **do**
3   Randomly sample L examples as a batch B_t;
4   Compute gradient $\mathbf{g}_t(x_i)$ for $\forall x_i \in B_t$;
5   Clip gradient $\bar{\mathbf{g}}_t(x_i) = \mathbf{g}_t(x_i)/\max(1, \frac{\|\mathbf{g}_t(x_i)\|_2}{C})$;
6   Add noise $\tilde{\mathbf{g}}_t = \frac{1}{L}(\sum_i \bar{\mathbf{g}}_t(x_i) + \mathcal{N}(0, \sigma^2 C^2 \mathbf{I}))$;
7      $\beta_1' = 1 - \beta_1, \beta_2' = 1 - \beta_2, u_t = \beta_1 \cdot u_{t-1} + \beta_1' \cdot \tilde{\mathbf{g}}_t, v_t = \beta_2 \cdot v_{t-1} + \beta_2' \cdot \tilde{\mathbf{g}}_t^2$;
8      $\bar{u}_t = u_t/\beta_1', \bar{v}_t = v_t/\beta_2'$;
9      $\Theta_{t+1} = \Theta_t - \alpha \cdot \bar{u}_t/(\sqrt{\bar{v}_t} + \gamma)$
10   **end**
11   Return Θ_T

4.1 Differentially Private Adam Optimizer

Adaptive moment estimation (Adam) [25] is an optimizer with a faster convergence rate than SGD. It consumes smaller memory when training over high-dimensional parameter spaces and representative datasets. We devise DP-Adam[2], for retaining Adam's advantages while training models for example-level privacy as in DP-SGD.

Algorithm 1 presents DP-Adam's pseudocode, including gradient computation, gradient clipping, noise addition, and Adam-like parameter update. In each step, we randomly sample L examples as a batch for estimating the gradient of $\mathcal{L}(\Theta)$. For every example in the batch, we compute its gradient $\mathbf{g}$ which is then clipped to $\mathbf{g}/\max\{1, \frac{\|\mathbf{g}\|_2}{C}\}$ with a clipping threshold C. Such clipping is also a popular ingredient of the non-private Adam for mitigating gradient explosion. Aggregating L clipped gradients thus has a bounded L_2-sensitivity $S_2(f) = C$. We then add i.i.d. noise drawn from a Gaussian distribution specified by C and the noise scale σ to the aggregated gradient. The remaining steps (Lines 7 to 9) stay the same as the raw Adam to update models using unbiased moment estimates but derived from noisy gradients.

4.2 Differentially Private Generative Adversarial Networks

As Algorithm 2, we build a DP version of GAN (DP-GAN) that can yield representative samples with example-level privacy. A GAN [21] contains a generative model G and a discriminative model D to solve the min-max optimization: $\min_G \max_D \mathbb{E}_{x \sim p_{\text{data}}(x)}[\log D(x)] + \mathbb{E}_{z \sim p_z(z)}[\log(1 - D(G(z)))]$, where p_{data} is a data generating distribution and p_z is a noise[3] prior. G maps noise samples z's

---

[2] Although it has been wrapped up in some privacy libraries, the research literature lacks a self-contained description.

[3] It is just an ingredient of GAN and not for any privacy purposes.

Algorithm 2: Training GAN with DP-Adam (Outline)

Input: A random subset $\{x_1, \ldots, x_L\}$ of d
Output: Updated model parameters Θ^D and Θ^G
1 Draw L noise samples $\{z_1, \ldots, z_L\} \sim p_z(z)$;
2 **for** $i \in [L]$ **do**
3     $\mathbf{g}_i^D = \nabla_{\Theta^D}[\log D(x_i) + \log(1 - D(G(z_i)))]$;
4     $\mathbf{g}_i^G = \nabla_{\Theta^G} \log(1 - D(G(z_i)))$;
5 **end**
6 Run DP-Adam (line 5 to 11) for all $\mathbf{g}^D$ to update Θ^D;
7 Run raw Adam for all $\mathbf{g}^G$ to update Θ^G

---

to data space $\mathsf{data_G}$, then D evaluates $\mathsf{data_G}$ and outputs the result set res of probabilities that samples (in $\mathsf{data_G}$) are real or fake. Given res, G outputs new $\mathsf{data_G}$ and repeats such an adversarial process with D.

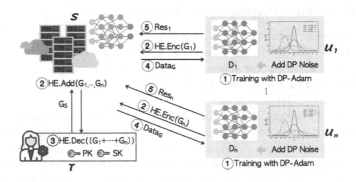

Fig. 1. A Hybrid Approach for Securing Federated GAN

We build DP-GAN atop DP-Adam and Adam to offer example-level privacy for a training dataset $\mathsf{d} = \{x_1, \ldots, x_N\}$. Algorithm 2 lists the pseudocode of one training step in DP-GAN. Since the private examples are only used for training D, we adopt DP-Adam to update model parameters Θ^D. The noise samples for training G do not need any protection; we can run the raw Adam to update parameters Θ^G.

5   Cryptographic Federated DP-GAN

In FL, we consider a central server $\mathcal{S}$ with n users $\mathcal{U}_1, \ldots, \mathcal{U}_n$, where $\mathcal{U}_i$ (for $i \in [1, n]$) has a private training dataset d_i of N examples. When training GANs in FL, $\mathcal{S}$ plays the role of G to generate more representative $\mathsf{data_G}$ (than in the centralized setting) based on $\{\mathsf{d}_1, \ldots, \mathsf{d}_n\}$, while $\mathcal{U}_i$ acts as D to evaluate $\mathsf{data_G}$.

5.1 Our Hybrid Approach

As in Fig. 1, our protocol involves interactions among $\mathcal{S}$ and $\mathcal{U}_{i \in [n]}$. For brevity and a modular discussion, our discussion starts with a trusted third party $\mathcal{T}$, which sets up the HE key pair.

Protocol 3 shows our hybrid approach using DP-Adam and CKKS for training federated GANs. It consists of two stages – aggregating HE-encrypted models in the first round (Lines 3 to 6) and DP tuning in following rounds (Lines 8 to 12).

Protocol 3: Cryptographic Federated DP-GAN

Input: Private datasets $d_1, \ldots, d_n$
Output: Model parameters Θ_i^D for $\mathcal{U}_i$ and Θ^G for $\mathcal{S}$
1 $\mathcal{T}$: Setup $(sk, pk, evk) \leftarrow$ CKKS.KeyGen(1^λ);
2 **for** $i \in [n]$ **do**
3 $\quad|\quad$ $\mathcal{U}_i$: Train Θ_i^D, Θ_i^G via DP-GAN; Send $[\![\eta_i \Theta_i^G]\!] \leftarrow$ CKKS.Enc$_{pk}(\eta_i \Theta_i^G)$ to $\mathcal{S}$;
4 **end**
5 $\mathcal{S}$: Aggregate $[\![\Theta^G]\!] \leftarrow \sum_i [\![\eta_i \Theta_i^G]\!]$ via CKKS.Add;
6 $\mathcal{T}$: Decrypt $\Theta^G =$ CKKS.Dec$_{sk}([\![\Theta^G]\!])$ for $\mathcal{S}$;
7 $\mathcal{S}$: Update G by Θ^G and output data$_G$ for $\mathcal{U}_{i \in [n]}$;
8 **while** *not converged* **do**
9 $\quad|\quad$ $\mathcal{U}_i$: Compute $\nabla \Theta_i^D$ to update Θ_i^D via DP-Adam;
10 $\quad|\quad$ $\mathcal{U}_i$: Run updated $D_i(\text{data}_G)$ and output res$_i$ to $\mathcal{S}$;
11 $\quad|\quad$ $\mathcal{S}$: Evaluate all res and update Θ^G via raw Adam;
12 $\quad|\quad$ $\mathcal{S}$: Run updated G to output new data$_G$ for $\mathcal{U}_{i \in [n]}$
13 **end**

Cryptographic Aggregation of Model Parameters. Although FL does not collect all the users' raw data, sensitive information (beyond the membership of an example) can still be recovered from noisy parameters (or gradients) used for updating global models at the server. For mitigation, we incorporate CKKS to encrypt local noisy models while allowing aggregation to be performed on ciphertexts. Except for one-time cryptographic aggregation, our solution does not transmit the global model or local models anymore.

As a one-time setup, $\mathcal{T}$ first sets up CKKS by running CKKS.KeyGen(1^λ). $\mathcal{T}$ keeps sk secret and publishes (evk, pk). Each $\mathcal{U}_{i \in [n]}$ runs DP-GAN locally over d_i and outputs (Θ_i^D, Θ_i^G), where Θ_i^G is scaled by a weight η_i. Then, it encrypts the weighted model by CKKS under pk and sends $[\![\eta_i \Theta_i^G]\!]$ to $\mathcal{S}$. From the additive homomorphism, $\mathcal{S}$ aggregates all the encrypted models using CKKS.Add$(\cdot)$. With $\mathcal{T}$, $\mathcal{S}$ can obtain decrypted and aggregated model parameters Θ^G for outputting data$_G$ to $\mathcal{U}_{i \in [n]}$.

Global Model Tuning via (DP-)Adam. With the aggregated model Θ^G, $\mathcal{S}$ outputs data$_G$ to be evaluated at the user side. Each $\mathcal{U}_i$ updates its Θ_i^D by DP-Adam since it takes private d_i (and non-private data$_G$) as input. It then runs updated D_i to evaluate $\mathsf{data_G}$, which returns a set of probabilities res_i to $\mathcal{S}$. (A higher probability indicates that $\mathsf{data_G}$ is more likely to be real.) $\mathcal{U}_i$ does not need to encrypt res_i (as in the cryptographic aggregation of Θ_i^G), assuming that non-sensitive information can be inferred from the probabilities. With all $\mathsf{res}_{i \in [n]}$, $\mathcal{S}$ updates Θ^G with the raw Adam and runs updated G to output new $\mathsf{data_G}$. The tuning of $D_{i \in [n]}$ and G repeats until convergence. Finally, $\mathcal{S}$ is able to learn the global statistics of $d_{i \in [n]}$ while providing example-level privacy for each user.

5.2   Generalization

Besides training GANs, FL with our hybrid protection can be generalized for "transferring knowledge" (*e.g.*, meta learning) from local models to a central one maintained by the server.

We consider adapting long short-term memory-based meta learning (LSTM-ML) [34] as an illustration of its empirical performance. Let $\mathcal{U}_{i \in [n]}$ first train a model pair – a meta learner and a learner classifier – over d_i for a specific task using DP-Adam. Each local meta learner is encrypted and then aggregated by $\mathcal{S}$ to a global one for capturing both short-term knowledge within a task and long-term knowledge common among all the tasks; each learner classifier is never revealed. In later model tuning, users update their local models (taking the server's feedback as input) by DP-Adam, while the server updates its global model with knowledge (*e.g.*, model architectures) from users. Finally, the server can output a classifier for even a new task.

We also remark that our design can naturally extend to more complicated aggregation (beyond summation) since CKKS supports multiplication on cipher-texts. Albeit CKKS serves as a performance baseline, our modular construction allows replacement with any future improvement of HE or secure aggregation.

6   Security Analysis

Security of our framework boils down to two parts – DP-Adam run by each user and initial models encrypted by HE.

6.1   Differential Privacy Analysis

The conventional approaches achieving (ϵ, δ)-DP are from the sensitivity method [10] that adds some noise sampled from Gaussian distributions, proportional to the sensitivity. We first follow the proof paradigms [3,11,25] for analyzing DP-Adam, and later apply composition [23] over it. Specific to DP-Adam, we restrict a real-value function $\mathbf{f}$ for simplifying analysis (used in [3]), and later explain the noise effect of Algorithm 1.

Lemma 1 (Sensitivity of DP-Adam). *Assuming data matrix $\mathbf{X}$ to be normalized, the sensitivity of mini-batch update computation is bounded by $S_f^2 \leq C^2$.*

Proof. To analyze the sensitivity of $\mathbf{f}$, we consider two data matrices $\mathbf{X}, \mathbf{X}'$ that differ only in a single example $\mathbf{x}_k$. The information of $\mathbf{x}_k$ is removed from $\mathbf{X}'$ by initializing $\mathbf{X}' = \mathbf{X}$ and setting its k-th column to be zero. Let Δ and Δ' be the updates computed on $\mathbf{X}$ and $\mathbf{X}'$, respectively. By the clipping step, DP-Adam guarantees $|\bar{\mathbf{g}}_k| \leq C$. By the normalization, we have $\|\mathbf{x}_k\| = 1$. The difference is $\|\Delta - \Delta'\| \leq \|\mathbf{x}_k\| \|\bar{\mathbf{g}}_k\| \leq C$. Therefore, the sensitivity of $\mathbf{f}$ is bounded by $S_f^2 := \max_{\mathbf{X}\backslash\mathbf{X}'=\mathbf{x}_k} \|\Delta - \Delta'\|^2 \leq C^2$.

Now, let's analyze parameter updating $\Theta_{t+1} = \Theta_t - \alpha \cdot \bar{u}_t/(\sqrt{\bar{v}_t} + \gamma)$ via DP-Adam. At time t, step size is $\Delta_{\mathsf{stp}} = |\alpha \cdot \bar{u}_t/(\sqrt{\bar{v}_t} + \gamma)|$ for updating Θ. Assuming $\gamma = 0$ (originally a very small value, 10^{-8}), then $\Delta'_{\mathsf{stp}} = |\alpha \cdot \bar{u}_t/\sqrt{\bar{v}_t}|$. Effective step size is bounded by $\Delta_{\mathsf{stp}} \leq \Delta'_{\mathsf{stp}} \leq \alpha \cdot \max((1 - \beta_2)/\sqrt{1 - \beta_2}, 1)$. Since $|\mathbb{E}[\mathbf{g}]/\sqrt{\mathbb{E}[\mathbf{g}^2]}| \leq 1$ and $|\bar{\mathbf{g}}_k| \leq C$, we know $\bar{u}_t/\sqrt{\bar{v}_t} \approx \pm 1$. Thus, effective step of DP-Adam is bounded by $\Delta_{\mathsf{stp}} \lesssim \Delta'_{\mathsf{stp}} \lesssim \alpha$ (whereas [3] bounded by ηC). $\square$

By Lemma 1 and composition theorem [15], models trained by DP-Adam satisfy $(O(q\epsilon\sqrt{T}), \delta)$-DP for $q = L/N$ at each client's side. The proof[4] directly follows from that of DP-SGD [3] due to the running of T sub-sampled GMs under the adaptive composition (with the only difference in post-processing for updating models).

6.2 Cryptographic Indistinguishability-Based Security

CKKS provides indistinguishability under chosen plaintext attack [9], or say, satisfying IND-secure as in Lemma 2. It means that any adversary (without secret key) can not efficiently distinguish the encryptions of two (adversarially chosen) messages given the encryption oracle for plaintexts.

Lemma 2. *CKKS is IND-secure [9].*

Lemma 3. *If an $\mathcal{M}$ over neighboring databases D, D' satisfies (ϵ, δ)-DP, D and D' are $(e^\epsilon/2^\lambda + \delta)$-indistinguishable by viewing $\mathcal{M}(D), \mathcal{M}(D')$.*

Proof. We first analyze an adversary $\mathcal{A}$'s advantage for $(\epsilon, 0)$-DP. Recall that λ represents the input length in cryptography. We label each D by a λ-bit string. By specifying a binary setting, we have $\Pr[X = D] = 1/2^\lambda$ for a random guess by an adversary $\mathcal{A}$. By DP's definition, we attain $e^{-\epsilon}/2^\lambda \leq \Pr[X = D|\mathcal{M}(X) = y] \leq e^\epsilon/2^\lambda$. When ϵ is very small, $\lim e^\epsilon \to 1 \Rightarrow \lim e^\epsilon/2^\lambda \to 1/2^\lambda$.

Now, let's take δ into consideration. The δ defines the probability of information accidentally being leaked. That is, the $\mathcal{A}$ can distinguish the outputs from D or D' since $Pr[\mathcal{M}(D) = y]$ and $\Pr[\mathcal{M}(D') = y]$ may be substantially different. Combining two $\mathcal{A}$'s advantages, we get $e^\epsilon/2^\lambda + \delta$. $\square$

By employing cryptographic indistinguishability, we combine security analysis of differential privacy and cryptography. For the whole federated training process, we can (math-)trivially sum all indistinguishable advantages (derived from Lemma 2 and Lemma 3) from an $\mathcal{A}$'s view.

---

[4] We remove this straightforward proof due to page limit.

7 Experiments

Our experiments are carried out on the commodity PC running CentOS with Intel Xeon Gold 5118 24-core CPU, 62.5 GiB RAM, and GeForce RTX 2080 Ti GPU. We implemented our algorithms/protocol using Python 3.7 supported in PyTorch. We also used PySyft [1] for FL and Microsoft SEAL [2] for CKKS [9].

To focus on the computational overhead introduced by our use of HE over DP-only approaches, or the computational saving by our minimized use of HE over cryptography-only or hybrid approaches, we consider the base case of two users with a server. It is true that the use of HE introduces communication overhead over the pure use of DP or secret sharing to hide the plaintext data for a single data item. Again, the former is known to be vulnerable to inference attacks, while secret-sharing-based designs require way more communication rounds, similar to existing cryptography-only or hybrid approaches. Repeating our experiment over an increasing number of users would just widen the performance gap due to the communication overheads for the holistic process.

To configure parameters for reliable results, we test a wide range from non-convergence to convergence, which we omit the exhaustive report here, especially under the page limit.

Datasets. We used three typical datasets: i) MNIST [26] has 60k training and 10k test images of handwritten digits with total 10 classes, ii) Fashion-MNIST [42] is MNIST-like with each example being a 28×28 grayscale image, and iii) Mini-ImageNet is a subset of ImageNet [12] with 100 classes. Each class has the same size as that of MNIST. We then randomly split each training dataset evenly for the two users.

Parameters configuration. In DP-Adam, we fix $\delta = 10^{-5}$ which is in $O(1/N)$ and use different σ's (representing different privacy levels). We set $\beta_1 = 0.9$, $\beta_2 = 0.999$, $\gamma = 1e - 8$, no weight decay, and the learning rate $\alpha = 0.0002$ for GANs and $\alpha = 0.001$ for LSTM-ML. For CKKS, we pick the ring dimensions as 4096 and 8192, governing two security levels.

Baselines. We implemented DP-SGD as a baseline with the same parameter setting from [3]. We also compare with BatchCrypt [44], a secure FL framework built atop only additively HE, and HybridAlpha [43], a hybrid construction (using DP and HE) similar to ours.

7.1 Privacy Versus Utility for GANs/LSTM-ML

The training loss of G and D (derived by binary cross entropy) can be used as an "indicator" for model efficacy: the lower the loss, the better the efficacy. Figure 2 shows the results of training-then-tuning via DP-Adam/SGD with two σ values.

Remark 1. As in the non-private case, DP-Adam converges faster and is more stable than DP-SGD. For similar efficacy, DP-SGD requires a larger learning rate or more training steps. As can be seen, the training loss of DP-Adam stays stable throughout the tuning at $\sigma = 8$, but using DP-SGD is initially unstable. Our results also show that models trained with DP-Adam outperform those with

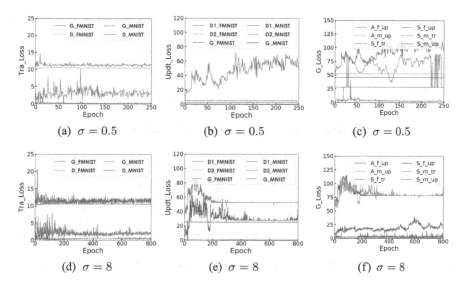

Fig. 2. The training loss of GANs over MNIST (A_/S_ = DP-Adam/-SGD, f_/m_ = FMNIST/MNIST, tr_/up_ = training/update)

DP-SGD when configuring the parameters identically. For example, we should set α of DP-SGD 20× larger than that of DP-Adam to generate visually similar fake images if the number of training epochs is 800.

Remark 2. Adding noise with larger σ increases the training instability, making the convergence slower; when σ is large enough, the training is not convergent anymore. For a single step, the training loss may not even decrease if drawing large noise. Training with more epochs often yields more accurate models but with higher privacy loss (due to the composition).

The variation of training loss becomes more irregular with increasing σ. One can also use more training epochs for better efficacy. The peak values of training loss in model tuning are larger than those of local training (Fig. 2a vs. 2b). When large noise is drawn, the training loss of G increases readily, e.g., the orange line at the 200-th epoch in Fig. 2f.

Remark 3. The noise "tolerance" depends on the parameter configuration, the choice of training datasets and/or models. For example, training GANs on Fashion-MNIST converges faster than on MNIST with the same setting; Fashion-MNIST is able to tolerate larger noise than MNIST. For LSTM-ML, Fig. 3 depicts the accuracy of the learner classifier at $\mathcal{U}$ and meta learner at $\mathcal{S}$ under different privacy levels. The baseline accuracy is ~60% [34].

We also evaluated LSTM-ML on a new task for classifying non-training data. Table 1 shows the accuracy with standard deviation (denoted by Δ) where the number of epochs is 100. The accuracy of using DP-Adam is roughly 10% better than using DP-SGD. More importantly, DP-SGD does not work well even in small-noise settings (e.g., $\sigma = 0.005$) due to its slow convergence or possibly falling into local optimums.

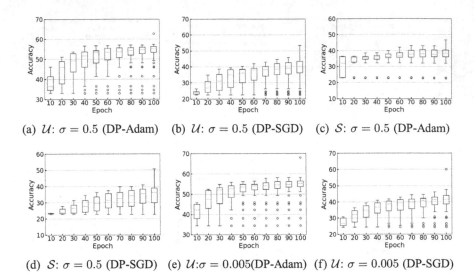

(a) $\mathcal{U}$: $\sigma = 0.5$ (DP-Adam)   (b) $\mathcal{U}$: $\sigma = 0.5$ (DP-SGD)   (c) $\mathcal{S}$: $\sigma = 0.5$ (DP-Adam)

(d) $\mathcal{S}$: $\sigma = 0.5$ (DP-SGD)   (e) $\mathcal{U}$:$\sigma = 0.005$(DP-Adam)   (f) $\mathcal{U}$: $\sigma = 0.005$ (DP-SGD)

Fig. 3. Averaged LSTM-ML accuracy

Table 1. Accuracy of LSTM-ML for a new task

σ	DP-Adam		DP-SGD	
	$\mathrm{Acc}_{\mathcal{U}}(\Delta)$	$\mathrm{Acc}_{\mathcal{S}}(\Delta)$	$\mathrm{Acc}_{\mathcal{U}}(\Delta)$	$\mathrm{Acc}_{\mathcal{S}}(\Delta)$
0.005	59.6($\pm$9)	45.8($\pm$8)	48.9($\pm$8)	39.5($\pm$8)
0.05	59.9($\pm$8)	43.2($\pm$9)	48.5($\pm$8)	39.4($\pm$7)
0.5	59.3($\pm$8)	43.7($\pm$9)	48.4($\pm$8)	38.7($\pm$8)
1	55.0($\pm$8)	37.2($\pm$8)	50.5($\pm$9)	39.3($\pm$8)

7.2   Computation and Communication Costs

The communication overheads between $\mathcal{S}$ and $\mathcal{U}$ are given in Table 2. We separately report the upload (*e.g.*, HE-encrypted parameters and res_i) and download (*e.g.*, $\mathsf{data_G}$) results. The ciphertext sizes (of a single parameter value) for GANs and LSTM-ML are 22.3 KB and 13.6 KB, respectively. Compared to BatchCrypt, we can achieve 24× uplink bandwidth saving.

At last, we present the time costs (dominated by HE operations) of aggregating parameters in Table 3. Each $\mathcal{U}$ encrypts parameter entries individually, and $\mathcal{S}$ performs homomorphic additions: the aggregation time is linear in the parameter size. As opposed to HybridAlpha aggregating parameters securely in each round, ours is just one-time, being 50× and 1.2× faster at $\mathcal{U}$ and $\mathcal{S}$, respectively.

Table 2. Communication costs (in gigabytes)

Model	Ours			BatchCrypt		
(epochs)	Up	Down	Total	Up	Down	Total
GAN (250)	66.23	7.54	73.77	203.66	7.54	211.20
GAN (800)	66.73	24.19	90.92	636.14	24.19	660.33
ML (100)	42.74	47.00	89.74	1042.92	47.00	1089.92
ML (10k)	42.74	47.00	89.74	1042.92	47.00	1089.92

Table 3. Time costs (in minutes) of parameter aggregation

Model	Ours		HybridAlpha		Model	Ours		HybridAlpha	
	$\mathcal{U}$	$\mathcal{S}$	$\mathcal{U}$	$\mathcal{S}$		$\mathcal{U}$	$\mathcal{S}$	$\mathcal{U}$	$\mathcal{S}$
GAN	660	14.0	33221	17.1	LSTM-ML	14	1.7	705	2.1

8   Conclusion

In retrospect, one might consider the current state of affairs in federated learning as a result of "ad hoc" improvements in addressing both the privacy and efficiency requirements. The premise starts with asking each contributor to contribute only the gradient for the apparent privacy benefits, which rules out an arguably more intuitive option of asking for a locally trained model (with obvious privacy implications). Nevertheless, gradients, being useful for training an aggregated model, still contain a lot of "residual" sensitive information. As a remedy, differential privacy has been incorporated to protect the gradients. Unfortunately, they are still "leaky" and remain exploitable by sophisticated membership inference attacks. Cryptography offers a general solution for processing encrypted data with a strong indistinguishability-based guarantee over all possible equal-length messages. Its great versatility, however, comes with significant overheads.

We revisit the "traditional wisdom" of federated learning of working over gradients. We put forth a new variant in the privacy-efficiency design space. On one hand, we allow the data contributors to directly share models locally trained from their training data. On the other, this part is now cryptographically protected for secure aggregation. In other words, cryptographic operations are confined to obtaining an initial global model. For the rest of the computation, we resort to a differentially private mechanism. This achieves the best of both worlds for selected applications of federate learning in which the subsequent computations are not exploited by any clever inference attacks. Particularly, we showcase our approach in generative adversarial networks and meta learning.

We hope this work can inspire further development in privacy-preserving federated learning. On the "destructive" side, our work motivates more analysis of the feasibility of innovative inference attacks from intermediate computation results beyond gradients or models. On the constructive side, it is interesting to explore alternative designs in the grand design space of privacy versus efficiency (versus utility). Furthermore, the potential of privacy-preserving designs tailored for specific federated learning applications remains open.

References

1. www.github.com/OpenMined/PySyft. Accessed 15 Oct 2023
2. Microsoft SEAL (release 3.3). www.github.com/Microsoft/SEAL. Accessed 15 Oct 2023
3. Abadi, M., et al.: Deep learning with differential privacy. In: CCS (2016)
4. Andrew, G., Thakkar, O., McMahan, B., Ramaswamy, S.: Differentially private learning with adaptive clipping. In: NeurIPS (2021)
5. Augenstein, S., et al.: Generative models for effective ML on private, decentralized datasets. In: ICLR (2020)
6. Bernstein, D.J., Hamburg, M., Krasnova, A., Lange, T.: Elligator: elliptic-curve points indistinguishable from uniform random strings. In: CCS (2013)
7. Bonawitz, K.A., et al.: Practical secure aggregation for privacy-preserving machine learning. In: CCS (2017)
8. Chen, D., Orekondy, T., Fritz, M.: GS-WGAN: a gradient-sanitized approach for learning differentially private generators. In: NeurIPS (2020)
9. Cheon, J.H., Kim, A., Kim, M., Song, Y.: Homomorphic encryption for arithmetic of approximate numbers. In: Takagi, T., Peyrin, T. (eds.) ASIACRYPT 2017. LNCS, vol. 10624, pp. 409–437. Springer, Cham (2017). https://doi.org/10.1007/978-3-319-70694-8_15
10. Cynthia, McSherry, F., Nissim, K., Smith, A.D.: Calibrating noise to sensitivity in private data analysis. In: TCC (2006)
11. Damaskinos, G., Mendler-Dünner, C., Guerraoui, R., Papandreou, N., Parnell, T.P.: Differentially private stochastic coordinate descent. In: AAAI (2021)
12. Deng, J., Dong, W., Socher, R., Li, L., Li, K., Fei-Fei, L.: ImageNet: a large-scale hierarchical image database. In: CVPR (2009)
13. Dong, Y., Chen, X., Jing, W., Li, K., Wang, W.: Meteor: improved secure 3-party neural network inference with reducing online communication costs. In: WWW (2023)
14. Dwork, C., McSherry, F., Nissim, K., Smith, A.D.: Calibrating noise to sensitivity in private data analysis. In: TCC (2006)
15. Dwork, C., Roth, A.: The algorithmic foundations of differential privacy. Found. Trends Theor. Comput. Sci. **9**(3–4), 211–407 (2014)
16. Dwork, C., Rothblum, G.N., Vadhan, S.P.: Boosting and differential privacy. In: FOCS (2010)
17. Fan, C., Liu, P.: Federated generative adversarial learning. In: PRCV (2020)
18. Fredrikson, M., Jha, S., Ristenpart, T.: Model inversion attacks that exploit confidence information and basic countermeasures. In: CCS (2015)
19. Gadotti, A., Houssiau, F., Annamalai, M.S.M.S., de Montjoye, Y.: Pool inference attacks on local differential privacy: quantifying the privacy guarantees of apple's count mean sketch in practice. In: USS (2022)
20. Gentry, C.: Fully homomorphic encryption using ideal lattices. In: STOC (2009)
21. Goodfellow, I.J., et al.: Generative adversarial nets. In: NeurIPS (2014)
22. Huang, Z., Hu, R., Guo, Y., Chan-Tin, E., Gong, Y.: DP-ADMM: ADMM-based distributed learning with differential privacy. IEEE Trans. Inf. Forensics Secur. **15**, 1002–1012 (2020)

23. Kairouz, P., Oh, S., Viswanath, P.: The composition theorem for differential privacy. In: ICML (2015)
24. Kim, M., Günlü, O., Schaefer, R.F.: Federated learning with local differential privacy: trade-offs between privacy, utility, and communication. In: ICASSP (2021)
25. Kingma, D.P., Ba, J.: Adam: a method for stochastic optimization. In: ICLR (2015)
26. LeCun, Y.: The MNIST database of handwritten digits (1998)
27. Li, Z., Huang, Z., Chen, C., Hong, C.: Quantification of the leakage in federated learning. In: NeurIPS Workshop on FL (2019)
28. Mandal, K., Gong, G.: PrivFL: practical privacy-preserving federated regressions on high-dimensional data over mobile networks. In: CCSW@CCS (2019)
29. McMahan, B., Moore, E., Ramage, D., Hampson, S., Arcas, B.A.: Communication-efficient learning of deep networks from decentralized data. In: AISTATS (2017)
30. McMahan, H.B., Ramage, D., Talwar, K., Zhang, L.: Learning differentially private recurrent language models. In: ICLR (2018)
31. Mironov, I.: Rényi differential privacy. In: CSF (2017)
32. Nasr, M., Shokri, R., Houmansadr, A.: Comprehensive privacy analysis of deep learning: passive and active white-box inference attacks against centralized and federated learning. In: S&P (2019)
33. Ng, L.K.L., Chow, S.S.M.: SoK: cryptographic neural-network computation. In: S&P (2023)
34. Ravi, S., Larochelle, H.: Optimization as a model for few-shot learning. In: ICLR (2017)
35. Shokri, R., Stronati, M., Song, C., Shmatikov, V.: Membership inference attacks against machine learning models. In: S&P (2017)
36. Song, W., Fu, C., Zheng, Y., Cao, L., Tie, M.: A practical medical image cryptosystem with parallel acceleration. J. Ambient. Intell. Humaniz. Comput. **14**, 9853–9867 (2022)
37. Stevens, T., Skalka, C., Vincent, C., Ring, J., Clark, S., Near, J.: Efficient differentially private secure aggregation for federated learning via hardness of learning with errors. In: USENIX Security (2022)
38. Sun, L., Qian, J., Chen, X.: LDP-FL: practical private aggregation in federated learning with local differential privacy. In: IJCAI (2021)
39. Truex, S., et al.: A hybrid approach to privacy-preserving federated learning. In: AISec@CCS (2019)
40. Wang, X., Ranellucci, S., Katz, J.: Authenticated garbling and efficient maliciously secure two-party computation. In: CCS (2017)
41. Wei, K., et al.: Federated learning with differential privacy: algorithms and performance analysis. IEEE Trans. Inf. Forensics Secur. **15**, 3454–3469 (2020)
42. Xiao, H., Rasul, K., Vollgraf, R.: Fashion-MNIST: a novel image dataset for benchmarking machine learning algorithms. arXiv:1708.07747 (2017)
43. Xu, R., Baracaldo, N., Zhou, Y., Anwar, A., Ludwig, H.: HybridAlpha: an efficient approach for privacy-preserving federated learning. In: AISec@CCS (2019)
44. Zhang, C., Li, S., Xia, J., Wang, W., Yan, F., Liu, Y.: BatchCrypt: efficient homomorphic encryption for cross-silo federated learning. In: USENIX ATC (2020)
45. Zhang, W., Tople, S., Ohrimenko, O.: Leakage of dataset properties in multi-party machine learning. In: USENIX Security (2021)
46. Zhang, W., Fu, C., Zheng, Y., Zhang, F., Zhao, Y., Sham, C.: HSNet: a hybrid semantic network for polyp segmentation. Comput. Biol. Med. **150**, 106173 (2022)

A Hessian-Based Federated Learning Approach to Tackle Statistical Heterogeneity

Adnan Ahmad[1]([✉])[ID], Wei Luo[1][ID], and Antonio Robles-Kelly[1,2][ID]

[1] School of Information Technology, Deakin University, Geelong, VIC 3220, Australia
{ahmadad,wei.luo,antonio.robles-kelly}@deakin.edu.au
[2] Defence Science and Technology Group, Edinburgh, SA 5111, Australia

Abstract. Federated learning (FL) involves collaboration between clients with limited data to produce a single optimal global model through consensus. One of the difficulties with FL is the differences in data statistics between local clients. Clients with statistically heterogeneous data deviate from the global target, resulting in a slower convergence rate and increased communication resource consumption. To address this problem, we propose a new approach, FedH, that maintains the proximity of local models to the global target while maximizing communication efficiency and computational resources. We use the Hessian matrix to constrain client updates that deviate from the global target. Our results demonstrate the superiority of FedH over FL baselines such as FedAvg, FedProx, and Fedcurv when applied to benchmark datasets such as MNIST, Fashion-MNIST, and CIFAR-10 across a range of statistical heterogeneity levels.

Keywords: Federated learning · model aggregation · client divergence · Hessian matrix

1 Introduction

Machine learning (ML) models require enormous amounts of data for training. Obtaining such a large amount of training data is difficult because the data is usually stored in silos on different edge devices. Therefore, large amounts of communication resources are required to transfer the data silos to the central server for training. In addition, this also violates the user's privacy. To solve this problem, several decentralized ML methods have been introduced in recent years. Federated Learning (FL) [1] is one of the decentralized approaches in which a model is trained in a decentralized manner while the training data is stored on edge devices. A typical FL process involves a central server that initiates training by transmitting initial parameter estimates to the participants. Participants then initialize their respective models with the received parameter estimates by performing some Stochastic Gradient Decent (SGD) steps on their local data. After training, participants communicate the model updates to the server. The server finally takes the weighted average of the received updates and sends them back to all the involved participants, completing an FL round of communication. The goal is to obtain a single global model that should be better than the models trained by the participants at the local level.

The Federated SGD (FedSGD) [1] is the first attempt to create a global model in a decentralized manner in the FL domain. In this approach, each participant transmits gradient updates to the server after each SGD step, and the central server considers these updates to refine the parameter estimates of the global model. However, FedSGD involves higher communication costs because updates must be exchanged after each SGD step, which can run into the thousands when heavily parameterized models are trained to achieve satisfactory performance levels. To address the communication cost issue, Federated Averaging (FedAvg) was introduced in [1]. It allows each participant to communicate their local model parameter estimates after multiple epochs of local training, reducing communication rounds and allowing participants to perform more local training before sending updates to the central server. This results in a significant reduction in communication overhead. As a result, FedAvg is considered the most commonly used aggregation method in FL.

FedAvg shows superior performance in an ideal environment where training data are independently and identically distributed (IID) between participants. This is because each local training data source has the same distribution and the local models converge to the same optimum. However, in practice, the data is often not IID distributed due to various factors, such as personalization and geography effects [2]. In a typical classification task, the training data and the class labels are unbalanced. For example, the local data on a given client may have unbalanced training patterns per class, or may even have no training pattern from a given class. In a non-IID environment, each local training data source has a different distribution that diverges the particular local model from the global objective. This client divergence in the presence of statistical heterogeneity is known in the literature as the "client drift" problem [3]. FedAvg is prone to client drift which leads to slower convergence. Therefore, merging local updates with a naive aggregation approach, such as FedAvg, may result in additional rounds of communication to achieve a satisfactory level of performance.

Several approaches to mitigate client drift have been proposed to ensure robust convergence, as discussed in detail in Sect. 2. Modifying the clients' local loss function is a common strategy to keep the local parameter estimates closer to the global ones [3,4]. However, existing approaches increase the communication cost within a single round of communication because they require additional components such as gradient information [3,5] to be communicated along with the parameter updates.

To address the aforementioned challenge, we propose a novel aggregation approach i.e., *FedH*, that not only ensures robust convergence but also consumes the same amount of communication and computational resources as FedAvg. Our proposed approach presents a modified version of the clients' local objectives to mitigate the clients' divergence problems. In particular, we add a regularization term to the clients' local loss function to prevent it from diverging from the global objective. The proposed approach uses the Hessian matrix, a well-known second-order metric for quantifying the curvature of the loss function. The diagonal elements of the Hessian matrix serve as scaling coefficients for the additional term we include in the local loss function. Rather than computing the exact Hessian, we use common techniques to approximate the Hessian [6,7] to avoid the computational complexity of the participating devices.

Our main contributions are summarized as follows:

- We propose a novel method of model aggregation to ensure robust convergence under statistically heterogeneous data distributions.
- We present an approach that efficiently uses the second-order information of the model loss function to quantify the degree of divergence between global and local objectives.
- Our method consists of approximating the Hessian matrix from first-order information, which acts as a scaling matrix in clients' local goals.
- Through extensive experiments, we demonstrate the superior performance of the proposed approach by training neural networks for a variety of image classification tasks.

The remainder of the paper is organised as follows. Section 2 discusses related work, while Sect. 3 presents the problem formulation. Section 4 presents the proposed method for solving the statistical heterogeneity problem along with preliminary remarks useful for understanding the main concept. Section 5 describes the experimental evaluation and the main results of this work. Finally, Sect. 5 discusses the conclusions and possible future work.

2   Related Work

This section presents some of the existing work on the convergence of FedAvg in a non-IID data distribution. In recent years, since the advent of the FedAvg algorithm, a large number of researchers have addressed the problem of statistical heterogeneity in FL. For example, Hsu et al. show. [8] the linear relationship between client divergence and the degree of statistical heterogeneity in the training data. They show that the greater the disparity between datasets, the more pronounced the performance degradation of FedAvg. Similarly, Zhao et al. measure the drop in accuracy using the Earth Mover Distance (EMD) between data distributions in their work [9]. They also propose a data-sharing strategy to minimize the differences in data distributions. However, it should be noted that data sharing is not always possible due to privacy concerns and contradicts the core principles of the FL algorithm.

In addition, ensemble learning and model distillation techniques have also been used in FL to aggregate models. In the work of Guha et al. [10], a single round of communication is used to combine local models using a model distillation technique. Lin et al. [11] extend this work by implementing the ensemble learning technique across multiple communication rounds. Meanwhile, FEDBE [12] approaches the model distillation process from a Bayesian perspective. The FEDBE approach, inspired by the stochastic weighting method in [13], models client updates using a Gaussian distribution and uses the Monte Carlo method to extract models from the distribution. However, these ensemble learning and model distillation methods may not be practical for certain FL tasks because they require that the server have access to some shared data in order to distil the ensemble model into a single global model.

In addition to the above techniques, modifying the FedAvg local likelihood function is a popular method to account for statistical heterogeneity. Several studies attempt to solve the client divergence problem by modifying the client loss function [3–5,14]. They do this by incorporating an isotropic regularization term into the local loss function, along with a hyperparameter that weights each parameter estimate equally. For example, FedProx [4] adds a proximal term to the local loss function to control gradient dissimilarity during training. FedDANE [14] takes a similar approach with an additional gradient correction term to handle client divergence. SCAFFOLD [3] reduces gradient divergence using a variance reduction technique in the local likelihood function.

To solve the problem of statistical heterogeneity among client data distributions, researchers have also explored second-order methods. FedCurv [5], uses the Fisher Information (FI) matrix to modify the proximal term introduced by FedProx [4]. Specifically, the proximal term is modified by including the diagonal of the FI matrix. This allows each parameter estimate to be regularized based on its importance to the performance of the global model. Islamov et al. [15] presented a communication-efficient Newton method for distributed optimization. While effective, it does not meet privacy requirements because the central server needs access to the clients' training samples to compute the second-order information, making it an unsuitable option in scenarios where privacy is a concern. Building on the idea presented in the work of Islamov [15], Randi et al. [16] introduced FedNL, a method that uses a global Hessian matrix to execute a Newton step on the server. In addition, the work of Qian et al. [17] presented different versions of FedNL that use matrix compression methods to improve the compressed Hessian matrix by changing the basis in the matrix space. Liu et al. [18] presented another Newton-based method using the L-BFGS algorithm [19] to overcome communication constraints. However, it should be noted that the above methods require additional terms to be exchanged between the server and the clients in each communication round, in addition to the parameter estimates. In contrast, the communication cost incurred by our method in each round is comparable to that of FedAvg.

3   Problem Formulation

In a typical FL process, a central server aims to train a global model with a distributed objective given by:

$$\min_{w \in \mathbb{R}^d} f(\boldsymbol{w}) \quad \text{where} \quad f(\boldsymbol{w}) = \sum_{k=1}^{K} \frac{n_k}{n} f(\boldsymbol{w}_k) \tag{1}$$

K is the total number of participants with local data $D_k(x,y)$ available for collaboration. $f(.)$ is the loss function. $\boldsymbol{w}$ are the parameter estimates of the global model. $\boldsymbol{w}_k$ are the parameter estimates of the kth local model. n_k represents the number of training samples at client k, and n represents the total number of training samples across all clients.

In a standard federated averaging process, such as FedAvg, each participant k locally trains its respective model with local data $D_k(x,y)$ for a designated number of epochs and communicates updates to the server. The client adopts the SGD method to update parameter estimates given by:

$$\boldsymbol{w}_k = \boldsymbol{w}_k - \eta \nabla f(\boldsymbol{w}_k) \tag{2}$$

where η is the learning rate. After the server receives updates from clients, it aggregates local updates using a weighted average method given by:

$$\boldsymbol{w}^{t+1} = \sum_{k=1}^{K} \frac{n_k}{n} \boldsymbol{w}_k \tag{3}$$

Here, $\boldsymbol{w}^{t+1}$ indicates the global parameter updates after the current training round.

The standard aggregation technique, FedAvg, demonstrates fast convergence under ideal conditions where each participant has an IID data distribution in their training data [20]. This is because an IID distribution results in all local models pursuing a similar learning trajectory to reach their local optimal solution, resulting in limited differences between the local parameter estimates. When these local updates are combined using a standard aggregation approach, it leads to a global model that attains the same optimal solution as the local models. However, in real-world scenarios, FedAvg has been found to have sub-par convergence due to the non-IID nature of real-world data, caused by factors such as geography effects and personalization [2]. Participants may have differing amounts of training samples or labels, leading to unique data distributions. Consequently, in a non-IID environment, each local model deviates from the global optimum and follows its own individual learning path to reach its own local optimum.

As previously discussed in Sect. 2, various methods have been proposed to modify the client's local objective (Eq. 2) in order to alleviate client divergence. Our approach employs different methods to regulate the local gradients in order to avoid negative impacts on the global model's performance. Our aim is to incorporate the Hessian matrix into the client's local loss functions, which will be described in more detail in the following section.

Algorithm 1. FedH

1: **Server Input:** Initial global model w^t, R: number of communication rounds, E : number of epochs, λ
2: **Output:** Final global model w^{t+1}
3: Let $t = 0$
4: **for** $r = 0$ to R **do**
5:     Communicate w^t to all clients
6:     **for** For each client $k \in K$ in parallel **do**
7:         **Initialize** local model $w_k \leftarrow w^t$
8:         **for** epoch $\leftarrow 0$ to E **do**
9:             **Compute Gradients:** $\nabla f(w_k)$
10:            **if** $epoch == 0$ **then**
11:                **Client Approx. :** $\nabla^2 f(w) = \nabla^2 f(w_k)$
12:            **end if**
13:            **Client Approx.:** $\nabla^2 f(w_k)$
14:            **Client update:** Equation 5
15:        **end for**
16:        **Communicate** updated w_k to the central server.
17:     **end for**
18:     Server aggregates weights: $w^{t+1} = \sum_{k=1}^{K} \frac{n_k}{n} w_k$
19: **end for**
20: **return** w^{t+1}

---

4   Proposed Method

In this section, we briefly review the preliminaries (Sect. 4.3), which are useful for understanding the proposed method. Next, we propose a local objective function (Sect. 4.1), and finally, we explain the implementation details of the proposed method (Sect. 4.2).

In this paper, we propose a novel method of model aggregation using the approximation of the Hessian matrix from the first-order derivatives. The Hessian matrix is a square matrix that describes the second-order partial derivatives of the loss function. It is used to quantify the local curvature of the function, which can be important in optimization problems because it determines whether a critical point is a local minimum, a local maximum, or a saddle point. In information geometry, the Hessian matrix plays a key role in characterizing the geometry of statistical models and their parameter spaces. The Hessian matrix provides information about the rate at which the function is extremized [6,7], or how fast it changes when the input variables change.

The Hessian is represented as $\mathbf{H} = \mathbf{J}(\nabla f(w))$, where $\mathbf{J}(\nabla f(w))$ is the Jacobian matrix of the gradient of the loss function with respect to the model weights w. The Hessian matrix is often used in second-order optimization methods such as Newton's method to improve the convergence properties of optimization methods [21]. However, computing the Hessian matrix in its exact form for high-dimensional models can be very computationally intensive due to the increased number of parameters [22]. Therefore, in practice, alternative methods such as approximation techniques or approximating the Hessian matrix by diagonal matrices or low rank matrices are often used to solve this problem. As a result, the Gauss-Newton optimization method [23,24] is often used as an alternative to the Newton method.

Therefore, in our approach, we adopt a similar technique that exploits the first-order information of the loss function to approximate the Hessian [25]. This is done following a best practice in Gauss-Newton algorithms [24], where the Hessian is approximated by squaring the Jacobian matrix [22,26]. The approximation of the Hessian for a given loss function, $f(w_k)$, is given as follows:

$$\tilde{\mathbf{H}} = \mathbf{J}\big(f(\boldsymbol{w})\big)^T \mathbf{J}\big(f(\boldsymbol{w})\big) \tag{4}$$

4.1   Proposed Local Objective

In this section, we discuss the proposed local objective to be solved by each participant. As mentioned in Sect. 3, our method proposes a modification of the local objective function (Eq. 2). Instead of optimizing the loss function $f(w_k)$, client k solves the following objective:

$$\boldsymbol{w}_k = \boldsymbol{w}_k - \eta\Big(\nabla f(\boldsymbol{w}_k) + \lambda\big(\boldsymbol{\tau}|\boldsymbol{w} - \boldsymbol{w}_k|\big)\Big), \tag{5}$$

where

$$\boldsymbol{\tau} = \frac{\nabla^2 f(\boldsymbol{w}) - \nabla^2 f(\boldsymbol{w}_k)}{\|\nabla^2 f(\boldsymbol{w}) - \nabla^2 f(\boldsymbol{w}_k)\|}$$

$\nabla^2 f(\boldsymbol{w})$ represents the diagonal approximation of the Hessian of the loss function with respect to the global parameter estimates $\boldsymbol{w}$ and $\nabla^2 f(\boldsymbol{w}_k)$ denotes the diagonal approximation of the Hessian of the local loss function with respect to the current parameter estimates $\boldsymbol{w}_k$. Note that the statistical heterogeneity can result in a deviation of $\boldsymbol{w}_k$ from the global parameter estimates $\boldsymbol{w}$. To prevent this deviation, we use $|\boldsymbol{w} - \boldsymbol{w}k|$, which helps maintain the proximity of the current estimates $\boldsymbol{w}_k$ to the global parameters $\boldsymbol{w}$. Similar terms have also been widely used in the literature [20,27]. However, the standalone usage of this term treats all parameter estimates as equally important, which can limit the effectiveness of the parameters of local updates that are not much important to the global objective. To address this, we scale the components of the regularization term, $|\boldsymbol{w} - \boldsymbol{w}_k|$, with the coefficients term, $\boldsymbol{\tau}$, originating from the Hessian diagonal, since the diagonal of the approximated Hessian can be used as a scaling matrix [28]. $\nabla^2 f(\boldsymbol{w})$ quantifies the tendency for the global model parameters to change when input data $D_k(x, y)$ are used. Similarly, $\nabla^2 f(\boldsymbol{w}_k)$ determines the influence of $D_k(x, y)$ on local model parameters. If a component of $\nabla^2 f(\boldsymbol{w}_k)$ is larger than $\nabla^2 f(\boldsymbol{w})$, then its normalized difference, $\boldsymbol{\tau}$, will penalize the corresponding component in $|\boldsymbol{w} - \boldsymbol{w}_k|$. Therefore, integrating this entire term controlled by a hyperparameter λ into the local loss function can act as a great remedy for client divergence problems, keeping $\boldsymbol{w}_k$ closer to $\boldsymbol{w}$ without limiting the effectiveness of updates for those parameters that are not critical to the global objective.

Table 1. Test accuracy as a function of communication rounds is reported after 50 communication rounds. The absolute best performance is reported in bold.

non-IID Type	Data	FedAvg	FedProx	FedCurv	FedH
non-IID(I)	MNIST	89.46 ± 1.332	89.27 ± 1.425	89.44 ± 1.421	$\mathbf{90.63 \pm 1.190}$
	Fashion-MNIST	74.41 ± 3.332	74.12 ± 3.474	72.55 ± 4.881	$\mathbf{76.48 \pm 3.120}$
	CIFAR10	60.86 ± 1.190	59.20 ± 1.055	62.03 ± 0.961	$\mathbf{62.38 \pm 1.630}$
non-IID(II)	MNIST	82.79 ± 1.542	82.45 ± 1.459	81.75 ± 3.579	$\mathbf{86.67 \pm 2.336}$
	Fashion-MNIST	64.21 ± 5.019	63.08 ± 5.058	55.24 ± 4.318	$\mathbf{66.85 \pm 5.828}$
	CIFAR10	49.82 ± 2.296	43.33 ± 1.415	50.25 ± 2.653	$\mathbf{52.59 \pm 3.334}$
Dir(0.1)	MNIST	94.52 ± 0.952	94.60 ± 1.041	94.56 ± 0.889	$\mathbf{95.70 \pm 0.780}$
	Fashion-MNIST	82.32 ± 1.639	82.16 ± 1.680	82.07 ± 1.869	$\mathbf{82.88 \pm 2.080}$
	CIFAR10	68.55 ± 0.247	69.00 ± 0.203	68.49 ± 0.254	$\mathbf{69.80 \pm 0.890}$
Dir(0.05)	MNIST	90.77 ± 3.851	90.82 ± 3.823	90.81 ± 3.671	$\mathbf{92.04 \pm 1.272}$
	Fashion-MNIST	77.74 ± 5.624	78.47 ± 5.263	78.68 ± 5.174	$\mathbf{79.78 \pm 1.448}$
	CIFAR10	62.47 ± 4.898	62.27 ± 4.202	63.30 ± 4.842	$\mathbf{63.62 \pm 4.109}$
Dir(0.01)	MNIST	69.63 ± 5.289	71.30 ± 7.934	67.75 ± 8.108	$\mathbf{79.34 \pm 3.504}$
	Fashion-MNIST	58.33 ± 5.094	62.22 ± 1.929	63.75 ± 0.913	$\mathbf{72.62 \pm 0.575}$
	CIFAR10	46.25 ± 7.230	38.75 ± 7.687	47.68 ± 6.748	$\mathbf{48.52 \pm 6.732}$

4.2 FedH Aggregation

Now we turn our attention to the implementation of the proposed method. As mentioned earlier, we approximate Hessian from the first-order information of the loss function. Our method requires the same communication and computation resources as that of FedAvg. Hessian can be easily approximated during local training without imposing an extra burden on clients. Algorithm 1 presents the proposed method which we named FedH. The server takes the initial model parameters w^t, the total number of passes(epochs) over local data E, and a hyperparameter λ as inputs and returns final model weights w^{t+1} as output after R communication rounds. In each communication round r, the server communicates the initial model parameters w^t to all participants. Each participant initializes their local models with current global parameters w^t and computes gradients with their local data $D_k(x, y)$ for the first epoch. We can approximate $\nabla^2 f(w)$ from local gradients in first epochs as we initialize local parameters with global ones. Therefore, there is no need to perform extra passes over the data. Each client then needs to approximate the $\nabla^2 f(w)$ for each epoch and update the local parameter estimates with Eq. 5.

As for as communication efficiency and privacy is concerned, note that our method is akin to FedAvg in terms of communication cost. The central server and participants only need to communicate parameter updates in each communication round. Therefore, Like FedAvg, our method also respects privacy and weights compression schemes and can be integrated with state-of-the-art algorithms.

4.3   Preliminaries

5   Experimental Evaluation

This section explains the datasets and the models (Sect. 5.1) used to perform the experiments to test the effectiveness of the proposed method. Section 5.2, the baseline is described, and finally, the obtained results are discussed (Sect. 5.3).

5.1   Datasets and Models

We consider three publicly available image classification datasets to evaluate the proposed method. These are MNIST [29], Fashion-MNIST [30], and CIFAR10 [31]. MNIST is a digit classification dataset consisting of 50, 000 training and 10, 000 test samples. Each sample consists of 28×28 single channel images with one target. Fashion-MNIST contains the same number of samples and image sizes, but these are images of fashion clothing from Zalando's database. CIFAR10 contains 50, 000 training samples and 10, 000 test samples, each sample consisting of 32×32 random images with three channels.

We use multilayer perception (MLP) to learn the MNIST and Fashion MNIST datasets. We use the same architecture used by the authors in [1]. MLP consists of two hidden layers, each consisting of 200 units with ReLU activation. We train a convolutional neural network (CNN) for the CIFAR10 image classification task. The CNN consists of three convolutional layers with 3×3 kernels (channel sizes 32, 64, and 128) followed by two fully connected layers. Note that our goal is not to achieve the highest accuracy in the datasets. Our main aim is to achieve better convergence compared to existing methods, even when working with non-IID data. Therefore, we use models that are widely used in the FL literature.

5.2   Baselines and Settings

We compare our method to the standard FedAvg [1] aggregation method. In addition, we also consider FedProx [4] and FedCurv [5] as additional baselines. The motivation for selecting these baselines is that these baselines also account for client divergence by modifying clients' local loss functions. FedProx adjusts clients' local objective functions with a proximal term, $\frac{\mu}{2}\|w - w_k\|_2$, to avoid clients' drift. FedCurv improves FedProx by using the FI matrix alongside the proximal term.

We run our experiments in a cross-silo FL environment [2] where we consider 10 active participants throughout the training process. We fit the hyperparameters for all models using cross-validation and the learning rate is set to 0.01 with 5 training epochs between communication rounds. We use the SGD optimizer with a weight decay of 0.003 and a server momentum of 0.9 on each client to optimize each model. For FedProx [4] and FedCurv [5], we use the same hyperparameter settings that the authors used in their experiments.

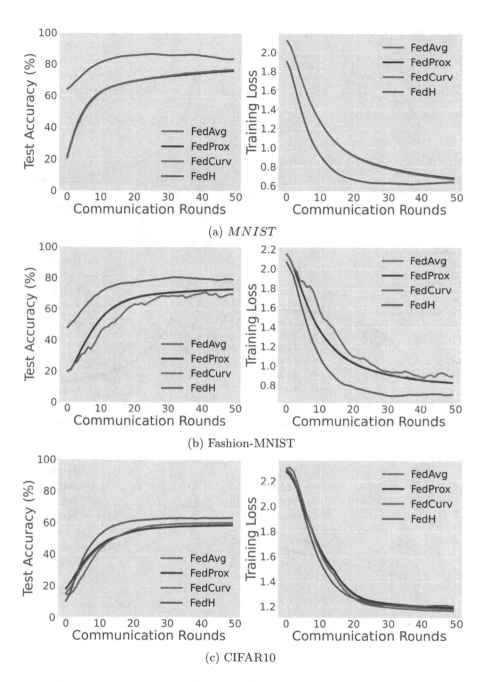

Fig. 1. Test accuracy and loss of the global model as a function of communication rounds when a non-IID (I) data distribution scheme is used to distribute the MNIST, Fashion MNIST, and CIFAR10 datasets between clients.

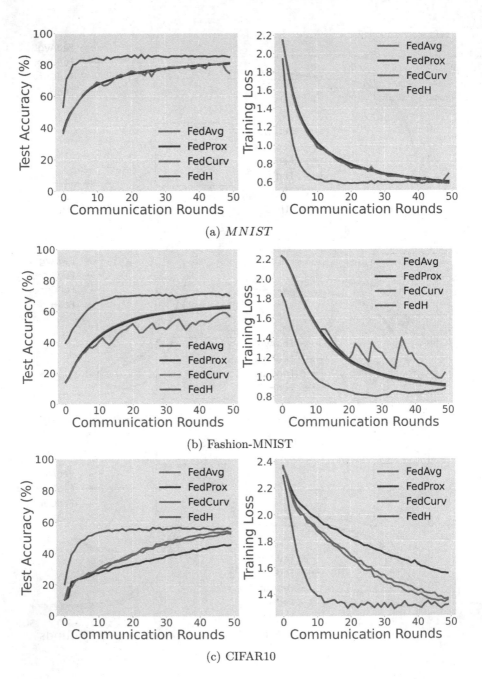

Fig. 2. Test accuracy and loss of the global model as a function of communication rounds when a non-IID data distribution scheme (II) is used to distribute MNIST, Fashion MNIST, and CIFAR10 datasets between clients.

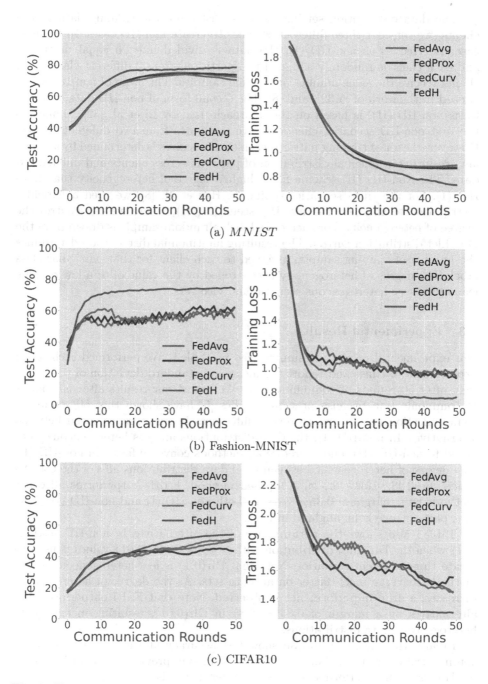

Fig. 3. Test accuracy and loss of the global model as a function of communication rounds when the MNIST, Fashion-MNIST and CIFAR10 datasets are distributed between clients with a Dirichlet distribution for $\alpha = 0.01$.

The data distribution settings used to distribute the training data to the clients, we consider three different schemes to create non-IID cases. In the first case, referred to as non-IID (I), the data is divided into 20 equal partitions and each client is randomly assigned two partitions from 2 different classes. All clients receive the same number of training samples. This type of non-IID setup is used in the work of [5,32], and [8]. The second form of non-IID data, referred to as non-IID (II), is based on the approach used by Li et al. [32]. Similar to the first non-IID scenario, clients receive two parts from two different classes. However, the set of training patterns each client receives is determined by a power law, resulting in unequal distribution of patterns across clients and unbalanced data. The Non-IID (II) scheme has a higher degree of heterogeneity than non-IID (I). For the third scenario, similar to Hsu et al. [8], we used a Dirichlet distribution to simulate the non-IID scenario (III), where we can control the degree of heterogeneity. For each class label, a random sample is drawn from the Dirichlet distribution $Dir(\alpha)$. The resulting multinomial distribution determines the number of training samples assigned to each client for that particular class label. The degree of heterogeneity is controlled by the value of α: a low α value leads to a very heterogeneous scenario and vice versa.

5.3   Experimental Results

Our experimental results are summarized in Table 1. We performed each set of experiments 10 times and report the mean and standard deviation of test accuracy after 50 rounds of communication. We report our results after 50 rounds of communication because previous studies [5] have shown that 50 rounds of communication is sufficient to reveal differences in convergence speed between alternatives. In non-IID (I), the data distributions are less heterogeneous compared to non-IID (II), and therefore all methods converge faster in non-IID (I). The degree of heterogeneity between local data distributions affects the convergence speed [8]. However, our proposed method, FedH, outperforms all other methods by a margin of $0.35 \sim 3.88$ for both non-IID (I) and non-IID (II). The best performance is highlighted in bold.

Table 1 also shows the performance of the alternatives in non-IID settings (III) when the Dirichlet distribution with $\alpha = \{0.1, 0.05, 0.01\}$ values is used to create three different scenarios. Note that $Dir(0.1)$ is less heterogeneous, so all methods converge much faster on all data sets. As the degree of heterogeneity increases, a drop in performance is observed. Note that FedH outperforms the alternatives by a margin of $0.84 \sim 8.87$ in $Dir(0.01)$, resulting in extremely heterogeneous data distributions.

Finally, the Figs. 1, 2 and 3b show test accuracy and loss as a function of communication rounds. These figures confirm our previous observations that FedH consistently performs better than other methods.

6   Conclusion and Future Work

This paper presents a novel aggregation approach in FL to aggregate local models trained on heterogeneous data distributions into a single global model. The proposed method addresses client model divergence by modifying the local loss function. The proposed method uses the Hessian matrix of loss functions to regularize clients' local objectives. We demonstrate the superior performance of the proposed method on a variety of image classification tasks learned under different levels of statistical heterogeneity. In future work, the proposed approach can be further improved by using the off-diagonal elements of the Hessian matrix. To this end, we plan to make efficient use of the eigenvalue decomposition of the Hessian matrix in the local objective functions.

References

1. McMahan, H.B., Moore, E., et al.: Communication-efficient learning of deep networks from decentralized data. In: AISTATS (2017)
2. Kairouz, P., McMahan, H.B., Avent, B., Bellet, A., Bennis, M., et al.: Advances and open problems in federated learning. Found. Trends Mach. Learn. **14**, 1–210 (2019)
3. Karimireddy, S., Kale, S., Mohri, M., et al.: SCAFFOLD: stochastic controlled averaging for federated learning. In: ICML, 13–18 Jul 2020, pp. 5132–5143 (2020)
4. Li, T., Sahu, A.K., et al.: Federated optimization in heterogeneous networks. In: Dhillon, I., Papailiopoulos, D., Sze, V. (eds.) Proceedings of Machine Learning and Systems, vol. 2, pp. 429–450 (2020)
5. Shoham, N., et al.: Overcoming forgetting in federated learning on non-IID data. In: FL-NeurIPS (2019). arXiv:1910.07796
6. Pennington, J., Bahri, Y.: Geometry of neural network loss surfaces via random matrix theory. In: ICML (2017)
7. Pennington, J., Worah, P.: The spectrum of the fisher information matrix of a single-hidden-layer neural network. In: Conference on Neural Information Processing Systems (2018)
8. Hsu, T., Qi, H., Brown, M.: Measuring the effects of non-identical data distribution for federated visual classification. arXiv:abs/1909.06335 (2019)
9. Zhao, Y., Li, M., Lai, L., et al.: Federated learning with non-IID data. arXiv:abs/1806.00582 (2018)
10. Guha, N., Talwalkar, A., Smith, V.: One-shot federated learning. CoRR, abs/1902.11175 (2019)
11. Lin, T., Kong, L., Stich, S.U., Jaggi, M.: Ensemble distillation for robust model fusion in federated learning. Adv. Neural. Inf. Process. Syst. **33**, 2351–2363 (2020)
12. Chen, H.-Y., Chao, W.-L.: FedBE: making Bayesian model ensemble applicable to federated learning. In: ICLR (2021)
13. Maddox, W., Garipov, T., Izmailov, P., et al.: A simple baseline for Bayesian uncertainty in deep learning. In: NeurIPS (2019)
14. Li, T., Sahu, A.K., Zaheer, M., et al.: Feddane: a federated newton-type method. In: 2019 53rd Asilomar Conference on Signals, Systems, and Computers, pp. 1227–1231 (2019)

15. Islamov, R., Qian, X., Richtárik, P.: Distributed second order methods with fast rates and compressed communication. In: ICML (2021)
16. Safaryan, M., Islamov, R., et al.: FedNL: making newton-type methods applicable to federated learning. In: ICML, ser. Workshop on Federated Learning for User Privacy and Data Confidentiality (2021)
17. Qian, X., et al.: Basis matters: better communication-efficient second order methods for federated learning. In: AISTATS (2022)
18. Liu, Y., Zhu, Y., James, J.: Resource-constrained federated learning with heterogeneous data: formulation and analysis. IEEE Trans. Network Sci. Eng. (2021)
19. Liu, D.C., Nocedal, J.: On the limited memory BFGs method for large scale optimization. Math. Program. **45**, 503–528 (1989)
20. Li, T., Sahu, A.K., Zaheer, M., Sanjabi, M., Talwalkar, A., Smith, V.: Federated optimization in heterogeneous networks. Proceedings of Machine Learning and Systems, vol. 2, pp. 429–450 (2020)
21. Nocedal, J., Wright, S.J.: Numerical Optimization. Springer, New York (1999). https://doi.org/10.1007/0-387-22742-3
22. LeCun,Y.A., Bottou, L., Orr, G.B., Müller, K.-R.: Efficient backprop. In: Neural Networks: Tricks of the Trade (2012)
23. Becker, S., LeCun, Y.: Improving the convergence of back-propagation learning with second-order methods. In: Technical Report CRG-TR-88-5 (1989)
24. Schraudolph, N.N.: Fast curvature matrix-vector products. In: ICANN (2001)
25. Chen, P.: Hessian matrix vs. gauss-newton hessian matrix. SIAM J. Numer. Anal. **49**, 1417–1435 (2011)
26. Nocedal, J., Wright, S.: Numerical Optimization. Springer, New York (2006). https://doi.org/10.1007/978-0-387-40065-5
27. Shamir, O., Srebro, N., Zhang, T.: Communication-efficient distributed optimization using an approximate newton-type method. In: International Conference on Machine Learning. PMLR 2014, pp. 1000–1008 (2014)
28. Oren, S.S., Luenberger, D.G.: Self-scaling variable metric (SSVM) algorithms. Part I: criteria and sufficient conditions for scaling a class of algorithms. Manage. Sci. **20**(5), 845–862 (1974)
29. Deng, L.: The mnist database of handwritten digit images for machine learning research [best of the web]. IEEE Signal Process. Mag. **29**(6), 141–142 (2012)
30. Cohen, G., Afshar, S., Tapson, J., Van Schaik, A.: Emnist: extending mnist to handwritten letters. In: International Joint Conference on Neural Networks (IJCNN). IEEE 2017, pp. 2921–2926 (2017)
31. Krizhevsky, A., Hinton, G.: Convolutional deep belief networks on cifar-10. Unpublished manuscript, 40(7), 1–9 (2010)
32. Li, X., Huang, K., et al.: On the convergence of fedavg on non-IID data. In: ICLR (2020)

Analyzing the Convergence of Federated Learning with Biased Client Participation

Lei Tan[1,2], Miao Hu[1,2], Yipeng Zhou[3], and Di Wu[1,2(✉)]

[1] School of Computer Science and Engineering, Sun Yat-sen University,
Guangzhou 510006, China
tanlei6@mail2.sysu.edu.cn, humiao5@mail.sysu.edu.cn,
wudi27@mail.sysu.edu.cn
[2] Guangdong Key Laboratory of Big Data Analysis and Processing,
Guangzhou 510006, China
[3] School of Computing, Faculty of Science and Engineering, Macquarie University,
Sydney, NSW 2109, Australia
yipeng.zhou@mq.edu.au

Abstract. Federated Learning (FL) is a promising decentralized machine learning framework that enables a massive number of clients (e.g., smartphones) to collaboratively train a global model over the Internet without sacrificing their privacy. Though FL's efficacy in non-convex problems is proven, its convergence amidst biased client participation lacks theoretical study. In this paper, we analyze the convergence of FedAvg on non-convex problems, which is the most renowned FL algorithm. We assume even data distribution but non-IID among clients, and elucidate the convergence rate of FedAvg in situations characterized by biased client participation. Our analysis reveals that biased client participation can significantly reduce the precision of the FL model. We validate this through trace-driven experiments, demonstrating that unbiased client participation results in 11% to 50% higher test accuracy compared to extremely biased client participation.

Keywords: Federated learning · Non-convex · Biased participation · Convergence analysis

1 Introduction

Federated Learning (FL), an emerging paradigm in the realm of decentralized machine learning (ML), enables multiple edge devices (e.g., smartphones, ipads) to jointly train an ML model without exposing their sensitive data [1,19,20], and thus prevents the leakage of user privacy. In the FL training, we need to conduct a series of global iterations, and a certain number of FL clients participate in each iteration by submitting their computational outcomes (e.g., parameters, gradients) to a central parameter server (PS). The PS, in turn, undertakes the task of aggregating the submitted model parameters, utilizing techniques such as FedAvg [7,20] or FedSGD [8,18,20]. Subsequent to the aggregation process, the aggregated model is subsequently disseminated back to the clients. As a

© The Author(s), under exclusive license to Springer Nature Switzerland AG 2023
X. Yang et al. (Eds.): ADMA 2023, LNAI 14177, pp. 423–439, 2023.
https://doi.org/10.1007/978-3-031-46664-9_29

novel distributed framework, FL has been widely used in the fields of Web-of-Things [11], healthcare [23], environmental sound detection [21] and so on.

FL differs from traditional distributed optimization [5] in the following two aspects: *First*, the assumption of *IID (identically and independently distributed)* distribution of training data does not hold in the FL scenarios. The IID assumption is commonly used in traditional ML models. In FL, the data on a client is generated by a user itself when using the device. Therefore, any local data fails to represent the global distribution. *Second,* the communication between the PS and clients is restricted and unreliable. For example, due to the dynamics of network conditions, mobile devices cannot stay online forever. It is highly possible that clients at different geographical locations cannot participate in the entire training process in an unbiased manner. As a result, some clients may participate in much more rounds of iterations than other clients [9,16], and the frequency of each client's participation in the FL training process is different [22].

McMahan *et al.* [20] empirically showed that FedAvg (Federated Averaging) performs well on non-convex optimization with non-IID data distribution. Subsequently, theoretical guarantees of the FL algorithm in strongly convex and non-convex cases are provided by the researchers [8,17]. However, these works all assume that client participation is unbiased. In fact, due to restricted and unreliable communication, the number of times clients participate in federated training is variable. There still lacks theoretical convergence analysis for non-convex optimization with non-IID data distribution and biased client participation.

Motivated by the above problem, our goal in this paper is to analyze the convergence guarantee of FedAvg for non-convex optimization on non-IID data with biased client participation. Despite the aforementioned research efforts, quite a few challenges remain. *i)*. The properties of convex functions cannot be applied to non-convex functions, and finding the optimal solution for non-convex problems is NP-hard in general, resulting in that non-convex optimization analysis being more complicated than convex optimization analysis. Furthermore, the unique characteristics of FL (e.g., non-IID training data, biased client participation) increase the complexity of the analysis. *ii)*. The current FL aggregation strategies do not take the bias of client participation into account. It is still unknown how biased client participation in the training process affects the effectiveness of the global FL model training.

In this paper, we conduct a theoretical analysis of decentralized FL for non-convex optimization and obtain the convergence rate of FedAvg within the context of biased client participation. To address the first challenge, the disparity existing between local distribution and global distribution is used to describe the feature of non-IID training data. To address the second challenge, we integrate the number of times each client participates in FL into its aggregation weight and get the corresponding aggregation formula. Through analyzing the deduced convergence rate, we observe that biased client participation in the training process exerts a detrimental impact on the convergence rate. Summarizing our key contributions:

- We conduct a convergence analysis of FL based on its natural properties and provide theoretical convergence guarantees for non-convex problems.

This theoretical analysis can be extended to the case with biased client participation by substituting the aggregation coefficient. Moreover, we reveal the relationship between the relevant variables and the convergence rate of FL for non-convex optimization on non-IID data.

- We further analyze the convergence rates by considering the extent of the bias of client participation in FL. Based on our analysis results, we find that the bias of client participation exerts a detrimental impact on the convergence of FedAvg, i.e., the more serious bias results in a lower precision global model.
- We conduct a series of real trace-driven experiments to affirm the correctness of our theoretic findings. The results point out that biased client participation incurs a decrease in the precision of the global FL model. The accuracy of the unbiased state is 11%–50% higher than that of the extreme state.

2 Related Work

FL can learn global models without centralized training data, and its mainstream optimization algorithm is FedAvg, which is an algorithm based on averaging local gradient updates. The analysis of FedAvg is far from trivial due to the natural characteristics of FL.

Khaled et al. [10] provided a convergence analysis of distributed gradient descent on heterogeneous data. Li et al. [17] presented a convergence analysis of FedAvg with a focus on non-IID data, successfully establishing a convergence rate applicable to strongly convex problems. Cho et al. [6] analyzed the convergence of FL for strategies involving biased client selection in strongly convex problems. Balakrishnan et al. [3] proposed to select clients with representative gradient information, and then send these updates to the PS. The authors also provided a convergence analysis on convex problems in the heterogeneous setting.

These latest works provided convergence guarantees on non-IID data for convex, however, its result is only applicable to strongly convex problems. Unfortunately, convex problems merely account for a part of what we have to solve in FL, and many more problems are non-convex. As an example, most neural networks widely utilized in artificial intelligence are non-convex. Yu et al. [26] proved the distributed SGD method for non-convex problems possesses the linear speedup property in the heterogeneous setting. Li et al. [15] introduced a framework called FedProx to solve statistical heterogeneity and provided convergence guarantees on non-IID data. By incorporating a regularization term in the local objective similar to that in FedProx, Li et al. [14] proposed a personalized FL framework, Ditto, and provided convergence guarantees for each local model. Haddadpour et al. [8] provided the convergence rates of the local GD/SGD algorithm to federated learning both in general non-convex and non-convex under Polyak-Lojasiewicz condition. Amiri et al. [2] studied the effect of a shared wireless medium with limited downlink bandwidth on the performance of FL, and further provided a convergence analysis of the simulated downlink method.

However, the existing theoretical analysis for either convex or non-convex problems did not consider biased client participation. In the practical FL environment, each client participates in the FL training process for a different number of rounds due to the influence of its network conditions and willingness to participate in FL. The objective of our paper is to analyze the impact of biased client participation on FL convergence for non-convex problems from a theoretical perspective.

3   Preliminaries of Federated Averaging

The FL framework is crafted to uphold clients' data privacy, wherein the PS assumes the role of orchestrating the training of the global model across the network. Assume that there are N clients with local datasets $\mathcal{D}_1, \mathcal{D}_2, ..., \mathcal{D}_N$. The global objective function is to minimize:

$$F(\omega) = \sum_{k=1}^{N} p_k F_k(\omega), \tag{1}$$

$F_k(\omega) = \frac{1}{n_k} \sum_{i \in \mathcal{D}_N} f(\omega, x_i)$ is the local objective function at each client k, p_k represents the weight attributed to the k-th client adhering to the condition $p_k \geq 0$ and $\sum_{k=1}^{N} p_k = 1$. $f(\omega, x_i)$ represents the loss function for the prediction on sample x_i using parameters ω. We define n_k as the size of dataset $\mathcal{D}_k$, denoted by $|\mathcal{D}_k|$.

In FedAvg, a subset $m \ll N$ of clients is selected to optimize the local objective F_k on client k at each iteration t. The total number of iterations is denoted as T. Generally, to minimize communication cost, client-server communication and parameter updates are conducted only once with several iterations [20]. E is defined as the count of local epochs that occur between two consecutive instances of client-server communication. The variable T_c is defined to indicate the total quantity of rounds encompassing client-server communications and parameter updates required to generate the global model. So we can get the relationship between T and T_c, that is, $T_c = \lfloor \frac{T}{E} \rfloor$. The set of m clients participating in the t-th iteration is symbolized by S_t, then $|S_t| = m$.

Within the FedAvg framework, as each client independently advances one gradient descent step on the prevailing model using its local data, we have

$$\omega_k^{t+1} = \omega^t - \eta_t \nabla F_k(\omega_k^t, \xi_k^t), \tag{2}$$

where ω^t signifies the present model parameters, η_t denotes the learning rate, and ξ_k^t and ω_k^{t+1} correspond to the local data and updated parameters of the k-th client, respectively.

The aggregation phase at the server entails computing the average of the locally updated outcomes, thereby yielding a fresh global model. When full client participation in the aggregation step of FedAvg, the global model is

$$\omega^{t+1} = \sum_{k=1}^{N} p_k \omega_k^{t+1}. \tag{3}$$

If solely a subset comprising m clients engages in the training process during each iteration t of FedAvg, Eq. (3) can be rewritten as [17]:

$$\omega^{t+1} = \frac{N}{m} \sum_{k \in S_t} p_k \omega_k^{t+1}. \tag{4}$$

Note that $\sum_{k=1}^{N} p_k = 1$ and $|S_t| = m$, therefore, $\mathbb{E} \frac{N}{m} \sum_{k \in S_t} p_k = 1$ can be proved easily [17]. When each client performs a single local update per round, we can obtain the global model parameters ω^{t+1} by combining Eq. (2) and Eq. (4).

$$\omega^{t+1} = \omega^t - \eta_t \frac{N}{m} \sum_{k \in S_t} p_k \nabla F_k(\omega_k^t, \xi_k^t). \tag{5}$$

4 Convergence Analysis of Federated Learning

In this section, we analyze the convergence of FL with or without biased client participation when the aggregation weight encompasses the count of times each client engages in the complete training process. Table 1 provides an overview of the principal symbols utilized within this paper along with their corresponding explanations.

Table 1. Catalog of symbols and their meanings

Notation	Description
k, N	the client index, the total count of clients
$\mathcal{D}_k$	the dataset specific to client k
n_k, n	the size of dataset $\mathcal{D}_k$, the total number of samples over all clients
t, T	the iteration index, the total count of iterations
E, T_c	the epochs of local computation, the total count of communication rounds
$F_k(\cdot), F(\cdot)$	client k's local objective function, the overarching global objective function
ω^t, ω	the global model parameter at iteration t, the parameters of global model
ω_k^t	the parameter of client k at iteration t
ξ_k^t	the sample selected uniformly from local data $\mathcal{D}_k$ at iteration t
S_t, m	the subset chosen at iteration t, the magnitude of the subset selected in each iteration
p_k, η_t	the aggregation weight of client k, learning rate at iteration t
G^2, δ^2	the upper bound of the expected squared norm of stochastic gradient, the upper bound of the variance in stochastic gradient for each client
M_k^t	the total count of times client k is involved in the training process

4.1 Notations and Assumptions

Similar to earlier studies [16,26], we posit the ensuing assumptions concerning the functions $F(\omega)$ and all $F_k(\omega)$.

Assumption 1 (Smoothness). $F(\omega)$ and $F_k(\omega)$ exhibit ρ-smoothness with respect to ω, indicating that for any ω^1 and ω^2:

$$F(\omega^1) \leq F(\omega^2) + \langle \nabla F(\omega^2), \omega^1 - \omega^2 \rangle + \frac{\rho}{2}\|\omega^1 - \omega^2\|^2, \tag{6}$$

Assumption 2 (Bounded gradient). The expected squared norm of stochastic gradients is uniformly constrained:

$$\mathbb{E}\|\nabla F_k(\omega_k^t, \xi_k^t)\|^2 \leq G^2 \quad \forall \, \omega, \, t. \tag{7}$$

Assumption 3 (Bounded variance). The stochastic gradient variance within each client is bounded:

$$\mathbb{E}\|\nabla F_k(\omega_k^t, \xi_k^t) - \nabla F_k(\omega_k^t)\|^2 \leq \delta^2 \quad \forall \, k, \, \omega, \, \xi, \tag{8}$$

where

$$\nabla F_k(\omega_k^t) = \mathbb{E}\big[\nabla F_k(\omega_k^t, \xi_k^t)\big] \quad \forall \, k, \, \omega, \, \xi. \tag{9}$$

Assumption 4 (Sample even distribution). Like the assumption in previous works [15, 17], this work also assumes that each client has the same number of samples:

$$n_k = \frac{n}{N}, \quad \forall \, k, \tag{10}$$

where $n = \sum_{k=1}^{N} n_k$ corresponds to the total training data size.

4.2 Convergence Results of FedAvg with or Without Bias

In the following, we perform a convergence analysis on FL with or without bias by considering the number of times each client participates in FL. Two definitions are first given below.

Definition 1. The characteristic function $I_k(t)$ records whether the client k participates in the global training process at iteration t:

$$I_k(t) = \begin{cases} 1 & k \in S_t \\ 0 & k \notin S_t \end{cases}, \tag{11}$$

where $k \in S_t$ indicates that the client k participated in the training process at the t-th iteration. When $k \in S_t$, we have $I_k(t) = 1$, otherwise $I_k(t) = 0$.

Definition 2. Biased client participation. Let M_k^t symbolize the cumulative count of client k's participation in the training process from the initial iteration up to the t-th iteration, then we have:

$$M_k^t = \sum_t I_k(t), \tag{12a}$$

$$s.t. \quad \sum_{k=1}^{N} M_k^t = mt, \tag{12b}$$

$$M_k^t \leq t. \tag{12c}$$

To show the influence of the bias defined in Definition 2 on FL training, we consider M_k^t as a factor in the weight p_k of FedAvg. Specifically, the manifestation of p_k is

$$p_k = \frac{n_k M_k^t}{\sum_{k=1}^{N} n_k M_k^t}. \tag{13}$$

We first demonstrate an important lemma to show the disparity between the local distribution and the global distribution.

Lemma 1. *Suppose Assumption 2 holds, we can obtain:*

$$\mathbb{E}\left\|\omega_k^t - \omega^t\right\|^2 \leq \eta_t^2 E^2 G^2. \tag{14}$$

Proof. The proof of Lemma 1 can be found in the appendix.

Lemma 1 shows that the gap between local distribution and global distribution is proportional to the square of the number of local computation epochs E and the local gradient variance G^2.

Then, we derive the following inequality for FL convergence rate with or without bias, through the incorporation of each client's participation frequency in the process.

Lemma 2. *Assuming that Assumptions 1 through 4 are satisfied, and choose $\eta_t = \frac{1}{\rho}\sqrt{\frac{1}{T}}$. We can obtain the inequality about convergence rate of FedAvg of FL with or without bias.*

$$\frac{1}{T}\sum_{t=0}^{T-1}\mathbb{E}\|\nabla F(\omega^t)\|^2 \leq \frac{2\rho\left[F(\omega^0) - F(\omega^*)\right]}{\sqrt{T}} + \left[\frac{N^2 E^2 G^2}{mT} + \frac{N^2 \delta^2}{m^2\sqrt{T}}\right]\frac{1}{T}\sum_{t=0}^{T-1}\sum_{k\in S_t}\left(\frac{M_k^t}{mt}\right)^2. \tag{15}$$

Proof. The proof of Lemma 2 can be found in the appendix.

According to Lemma 2, the convergence rate of FL is not only related to T but also affected by M_k^t the number of times client k participate in FL.

Next, we will discuss the effect of the variable M_k^t on the convergence rate. Let

$$\tau = \frac{1}{T}\sum_{t=0}^{T-1}\sum_{k\in S_t}\left(\frac{M_k^t}{mt}\right)^2. \tag{16}$$

Finally, we demonstrate the respective convergence rates of FL under conditions of unbiased client participation and biased client participation.

The Convergence Rate of Unbiased State. It is an ideally *unbiased state* that reflects each client has the same frequency to participate in the global training process. As $\sum_{k=1}^{N} M_k^t = mt$, if and only if $M_k^t = \frac{mt}{N}$ for all clients, and $\frac{mt}{N}$ is the expectation that each client participates in the update process at iteration t. We have

$$\tau = \frac{1}{T} \sum_{t=0}^{T-1} \sum_{k \in S_t} \left(\frac{M_k^t}{mt}\right)^2 = \frac{m}{N^2}. \tag{17}$$

Theorem 1. *Assuming that Assumptions 1 through 4 are satisfied, and choose $\eta_t = \frac{1}{\rho}\sqrt{\frac{1}{T}}$. We can obtain a convergence rate of the unbiased state of FedAvg as:*

$$\frac{1}{T} \sum_{t=0}^{T-1} \mathbb{E}\|\nabla F(\omega^t)\|^2 \leq \frac{2\rho\left(F(\omega^0) - F(\omega^*)\right)}{\sqrt{T}} + \frac{E^2 G^2}{T} + \frac{\delta^2}{m\sqrt{T}}. \tag{18}$$

Proof. By substituting Eq. (17) into Lemma 2, we can obtain the inequality.

The Convergence Rate of Extremely Biased State. The *extremely biased state* is defined as each iteration of the update process is participated by constant m clients. That is, the constant m clients participate in the entire training process, while the other $N - m$ clients are not available. Since $M_k^t \leq t$, we can obtain $\left(\frac{M_k^t}{mt}\right)^2 \leq \frac{1}{m^2}$. When $M_k^t = t$, we have

$$\tau = \frac{1}{T} \sum_{t=0}^{T-1} \sum_{k \in S_t} \left(\frac{M_k^t}{mt}\right)^2 = \frac{1}{m}. \tag{19}$$

Theorem 2. *Given that Assumptions 1 through 4 are met, and selecting $\eta_t = \frac{1}{\rho}\sqrt{\frac{1}{T}}$, the convergence rate of the extremely biased state of FedAvg can be deduced using Lemma 2.*

$$\frac{1}{T} \sum_{t=0}^{T-1} \mathbb{E}\|\nabla F(\omega^t)\|^2 \leq \frac{2\rho\left(F(\omega^0) - F(\omega^*)\right)}{\sqrt{T}} + \frac{N^2 E^2 G^2}{m^2 T} + \frac{N^2 \delta^2}{m^3 \sqrt{T}}. \tag{20}$$

Proof. By substituting Eq. (19) into Lemma 2, we can obtain the inequality.

Discussion. Using Theorem 1 and Theorem 2, we can draw the following conclusions. Firstly, it's highlighted that the convergence rate of FL depends on the values of T, m, and E for a specific optimization objective. Here, T represents the overall count of iterations, m is the chosen subset size per iteration, and E indicates the number of local computation epochs. Secondly, the values on the right-hand side of the inequalities in Theorem 1 and Theorem 2 determine the precision of the solution produced by the FL algorithm. A smaller value signifies higher solution precision. We observe that as the number of iterations T increases, the right side of the inequalities approaches 0, implying that FedAvg can ultimately achieve convergence. Thirdly, Theorem 1 and Theorem 2 also reveal the correlation between the quantity of clients chosen per iteration and the convergence of FedAvg. A larger m leads to faster convergence. Lastly, as $T_c = \lfloor \frac{T}{E} \rfloor$, we derive the relationship between FedAvg's convergence and E as: $\frac{1}{T} \sum_{t=0}^{T-1} \mathbb{E}\|\nabla F(\omega^t)\|^2 \propto \frac{1}{\sqrt{T_c E}} + \frac{E}{T_c} + \frac{1}{\sqrt{T_c E}}$, implying that the convergence rate of FedAvg is a function of E, exhibiting an initial decrease followed by an eventual increase.

By comparing convergence rates in Theorem 1 and Theorem 2, we can answer the questions raised in the second challenge. In FL, a subset of clients, where $m \ll N$, is chosen for local model training in each iteration, and we have $\frac{N^2}{m^2} \gg 1$. Upon contrasting the values situated on the right-hand side of the inequality in Eq. (18) from Theorem 1 and Eq. (20) from Theorem 2, we can derive

$$\frac{2\rho\left(F(\omega^0) - F(\omega^*)\right)}{\sqrt{T}} + \frac{E^2 G^2}{T} + \frac{\delta^2}{m\sqrt{T}} \leq \frac{2\rho\left(F(\omega^0) - F(\omega^*)\right)}{\sqrt{T}} + \frac{N^2 F^2 G^2}{m^2 T} + \frac{N^2 \delta^2}{m^3 \sqrt{T}}. \tag{21}$$

Since the value on the right-hand side of the convergence rate inequality indicates the precision of the solution of the FL algorithm, we can conclude that the precision of the solution of FedAvg in the unbiased state is higher than that in the extremely biased state. In the extreme case, the distribution of the other $N - m$ clients cannot be represented well by the global model. Each iteration only has constant m clients participating in the training procedure, resulting in the precision of FedAvg in the extremely biased state being lower than that in the unbiased state. So, we can conclude that the bias of client participation will have a negative impact on the precision of the global model in FL.

5   Experimental Validation

Within this section, we perform a series of experiments to verify the theoretical analysis.

5.1   Experimental Settings

Datasets. This section validates our theoretical findings through experiments on three real-world datasets: MNIST [4], CIFAR-10 [12], and CIFAR-100 [12]. MNIST includes 10 categories of handwritten digits, with samples from 250 individuals, 60,000 in the training set, and 10,000 in the testing set. CIFAR-10 and CIFAR-100 stem from labeled subsets of 80 million small images. CIFAR-10 comprises 60,000 images across 10 classes, with 6,000 images per class. It's split into 50,000 training images and 10,000 test images. CIFAR-100 has 100 classes, each with 600 images. For each class, there are 500 training images and 100 testing images.

Models. Referring to [20,25], we employed a simplified Convolutional Neural Network (CNN) model for classifying MNIST, CIFAR-10, and CIFAR-100 datasets. In MNIST, the CNN consists of two 5×5 convolutional layers followed by max-pooling and ReLU activation. Two fully connected layers with 50 and 10 units, and a softmax-equipped output layer complete the network. The CNN models for CIFAR-10 and CIFAR-100 are different. They have three fully connected layers with 384, 192, and 10 units for CIFAR-10, and 384, 192, and 100 units for CIFAR-100.

Implementation Details. Similar to [20], MNIST and CIFAR-10 datasets are split among $N = 100$ clients in a non-IID manner. This ensures each client has only 2 categories. CIFAR-100 is similarly split among $N = 100$ clients, each with up to 10 classes. η_t uses a step decay with 0.995 rate, starting at $\eta_0 = 0.001$. Each round involves $m = 20$ clients for training, with $E = 5$ local updates per client. Moreover, we verify our theory with other FL algorithms like FedProx and Ditto.

Moreover, we conduct experiments covering various degrees of non-IID. Referring to the definition in [13], non-IID$(\lambda)(1 \leq \lambda \leq 10)$ denotes that each client possesses samples from a randomly selected subset of λ out of the total 10 classes. In an intuitive sense, the degree of non-IID is inversely proportional to the value of λ. Accordingly, the above statistical degree of heterogeneous training data on MNIST dataset is non-IID(2).

We assess the performance of three FL algorithms in two scenarios. One is the unbiased state, where m clients are selected from the entire client pool for training in each iteration, as per Theorem 1. In the unbiased state, all clients have equal participation throughout the FL training. The other scenario is the extreme state, where a constant set of m clients are engaged in the FL process (with the remaining $N - m$ clients not participating). This corresponds to the extremely biased state outlined in Theorem 2.

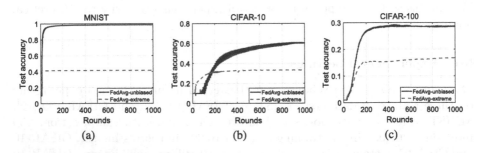

Fig. 1. The test accuracy of FedAvg on three datasets.

5.2   Experimental Results

We verify the correctness of our theoretical analysis in different FL algorithms on three real-world datasets. Each figure and table shows the mean performance over three independent runs.

Table 2. The test accuracy exhibited by various FL algorithms under two conditions across the three datasets.

Algorithm and state	Test Accuracy (%)		
	MNIST	CIFAR-10	CIFAR-100
FedAvg-unbiased	**98.57**	**60.83**	**28.73**
FedAvg-extreme	41.48	33.69	16.73
FedProx-unbiased	**98.60**	**60.26**	**29.06**
FedProx-extreme	41.47	33.58	16.97
Ditto-unbiased	**98.59**	**60.51**	**28.67**
Ditto-extreme	41.45	34.15	16.74

Performance of Different FL Algorithms in the Extreme and Unbiased States. Figure 1 shows the performance of FedAvg in the extreme state and the unbiased state on three datasets. From this figure, it can be found that FedAvg show higher test accuracy on all three datasets in the unbiased state. Since only the data distribution of fixed m clients is learned, the accuracy of the extreme state improves relatively quickly at the beginning of training, but then falls into a bottleneck, and its final test accuracy is lower than that in the unbiased state. In detail, Table 2 shows the accuracy results of FedAvg, FedProx, and Ditto on the three datasets under unbiased and extreme states, respectively. Evidently, biased client participation adversely affects the performance of global models across various FL algorithms on the three distinct datasets. Specifically, FedAvg in the extreme state is far from reaching the same accuracy as that in the unbiased state. For MNIST, CIFAR-10, and CIFAR-100, the highest accuracy of FedAvg in the unbiased state exceeds 95%, 60%, and 28%, respectively. While in the extreme state, it is less than 45%, 35%, and 17%.

Similarly, as shown in Table 2, for different FL algorithms FedProx and Ditto, the biased client participation also affects the performances of their global models, for example, the accuracy of the global model decreases. Through experiments in different FL algorithms, it is verified that our theoretical analysis in Sect. 4.2 is correct, that is, biased client participation reduces the precision of the global model of the FL algorithm.

Impact of Varying Degrees of Non-IID Characteristics. We establish distinct non-IID degrees and proceed to conduct experiments utilizing various FL algorithms on the MNIST dataset. Specifically, we distribute the MNIST dataset to 100 clients using three different non-IID degrees. These three data distribution approaches correspond to instances where each client possesses only 2, 4, or 6 distinct label types. We employ the notations non-IID(2), non-IID(4), and non-IID(6) to signify these varying non-IID degrees. It can be seen from the settings that Table 2 in Sect. 5.2 shows the results of MNIST dataset with the degree of non-IID(2). In this subsection, We proceed to perform additional experiments

Table 3. The test accuracy of distinct FL algorithms is evaluated under two states, considering varying degrees of non-IID data distribution, using the MNIST dataset.

Algorithm and state	Test Accuracy (%)		
	non-IID(2)	*non-IID(4)*	*non-IID(6)*
FedAvg-unbiased	**98.57**	**98.89**	**98.91**
FedAvg-extreme	41.48	69.31	79.39
FedProx-unbiased	**98.60**	**98.93**	**98.89**
FedProx-extreme	41.47	69.32	79.41
Ditto-unbiased	**98.59**	**98.92**	**98.92**
Ditto-extreme	41.45	69.31	79.37

on the MNIST dataset, encompassing the non-IID(4) and non-IID(6) scenarios. From Table 3, it can be found that for different non-IID degrees data, biased client participation affects the performance of FL's global model.

In summary, through the above experimental results, it is obvious that the performances of FL algorithms in unbiased and extreme states are quite different. Furthermore, we can found that extreme client participation has a huge negative impact on the performance of the global model in the FL task, FL algorithms cannot achieve satisfactory performance in the extreme state. These experimental results verify the above discussion, and we can conclude that the bias of client participation significantly impacts the performance of the FL's global model.

6   Conclusion

FL has emerged as the predominant paradigm for collaborative optimization among multiple parties while upholding privacy protection. In this paper, we conducted the convergence rate of FL by considering the number of times each client participates in FL. We analyzed the convergence of the FL algorithm with unbiased client participation and biased client participation, respectively. Finally, we validated the theoretical findings using experiments on three real-world datasets.

Through theoretical analysis and experimental verification, we proved that biased client participation will have a negative effect on the global model's precision of FL. This can provide a guide for future research. The adverse impact of biased client participation on FL can potentially be mitigated through the design of unbiased aggregation algorithms or equitable client selection strategies. In the future, we plan to design practical unbiased optimization algorithms and fair client selection algorithms for FL.

Acknowledgements. This study received support from the National Natural Science Foundation of China through Grants U1911201 and U2001209, the Natural Science Foundation of Guangdong under Grant 2021A1515011369, and the Science and Technology Program of Guangzhou under Grant 2023A04J2029.

Appendiex

Within this section, we will provide proofs for Lemma 1 and Lemma 2.

Proof of Lemma 1

For any $t \geq 0$, there exits a $t - t_0 \leq E$, and $\omega_k^{t_0} = \omega^{t_0}$ for all $k = 1, 2, ..., N$. Similar to previous work [17], we have

$$
\begin{aligned}
\mathbb{E}\left\|\omega_k^t - \omega^t\right\|^2 &= \mathbb{E}\left\|(\omega_k^t - \omega^{t_0}) - (\omega^t - \omega^{t_0})\right\|^2 \\
&\leq \mathbb{E}\left\|\omega_k^t - \omega^{t_0}\right\|^2 \\
&\leq \eta_t^2 E \sum_{i=t_0}^t \mathbb{E}\left\|\nabla F_k(\omega_k^i, \xi_k^i)\right\|^2 \\
&\leq \eta_t^2 E^2 G^2.
\end{aligned}
$$

Proof of Lemma 2

Since $n_k = \frac{n}{N}$ for all n_k, $\sum_{k=1}^N M_k^t = mt$, we can derive that

$$
\sum_{k \in S_t} p_k^2 = \sum_{k \in S_t} \left(\frac{\frac{n}{N} M_k^t}{\sum_{k=1}^N \frac{n}{N} M_k^t}\right)^2 = \sum_{k \in S_t} \left(\frac{M_k^t}{mt}\right)^2. \tag{22}
$$

Utilizing the ρ-smoothness property of $F(\omega)$, the subsequent inequality can be derived:

$$
\mathbb{E}F(\omega^{t+1}) \leq \mathbb{E}F(\omega^t) + \mathbb{E}\left\langle \nabla F(\omega^t), \omega^{t+1} - \omega^t \right\rangle + \frac{\rho}{2}\mathbb{E}\left\|\omega^{t+1} - \omega^t\right\|^2. \tag{23}
$$

By applying the fact: $\mathbb{E}\|x\|^2 = \mathbb{E}\left[\|x - \mathbb{E}x\|^2\right] + \|\mathbb{E}x\|^2$, we can obtain

$$
\begin{aligned}
\mathbb{E}\left\|\omega^{t+1} - \omega^t\right\|^2 &= \eta_t^2 \mathbb{E}\left\|\frac{N}{m}\sum_{k \in S_t} p_k \nabla F_k(\omega_k^t, \xi_k^t)\right\|^2 \\
&= \eta_t^2 \mathbb{E}\left\|\frac{N}{m}\sum_{k \in S_t} p_k \left[\nabla F_k(\omega_k^t, \xi_k^t) - \nabla F_k(\omega_k^t)\right]\right\|^2 + \eta_t^2 \mathbb{E}\left\|\frac{N}{m}\sum_{k \in S_t} p_k \nabla F_k(\omega_k^t)\right\|^2.
\end{aligned} \tag{24}
$$

Since each client works in parallel and independently and according to Assumption 3, we have

$$
\begin{aligned}
\mathbb{E}\left\|\omega^{t+1} - \omega^t\right\|^2 &= \frac{\eta_t^2 N^2}{m^2}\sum_{k \in S_t} p_k^2 \mathbb{E}\left\|\nabla F_k(\omega_k^t, \xi_k^t) - \nabla F_k(\omega_k^t)\right\|^2 \\
&\quad + \eta_t^2 \mathbb{E}\left\|\frac{N}{m}\sum_{k \in S_t} p_k \nabla F_k(\omega_k^t)\right\|^2 \\
&\leq \frac{\eta_t^2 N^2 \delta^2}{m^2}\sum_{k \in S_t} p_k^2 + \eta_t^2 \mathbb{E}\left\|\frac{N}{m}\sum_{k \in S_t} p_k \nabla F_k(\omega_k^t)\right\|^2.
\end{aligned} \tag{25}
$$

We further note that

$$
\begin{aligned}
\mathbb{E}\left\langle \nabla F(\omega^t), \omega^{t+1} - \omega^t \right\rangle &= -\eta_t \mathbb{E}\left\langle \nabla F(\omega^t), \frac{N}{m} \sum_{k \in S_t} p_k \nabla F_k(\omega^t, \xi_k^t) \right\rangle \\
&= -\eta_t \mathbb{E}\left[\mathbb{E}\left[\left\langle \nabla F(\omega^t), \frac{N}{m} \sum_{k \in S_t} p_k \nabla F_k(\omega^t, \xi_k^t) \right\rangle \middle| \xi^t \right] \right] \\
&= -\eta_t \mathbb{E}\left\langle \nabla F(\omega^t), \mathbb{E}\left[\frac{N}{m} \sum_{k \in S_t} p_k \nabla F_k(\omega^t, \xi_k^t) \middle| \xi^t \right] \right\rangle \\
&= \underbrace{-\eta_t \mathbb{E}\left\langle \nabla F(\omega^t), \frac{N}{m} \sum_{k \in S_t} p_k \nabla F_k(\omega^t) \right\rangle}_{A1}.
\end{aligned}
\tag{26}
$$

Firstly, for bound A1, we can obtain

$$
A1 = -\frac{\eta_t}{2}\mathbb{E}\|\nabla F(\omega^t)\|^2 - \frac{\eta_t}{2}\mathbb{E}\left\| \frac{N}{m} \sum_{k \in S_t} p_k \nabla F_k(\omega_k^t) \right\|^2 + \underbrace{\frac{\eta_t}{2}\mathbb{E}\left\| \frac{N}{m} \sum_{k \in S_t} p_k \nabla F_k(\omega_k^t) - \nabla F(\omega^t) \right\|^2}_{A2}.
\tag{27}
$$

Secondly, $\omega^{t+1} = \frac{N}{m}\sum_{k \in S_t} p_k \omega_k^{t+1}$ according to Eq. (4), therefore we can obtain $\nabla F(\omega^{t+1}) = \frac{N}{m}\sum_{k \in S_{t+1}} p_k \nabla F_k(\omega^{t+1})$ [24]. For bound A2, we can obtain

$$
\begin{aligned}
A2 &= \frac{\eta_t}{2}\mathbb{E}\left\| \frac{N}{m} \sum_{k \in S_t} p_k \nabla F_k(\omega_k^t) - \frac{N}{m} \sum_{k \in S_t} p_k \nabla F_k(\omega^t) \right\|^2 \\
&= \frac{\eta_t}{2}\mathbb{E}\left\| \frac{N}{m} \sum_{k \in S_t} p_k \left[\nabla F_k(\omega_k^t) - \nabla F_k(\omega^t) \right] \right\|^2.
\end{aligned}
\tag{28}
$$

According to the Cauchy-Buniakowsky-Schwarz inequality, we have

$$
A2 \leq \frac{\eta_t N^2}{2m^2} \sum_{k \in S_t} p_k^2 \sum_{k \in S_t} \mathbb{E}\left\| \nabla F_k(\omega_k^t) - \nabla F_k(\omega^t) \right\|^2.
\tag{29}
$$

By using Assumption 1, we can obtain

$$
A2 \leq \frac{\eta_t \rho^2 N^2}{2m^2} \sum_{k \in S_t} p_k^2 \sum_{k \in S_t} \mathbb{E}\left\| \omega_k^t - \omega^t \right\|^2.
\tag{30}
$$

By using Lemma 1, we can derive the bound of A2 as

$$
A2 \leq \frac{\eta_t^3 \rho^2 N^2 E^2 G^2}{2m} \sum_{k \in S_t} p_k^2.
\tag{31}
$$

Upon substituting Eq. (31) into Eq. (27), we arrive at the upper bound for A1 as follows:

$$A1 \leq -\frac{\eta_t}{2}\mathbb{E}\big\|\nabla F(\omega^t)\big\|^2 - \frac{\eta_t}{2}\mathbb{E}\left\|\frac{N}{m}\sum_{k \in S_t} p_k \nabla F_k(\omega_k^t)\right\|^2 \tag{32}$$
$$+ \frac{\eta_t^3 \rho^2 N^2 E^2 G^2}{2m}\sum_{k \in S_t} p_k^2.$$

By combining the results of Eq. (25), Eq. (26) and Eq. (32), we can obtain

$$F(\omega^{t+1}) \leq F(\omega^t) - \frac{\eta_t}{2}\mathbb{E}\big\|\nabla F(\omega^t)\big\|^2$$
$$- \frac{\eta_t - \eta_t^2 \rho}{2}\mathbb{E}\left\|\frac{N}{m}\sum_{k \in S_t} p_k \nabla F_k(\omega_k^t)\right\|^2 \tag{33}$$
$$+ \frac{\eta_t^3 \rho^2 N^2 E^2 G^2}{2m}\sum_{k \in S_t} p_k^2 + \frac{\eta_t^2 \rho N^2 \delta^2}{2m^2}\sum_{k \in S_t} p_k^2.$$

The conclusion that $0 \leq \eta_t \leq \frac{1}{\rho}$ can be obtained from the setting $\eta_t = \frac{1}{\rho}\sqrt{\frac{1}{T}}$, we can obtain

$$\frac{\eta_t}{2}\mathbb{E}\big\|\nabla F(\omega^t)\big\|^2 \leq F(\omega^t) - F(\omega^{t+1}) + \frac{m\eta_t^3 \rho^2 N^2 E^2 G^2 + \eta_t^2 \rho N^2 \delta^2}{2m^2}\sum_{k \in S_t} p_k^2. \tag{34}$$

By dividing both the left side and the right side by $\frac{\eta_t}{2}$, we have

$$\mathbb{E}\big\|\nabla F(\omega^t)\big\|^2 \leq \frac{2\left[F(\omega^t) - F(\omega^{t+1})\right]}{\eta_t} + \frac{m\eta_t^2 \rho^2 N^2 E^2 G^2 + \eta_t \rho N^2 \delta^2}{m^2}\sum_{k \in S_t} p_k^2. \tag{35}$$

According to Eq. (13), we have $\sum_{k \in S_t} p_k^2 = \sum_{k \in S_t} \left(\frac{M_k^t}{mt}\right)^2$. As $\eta_t = \frac{1}{\rho}\sqrt{\frac{1}{T}}$, we can sum Eq. (35) from $t = 0$ to $T - 1$ and obtain

$$\frac{1}{T}\sum_{t=0}^{T-1}\mathbb{E}\big\|\nabla F(\omega^t)\big\|^2 \leq \frac{2\rho\left[F(\omega^0) - F(\omega^*)\right]}{\sqrt{T}} + \left[\frac{N^2 E^2 G^2}{mT} + \frac{N^2 \delta^2}{m^2 \sqrt{T}}\right]\frac{1}{T}\sum_{t=0}^{T-1}\sum_{k \in S_t}\left(\frac{M_k^t}{mt}\right)^2.$$

where ω^* is the optimal solution.

References

1. Abay, A., Zhou, Y., Baracaldo, N., Rajamoni, S., Chuba, E., Ludwig, H.: Mitigating bias in federated learning. arXiv preprint arXiv:2012.02447 (2020). https://doi.org/10.48550/arXiv.2012.02447
2. Amiri, M.M., Gündüz, D., Kulkarni, S.R., Poor, H.V.: Convergence of federated learning over a noisy downlink. IEEE Trans. Wireless Commun. **21**(3), 1422–1437 (2021). https://doi.org/10.1109/TWC.2021.3103874

3. Balakrishnan, R., Li, T., Zhou, T., Himayat, N., Smith, V., Bilmes, J.: Diverse client selection for federated learning via submodular maximization. In: International Conference on Learning Representations (ICLR) (2021)

4. Chen, F., Chen, N., Mao, H., Hu, H.: Assessing four neural networks on handwritten digit recognition dataset (MNIST). arXiv preprint arXiv:1811.08278 (2018). https://doi.org/10.48550/ARXIV.1811.08278

5. Chilimbi, T., Suzue, Y., Apacible, J., Kalyanaraman, K.: Project adam: building an efficient and scalable deep learning training system. In: Proceedings of the 11th USENIX conference on Operating Systems Design and Implementation (OSDI), pp. 571–582 (2014)

6. Cho, Y.J., Wang, J., Joshi, G.: Client selection in federated learning: convergence analysis and power-of-choice selection strategies. arXiv preprint arXiv:2010.01243 (2020). https://doi.org/10.48550/arXiv.2010.01243

7. Duan, M., Liu, D., Chen, X., Liu, R., Tan, Y., Liang, L.: Self-balancing federated learning with global imbalanced data in mobile systems. IEEE Trans. Parallel Distrib. Syst. **32**(1), 59–71 (2020). https://doi.org/10.1109/TPDS.2020.3009406

8. Haddadpour, F., Mahdavi, M.: On the convergence of local descent methods in federated learning. arXiv preprint arXiv:1910.14425 (2019). https://doi.org/10.48550/arXiv.1910.14425

9. Kairouz, P., et al.: Advances and open problems in federated learning. Found. Trends® Mach. Learn. **14**(1–2), 1–210 (2021)

10. Khaled, A., Mishchenko, K., Richtárik, P.: First analysis of local GD on heterogeneous data. arXiv preprint arXiv:1909.04715 (2019). https://doi.org/10.48550/ARXIV.1909.04715

11. Khan, L.U., Saad, W., Han, Z., Hossain, E., Hong, C.S.: Federated learning for internet of things: recent advances, taxonomy, and open challenges. IEEE Commun. Surv. Tutor. (2021). https://doi.org/10.1109/COMST.2021.3090430

12. Krizhevsky, A.: Learning Multiple Layers of Features From Tiny Images. University of Toronto, Toronto (2012)

13. Li, A., Zhang, L., Tan, J., Qin, Y., Wang, J., Li, X.Y.: Sample-level data selection for federated learning. In: IEEE Conference on Computer Communications (INFOCOM), pp. 1–10 (2021). https://doi.org/10.1109/INFOCOM42981.2021.9488723

14. Li, T., Hu, S., Beirami, A., Smith, V.: Ditto: Fair and robust federated learning through personalization. In: Proceedings of the 38th International Conference on Machine Learning (ICML), pp. 6357–6368. PMLR (2021)

15. Li, T., Sahu, A.K., Zaheer, M., Sanjabi, M., Talwalkar, A., Smith, V.: Federated optimization in heterogeneous networks. Proc. Mach. Learn. Syst. **2**, 429–450 (2020)

16. Li, T., Sanjabi, M., Smith, V.: Fair resource allocation in federated learning. In: International Conference on Learning Representations (ICLR) (2020)

17. Li, X., Huang, K., Yang, W., Wang, S., Zhang, Z.: On the convergence of Fedavg on non-IID data. In: Eighth International Conference on Learning Representations (ICLR) (2020)

18. Liu, R., Cao, Y., Yoshikawa, M., Chen, H.: Fedsel: Federated SGD under local differential privacy with top-k dimension selection. In: DASFAA (2020)

19. Ma, J., Xie, M., Long, G.: Personalized federated learning with robust clustering against model poisoning. In: Chen, W., Yao, L., Cai, T., Pan, S., Shen, T., Li, X. (eds.) ADMA 2022. LNCS, vol. 13726, pp. 238–252. Springer, Cham (2022). https://doi.org/10.1007/978-3-031-22137-8_18

20. McMahan, B., Moore, E., Ramage, D., Hampson, S., Arcas, B.A.: Communication-efficient learning of deep networks from decentralized data. In: Artificial Intelligence and Statistics (AISTATS), pp. 1273–1282 (2017)
21. Segarceanu, S., Gavat, I., Suciu, G.: Evaluation of deep learning techniques for acoustic environmental events detection. Romanian J. Technical Sci. Appl. Mech. **66**(1), 19–37 (2021)
22. Tan, L., et al.: Adafed: optimizing participation-aware federated learning with adaptive aggregation weights. IEEE Trans. Network Sci. Eng. **9**(4), 2708–2720 (2022). https://doi.org/10.1109/TNSE.2022.3168969
23. Xu, J., Glicksberg, B.S., Su, C., Walker, P., Bian, J., Wang, F.: Federated learning for healthcare informatics. J. Healthcare Inform. Res. **5**(1), 1–19 (2021)
24. Yang, H., Fang, M., Liu, J.: Achieving linear speedup with partial worker participation in non-IID federated learning. In: International Conference on Learning Representations (ICLR) (2021)
25. Yang, W., et al.: Gain without pain: Offsetting DP-injected Nosics stealthily in cross-device federated learning. IEEE Internet Things J. **9**(22), 22147–22157 (2021). https://doi.org/10.1109/JIOT.2021.3102030
26. Yu, H., Jin, R., Yang, S.: On the linear speedup analysis of communication efficient momentum SGD for distributed non-convex optimization. In: International Conference on Machine Learning (ICML), pp. 7184–7193 (2019)

Privacy Lost in Online Education: Analysis of Web Tracking Evolution

Zhan Su[1], Rasmus Helles[1(✉)], Ali Al-Laith[1], Antti Veilahti[1], Akrati Saxena[2], and Jakob Grue Simonsen[1]

[1] University of Copenhagen, Copenhagen, Denmark
{zhan.su,alal,simonsen}@di.ku.dk,
{rashel,antti}@hum.ku.dk
[2] Leiden Institute of Advanced Computer Science, Leiden University, Leiden, The Netherlands
a.saxena@liacs.leidenuniv.nl

Abstract. Digital tracking poses a significant and multifaceted threat to personal privacy and integrity. Tracking techniques, such as the use of cookies and scripts, are widespread on the World Wide Web and have become more pervasive in the past decade. This paper focuses on the historical analysis of tracking practices specifically on educational websites, which require particular attention due to their often mandatory usage by users, including young individuals who may not adequately assess privacy implications. The paper proposes a framework for comparing tracking activities on a specific domain of websites by contrasting a sample of these sites with a control group consisting of sites with comparable traffic levels, but without a specific functional purpose. This comparative analysis allows us to evaluate the distinctive evolution of tracking on educational platforms against a standard benchmark. Our findings reveal that although educational websites initially demonstrated lower levels of tracking, their growth rate from 2012 to 2021 has exceeded that of the control group, resulting in higher levels of tracking at present. Through our investigation into the expansion of various types of trackers, we suggest that the accelerated growth of tracking on educational websites is partly attributable to the increased use of interactive features, facilitated by third-party services that enable the collection of user data. The paper concludes by proposing ways in which web developers can safeguard their design choices to mitigate user exposure to tracking.

Keywords: Web-tracking · Information Security · Privacy · Online Education

1 Introduction

Privacy lost occurs when an individual's personal information or data is disclosed, shared, or accessed by others without their permission, which can result in various negative consequences, such as identity theft, financial fraud, damage to reputation, and discrimination. To investigate these issues, researchers may examine the historical practice of third-party web tracking, as described by [17]. Third-party web tracking involves third-party entities, such as advertisers, social media widgets, and website analytics engines, that are embedded in the first-party sites that users directly visit and are capable of re-identifying users across domains while they browse the web. The proliferation of web tracking has spurred a growing body of research in the computer security and privacy community, which seeks to understand, quantify, and counteract these privacy risks posed by tracking companies compiling lists of websites that users have visited [2,3,17].

As the education industry transitions from traditional offline models to online or hybrid models, the need for privacy protection on educational websites is becoming increasingly prominent. This issue is crucial because the loss of privacy on educational websites can undermine the fundamental principles of privacy and security that are essential for individuals to feel safe and empowered while using the internet for educational purposes. By protecting users' privacy, educational websites can promote trust, openness, and responsibility, which are essential for fostering a positive and inclusive online learning experience. Therefore, several researchers have started studying the practice of web tracking in educational websites [11,12,21,24].

To deepen our comprehension of the nature and progression of tracking on educational websites, we propose an analytical framework that enables a comparative analysis of tracking on a specific type of site (in this case, education) in relation to a control group of sites with comparable traffic levels but of different types. The framework involves three steps: we construct a sample of educational websites, and a control sample of non-educational websites that have similar levels of traffic (Sect. 3.1). We then retrieve the historical websites from the Internet Archive's Wayback Machine[1] for both samples. Third, we scan the HTML file snapshots of the collected websites using the Wayback Machine (Sect. 3.2), and extract third-party trackers embedded in the HTML files (Sect. 3.3).

We aim to answer the following research questions, which we present along with our main findings:

RQ1: How has the use of trackers on educational websites evolved from 2012 to 2021?
In Sect. 4.1, we examine the average number of trackers from 2012 to 2021 and observe a general trend of tracker growth. Until 2018, both educational and non-educational sites sees substantial growth, but they diverge around the time of the introduction of the GDPR in 2018: at this point there is a minor drop in tracking on non-educational sites, which is not seen on educational sites, where the development merely stagnates.

RQ2: How does the evolution of the use of trackers differ between educational and non-educational websites?
Section 4.1 also addresses differences between educational and non-educational websites in the evolution of tracking between 2012 and 2021. The results show that despite the similarity of the underlying trend, the intensity of tracking has

---

[1] https://archive.org/.

grown relatively more on educational sites and that the growth has not similarly reverted as on non-educational websites after the introduction of the GDPR. The results are further supported by a Wilcoxon signed rank (WS) test conducted in Sect. 4.2, demonstrating that the intensity of tracking on educational sites surpassed that of non-educational sites in 2017.

RQ3: Is there a qualitative difference in what kind of trackers are used on educational and non-educational websites?
The quantitative difference between tracking on educational and non-educational sites that we find in the average number of trackers also shows up in the different compositions of the portfolios of trackers found at the two types of sites. We substantiate this statistically by using the Kolmogorov-Smirnov test (KS) to compare the distribution of trackers in these two groups of sites. To investigate the source of these differences, Sect. 4.3 examines the occurrence of some of the most popular trackers, demonstrating that the use of Twitter, Youtube, and Facebook has evolved very differently between educational and non-educational websites. In addition, Sect. 4.5 compares the presence of trackers presenting particular categories, demonstrating that tracking related to enhancing customer interaction in particular seems to have become relatively more common on educational websites over the past few years.

Our contributions can be summarized in two main points: (I) We develop a list of both educational and non-educational websites to investigate the issue of *privacy lost* in online education. The complete code and dataset we compiled can be accessed at[2]. (II) We conduct a quantitative and qualitative analysis of third-party tracking on educational websites, focusing on third-party services from 2012 to 2021. Our findings highlight potential concerns regarding the autonomy and fairness of education.

2   Related Work

Tracking through third-party cookies and scripts has been extensively studied from various perspectives. A significant portion of this research has focused on mapping the prevalence of trackers across samples of websites, such as those found on the Alexa top lists [1,9,19]. Other studies have investigated tracking on different platforms, including the mobile ecosystem [5,6,16].

Karaj et al. [13] proposed a method for measuring web tracking using a browser extension, resulting in a dataset covering 1.5 billion page loads collected over 12 months period from real users. Krishnamurthy and Wills [9] presented a dataset on tracking based on a crawl of the top 1 million websites. They developed an open-source web privacy measurement tool called OpenWPM, which allows researchers to detect, quantify, and characterize emerging online tracking behaviors. Our work is related to several general areas:

---

[2] https://github.com/shuishen112/Privacy_Lost.git.

Historical Web Tracking. Krishnamurthy and Wills provided early insights into web tracking, demonstrating the evolution of third-party organizations between 2005 and 2008 [15]. Lerner et al. presented longitudinal measurements of third-party web tracking behaviors from 1996–2016 [17]. Karaj et al. conducted a large-scale and long-term measurement of online tracking based on real users [13]. Agarwal and Sastr analyzed the top 100 Alexa websites over 25 years using data from the Internet Archive, studying changes in website popularity and examining different categories of websites and their popularity trends over time [2]. Amos et al. curated a dataset of 1,071,488 English language privacy policies spanning over two decades and encompassing more than 130,000 different websites [3].

Web Tracking after GDPR. Numerous studies have investigated web tracking following the implementation of the GDPR (General Data Protection Regulation) in the EU in May 2018, which imposed constraints on online data collection. These studies generally indicate a pattern of diminished tracking activity [7,20,22], but they also reveal that most sites appear unable or unwilling to fully comply with regulations [10,14,23], and tracking companies can still likely monitor user behavior [20].

Web Tracking in Educational Websites. A body of research focuses explicitly on educational websites, which are known to have a higher incidence of tracking technology than sites aimed at minors [24]. In particular, university websites exhibit a substantial prevalence of major tracking companies (e.g., Google, Facebook) [12]. While several recent papers discuss the implications of tracking on educational websites, there seems to be a lack of studies investigating third-party tracking on substantial samples of educational websites post-2018 or examining the development of tracking over time for these websites.

3   Data Collection

We provide a concise overview of our data collection framework, which comprises three main components. Firstly, we discuss the process of gathering educational and non-educational websites, as detailed in Sect. 3.1. Secondly, we present the methodology for scanning historical snapshots from Internet Archive's Wayback machine, which is described in Sect. 3.2. Finally, we discuss the approach for extracting third-party trackers from HTML files, which is outlined in Sect. 3.3.

3.1   Collecting Websites

To understand the evolution of web tracking in educational websites, we compare them to a control set of non-educational websites to see whether there are any changes related to education in particular. The comparison set is explicitly controlled for popularity so that the two sets consisting of educational and non-educational websites have equal rank distribution. The studied websites must also have available historical data stored in internet archives.

We construct the two rank-matched sets of educational and non-educational websites as follows:

Step 1. We extract the educational websites from DMOZ[3]. DMOZ is a large communally maintained open directory that categorizes websites based on web-page content, and we use the DMOZ classification of educational websites. There are 146,941 websites in the DMOZ database labeled as educational websites.

Step 2. Next, we limit the set of educational websites to those occurring on the Open PageRank Initiative[4], which maintains a list of the top 10 million websites ranked based on their Open PageRank. There are 55,390 educational websites present among the top 10 million. This filtering is done so that we can create a comparable control set.

Step 3. We use the Internet Archive's Wayback Machine[5] for archived data. Therefore, the set of educational websites is further limited to those with at least one snapshot per year in every year from 2012 through 2021 to ensure that annual comparisons are balanced. This results in 17,975 educational websites altogether.

Step 4. Based on the list of educational websites from Step 3, we construct a set of non-educational websites with rank (Open Pagerank) distribution matching educational websites. Starting with the educational website of the highest rank, this is done recursively by choosing for each educational website a non-educational website that satisfies the following three conditions:

(a) The website has the lowest possible rank below the matching educational website.
(b) The historical data of the website is available on Internet Archive's Wayback Machine.
(c) The website has not already been added to the control set of non-educational websites.

For instance, if there are two educational websites of ranks 19 and 20, then ranks 21 and 22 would be chosen to the control set of non-educational sites, provided that they are not educational websites and have archived versions available. We study the rank gap (The rank of non-educational websites minus the rank of matching educational websites) distribution. The mean rank gap is 2.86, while the maximum gap is 255. We also find that 99% of the rank gap is below 16.

---

[3] https://dataverse.harvard.edu/dataset.xhtml?persistentId=doi:10.7910/DVN/OMV93V.
[4] https://www.domcop.com/top-10-million-websites.
[5] https://archive.org.

3.2   Scanning the Historical Snapshot

There are two primary methods for scanning historical snapshots: the Way-back CDX Server[6] and the waybackpy Python library[7]. The Wayback CDX Server is a standalone HTTP servlet that serves the index used by the Wayback Machine to search for captures. The second method involves using the Wayback-MachineCDXServerAPI provided by the waybackpy library to retrieve historical snapshots at specific times. For our research, we opted to use the Wayback CDX Server as our scanning method.

3.3   Extraction of Third-Party Trackers

Each website is examined for requests to other URLs initiated during the web-site's loading. These requests will always be embedded in three HTML-elements: *"iframe"*, *"script"* and *"img"*. We only consider the requests generated automat-ically without user action. That is why we omitted the *"a"*-element[8].

The list of third-party services (TPSs) was compiled by extracting all URLs found in the three HTML elements mentioned earlier across the entire dataset. For each website and URL, we checked whether the main domain of the linked URL (e.g., 'google' in 'www.google.dk') differed from the main domain of the website. If the domains were different, the URL was considered a 'third party' and the domain (e.g., 'google') along with the suffix (e.g., 'dk') were added to the list of TPSs.

To clarify our terminology, we will use the term 'trackers' instead of 'third-party services' for the remainder of this paper. While many third-party services serve various functions, such as providing weather data or chat services, some are solely designed for tracking and provide data that is used for personalized ban-ner ads. However, even third-party services that seemingly provide non-tracking functionality have the potential to gather valuable data from users, such as their timestamped IP addresses and the websites they visit when the third-party ser-vice is activated. This information may be used by the third-party provider or sold to data brokers, or both. As all third-party services can track users, we refer to all such services invoked through websites as trackers [18].

We utilized the trackers list[9], which covers the period between May 2017 and August 2022 [13]. The trackers on the list are ranked according to their *tracker reach*, which is a metric defined in the aforementioned paper [13]. It should be noted that each tracker corresponds to multiple tracker domains. For

---

[6] https://github.com/internetarchive/wayback/tree/master/wayback-cdx-server.

[7] https://akamhy.github.io/waybackpy/.

[8] A "ping"-attribute in the "a"-element allows requests to be made to multiple URLs without the user being aware of this, but there were no ping-attributes used in the data used in this study.

[9] https://whotracks.me/trackers.html.

instance, Doubleclick is associated with three tracker domains: '2mdn.net', 'doubleclick.net', and 'invitemedia.com'. In total, the tracker list comprises 1,285 tracker domains.

4   Analysis and Discussion

Our analysis focuses on the changes in web tracking between 2012 and 2021, with a particular emphasis on the qualitative and quantitative differences in tracker usage between educational and non-educational websites.

4.1   Evolution of Tracker Domains per Website

To begin our analysis, we computed the average number of trackers per website for each year. Figure 1 displays the trends in tracker usage on educational and non-educational websites between 2012 and 2021. In general, a striking 94.5% increase in the average number of trackers on educational websites was observed, while the control group experienced a comparatively modest 31.3% increase from 2012–2021. Specifically, when observing the trends of

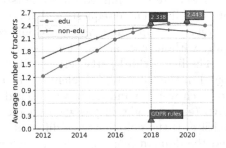

Fig. 1. Evolution of the average number of tracker domains per webiste.

growth each year, we notice a plateau or slight reversal in growth occurring after 2017. Notably, the vertical line in Fig. 1 represents the formal implementation of GDPR in 2018. It is interesting to observe that the number of trackers on non-educational sites experienced a slight decline, whereas tracker usage on educational sites appeared to taper off around the same time.

Figure 2a shows a box plot of the number of trackers per year for educational websites. The plot suggests that the evolution in the average number of trackers after 2018 observed in Fig. 1 is driven by an increased dispersion in tracking across distinct educational sites, as the third quartile increases from 2017–2018, while the median (the horizontal line in each box) remains almost the same in the period 2017–2021. As a comparison, Fig. 2b shows a boxplot for non-educational websites. The plot also suggests the evolution in the average number of trackers in Fig. 1. The third quartile increased from 2017–2018 but has dropped since 2019. Especially, the first quartile decrease in 2021.

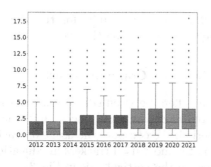

(a) educational websites.

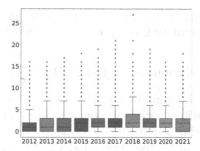

(b) non-educational websites.

Fig. 2. Number of trackers in each year for educational websites and non-educational websites.

According to the findings, it appears that users who browse educational websites are at a higher risk of having their online behavior information collected and potentially utilized by various services and websites. This disparity between educational and non-educational websites highlights the potential inadequacy of the GDPR in addressing privacy concerns specific to educational sites.

4.2 The Number of Trackers on Educational and Non-educational Websites

The tracking trends presented in Sect. 4.1 are purely descriptive, hence we conducted a statistical test to determine if there is a significant difference between educational and non-educational websites. Specifically, we performed a matched-pairs Wilcoxson signed rank(WS)[10] test to evaluate if the medians of the educational and non-educational samples are different for each year. This test is appropriate for paired data, which is the case for our study due to the rank-based construction of the data, and does not make any assumptions about the underlying distributions, making it a non-parametric test.

The input of the WS test is the number of tracker domains for each educational and non-educational website. The results of the tests are summarized in Table 1; as usual, small p-values indicate statistical significance; for all years, except 2017 $p < 0.01$

Table 1. Summary of the WS-test showing differences in each year, $N = 17975$.

Year	WS-test	
	p-value	Z
2012	5.2×10^{-86}	−19.66
2013	6.8×10^{-55}	−15.6
2014	1.6×10^{-46}	−14.32
2015	6.7×10^{-28}	−10.95
2016	7.1×10^{-11}	−6.52
2017	6.6×10^{-2}	−1.84
2018	1.5×10^{-5}	−4.33
2019	8.7×10^{-18}	−8.59
2020	3.2×10^{-20}	−9.21
2021	2.3×10^{-30}	−11.45

---

[10] http://www.biostathandbook.com/wilcoxonsignedrank.html.

for both tests. The sole exception is 2017, where $p > 0.05$ for the WS-test, consistent with the prevalence curves (see Fig. 1) crossing that year.[11]

4.3 Evolution of Usage Rate for the Most Common Trackers

To understand how tracking development differs between educational and non-educational sites, we compare how the ten most commonly occurring trackers have changed during the measurement period. We compute the usage of trackers based on the usage rate and select the top ten most used trackers in educational websites in 2012. The top ten trackers are on the vertical axis in Fig. 3.

We define the *usage rate* of each tracker as $f(t) = \frac{N(t)}{N(w)}$ where N(t) is the total number of websites where tracker t occurs, and $N(w)$ is the total number of websites in the sample. We calculate the relative increase I in usage rates of the top ten trackers most common on educational websites from 2012 to 2012 as $I = \frac{f(t)_{2021} - f(t)_{2012}}{f(t)_{2012}}$. The relative change of usage rate is shown in Fig. 3. We observe that the overall usage rate increased for the five top trackers on educational websites, including the social media sites Twitter and Facebook, and Youtube. It also increased for two Google-related trackers.

It decreased for five trackers, including Twimg (operated by Twitter) decrease by 66.4%, Addthis (−46.9%), Google-analytics (−32.5%), Adobe (−31.6%) and Googlesyndication (−1.1%) on educational websites. For non-educational websites, the usage rate increased only for the three Alphabet-operated trackers (Youtube, Googleapis and Google) and saw the largest decrease for Twimg, by 78.4%.

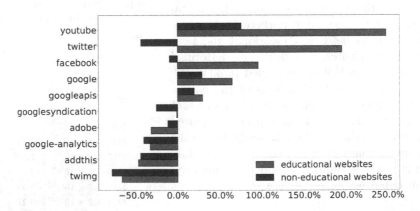

Fig. 3. Top usage rate change of trackers. X-axis is the percentage change of usage rate. Y-axis is the tracker name.

---

[11] Note that the level of statistical significance in the test means that correcting the alpha level for multiple comparisons does not alter the finding. This also holds for Table 2.

Notably, the use of Twitter, Youtube, and Facebook has evolved very differently between educational and non-educational websites. All three have an increased presence on educational sites, whereas their use has declined (Twitter and Facebook) or grown much slower (Youtube) on non-educational sites. The presence of trackers from these companies on educational sites helps target ads at the users when they visit the platforms and can also help educational sites to advertise their services to new, potential users with profiles similar to their existing users.

4.4   Distribution of Trackers in Educational and Non-educational Sites

The results in Sect. 4.1 show that the level of tracking differs significantly between educational and non-educational sites. We will investigate if the difference also relates to the composition of trackers used on the two groups of sites and differences in intensity. As stated in 3.3, all third-party services may collect data that can be used for tracking. However, the value proposition to the website owner differs between different services since they provide various functionalities to the site. Therefore, an analysis of the other functionalities also indicates what the site owner has sought to gain from embedding the service (or tracker), irrespective of the tracking of user behavior it enables. This analysis will, in turn, show if educational sites have followed a different path in integrating trackers than other sites.

For each year, we employ the two-sample Kolmogorov-Smirnov(KS) test–the standard non-parametric test for comparing distributions– to test whether the educational, resp. non-educational samples are drawn from the same underlying distribution. The test is suitable for paired data, similar to the WS test reported before. As indicated in Table 2, there is a significant difference in the distribution of trackers found on the two groups of sites in each year between 2012 and 2021. This indicates that in addition to the different quantitative trends, there appears to be a qualitative difference in the kind of trackers used on educational and non-educational websites.

Table 2. Summary of the KS-test showing differences in each year.

Year	KS-test	
	p-value	Statistic
2012	4.7×10^{-6}	0.25
2013	6.0×10^{-5}	0.22
2014	4.1×10^{-5}	0.23
2015	2.7×10^{-3}	0.18
2016	1.4×10^{-3}	0.19
2017	1.4×10^{-3}	0.19
2018	2.8×10^{-4}	0.21
2019	1.2×10^{-3}	0.19
2020	4.4×10^{-4}	0.2
2021	3.3×10^{-3}	0.17

4.5   Evolution of Different Categories of Trackers

To understand how the overall differences in tracker distribution identified in the previous section relate more closely to different purposes of web functionality, we look at the changes across different types of trackers. Since no exhaustive categorization of trackers exists, we use the tracker typology made available by the WhoTracksMe initiative in June 2022, which to our knowledge, is the most comprehensive and up-to-date list, that is openly available[12].

This typology matches 1285 trackers in our dataset. While this represents only a subset of the total number of trackers, the list coincides with 213 of the 1285 most common trackers in our analysis. In the following, we examine the distribution of trackers across categories but only do so for the subset of the most common trackers. The WhoTracksMe list categorizes most common trackers into one of *Site Analytics, Customer Interaction, Advertising, Cdn, Social Media, Audio Video Player, Essential, Misc.* A more detailed explanation of the eight tracker categories is found in Appendix A.

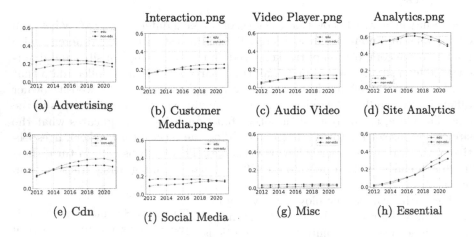

Fig. 4. Evolution of different categories of trackers from 2012 to 2021.

For each category c of trackers, we calculate usage rate $f(c)$ as $f(c) = \frac{N(c)}{N(w)}$ where $N(c)$ is the number of websites this type of trackers occur on, and $N(w)$ is the total number of websites in the sample. We calculated the usage rate $f(c)$ for each category c of trackers as $f(c) = \frac{N(c)}{N(w)}$, where $N(c)$ is the number of websites on which this type of tracker occurs, and $N(w)$ is the total number of websites in the sample. The evolution of different tracker categories in Fig. 4 indicates that, when comparing the levels in 2012 and 2021, educational sites have increased their use of all trackers except for *Site Analytics*. Even though *Advertising* and *Cdn* trackers have dropped slightly since their peak, they are still at a higher level than in 2012. In contrast, the usage rate for *Site Analytics, Advertising,* and *Social Media* is lower in 2021 than in 2012 for non-educational sites.

When comparing the two groups, two significant trends are apparent. Firstly, educational sites exhibit higher growth in the use rate of trackers related to interactive site features and audio-visual content. For instance, the *Audio Video Player* category witnessed a growth of 225.0% on educational websites from

---

[12] https://whotracks.me/trackers.html.

2012–2021 compared to 73.1% for non-educational websites. While the increase slowed down in 2018 or was even reverted for non-educational websites, it has increased again since 2019 and decreased since 2020. Additionally, *Cdn* services and *Customer Interaction* trackers have become more commonly used and grown more on educational sites than non-educational sites during the period. The increase from 2012–2021 was 63.7% for educational websites and 32.5% for non-educational websites in the case of *Customer Interaction* trackers. This growth in interactive features and audio-visual content on educational sites is consistent with the evolution of online learning, which has become more interactive and audio-visually engaging over the years [8]. Moreover, these trends make the sites more bandwidth-consuming, which is also in line with the growth of *Cdn* services.

Second, the use of *Social Media* (increased by 66.2%) and *Advertising* (18.0%) related trackers grow for educational sites but display a net drop for non-educational sites. Compared to the level in 2012, the use of both trackers is higher in 2021, whereas both are lower on non-educational sites. For both categories, the educational sites begin at a lower level than the non-educational sites. In both categories, the difference becomes less pronounced over time, and for *Social Media* ends up at the same level. This indicates that purely commercial tracking on educational sites has evolved from being comparatively less common than on other types of sites to be similar. For both types of sites, the use of trackers for *Advertising* has gone down in recent years, but for educational sites, the peak is more recent (2018) than for non-educational sites (2014). For non-educational sites, this is consistent with the overall development in commercial tracking, which has seen a general trend toward concentration around a few major players. In 2012, the market for commercial tracking was less dominated by monopolies such as Alphabet and Meta than it has since become [4]. The market domination of fewer players is consistent with the falling trend in the use rate. For educational sites, the continued growth is consistent with them catching up to the market standard for commercial tracking, which is also suggested by the strong growth of trackers from the top market players observed in Sect. 4.3.

4.6   Discussion

The evolution of web tracking over time aligns with predictions surrounding the implementation of GDPR in the EU, although its impact on educational sites has been less significant than on non-educational sites. While non-educational sites have experienced a decline in the use of trackers since the introduction of GDPR, the usage of trackers on educational sites has increased and remained stable at a higher level. Moreover, the prevalence of purely commercial tracking, such as advertising and social media tracking, has grown on educational sites, approaching similar levels as on non-educational sites.

The tracking via third-party services outside the commercial categories also serves other purposes (e.g., making it possible to embed a chat function on a web page). Any third-party service with a substantial use rate across the web gives the third-party company that operates the service the opportunity to collect information about the end user. The overall increase in tracking also means that users of educational sites have become more exposed to having information about their online behavior collected and (potentially) exploited across a range of different services and sites. This development is interesting from a normative perspective: tracking on educational sites comes with specific privacy concerns since using these sites is not necessarily voluntary, and user consent to tracking is, therefore, less meaningful. Use of these types of sites happens both at different levels of the educational systems (schools and universities), and in the private sector, for example, as training of employees. The increased use of tracking, both through the inclusion of purely commercial trackers and by embedding other third-party services, suggests that learning activities are increasingly open for commercially oriented analytical exploitation.

The trends of tracking we have identified also suggest that the associated business model(s) remain active and are of increasing relevance in the online education sector. Our paper particularly raises the concern whether the GDPR in its current form suitably addresses privacy issues related to websites whose use is not predominantly voluntary, such as educational websites, but also many other sites like public websites, where the trends of tracking form an important research question on its right and should be addressed in studies in the future.

Reviewing the results, we do not find convincing evidence that tracking on educational websites has been substantially impacted by the COVID-19 pandemic. Despite the fact that additional traffic to these types of sites during the lock-down periods would represent a valuable asset for site owners, no trends in the data meaningfully relate to this. This may be related to the fact that educational sites had already adjusted their portfolio of commercial tracking in particular to facilitate monetization of increased traffic, e.g. through re-targeting of potential students on social media.

5   Conclusions

In this paper, we present a framework for examining historical web tracking within a defined set of sites, and apply it to a sample of educational websites. We constructed a sample comprising educational websites and a control group of non-educational websites that shared similar ranking positions. Utilizing the Internet Archive's Wayback Machine, we gathered historical data on third-party trackers. Our analysis involved 17,975 pairs of websites and their corresponding controls, spanning the period from 2012 to 2021. We observed a notable overall rise in tracking activities on educational sites.

Then we conduct a quantitative and qualitative analysis of third-party tracking on educational websites. We discover that the growth rate of educational websites has surpassed the control group from 2012–2021. Our investigation into the relative expansion of various tracker types suggests that the accelerated growth of tracking on educational websites may be attributed to the rising use of customer interaction, audio-visual content, and social media integration within these platforms.

Our analysis raises concerns about privacy and independence in education. Privacy issues in educational websites should be prioritized, as they can lead to unauthorized disclosure of confidential information, loss of trust, legal consequences, and intellectual property compromise. Furthermore, researchers may wish to analyze privacy lost in other areas, such as news or sports, from a historical perspective. Our framework offers a convenient solution for creating comparable websites and collecting historical third-party tracker data in these domains.

Appendix

A   Tracker Categories

Trackers differ both in the technologies they use, and the purpose they serve. Based on the service they provide to the site owner, we have categorized the trackers in the following:

Advertising. Provides advertising or advertising-related services such as data collection, behavioral analysis or re-targeting.

Customer Interaction. Includes chat, email messaging, customer support, and other interaction tools

Essential. Includes tag managers, privacy notices, and technologies that are critical to the functionality of a website

Site Analytics. Collects and analyzes data related to site usage and performance. Social Media Integrates features related to social media sites

Audio Video Player. Enables websites to publish, distribute, and optimize video and audio content

CDN (Content Delivery Network). Content delivery network that delivers resources for different site utilities and usually for many different customers.

Misc (Miscellaneous). This tracker does not fit in other categories.

Essential. Includes tag managers, privacy notices, and technologies that are critical to the functionality of a website

References

1. Acar, G., Eubank, C., Englehardt, S., Juarez, M., Narayanan, A., Diaz, C.: The web never forgets: persistent tracking mechanisms in the wild. In: Proceedings of the 2014 ACM SIGSAC Conference on Computer and Communications Security, pp. 674–689 (2014)
2. Agarwal, V., Sastry, N.: Way back then: a data-driven view of 25+ years of web evolution. In: Proceedings of the ACM Web Conference 2022, pp. 3471–3479 (2022)
3. Amos, R., Acar, G., Lucherini, E., Kshirsagar, M., Narayanan, A., Mayer, J.: Privacy policies over time: Curation and analysis of a million-document dataset. In: Proceedings of the Web Conference 2021, pp. 2165–2176 (2021)
4. Bilić, P., Prug, T.: The Political Economy of Digital Monopolies: Contradictions and Alternatives to Data Commodification. Policy Press, Bristol (2021)

5. Binns, R., Lyngs, U., Van Kleek, M., Zhao, J., Libert, T., Shadbolt, N.: Third party tracking in the mobile ecosystem. In: Proceedings of the 10th ACM Conference on Web Science, pp. 23–31 (2018)
6. Binns, R., Zhao, J., Kleek, M.V., Shadbolt, N.: Measuring third-party tracker power across web and mobile. ACM Trans. Internet Technol. (TOIT) **18**(4), 1–22 (2018)
7. Dabrowski, A., Merzdovnik, G., Ullrich, J., Sendera, G., Weippl, E.: Measuring cookies and web privacy in a post-GDPR world. In: Choffnes, D., Barcellos, M. (eds.) PAM 2019. LNCS, vol. 11419, pp. 258–270. Springer, Cham (2019). https://doi.org/10.1007/978-3-030-15986-3_17
8. Elisabeta, P.M., Alexandru, M.R.: Comparative analysis of e-learning platforms on the market. In: 2018 10th International Conference on Electronics, Computers and Artificial Intelligence (ECAI), pp. 1–4. IEEE (2018)
9. Englehardt, S., Narayanan, A.: Online tracking: A 1-million-site measurement and analysis. In: Proceedings of the 2016 ACM SIGSAC Conference on Computer and Communications Security, pp. 1388–1401 (2016)
10. Hu, X., Sastry, N.: Characterising third party cookie usage in the EU after GDPR. In: Proceedings of the 10th ACM Conference on Web Science, pp. 137–141 (2019)
11. Jarke, J., Breiter, A.: The datafication of education. Learn. Media Technol. **44**(1), 1–6 (2019)
12. Jordan, K.: Degrees of intrusion? a survey of cookies used by UK higher education institutional websites and their implications. A Survey of Cookies Used by UK Higher Education Institutional Websites and Their Implications (March 16, 2018) (2018)
13. Karaj, A., Macbeth, S., Berson, R., Pujol, J.M.: Whotracks. me: shedding light on the opaque world of online tracking. arXiv preprint arXiv:1804.08959 (2018)
14. Kretschmer, M., Pennekamp, J., Wehrle, K.: Cookie banners and privacy policies: Measuring the impact of the GDPR on the web. ACM Trans. Web (TWEB) **15**(4), 1–42 (2021)
15. Krishnamurthy, B., Wills, C.: Privacy diffusion on the web: a longitudinal perspective. In: Proceedings of the 18th International Conference on World Wide Web, pp. 541–550 (2009)
16. Krupp, B., Hadden, J., Matthews, M.: An analysis of web tracking domains in mobile applications. In: Hooper, C., Weber, M., Weller, K., Hall, W., Contractor, N., Tang, J. (eds.) WebSci 2021: 13th ACM Web Science Conference 2021, Virtual Event, United Kingdom, 21–25 June 2021, pp. 291–298. ACM (2021)
17. Lerner, A., Simpson, A.K., Kohno, T., Roesner, F.: Internet jones and the raiders of the lost trackers: an archaeological study of web tracking from 1996 to 2016. In: 25th USENIX Security Symposium (USENIX Security 16) (2016)
18. Libert, T.: Exposing the hidden web: an analysis of third-party http requests on 1 million websites. arXiv preprint arXiv:1511.00619 (2015)
19. Mathur, A., et al.: Dark patterns at scale: Findings from a crawl of 11k shopping websites. In: Proceedings of the ACM on Human-Computer Interaction 3(CSCW), pp. 1–32 (2019)
20. Sanchez-Rola, I., et al.: Can i opt out yet? GDPR and the global illusion of cookie control. In: Proceedings of the 2019 ACM Asia Conference on Computer and Communications Security, pp. 340–351 (2019)
21. Saxena, A., Saxena, P., Reddy, H., Gera, R.: A survey on studying the social networks of students. arXiv preprint arXiv:1909.05079 (2019)

22. Sørensen, J., Kosta, S.: Before and after GDPR: the changes in third party presence at public and private European websites. In: The World Wide Web Conference, pp. 1590–1600 (2019)
23. Urban, T., Tatang, D., Degeling, M., Holz, T., Pohlmann, N.: Measuring the impact of the GDPR on data sharing in ad networks. In: Proceedings of the 15th ACM Asia Conference on Computer and Communications Security, pp. 222–235 (2020)
24. Vlajic, N., El Masri, M., Riva, G.M., Barry, M., Doran, D.: Online tracking of kids and teens by means of invisible images: COPPA vs. GDPR. In: Proceedings of the 2nd International Workshop on Multimedia Privacy and Security, pp. 96–103 (2018)

DANAA: Towards Transferable Attacks with Double Adversarial Neuron Attribution

Zhibo Jin[1] , Zhiyu Zhu[1] , Xinyi Wang[2] , Jiayu Zhang[3] , Jun Shen[4] ,
and Huaming Chen[1(✉)]

[1] The University of Sydney, Camperdown, Australia
{zjin0915,zzhu2018}@uni.sydney.edu.au, huaming.chen@sydney.edu.au
[2] Jiangsu University, Zhenjiang, China
[3] Suzhou Yierqi, Changshu, China
[4] University of Wollongong, Wollongong, Australia

Abstract. While deep neural networks have excellent results in many fields, they are susceptible to interference from attacking samples resulting in erroneous judgments. Feature-level attacks are one of the effective attack types, which target the learned features in the hidden layers to improve their transferability across different models. Yet it is observed that the transferability has been largely impacted by the neuron importance estimation results. In this paper, a double adversarial neuron attribution attack method, termed 'DANAA', is proposed to obtain more accurate feature importance estimation. In our method, the model outputs are attributed to the middle layer based on an adversarial non-linear path. The goal is to measure the weight of individual neurons and retain the features that are more important toward transferability. We have conducted extensive experiments on the benchmark datasets to demonstrate the state-of-the-art performance of our method. Our code is available at: https://github.com/Davidjinzb/DANAA.

Keywords: Transferability · Adversarial attack · Attribution-based attack

1 Introduction

Deep neural networks (DNNs) have been used in a wide range of applications in different fields, such as face recognition [6], voice recognition [1] and sentiment analysis [30]. DNNs can also achieve state-of-the-art performance in tasks such as security verification in unconstrained environments where very low false positive rate metrics are required [6]. However, deep learning models are shown to be vulnerable to interference from adversarial samples. Attackers can manipulate the model outcome by deliberately adding the perturbations to the original samples to attack the models [28].

In general, the current approaches to attack models can be categorised into two types: white-box attack [12] and black-box attack [22]. For white-box attacks, the attacker knows the relevant parameters of the target model and can formulate

© The Author(s), under exclusive license to Springer Nature Switzerland AG 2023
X. Yang et al. (Eds.): ADMA 2023, LNAI 14177, pp. 456–470, 2023.
https://doi.org/10.1007/978-3-031-46664-9_31

the most suitable attack method. For black-box attacks, on the other hand, the attacker does not have access to the model parameters. In terms of the characteristics of the white-box and black-box attack methods, the black-box attack provides the adversarial performance of the attacking samples, which is useful for improving the robustness of deep learning models in real-world scenarios. Specifically, the black-box attack methods have three types, including query-based method [14], transfer-based method [8] and hybrid method [9].

The objective of the query-based method is to interrogate the model to extract pertinent input or output information, and subsequently utilize this limited information to iteratively generate optimal adversarial samples. However, such method is subject to restrictions imposed by access permissions and often require multiple queries to obtain excellent adversarial samples. The transfer-based method aims to train and generate adversarial samples on a known-information local surrogate model, which are then transferred and tested on the target black-box model for the attack success rate. Compared to query-based methods, transfer-based methods do not require additional access to the model and can bypass certain adversarial defense mechanisms aimed at queries. The hybrid method combines the principles of query and transfer approaches. Although it can achieve sufficiently high attack success rate, it also implies that it is susceptible to adversarial defense mechanisms targeting both queries and transfers. Therefore, in this paper, we focus on transfer-based method.

As a common approach of transfer-based attack, feature-level attack attempts to maximise the internal feature loss by attacking intermediate layers' features to improve the transferability of the attack [32]. The aim is to increase the weight of negative features in the middle layer of the model while decreasing the weight of positive features. More negative features will be retained to assist the diversion of the model's predictions. However, it is still challenging to harmoniously differentiate the middle-level features via feature-level attack method, which is also prone to its local optimum [32]. Moreover, it is well-known that the effectiveness of transfer-based black-box attacks is influenced by the overfitting on surrogate models and specific adversarial defenses. To address these challenges, we propose to utilise the information of neuron importance estimation for the middle layer to identify the adversarial features more accurately. In addition, we also evaluate the transferability of our proposed method on adversarially trained models, which will be specifically discussed in Sect. 4. The results demonstrate that our method achieves favorable attack success rates even on target models protected by adversarial defenses.

To obtain adversarial samples with higher transferability, this paper presents a double adversarial neuron attribution attack (DANAA). DANAA method attributes the model outputs to the middle layer neurons, thus measuring individual neuron weights and retaining features that are more important towards transferability. We use adversarial non-linear path selection to enrich the attacking points, which improves the attribution results. Extensive experiments on the benchmarking datasets following the literature methods have been conducted. The results show that, DANAA can achieve the best performance for the adversarial attacks. We anticipate this work will contribute to the attribution-based neuron importance estimation and provides a novel approach for transfer-based black-box attack. Our contributions are summarised as follows:

- We propose DANAA, an innovative method of non-linear gradient update paths to achieve a more accurate neuron importance estimation, for a more in-depth study of the route to attribution method.
- We present both theoretical and empirical investigation details for the attribution algorithm in DANAA, which is a core part of the method, in Sect. 3.
- A comprehensive statistical analysis is performed based on our benchmarking experiments on different datasets and adversarial attacks. The results in Sect. 4 demonstrates the state-of-the-art performance of DANAA method.

2   Related Work

In this section, we review the literature on white-box attacks, query-based black-box attacks, transfer-based black-box attacks, and hybrid black-box attacks.

2.1   Common White-Box Attacks

Previous work has demonstrated that neural networks are highly susceptible to misclassification by pre-addition of perturbed test samples. Such processed samples are called adversarial samples. The emergence of adversarial samples has led to the development of a range of adversarial defences to ensure the model performance [16,28,29].

Currently, adversarial attacks can be divided into white-box attacks and black-box attacks depending on the level of available information for the model being attacked. There are various approaches for white-box attacks, such as gradient-based and GAN-based. Gradient-based white-box attacks include FGSM [12], I-FGSM [16], PGD [20] and C&W [3]. Some recent GAN-based white-box attack methods are AdvGAN [33], GMI [37], KED-MI [4] and Plug&Play [24]. While white-box attacks are effective in measuring the robustness of a model under attack, in real-world scenario, the parameters of the model are often not accessible, leading to the development of black-box attacks.

2.2   Query-Based Black-Box Attacks

Query-based attacks are a branch of black-box attacks aiming to train an effective adversarial sample by performing a small-scale attack on the target model to query the model parameters, such as the model labels and confidence levels. These parameters can be used as part of the dataset to assist in training the migration algorithm to verify the migration of the black-box model. Ilyas et al. [14] were the first to propose a query-based black-box attack approach. Following, they proposed combining prior and gradient estimation of historical queries and data structures based on Bandit Optimization, which greatly reduces the number of queries [15]. Li et al. [17] proposed a query-efficient boundary-based black box attack method (QEBA). It proved that the gradient estimation of the boundary-based attack over the entire gradient space is invalid in terms of the number of queries. Andriushchenko et al. [2] proposed the square search attack method, which selects local square blocks at random locations in the image to search and update the direction of the attack.

2.3   Transfer-Based Black-Box Attacks

The transferability of adversarial attacks refers to the applicability of the adversarial samples generated by the local model to the target model for attack. The attacker firstly uses the parameters obtained from the attack on the local model to train the adversarial samples, then uses these samples to perform a black-box attack on the target model to verify the success rate.

There are three main categories of transfer-based black-box attacks, namely gradient calculation methods, input transformation methods and feature-level attack methods. Gradient calculation methods such as MIM [7], VMI-FGSM [31] and SVRE [35] improve transferability by designing new gradient updates. Input transformation methods such as DIM [34], PIM [11] and SSA [18] boost the transferability by using input transformations to simulate the ensemble process of the model, while feature-level attacks focus on the middle-layer features.

Some state-of-the-art feature-level attack methods include NRDM, FDA, FIA and NAA, etc. NRDM [21] attempts to maximise the degree of distortion between neurons, but it does not take into account the role of positive and negative features in the attack. FDA [10] averages the neuronal activation values to obtain an estimate of the importance of a neuron. However, this method does not distinguish the degree of each neuron's importance and the discrimination between positive and negative features is still too low. FIA [32] multiplies the activation values of neurons and back-propagation gradients for estimation, but its effect on the original input is affected by over-fitting and the results are not accurate. NAA [36] effectively improves the transferability of the model and reduces computational complexity by attributing the model's output to an intermediate layer to obtain a more accurate importance estimation. However, its attribution method focuses more on the gradient iteration process considering linear path, and there is still room for improvement in the non-linear path condition.

2.4   Hybrid Black-Box Attacks

Hybrid method is a combination of query-based method and transfer-based method. It not only considers the priori nature of the transfer but also utilizes the gradient information obtained from the query, which resolves the challenges of high access cost for the query attack and low accuracy for transfer attack.

Dong et al. [5] proposed a hybrid method named P-RGF, which used the gradient of surrogate model as prior knowledge to guide the query direction of RGF and obtained the same success rate as RGF with fewer queries. Fu et al. [9] train Meta Adversarial Perturbation (MAP) on an surrogate model and perform black-box attacks by estimating the gradient of the model, which has good transferability and generalizability. Ma et al. [19] introduced Meta Simulator to black-box attacks based on the idea of meta-learning. By combining query and transfer based attacks, the researchers not only significantly reduce the number of queries, but also reduce the complexity of queries by transferring the adversarial samples trained on the surrogate model to the target model.

While there are different types of black-box attack methods, transfer-based attacks is considered as the most convenient method which doesn't require additional information queries for the model. However, it poses the challenge of a good transferability for the adversarial samples. Therefore, in this work, we target the transfer-based attack methods. Especially, we introduce the attribution method for the middle-layer feature estimation, which shows a promising performance with our experiments.

3   Method

3.1   Preliminaries

When an adversarial attack to the target model can be successfully launched given an adversarial samples trained with a local DNN model, we consider there is a strong transferability relationship between these two models. Formally, with a deep learning network $N : R^n \rightarrow R^c$ and original image sample $x^0 \in R^n$, whose true label is t, if the imperceptible perturbation $\sum_{k=0}^{t-1} \triangle x^k$ is applied on the original sample x^0, we may mislead the network N with the manipulated input $x^t = x^0 + \sum_{k=0}^{t-1} \triangle x^k$ to the label of m, which can also be denoted as x^{adv}. Assuming the output of the sample x as $N(x)$, the optimization goal will be:

$$\left\| x^t - x^0 \right\|_n < \epsilon \quad subject\ to \quad N(x^t) \neq N(x^0) \tag{1}$$

where $\|\cdot\|_n$ represents the n-norm distance. Considering the activation values in the middle layers of network N, we denote the activation value of y-th layer as y and the activation value of j-th neuron as y_j.

3.2   Non-linear Path-Based Attribution

Inspired by [25] and [36], we define the attribution results of input image x^t(with $n \times n$ pixels) as

$$A := \sum_{i=1}^{n^2} \int \triangle x_i^t \frac{\partial N(x^t)}{\partial x^t} dt \tag{2}$$

As shown in Fig. 1, different from the NAA algorithm [36], our paper proposes a new attribution idea that uses a non-linear gradient update path instead of the original linear path, which allows the model to find the optimal path against the attack itself. In Eq. 2, the gradient of N iterates along the non-linear path $x^t = x^0 + \sum_{k=0}^{t-1} \triangle x^k$, in which $\frac{\partial N}{\partial x^t}(\cdot)$ is the partial derivative of N to the i-th

Fig. 1. Non-linear gradient update path diagram

pixel. For each iteration, $\triangle x^t = sign(\frac{\partial N(x^t)}{\partial x^t}) + N(0, \sigma)$. We further apply the learning rate and Gaussian noise to update the perturbation.

Afterwards, we can approximate A as $N(x)$ depending on basic advanced mathematics and extend the attribution results to each layer. The formula of attribution can then be expressed as:

$$A_{y_j} := \sum_{i=1}^{n^2} \int \triangle x_i^t \frac{\partial N(x^t)}{\partial y_j(x^t)} \frac{\partial y_j(x^t)}{\partial x^t} dt \qquad (3)$$

where A_{y_j} represents the attribution of j-th neuron in the layer y, $\sum A_{y_j} = A$. We provide the relevant proof of our non-linear path-based attribution in following section.

3.3 Proof of Non-linear Path-Based Attribution

Since we now have A_{y_j} as Eq. 3, assuming that the neurons on the middle layer of the deep neural network are independent from each other, A_{y_j} can be expressed as

$$A_{y_j} := \int \frac{\partial N(x^t)}{\partial y_j(x^t)} \sum_{i=1}^{n^2} \triangle x_i^t \frac{\partial y_j(x^t)}{\partial x^t} dt \qquad (4)$$

where $\frac{\partial N(x^t)}{\partial y(x^t)}$ is the gradient of $N(x^t)$ to the j-th neuron, $\sum_{i=1}^{n^2} \triangle x_i^t \frac{\partial y_j(x^t)}{\partial x^t}$ is the sum of the gradient of y_j to each pixel on $x^t (x^t \in R^n)$. Since the two gradient sequences are zero covariance, we then convert Eq. 4 into:

$$A_{y_j} := \int \frac{\partial N(x^t)}{\partial y_j(x^t)} dt \cdot \int \sum_{i=1}^{n^2} \triangle x_i^t \frac{\partial y_j(x^t)}{\partial x^t} dt \qquad (5)$$

Combining the principles of calculus, we can prove that

$$\int \sum_{i=1}^{n^2} \triangle x_i^t \frac{\partial y_j(x^t)}{\partial x^t} dt = y_j^t - y_j^0 \qquad (6)$$

then we denote $y_j^t - y_j^0$ as $\triangle y_j^t$, Eq. 5 can be converted into

$$A_{y_j} := \triangle y_j^t \int \frac{\partial N(x^t)}{\partial y_j(x^t)} dt \qquad (7)$$

Denoting $\int \frac{\partial N(x^t)}{\partial y_j(x^t)} dt$ as $\gamma(y_j)$, which means the gradient of network N along our non-linear path with attention to the j-th neuron. Afterwards, we can get $A_{y_j} = \triangle y_j^t \cdot \gamma(y_j)$. Since the neuron y_j is on the middle layer y, finally the attribution result of the layer y can be expressed as

$$A_y = \sum_{y_j \in y} A_{y_j} = \sum_{y_j \in y} \triangle y_j^t \cdot \gamma(y_j) = \triangle y^t \cdot \gamma(y) \tag{8}$$

Algorithm 1. Double Adversarial Neuron Attribution Attack

Require: Deep network N, target layer y
Require: Manipulated input x^t with label m
Require: Perturbation budget ϵ and iteration number T
Require: Original input x^0 and integrated step τ
1: $\alpha = \frac{\epsilon}{T}$, $\gamma(y_j) = 0$, $g_0 = 0$, $\mu = 1$, $x_0^{adv} = x^t$
2: **for** $t = 0 \leftarrow \tau$ **do**
3:     $x^{t+1} = clip_x^\epsilon \{x^t + lr \cdot sign(\frac{\partial N(x^t)}{\partial x^t}) + N(0, \sigma)\}$
4:     $\gamma(y_j) = \gamma(y_j) + \nabla_{y(x^t)} N(x^t)$
5: **end for**
6: **for** $s = 0 \leftarrow T - 1$ **do**
7:     $A_y = \triangle y^t \cdot \gamma(y)$
8:     $g_{s+1} = \mu \cdot g_s + \frac{\nabla_{x^t} A_y}{\|\nabla_{x^t} A_y\|_1}$
9:     $x_{s+1}^{adv} = Clip_{x^t}^\epsilon \{x_{s+1}^{adv} + \alpha \cdot sign(g_{s+1})\}$
10: **end for**

Algorithm 1 shows the specific pseudocode structure of our DANAA algorithm with Non-linear Path-based attribution.

4   Experiments

Extensive experiments have been conducted to demonstrate the efficiency of our method. Following sections cover the topic of leveraged datasets, benchmarking models and incorporated metrics. We also provide the experimental settings. We performed five rounds of benchmarking experiments to compare our algorithm with other methods, demonstrating the superiority of our approach to the baselines in terms of transferability for adversarial attacks. Moreover, we conducted the ablation study to investigate our approach, focusing on the impact of various learning rates and noise deviation on attack transferability.

4.1   Dataset

Following other literature methods, the widely-used datasets from NAA work [36] are considered in this paper. The datasets consist 1000 images of different categories randomly selected from the ILSVRC 2012 validation set [23], which we called a multiple random sampling(MRS) dataset.

4.2   Model

We include four widely-used models for image classification tasks, namely Inception-v3 (Inc-v3) [27], Inception-v4 (Inc-v4) [26], Inception-ResNet-v2 (IncRes-v2) [26], and ResNet-v2-152 (Res152-v2) [13], as source models for assessing the attacking performance of our algorithm. We start with four pretrained models without adversarial learning, which include Inc-v3, Inc-v4, IncRes-v2, and Res152-v2. Later on, we construct more robust models for a in-depth comparison, such as including adversarial training for the pretrained models. This results in two adversarial trained models, including Inception-v3(Inc-v3-adv) and Inception-Resnet-v2 (IncRes-v2-adv) [16]. The remaining three models are based on the ensemble models: the ensemble of three adversarial trained Inception-v3(Inc-v3-adv-3), the ensemble of four adversarial trained Inception-v3 (Inc-v3-adv-4), and the ensemble of three adversarial trained Inception-Resnet-v2 (IncRes-v2-adv-3), following the work from [29]. In [29], the models are combined by training the sub-models of the corresponding model independently and finally weighting the results of each sub-model to increase the accuracy and robustness of the model.

4.3   Evaluation Metrics

The attack success rate is selected as the metric to evaluate the performance. It measures the proportion of the dataset where our method produces incorrect label predictions after attacking. Hence, a higher success rate indicates improved performance of the attack method.

4.4   Baseline Methods

For comparison in our experiment, we selected five state-of-the-art attack methods as the baseline, including MIM [7], NRDM [21], FDA [10], FIA [32], and NAA [36]. Furthermore, to test the effect of each model after combining input transformation methods and to verify the superiority of our algorithm, we apply both DIM and PIM to the attack methods. The implementation details can be found in the open source repository. Consequently, we extend the model comparison set with MIM-PD, NRDM-PD, FDA-PD, FIA-PD, NAA-PD and DANAA-PD, respectively.

4.5   Parameter Setting

In the experiment, we set the parameters as following: the learning rate (lr) is 0.0025; the noise deviation is 0.25; and the maximum perturbation rate is 16, which is derived from the number of iterations (15) and the step size (1.07). The batch size is 10, and the momentum of the optimization process is 1. Since we introduced the DIM and PIM algorithms to verify the superiority of our model when combining input transformation methods, we set the transformation probability of DIM to 0.7, and the amplification factor and kernel size of PIM are 2.5

and 3, respectively. For the target layer of the attack, we choose the same layer as in NAA. Specifically, we attack InceptionV3/InceptionV3/Mixed_5b/concat layer for Inc-v3; InceptionV4/InceptionV4/Mixed_5e/concat layer for Inc-v4; InceptionResnetV2/Ince-ptionResnetV2/Conv2d_4a_3x3/Relu layer for IncRes-v2; the ResNet-v2-152/blo-ck2/unit_8/bottleneck_v2/add layer of Res152-v2 [36].

4.6   Result

All the experiments are carried out with the hardware of RTX 2080Ti card. A detailed replication package can be found in the open source repository at https://github.com/Davidjinzb/DANAA. We subsequently compile the results of all the attack methods without and with the input transformation methods (ending with PD) in Table 1.

In Table 1, we can see that, DANAA has retained a strong and robust performance across all the models, in comparison with other attack methods. Especially, DANAA demonstrated notable improvements on five models that are adversarial trained. We can observe a largest improvement of the attacking performance is between our method and NAA method [36], which is the generally second best attacking method in the comparison experiments. The ratio of improvement is 9.0%. Across all local models, our approach demonstrated an overall average improvement of 7.1% as compared to NAA on the adversarial trained models. By introducing the PD concept, our method achieves a maximum improvement of 9.8% over NAA-PD and an overall average improvement of 7.3% on the adversarial trained models.

4.7   Ablation Study

In this section, we investigate the impact of the learning rate and Gaussian noise deviations on the performance of the proposed method.

The Impact of Learning Rates. Experiments are conducted using different scales of learning rates, which are 0.25, 0.025, 0.0025 and 0.00025. In Fig. 2, the DANAA method exhibits the highest attack success rate for nearly all models when the selected learning rate was 0.0025. In Fig. 3, the highest attack success rates are achieved on most models for DANAA-PD method.

Notably, when using Inception-ResNet-v2 as the source model, although at a learning rate of 0.0025 DANAA-PD ranked second best in attack success rate on the models without adversarial training, its effectiveness on the model with adversarial training is still much higher than those at other learning rates.

The Impact of Gaussian Noise Deviation (Scale). To verify the effect of adding Gaussian noise to the gradient update on model transferability in this paper, we selected different noise deviations for testing in this subsection. As shown in Fig. 4 and Fig. 5, five different scales of the Gaussian noise deviation ranging from 0.2 to 0.4 are used in this experiment. In general, higher value of scale tends to have more superior results for the normal training model while sacrificing performance for the more robust one. Conversely, a lower value of scale results in less improvements for the normal trained model but better performance for the adversarial trained model. Accordingly, the scale value of 0.25 is selected for the optimal performance in this paper.

Table 1. Attack success rate of multiple methods on different models

Model	Attack method	Inc-v3	Inc-v4	IncRes-v2	Res152-v2	Inc-v3-adv	IncRes-v2-adv	Inc-v3-adv-3	Inc-v3-adv-4	IncRes-v2-adv-3
Inc-v3	MIM	100	41.9	39.7	32.8	22.1	18.4	14.9	15.7	8.2
	NRDM	90.4	61.4	52.5	49.9	26.1	19.2	9.5	12.9	4.7
	FDA	81.7	42.9	37.1	35.1	19.4	12.6	9.3	12.2	5.0
	FIA	96.5	79.1	77.8	71.8	54.8	53.9	43.1	44.2	23.2
	NAA	97.0	83.0	80.6	74.7	56.2	59.4	49.5	50.4	31.5
	DANAA	98.1	86.8	84.8	80.3	64.4	68.4	55.4	56.5	33.1
Inc-v4	MIM	58.2	99.9	45	40.4	23.5	20.4	17.7	20.3	9.7
	NRDM	78.0	96.4	62.8	62.3	26.1	25	17.3	16.6	6.8
	FDA	84.6	99.6	71.8	68.8	28.2	26.1	17.4	17.1	7.0
	FIA	74.6	91.0	69.6	65.7	43.5	47.3	39.3	39.9	23.5
	NAA	83.3	95.8	77.9	73.3	49.5	53.2	48.0	46.5	31.4
	DANAA	86.8	97.2	82.4	76.9	54.9	61	53.8	53.5	35
IncRes-v2	MIM	60	51.9	99.2	42.2	25.9	30.5	21.7	23.3	12.3
	NRDM	72.8	67.9	77.9	59.7	35.7	30.8	16.4	17.1	7.3
	FDA	69.0	68.0	78.2	56.2	34.5	29.7	16.2	15.4	7.7
	FIA	71.0	68.2	78.8	63.9	53.8	56.4	47.4	45.8	37.6
	NAA	79.5	76.4	89.3	71.1	60.3	64.8	56.9	55.0	47.3
	DANAA	82.7	80.4	91.5	77.7	66.3	72.2	64.7	60.8	56
Res152-v2	MIM	52.9	47.3	44.9	99.4	26.6	25.1	24.3	24.4	13.3
	NRDM	72.7	68.8	59.5	89.9	39.1	31.0	20.3	18.1	9.3
	FDA	15.7	9.2	8.3	26.2	13.1	6.8	9.3	9.7	4.0
	FIA	80.7	78.2	77.5	98.0	58.5	58.2	53.0	48.4	34.4
	NAA	84.7	83.5	82.3	97.6	61.8	67.0	59.1	58.1	46.1
	DANAA	86.4	86.8	85.9	98.8	68.1	71.7	65.1	62.0	48.4
Inc-v3	MIM-PD	99.7	72.8	66.9	54.1	31.7	29.1	20.2	21.7	9.7
	NRDM-PD	86.3	68.6	64.3	58.0	31.1	22.6	10.6	13.8	5.9
	FDA-PD	74.7	49.3	46.5	40.9	23.7	15.4	10.5	13.1	6.2
	FIA-PD	96.9	83.5	82.7	79.8	61.4	62.1	47.0	48.2	27.5
	NAA-PD	97.2	87.0	85.6	81.1	64.9	65.8	53.4	51.6	33.6
	DANAA-PD	97.9	89.4	89.4	84.8	70.6	72.3	61.7	60.9	40.1
Inc-v4	MIM-PD	81.3	99.4	71.0	59.7	31.6	28.0	22.9	23.3	12.7
	NRDM-PD	90.3	97.0	79.5	76.8	34.1	34.4	21.1	19.7	8.6
	FDA-PD	93.2	99.2	86.4	82.4	36.7	37.4	20.3	21.1	10.0
	FIA-PD	84.0	92.4	81.2	77.1	55.2	58.6	48.9	47.5	29.3
	NAA-PD	90.5	96.9	87.6	83.9	58.4	64.3	54.0	53.4	34.6
	DANAA-PD	90.4	96.5	87.9	84.9	63.9	71.0	61.9	60.1	42.7
IncRes-v2	MIM-PD	80.7	76.5	98.0	65.8	36.9	42.7	29.4	28.6	17.1
	NRDM-PD	76.4	74.1	78.7	64.1	40.7	32.4	17.5	18.8	6.7
	FDA-PD	78.1	76.2	80.7	66.5	41.3	35.6	18.4	17.0	7.6
	FIA-PD	76.5	73.4	81.7	71.1	60.0	62.5	50.3	47.0	36.4
	NAA-PD	81.4	78.2	89.9	76.4	65.2	67.7	59.9	57.1	46.0
	DANAA-PD	83.7	80.4	89.8	80.6	70.3	73	65.8	63.1	55.8
Res152-v2	MIM-PD	81.5	77.5	76.2	99.4	41.5	44.5	34.8	33.6	18.4
	NRDM-PD	84.1	82.1	73.1	90.1	51.6	43.5	28.3	22.5	11.2
	FDA-PD	22.1	12.7	11.4	23.4	19.6	10.4	9.9	11.7	5.4
	FIA-PD	88.6	86.1	87.0	98.3	70.9	71.0	63.6	58.6	43.4
	NAA-PD	90.2	88.5	89.0	98.0	73.5	76.1	70.3	66.3	52.2
	DANAA-PD	92.0	91.7	91.8	98.7	79.3	82.1	76.1	73.4	60.8

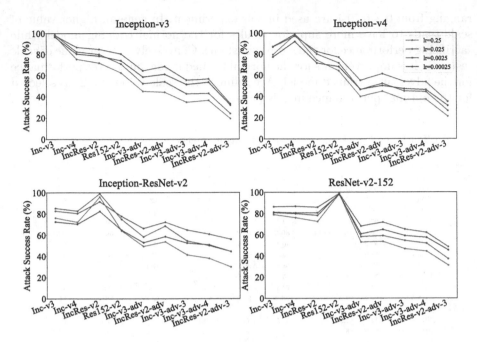

Fig. 2. DANAA attack success rate performance at different learning rates

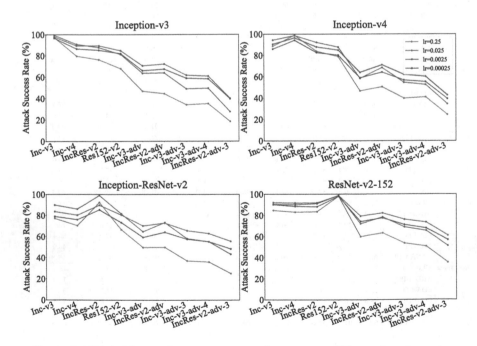

Fig. 3. DANAA-PD attack success rate performance at different learning rates

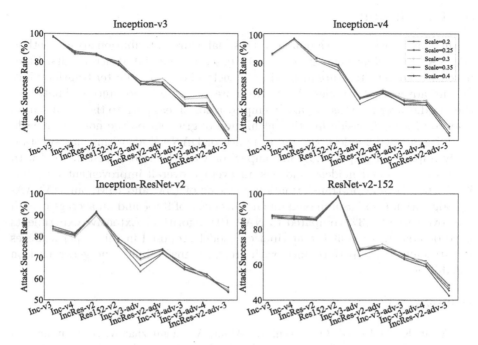

Fig. 4. DANAA attack success rate performance at different noise deviation

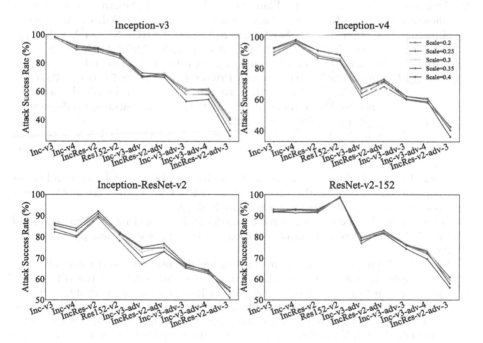

Fig. 5. DANAA-PD attack success rate performance at different noise deviation

5   Conclusion

In this paper, we propose a double adversarial neuron attribution attack method (DANAA) to achieve enhanced transfer-based adversarial attack results. Compared with other literature methods, our method obtains a better transferability for the adversarial samples. To derive more accurate importance estimates for the middle layer neurons, we firstly employ a non-linear path to the perturbation update process. Considering the calculation of gradient on the non-linear path, for all examined models, the performance of DANAA algorithm has substantially improved by up to 9.0% in comparison with the second best method with adversarial trained models, and has an average overall improvement by 7.1%. With the information transformation methods of DIM and PIM, our DANAA-PD algorithm also has a maximum enhancement of 9.8% and an average overall improvement of 7.3% compared to NAA-PD algorithm. Extensive experiments have demonstrated that the attribution model proposed in this paper achieves the state-of-the-art performance, with greater transferability and generalisation capabilities.

References

1. Aizat, K., Mohamed, O., Orken, M., Ainur, A., Zhumazhanov, B.: Identification and authentication of user voice using DNN features and I-vector. Cogent Eng. **7**(1), 1751557 (2020)
2. Andriushchenko, M., Croce, F., Flammarion, N., Hein, M.: Square attack: a query-efficient black-box adversarial attack via random search. In: Vedaldi, A., Bischof, H., Brox, T., Frahm, J.-M. (eds.) ECCV 2020, Part XXIII. LNCS, vol. 12368, pp. 484–501. Springer, Cham (2020). https://doi.org/10.1007/978-3-030-58592-1_29
3. Carlini, N., Wagner, D.: Towards evaluating the robustness of neural networks. In: 2017 IEEE Symposium on Security and Privacy (SP), pp. 39–57. IEEE (2017)
4. Chen, S., Kahla, M., Jia, R., Qi, G.J.: Knowledge-enriched distributional model inversion attacks. In: Proceedings of the IEEE/CVF International Conference on Computer Vision, pp. 16178–16187 (2021)
5. Cheng, S., Dong, Y., Pang, T., Su, H., Zhu, J.: Improving black-box adversarial attacks with a transfer-based prior. In: Advances in Neural Information Processing Systems, vol. 32 (2019)
6. Deng, J., Guo, J., Xue, N., Zafeiriou, S.: Arcface: additive angular margin loss for deep face recognition. In: Proceedings of the IEEE/CVF Conference on Computer Vision and Pattern Recognition, pp. 4690–4699 (2019)
7. Dong, Y., et al.: Boosting adversarial attacks with momentum. In: Proceedings of the IEEE Conference on Computer Vision and Pattern Recognition, pp. 9185–9193 (2018)
8. Dong, Y., Pang, T., Su, H., Zhu, J.: Evading defenses to transferable adversarial examples by translation-invariant attacks. In: Proceedings of the IEEE/CVF Conference on Computer Vision and Pattern Recognition, pp. 4312–4321 (2019)
9. Fu, J., Sun, J., Wang, G.: Boosting black-box adversarial attacks with meta learning. In: 2022 41st Chinese Control Conference (CCC), pp. 7308–7313. IEEE (2022)
10. Ganeshan, A., BS, V., Babu, R.V.: FDA: feature disruptive attack. In: Proceedings of the IEEE/CVF International Conference on Computer Vision, pp. 8069–8079 (2019)

11. Gao, L., Zhang, Q., Song, J., Liu, X., Shen, H.T.: Patch-wise attack for fooling deep neural network. In: Vedaldi, A., Bischof, H., Brox, T., Frahm, J.-M. (eds.) ECCV 2020, XXVIII. LNCS, vol. 12373, pp. 307–322. Springer, Cham (2020). https://doi.org/10.1007/978-3-030-58604-1_19

12. Goodfellow, I.J., Shlens, J., Szegedy, C.: Explaining and harnessing adversarial examples. arXiv preprint arXiv:1412.6572 (2014)

13. He, K., Zhang, X., Ren, S., Sun, J.: Deep residual learning for image recognition. In: Proceedings of the IEEE Conference on Computer Vision and Pattern Recognition, pp. 770–778 (2016)

14. Ilyas, A., Engstrom, L., Athalye, A., Lin, J.: Black-box adversarial attacks with limited queries and information. In: International Conference on Machine Learning, pp. 2137–2146. PMLR (2018)

15. Ilyas, A., Engstrom, L., Madry, A.: Prior convictions: black-box adversarial attacks with bandits and priors. arXiv preprint arXiv:1807.07978 (2018)

16. Kurakin, A., Goodfellow, I., Bengio, S.: Adversarial machine learning at scale. arXiv preprint arXiv:1611.01236 (2016)

17. Li, H., Xu, X., Zhang, X., Yang, S., Li, B.: QEBA: query-efficient boundary-based blackbox attack. In: Proceedings of the IEEE/CVF Conference on Computer Vision and Pattern Recognition, pp. 1221–1230 (2020)

18. Long, Y., et al.: Frequency domain model augmentation for adversarial attack. In: Avidan, S., Brostow, G., Cissé, M., Farinella, G.M., Hassner, T. (eds.) ECCV 2022, Part IV. LNCS, vol. 13664, pp. 549–566. Springer, Cham (2022)

19. Ma, C., Chen, L., Yong, J.H.: Simulating unknown target models for query-efficient black-box attacks. In: Proceedings of the IEEE/CVF Conference on Computer Vision and Pattern Recognition, pp. 11835–11844 (2021)

20. Madry, A., Makelov, A., Schmidt, L., Tsipras, D., Vladu, A.: Towards deep learning models resistant to adversarial attacks. arXiv preprint arXiv:1706.06083 (2017)

21. Naseer, M., Khan, S.H., Rahman, S., Porikli, F.: Task-generalizable adversarial attack based on perceptual metric. arXiv preprint arXiv:1811.09020 (2018)

22. Papernot, N., McDaniel, P., Goodfellow, I., Jha, S., Celik, Z.B., Swami, A.: Practical black-box attacks against machine learning. In: Proceedings of the 2017 ACM on Asia Conference on Computer and Communications Security, pp. 506–519 (2017)

23. Russakovsky, O., et al.: Imagenet large scale visual recognition challenge. Int. J. Comput. Vision 115, 211–252 (2015)

24. Struppek, L., Hintersdorf, D., Correia, A.D.A., Adler, A., Kersting, K.: Plug & play attacks: towards robust and flexible model inversion attacks. arXiv preprint arXiv:2201.12179 (2022)

25. Sundararajan, M., Taly, A., Yan, Q.: Axiomatic attribution for deep networks. In: International Conference on Machine Learning, pp. 3319–3328. PMLR (2017)

26. Szegedy, C., Ioffe, S., Vanhoucke, V., Alemi, A.: Inception-v4, inception-resnet and the impact of residual connections on learning. In: Proceedings of the AAAI Conference on Artificial Intelligence, vol. 31 (2017)

27. Szegedy, C., Vanhoucke, V., Ioffe, S., Shlens, J., Wojna, Z.: Rethinking the inception architecture for computer vision. In: Proceedings of the IEEE Conference on Computer Vision and Pattern Recognition, pp. 2818–2826 (2016)

28. Szegedy, C., et al.: Intriguing properties of neural networks. arXiv preprint arXiv:1312.6199 (2013)

29. Tramèr, F., Kurakin, A., Papernot, N., Goodfellow, I., Boneh, D., McDaniel, P.: Ensemble adversarial training: attacks and defenses. arXiv preprint arXiv:1705.07204 (2017)

30. Wadawadagi, R., Pagi, V.: Sentiment analysis with deep neural networks: comparative study and performance assessment. Artif. Intell. Rev. **53**(8), 6155–6195 (2020)
31. Wang, X., He, K.: Enhancing the transferability of adversarial attacks through variance tuning. In: Proceedings of the IEEE/CVF Conference on Computer Vision and Pattern Recognition, pp. 1924–1933 (2021)
32. Wang, Z., Guo, H., Zhang, Z., Liu, W., Qin, Z., Ren, K.: Feature importance-aware transferable adversarial attacks. In: Proceedings of the IEEE/CVF International Conference on Computer Vision, pp. 7639–7648 (2021)
33. Xiao, C., Li, B., Zhu, J.Y., He, W., Liu, M., Song, D.: Generating adversarial examples with adversarial networks. arXiv preprint arXiv:1801.02610 (2018)
34. Xie, C., et al.: Improving transferability of adversarial examples with input diversity. In: Proceedings of the IEEE/CVF Conference on Computer Vision and Pattern Recognition, pp. 2730–2739 (2019)
35. Xiong, Y., Lin, J., Zhang, M., Hopcroft, J.E., He, K.: Stochastic variance reduced ensemble adversarial attack for boosting the adversarial transferability. In: Proceedings of the IEEE/CVF Conference on Computer Vision and Pattern Recognition, pp. 14983–14992 (2022)
36. Zhang, J., et al.: Improving adversarial transferability via neuron attribution-based attacks. In: Proceedings of the IEEE/CVF Conference on Computer Vision and Pattern Recognition, pp. 14993–15002 (2022)
37. Zhang, Y., Jia, R., Pei, H., Wang, W., Li, B., Song, D.: The secret revealer: generative model-inversion attacks against deep neural networks. In: Proceedings of the IEEE/CVF Conference on Computer Vision and Pattern Recognition, pp. 253–261 (2020)

CRNN-SA: A Network Intrusion Detection Method Based on Deep Learning

Wanxiao Liu[ID], Jue Chen[✉][ID], and Xihe Qiu[ID]

Shanghai University of Engineering Science, Shanghai 201620, China
jadeschen@sues.edu.cn

Abstract. Network Intrusion Detection System (IDS) is crucial in defending the target network from intrusions. However, due to information loss and insufficient feature dimensions during feature extraction, the majority of existing detection algorithms are unable to fully utilize the data present in the original network. To address the aforementioned issues, this study examines the presence of temporal and spatial characteristics in network traffic data and proposes a new intrusion detection model named CRNN-SA which combines hierarchical Convolutional Neural Network (CNN), Recurrent Neural Network (RNN) and Self-Attention. This model extracts spatial features and temporal features by using CNN and RNN, respectively, and "connects" the features extracted by CNN and RNN to obtain fusion features. In order to express useful input information better, Self-Attention is utilized to allocate distinct weights to the combined characteristics. This model can effectively extract spatial and temporal features of data by increasing the granularity of synchronized input data. To ensure the accuracy of the model, it undergoes evaluation using the UNSW-NB15 dataset. The Accuracy and F1-score of the CRNN-SA model under the binary classification are 90.4% and 91.3%, respectively, and the metrics under the multi-class classification are 89.9% and 77.5%, respectively. Through experiments, it has been demonstrated that the combination of feature selection and deep learning models can significantly enhance the detection capability, resulting in a substantial decrease in the false positive rate.

Keywords: Network intrusion detection · Convolutional neural network · Recurrent neural network · Self-Attention mechanism

1 Introduction

Over the past few decades, web technologies have evolved rapidly and gained wide application in many fields, having a profound impact on social development. Similarly, attacks on network systems are becoming increasingly serious, with a wider and wider range of attacks and a variety of new attack tools and methods, which not only cause economic losses but also threaten national security. Therefore, an effective network intrusion detection solution is ultimately

© The Author(s), under exclusive license to Springer Nature Switzerland AG 2023
X. Yang et al. (Eds.): ADMA 2023, LNAI 14177, pp. 471–485, 2023.
https://doi.org/10.1007/978-3-031-46664-9_32

crucial for modern society. IDS [14] is a commonly utilized device for network security that has the capability to monitor real-time network activities and identify potential threats. A Network IDS (NIDS) is a type of IDS that is installed on network nodes and is capable of identifying attacks by directly examining the traffic on the network.

The NIDS system comprises of both misuse detection and anomaly detection [6]. The misuse detection method, which is widely employed in practical deployment, utilizes a set of rules based on expert knowledge to identify malicious activities. It detects attacks by comparing network traffic line by line, enabling quick detection with a minimal false positive rate [4]. Nevertheless, assailants consistently enhance their assault instruments and tactics, whereas misuse detection techniques are incapable of identifying these unfamiliar attacks. In recent years, the focus of intrusion detection has been on detecting unknown attacks, a task that can be accomplished using the anomaly detection method. Machine Learning and Deep Learning are the main technologies of intrusion detection [2].

The machine learning-based intrusion detection technique initially extracts characteristics from the initial network traffic using feature engineering, followed by model training to identify abnormalities. Models utilized to identify intrusions often, include Random Forest [5] and SVM [8]. While machine learning-based methods have attained comparatively high performance, as data complexity and diversity increase, they become increasingly dependent on features extracted through complex feature engineering (for example, packet lengths), whose design requires specialized knowledge and results in information loss, becoming bottlenecks for machine learning methods. By bypassing the limitations of machine learning techniques and automatically extracting features from the original data, the IDS method based on deep learning solves the aforementioned issues. Deep learning technology has advanced quickly and produced amazing results in a variety of applications, including intrusion detection, thanks to the growth of hardware and the production of vast amounts of data. CNN [18], RNN, LSTM [18], Transformer, and GAN [3] are examples of deep learning techniques that learn characteristics from several perspectives. For instance, LSTM learns the temporal characteristics of network traffic, CNN learns the spatial characteristics.

Although the existing methods based on deep learning have achieved high performance, they still have the following shortcomings:

- Packet header and packet payload play a key role in intrusion detection, nevertheless, most methods based on deep learning treat them as a whole at the same time, which makes it impossible to learn more concentrated features in the model.
- The number and size of data packets contained in the session are not fixed. Current solutions include explicit truncation and patching to a predetermined length, but the result is inevitably loss of information as the shortened component is unusable.
- Data packet intervals are not taken into consideration. Just as a phrase may be seen as a series of several words, a conversation can be seen as a collection of

various data packets. Contrary to sentence patterns, however, conversational parts are relatively far apart from one another. The common sequence method now in use is ineffective because processing time information is lost.

Aiming at the above problems, this paper constructs a new intrusion detection model named CRNN-SA which is based on CNN, RNN and Self-Attention. This model uses one-dimensional CNN and RNN to extract spatial features and temporal features respectively, "connects" the features extracted by them to obtain fusion features, introduces Self-Attention to extract the features, and further extracts features to select important features. The contributions are as follows:

- A new intrusion detection model named CRNN-SA is proposed, which takes into account the attributes of every element in the initial network traffic. This model has the ability to hierarchically learn the spatial- temporal characteristics of the network traffic.
- To allocate distinct weights to the combined characteristics, the Self-Attention mechanism is employed, and the secondary feature extraction is carried out to select important feature information.
- The model is evaluated on UNSW-NB15 dataset, and the experimental results show that the model is effective and robust. The prediction accuracy of binary and multi-class experiments reached 90.4% and 89.9%, respectively.

2 Related Works

IDS is a crucial security technology that may collect and filter network traffic to look for signs of network damage. Numerous ML-based intrusion/anomaly detection techniques have been researched. For instance, Marteau et al. [11] constructed an intrusion detection model using the forest of binary partition trees in order to detect point-by-point and collective anomalies. Abdelmoumin et al. [1] proposed an ensemble learning technique based on triple stacking to detect anomalies in the Internet of Things, which integrated PCA, first-level SVM and second-level neural network to aggregate predictions. Khan et al. [9] put forward an integrated IDS model, which uses AutoML based on soft voting method in network environment. The Internet of Things may experience high delays and resource usage due to ensemble learning's high training, testing, and computing overhead costs. Qi et al. [16] integrated PCA, isolated forests, and locally sensitive hashing (LSH) techniques to effectively and accurately detect anomalies in Industry 4.0. However, machine learning techniques can limit their learning ability and make it challenging to capture deep feature relationships between data.

Because of its strong capacity for learning and independence from any feature engineering, deep learning has garnered a great deal of interest in intrusion/anomaly detection. In IDS, some cutting-edge deep learning techniques have been applied. For instance, IDS with Variable LSTM was developed by Zhou et al. [19] to address the complexity of data compression. Park et al. [15]

created an IDS based on boundary balanced GAN (BEGAN) to detect networks in order to address the issue of data imbalance. To enhance the detection performance, they use features from trained encoders and apply supervised classifiers like DNN, CNN, and LSTM. These deep learning approaches, however, have a substantial computational cost. An detection system based on LSTM AE was created by Li et al. [10] and leverages edge computing to identify potential intrusions into the Internet of Things network.

3   Proposed CRNN-SA Approach for Intrusion Detection

RNN targets temporal features, whereas CNN targets spatial features. As seen in Fig. 1, the current HAST-IDS [17] design only connects CNN and RNN in series. When learning along a multi-layer CNN hierarchy, the learning effectiveness of the successor RNN (LSTM) is severely constrained and temporal features may be lost.

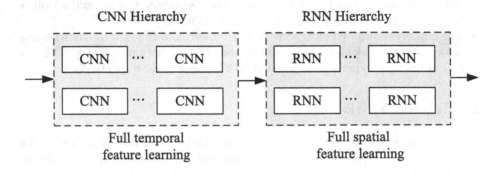

Fig. 1. HAST-IDS.

In response to the above, this paper mixes CNN and RNN subnets, adds Self-Attention, and finally learns synchronously in multiple steps, as shown in Fig. 2. Due to its ability to extract sophisticated characteristics from extensive data, this research paper prioritizes CNN over RNN. Consequently, the time information preserved in the output of CNN will be captured by RNN. Finally, Self-Attention can use attention mechanism to dynamically generate the weights of different connections, so it can be used to process variable-length input sequences. With the next step of data processing, the learning granularity becomes finer. CNN, RNN and Self-Attention can fully learn without interfering with each other.

3.1   CNN

Convolution and pooling are the two basic operations that make up CNN. A feature map is the common name for the output of convolution, which highlights the properties of the input data using a set of kernels. The activation function

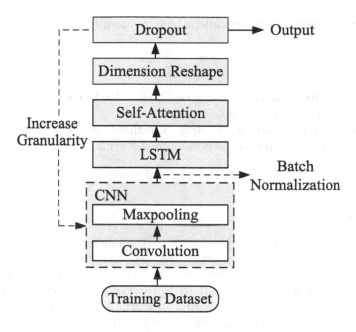

Fig. 2. CRNN-SA.

is used to further process the convolution output, and then pooling is applied to down-sample and eliminate irrelevant data. In addition, pooling aids in the removal of errors in data, thereby enhancing the learning process of subsequent layers. Its output feature map may accurately represent the original input data, CNN learns the input data by automatically modifying the filter in successive learning rounds. Since network packets exist in one-dimensional form, then one-dimensional convolution is enough, which is expressed as follows:

$$(F * G)(i) = \sum_{j=1}^{m} G(j) \cdot F(i - j + m/2) \tag{1}$$

Among them, F is the filter with the size of m, this paper chooses ReLU as the activation function.

3.2 Batch Normalization (BN)

During training, the range of input values changes dynamically layer by layer, a phenomenon known as covariance shift. An unstable learning outcome is produced by covariance shift, which makes the learning effectiveness of one layer dependent on other layers. Furthermore, because of covariance shift, it may be necessary to restrict the learning rate to a lower value in order to effectively learn the data from various input ranges, consequently decelerating the learning process. This paper uses Batch Normalization to solve the above problems and adjust CNN output in RNN module, which is shown below:

$$\hat{x} = \frac{x - \mu_B}{\sqrt{\delta_B^2 + \varepsilon}} \tag{2}$$

x is the value entered in batch, ε is a negligible value to ensure that the denominator in the formula is not zero.

The output $\hat{y}$ as shown in Eq. (3) is normalized, in which both φ and ϕ are trained in the learning process to obtain great learning results.

$$\hat{y} = \varphi \hat{x} + \phi \tag{3}$$

3.3   LSTM

In contrast to CNN, which learns information from individual data records, RNN is capable of establishing connections between data records by incorporating previous learning into current learning. This enables RNN to capture the time characteristics present in the input data. However, traditional RNNs provide basic feedback, which has the potential to accrue learning errors over time. If this happens, the final learning results may be invalidated. LSTM is a kind of gated recursive neural network, which can alleviate this kind of problem, and control the feedback through a set of gate functions, so that short-term errors will be eliminated eventually, leaving only lasting features. Therefore, this paper uses LSTM for RNN. Four sub-networks, along with a set of control gates and a memory unit. The figure contains input and output values that are vectors of equal size, determined by the input. As shown below:

$$b + U \times x(t) + W \times h(t-1) \tag{4}$$

After four subnets, two types of control gates ($\alpha, \tanh$) are used to determine the previous learning and current output $h(t)$ of feedback $s(t)$, which are specifically expressed as follows:

$$s(t) = \sigma(f(t)) * s(t-1) + \sigma(p(t)) * \tanh g(t) \tag{5}$$

$$h(t) = \tanh s(t) * \sigma(q(t)) \tag{6}$$

The input is learned by LSTM through the modification of weights and σ values in these networks, enabling the output to effectively capture the time characteristics of the input data (Fig. 3).

3.4   Self-Attention(SA)

Combining CNN with LSTM, we can use LSTM for sequence prediction while CNN for feature extraction. By introducing attention mechanism, more important parts can be selected from a large amount of information, which is conducive to improving the accuracy of intrusion detection. As shown in Fig. 4, SA can use

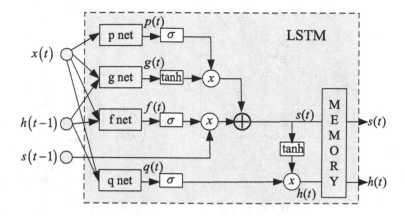

Fig. 3. LSTM data processing graph.

attention mechanism to dynamically generate the weights of different connections, so it can be used to process variable-length input sequences and can be used as a layer of neural network, as shown below:

$$Attention(Q, K, V) = soft \max \frac{K^T Q}{\sqrt{F_k}} V \tag{7}$$

where Q represents a set of query vector moments, K represents a set of key vector matrices, and V represents a vector matrix. Firstly, the point multiplication of Q and K is calculated, and divided by $\sqrt{F_k}$ to prevent the result from being too large, then the result is normalized to probability distribution by Softmax operation, and finally the weight is obtained by multiplying it by matrix V.

3.5 Output

Due to the variation in learning granularity between different levels of CNN+RNN, the output size of a level may not match the expected input size of the subsequent level. As a result, the data of the following module is reshaped using a Dimension Reshape layer. Additionally, over-fitting is a common issue when learning with deep neural networks. Dropout is used in this research to address over-fitting by randomly removing some connections from the DNN. In conclusion, an extra convolutional layer and a Global Average Pooling (GAP) layer are employed to extract additional spatio-temporal characteristics, and the ultimate learning outcome is produced through the final fully connected layer.

4   Dataset

The assessment of neural network architecture is intricately linked to the dataset employed. The evaluation results of numerous data sets collected for NID are deemed unreliable [7] due to the presence of excessive redundant data [12], which

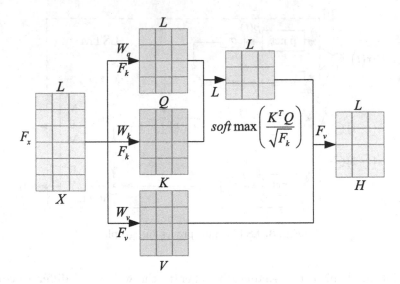

Fig. 4. Self-Attention Structure.

makes the evaluation results unreliable. In order to ensure the effectiveness of the evaluation, this paper chooses the non-redundant data set UNSW-NB15.

UNSW-NB15 data set [13] was generated by Australian Network Security Center (ACCS) in 2015. The attack samples of the data set were collected from three websites: Common Vulnerability and Exposures (CVE), Symantec Corporation (BID) and Microsoft Security Bulletin (MSD). Subsequently, the laboratory environment was used to simulate the sample attacks and create the data set. The UNSW-NB15 dataset consists of nine categories of attacks.

5 Settings

In order to evaluate the design of this paper, which implement CRNN-SA with PyTorch. For comparison, this paper also implements a set of most advanced machine learning algorithms. The description, results and discussion of the experiment are described as follows. In this experiment, RMSprop is used to optimize the weight and bias of CRNN-SA model training, and the dropout is set to be 0.5.

5.1 Data Preprocessing

In this paper, the UNSW-NB15 dataset is split according to the ratio of 7:3, and 70% of the data is used as the training dataset. The remaining 30% data is used as a test data set to evaluate the trained network intrusion detection model.

We use both binary classification and multi-classification to evaluate data sets. In the first set of experiments, there are only two kinds of results in binary classification, normal or attack. In the second set of experiments, the multi-category classification label has a range of values that can be assigned based on the attack type. Before evaluating these models, the data set is preprocessed through the following three steps.

1. Transform classification features: To ensure the experiment's effectiveness, the data must adhere to the input format specified by the neural network. The initial network data contains certain categorical features that need to be transformed. Learning algorithms are unable to process textual information and require conversion into numerical values.
2. Standardization: The input data may have different distributions of mean and standard derivative, which may affect the learning efficiency. In this paper, the input data are scaled using standardization methods so that the mean is 0 and the standard deviation is 1.
3. Hierarchical K-fold cross-validation: The UNSW-NB15 dataset used in this paper contains a total of 257,673 samples. In order to use a large amount of non-redundant data for training and validation, this paper adopts a hierarchical K-fold cross-validation strategy. In this scheme, all the samples in the data set are divided into K groups, in which K1 group is used as a whole for training and the other group is used for verification.

5.2   Evaluation Metrics

This paper evaluates CRNN-SA according to verification Accuracy (ACC), DR (Detection Rate) and FPR. ACC measures CRNN-SA's ability to correctly predict normal traffic and attacked traffic, while DR indicates the ability to predict only attacks, as shown below:

$$ACC = \frac{TP + TN}{TP + TN + FP + FN} \tag{8}$$

$$Precision = \frac{TP}{TP + FP} \tag{9}$$

$$Recall = \frac{TP}{TP + FN} \tag{10}$$

$$F1 - score = 2 \times \left(\frac{Precision \times Recall}{Precision + Recall} \right) \tag{11}$$

where TP is the number of attacks, TN is the number of normal traffic correctly classified, FP is the number of attacks wrongly classified as normal traffic, and FN is the number of attacks wrongly classified as normal traffic.

5.3   Analysis of Experimental Results

Firstly, this paper measures the performance of CRNN-SA model according to two situations: (1) Two classification, that is, CRNN-SA model predicts a data packet in only two situations: attack or normal traffic; (2) Classification with multiple classes, that is, the CRNN-SA model identifies a data packet as normal or an attack type given in the attack model of UNSW-NB15 data set (10 classes). The experimental results are described as follows:

Binary Classification. The detection results of CRNN-SA algorithm in this paper under the binary classification including Accuracy and F1-score are shown in Table 1. It can be observed that the accuracy of other comparison algorithms is between 71.6% and 88.5%. In contrast, the CRNN-SA algorithm proposed in this paper shows the best detection performance among all the comparison model methods. The Accuracy of CRNN-SA is 90.4%, and the F1-score is 91.3%.

Table 1. Result of Binary Classification.

Methods	Accuracy	F1-score	Methods	Accuracy	F1-score
LR	0.753	0.792	RF	0.877	0.912
GNB	0.716	0.818	CNN-LSTM	0.835	0.889
KNN	0.829	0.869	LSTM	0.767	0.798
DT	0.885	0.91	GRU	0.777	0.818
AdaB	0.839	0.884	DNN	0.827	0.879
CRNN-SA	**0.904**	**0.913**			

Classification with Multiple Classes. The results of classification with multiple classes are displayed in Table 2. It can be observed that the prediction accuracy of the proposed CRNN-SA algorithm for UNSW-NB15 data set reaches 89.9%, while other algorithms are in the range of 8.5% to 73.6%. In conclusion, For UNSW-NB15 data set, the CRNN-SA algorithm proposed in this paper shows the best performance compared with the RF and DT methods of multi-classification. Moreover, the F1-score of CRNN-SA is 77.5%.

Table 2. Result of Multi-class Classification.

Methods	Accuracy	F1-score	Methods	Accuracy	F1-score
LR	0.561	0.428	RF	0.736	0.695
GNB	0.085	0.130	CNN-LSTM	0.680	0.615
KNN	0.652	0.638	LSTM	0.661	0.598
DT	0.735	0.718	GRU	0.665	0.608
AdaB	0.631	0.557	DNN	0.663	0.608
CRNN-SA	**0.899**	**0.775**			

Ablation Experiment. To further investigate the efficacy of the CRNN-SA model introduced in this research, the ablation experiment was examined. The experimental results of CRNN-SA model and ablation model are shown in Table 3 and 4. As can be seen from Table 3 and 4, the detection performance of CNN model and RNN model alone is poor. On the contrary, the Accuracy and F1-score of the CRNN-SA model under the binary classification are 90.4% and 91.3%, respectively, and the metrics under the multi-class classification are 89.9% and 77.5%, respectively. In a conclusion, the ablation experiment shows that the improved CRNN-SA model has better detection performance than CNN and RNN models, and proves the effectiveness and accuracy of the CRNN-SA model.

Table 3. Binary Classification ablation experiment.

Methods	Accuracy	F1-score
CNN	0.856	0.897
RNN	0.807	0.867
CRNN-SA	**0.904**	**0.913**

Table 4. Multi-classification ablation experiment.

Methods	Accuracy	F1-score
CNN	0.684	0.627
RNN	0.662	0.587
CRNN-SA	**0.899**	**0.775**

In order to further discuss the results, this paper visualizes the Binary Classification indexes of CRNN-SA algorithm, including Accuracy, FPR, Recall and F1-score curves, for the UNSW-NB15 network attack dataset, as shown in Fig. 5, Fig. 6 and Fig. 7, respectively.

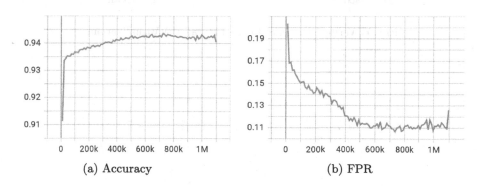

(a) Accuracy   (b) FPR

Fig. 5. Visualization diagram of Accuracy (a) and FPR (b) of training.

In this paper, the multi-classification indexes of CRNN-SA, including Accuracy, FPR, Recall and Loss curves, are visualized in the UNSW-NB15 network attack data set, as shown in Fig. 8 and Fig. 9. Additionally, the CRNN-SA Confusion Matrix Result for the UNSW-NB15 dataset is illustrated in Fig. 10.

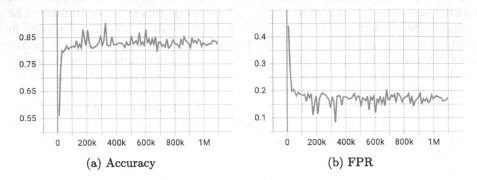

(a) Accuracy   (b) FPR

Fig. 6. Verified Accuracy (a) and FPR (b) Visualization.

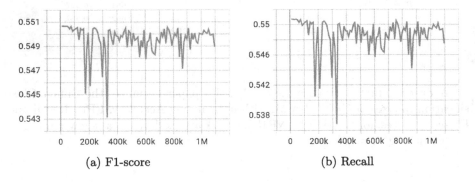

(a) F1-score   (b) Recall

Fig. 7. Verified F1-score (a) and Recall (b) Visualization.

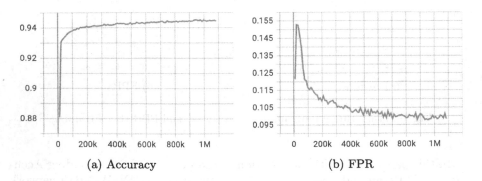

(a) Accuracy   (b) FPR

Fig. 8. Visualization diagram of Accuracy (a) and FPR (b) of training.

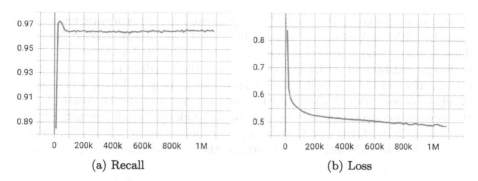

(a) Recall   (b) Loss

Fig. 9. Visualization of Recall (a) and Loss (b) of training.

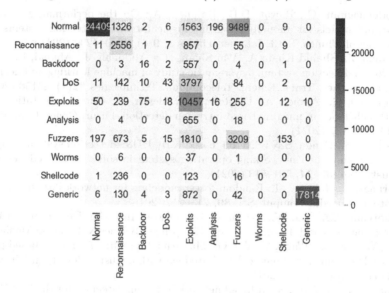

Fig. 10. The confusion matrix result for the UNSW-NB15 dataset is provided by CRNN-SA.

6 Conclusion

In order to identify intrusions on large-scale networks, a CRNN-SA based on DNN architecture is developed in this research. CNN and LSTM are used by CRNN-SA to learn the spatial and temporal properties of network traffic data, respectively. This work investigates the input data at the same granularity by synchronizing CNN and RNN in order to prevent the information loss brought on by the differing learning emphases of CNN and RNN. In order to express useful input information better, Self-Attention is presented to assign different weights to the fused features, so that the spatial and temporal features of data can be extracted effectively. In addition, in order to strengthen learning, batch normalization is added to the model. A various of experiments are carried out on the non-redundant data set UNSW-NB15, and the experimental results show that the CRNN-SA model can effectively utilize CNN, LSTM and Self-Attention. The Accuracy and F1-score of the CRNN-SA model under the binary classification are 90.4% and 91.3%, respectively, and the metrics under the multi-class classification are 89.9% and 77.5%, respectively. Compared with other optimal baseline models, the CRNN-SA model proposed in this paper can greatly improve the verification accuracy and decrease the false alarm rate of network intrusion detection. In the future work, this paper will measure the training and testing time of the selected methods to determine the best performance in efficiency.

References

1. Abdelmoumin, G., Rawat, D.B., Rahman, A.: On the performance of machine learning models for anomaly-based intelligent intrusion detection systems for the Internet of Things. IEEE Internet Things J. **9**(6), 4280–4290 (2021)
2. Ahmad, Z., Shahid Khan, A., Wai Shiang, C., Abdullah, J., Ahmad, F.: Network intrusion detection system: a systematic study of machine learning and deep learning approaches. Trans. Emerg. Telecommun. Technologies **32**(1), e4150 (2021)
3. Andresini, G., Appice, A., De Rose, L., Malerba, D.: Gan augmentation to deal with imbalance in imaging-based intrusion detection. Futur. Gener. Comput. Syst. **123**, 108–127 (2021)
4. Badotra, S., Panda, S.N.: SNORT based early DDoS detection system using open-daylight and open networking operating system in software defined networking. Clust. Comput. **24**, 501–513 (2021)
5. Farnaaz, N., Jabbar, M.: Random forest modeling for network intrusion detection system. Procedia Comput. Sci. **89**, 213–217 (2016)
6. Ghorbani, A.A., Lu, W., Tavallaee, M.: Network Intrusion Detection and Prevention: Concepts and Techniques, vol. 47. Springer Science & Business Media (2009)
7. Hu, W., Gao, J., Wang, Y., Wu, O., Maybank, S.: Online Adaboost-based parameterized methods for dynamic distributed network intrusion detection. IEEE Trans. Cybern. **44**(1), 66–82 (2013)
8. Jing, D., Chen, H.B.: SVM based network intrusion detection for the UNSW-NB15 dataset. In: 2019 IEEE 13th International Conference on ASIC (ASICON), pp. 1–4. IEEE (2019)
9. Khan, M.A., Iqbal, N., Jamil, H., Kim, D.H., et al.: An optimized ensemble prediction model using autoML based on soft voting classifier for network intrusion detection. J. Netw. Comput. Appl. **212**, 103560 (2023)
10. Li, R., Li, Q., Zhou, J., Jiang, Y.: ADRIoT: an edge-assisted anomaly detection framework against IoT-based network attacks. IEEE Internet Things J. **9**(13), 10576–10587 (2021)
11. Marteau, P.F.: Random partitioning forest for point-wise and collective anomaly detection-application to network intrusion detection. IEEE Trans. Inf. Forensics Secur. **16**, 2157–2172 (2021)
12. McHugh, J.: Testing intrusion detection systems: a critique of the 1998 and 1999 DARPA intrusion detection system evaluations as performed by Lincoln laboratory. ACM Trans. Inf. Syst. Secur. (TISSEC) **3**(4), 262–294 (2000)
13. Moustafa, N., Slay, J.: The evaluation of network anomaly detection systems: statistical analysis of the UNSW-NB15 data set and the comparison with the KDD99 data set. Inf. Secu. J. Glob. Perspect. **25**(1–3), 18–31 (2016)

14. Mukherjee, B., Heberlein, L., Levitt, K.: Network intrusion detection. IEEE Netw. **8**(3), 26–41 (1994). https://doi.org/10.1109/65.283931
15. Park, C., Lee, J., Kim, Y., Park, J.G., Kim, H., Hong, D.: An enhanced AI-based network intrusion detection system using generative adversarial networks. IEEE Internet Things J. **10**(3), 2330–2345 (2022)
16. Qi, L., Yang, Y., Zhou, X., Rafique, W., Ma, J.: Fast anomaly identification based on multiaspect data streams for intelligent intrusion detection toward secure industry 4.0. IEEE Trans. Ind. Inf. **18**(9), 6503–6511 (2021)
17. Wang, W., et al.: HAST-IDS: learning hierarchical spatial-temporal features using deep neural networks to improve intrusion detection. IEEE Access **6**, 1792–1806 (2017)
18. Yu, L., et al.: PBCNN: packet bytes-based convolutional neural network for network intrusion detection. Comput. Netw. **194**, 108117 (2021)
19. Zhou, X., Hu, Y., Liang, W., Ma, J., Jin, Q.: Variational LSTM enhanced anomaly detection for industrial big data. IEEE Trans. Industr. Inf. **17**(5), 3469–3477 (2020)

Applications

Conspiracy Spoofing Orders Detection with Transformer-Based Deep Graph Learning

Le Kang[1,2(✉)], Tai-Jiang Mu[1], and Xiaodong Ning[2]

[1] Department of Computer Science and Technology, Tsinghua University, Beijing, China
kangl18@mails.tsinghua.edu.cn, taijiang@tsinghua.edu.cn
[2] Zhengzhou Commodity Exchange, Zhengzhou, China
xdning@czce.com.cn

Abstract. Spoofing behaviors profits from other participants via placing large amounts of deceptive orders to create misleading signals with the help of advanced electronic trading techniques, severely harming the healthy and sustainable development of modern financial markets. Existing works focus mostly on detecting individual spoofing issuers or orders, which ignore the capability of mining hidden conspiracy fraud patterns. Consequently, they face significant challenges in tracking organized spoofing behaviors. To this end, in this paper, we proposed a novel transformer-based graph neural network to learn the temporal interconnected transaction representations for conspiracy spoofing behavior detection. In particular, we model user orders in the financial market by the temporal transaction graph layer first, which produces temporal and relational encodings of the user's order behaviors. Then, we devise a transformer-based neural network to automatically learn the deeper sequential and inter-connected representations, which could perceive a boarder view of the user's transaction patterns. Finally, the temporal and relational features are jointly optimized by the detection network in an end-to-end manner. We conduct extensive experiments on a real-world dataset from one of the largest financial exchange markets in East Asia. The result strongly proves the superior performance of our proposed method in fighting against conspiracy spoofing behaviors with higher accuracy without handcraft feature engineering, compared with ten state-of-the-art baselines. To the best of our knowledge, this is the first work that proposed a transformer-based deep graph learning approach for spoofing detection challenges and demonstrates its practical ability in real-world applications.

Keywords: Spoofing Detection · Graph Neural Network · Transformer Neural Network

1 Introduction

Spoofing [26], also known as "bait and switch", refers to serious market manipulation in the financial market such as stock markets, futures markets, bond

X. Yang et al. (Eds.): ADMA 2023, LNAI 14177, pp. 489–503, 2023.
https://doi.org/10.1007/978-3-031-46664-9_33

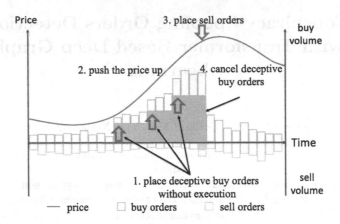

Fig. 1. A typical procedure of spoofing transactions. Fraudsters create lots of fake LOB balances and cancel them on the bait side to gain illegal profit from the financial market.

markets, etc. It is a federal felony in the United States [27] and forbidden by all major financial markets worldwide [23]. With the rapid development of electronic trading techniques, the occurrence of spoofing orders [25] shows a growth trend and draws greater attention from financial regulation in many countries. For example, in the past few years, the U.S. Commodity Trading Futures Commission (CFTC) has dealt with an increasing number of spoofing cases, which punished the violator over $900 million for a single case in the year of 2020[1]. There is no doubt that spoofing orders can have devastating effects on healthy financial markets, social well-being, as well as sustainable economic development. Therefore, it is of great significance to detect spoofing orders, which has attracted the great attention of academia and industry.

Spoofing is a market abuse behavior where a trader submits a number of orders at different price levels on the bait side with no intention of ever seeing them executed, aiming to push the transaction price in a favorable direction. Subsequently, the trader will switch to the opposite side and trade at a particular price range. Finally, the majority of the virtual orders on the bait side will be withdrawn. Via this method, the trader can buy or sell an asset at a better price, artificially inflate the supply and demand of an asset, and gain illegal profit. Figure 1 illustrates a typical procedure of spoofing, the trader places orders on the sell side (green) and places a series of orders (red) at a different price on the buy side, aiming to give the impression of substantial supply/demand and drive the prices up. After the genuine order trades, the multiple buy orders will be rapidly withdrawn. By spoofing, the trader can buy or sell an asset at a better price and gain illegal profit. This spoofing strategy creates a lot of fake limited order book (LOB) balances, which bring significant threats to the fair

---

[1] https://www.cftc.gov/PressRoom/PressReleases/8260-20.

systematical, and healthy of the financial market by harming the interest of a broad range of innocent mid-and-small investors.

The financial industry has developed spoofing detection since the early 1970s [22], from the statistical compliance rules to conventional machine learning approaches [5]. Recently, deep and sequential learning techniques have been introduced for spoofing detection [28]. For example, most researchers leverage time series learning methods to quantify the impact and cost of spoof activity for better regulation policy [9]. [21] tried to detect occurrences of spoofing activities by correlation analysis. However, an increasing number of criminals have been organized like enterprises recently, which can be far-reaching and move quickly from one asset to another. To fight against these human brain-armed criminal behaviors, most existing spoofing detection methods still focus on mining sequential patterns from historical orders individually, facing significant challenges in learning meaningful representations from temporally inter-connected conspiracy patterns.

Therefore, in this paper, we present a novel transformer-based deep graph learning method, named RTG-Trans, to detect the conspiracy spoofing trading behaviors by jointly learning the inter-connected relational feature and complicated temporal features. In particular, the relational features are learned by the proposed graph neural layer, which directly models the user's order relations from the constructed transaction graph. The temporal features signify the temporal characteristics inherent in transaction actions, which are generated by concatenating the features of each transaction action. Then, we model the encodings of the users' sequential transactions by transformer-based sequential learning into temporal representations. Finally, the inter-connected relational representations and complicated temporal features are jointly optimized by an end-to-end detection network so that the model can effectively capture the complicated (temporally inter-connected) spoofing patterns. Extensive experiments on a real-world dataset from one of the largest financial exchange markets in East Asia strongly demonstrate the effectiveness of our proposed method, compared with ten state-of-the-art baseline methods. Due to its superior performance and practical capability, the proposed approach is now on the way to becoming a fundamental component in the real-world spoofing detection system. In brief, our main contributions are summarized as follows:

- We propose a novel transformer-based deep graph learning method for spoofing detection by jointly capturing relational features and temporal features. To the best of our knowledge, this is the first work that addresses the spoofing detection problem by transformer-based deep graph learning.
- To effectively capture the temporally inter-connected spoofing patterns, we propose the deep graph neural layer for relational pattern learning and transformer-based sequential layer for temporal feature learning, which are then jointly optimized by an end-to-end detection network.
- Our approach is extensively evaluated in the analytic system of one of the largest financial exchange markets in East Asia. The result strongly demon-

strates the effectiveness and the superiority of our proposed method, compared with ten state-of-the-art baselines.

2 Related Works

2.1 Spoofing Detection

In recent years, spoofing [1] has been studied by many researchers. [19] investigated two popular scenarios of stock price manipulations: pump-and-dump and spoofing trading. They utilized statistical methods to define the two kinds of trading and tried to use a neural network to model them. The experimental results show that the model achieves 88.28% accuracy for detecting pump-and-dump, but it fails to model spoof trading effectively. [33] presented an agent-based model of manipulating prices in financial markets, realizing the spoofing trading mechanism. [21] utilized the rule-based algorithm and set the corresponding parameters to detect potential spoofing cases in ten stocks from the Ibovespa index. [5] presented a model that simulates the trading tactics of an investor who engages in spoofing of the limit order book with the goal of boosting profits generated from selling a security. [28] offered a micro-structural examination of spoofing in a simple static scenario. They introduce a multilevel imbalance that impacts the resulting price action and proceed to outline the optimization strategy utilized by a potential spoofer. In addition to the above work, some researchers focused on utilizing Markov Decision Model (MDP) based methods to model the different trading strategies. For example, [35] modeled the trading on the futures market as a MDP, and proposed the IRL algorithm based on linear programming to characterize the behaviors of high-frequency traders and identify the manipulative spoofing strategy. [4] proposed an adaptive hidden Markov model with anomaly states for modeling price manipulation activities. However, the series of MDP-based methods pay more attention to strategy modeling instead of spoofing detection. Furthermore, few existing works focus on the effective learning of temporal and inter-connected features of spoofing order detection.

2.2 Deep Feature Learning on Financial Data

With the continuous development of artificial intelligence technology [24], graph deep learning has gained much attention. Machine learning techniques [2] based on graphs are not only widely used in the fields of image, natural language processing, knowledge graph, and network security, but it has also proven to be extremely effective and reliable in financial tasks [3]. [30] put forward a fresh framework called the HITS-based fraudulent phone call detection system, which uses data mining of users' telecommunication records to uncover fraudulent phone calls. They investigated duration-relatedness and frequency-relatedness by leveraging telecommunications to extract descriptive features that can be used to weigh the Common Phone Graph (CPG) and Unique Phone Graph (UPG).

[12] introduced a fraud detection framework based on Convolutional Neural Networks, which is designed to capture the inherent patterns of fraudulent behaviors learned from labeled data. A wealth of transaction data is expressed as a feature matrix, over which a CNN is employed to identify a collection of underlying patterns for each sample. [9] proposed a temporal attention-based graph network (STAGN) for credit card fraud detection. In particular, they first learned temporal and location-based transaction graph features by graph neural networks. [8] first attempted to provide loan risk assessment by adding a high-order graph attention representation to predict loan delinquencies and warn of domino effect loan crises. To enhance GNN-based fraud detectors against fraudsters' feature disguise and relationship disguise, [11] proposed a label-aware similarity metric and similarity-aware neighbor selector using reinforcement learning. To solve the sequence-based fraud detection problem, [40] proposed a Hierarchical Interpretable Network to model the user's behavioral sequences, which not only improves the performance of fraud detection but also gives a reasonable interpretation of the prediction results. Transformer models also have achieved impressive success in finance, such as [10] proposed a gaussian transformer to predict stock movement and [34] utilized transformer for volatility prediction. Though existing methods have achieved a decent performance, little research has been done on spoofing transaction detection by jointly learning the temporal and relational features. To the best of our knowledge, this is the first work that learn the potential relational-temporal representations of transaction behavior to automatically detect spoofing activities with a transformer-based deep graph neural network.

3   The Proposed Method

3.1   Preliminaries

Transaction Record. For a set of transaction records $R = \{r_1, r_2, \cdots, r_n\}$, each transaction record r can be defined as a tuple of attributes $r = \{u, a, f\}$, composed of a user ID u, the set of transaction actions $a = \{a_1, a_2, \cdots, a_m\}$ of u, and the features $f = \{f_1, f_2, \cdots, f_m\}$, where m is the number of actions, and $f_i = (f_i^1, f_i^2, \cdots, f_i^k)$, is the k-dimensional feature dimension of each action. In the preprocessing, we normalize all features of all users.

Problem Definition. A spoofing event d in this paper is an illegal transaction that distorts the shape of the limit order book for the purpose of misleading others into trading. Thus, the spoofing event is a special type of transaction, which means that it also retains the $\{u, a, f\}$ attributes. The complete real-world spoofing event data provided by our partner organizations provide us with a unique opportunity to solve the fraud detection problem. In summary, we now formalize our spoofing detection problem as follows. Given a set of transaction records $R = \{U, A, F\}$, a set of spoofing events D, which are subsets of the set of transactions $\{D|D \subset R\}$. For each transaction, we want to infer the probability

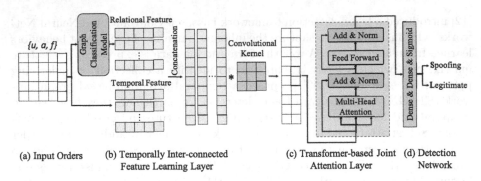

Fig. 2. Overall architecture of the RTG-Trans model. The model contains four modules: the input orders, the temporally inter-connected feature learning layer, the transformer-based joint attention layer, and the detection network.

of whether it is a spoofing event based on the transaction records and features. The goal is to achieve high accuracy in spoofing prediction, as well as to explore spoofing patterns in transactions.

3.2   Model Architecture

Figure 2 shows the network architecture of our proposed model RTG-Trans. First, our model takes the user's transaction records as input and feeds them into two different higher-order tensor spaces to capture both temporal features and relational features. Next, we use the convolutional layer to obtain the fused transaction representation vector, since the convolutional layer helps model the hidden patterns of transactions. Then, we use the Transformer to extract information by strengthening the important features and weakening the unimportant ones. Finally, we reshape the learned feature representation from tenors to vectors for the spoofing detection task by a detection network. We will first introduce the components of the model, and then discuss the setup and optimization of the detection layer in the following sections.

3.3   Temporally Inter-connected Feature Learning Layer

Temporal Feature. The transaction actions are temporal sequence data, which has temporal information. To reflect the temporal characteristics of transaction actions, here we construct temporal features in a simple way, i.e. stacking the features of each action f_i in chronological order as temporal features. We represent the temporal characteristics of each transaction record of a user as fea_{tem}, where $fea_{tem} = [f_1, f_2, \cdots, f_m]^T$ and $f_i \in \mathbb{R}^k$, $fea_{tem} \in \mathbb{R}^{m \times k}$. In this paper, for each trading action, we extract features including buy or sell, open or close, quantity, price, level and volume of the order, and volume at the 5 best buy prices and 5 best sell prices on the LOB.

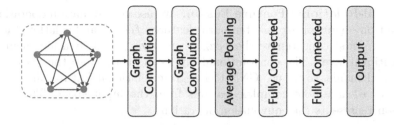

Fig. 3. The graph classification model for spoofing detection.

Relational Feature. GCN [7] is a graph neural network structure that has become popular in recent years, which is a natural extension of the convolutional neural network in the graph domain. In our method, all transaction records of each user constitute a graph. Each node in the graph is a transaction record, and the representation of nodes is represented by the data in the transaction records, and the relationship between nodes is the sequential relationship of transaction records, so the graph we construct is a one-way graph. To determine whether there is spoofing in all transaction records of a user to perform graph binary classification on the graph composed of all transaction records. The binary classification model of the graph we constructed is shown in Fig. 3, where the input is the graph represented by its adjacency matrix and node feature matrix. The first two layers are graphical convolution with multiple neural units and ReLU activation functions per layer. The next layer is the Average Pooling layer, where the learned node representations are aggregated to create graph representations. The graph representation is fed into two fully connected layers, each of which also has multiple neural units and ReLU activation functions. The final layer is the output layer with one unit and a sigmoid activation function.

The binary classification of a graph is defined as follows. We use A to denote the adjacency matrix of a graph, m is the number of nodes, and each node has a k-dimensional feature vector. We use $X \in \mathbb{R}^{m \times k}$ to denote the node information matrix of the graph, with each row representing a node. Given a graph A and its node information matrix X, our graph convolutional layer takes the form $Z = f(\widetilde{D}^{-1}\widetilde{A}XW)$ where $\widetilde{A} = A + I$ is the adjacency matrix of the graph with added self-loops, $\widetilde{D}$ is its diagonal degree matrix with $\widetilde{D}_{ii} = \sum_j \widetilde{A}_{ij}$, $W \in \mathbb{R}^{k \times k'}$ is a matrix of trainable graph convolution parameters, f is a nonlinear activation function, and $Z \in \mathbb{R}^{m \times k'}$ is the output activation matrix. In this Average Pooling layer, the input is the output of the GCN layer Z_{gcn}, where each row is a vertex's feature descriptor and each column is a featured channel. The Average Pooling outputs a tensor Z_{mp}. After Average Pooling, we add a fully-connected layer followed by a softmax layer. When the graph classification model is trained, we use the output of the second fully connected layer as a representation of each node. The representations of these nodes capture the node relationship information, which we denote as $fea_{gcn} \in \mathbb{R}^{m \times k}$.

Temporal-Relational Feature Fusion. We use concatenation operation and convolution operation to fuse temporal features fea_{tem} and relational features fea_{gcn}. We first concatenate the fea_{tem} and fea_{gcn} and then fed them to the convolutional layer. The convolutional layer can clearly fuse the features from different domains, so we use CNN to integrate relational features and temporal features to learn more complex features from the input space. The following equation represents the convolution operation:

$$fea_j^c = \sum_j fea^{c-1}(c_t - c_t', c_s - c_s')\omega_i^c(c_t', c_s') \tag{1}$$

where ω_i^c is the kernel in the cth layer and i^{th} kernel which convolves over the feature fea^{c-1}. The first layer of fea^{c-1} is the splicing of fea_{tem} and fea_{gcn}. c_t and c_s are the dimension of fea_{tem} and fea_{gcn}, which are equal to m and k of the first convolutional layer. ω_i^c is the element-wise weight in the convolutional kernel (c_t', c_s'). Thus, the output feature fea_{cnn} is calculated by $fea_{cnn} = \sigma(\sum_j fea_j^c + b^c)$, where b^c is the bias parameter and σ denotes the sigmoid function. The output feature fea_{cnn} is then resized and fed to the Transformer.

3.4　Transformer-Based Joint Attention Layer

Transformer [32] is a loop-avoiding model structure that relies entirely on the attention mechanism to model the global dependencies of inputs. Because the modeling of dependencies relies entirely on the attention mechanism, the attention mechanism [6] used by Transformer is called self-attention [39]. In this paper, we use self-attention to enhance the important features and weaken the unimportant features among relational features and temporal features. The self-attention can be described as:

$$\text{Attention}(Q, K, V) = \text{softmax}(\frac{QK^T}{\sqrt{d}})V, \tag{2}$$

where $Q \in \mathbb{R}^{m \times k}$, $K \in \mathbb{R}^{m \times k}$, and $V \in \mathbb{R}^{m \times k}$ denote the query matrix, key matrix, and value matrix respectively. We set $Q = K = V$, and set d equal to k. The multi-head attention first linearly projects the queries, keys, and values h times via using different linear projections, and the scaled dot-product attention is performed on h attention layers in parallel. Finally, the results of scaled dot-product attention are concatenated and once again projected to get a new representation. Formally, multi-head attention can be defined as follows:

$$head_i = \text{Attention}(QW_i^Q, KW_i^K, VW_i^V) \tag{3}$$

$$fea_{trans} = \text{Concat}(head_1, \cdots, head_h)W_o \tag{4}$$

where $W_i^Q \in \mathbb{R}^{d_h \times d_k}$, $W_i^K \in \mathbb{R}^{d_h \times d_k}$, $W_i^V \in \mathbb{R}^{d_h \times d_k}$ and $W_o \in \mathbb{R}^{2d_h \times 2d_h}$ are trainable projection parameters, d_h is the dimension of values and d_k is the dimension of queries or keys in attention, and $d_k = d_h$. The representations obtained after using the self-attention mechanism are denoted as fea_{tra}.

3.5   The Detection Network and Optimization

In the detection network, the input is the fea_{tra}, which is learned by the convolutional layer and Transformer. The detection network aims to learn the probability of whether it is a spoofing. Hence, we employ a simple two-layer fully connected network with sigmoid activation as the detection network. The loss function is the likelihood defined as follows:

$$Loss = -\frac{1}{n} \sum_{i=1}^{n} [y_i \log detect(fea_{tra}) + (1 - y_i) \log(1 - detect(fea_{tra}))] \quad (5)$$

where n is the number of the transaction record, y_i denotes the label of i-th records, which is 1 if the record is spoofing and 0 otherwise. $detect(fea_{tra})$ is the detection function that maps fea_{tra} to a real-valued score, indicating the probability of whether the current transaction is spoofing. In this paper, our model can be optimized by standard SGD-based algorithms, and we use Adam Optimizer to learn the parameters.

4   Experiments

4.1   Experimental Settings

Dataset. To the best of our knowledge, there does not exist any available public spoofing detection dataset with ground-truth labels. Therefore, we collect large-scale labeled spoofing records from one of the largest commodity futures exchange markets in East Asia. The dataset records real-world users' transactions spanning from January 1, 2019 to June 5, 2020. We chose historical records from 58 contracts of 17 futures, as these futures have the most actively traded futures contracts that covered a wide range of traders. The trading data contains order data, transaction data, and order book data. The order data reflects the interest of the trader to buy, sell or cancel the asset at a particular price, the transaction data reflects the matching result of buy and sell orders, and the order books show the price and volume of the currently untraded buy and sell orders, which clearly reflects the predominating market sentiment. It should be noted that the order books provide five levels of depth, i.e., the price and volume on the 5 best buy prices and the 5 best sell prices. Experienced supervisors have labeled the spoofing traders, including the spoofing trading time and the corresponding orders. We conducted statistical analysis on the labeled spoofing trading data and found that the vast majority of spoofing activities contain no more than 50 orders. We thus treat trading records as a sequence and the length of each sequence is set to 50. In order to model the spoofing characteristics, we extract 16-dimensional features including order-related information and corresponding order book information. The final feature contains buy or sell, open or close, quantity, price, level and volume of the order, and volume at the 5 best Buy prices and 5 best Sell prices on the limited order books. The dataset consists of a total of 11,790 records, and we randomly split the dataset into 66%

for training and 34% for testing. According to the strict confidentiality policy and non-disclosure agreement, we cannot directly release this dataset due to the user's privacy protection. But we are working on an agreement to simulate an artificial dataset by deep models, such as GraphRNN [36]. Thus, the simulated data could preserve a similar distribution to this real-world dataset. We plan to release them publicly after passing the legal process. We believe it is very valuable for the research community and could contribute a lot to future work in the literature.

Baselines. For comparison, we tested both the traditional machine learning methods and deep learning models on our benchmark dataset to demonstrate the superiority of our proposed RTG-Trans. **Logistic Regression (LR)** [13], which uses logistic regression to detect spoofing, taking transaction records information as features. **Random Forest (RF)** [38], a decision tree-based ensemble learning method that could be employed for spoofing detection. **Adaboost** [17], which is an iterative algorithm that trains different weak classifiers for the same training set, and then aggregates these weak classifiers to form a stronger final strong classifiers for detecting spoofing. **Gradient Boosting Decision Tree (GBDT)** [37], which is a gradient boosting method for optimizing classification metrics and efficiently handling mixed types of data. **Multilayer Perceptron (MLP)** [15], which is a feed-forward network with two hidden layers for detecting spoofing. **Long Short Term Memory (LSTM)** [14], which encodes transaction records information with an LSTM neural network to detect spoofing. **EigenGCN** [20], which learns the correlations between different transaction records with a GCN neural network for detecting spoofing. **BiTransformer** [29], which is a loop-avoiding model structure that relies entirely on attention mechanisms to model the global dependencies of inputs and outputs. **RetaGNN** [16], which presents a novel Relational Temporal Attentive Graph Neural Network model to learn the mapping from a local graph of the given user-item pair to their interaction score and to train the learnable relation weight matrices. **GRU-DM** [31], which is based on a highly extendable Gated Recurrent Unit model and allows the inclusion of market variables that can explain spoofing and potentially other illicit activities.

Evaluation Metrics. Spoofing detection is a binary classification task. In real futures trading, to detect the presence of spoofing behavior for the full volume of transactions, and in addition to the Accuracy indicator, the AUC indicator is added to evaluate the classification ability of the model, the Precision indicator evaluates the accuracy of the model prediction, the Recall indicator evaluates the detection ability of the model for spoofing behavior, and the F1 indicator is the combination of Precision and Recall.

Parameter Setting. We use the Adam [18] method for optimization, with the learning rate set to 0.001 and the training count set to 100. The LR, RF,

Table 1. Performance comparison with baselines. The best number is in boldface and the second best is underlined.

Method	AUC	Accuracy	Precision	Recall	F1	Model Size
LR	0.7087	0.6925	0.8168	0.5952	0.6886	201
RF	0.7339	0.7165	0.8498	0.6118	0.7115	12,120
Adaboost	0.7642	0.7551	0.8447	0.6998	0.7654	7,272
GBDT	0.7587	0.7368	<u>0.9026</u>	0.6044	0.7240	13,635
MLP	0.7796	0.7228	0.8499	0.7313	0.7862	15,201
LSTM	0.8264	0.8167	**0.9050**	0.7589	0.8255	73,601
EigenGCN	0.8292	0.8308	0.8605	0.8398	0.8501	65,793
BiTransformer	0.8091	0.8120	0.8396	0.8293	0.8344	95,714
RetaGNN	<u>0.8398</u>	<u>0.8332</u>	0.8463	<u>0.8792</u>	<u>0.8624</u>	65,203
GRU-DM	0.8320	0.8282	0.8514	0.8551	0.8533	56,301
RTG-Trans	**0.8469**	**0.8523**	0.8607	**0.8845**	**0.8724**	154,937

Adaboost and GBDT are implemented using scikit-learn library. In LR, the parameter C is set to 0.5. In RF, Adaboost and GBDT, the parameter max_depth is set to 5 and the parameter n_estimators is set to 10. MLP, LSTM, EigenGCN, BiTransformer, RetaGNN, GRU-based and RTG-Trans are implemented using PyTorch, and the batch size is set to 128. The MLP model has two layers, the number of neurons in the first layer is 200, and the number of neurons in the second layer is 50. The dimension of the hidden layer in LSTM is set to 100 and the layer of LSTM is two. The number of layers of the GRU-based model is three layers, the first two layers are GRU, the last layer is a fully connected layer, and the dimensions are 128, 64, and 50 respectively. In EigenGCN, the first two layers are graph convolution with 8 units and ReLU activation per layer, and the last two fully connected layers have (8, 8) units and ReLU activation, respectively. The final layer is the output layer with a single unit and sigmoid activation. In the BiTransformer and RTG-Trans, the number of attention headers is set to 4, and the dimensions of key, query, value are 768. The parameter sizes of all models are shown in the last column of Table 1.

4.2 Results and Analysis

We evaluate the performance of different models in spoofing detection tasks. We repeated the experiments 10 times and the average experimental results are shown in Table 1. As can be seen from the table, in general, our method outperforms all competitors, including recent methods such as RetaGNN and GRU-based. Our method achieves the best results in most of the metrics, except Precision, indicating that our method is very effective in spoofing detection. In all baselines, the deep learning-based approach (MLP, LSTM, EigenGCN, BiTrans-former, RetaGNN, GRU-based and RTG-Trans) outperforms the machine-

Table 2. Ablation Studies. "RTG-Trans-no-X" means removing module X from our full model RTG-Trans.

Method	AUC	Accuracy	Precision	Recall	F1
RTG-Trans	**0.8469**	**0.8523**	**0.8607**	**0.8845**	**0.8724**
RTG-Trans-no-TEMP	0.7805	0.7973	0.8566	0.7396	0.7938
RTG-Trans-no-GCN	0.7458	0.7418	0.8090	0.7173	0.7604
RTG-Trans-no-CNN	0.8306	0.8380	0.8415	0.8827	0.8616
RTG-Trans-no-Trans	0.7802	0.7874	0.8583	0.7370	0.7930

learning-based approach (LR, RF, Adaboost and GBDT), demonstrating the need for deep models for spoofing detection. The results of EigenGCN are slightly better than LSTM, probably due to the fact that the information about the relationship between orders can be extracted by EigenGCN, and this information is more useful for spoofing detection. Although BiTransformer works better than LSTM in tasks concerning text information processing, in our experiments, LSTM gives better results than BiTransformer. The possible reason is that our feature dimension is relatively small, only 16 dimensions, which is much less than the hundreds of dimensions in text processing tasks, and may be overfitting when using BiTransformer to extract features in our task.

We also counted and compared the training time of the model, LR and RF are about 15 min, Adaboost, GBDT and MLP are about 40 min, LSTM is about 50 min, EigenGCN is about 60 min, BiTransformer is about 68 min, RetaGNN is 75 min, GRU-DM is about 65 min, RTG-Trans is about 70 min. In the inference stage, all models are at the second level, and there is not much difference. Other models are around 0.5 s, our model is around 0.8 s. Although the training time of the model in this paper has been lengthened, the prediction accuracy is the best.

4.3   Ablation Study and Case Study

To show the effectiveness of each module in our model, we conducted ablation experiments. We removed each core module from our full model separately, and the experimental results are shown in Table 2. When the module GCN is removed, the accuracy decreases the most, indicating that the information about the relationship between orders extracted from the GCN module is very important. When the Temporal or Transformer module is removed, the accuracy decreases as well, indicating that these two modules are also influential. Overall, the removal of different modules from our proposed model leads to different degrees of performance degradation, which indicates that each module in our proposed model contributes to the overall spoofing detection performance.

We also give an example of spoofing, as shown in Fig. 4a. The client submitted 800 lots of buy and open orders at 10:25:03, 04, 09, 11, etc., and at the same time made a sell and open order in the opposite direction at 10:25:14 and completed the transaction, and canceled the order immediately after the transaction. This

Time	Trade Type	Mount	Price	Is Deal?
10:25:03	BO	200	2046	no
10:25:04	BO	200	2046	no
10:25:09	BO	200	2048	no
10:25:11	Withdrawal(BO)	200	–	yes
10:25:11	BO	100	2050	no
10:25:11	BO	100	2050	no
...				
10:25:14	SO	600	2054	yes
10:25:15	Withdrawal(BO)	800	–	no

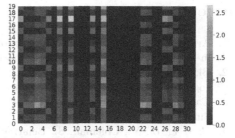

(a) The plot of the data sample.   (b) The visualization of the extracted features of the model.

Fig. 4. The pictures of the case study.

behavior has the typical characteristics of spoofing. The purpose of the buyer's order is to promote the price increase, which is conducive to the customer's order in the opposite direction and the transaction. We put these data into the model for inference, and visualize the features of these features in the penultimate layer of the model, as shown in Fig. 4b. The horizontal coordinate is the feature dimension of the data, and the vertical coordinate is the number of entries of the data. As can be seen in the subplots, there are several rows of features that are clearly different from the other rows, and these rows correspond to the label of spoofing. This is a good indication that the features extracted by our method are very effective in detecting whether they are spoofing or not.

5   Conclusion

In this paper, we propose a novel transformer-based deep graph learning approach to address the inter-connected temporal pattern representation challenges in real-world conspiracy spoofing detection. We achieve this goal by devising a graph neural layer to directly model the user's order relations from the constructed transaction graph and proposing transformer-based sequential learning to learn users' sequential transaction representations. The relational representations and temporal features are jointly optimized by an end-to-end detection network so that the model can effectively capture the inter-connected temporal pattern spoofing patterns. We conduct extensive experiments on a real-world dataset from one of the largest financial exchange markets in East Asia. The result shows that our method achieves superior performance compared to ten state-of-the-art baseline methods. Now, the proposed approach is on the way to becoming a fundamental component in the deployed online spoofing detection system. To the best of our knowledge, this is the first work to solve the spoofing detection problem by a transformer-based deep graph neural network. We are willing to discuss and share the research results and source codes with the research community and regulatory agencies. We believe it is very valuable and could inspire more works in the literature to protect the health of modern financial markets and the interest of a broad range of mid-and-small investors.

References

1. Arora, S., Bhatia, M.: Fingerprint spoofing detection to improve customer security in mobile financial applications using deep learning. AJSE, pp. 2847–2863 (2020)
2. Bkassiny, M., Li, Y., Jayaweera, S.K.: A survey on machine-learning techniques in cognitive radios. IEEE Commun. Surv. Tutorials, 1136–1159 (2012)
3. Cao, L.: Ai in finance: challenges, techniques, and opportunities. ACM Comput. Surv. (CSUR) **55**(3), 1–38 (2022)
4. Cao, Y., Li, Y., Coleman, S., Belatreche, A., McGinnity, T.M.: Adaptive hidden Markov model with anomaly states for price manipulation detection. TNNLS, pp. 318–330 (2014)
5. Cartea, Á., Jaimungal, S., Wang, Y.: Spoofing and price manipulation in order-driven markets. Appl. Math. Finance **27**(1–2), 67–98 (2020)
6. Chen, J., Zhuang, F., Hong, X., Ao, X., Xie, X., He, Q.: Attention-driven factor model for explainable personalized recommendation. In: SIGIR, pp. 909–912 (2018)
7. Cheng, D., Niu, Z., Tu, Y., Zhang, L.: Prediction defaults for networked-guarantee loans. In: ICPR, pp. 361–366 (2018)
8. Cheng, D., Tu, Y., Ma, Z.W., Niu, Z., Zhang, L.: Risk assessment for networked-guarantee loans using high-order graph attention representation. In: IJCAI, pp. 5822–5828 (2019)
9. Cheng, D., Wang, X., Zhang, Y., Zhang, L.: Graph neural network for fraud detection via spatial-temporal attention. TKDE, pp. 1–13 (2020)
10. Ding, Q., Wu, S., Sun, H., Guo, J., Guo, J.: Hierarchical multi-scale gaussian transformer for stock movement prediction. In: IJCAI, pp. 4640–4646 (2020)
11. Dou, Y., Liu, Z., Sun, L., Deng, Y., Peng, H., Yu, P.S.: Enhancing graph neural network-based fraud detectors against camouflaged fraudsters. In: CIKM, pp. 315–324 (2020)
12. Fu, K., Cheng, D., Tu, Y., Zhang, L.: Credit card fraud detection using convolutional neural networks. In: ICONIP, pp. 483–490 (2016)
13. Gong, J., Sun, S.: A new approach of stock price prediction based on logistic regression model. In: NISS, pp. 1366–1371 (2009)
14. Graves, A., Fernández, S., Schmidhuber, J.: Bidirectional lstm networks for improved phoneme classification and recognition. In: ICANN, pp. 799–804 (2005)
15. Heidari, A.A., Faris, H., Aljarah, I., Mirjalili, S.: An efficient hybrid multilayer perceptron neural network with grasshopper optimization. Soft Computing, pp. 7941–7958 (2019)
16. Hsu, C., Li, C.T.: Retagnn: Relational temporal attentive graph neural networks for holistic sequential recommendationn. In: WWW, pp. 2968–2979 (2021)
17. Khairy, R.S., Hussein, A., ALRikabi, H.S.: The detection of counterfeit banknotes using ensemble learning techniques of adaboost and voting. IJISAE, pp. 326–339 (2021)
18. Kingma, D.P., Ba, J.: Adam: A method for stochastic optimization. In: ICLR, pp. 1–15 (2015)
19. Leangarun, T., Tangamchit, P., Thajchayapong, S.: Stock price manipulation detection based on mathematical models. IJTEF, pp. 81–88 (2016)
20. Ma, Y., Wang, S., Aggarwal, C.C., Tang, J.: Graph convolutional networks with eigenpooling. In: KDD, pp. 723–731 (2019)
21. Mendonça, L., De Genaro, A.: Detection and analysis of occurrences of spoofing in the Brazilian capital market. J. Financial Regulation Compliance (2020)

22. Oghenerukevbe, E.A.: Customers perception of security indicators in online banking sites in Nigeria. J. Internet Banking Commerce **13**(3), 1–14 (1970)
23. Olchyk, A.: A spoof of justice: double jeopardy implications for convictions of both spoofing and commodities fraud for the same transaction. Am. UL Rev. **65**, 239 (2015)
24. Park, S.H., Han, K.: Methodologic guide for evaluating clinical performance and effect of artificial intelligence technology for medical diagnosis and prediction. Radiology, pp. 800–809 (2018)
25. Sadineni, P.K.: Detection of fraudulent transactions in credit card using machine learning algorithms. In: I-SMAC, pp. 659–660 (2020)
26. Scopino, G.: Preventing spoofing: From criminal prosecution to social norms. U. Cin. L. Rev., 1–8 (2016)
27. Sinrod, E.J., Reilly, W.P.: Cyber-crimes: A practical approach to the application of federal computer crime laws. Santa Clara Computer & High Tech. LJ **16**, 177 (2000)
28. Tao, X., Day, A., Ling, L., Drapeau, S.: On detecting spoofing strategies in high-frequency trading. Quantitative Finance **22**(8), 1405–1425 (2022)
29. Tezgider, M., Yildiz, B., Aydin, G.: Text classification using improved bidirectional transformer. Concurrency and Computation: Practice and Experience, pp. 1–12 (2021)
30. Tseng, V.S., Ying, J.C., Huang, C.W., Kao, Y., Chen, K.T.: Fraudetector: a graph-mining-based framework for fraudulent phone call detection. In: KDD, pp. 2157–2166 (2015)
31. Tuccella, J.N., Nadler, P., Şerban, O.: Protecting retail investors from order book spoofing using a gru-based detection model. arXiv preprint, pp. 1–13 (2021)
32. Vaswani, A., et al.: Attention is all you need. In: NIPS, pp. 5998–6008 (2017)
33. Wang, X., Wellman, M.P.: Spoofing the limit order book: an agent-based model. In: AAAI, pp. 651–659 (2017)
34. Yang, L., Ng, T.L.J., Smyth, B., Dong, R.: Html: Hierarchical transformer-based multi-task learning for volatility prediction. In: Proceedings of The Web Conference 2020, pp. 441–451 (2020)
35. Yang, S., Paddrik, M., Hayes, R., Todd, A., Kirilenko, A., Beling, P., Scherer, W.: Behavior based learning in identifying high frequency trading strategies. In: CIFEr, pp. 1–8 (2012)
36. You, J., Ying, R., Ren, X., Hamilton, W., Leskovec, J.: Graphrnn: generating realistic graphs with deep auto-regressive models. In: International Conference on Machine Learning. pp. 5708–5717. PMLR (2018)
37. Zhang, T., He, W., Zheng, H., Cui, Y., Song, H., Fu, S.: Satellite-based ground pm2. 5 estimation using a gradient boosting decision tree. Chemosphere **268**, 1–45 (2021)
38. Zhang, W., Wu, C., Zhong, H., Li, Y., Wang, L.: Prediction of undrained shear strength using extreme gradient boosting and random forest based on bayesian optimization. Geoscience Frontiers, pp. 469–477 (2021)
39. Zhu, P., Cheng, D., Yang, F., Luo, Y., Qian, W., Zhou, A.: Zh-ner: Chinese named entity recognition with adversarial multi-task learning and self-attentions. In: DAS-FAA, pp. 603–611 (2021)
40. Zhu, Y., Xi, D., Song, B., Zhuang, F., Chen, S., Gu, X., He, Q.: Modeling users' behavior sequences with hierarchical explainable network for cross-domain fraud detection. In: WWW, pp. 928–938 (2020)

A Novel Explainable Rumor Detection Model with Fusing Objective Information

Junlong Wang(✉), Dechang Pi(D), Mingtian Ping, and Zhiwei Chen

College of Computer Science and Technology, Nanjing University of Aeronautics and Astronautics, Nanjing, China
dragoner@nuaa.edu.cn

Abstract. Amidst the dynamic expansion of social networks, the dissemination of rumors has accelerated, rendering rumor detection an imperative and formidable endeavor in the realm of online environment governance. Traditional rumor detection methodologies have predominantly neglected the significance of interpretability. To rectify this deficiency, we introduce a sophisticated and interpretable rumor detection model, denoted as FOEGCN. This avant-garde model discerns objective information from an extensive database predicated on subjective data, subsequently employing a graph neural network to classify rumors based on a fusion of objective and subjective intelligence. Concurrently, FOEGCN elucidates the detection results via a visually compelling interpretation. Rigorous experiments conducted on a pair of publicly accessible datasets substantiate that our proposed model surpasses existing baseline methods in both rumor and early rumor detection assignments. The FOEGCN model enhances performance by 1% and 1.6% in terms of accuracy metrics. A comprehensive case study further accentuates the model's superior interpretability, making it an exemplary solution for tackling the challenges of rumor detection.

Keywords: rumor detection · graph neural network · interpretability

1 Introduction

With the growing prevalence of social media, users are now able to access a broader range of information quickly. However, this convenience has come with the downside of widespread rumors [1]. Rumors are defined as information spread on the internet without being proven true. They have now become a significant social problem that poses a threat to public security and national stability. For instance, during the COVID-19 pandemic, the rumor that "5G can spread the coronavirus" spread like wildfire, causing a crisis of trust and disrupting social order. This caused significant harm to society, resulting in considerable economic losses. In the face of the panic and threats that rumors may cause, it has become crucial to identify an effective and early detection method for identifying rumors on social media.

Current research on rumor detection primarily focuses on treating it as a natural language classification task. However, merely determining whether a piece of text is

© The Author(s), under exclusive license to Springer Nature Switzerland AG 2023
X. Yang et al. (Eds.): ADMA 2023, LNAI 14177, pp. 504–517, 2023.
https://doi.org/10.1007/978-3-031-46664-9_34

a rumor is insufficient to make people understand and trust the model's classification results. A robust rumor detection system should possess interpretability, which encompasses two critical functions: rumor identification and evidence provision. The provision of evidence is essential for supporting judgments based on rumor detection results.

Previous research on interpretability can be broadly classified into two categories. The first category involves using deep neural network-based approaches to learn credibility metrics from subjective information, such as source tweets and comments [2]. While these methods are effective, they lack transparency, making it difficult to explain their underlying rationale. To address this weakness, the second category of research explores evidence-based verification solutions that acquire objective information from reliable sources through appropriate deep learning models [3, 4]. For example, Thorne et al. [3] developed a multi-task learning model that extracts evidence from Wikipedia and synthesizes information from multiple documents to verify statements. However, this approach neglects the fact that subjective information in rumors can also serve as evidence.

To overcome these limitations, we propose a fused objective and subjective information graph convolutional network (FOEGCN) to address the above challenges in this paper. Our model fuses subjective information with objective information by introducing objective information from document databases, such as Wikipedia, as evidence for rumor detection. Additionally, our model uses a graph structure to represent the propagation path according to the propagation structure of rumors and employs an improved graph convolutional neural network for rumor detection. Our model is based on the improved BiGCN model, and achieves an accuracy of 86.7% and 83.2% on the open-source datasets PHEME16 and PHEME18, respectively. The experimental results confirm the effectiveness and feasibility of our proposed method and provide interpretability for rumor detection.

2 Related Work

2.1 Rumor Detection

The content-based method for rumor detection utilizes text information within articles. Ma et al. [5] first introduced deep learning ideas to the field of rumor detection by feeding text into a recurrent neural network and representing the text information using the hidden vector of the recurrent neural network. Due to the gradient explosion and vanishing problem of recurrent neural networks, Yu et al. [6] proposed to extract text features using convolutional neural networks. Cheng et al. [7] used a variational self-encoder VAE to self-encode text information to obtain an embedding representation of news text.

The structure-based method focuses on the unique propagation structure of rumors. Liu et al. [8] considered the rumor propagation process as a time series, arranged the source tweets and replies in chronological order, and modeled them separately using recurrent neural networks and convolutional neural networks. Tian et al. [9] defined the news dissemination process as a graph that uses top-down graphs to express spreading rumors and bottom-up graphs to express diffused information. Wei et al. [10] argued that unreliable edge relations in the propagation structure can trigger uncertainty.

The objective information-based method uses known information to assist in determining the veracity of a claim. Li et al. [11] queried factual evidence in the objective information corpus using a pre-trained fact-checking model. Then the evidence and objective information are constructed as a graph, and the information is fused using a graph convolutional neural network. Hu et al. [12] constructed the textual information into a directed heterogeneous graph and extracted the entity information in the tweets through the heterogeneous graph. Lu et al. [13] used convolutional neural networks and recurrent neural networks to learn rumor propagation representations based on user features. They used graph convolutional neural networks to construct interactions between users.

2.2 Interpretability Research

Interpretability research can generally be divided into intrinsic interpretability [14] and late interpretability [15] Intrinsic interpretability occurs by constructing its explanatory model, incorporating interpretability directly into the model structure. Wu [14] provided a reasonable explanation for relationship prediction using a capsule network based on its unique characteristics. Post-hoc Interpretability requires designing a second model that explains the existing model. Bojan [15] explained machine learning methods based on probability theory using Bayesian networks.

3   The Proposed Models

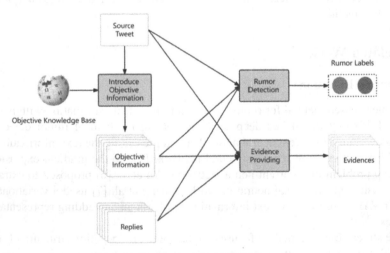

Fig. 1. Structure of the FOEGCN model

This paper proposes a subjective and objective information graph convolutional network model that integrates objective information to solve the problem of rumor detection. The model first uses the objective information introduction module to obtain objective information from the Wikipedia database. Put the obtained objective information and its subjective information into the rumor detection module to judge the accuracy of the rumor. At the same time, the evidence providing module is used to give the factual basis for the judgment of rumors. The overall framework of the model is shown in Fig. 1.

3.1 Objective Information Introduction

Using method [4] similar to the previous work, the target information introduction module has two steps: document search and sentence search. In the document search, the entity information of the source tweet is loaded into the Wikipedia database, and the keyword match algorithm is used to find the relevant documents. In sentence retrieval, we match the source tweets with all sentences in the relevant documents obtained through document retrieval and identify the k most relevant sentences as objective information.

3.2 Rumor Detection

The overall architecture of the rumor detection module appears in Fig. 2.

A. Text Embedding

We must find a way to embed textual content into a low-dimensional space to obtain sentence vectors. A common approach is to use the BERT pre-trained model [23, 24]. However, the size of the benchmark dataset is too small for fine-tuning and the model is very time-consuming and requires a lot of memory, so we compute the word vector for each word by TF-IDF and then use the average vector as a representative of the whole tweets. To make the vectors accessible to the contextual relationships in the text and to extract the hidden information in the text using the improved BiLSTM model, as shown in Fig. 2, A. For the improved BiLSTM model, to overcome the gradient vanishing problem of LSTM, a new activation function RTLU is proposed in this paper, due to replace the tanh function in LSTM, and its function equation is shown in (1).

$$f(x) = \left\{ \begin{array}{l} x, x \geq 0 \\ \tanh x . x < 0 \end{array} \right\} \tag{1}$$

The text is converted into hidden nodes by converting the sentence embedding as described above, and the conversion Eq. (2) is as follows.

$$[r, x_1, \ldots, x_j] \rightarrow [h_{x_0}^0, h_{x_1}^0, h_{x_j}^0]$$
$$[r, s_1, \ldots, s_j] \rightarrow [h_{s_0}^0, h_{s_1}^0, h_{s_j}^0] \tag{2}$$

where r is the source tweet text, x_i is the reply text, and s_i is the objective information text. Where h_x^0 and h_s^0 serve as the initial hidden states of source-tweet-reply and source-tweet-objective messages.

B. Figure Configuration and Update

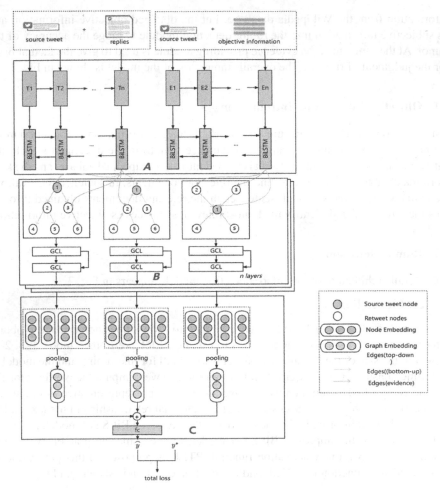

Fig. 2. Rumor detection module

For a post c^i, we construct two graph structures G_i^{td}, G_i^{bu} for its subjective information, top-down and bottom-up. Because it can extract both diffusion and propagation feature information [9]. These two graphs differ in that they have different adjacency matrices but the same identity matrix. That is $A_i^{td} = A_i$, $A_i^{td} = A_i$. For the objective information of the posts, we construct a star graph G_i^{ev}, as shown in Fig. 2, B. To better demonstrate the approach in this paper, we ignore the superscript in the following description.

The iterative formula is shown in Eq. (3–4), continuously iterating the hidden states of both.

$$h_{x_{N(v)}}^{k} \leftarrow AGG_k^{pool}(\{h_{p_u}^{k-1}, \forall u \in N(v)\})$$
$$h_{x_v}^{k} \leftarrow \sigma(W_x^k \cdot CON(h_{x_v}^{k-1}, h_{x_{N(v)}}^{k})) \tag{3}$$

$$h^k_{s_{N(v)}} \leftarrow AGG^{pool}_k(\{h^{k-1}_{s_u}, \forall u \in N(v)\})$$
$$h^k_{s_v} \leftarrow \sigma(W^k_s \cdot CON(h^{k-1}_{s_v}, h^k_{s_{N(v)}}))$$

(4)

where $h^k_{x_{N(v)}}$ and $h^k_{s_{N(v)}}$ denote the neighborhood vectors of replies and objective information, k denotes the number of iterations elapsed, $N(v)$ denotes the nearest neighbors of v nodes, CON denotes the connectivity function, σ denotes the nonlinear activation function, and AGG^{pool}_k is the aggregation function. $h^k_{x_v}$, $h^k_{s_v}$ are used as the final representations of source-tweet replies and source-tweet-objective information.

In this article, we choose the max pooling aggregator. The formula is shown in Eq. (5).

$$AGG^{pool}_k = \max(\{\sigma(W_{pool} \cdot h^k_{graph_{N(v)}} + b_{pool}), \forall u_i \in N(v)\})$$

(5)

After k iterations of information transfer based on dialogue structure and star structure, we obtained the final representation of dialogue embedding results and evidence embedding results. The formula is shown in (6):

$$H^{td} \leftarrow h^k_{x_v}, \forall v \in V_x$$
$$H^{bu} \leftarrow h^k_{x_v}, \forall v \in V_x$$
$$H^{ev} \leftarrow h^k_{s_v}, \forall v \in V_s$$

(6)

where H^{td}, H^{bu}, H^{ev} denote the hidden feature vectors after k iterations of the self-directed lower graph, self-directed upper graph, and objective information graph, respectively.

C. Classification

We consider the rumor detection task as a graph classification problem. As shown in Fig. 2, C, to aggregate the node representations in the graph, this paper first performs the maximum pooling operation on the node representations in the propagation graph H^{td}, the node representations in the diffusion graph H^{bu}, and the node representations in the objective info graph H^{ev}. Equation (7) shows the set operation of the nodes.

$$C^{td} = maxpooling(H^{td})$$
$$C^{bu} = maxpooling(H^{bu})$$
$$C^{ev} = maxpooling(H^{ev})$$

(7)

where $maxpooling$ is the maximum pooling operation, C is the node after pooling.

The nodes that complete the pooling operation connect. The prediction results are represented by a network consisting of a fully connected layer and a softmax layer. Equation (8) shows the specific prediction method.

$$\hat{y} = soft\max(W_c[C^{td}; C^{bu}; C^{ev}] + b_c)$$

(8)

where W_c and b_c are the learnable parameter matrices.

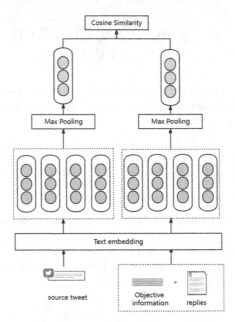

Fig. 3. Evidence Provision Module

3.3 Evidence Provided

The evidence-providing module appears in Fig. 3.

Source tweets and objective information responses were used as two sets of inputs, and word vectors were generated by employing the same text embedding as in the rumor detection module.

The pre-trained BERT model is used to perform word embedding on the processed clauses. Equation (9) shows the specific computation.

$$V_s = BERT(W_s)$$
$$V_I = BERT(W_I)$$
(9)

After max-pooling the obtained word vectors, we use the vectors for cosine calculation of the similarity measure. Equation (10) shows the specific calculation method.

$$\text{score} = \cos(\theta) = \frac{V_s \bullet V_I}{||V_s|| \times ||V_I||} = \frac{\sum_{i=1}^{n} (V_S^i \times V_I^i)}{\sqrt{\sum_{i=1}^{n} (V_S^i)^2} \times \sqrt{\sum_{i=1}^{n} (V_I^i)^2}}$$
(10)

where W_s and W_I are the word groups of the source tweets and objective information-responses after the completion of word splitting, V_s and V_I are the word vectors of the source tweets and objective information responses after the completion of text embedding, score indicates the similarity of the source tweet to a set of objective message – replies.

4 Experiments and Analysis

4.1 Dataset and Comparison Methods

Our paper evaluates the model for the rumor detection module on two real-world benchmark datasets: PHEME16 and PHEME18 [16]. Table 1 summarizes the statistics of the dataset. Non-existent or anonymous statistics marked with a "-".

Table 1. Statistics of Rumor Dataset

Dataset	PHEME16	PHEME18
# of users	49345	50593
# of events	5802	6425
# of events	103212	105453
# of false rumors	3830	1067
# of true rumors	1972	638
# of unverified rumors	–	698
# of non-rumors	–	4022

We compare our proposed method with the most commonly used machine learning methods and several recent deep learning methods to demonstrate the effectiveness of our proposed model. The specific baseline methods used are as follows.

DTC [17]: A decision tree model built on manual features.

SVM-TK [18]: A SVM classifier based on propagation tree kernel structure.

RvNN [19]: A tree structure-based recurrent neural network.

GCN [20]: A rumor detection model based on graph convolutional neural network.

GraphSAGE [21]: A rumor detection model for neural networks based on aggregator learning with graph convolutional neural networks.

GAT [22]: A rumor detection model for neural networks based on the attention mechanism of graph convolutional neural network.

BiGCN [9]: A bi-directional propagation graph convolutional neural network rumor detection model.

4.2 Implementation Details and Evaluation Metrics

We focus on comparing the rumor detection module with state-of-the-art models, where experimental results for DTC, RvNN, and SVM-TK come from the literature [22]. For the other baseline models, we replicated the above methods using PyTorch. For a fair comparison, we randomly divide the data set into five parts for five-fold cross-validation and average the obtained results as the final result. In the experimental process, we used a rare word deletion method to reduce the noise in the data, i.e., removing words with less than 2 occurrences. This method may lead to null values in the samples, so the null values are removed from this paper. We used stochastic gradient descent and Adam's algorithm to update the graph parameters. The dimensionality of the hidden feature vector for each node is 64. The dropout rate in DropEdge is 0.2 and the dropout rate is 0.5. The training process is iterated after 200 epochs and an early stop is used when the validation loss stops decreasing before 10 epochs. To make a fair comparison, this paper follows previous work [9, 10] and uses precision, recall, and F1 score as metrics to evaluate the model's overall performance.

4.3 Experimental Results

Our paper compares FOEGCN with the model presented in the comparison algorithm. The results for PHEME18 and PHEME16 appear in Table 2 and Table 3.

Table 2. PHEME18 results

PHEME18					
Method	Acc	NR	FR	TR	UR
		F1	F1	F1	F1
GCN	0.806	0.878	0.550	0.586	0.401
SAGE	0.808	0.881	0.543	0.589	0.403
GAT	0.806	0.880	0.539	0.587	0.412
BiGCN	0.816	0.891	0.742	0.679	0.581
FOEGCN	**0.832**	**0.900**	0.569	0.666	**0.667**

All the compared methods fall into three groups. The first group is the artificial feature-based methods, including DTC and SVM-TK; the second refers to traditional deep learning methods, including RvNN; and the last consists of graph neural network-based methods.

We can deduce the following from Table 2 and Table 3.

1. The model proposed in this paper consistently outperforms other comparative methods on all datasets. In PHEME16, the rumor detection accuracy of FOEGCN outperforms the most advanced model by 1.6%. In PHEME18, the rumor detection accuracy of FOEGCN is 1% higher than that of the most advanced model.
2. The graph neural network-based approach is superior to the artificial feature-based and traditional deep learning approaches. It demonstrates that graph structure can better represent the textual features in the rumor detection domain.
3. The traditional neural network-based approach is significantly higher than the manual feature-based approach, which indicates that in the field of rumor detection, it is challenging to find rumor features that can be generalized through manual feature extraction. A deep learning-based approach is required to better extract text features from them.

Table 3. PHEME16 results

PHEME16					
Method	Acc	Class	Pre	Rec	F1
DTC	0.670	FR	–	–	0.755
		TR	–	–	0.494
SVM-TK	0.785	FR	–	–	0.839
		TR	–	–	0.677
GRU-RNN	0.829	FR	–	–	0.873
		TR	–	–	0.658
GCN	0.854	FR	0.865	0.923	0.888
		TR	0.825	0.737	0.746
GraphSAGE	0.851	FR	0.865	0.918	0.886
		TR	0.815	0.722	0.745
GAT	0.857	FR	0.873	0.917	0.890
		TR	0.819	0.737	0.756
BiGCN	0.857	FR	0.888	0.897	0.888
		TR	0.793	0.777	0.766
FOEGCN	**0.867**	FR	0.892	0.908	**0.899**
		TR	0.810	0.783	**0.794**

4.4 Early Rumor Detection

Early rumor detection means detecting a rumor at an early stage before it spreads widely on social media so that people can take appropriate action earlier. It is crucial for real-time rumor detection systems. To evaluate the performance of early rumor detection, we adopt the method proposed in the literature [9] to test the period of rumor detection by controlling the detection period or the number of tweets after the source tweet information publication. The earlier the detection deadline, or the lower the number of tweets, the less information is available for dissemination. The performance of early rumor detection is shown in Fig. 4.

There are some conclusions we can draw from Fig. 4:

1. All models climb as detection deadlines pass or tweet counts increase. In particular, FOEGCN achieves higher accuracy scores on tweet counts than similar models.
2. Compared with RvNN, a traditional deep learning-based approach outperforms GCN, SAGE, GAT, BiGCN, and FOEGCN. It indicates that the rich structural features of neural graph networks have a crucial role in rumor detection and can better represent rumor information.
3. FOEGCN achieves better results in early detection compared with other models. The results show that the model can assist in judging rumors by objective information so

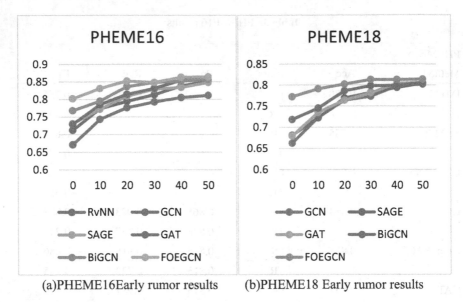

(a)PHEME16Early rumor results   (b)PHEME18 Early rumor results

Fig. 4. Early rumor results

that the model no longer relies only on subjectivity and enhances the robustness of early rumor detection.

In summary, FOEGCN has better performance in long-term rumor detection and exemplary performance in early rumor detection.

4.5   Case Study

In this section, we conduct a case study to show the interpretability of rumor detection. We randomly selected a rumor from PHEME16, as shown in Fig. 5. Tweets as nodes and relationships are modeled as edges in the graph. In the graph, red nodes indicate source tweets, replies 1–4 indicate follow-up replies and messages 1–3 indicate the objective information introduced. As seen in the figure, our proposed model can generate evidence for visualizing the accuracy of rumors based on both objective and subjective information. In this paper, k is set to 3, so that the three most relevant pieces of evidence are generated according to their relevance ranking.

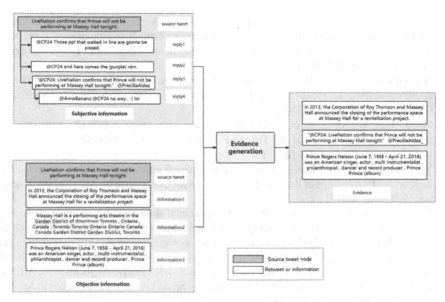

Fig. 5. Case study

5 Conclusion and Future Study

This paper proposes a new interpretation model FOEGCN. We retrieve objective information from Wikipedia and use objective information and subjective information as features to judge the authenticity of the text by updating the graph convolutional network with edge weights. After completing the judgment of text authenticity, a rumor judgment basis based on subjective and objective information is generated. The experimental results show that the model proposed in this paper can extract entities from objective information to improve the performance of rumor detection. At the same time, this paper also conducts ablation experiments to verify the effectiveness of each key module in this model and examines the effect of model interpretation and edge weight update through case studies. In the future, we will consider incorporating more information, such as time and comment sentiment, and try to use knowledge graphs to import objective information.

Acknowledgment. This work was supported by National Science and Technology Innovation 2030-Key Project of "New Generation Artificial Intelligence" under Grant 2021ZD0113103.

References

1. Qiao, D., Qiu, L., Sun, C.: An ICDRI rumor propagation model based on the individual discrepancy theory in social networks. Int. J. Mod. Phys. C **34**(01), 235–250 (2023)
2. Volkova, S., Shaffer, K., Jang, J., Hodas, N.: Separating facts from fiction: Linguistic models to classify suspicious and trusted news posts on twitter. In: Proceedings of the 55th Annual Meeting of the Association for Computational Linguistics (Volume 2: Short papers), pp. 647–653 (2017)

516   J. Wang et al.

3. Thorne, J., Vlachos, A., Christodoulopoulos, C., Mittal, A.: FEVER: a Large-scale Dataset for Fact Extraction and VERification. In: Proceedings of the 2018 Conference of the North American Chapter of the Association for Computational Linguistics: Human Language Technologies, pp. 809–819 (2018)
4. Hanselowski, A., Zhang, H., Li, Z., et al.: UKP-Athene: multi-sentence textual entailment for claim verification. In: Proceedings of the First Workshop on Fact Extraction and Verification (FEVER), pp. 103–108 (2018)
5. Ma, J., Gao, W., Mitra, P., et al.: Detecting rumors from microblogs with recurrent neural networks. In: Proceedings of the Twenty-Fifth International Joint Conference on Artificial Intelligence, pp. 3818–3824 (2016)
6. Yu, F., Liu, Q., Wu, S., Wang, L., Tan, T.: A convolutional approach for misinformation identification. In: Proceedings of the 26th International Joint Conference on Artificial Intelligence, pp. 3901–3907 (2017)
7. Cheng, M., Nazarian, S., Bogdan, P.: VRoC: variational autoencoder-aided multi-task rumor classifier based on text. In: Proceedings of the Web Conference 2020, pp. 2892–2898 (2020)
8. Liu, Y., Wu, Y-FB.: Early detection of fake news on social media through propagation path classification with recurrent and convolutional networks. In: Proceedings of the Thirty-Second AAAI Conference on Artificial Intelligence and Thirtieth Innovative Applications of Artificial Intelligence Conference and Eighth AAAI Symposium on Educational Advances in Artificial Intelligence, pp. Article 44 (2018)
9. Bian, T., et al.: Rumor detection on social media with bi-directional graph convolutional networks. In: Proceedings of the AAAI Conference on Artificial Intelligence, pp. 549–556 (2020)
10. Wei, L., Hu, D., Zhou, W., Yue, Z., Hu, S.: Towards propagation uncertainty: edge-enhanced Bayesian graph convolutional networks for rumor detection. In: Proceedings of the 59th Annual Meeting of the Association for Computational Linguistics and the 11th International Joint Conference on Natural Language Processing (Volume 1: Long Papers), pp. 3845–3854 (2021)
11. Li, J., Ni, S., Kao, H-Y.: Meet the truth: leverage objective facts and subjective views for interpretable rumor detection. In: Findings of the Association for Computational Linguistics: ACL-IJCNLP 2021, pp. 705–715 (2021)
12. Hu, L., et al.: Compare to the knowledge: graph neural fake news detection with external knowledge. In: Proceedings of the 59th Annual Meeting of the Association for Computational Linguistics and the 11th International Joint Conference on Natural Language Processing, pp. 754–763 (2021)
13. Lu, Y-J., Li, C-T.: GCAN: graph-aware co-attention networks for explainable fake news detection on social media. In: Proceedings of the 58th Annual Meeting of the Association for Computational Linguistics, pp. 505–514 (2020)
14. Wu, J., Mai, S., Hu, H.: Contextual relation embedding and interpretable triplet capsule for inductive relation prediction. Neurocomputing 505, 80–91 (2022)
15. Mihaljević, B., Bielza, C., Larrañaga, P.: Bayesian networks for interpretable machine learning and optimization. Neurocomputing 456, 648–665 (2021)
16. Kochkina, E., Liakata, M., Zubiaga, A.: All-in-one: multi-task learning for rumour verification. In: Proceedings of the 27th International Conference on Computational Linguistics, pp. 3402–3413 (2018)
17. Castillo, C., Mendoza, M., Poblete, B.: Information credibility on twitter. In: The Web Conference (2011)
18. Ma, J., Gao, W., Wong, K-F.: Detect rumors in microblog posts using propagation structure via kernel learning. In: Proceedings of the 55th Annual Meeting of the Association for Computational Linguistics (Volume 1: Long Papers), pp. 708–717 (2017)

19. Ma, J., Gao, W., Wong, K-F.: Rumor detection on twitter with tree-structured recursive neural networks. In: Proceedings of the 56th Annual Meeting of the Association for Computational Linguistics (Volume 1: Long Papers), pp. 1980–1989 (2018)
20. Yao, L., Mao, C., Luo, Y.: Graph convolutional networks for text classification. In: Proceedings of the Thirty-Third AAAI Conference on Artificial Intelligence and Thirty-First Innovative Applications of Artificial Intelligence Conference and Ninth AAAI Symposium on Educational Advances in Artificial Intelligence, pp. Article 905 (2019)
21. Vu, D.T., Jung, J.J.: Rumor detection by propagation embedding based on graph convolutional network. Int. J. Comput. Intell. Syst. **14**, 1053–1065 (2021)
22. Lin, H., Ma, J., Cheng, M., Yang, Z., Chen, L., Chen, G.: Rumor detection on twitter with claim-guided hierarchical graph attention networks. In: Proceedings of the 2021 Conference on Empirical Methods in Natural Language Processing, pp. 10035–10047 (2021)
23. Li, Z., Ma, H., Lv, Y., Hao, S.: Joint extraction of entities and relations in the news domain. In: Advanced Data Mining and Applications: 18th International Conference, pp. 28–30 (2022)
24. Ranganathan, J., Tsahai, T.: Sentiment analysis of tweets using deep learning. In: Chen, W., Yao, L., Cai, T., Pan, S., Shen, T., Li, X. (eds.) Advanced Data Mining and Applications. Lecture Notes in Computer Science(), vol. 13725, pp. 106--117. Springer, Cham (2022). https://doi.org/10.1007/978-3-031-22064-7_9

VLS: A Reinforcement Learning-Based Value Lookahead Strategy for Multi-product Order Fulfillment

Ryan Wickman[1], Junxuan Li[2], and Xiaofei Zhang[1(✉)]

[1] The University of Memphis, Memphis, USA
{rwickman,xiaofei.zhang}@memphis.edu
[2] Microsoft, Redmond, USA
junxuanli@microsoft.com

Abstract. The fast-paced growth of online ordering comes with additional challenges that are not as prevalent in the traditional brick-and-mortar retail sector. For one, e-commerce retailers have the challenge and flexibility of selecting which warehouses and stores should fulfill online orders. This is known as the omnichannel order fulfillment problem. In this work, we combine a lookahead search tree strategy with a reinforcement learning-based cost-to-go estimator to produce an effective cost-saving order fulfillment strategy, named the *value lookahead strategy (VLS)*. Furthermore, we design and implement a simulator with the capabilities to simulate a wide variety of order fulfillment scenarios which can allow for developing, training, and evaluating order fulfillment strategies, even in the presence of limited data. We show that using these in conjunction can produce an order fulfillment strategy with lower total fulfillment costs than all other order fulfillment strategies we compared against.

Keywords: Reinforcement learning · Multi-product Order Fulfillment

1 Introduction

There has been a steady growth of e-commerce volume in social-economy activities during the past few years. For example, in February 2022, the US Department of Commerce reported total e-commerce sales for 2021 at $870.8 billion with an increase of 14.2% from 2020. E-commerce sales in 2021 accounted for 13.2% of total retailer sales in the US [25]. With the continuously growing scale of online retail [23,25], the omnichannel order fulfillment scenario, becomes increasingly popular. In this scenario, a variety of fulfillment nodes (e.g., store, warehouse, etc.) can fulfill orders. Compared to traditional brick-and-mortar retailers, an e-commerce retailer (e-retailer) has the flexibility to choose where to fulfill the orders, as illustrated in Fig. 1. This flexibility has several benefits: (1) Given a target performance metric, e.g., shipping cost, lead time, turnover, it enables an e-retailer to minimize this metric (or a portfolio of metrics). (2) Given the current

X. Yang et al. (Eds.): ADMA 2023, LNAI 14177, pp. 518–534, 2023.
https://doi.org/10.1007/978-3-031-46664-9_35

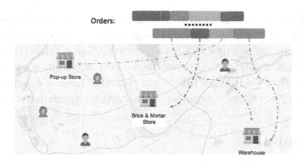

Fig. 1. A visual example of the multi-product fulfillment problem under the omnichannel setting. Orders from geographically distributed users are fulfilled by multiple stores of various types and locations. Particularly, different parts of an order can be sent to different stores with the aim of reducing the overall cost defined by retail companies (e.g., operation/deliver/storage cost, etc.)

state of a fulfillment network, it allows an e-retailer to avoid over-/under-stocking products at all warehouses. This can allow for balancing the global inventory levels, and hence, optimizing the service periods for all products. However, the optimal order fulfillment under this setting is non-trivial considering the exponential search space. Thus, algorithms that can reach near-optimality in solving this problem are of increasing interest to companies that want to minimize their fulfillment costs.

Surprisingly, to our best knowledge, there has been little research effort towards this problem from the academic world, and the widely adopted fulfillment strategies are rather naive solutions. An academic literature review [27] reported that many e-retailers are only applying "myopic" policies to calculate their fulfillment plans, whereby a batch of orders are accumulated over a predefined time window (often in hours), and fulfillment plans are calculated by simply minimizing the existing cost or shipping distances. This greedy and myopic optimization is proven to be sub-optimal both theoretically and practically, see [1,3,8]. Moreover, the traditional batch-processing decision mode is asynchronous, which leads to lower supply chain visibility and degraded operational agility.

In this work, we propose and develop an intelligent, data-driven order fulfillment strategy that can learn the most cost-effective, near-optimal way to fulfill any order in any multi-product order fulfillment scenario. This led us to build a simulator that is configurable using historical transaction data to simulate various order fulfillment scenarios. The simulator will produce synthetic orders that are processed by our deep reinforcement learning strategy, which will in turn train a cost-to-go model. The proposed *value lookahead strategy (VLS)* combines a simple lookahead search tree with a cost-to-go model with the aim to minimize the total fulfillment costs in the long term. The lookahead search tree will be used to evaluate the different ways to fulfill an order and the cost-to-go model will provide an estimate of the future cost of the order fulfillment. The output of VLS assists an e-retailer in selecting cost-efficient fulfillment plans, which are consumed by the fulfillment nodes to generate shipment/pickup operations. To the best of our knowledge, VLS is the first reinforcement learning empowered data-driven approach for this problem. And our simulator could serve as a general-purpose testbed for any future effort in this field. Experiments over both real-world datasets and the simulator validate the effectiveness of our solution.

2   Background

2.1   Problem Definition

We consider an e-retailing order fulfillment problem with finite inventory, without inventory replenishment, with orders that are multi-product, and where products may be requested multiple times. An e-retailing sales order consists of product information, shipping address, and ordering quantities for each ordered product. A fulfillment node is defined as a location carrying inventories, from where inventories can be committed to a sales order. Such a fulfillment node could be a warehouse or a fulfillment center, or a physical store if the retailing network is omnichannel. Once an order is validated, an order fulfillment optimization problem tries to answer the following questions: 1) What fulfillment nodes should we select to fulfill this order? 2) From each fulfillment node, what combination of products should be fulfilled? 3) From each fulfillment node, how many products should be fulfilled?

We denote an order received at time t as order o_t, and introduce key indices of parameters/variables used to formulate an order fulfillment optimization problem. Let $i \in I(t)$ be the product index, $I(t)$ be the set of products included in order t, $j \in J$ be the fulfillment node index, J be the set of all fulfillment nodes, $k \in K$ be the demand region index, K be the set of all demand regions, and $k(t)$ denotes the demand region of order t. For each product i, let o_{ti} be the order quantity in the order at time t, $o_t = \{o_{ti}\}_{i \in I(t)}$ be the order quantities of all products, and $n_t = \sum_{i \in I(t)} o_{ti}$ be the order size. For each product i and node j, let s_{tij} be the onhand inventory level, $s_t = \{s_{tij}\}_{i \in I(t), j \in J}$ be the onhand inventory levels in the fulfillment network, and $m_{tj} = \sum_{j \in J, i \in I(t)} s_{tij}$ be the inventory size. Let p_t be the demand distribution of each product. That is, the likelihood of each product being contained in an order, independent of all other products in the order. Then, at time t the fulfillment network can be described as state (o_t, s_t, p_t).

An action for order t at state (o_t, s_t, p_t) is a fulfillment plan, denoted as $x_t = \{x_{tij}\}_{i \in I(t), j \in J}$. x_{tij} is the quantity of product i fulfilled from node j. The action space of x_t is characterized by both demand and inventory constraints:

- For each product i, the fulfillment quantities from all nodes should sum to the ordered quantity. For $i \in I(t) : \sum_{j \in J} x_{tij} = o_{ti}$.
- For each product i and node j, the fulfillment quantity should not exceed the onhand inventory. For $i \in I(t), j \in J : x_{tij} \leq s_{tij}$.

Once inventories are committed to an order, they are reserved for picking and shipping (thus removed from onhand inventory). At the next decision epoch (when order $t + 1$ occurs), the inventory dynamic is $s_{t+1,ij} = s_{tij} - x_{tij}$.

2.2   Heuristic Cost Estimation

In practice, a fulfillment cost is often hard to estimate with an exact dollar amount, and thus, this cost is often approximated as a function of fulfilling distance. In this work, we designed a heuristic cost function that will be calculated each time an ordered product is allocated to a fulfillment node. This is important for our multi-product setting as an order can be fulfilled by several different fulfillment nodes. Furthermore, our method of fulfillment benefits from a fine-grain cost analysis, which we will discuss in the Method section. We use an exponential decaying marginal cost of fulfilling one additional unit of product. Let $d_{jk(t)}$ be the distance between the node j and demand region $k(t)$. Let $q(j)$ be the number of ordered products already allocated to node j (regardless of their types), then the marginal cost of allocating an additional product to node j is $d_{jk(t)} \cdot \alpha^{q(j)}$, which is decaying in $q(j)$ by a discount factor $\alpha \in [0, 1)$. This marginal decaying property encourages orders to be fulfilled from the same fulfillment node. Let $c(x_t)$ be an cost accrued when fulfillment plan x_t is executed. Thus the total cost of executing a fulfillment plan x_t is

$$c(x_t) = \sum_{j \in J} d_{jk(t)} \cdot \left[1 + \alpha + \cdots + \alpha^{\sum_{i \in I(t)} x_{tij} - 1} \right] \qquad (1)$$

The cost modeling can be viewed as breaking the cost down into unit product and node pairs where attention to unit product fulfillment decisions is measured. The objective of this work is to design a strategy to minimize a long-term order fulfillment cost which consists of costs of a series of orders, while future order information is not known a priori. So, this heuristic implies this corresponds to jointly minimizing the distance between the fulfillment nodes and regions, and minimizing the number of fulfillment nodes assigned to fulfill an order.

2.3   MDP Formulation

A real-time order fulfillment problem can be formulated as a discounted cost infinite horizon Markov Decision Process (MDP), $v = Hv$, where operator H is defined as

$$Hv(o_t, s_t, p_t) = c(x_t) + \gamma \mathbb{E} \left[v(o_{t+1}, s_{t+1}, p_{t+1}) \right]$$

$$s.t. \sum_{j \in J} x_{tij} = o_{ti}, \quad \forall i \in I(t), \qquad (2)$$

$$x_{tij} \leq s_{tij}, \quad \forall i \in I(t), j \in J$$

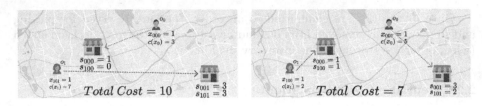

Fig. 2. A motivating example with 2 orders and 2 fulfillment nodes. The oracle (R) achieves a lower total cost than the naive strategy (L).

where $\gamma \in [0, 1)$ is a discount factor and $\mathbb{E}\left[v(o_{t+1}, s_{t+1}, p_{t+1})\right]$ is the expected cost-to-go, and v values zero when inventory is insufficient.

Given the fact that s_t is finite for all t, x_t is a finite space for all t, due to the second constraint. In this way, $c(x_t)$ is a bounded single period cost, and according to results in [18], there exists an optimal fulfillment policy which is the fixed point of Eq. (2).

As this problem can be formulated by an MDP, it can be solved using existing methods in reinforcement learning (RL) [24]. In RL, an agent acts according to a policy π_θ by choosing an action based on the current state given by the environment. The environment will respond with a reward r_t and a next state. The goal of the RL agent is to learn a policy that can maximize the sum of discounted rewards, or the *return* $G_t = \sum_{i=t}^{T} \gamma^{i-t} r_i$.

We can easily shape our current problem definition into this setting by defining the reward as $r_t = -c(x_t)$ and the cost-to-go as the return G_t, respectively. Thus, the overall objective is to find a policy that will minimize the sum of discounted costs.

A Motivating Example. In the initial stages of designing our approach, we kept in mind that this is a risk-sensitive scenario where epistemic and aleatoric uncertainty could lead to real costs in production. Furthermore, before a fulfillment decision is made, evaluating the different possible fulfillment plans on their cost and true, future costs can achieve a lower bound on the total costs. A motivating example with two fulfillment nodes, two orders, and one unit product per order is given in Fig. 2. We compare two strategies: naive (left) and oracle (right). The naive strategy always fulfills the order based on the immediate order cost, but the oracle strategy has access to the true cost-to-go function.

In this example, we use the standard euclidean metric for the cost, and thus the naive strategy always chooses the nearest fulfillment node with non-zero inventory (i.e., $s_{tij} \neq 0$). When the first order o_0 arrives at $t = 0$ it sets $x_{000} = 1$ indicating it wants fulfillment node $j = 0$ to fulfill this order. However, that leads to a problem when o_1 arrives at $t = 1$, because the inventory at node $j = 0$ will be empty, that is $s_{100} = 0$, and thus it is required to set $x_{101} = 1$ and fulfill from the farthest fulfillment node.

Alternatively, the oracle strategy can rectify this issue. Assume the oracle has access to a simple myopic cost-to-go model that can evaluate the true future costs of fulfilling an order, based on orders that arrive after the current timestep.

When o_0 arrives it can utilize the cost-to-go model, which will allow it to know fulfilling from node $j = 1$, that has a slightly higher immediate cost, will lead to lower cost in the long term and thus sets $x_{001} = 1$. This allows order o_1 to be fulfilled from the nearest fulfillment node $j = 0$, setting $x_{100} = 1$, and leads to an overall lower total cost.

It is worth noting that both fulfillment nodes fulfilled the same number of orders and products in both strategies. However, the order in which they are fulfilled led to a difference in the total cost, even though they both end up in the same, final state. This evidences the importance of effectively approximating the cost-to-go function.

3   Method

In this section, we will start out by describing our order fulfillment strategy. Then, we will provide a simple solution to reduce the time complexity of our strategy. Finally, we will go into depth about the design, organization, and processes of the simulator we implemented.

3.1   Order Fulfillment Strategy

In Fig. 2 we provided an example that reveals the effectiveness of utilizing the cost-to-go function in an order fulfillment strategy. However, the cost-to-go model only evaluates the future cost of a fulfillment plan. The control part of the strategy, which constructs the fulfillment plan for each order, is managed by a lookahead search tree algorithm.

This algorithm defines actions taken by the policy through exhaustively searching over every possible way to fulfill an order and keeping track of the cost of each. Once all the products in an order have been assigned to fulfillment nodes, the order fulfillment plan x_t and its associated cost $c(x_t)$ is produced. Then, the state is updated by x_t and the cost-to-go model will provide an estimate of the cost-to-go. The values produced by both parts are used to evaluate a fulfillment decision: $v_t = r_t + \gamma v_{t+1}$, where r_t is the negative cost of the fulfillment plan, v_{t+1} is the future cost-to-go, v_t is the value of the current order fulfillment decision, and $\gamma \in [0, 1)$ is a discount factor on the cost-to-go. Once v_t has been produced for every possible fulfillment plan, the one that maximizes this value, and thus minimizes the immediate and future cost is returned. We call this approach the *value lookahead strategy (VLS)* and outline it in Algorithm 1.

The discount factor on the cost-to-go estimate inherently allows for a trivial interpolation between a myopic order fulfillment strategy when $\gamma = 0.0$ and a farsighted order fulfillment strategy when $\gamma = 0.99$. This is important as it allows for managing the risk of deploying this in the real world. For example, when increasing the value of γ, the performance of the strategy can be carefully monitored and the accrued cost of the strategy can be tested against the expected cost of the myopic strategy.

Algorithm 1: Value Lookahead Strategy

---

input : o_t, s_t, and p_t
output: x_t the fulfillment plan

$X_t \leftarrow$ get set of all possible fulfillments on o_t;
$c_t, x_t \leftarrow \infty, \emptyset$;
for $x'_t \in X_t$ **do**
    $c'_t \leftarrow$ get cost of x'_t;
    // Add discounted cost-to-go
    $c_t \leftarrow c_t + \gamma v(s_t - x'_t, p_t)$;
    if $c'_t \leq c_t$ **then**
        $c_t, x_t \leftarrow c'_t, x'_t$;
return x_t;

3.2 Cost-to-Go Model

The value v_t is the standard state-value function used in RL, so standard methodologies can be applied to learn it. We use a state-value version of DQN [16] to represent the cost-to-go estimator, which is parameterized by a deep neural network. The model is trained over sampled batches of trajectories, sampled using prioritized replay [22]. Our model architecture is shown in Fig. 3. The input into the model consists of the fulfillment nodes' inventory s_t, fulfillment nodes' geolocation, and the demand distribution p_t. The model consists of a node encoder, a demand encoder, a transformer encoder [26], and a state-value function head. The node encoder takes in a fulfillment node and produces an embedding for each node. The demand encoder takes in the demand distribution and produces a demand embedding. These embeddings are concatenated and fed into the transformer encoder which performs multi-head attention over the embeddings. Finally, the state-value function head will produce the state value which is equivalent to the estimate of the discounted cost-to-go estimate.

The node encoder and demand encoder both consists of 4 1D convolutional layers with kernel and stride 5, and channel sizes 64, 128, 256, and 512 each followed by GELU activation [7]. These are used to downsample the inventory and demand distribution down into a 512-dimensional embedding. The transformer encoder consists of 6 transformer encoder layers with a hidden size of 1024 and the state-value head consists of a single fully-connected layer with a scalar output for the estimate.

3.3 Depth-Limited Search

While our approach seems promising, there is one issue we have yet to discuss: its ability to scale with the size of orders. In truth, searching for the optimal order in a multi-product setting is NP-hard [8]. The time-complexity of the lookahead search tree algorithm grows exponentially as the order size and the number of fulfillment nodes increase, i.e., $O(|J|^n)$ where $|J|$ is the search tree branch factor that is the number of fulfillment nodes and n is the maximum depth, equivalent to the order size.

One insight that can be made about our approach is that the cost-to-go model does not consider the current order it is attempting to fulfill, but only its effect on the inventory levels at the fulfillment nodes. This means the lookahead search tree algorithm can be terminated early before a complete fulfillment plan is constructed, and a partial order fulfillment plan can be evaluated. A simple and common solution to handling this type of problem is to perform a depth-limited search where the depth limit ℓ is enforced [21].

Due to the depth limitation, our approach must be augmented by splitting the orders into partial orders of size ℓ, applying VLS on each partial order, and then aggregating the fulfillment plans on these partial orders. In other words, the order can be split up into mutually exclusive subsets of products where the sum of all products' quantities in a partial order are less than ℓ. Let $z \in Z(t)$ be the index of a partial and assume each of the unit products is unique, then the following is true about the partial orders: 1) all of them will make up the order, $o_t = \bigcup_{z=1}^{|Z(t)|} o_t^z$; 2) each will be a unique subset of products, $\bigcap_{z=1}^{|Z(t)|} o_t^z = \emptyset$; and 3) order size will not exceed the depth-limit, $\forall_z |o_t^z| \le \ell$.

As each of these partial orders is sequentially processed by VLS, the inventory of fulfillment nodes and the fulfillment plan x_t are updated after each partial order. Let x_t^z be the fulfillment plan for o_t^z, then the complete fulfillment plan for an order is $x_t = \sum_{z=1}^{|Z(t)|} x_t^z$. Overall, doing a depth-limited search will significantly reduce the time complexity down to $O(|J|^\ell)$, where $\ell \ll n$ usually holds in practice.

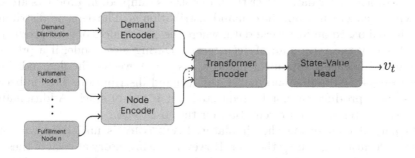

Fig. 3. The VLS model architecture.

3.4 Simulator

Due to the lack of publicly available datasets in this field of research[1], we design and implement a highly configurable order generate-fulfill simulator that can

---

[1] Most existing datasets are not publicly available and tend to based on collaboration with companies [1,3,19].

simulate discrete-event order fulfillment under various settings of order generation. We use this simulator to generate data for training the cost-to-go model, as well as a testbed for comparing a variety of order fulfillment strategies.

In our design, we simulate three entities: ORDER, POLICY, and INVENTORY, each can be configured independently. INVENTORY (I) models fulfillment nodes that carry out the order fulfillment action. Their capacity, location, as well as stock level can be configured and updated on demand. ORDER (O) generates orders following a given configuration, where multiple factors can be specified, i.e., volume, frequency, and demand distribution, to name a few. It empowers our simulator to generate any type of order demand. POLICY (P) governs the strategy to fulfill orders based on both the potential order pattern and the latest inventory status.

We summarize the simulator's workflow in Algorithm 2. At the beginning of the simulation, we first apply the configurations to ORDER and INVENTORY. We assume that an order must not request a product that does not exist in INVENTORY. Therefore, when we initialize INVENTORY, the product catalog is determined. When we start the order generate-fulfill process, we iterate over T timesteps (which can be user-specified), where an order o_t will be generated at each timestep. Following previous literature on RL, we call this T timestep order generate-fulfill process an episode.

o_t is generated by sampling from three distributions that are controlled by ORDER: the distribution of geolocations of order request, the order size distribution, and the demand distribution. The geolocation is solely based on historical data where a user's geolocation will be randomly sampled from a set of historical user locations. The order size uses a Poisson distribution that is parameterized using historical order data. After the order size is sampled, n_t products are sampled with replacement from the demand distribution p_t. The demand distribution is configured based on historical data, where the occurrences of each product are normalized by the total sum of occurrences. During an episode, if a product i is exhausted from all fulfillment node, $\forall_{j \in J} s_{tj} = 0$, we set the demand probability to 0, $p_{ti} = 0$ and re-normalize the demand distribution. This is because we assume a product cannot be requested if it is out of stock. Additionally, it provides a better state representation for the cost-to-go.

As part of configuring the simulator, INVENTORY is initialized with fixed inventory locations given by the user. However, the inventory at each of these fulfillment nodes is randomized before each order generate-fulfill process. After the marginal distribution is initialized, each inventory node will sample its inventory size m_j uniformly from a configurable range $[m_{min}, m_{max}]$. Then, m_j products will be sampled from the demand distribution to define the initial inventory of the fulfillment nodes.

The distributions that control the simulator are to provide randomness in the synthetic order data that is generated, randomness in the inventory, and to realistically capture various types of order fulfillment scenarios. Furthermore, we optionally provide additional randomness to the demand distribution, as we want to prevent overfitting on our cost-to-go model and show more robust

---

Algorithm 2: Simulator Workflow

---

  input : O, I, and P
  output: $\text{Cost}(C)$ on fulfillment
  $O.init(O\_conf)$; $I.init(I\_conf)$
  for $t = 1$ **to** T **do**
    Order $o_t \leftarrow O.generate()$;
    Fulfillment $x_t \leftarrow P(o_t, I.state(t))$;
    $I.update(t)$;
    Record the cost $C(x_t)$;
  return C

---

results. Before each order generate-fulfill process, the historical product counts are shuffled and added with random noise generated from a beta-binomial distribution. In the pursuit to capture various types of order fulfillment beyond what the data that populated the simulator provides, we sought to randomize the moments of the demand distribution. Therefore, we randomly parameterize the beta-binomial distribution, which is explained more in the experiment setup.

When POLICY receives o_t, it will make a decision on what part of the order will be fulfilled by which fulfillment node by considering a few factors: the order, the latest inventory stock states, the location of the order, fulfillment nodes, and the latest demand distribution. While not every strategy has to fully utilize this information, it is provided in the interface. However, every POLICY is required to return the fulfillment plan x_t.

While we define randomness for ORDER and INVENTORY, these are completely optional. We also provide a mode of the simulator for backtesting POLICY on historical orders and fulfillment nodes. Thus, we make sure to delineate which mode we are referring to when we discuss our results in the experiment section.

4 Experiments

First, we outline our experiment setup in detail followed by the empirical study of VLS against the baseline methods, where the results demonstrate the superior performance of VLS. Then, we discuss the findings of several ablation studies where we evaluate different components of VLS.

4.1 Experiment Setup

Dataset. As this field is lacking in order fulfillment datasets, we utilized the Olist dataset [17] with some modifications to better fit our problem formulation. The Olist dataset is a public Brazilian e-commerce dataset that contains $100k$ historical order transactions made at the Olist store. In this dataset, there were a lot of products that were rarely used, so we decided to filter out products that were not used at least 10 times. This forces the order fulfillment strategy to learn

Table 1. Olist Dataset

Statistic	Value After Filtering
Number of Orders	44,303
Number of Products	1,919
Mean Order Size	1.13
Median Order Size	1
Max Order Size	20
Mean Product Occurrence	26.15
Median Product Occurrence	16
Max Product Occurrence	527

how to actively manage the inventory knowing that a different order may request the same product. A summary of the dataset is given in Table 1. Additionally in the Olist dataset, there are 3095 unique sellers that each fulfills only a few orders. We decided to aggregate the sellers by using K-Means clustering where we found 11 clusters to provide the highest Silhouette score. During the evaluation, the products that were fulfilled by that seller are then used as the initial inventory at the corresponding fulfillment node that will be centered at the cluster means. Finally, we used a 80%/10%/10% training/validation/test split on the orders; respectively.

Simulator Training. We used the simulator to train VLS, where the training dataset was used to configure it. The simulator was reset and reconfigured after a maximum of 128 orders were fulfilled, as we wanted the policy to generalize to various order fulfillment scenarios with varying demand distribution and fulfillment node inventory. The product occurrences were added with random integer noise sampled from a beta-binomial distribution with parameters n_{beta} for the number of trails and shape parameters α_{beta} and β_{beta}. Prior to each episode, theses parameters were uniformly and randomly sampled, where $n_{beta} \in [5, 15]$, $\alpha_{beta} \in [0.1, 2]$, and $\beta_{beta} \in [0.1, 6]$. We choose these ranges as they are likely to produce a right-skewed distribution and tended to work well in practice by preventing the cost-to-go model from overfitting. After the random noise was added, the product occurrences were normalized to produce the demand distribution. We used $\alpha = 0.5$ for the discount factor on the cost. Additionally, we divided the per-order costs by the maximum geolocation distance (≈ 142.28) as we wanted all the per-order costs to have a magnitude close to 1. The costs we show in the results are divided by this same constant, as we believe it gives cleaner-looking results.

For the prioritized replay, we set the priority exponent to 0.9 and the importance-sampling exponent to 0.4, $\gamma = 0.99$ for the discounted return, use a batch size of 32, and initially set $\epsilon = 0.99$ for ϵ-greedy exploration and linearly decay it to 0.001 over the first 32k policy update steps. We use a 10-step return, the Adam optimizer [10] with a learning rate of $2e^{-4}$, and the target-value network that was updated every 512 policy update steps. We trained the model on ≈ 750k orders generated from the simulator, where after every 4 orders were fulfilled we updated the cost-to-go model. Figure 4 shows the learning curves of VLS. The loss and the avg. order costs are both decreasing over time which shows the model's fulfillment decisions keep improving.

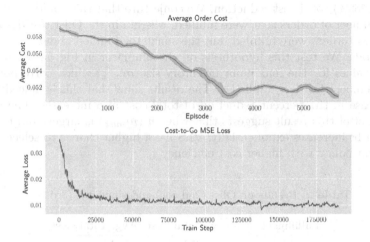

Fig. 4. Learning curves for VLS on episodes generated from the simulator.

Baseline Methods. We created two baseline methods: naive and random. The naive fulfillment algorithm fulfills orders greedily, where each unit product in an order is fulfilled based on what gives the lowest cost. The random fulfillment algorithm randomly fulfills every unit product in an order. We also tested against an existing linear programming approach based on prior research work that uses the primal-dual schema [3] and VLS with only the lookahead tree search (i.e., $\gamma = 0$). The source code for this project can be found at https://github.com/rwickman/IOM.

4.2   Order Fulfillment Results

The results from the Olist testing dataset are given in Table 2. We can see the advantage of considering the future cost of fulfilling an order as the minimum total cost is achieved with VLS when $\gamma = 0.99$. VLS with $\gamma = 0$ does better than all other baseline methods which show the importance of minimizing the cost of fulfilling each order, even if the future cost is not considered. As we use a simulator to train VLS, it is important to evaluate if the order fulfillment strategies' performance on the Olist testing dataset correlates to how they perform on the simulator. We ran a total of 16 episodes, where for each episode the demand distribution was randomly sampled and the maximum inventory of each inventory node was sampled between $[1, 2000]$. We averaged the total costs of each episode and provide the results in Table 2 (3rd col.). We observe the same ranking among different strategies, but with a slightly amplified scale for the total cost. This supports that our simulator allows for synthetic orders that can be used to approximate real-world order fulfillment scenarios.

From the results, we observe that VLS has a 0.592% total cost reduction on Olist compared to the naive strategy. While the simulator derives a more significant (4.269%) total cost reduction. We conjecture that this could be attributed to the number of orders that were fulfilled, wherein in the Olist test dataset only about 4000 orders were fulfilled, but the simulator fulfilled about 10, 000 orders per episode. We test this hypothesis on the simulator in Fig. 5a where we test various values for the maximum inventory size m_{max}, which corresponds to a larger number of orders fulfilled n. The results show that this is indeed true. As we increase n, the percent reduction of total cost also increases. The managerial insight of this result suggests that when n (m_{max}) is larger, and the size of event paths increases exponentially, VLS has a higher chance to select an order fulfillment policy with higher cost efficiency.

Table 2. Results on Olist (2nd col.) and simulator (3rd col.).

Fulfillment Strategy	Total Cost	Avg. Total Cost
Random	279.32	479.16
Naive	251.57	373.08
Primal-Dual	252.37	374.09
VLS ($\gamma = 0$)	251.49	363.16
VLS ($\gamma = 0.99$)	**250.08**	**357.15**

4.3   Interpolating Lookahead Strategies

Our strategy design allows us to easily weigh the importance we put on the future cost. This can also be viewed as a form of backtesting, where the optimal γ could be found without requiring prior deployment into the real world. Thus, we test over the testing set where results are given in Fig. 5b. In the figure, we can see that as the value of γ increases, the total cost is reduced. This shows that using future cost estimation assists in making more cost-effective order fulfillment decisions. Additionally, we only trained the model on $\gamma = 0.99$. So, these results show that the fulfillment strategy can adapt to different cost-to-go discount factors even without being explicitly trained with them.

4.4   Evaluating Depth-Limited Search

The results given in Table 2 assume that $\ell = n$, however in practice this could lead to a long delay for each order fulfillment decision. Therefore, we test various

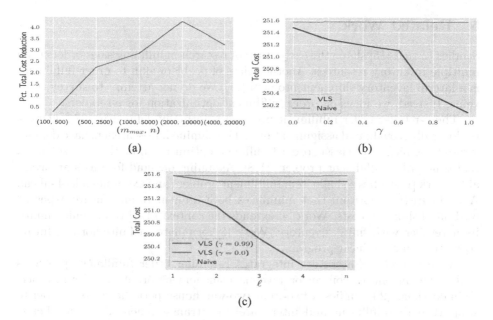

Fig. 5. (a) Results with increasing the size of the number of orders fulfilled on the simulator. In this figure, we can see that as the number of orders increases, the percentage increase compared to the naive policy increases. (b) Results with varying γ. In the figure, we can see that as γ increases, the total cost is reduced. (c) Results with varying ℓ. In the figure, we can see that as ℓ increases, the total cost is reduced for both cases when $\gamma = 0.99$ and $\gamma = 0.0$.

ℓ values to explore the time vs fulfillment cost trade-off on the depth-limited search algorithm. The results of this experiment on the Olist testing dataset are given in Fig. 5c. As expected, the total cost decreases as the depth limit increases. Interestingly, there seems to be a point at which increasing ℓ only results in zero or negligible gains for both cases of $\gamma = 0.99$ and $\gamma = 0.0$. This can be partially explained by the order sizes in the test dataset where 90% of the orders only contain one product and 98% of the orders contain less than 5 products.

Another thing to point out is that when $\ell = 1$ and $\gamma = 0.99$, VLS does better than when $\gamma = 0$ and $\ell = n$. In other words, the cost-to-go model estimate is effective enough that it performs better than a simple, myopic lookahead strategy. This is important because if $\ell = 1$, this effectively removes the constraint of VLS being NP-hard and reduces its exponential time complexity of $O(|J|^\ell)$ to linear time $O(|J|)$.

Regardless, these results show evidence of an apparent trade-off between time and cost, where increasing ℓ can lead to lower cost, but a longer delay and vice versa. Also, with an effective cost-to-go estimator, we can limit the depth size and still outperform all other strategies we tested against.

5   Related Work

Studies [2] and [15] shape the early stage of research in fulfilling sales orders from multiple inventory resources. Another class of problem akin to order fulfillment is a lateral inventory shipment problem [11]. We do not attempt to survey these veins. Instead, we focus on order fulfillment optimization for e-retailers.

The e-retailer order fulfillment problem is formally defined in [27], where a periodically reevaluated assignment model is examined. Researchers have demonstrated the pitfalls of classic order fulfillment optimization methods in solving a real-time order fulfillment problem [1,8]. Assuming demand forecasts are available, work [1] studies a single item fulfillment problem and examines a look-ahead ADP heuristic algorithm that minimizes both immediate and future expected outbound shipping costs. Work [8] extends the context of [1] to an online multi-item retailer with finite inventory. Work [8] formulates a finite-horizon future cost function as a linear program.

A large body of literature integrates E-retailer order fulfillment problems with other inventory control or revenue management problems. For instance, Ramakrishnar [20] studies a two-item two-warehouse periodic review model to support both fulfillment and inter-warehouse transshipment decisions. Lei *et al.* [12] considers an e-retailer with the capability to select product prices during a finite selling season. The e-retailer maximizes its profit by jointly optimizing price selections and fulfillment plans. Work [13] proposes a multi-stage stochastic optimization model to solve inventory replenishment while considering reactive order fulfillment. We point readers to a recent paper [9] for a comprehensive review on problems at the crossroads of order fulfillment and other supply chain management problems.

6   Conclusion

We provided a new order fulfillment strategy called value lookahead strategy (VLS) and a simulator to train and test several order fulfillment strategies. We showed that VLS provides a cost-effective strategy by outperforming all other order fulfillment strategies we compared against. We discussed and explored VLS's ability to interpolate between a greedy order fulfillment strategy and one that utilizes the cost-to-go estimation by configuring γ. Additionally, we provided a solution using depth-limited search for VLS to scale to larger order sizes and showed that it can effectively reduce the search time complexity down to linear time. Finally, we showed the simulator provides a reasonable testbed for training and testing order fulfillment strategies even when data is sparse.

In future work, we would like to extend our order fulfillment scenario to the cases where inventory replenishment is allowed. Additionally, we want to explore potential improvements for VLS by incorporating recent advances in distributional reinforcement learning [4–6,14,28] to enable a better risk-sensitive strategy and explore offline reinforcement learning to train on historical data without the need for a simulator.

Acknowledgement. This work is partially funded by the NSF grant CCF-2217076.

References

1. Acimovic, J., Graves, S.C.: Making better fulfillment decisions on the fly in an online retail environment. Manuf. Serv. Oper. Manag. **17**(1), 34–51 (2015)
2. Alptekinoglu, A., Tang, C.: A model for analyzing multi-channel distribution systems. Eur. J. Oper. Res. **163**(3), 802–824 (2005)
3. Andrews, J.M., Farias, V.F., Khojandi, A.I., Yan, C.M.: Primal-dual algorithms for order fulfillment at urban outfitters, Inc. INFORMS J. Appl. Anal. **49**(5), 355–370 (2019)
4. Bellemare, M.G., Dabney, W., Munos, R.: A distributional perspective on reinforcement learning. In: International Conference on Machine Learning, pp. 449–458. PMLR (2017)
5. Dabney, W., Ostrovski, G., Silver, D., Munos, R.: Implicit quantile networks for distributional reinforcement learning. In: International Conference on Machine Learning, pp. 1096–1105. PMLR (2018)
6. Dabney, W., Rowland, M., Bellemare, M., Munos, R.: Distributional reinforcement learning with quantile regression. In: Proceedings of the AAAI Conference on Artificial Intelligence, vol. 32 (2018)
7. Hendrycks, D., Gimpel, K.: Gaussian error linear units (GELUs). arXiv preprint arXiv:1606.08415 (2016)
8. Jasin, S., Sinha, A.: An LP-based correlated rounding scheme for multi-item ecommerce order fulfillment. Oper. Res. **63**(6), 1336–1351 (2015)
9. Jiang, D., Li, X., Aneja, Y., Wang, W., Tian, P.: Integrating order delivery and return operations for order fulfillment in an online retail environment. Comput. Oper. Res. **143**, 105749 (2022)
10. Kingma, D., Ba, J.: Adam: a method for stochastic optimization. In: International Conference on Learning Representations (2014)
11. Lee, H.L.: A multi-echelon inventory model for repairable items with emergency lateral transshipments. Manage. Sci. **33**(10), 1302–1316 (1987)
12. Lei, Y.M., Jasin, S., Sinha, A.: Joint dynamic pricing and order fulfillment for e-commerce retailers. Manuf. Serv. Oper. Manag. **20**(2), 269–284 (2018)
13. Lim, Y.F., Jiu, S., Ang, M.: Integrating anticipative replenishment allocation with reactive fulfillment for online retailing using robust optimization. Manuf. Serv. Oper. Manag. **23**(6), 1616–1633 (2021)
14. Ma, Y., Jayaraman, D., Bastani, O.: Conservative offline distributional reinforcement learning. In: Advances in Neural Information Processing Systems, vol. 34 (2021)
15. Mahar, S., Bretthauer, K.M., Venkataramanan, M.: The value of virtual pooling in dual sales channel supply chains. Eur. J. Oper. Res. **192**(2), 561–575 (2009)
16. Mnih, V., et al.: Playing atari with deep reinforcement learning. arXiv preprint arXiv:1312.5602 (2013)
17. Olist, D., Sionek, A.: Brazilian e-commerce public dataset by Olist (2018). https://www.kaggle.com/dsv/195341
18. Puterman, M.L.: Markov Decision Processes: Discrete Stochastic Dynamic Programming, 1st edn. Wiley, Hoboken (1994)
19. Quanz, B., Deshpande, A., Xing, D., Liu, X.: Learning to shortcut and shortlist order fulfillment deciding. arXiv preprint arXiv:2110.01668 (2021)

20. Ramakrishna, K.S.: A two-item two-warehouse periodic review inventory model with transshipment (2020). https://researchdata.smu.edu.sg/articles/thesis/A_two-item_two-warehouse_periodic_review_inventory_model_with_transshipment/12291293

21. Russell, S., Norvig, P.: Artificial Intelligence: A Modern Approach, 3rd edn. Prentice Hall, Hoboken (2010)

22. Schaul, T., Quan, J., Antonoglou, I., Silver, D.: Prioritized experience replay. arXiv preprint arXiv:1511.05952 (2015)

23. Sopadjieva, E., Dholakia, U.M., Benjamin, B.: A study of 46,000 shoppers shows that omnichannel retailing works. Harv. Bus. Rev. 3, 1–2 (2017)

24. Sutton, R.S., Barto, A.G.: Reinforcement Learning: An Introduction, 2nd edn. The MIT Press, Cambridge (2018). https://incompleteideas.net/book/the-book-2nd.html

25. US Department of Commerce: Us census bureau news (2022). https://www2.census.gov/retail/releases/historical/ecomm/21q4.pdf

26. Vaswani, A., et al.: Attention is all you need. In: Advances in Neural Information Processing Systems, vol. 30. Curran Associates, Inc. (2017)

27. Xu, P.J., Allgor, R., Graves, S.C.: Benefits of reevaluating real-time order fulfillment decisions. Manuf. Serv. Oper. Manag. 11(2), 340–355 (2009)

28. Yang, D., Zhao, L., Lin, Z., Qin, T., Bian, J., Liu, T.Y.: Fully parameterized quantile function for distributional reinforcement learning. In: Advances in Neural Information Processing Systems, vol. 32 (2019)

Deep Reinforcement Learning for Stock Trading with Behavioral Finance Strategy

Shilong Deng[1], Zetao Zheng[1], Hongcai He[1], and Jie Shao[1,2(✉)]

[1] University of Electronic Science and Technology of China, Chengdu 611731, China
{dengshilon,ztzheng,hehongcai}@std.uestc.edu.cn, shaojie@uestc.edu.cn
[2] Sichuan Artificial Intelligence Research Institute, Yibin 644000, China

Abstract. Stock trading is a challenging task and has attracted extensive attention from artificial intelligence researchers. Deep Reinforcement Learning (DRL) approaches, which directly generate trading decisions by maximizing the expected return, have made more breakthroughs than supervised learning approaches. However, most existing DRL approaches ignore essential knowledge in the finance field (e.g., the momentum investment strategy) and fail to generate sophisticated trading decisions. Specifically, the momentum investment strategy, which is popular among professional investors for its profitability, aims to buy stocks that were past winners (with above-average performance) and sell past losers (with below-average performance). Inspired by this concept, we propose Momentum Investment Twin Delay Deep Deterministic (MITD3) policy gradient algorithm based on TD3. At the core of MITD3, a Momentum Investing critic (MI-critic) computes the Q-value according to the state-action pair and advantage historical performance of the actor, and gives a higher Q-value to encourage past winners and a lower value to penalize past losers. In addition, we devise the cross-time module in MITD3, in which the current actor interacts with the environment by receiving past states and making trading decisions, to evaluate the advantage historical performance of the actor. Experimental results on real market data show that our proposed MITD3 outperforms state-of-the-art DRL approaches and generates intuitively explainable stock trading decisions.

Keywords: Deep reinforcement learning · Stock trading · Behavioral finance · Momentum investment strategy

1 Introduction

The stock market, a worldwide platform for capital to flow between businesses and investors, plays a crucial role in the modern financial system. Investment companies and hedge funds are making great efforts to design a profitable stock trading strategy. It has become an attractive yet challenging problem for researchers to apply Artificial Intelligence (AI) approaches in the stock trading task. Existing studies in AI for stock trading can be roughly categorized into three types: classic Machine Learning (ML), Deep Learning (DL), and Deep Reinforcement Learning (DRL) approaches.

© The Author(s), under exclusive license to Springer Nature Switzerland AG 2023
X. Yang et al. (Eds.): ADMA 2023, LNAI 14177, pp. 535–549, 2023.
https://doi.org/10.1007/978-3-031-46664-9_36

Mainstream supervised ML and DL approaches model the stock trading task into two steps: predicting and trading. In the predicting step, they make the prediction of future price values [24] or price trends [18]. In the trading step, they make trading decisions designed by humans, e.g., selling all stocks whose prices are predicted to fall and buying specific shares of top-k stocks whose prices are predicted to rise. These approaches have longstanding challenges: (1) the prediction of future price values or price trends is not equivalent to profitability. Profitability is affected by a lot of factors, such as model deviation, trading decisions, simulation-to-reality gap, etc. (2) handcraft trading decisions suffer from little theoretical guarantee, poor generalization ability, and high human resources cost.

DRL approaches, which directly generate trading decisions by maximizing the expected return [11,22,23], do not encounter the above problems in supervised learning approaches. Moreover, DRL is powerful in solving dynamic decision making problems by interacting with the environment in a trial-and-error manner. Figure 1 shows the different stock trading frameworks in supervised learning and deep reinforcement learning.

However, most existing DRL approaches ignore introducing the essential knowledge in the financial field (e.g., the momentum investment strategy [1,6]) to generate sophisticated trading decisions. Integrating finance theory with DRL structure is a promising but unexplored prospect to make improvements. In this paper, we integrate momentum investment strategy into Twin Delay Deep Deterministic policy gradient algorithm (TD3) [5], and propose Momentum Investment TD3 (MITD3) for the stock trading task. The momentum investment strategy, which is proposed by behavioral finance researchers, has been implemented by many professional investors and hedge funds for its profitability. It aims to buy stocks that were past winners (with above-average performances) and sell past losers (with below-average performances). We utilize the historical performances of the market index to represent the average performances, and an actor with higher (lower) historical performances than the average performances is defined as a past winner (loser). Besides, a positive signal is released when momentum investing investors buy stocks that were past winners. In other words, the behaviors of past winners are encouraged, and vice versa. Inspired by this concept, we devise a Momentum Investing critic (MI-critic) at the core of MITD3. Our MI-critic, which computes the Q-value according to the state-action pair and advantage historical performance of the actor, would give a higher Q-value to encourage the actor that was the past winner and a lower value to penalize the past loser. The advantage historical performance of the actor is evaluated by the cross-time module, in which the current actor interacts with the environment by receiving past states and making trading decisions. The main contributions of our work are as follows:

- We integrate momentum investment strategy in behavioral finance into the DRL structure, and propose Momentum Investment TD3 (MITD3) for the stock trading task. To our best knowledge, we are the first to integrate finance theory with DRL for the stock trading task.

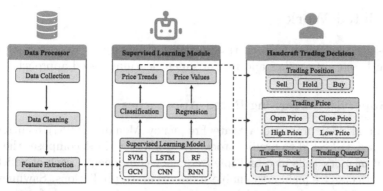

(a) Overview of stock trading framework in supervised learning.

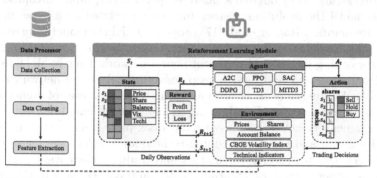

(b) Overview of stock trading framework in deep reinforcement learning.

Fig. 1. Comparison of stock trading frameworks between supervised learning and deep reinforcement learning. DRL approaches directly generate trading decisions, while supervised learning approaches need to handcraft them.

- Trading decisions made by MITD3 are intuitively explainable compared with existing DRL approaches. The actor optimizes trading decisions by maximizing the Q-value given by the critic, and the momentum investing critic encourages the actor that was the past winner by giving a higher Q-value, and vice versa. In this manner, trading decisions become intuitively explainable.
- We run our experiments on real-world stock market data and the proposed MITD3 achieves significant improvements against state-of-the-art DRL approaches.

The remainder of this paper is organized as follows. Section 2 introduces the related works. Section 3 describes preliminary knowledge in DRL for the stock trading task. Section 4 presents our proposed model. Section 5 demonstrates experiment settings and results. Section 6 makes the conclusion.

2   Related Work

We classify the existing stock trading studies into two categories: supervised learning based approaches and reinforcement learning based approaches.

2.1   Supervised Learning Based Approaches

Zhang et al. [24] proposed a State Frequency Memory (SFM) recurrent network based on LSTM to predict stock price values. SFM decomposes the hidden states of memory cells into multiple frequency components, each of which models a particular frequency to enable more accurate predictions. Sawhney et al. [18] proposed spatio-temporal hypergraph convolution network for stock price trends forecasting. They devised a gated temporal hypergraph convolution mechanism to model the evolution of stock movements related by a hypergraph in a time-aware manner. Rather et al. [17] proposed a hybrid model to predict stock returns. This model is constituted of autoregressive moving average model, exponential smoothing model, and recurrent neural network. Ding et al. [4] extracted events from news text and used a deep convolutional neural network to model the combined influence of short-term and long-term influences of events on stock price movements. Long et al. [16] proposed an attention-based bidirectional long short-term memory network to predict the stock price trends. They utilized the knowledge graph and graph embedding techniques to select the relevant stocks of the target for constructing the market and trading information.

However, these supervised learning based approaches have to design handcraft trading decisions based on the prediction results, while reinforcement learning based approaches directly generate trading decisions by algorithms.

2.2   Reinforcement Learning Based Approaches

Yang et al. [23] proposed an ensemble trading strategy using three actor-critic based reinforcement learning algorithms: Proximal Policy Optimization (PPO), Advantage Actor Critic (A2C), and Deep Deterministic Policy Gradient (DDPG). This ensemble strategy automatically selects the best performing agent to trade based on the Sharpe ratio in a period. Li et al. [11] extended the Deep Q-network (DQN) and the Asynchronous Advantage Actor Critic (A3C) for better adapting to the trading market. They utilized the Stacked Denoising Autoencoders (SDAEs) and LSTM as parts of the function approximator to extract robust market representations and resolve the financial time series dependence, respectively. Wu et al. [22] proposed Gated Deep Q-learning trading strategy (GDQN) and Gated Deterministic Policy Gradient trading strategy (GDPG) to ensure stable returns in different market conditions. They designed the reward function with risk-adjusted ratio and applied the Gated Recurrent Unit (GRU) to extract informative financial features. Li et al. [10] proposed an actor-critic based trading framework, where LSTM is shared by the value network and policy network to extract features. Not only the actor but also the critic are considered to make the final trading decisions during testing. Chen and

Table 1. Notations in problem formulation.

Notation	Description
s_t	State at time step t
a_t	Action at time step t
r_t	Reward given by the environment at time step t
π	Policy generated by the agent
$P(s_{t+1}\|s_t, a_t)$	Transition probability from state s_t to state s_{t+1} taking the action a_t
γ	Discount factor
b_t	Balance of account at time step t
p_t	Closing price of stocks at time step t
sh_t	Shares for each stock at time step t
vix_t	CBOE Volatility Index
$techi_t$	Technical indicators
v_t	Total value of assets at time step t
$Q^\pi(s, a)$	State-action value (Q-value) function, which is the expected return starting in state s, taking action a, and acting according to policy π until the terminal state

Gao [2] proposed Deep Recurrent Q-network (DRQN) by replacing the fully connected layer in DQN with a recurrent LSTM layer. They stated that introducing recurrence can improve the performance of DQN by capturing useful information in sequential data. Dang [3] applied Deep Q-Network (DQN), double DQN [8] and dueling DQN [21] to make trading decisions on more than 7,000 US-based stocks. This model does not use any external information but only the historical data of stock prices for trading.

However, these approaches ignore essential knowledge in the finance field (e.g., the momentum investment strategy) and fail to generate sophisticated trading decisions. To this end, we propose MITD3 by integrating momentum investment strategy in behavioral finance with the DRL structure to make improvements. The proposed MITD3 model outperforms state-of-the-art DRL approaches and generates intuitively explainable stock trading decisions.

3   Problem Formulation

Stock trading aims to buy and sell shares in companies to maximize the investment profit and minimize the risk. DRL agents, whose objective is to maximize the expected return, generate trading decisions by interacting with the stock market environment over a sequence of length T. Notations in this section are summarized in Table 1.

We formulate the stock trading as a Markov Decision Process (MDP). An MDP can be defined as a 5-tuple $< \mathcal{S}, \mathcal{A}, \mathcal{R}, \mathcal{P}, \gamma >$, where $\mathcal{S} = \{s_1, \ldots, s_T\}$ is a finite set of valid states, $\mathcal{A} = \{a_1, \ldots, a_T\}$ is a set of valid actions, $R : \mathcal{S} \times \mathcal{A} \times \mathcal{S} \to \mathbb{R}$ is the reward function with $r_t = R(s_t, a_t, s_{t+1})$, $\mathcal{P} : S \times A \to \mathbb{P}$ is the transition probability function with $P(s_{t+1}|s_t, a_t)$ being the transition probability from state s_t to state s_{t+1} taking the action a_t, and $\gamma \in (0, 1]$ is the discount factor. Following FinRL [13–15], a DRL library for stock trading, an MDP in the context of stocking trading can be described as follows:

State $\mathcal{S}$. The state contains a set of features that describe the state of trading stocks. In general, different types of information such as historical prices, trading volumes, sentiment scores, and financial statements can be used as the state of the stocks. The state s_t contains the following features:

- Balance $b_t \in \mathbb{R}_+$: the amount of cash left in the account at time step t.
- Closing price $p_t \in \mathbb{R}_+^m$: one of the most commonly used features. It is defined as a vector $p_t = [p_{t,1}, p_{t,2}, \ldots, p_{t,m}]^T$, where m represents the number of stocks.
- Share owned $sh_t \in \mathbb{Z}_+^m$: shares for each stock. It is defined as a vector $sh_t = [sh_{t,1}, sh_{t,2}, \ldots, sh_{t,m}]^T$.
- CBOE Volatility Index vix_t: an important real-time index that represents the market's expectations for volatility and provides a quantifiable measure of market risk and investors' sentiments.
- Technical indicators $techi_t$: including Moving Average Convergence Divergence (MACD), Bolling Bands (BOLL), Relative Strength Index (RSI), Commodity Channel Index (CCI), Directional Movement Index (DMI), and Simple Moving Average (SMA).

Action $\mathcal{A}$. The action, which represents the trading decision made by the agent that interacts with the stock market environment, is specified by a vector $a_t = [a_{t,1}, a_{t,2}, \ldots, a_{t,m}]$. $a_{t,i} \in \{-k, \ldots, -1, 0, 1, \ldots, k\}$ represents the number of shares traded by the agent, e.g., "buy 2 shares of MSFT" or "sell 2 shares of MSFT" are 2 or -2, respectively.

Reward $\mathcal{R}$. The reward received from the environment is the incentive for the agent to learn a better action. Intelligence, and its associated abilities, can be understood as subserving the maximization of reward by an agent acting in its environment [20]. The reward is defined as the change in total value of assets:

$$r(s_t, a_t, s_{t+1}) = ||v_{t+1}||_1 - ||v_t||_1, \tag{1}$$

$$v_t = p_t^T \cdot sh_t + b_t. \tag{2}$$

Policy π. The policy, which denotes the trading strategy, is essentially a network that outputs a probability distribution of action a given the current state s as the input. The agent's goal is to optimize the policy to maximize the expected return.

State-Action Value Function $Q^\pi(s, a)$. The state-action value function, also the Q-value function, is the expected return starting in state s, taking action a, and acting according to policy π until the terminal state. The Bellman equations for the state-action value (Q-value) functions are defined in Eq. (3):

$$Q^\pi(s, a) = \mathbb{E}[r(s, a, s') + \gamma \mathbb{E}[Q^\pi(s', a')]], \tag{3}$$

where s' is the next state of s, and a' is the action taken at state s', according to policy π.

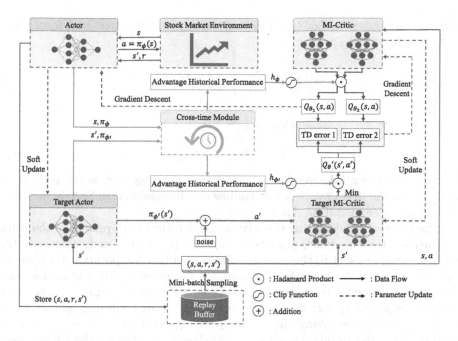

Fig. 2. Illustration of our proposed model. Cross-time module evaluates the advantage historical performance of the actor and target actor. The momentum investment strategy is implemented by the critic and target critic network, which utilizes advantage historical performance to encourage or penalize the behaviors of actor.

4   Method

Most existing DRL approaches ignore essential knowledge in the finance field (e.g., the momentum investment strategy) and fail to generate sophisticated trading decisions. To this end, we integrate momentum investment strategy into Twin Delay Deep Deterministic (TD3) policy gradient algorithm [5] and propose Momentum Investment TD3 (MITD3) for the stock trading task. In this section, we first describe the cross-time module, and then describe the Momentum Investing critic (MI-critic), and finally we summarize the detailed process to train the MITD3 agent. Our proposed MITD3 model is illustrated in Fig. 2. The detailed process to train the MITD3 agent is summarized in Algorithm 1.

4.1   Cross-Time Module

The advantage historical performance of the actor is evaluated by the cross-time module, in which the current actor interacts with the environment by receiving past states and making trading decisions. We utilize the Annualized Return (AR) over a window of past days to measure the historical performance, and AR of the Dow Jones Index (DJI) to represent the average historical performance. The advantage historical performance is defined as the gap of AR between the actor and DJI, as shown in Eq. (6). The actor is a past winner or past loser when $h_\phi > 0$ or $h_\phi < 0$, respectively.

$$AR_\phi = (\frac{\sum_{i=t-w}^{t-1} r_i}{v_0})^{\frac{365}{w-1}} - 1, \tag{4}$$

$$AR_{dji} = (\frac{p_{t-1}}{p_{t-w}})^{\frac{365}{w-1}} - 1, \tag{5}$$

$$h_\phi = AR_\phi - AR_{dji}, \tag{6}$$

where r_i is the reward given by the environment on day i, v_0 is the total value of assets on the first day, w is the window size investigated in Sect. 5, and p_i is the closing price of DJI on day i.

The process of calculating the advantage historical performance of the actor via the cross-time module is described as follows: (1) State s_t and policy π_ϕ of the actor network are sent to the cross-time module; (2) A temporary actor network and a temporary stock market environment are created; (3) The initial state of the environment is reset to s_{t-w}, where w is the window size; (4) The actor interacts sequentially with the environment by receiving state s_{t-w}, giving action $a_{t-w} = \pi_\phi(s_{t-w})$, receiving state s_{t-w+1}, giving action $a_{t-w+1} = \pi_\phi(s_{t-w+1})$, until reaching the terminal state s_{t-1}; (5) Reward r given by the environment at each time step is recorded to calculate the advantage historical performance h_ϕ in Eq. (6).

For the target actor, state s_{t+1} and target policy $\pi_{\phi'}$ are sent to the cross-time module, and the other steps are similar.

4.2 Momentum Investing Critic

As defined in Sect. 4.1, an actor with an advantage historical performance $h_\phi > 0$ or $h_\phi < 0$ is the past winner or past loser, respectively. The momentum investment strategy aims to buy stocks that were past winners and sell past losers [1,6]. A positive signal is released when momentum investing investors buy stocks that were past winners. In other words, the behaviors of past winners are encouraged, and vice versa. Inspired by this concept, we devise a Momentum Investing critic (MI-critic) at the core of MITD3. Our MI-critic, which computes the Q-value according to the state-action pair and advantage historical performance of the actor, would give a higher Q-value to encourage the actor that was the past winner and a lower value to penalize the past loser. The state-action value (Q-value) functions of the critic and target critic are shown in Eq. (7) and Eq. (8), respectively.

$$Q_{\theta,i}(s,a) \leftarrow Q_{\theta,i}(s,a) \odot \text{clip}((h_\phi, -c_2, c_2) + 1), \tag{7}$$

$$Q'_{\theta'}(s,a) \leftarrow Q'_{\theta'}(s,a) \odot \text{clip}((h'_\phi, -c_2, c_2) + 1), \tag{8}$$

where $Q_{\theta,i}(s,a)$ and $Q'_{\theta'}(s,a)$ on the right side of the arrows are calculated following Eq. (3), c_2 is the clip boundary value of the MI-critic investigated in

Sect. 5, h_ϕ and h'_ϕ are the advantage historical performance of the actor and target actor, respectively. In practice, the networks are usually optimized using mini-batch stochastic gradient descent, which is the reason we use the Hadamard product operator here.

As illustrated in Fig. 2, MITD3 can be divided into four-fold: (1) the simulated stock market environment and replay buffer for experience collection and off-policy training; (2) the (target) actor network for generating stock trading decisions; (3) the cross-time module for computing advantage historical performances of the (target) actor network; (4) the (target) MI-critic network for evaluating the profitability (measured by Q-value) of the (target) actor, and encouraging past winners or penalizing past losers. The actor network optimizes its profitability by maximizing the Q-value, and the MI-critic network optimizes its evaluation by minimizing the Temporal-Difference (TD) error.

4.3   Training Process of MITD3

The detailed process to train the MITD3 agent is summarized in Algorithm 1. In lines 1–2, we initialize the parameters of networks and create an empty replay buffer. From line 3 to line 16, the following procedures are repeated for the total T training step: (1) in lines 4–6, the MITD3 agent explores the environment and the transition tuples (s, a, r, s') are stored in the replay buffer, and N transitions are randomly sampled from the replay buffer during training; (2) in line 7, state s' is sent to the target actor, and a small amount of random noise is added to the target policy to generate the target action a'; (3) in line 8, the advantage historical performances of the actor and target actor are computed by the cross-time module, where c_2 is the clip boundary value investigated in Sect. 5; (4) in line 9, new Q-values of MI-critic and target MI-critic are computed according to the raw Q-values, h_ϕ and $h_{\phi'}$, as shown in Eq. (7) and Eq. (8); (5) in line 10, the target value of the MI-critic is computed according to the new Q-value and reward r, where d is 1 when s' is the terminal state, and d is 0 otherwise; (6) in line 11, the MI-critic network is updated by gradient descent with the Adam optimizer [9], to minimize the Temporal-Difference (TD) error; (7) in lines 12–15, after a fixed number of updates k to the critic, the actor network is updated by gradient descent with the Adam optimizer, to maximize the new Q-value given by the first MI-critic network, and the target actor and target MI-critic are updated using soft update.

5   Experiments

In this section, we conduct experiments to test the proposed MITD3 model. We first provide the details of the experimental settings. Then, the comparison between the MITD3 and baselines is presented. Finally, we verify the effectiveness of MI-critic and discuss the influences of clip boundary value and window size on MITD3.

Algorithm 1 Momentum Investment TD3

1: Initialize actor network ϕ, MI-critic networks θ_1, θ_2, and empty replay buffer $\mathcal{D}$
2: Initialize target networks $\phi' \leftarrow \phi, \theta'_1 \leftarrow \theta_1, \theta'_2 \leftarrow \theta_2$
3: **for** t = 1 to T **do**
4:     Observe state s, select action $a \sim \pi_\phi(s) + \epsilon$,
        $\epsilon \sim \mathcal{N}(0, \sigma)$, observe reward r and new state s'
5:     Store transition tuple (s, a, r, s') in $\mathcal{D}$
6:     Randomly sample mini-batch of N transitions from $\mathcal{D}$
7:     Compute target actions:
        $a' \leftarrow \pi_{\phi'}(s') + \epsilon, \quad \epsilon \sim \text{clip}(\mathcal{N}(0, \sigma), -c_1, c_1)$
8:     Compute actor's advantage historical performances:
        $h_\phi = AR_\phi - AR_{dji}, \quad h_{\phi'} = AR_{\phi'} - AR_{dji}$
9:     Compute new Q-values of (target) MI-critic:
        $Q'_{\theta'_i}(s', a') \leftarrow Q'_{\theta'_i}(s', a') \odot \text{clip}((h_{\phi'}, -c_2, c_2) + 1)$
        $Q_{\theta_i}(s, a) \leftarrow Q_{\theta_i}(s, a) \odot \text{clip}((h_\phi, -c_2, c_2) + 1)$
10:    Compute target value with new Q-values (i=1,2):
        $y \leftarrow r + \gamma(1 - d) \min_i [Q'_{\theta'_i}(s', a')]$
11:    Update MI-critic networks by gradient descent (i=1,2):
        $\nabla_{\theta_i} \frac{1}{N} \sum [Q_{\theta_i}(s, a) - y]^2$
12:    **if** t mod k = 0 **then**
13:        Update actor network by gradient descent:
            $\nabla_\phi \frac{1}{N} \sum Q_{\theta_1}(s, a)$
14:        Soft update the target networks (i=1,2):
            $\theta'_i \leftarrow \tau\theta'_i + (1 - \tau)\theta_i, \quad \phi' \leftarrow \tau\phi' + (1 - \tau)\phi$
15:    **end if**
16: **end for**

5.1   Experimental Settings

Datasets. The experiment datasets are from the real-world market to comprehensively evaluate our model. The stock data used in our experiments are from 30 constituent stocks of Dow Jones Index (DJI). Specifically, we use the daily closing price of stocks and the CBOE Volatility Index from 2009-01-01 to 2020-12-31 as the training set, and the daily closing price from 2021-01-01 to 2021-12-31 as the testing set. The technical indicators are calculated based on the closing price of stocks. Balance values and stock share values are given by the simulated stock market environment. All the stock data are downloaded from Yahoo Finance[1].

Baselines. We compare MITD3 with state-of-the-art DRL approaches: PPO [19], SAC [7], DDPG [12] and TD3 [5]. We run the above baselines and our proposed model on an open-source library FinRL [13–15].

Evaluation Metrics. We evaluate the performance of baselines and our model by the following five metrics: (1) Annualized Return (AR), (2) Cumulative Return (CR), (3) Annualized Volatility (AV), (4) Sharpe Ratio (SR) and (5) Maximum Drawdown (MDD). From the view of measuring trading performance, we prefer higher, higher, lower, higher and higher values for the five metrics, respectively.

Implementation Details. We implement our model via PyTorch. We adopt the Adam optimizer [9] on the training process with a single NVIDIA 3090 GPU. The training step is 10^6, and the batch size is 2048. We set the learning rate

---

[1] https://finance.yahoo.com.

to 3×10^{-5} and γ to 0.985. The dimension of hidden layers in actor and critic network is set to 512. The clip boundary c_2 in MI-critic is set to 0.05, and the window size w in cross-time module is set to 30. Results are averaged over 5 runs with random initialization seeds for all models.

Table 2. Performance comparison of different models over 5 different runs. All metrics are averaged values.

Models	Initial value	Final value↑	AR↑	CR↑	AV↓	SR↑	MDD↑
PPO	1000000	1210663	21.06%	21.06%	15.18%	1.33	−8.03%
SAC	1000000	1231967	23.19%	23.19%	**13.69%**	1.59	−6.51%
DDPG	1000000	1223036	23.32%	23.32%	14.62%	1.45	−7.99%
TD3	1000000	1234058	23.40%	23.40%	13.93%	1.59	−6.35%
MITD3 (N)	1000000	1171284	17.12%	17.12%	14.18%	1.24	−8.22%
MITD3 (ours)	1000000	**1283024**	**28.30%**	**28.30%**	13.74%	**1.89**	**−5.59%**

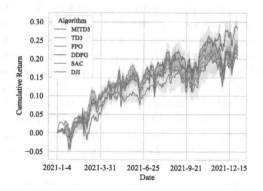

Fig. 3. Cumulative return of MITD3, baselines and market index, over 5 different runs. We run the experiment on the constituent stocks of Dow Jones Index (DJI).

5.2 Performance Comparisons

The performance comparison of the proposed MITD3 model with baselines is reported in Table 2, where MITD3 (N) is studied in Sect. 5.3. The compared models including PPO, SAC, DDPG, TD3 and our proposed MITD3 are evaluated by computing the averaged metrics over 5 runs. According to Table 2, we can observe that: (1) the TD3 model, which beats SAC slightly, outperforms the other baselines in terms of most metrics; (2) overall, our MITD3 outperforms all baselines in most metrics. More specifically, our model exceeds the best baseline TD3 by 20.94% and 18.86% in terms of AR and SR, respectively. Furthermore,

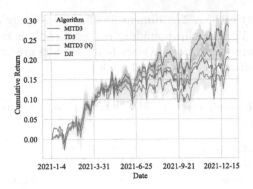

Fig. 4. Cumulative return of MITD3, TD3, MITD3 (N) and market index, over 5 different runs. We run the experiment on the constituent stocks of Dow Jones Index (DJI).

with an SR of 1.89 and an MDD of -5.59%, our proposed MITD3 model enjoys higher profit at lower risk.

Figure 3 shows the cumulative return of all algorithms in the U.S. stock market during the backtesting. Results are over 5 runs with random initialization seeds for all models. We include Dow Jones Index (DJI) as the market index to represent the buy-and-hold strategy. From the tendency of cumulative returns, we have the following observations. (1) The market index rises with volatility, and the testing period is a bull market generally. (2) In the early stage, many algorithms perform similarly and are beat by DJI. However, as time goes on, all algorithms perform better and better and beat DJI eventually. (3) Our proposed model outperforms all the other algorithms since the middle stage, which indicates the trading policy gains benefits from the guidance of the MI-critic. More specifically, the trading agent takes advantage of the momentum investment strategy and generates a sophisticated policy for the stock trading task. The trading decisions become intuitively explainable by integrating behavioral finance strategy with the DRL structure.

5.3   Ablation Studies

In this section, we study (i) whether MI-critic is contributing to the impressive performance of MITD3; (ii) the impact of clip boundary value c_2 on the performance; (iii) the impact of window size w on the performance.

(i) We evaluate the effect of MI-critic by replacing the advantage historical performances h_ϕ and $h_{\phi'}$ in MI-critic with random noise ϵ, where $\epsilon \sim \mathcal{N}(0, \sigma)$. MITD3 with noise is denoted as MITD3 (N), and the performance comparison is shown in Table 2 and Fig. 4. We can observe that MITD3 (N) is beat by all competitors (including the market index DJI) during the backtesting. Even worse, MITD3 (N) fails to achieve any improvement over TD3, let alone MITD3. Therefore, MI-critic is contributing to the impressive performance of MITD3 and

Table 3. Performance of MITD3 under different clip boundary values of MITD3, over 5 different runs. The window size is set to 30. All metrics are averaged values.

Clip boundary	AR↑	SR↑
$c_2 = 0.01$	22.41%	1.63
$c_2 = 0.025$	23.02%	1.71
$c_2 = 0.05$	**28.30%**	1.89
$c_2 = 0.10$	26.04%	**1.92**
$c_2 = 0.20$	21.52%	1.34
$c_2 = 0.25$	19.87%	1.33
$c_2 = 0.50$	20.69%	1.42

Table 4. Performance of MITD3 under different window sizes in the cross-time module, over 5 different runs. The clip boundary value is set to 0.05. All metrics are averaged values.

Window size	AR↑	SR↑
$w = 10$	17.09%	1.24
$w = 20$	23.46%	1.53
$w = 30$	28.30%	**1.89**
$w = 60$	28.74%	1.86
$w = 120$	**29.23%**	1.84

the momentum investment strategy is helpful to generate a sophisticated policy for the stock trading task.

(ii) Impact of clip boundary value. The boundary value c_2 in Eq. (7) and Eq. (8) is an important hyperparameter in MI-critic. To evaluate the influence of the boundary value c_2 in the clip function in MI-critic, we conduct experiments under different boundary values, over 5 different runs. The window size w in the cross-time module is set to 30. The results are reported in Table 3. From Table 3 we can see that: (1) MITD3 performs best on profitability (with an AR of 28.30%) when the clip boundary c_2 is set to 0.05; (2) MITD3 performs best with an SR of 1.92 at the clip boundary value of 0.1, with respect to return on risk; (3) with an AR of 28.30% and an SR of 1.89, 0.05 is the most suitable value for the clip boundary; (4) the performance of MITD3 is relatively stable when $c_2 \geq 0.20$, which is probably because the advantage historical performance is not usually larger than 0.20.

(iii) Impact of window size. The window size w in Eq. (4) and Eq. (5) is an important hyperparameter in the cross-time module. To evaluate its influence, we conduct experiments under different window sizes, over 5 different runs. The clip boundary value c_2 is set to 0.05. The results are reported in Table 4. From Table 4 we can find that using about 30 days of sequence length window for training leads to relatively higher profit and lower risk. A larger value of window size for the cross-time module can result in better performance but not obvious.

6 Conclusion

In this paper, we have explored the potential of integrating finance theory with the deep reinforcement learning structure. We integrate the momentum investment strategy in behavioral finance into TD3, and propose Momentum Investment TD3 (MITD3) for the stock trading task. The experimental results on real-world stock market data show that: (1) our proposed model outperforms state-of-the-art DRL models and market index DJI; (2) the trading agent takes advantage of the momentum investment strategy and generates a sophisticated policy for the stock trading task, and the trading decisions are intuitively explainable compared with existing DRL approaches.

Acknowledgements. This work is supported by the National Natural Science Foundation of China (No. 62276047).

References

1. Chan, L.K.C., Jegadeesh, N., Lakonishok, J.: Momentum strategies. J. Finance **51**(5), 1681–1713 (1996)
2. Chen, L., Gao, Q.: Application of deep reinforcement learning on automated stock trading. In: IEEE 10th International Conference on Software Engineering and Service Science, ICSESS 2019, Beijing, China, 18–20 October 2019, pp. 29–33 (2019)
3. Dang, Q.: Reinforcement learning in stock trading. In: Advanced Computational Methods for Knowledge Engineering - Proceedings of the 6th International Conference on Computer Science, Applied Mathematics and Applications, ICCSAMA 2019, Hanoi, Vietnam, 19–20 December 2019, pp. 311–322 (2019)
4. Ding, X., Zhang, Y., Liu, T., Duan, J.: Deep learning for event-driven stock prediction. In: Proceedings of the Twenty-Fourth International Joint Conference on Artificial Intelligence, IJCAI 2015, Buenos Aires, Argentina, 25–31 July 2015, pp. 2327–2333 (2015)
5. Fujimoto, S., van Hoof, H., Meger, D.: Addressing function approximation error in actor-critic methods. In: Proceedings of the 35th International Conference on Machine Learning, ICML 2018, Stockholmsmässan, Stockholm, Sweden, 10–15 July 2018, pp. 1582–1591 (2018)
6. Grinblatt, M., Titman, S., Wermers, R.: Momentum investment strategies, portfolio performance, and herding: a study of mutual fund behavior. Am. Econ. Rev. **85**(5), 1088–1105 (1995)
7. Haarnoja, T., Zhou, A., Abbeel, P., Levine, S.: Soft actor-critic: off-policy maximum entropy deep reinforcement learning with a stochastic actor. In: Proceedings of the 35th International Conference on Machine Learning, ICML 2018, Stockholmsmässan, Stockholm, Sweden, 10–15 July 2018, pp. 1856–1865 (2018)
8. van Hasselt, H., Guez, A., Silver, D.: Deep reinforcement learning with double q-learning. In: Proceedings of the Thirtieth AAAI Conference on Artificial Intelligence, 12–17 February 2016, Phoenix, Arizona, USA, pp. 2094–2100 (2016)

9. Kingma, D.P., Ba, J.: Adam: a method for stochastic optimization. In: 3rd International Conference on Learning Representations, ICLR 2015, San Diego, CA, USA, 7–9 May 2015, Conference Track Proceedings (2015)
10. Li, J., Rao, R., Shi, J.: Learning to trade with deep actor critic methods. In: 11th International Symposium on Computational Intelligence and Design, ISCID 2018, Hangzhou, China, 8–9 December 2018, vol. 2, pp. 66–71 (2018)
11. Li, Y., Zheng, W., Zheng, Z.: Deep robust reinforcement learning for practical algorithmic trading. IEEE Access **7**, 108014–108022 (2019)
12. Lillicrap, T.P., et al.: Continuous control with deep reinforcement learning. In: 4th International Conference on Learning Representations, ICLR 2016, San Juan, Puerto Rico, 2–4 May 2016, Conference Track Proceedings (2016)
13. Liu, X., et al.: FinRL-meta: market environments and benchmarks for data-driven financial reinforcement learning. In: Thirty-Sixth Conference on Neural Information Processing Systems Datasets and Benchmarks Track (2022)
14. Liu, X., et al.: FinRL: a deep reinforcement learning library for automated stock trading in quantitative finance. In: Deep RL Workshop, NeurIPS 2020 (2020)
15. Liu, X., Yang, H., Gao, J., Wang, C.D.: FinRL: deep reinforcement learning framework to automate trading in quantitative finance. In: ICAIF 2021: 2nd ACM International Conference on AI in Finance, Virtual Event, 3–5 November 2021, pp. 1:1–1:9 (2021)
16. Long, J., Chen, Z., He, W., Wu, T., Ren, J.: An integrated framework of deep learning and knowledge graph for prediction of stock price trend: an application in Chinese stock exchange market. Appl. Soft Comput. **91**, 106205 (2020)
17. Rather, A.M., Agarwal, A., Sastry, V.N.: Recurrent neural network and a hybrid model for prediction of stock returns. Expert Syst. Appl. **42**(6), 3234–3241 (2015)
18. Sawhney, R., Agarwal, S., Wadhwa, A., Shah, R.R.: Spatiotemporal hypergraph convolution network for stock movement forecasting. In: 20th IEEE International Conference on Data Mining, ICDM 2020, Sorrento, Italy, 17–20 November 2020, pp. 482–491 (2020)
19. Schulman, J., Wolski, F., Dhariwal, P., Radford, A., Klimov, O.: Proximal policy optimization algorithms. CoRR abs/1707.06347 (2017)
20. Silver, D., Singh, S., Precup, D., Sutton, R.S.: Reward is enough. Artif. Intell. **299**, 103535 (2021)
21. Wang, Z., Schaul, T., Hessel, M., van Hasselt, H., Lanctot, M., de Freitas, N.: Dueling network architectures for deep reinforcement learning. In: Proceedings of the 33nd International Conference on Machine Learning, ICML 2016, New York City, NY, USA, 19–24 June 2016, pp. 1995–2003 (2016)
22. Wu, X., Chen, H., Wang, J., Troiano, L., Loia, V., Fujita, H.: Adaptive stock trading strategies with deep reinforcement learning methods. Inf. Sci. **538**, 142–158 (2020)
23. Yang, H., Liu, X., Zhong, S., Walid, A.: Deep reinforcement learning for automated stock trading: an ensemble strategy. In: ICAIF 2020: The First ACM International Conference on AI in Finance, New York, NY, USA, 15–16 October 2020, pp. 31:1–31:8 (2020)
24. Zhang, L., Aggarwal, C.C., Qi, G.: Stock price prediction via discovering multi-frequency trading patterns. In: Proceedings of the 23rd ACM SIGKDD International Conference on Knowledge Discovery and Data Mining, Halifax, NS, Canada, 13–17 August 2017, pp. 2141–2149 (2017)

Ensemble Learning Based Employment Recommendation Under Interaction Sparsity for College Students

Haiping Zhu, Yifei Zhao, Yuchen Wu, Yan Chen, Wenhao Li, Qinghua Zheng, and Feng Tian[✉]

Department of Computer Science and Technology, Xi'an Jiaotong University, Shaanxi Provincial Key Laboratory of Big Data Knowledge Engineering, Xi'an, China
fengtian@mail.xjtu.edu.cn

Abstract. Recommendation systems play a crucial role in helping college students find job opportunities. However, the sparsity of interactions in employment recommendation for college students poses a challenge for models based on historical user preferences. To address this issue, we propose a novel model called Ensemble Learning based Employment Recommendation under Interaction Sparsity for College Students (EERIS). The model comprises two components: a similarity information component that uses pooled users to determine the nearest neighbor in user similarity measurement, and a global interaction component that uses interaction vectors of user groups to enhance interactions. To evaluate the missing interactions, we propose a loss function called CellLoss. These components are combined based on ensemble learning to improve the model's generalization and scalability. Our experiments on two real-world datasets demonstrate the superior performance of the EERIS model. Ablation experiments further confirm that each component positively contributes to the model's performance. Additionally, we design a revised metric for better model testing. Overall, the proposed EERIS model effectively addresses the interaction sparsity in employment recommendation for college students and provides satisfactory recommendations to students.

Keywords: Recommendation system · Employment recommendation · Interaction sparsity · Ensemble learning

1 Introduction

In recent years, the development and wide application of recommendation systems have led to the gradual emergence of employment recommendation systems that play a crucial role in assisting college students in their job-hunting process [3]. Historical behavior can largely reflect users' interests and preferences [19], making it a vital part of achieving personalized recommendations. However, due to personal reasons, there may be situations where there are not many user-item

X. Yang et al. (Eds.): ADMA 2023, LNAI 14177, pp. 550–564, 2023.
https://doi.org/10.1007/978-3-031-46664-9_37

interactions (with item referring to a job in this paper) taking place. The sparsity of interactions makes it difficult to comprehensively track and record user-item interactions during the job-hunting process. As a result, it becomes challenging to make satisfactory recommendations simply based on historical user-item interactions.

Classical recommendation models have limitations when faced with interaction sparsity. For example, the collaborative filtering algorithm (CF) [4], which uses similar historical users' interactions to recommend items, struggles to determine the nearest neighbor when there is a lack of common interactions between users. Additionally, some job websites [18] mainly extract a large number of features from users' resumes and job descriptions using natural language processing and deep learning technology to achieve person-job fit. However, these deep learning-based recommendation models require a large amount of data, such as side information and interactions, to support their training [6]. Sparse interactions can result in insufficient samples, leading to problems in model convergence and early stopping. Furthermore, most existing employment recommendation models for college students [20] are single models, which limits their generalization and scalability when dealing with interaction sparsity scenarios.

To address the aforementioned issues, we propose a model called Ensemble Learning based Employment Recommendation under Interaction Sparsity for College Students(EERIS), which consists of two components from two different perspectives: user similarity and global interaction. Firstly, to address interaction sparsity and determine the nearest neighbor in user similarity measurement, we construct a similarity information component (Sim Module) based on an improved CF algorithm that considers group similarity. This module introduces the concept of pooled user to capture information about specific employment groups, and considers both individual users and user groups when measuring user similarity. Secondly, we construct a global interaction component (Ginter Module) based on an improved autoencoder [13] to process the user-item interactions. To mitigate the impact of interaction sparsity on the autoencoder, we propose interaction vectors of user groups to enhance interactions, which allows the model to converge better. We also define a loss function called CellLoss to evaluate the missing interactions on the validation set, so that the model can have an appropriate early stopping. Finally, we combine the above components to obtain EERIS based on the idea of ensemble learning [2], which shows that combining multiple basic models can result in better generalization and accuracy compared with a single model.

The main contributions of this paper are as follows:

(1) We propose the pooled user to capture information about user groups. This approach supports the determination of the nearest neighbor in user similarity measurement when they cannot be classified as caused by interaction sparsity.

(2) We propose the interaction vectors of user groups and the loss function Cell-Loss in Ginter Module to mitigate the impact of interaction sparsity on the autoencoder, which enable the model to converge more effectively and facilitate early stopping.

(3) By utilizing ensemble learning, we combine the aforementioned components to enhance the model's generalization and scalability. The experimental results demonstrate that EERIS outperforms the baseline models.

2   Related Work

2.1   Interaction Sparsity

User-item interactions are considered one of the most important inputs of the recommendation model, but the sparsity problem of interactions, such as rating, evaluation and etc., always exists, which has a negative impact on the recommendation model. To solve this problem, many researchers propose the recommendation model based on matrix factorization by introducing a hidden vector [7]. Some studies introduce extra side information to alleviate this problem [10]. The cross-domain recommendation and the multi-task mechanism have also been found to perform better than an independent model on a single task [16]. They implicitly introduce more data to make the model more fully trained and are used as a solution to address the issue of interaction sparsity. For example, Tang et al. [15] propose Progressive Layered Extraction with a novel sharing structure to reuse data as much as possible.

Introducing more kinds of supporting data is an effective way to solve interaction sparsity, but the context-specific information often reduces the applicability of the model and narrows the model to the small scope of the specific domain. There may also be some difficulties in collecting and using them due to the limitation of enabling technology, policy and privacy. So, we construct Ginter Module with the interaction vectors of user groups and a loss function named CellLoss to improve the ability of the model to deal with interaction sparsity without introducing any additional information.

2.2   Employment Recommendation

Employment recommendation has attracted a lot of attention in the field of social and academic research. The main idea of the existing employment recommendation systems is the CF algorithm. The CF mainly uses the target user's interaction history to find similar historical users to refer to and make recommendations [17]. Due to the lower degree of intellectualization of education departments and factors from the users themselves such as time and energy, there are not adequate user-item interactions being collected to serve for the employment recommendation. So, some works improve the measurement method for user similarity with side information in order to avoid over-reliance on user-item interactions [18]. For example, Liu et al. [8] propose a concept of a student portrait based on records of students during the school period to calculate user similarity.

In summary, most of the existing employment recommendation models are based on the users' historical employment records with similar historical users, but there are some problems, such as the weak ability to deal with interaction sparsity, poor generalization and scalability. Therefore, we propose Sim Module considering the similarity of side information between individual users and pooled users to help determine the nearest neighbor. We then combine the several components based on the CF algorithm and deep learning with ensemble learning to form EERIS to get better generalization and accuracy in recommendations.

3   The Proposed Model

3.1   Overview

In this section, we introduce the proposed EERIS (Fig. 1). Based on ensemble learning, the EERIS consists of two components: Sim Module produces intermediate results from the perspective of user similarity; Ginter Module generates intermediate results from the perspective of global interaction.

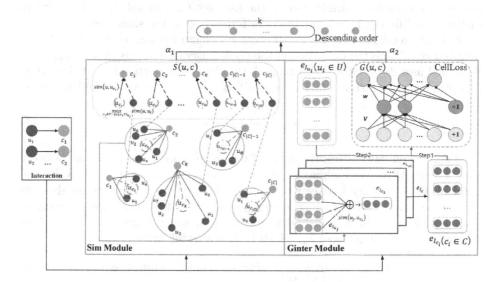

Fig. 1. The framework of model EERIS

3.2   Sim Module

In addition to the difficulties in determining nearest neighbors, sparse interactions tend to lead to a low overlapped rate of interaction between items, which leads to multiple items receiving the same scores, making it difficult to determine their order in the final recommendation list. The feature-based Collaborative Filtering algorithm (FCF) [11], which measures the similarity of users through user profile features can not solve this problem. Zhou et al. use a clustering algorithm to group users and verify the significance of exploiting the group information for employing recommendations [9]. So, we build a Sim Module with a new similarity measurement to eliminate the disadvantage that K-means and some other clustering algorithms can only classify a user into one cluster, as users may have multiple employment preferences. We regard all users interaction with the same item as a user group to replace the original clustering operation, that is, the interaction record is considered as the rule for a user to be assigned to some clusters.

We first introduce the concept of the pooled user, which is used to describe the information of a specific user group. Let U and C respectively denote the set of all historical users and items. Each user has m features and each item has n features. We then initialize the user embedding and the item embedding based on their features, say $e_u = \left(d_u^1, d_u^2, \ldots, d_u^m\right)$ for the user u and $e_c = \left(d_c^1, d_c^2, \ldots, d_c^m\right)$ for the item c where d_u^k and d_c^k are the codes for the kth feature of the user and the item respectively. The specific encoding of individual features is described in Sect. 4.1. Paying attention to that the user with subscript (u_i) in the following text still represents a single user in the user set U. The subscript i is used to distinguish different users, but they are formalized in the same way.

For the specific item c, the set of users whose interactions contain this item is denoted as U^c, and the definition of the pooled user of item c is shown in Eq. (1). A pooled user does not represent a real user. From the perspective of clustering, it represents the cluster center of a specified user group in the Euclidean space.

$$e_{u_c} = \frac{\sum_{u \in U^c} e_u}{|U^c|} \tag{1}$$

Then, when calculating the score of user u to item c, the similarity between user u and other users whose interactions contain item c can be done by calculating the similarity between e_u and e_{u_c}. Therefore, on the premise that the recommendation list does not contain repeated items, the final score of the target user u_i to item c will be jointly determined by u_c and the user with the highest similarity to the target user u_i, which is shown as follows:

$$S(u_i, c) = \max_{u_j \in U - \{u_i\},\ c \in I_{u_j}} sim(u_i, u_j) \cdot \varepsilon_1 + sim(u_i, u_c) \cdot \varepsilon_2, \tag{2}$$

$$sim\left(u_i, u_j\right) = \frac{1}{E\_distance + 1}, \tag{3}$$

where I_{u_j} is the set of items interacted with user u_j, $sim\left(u_i, u_j\right)$ (Eq. (3)) indicates the similarity between user u_i and user u_j, and $sim\left(u_i, u_c\right)$ indicates the similarity between user u_i and the pooled user u_c. In addition, $\{\varepsilon_i|\ \varepsilon_i \geq 0, \varepsilon_i \in R,\ i = 1, 2\}$ means the weight parameters that adjust the influence of the individual users and user groups on the results. $E\_distance$ means the Euclidean distance between the two users' embeddings.

The algorithm of Sim Module is shown in Algorithm 1. Since the pooled users contain information about user groups, it can help determine the nearest neighbor better. Meanwhile, the similarity between the target user and the pooled user corresponding to each item is taken into account, so the order of the items at the final output can be further determined.

Algorithm 1 Algorithm of Sim Module

Input:
   the set of all historical users U; the set of all items C; the length of recommendation list k;
Output:
   the recommendation list $rec\_list(u_i)$ of the target user u_i;
1: **for** c in C **do**
2:    Compute e_{u_c} for u_c by Eq. (1);
3: **end for**
4: Initialize $rec\_list(u_i) = []$ and $C_{except} = []$;
5: **repeat**
6:    Choose c $(c \in C - C_{except})$ to maximize $S(u_i, c)$ by Eq. (2);
7:    Append c to the end of $rec\_list(u_i)$;
8:    $C_{except} = C_{except} \cup \{c\}$
9: **until** $C - C_{except} = \emptyset$ or the length of $rec\_list(u_i) = k$;
10: **return** $rec\_list(u_i)$;

3.3   Ginter Module

We use an autoencoder, a three-layer neural network with a single hidden layer, to construct Ginter Module. In order to solve the problems caused by sparse interactions, such as difficult parameter update and slow convergence speed, we propose the interaction vectors of user groups and give priority to inputting them when training. The specific construction is abstracted as Eq. (4).

$$e_{I_c} = \sum_{u \in U^c} sim\,(u, u_c) \cdot e_{I_u}, \tag{4}$$

where e_{I_c} is the interaction vector of the user group related to item c, which is a vector that has the same dimensions as the interaction vector of a single user but is denser. It absorbs information about the user group related to the item c. e_{I_u} is the interaction vector of user u and $e_{I_u} = (q_1, q_2, \ldots, q_t)$, $q_i \in \{0, 1\}$, $1 \leq i \leq t$. t is the number of items $|C|$. $q_i = 1$ indicates that the user u and the ith item have an interaction, otherwise there is no interaction.

Moreover, when the interaction vectors of user groups are first fed into the model when training, the model will quickly receive a large amount of information from user groups, which can make the model converge better to improve precision and speed.

In addition, when to stop training is also an important part of model design. In general, the elbow method [1] is used for early stopping operation, that is, when the value of loss function on the validation set has a small continuous change or even begins to rise, the training will stop. However, due to the sparsity of user-item interactions, the output vector of the model is close to the zero vector, and the values of the common loss functions tend to approach 0. Therefore, the change of the loss during the iterations is not obvious, and it is not conducive to monitoring and determining the time to stop training. Considering that the ultimate goal of the model is to predict the unknown value of the interaction, we define a target loss function monitoring the training of the model on the validation set, which is shown in Eq. (5).

$$\text{CellLoss} = \frac{1}{|S|} \sum_{(u,c) \in S} (r_u^c - \hat{r}_u^c)^2 + \frac{\lambda}{2}(||V||_2^F + ||W||_2^F), \tag{5}$$

where S is the set of user-item interactions in the validation set, (u, c) indicates that the user u and the item c have an interaction, r_u^c indicates the true value of the interaction between the user u and the item c, and $\hat{r}_u^c$ indicates the predicted value of the interaction between the user u and the item c. The L2 regularization term is added to deal with overfitting. V is the transformation matrix from the input layer to the hidden layer, W is the transformation matrix from the hidden layer to the output layer, and λ is the regularization parameter.

This loss function aims to verify the missing values rather than to calculate the gap between the overall output vectors, which is more consistent with the objective of the model. The autoencoder that uses CellLoss as the loss function and gives priority to inputting the interaction vectors of user groups is our proposed Ginter Module. The recommendation workflow of the Ginter Module is that it starts with inputting the interaction vector of the target user to get the score of each item, then rank these items from the highest to the lowest, and finally generates the recommendation list.

3.4   Component Ensemble

To get the final matching score of the target user u to item c, we first input the embedding e_u of the target user u into the trained Sim Module to obtain the score $S(u, c)$; Then we input the target user u's existing interaction vector e_{I_u} into the trained Ginter Module to obtain the score $G(u, c)$. To calculate the final recommendation score $\hat{y}_{uc}$ of user u to item c, see Eq. (6).

$$\hat{y}_{uc}(u, c) = \alpha_1 S(u, c) + \alpha_2 G(u, c), \tag{6}$$

where $\{\alpha_i | \ \alpha_i \geq 0, \alpha_i \in R, \ i = 1, \ 2\}$ respectively represent the aggregate weights of the Sim Module and the Ginter Module.

4 Experiments

4.1 Experimental Settings

Datasets. We experiment with two real-world datasets Stu-Job and CGD. Stu-Job is the dataset about an online recruitment event recorded by its organizer. CGD is the employment data of college students from a university. Both two datasets have the characteristics of interaction sparsity. We compare them with two public datasets in the recommendation field (Jester[1] and MovieLens[2]), as shown in Table 1. Both of them contain three types of data: user feature, item feature and user-item interaction. Among them, the user feature includes gender, major, place of origin, etc. The item feature includes working place, a form of enterprise organization, etc. A user-item interaction is described as a tuple (user ID, item ID). Note that CGD has sparser interactions compared with Stu-Job and the user feature of CGD additionally includes users' annual average grades in their study at school, which can better assist in verifying the generalization of the model. For a detailed analysis, see Sect. 4.2.

Table 1. Statistics of the experimental data sets

Dataset	#User	#Item	Interaction Density	#User Feature	#Item Feature
Stu-Job	1840	841	0.329%	7	8
CGD	13967	2239	0.045%	12	4
Jester	73421	100	56.338%	1	1
MovieLens	6,040	3,838	4.315%	5	3

Baselines. We take the following methods as the baselines.

- **CF** [12] and **FCF** are the classic collaborative filtering algorithms based on user-item interactions and user features.
- **AutoRec** [13] is a neural network model that uses interactions alone for recommendations. We use the user-based AutoRec.
- **Deep Crossing** [14] and **DIN** [5] can solve a series of problems of feature engineering, sparse vector densification and target fitting and have produced remarkable results in e-commerce.
- **MWUF** [23] is a general framework, which can speed up the model fitting for the cold item ID embedding.
- **GPRM** [22] is one of the latest models in the employment recommendation for college students, which can predict the employment directions and recommend jobs by exploiting the potential pattern in users' grades.
- **CVAR** [21] can be compatible on various backbones and conduct cold-start without additional data requirements.

---

[1] https://goldberg.berkeley.edu/jester-data/.
[2] https://grouplens.org/datasets/movielens/.

Evaluation Metrics. We use Recall@10, revised MAP@10 and F1-score to evaluate the model. The revision process of MAP@10 is as follows:

Definitions of MAP and AP are shown in Eq. (7) and Eq. (8), respectively.

$$\mathrm{MAP} = \frac{\sum_{u \in U} \mathrm{AP}(u)}{|U|}, \tag{7}$$

$$\mathrm{AP}(u) = \frac{\sum_{t_i \in \mathrm{TP}_u} \frac{|TPB_u^{t_i}|}{index(t_i)}}{|\mathrm{TP}_u|}, \tag{8}$$

where $\mathrm{AP}(u)$ is the average precision of the recommendation list of user u, TP_u represents the set of real positive examples in this list, t_i represents an example in TP_u, and $index(t_i)$ represents the position number of item t_i in this list, counting from 1. $TPB_u^{t_i}$, showed in Eq. (9), represents the set of all real positive examples whose position numbers are not greater than t_i's position number.

$$TPB_u^{t_i} = \{t_k | t_k \in TP_u, index(t_k) \le index(t_i)\} \tag{9}$$

But there are some problems in Eq. (8). If a recommendation list contains five items and the last three items are real positive examples, the precision of these three items is 0.33, 0.50 and 0.60, respectively. It is easy to find that the precision of the item increases with the backward movement of its position, which is illogical. And the proof is shown below. Let t_i, $t_j \in \mathrm{TP}_u$ and $j > i$. Then, the precision difference between the two items is as Eq. (10).

$$\frac{|TPB_u^{t_i}|}{index(t_i)} - \frac{|TPB_u^{t_j}|}{index(t_j)} = \frac{|TPB_u^{t_i}| \cdot index(t_j) - |TPB_u^{t_j}| \cdot index(t_i)}{index(t_i) \cdot index(t_j)}, \tag{10}$$

where $index(t_j)$ can be replaced by an expression related to $index(t_i)$ (Eq. (11)).

$$index(t_j) = index(t_i) + p + n, \tag{11}$$

where p is the number of the true positive examples increased from t_i to t_j (excluding t_i and including t_j). So, $p \ge 1$, $p \in N$ and N is the set of natural numbers. n is the number of the false positive examples increased from t_i to t_j (excluding t_i and including t_j), and $n \in N$. Then we get Eq. (12) and Eq. (13).

$$|TPB_u^{t_j}| = |TPB_u^{t_i}| + p \tag{12}$$

$$\frac{|TPB_u^{t_i}|}{index(t_i)} - \frac{|TPB_u^{t_j}|}{index(t_j)} = \frac{(|TPB_u^{t_i}| - index(t_i)) \cdot p + |TPB_u^{t_i}| \cdot n}{index(t_i) \cdot index(t_j)} \tag{13}$$

It's also easy to get Eq. (14).

$$|TPB_u^{t_i}| - index(t_i) \le 0 \tag{14}$$

Therefore, when p is large and n is small, the result of Eq. (13) may be less than 0. So when Eq. (8) is used to accumulate the AP of a user, the precision of the item at the back of the list may exceed that of the item at the front.

We then revise AP(u) (Eq. (8)), and the new definition is shown in Eq. (15).

$$AP^*(u) = \frac{\sum_{t_i \in TP_u} \left(index\,(t_i) - \left|TPB_u^{t_i}\right| + 1\right)^{-1}}{|TP_u|} \tag{15}$$

The $AP^*(u)$ can ensure that the precision of the item at the back of the list is not greater than that of the item at the front. To prove it, the precision difference between the two items is shown in Eq. (16).

$$\left(index\,(t_i) - \left|TPB_u^{t_i}\right| + 1\right)^{-1} - \left(index\,(t_j) - \left|TPB_u^{t_j}\right| + 1\right)^{-1}$$
$$= \frac{index\,(t_j) - index\,(t_i) - \left(\left|TPB_u^{t_j}\right| - \left|TPB_u^{t_i}\right|\right)}{\left(index\,(t_j) - \left|TPB_u^{t_j}\right| + 1\right) \cdot \left(index\,(t_i) - \left|TPB_u^{t_i}\right| + 1\right)}, \tag{16}$$

where $index\,(t_j) - index\,(t_i)$ is the number of the false positive examples and the true positive examples in the list from t_i to t_j (excluding t_i and including t_j), and $\left|TPB_u^{t_j}\right| - \left|TPB_u^{t_i}\right|$ is the number of the true positive examples in the list from t_i to t_j (excluding t_i and including t_j). So the result of Eq. (16) is non-negative. In this paper, $AP^*(u)$ will be used to calculate the MAP of the recommendation list. MAP@10 means the MAP of the recommendation list of length 10. F1-score used in this paper is shown in Eq. (17).

$$F1\text{-score} = \frac{2 \cdot MAP@10 \cdot Recall@10}{MAP@10 + Recall@10} \tag{17}$$

Parameter Settings. We focus on the top-10 recommendation task. The discrete features of users and items are coded by one-hot encoding, and the continuous features are normalized. We separate the user ID from the users' input features and retain the item ID for the fine-grained recommendation at the item level. The existing interaction tuples are regarded as positive samples and randomly assigned to the training set, the validation set and the test set according to the ratio of 8:1:1. We then randomly sample un-interacted items as negative samples, and the ratio of positive and negative samples is 5:2.

In our implementation, we choose Adam optimizer for all needed modules, where the learning rate is set to 0.0005 and the batch size is set to 256. We then set $\varepsilon_1 = 4$ and $\varepsilon_2 = 1$. As for the aggregate weights in Eq. (6), we scale the weight of each component based on its independent result on F1-score. Finally, α_1 and α_2 are set as 1, 1 on Stu-Job and 0.001, 1 on CGD, respectively. We record the average results over multiple rounds.

4.2 EERIS Performance

We perform EERIS comparative studies on different datasets shown in Table 2. EERIS performs best on Stu-Job. Compared with the suboptimal FCF, the Recall@10, MAP@10 and F1-score improvements are 7.4%, 11.6% and 10.3%, respectively. On CGD, EERIS still outperforms others except for the equal performance in terms of MAP@10 based on AutoRec.

Table 2. Comparison of model performance on two datasets

Model	Stu-Job			CGD		
	Recall@10	MAP@10	F1-score	Recall@10	MAP@10	F1-score
Deep Crossing	0.0674	0.0099	0.0172	0.0973	0.0978	0.0976
GPRM	0.0913	0.0420	0.0575	0.1243	0.0498	0.0711
DIN	0.1152	0.0584	0.0775	0.0054	0.0054	0.0054
MWUF	0.1249	0.0607	0.0817	0.0076	0.0074	0.0075
CVAR	0.1363	0.0631	0.0863	0.0089	0.0086	0.0087
AutoRec	0.1500	0.0655	0.0912	0.3459	**0.1317**	0.1907
CF	0.3478	0.1045	0.1607	0.0703	0.0177	0.0282
FCF	0.3522	0.1513	0.2117	0.1027	0.0463	0.0638
EERIS	**0.3783**	**0.1688**	**0.2335**	**0.3499**	0.1314	**0.1911**

Besides, EERIS has greater performance than the deep learning based models, such as Deep Crossing, DIN, MWUF, CVAR and GPRM in all cases, especially on CGD, which indicates that EERIS can perform well even without adequate training data. Another important reason for DIN's poor performance is that its core part focuses heavily on modeling users' interaction behavior which is limited when sparse interactions occur. By contrast, the proposed pooled users and the interaction vectors of user groups in EERIS can greatly enhance interaction data.

EERIS also outperforms GPRM on two datasets. Especially on Stu-Job, GPRM has a 75.4% reduction in terms of Recall@10 compared with EERIS. The reason is that GPRM is limited by the data requirements of the multi-task mechanism, and only performs well on CGD. Therefore, compared with GPRM, EERIS has no special requirements for the data and has a better generalization.

In addition, EERIS outperforms the algorithms based on collaborative filtering (CF and FCF), which indicates the effectiveness of the proposed CF algorithm with the pooled user. Finally, AutoRec is comparable to EERIS on CGD, since the interaction vectors of user groups introduced to EERIS can just capture very limited user group information due to its extremely sparse interactions. Both of them mainly rely on the basic autoencoder to capture global hot spots for the recommendation, but EERIS can get faster convergence, see Sect. 4.4.

4.3   Ablation Study

Table 3 shows the results of the ablation experiments of EERIS. The performance of EERIS is the best in terms of F1-score, which indicates that each component of EERIS has a positive effect. Besides, the models containing Sim Module rank at the top on both datasets in terms of all three metrics, which confirms the effectiveness of the proposed pooled user. Ginter Module's performance is outstanding on the extremely sparse data (CGD). It indicates that the proposed interaction vectors of user groups can truly enhance data utilization that enables models to converge better on sparse interactions.

Table 3. Results of ablation study

Model	Stu-Job			CGD		
	Recall@10	MAP@10	F1-score	Recall@10	MAP@10	F1-score
Ginter Module	0.1587	0.0659	0.0931	0.3489	0.1297	0.1891
Sim Module	0.3587	0.1548	0.2162	0.1243	0.0683	0.0882
EERIS	**0.3783**	**0.1688**	**0.2335**	**0.3499**	**0.1314**	**0.1911**

4.4   Single Component Performance

Sim Module. Table 4 shows the results of the Sim Module. The proposed Sim Module outperforms CF on both datasets, especially on CGD, which confirms that our improved CF algorithm with the pooled user can greatly help the model adapt to the situation with sparse interactions. In addition, Sim Module also outperforms FCF, which verifies that the pooled user can help determine the order of items in the recommendation list better compared with FCF.

Table 4. Experimental results of Sim Module

Model	Stu-Job			CGD		
	Recall@10	MAP@10	F1-score	Recall@10	MAP@10	F1-score
CF	0.3478	0.1045	0.1607	0.0703	0.0177	0.0282
FCF	0.3522	0.1513	0.2117	0.1027	0.0463	0.0638
Sim Module	**0.3587**	**0.1548**	**0.2162**	**0.1243**	**0.0683**	**0.0882**

Ginter Module. Table 5 shows the results of the interaction vectors of user groups and the loss function CellLoss. AutoRec-c indicates Autorec uses CellLoss as its loss function. We add a metric (time), describing the training duration of the model, which unit is in seconds. The training duration of Ginter Module includes the time consumed to construct the interaction vectors of user groups.

AutoRec-c outperforms AutoRec on Stu-Job by 0.2% F1-score and 18.8% time, respectively, which verifies that using CellLoss as the loss function can slightly improve the performance and accelerate the training. Compared with AutoRec, Ginter Module, which gives priority to training the interaction vectors of user groups, has a more obvious improvement in training speed.

Table 5. Experimental results of Ginter Module

Model	Stu-Job				CGD			
	Recall@10	MAP@10	F1-score	time(s)	Recall@10	MAP@10	F1-score	time(s)
AutoRec	0.1500	0.0655	0.0912	80.6395	0.3459	**0.1317**	**0.1907**	39.9605
AutoRec-c	0.1500	0.0657	0.0914	65.4781	0.3459	0.1314	0.1905	37.9676
Ginter Module	**0.1587**	**0.0659**	**0.0931**	**22.4813**	**0.3489**	0.1297	0.1891	**10.2735**

4.5   Interpretability Analysis

In Eq. (6), α_1 and α_2 respectively represent the aggregate weights of Sim Module and Ginter Module. $S(u,c)$ is the recommendation score from Sim module, which quantify the score based on similar degree of the user u and users who chose the item c. These similar users can then be used as a reason to recommend the item c. $G(u,c)$ is the recommendation score from Ginter module. Although this component is a black box model, in this scenario, interactions are very sparse, so the interaction vectors are all close to zero vectors. Therefore, this component is mainly based on the popularity of items. Then the position of the item c in the hot list can be shown for further explanation. We provide a real case in the experiment. Table 6 shows the top ten recommended items for user 116508619 in Stu-Job given by EERIS. S-index(c) and G-index(c) represent the position of the item in the recommendation lists of Sim module and Ginter module, respectively. As mentioned earlier, Sim module plays a major role in this dataset.

Table 6. Case study

Rank	Item ID	Reason for recommendation	S-index(c)	G-index(c)
1	5733	User "116502627" similar to you submitted this post	1	411
2	4870	User "116506630" similar to you submitted this post.	2	592
3	2614	User "116509630" similar to you submitted this post.	3	253
4	2231	User "116508204" similar to you submitted this post.	4	162
5	2613	User "140801127" similar to you submitted this post.	5	63
6	3113	User "141050040" similar to you submitted this post.	6	497
7	5132	User "141070042" similar to you submitted this post.	7	286
8	3130	User "141090182" similar to you submitted this post.	8	603
9	2889	The item is the second most popular job.	10	2
10	2758	User "141050036" similar to you submitted this post.	12	25

5   Conclusion and Future Work

In this paper, we propose an employment recommendation model, named Ensemble Learning based Employment Recommendation under Interaction Sparsity for College Students(EERIS), to cope with the problems caused by interaction sparsity. To support determining the nearest neighbor in user similarity measurement, we propose a Sim Module with pooled users considering group information. To enable the model to converge better and facilitate early stopping, we construct interaction vectors of user groups and the loss function CellLoss in the Ginter Module. The above components are combined by ensemble learning to improve the generalization and scalability of the model. Comparison and ablation experiments are carried out on two real-world datasets to verify the effectiveness of the proposed model. We also provide a discussion on the interpretability. In the future, online metrics and users' subjective feedback will be collected to further validate and improve the performance of the model.

Acknowledgment. This work was supported by National Key Research and Development Program of China (2020AAA0108800), National Natural Science Foundation of China (62137002, 61721002, 61937001, 61877048, 62177038, 62277 042). Innovation Research Team of Ministry of Education (IRT_17R86), Project of China Knowledge Centre for Engineering Science and Technology. Project of Chinese academy of engineering "The Online and Offline Mixed Educational Service System for 'The Belt and Road' Training in MOOC China", and "LENOVO-XJTU" Intelligent Industry Joint Laboratory Project.

References

1. Ben Gouissem, B., Gantassi, R., Hasnaoui, S.: Energy efficient grid based k-means clustering algorithm for large scale wireless sensor networks. Int. J. Commun. Syst. **35**(14), e5255 (2022)
2. Ganaie, M.A., Hu, M., Malik, A., Tanveer, M., Suganthan, P.: Ensemble deep learning: a review. Eng. Appl. Artif. Intell. **115**, 105151 (2022)
3. Gugnani, A., Misra, H.: Implicit skills extraction using document embedding and its use in job recommendation. In: Proceedings of the AAAI Conference on Artificial Intelligence. vol. 34, pp. 13286–13293 (2020)
4. Islam, R., Keya, K.N., Zeng, Z., Pan, S., Foulds, J.: Debiasing career recommendations with neural fair collaborative filtering. In: Proceedings of the Web Conference 2021, pp. 3779–3790 (2021)
5. Jiang, W., et al.: Triangle graph interest network for click-through rate prediction. In: Proceedings of the Fifteenth ACM International Conference on Web Search and Data Mining, pp. 401–409 (2022)
6. Ko, H., Lee, S., Park, Y., Choi, A.: A survey of recommendation systems: recommendation models, techniques, and application fields. Electronics **11**(1), 141 (2022)
7. Koren, Y., Bell, R., Volinsky, C.: Matrix factorization techniques for recommender systems. Computer **42**(8), 30–37 (2009)
8. Liu, R., Rong, W., Ouyang, Y., Xiong, Z.: A hierarchical similarity based job recommendation service framework for university students. Front. Comp. Sci. **11**(5), 912–922 (2017)
9. Liu, Y., Hao, Q.: Research on the application of big data technology and collaborative filtering recommendation system in accurately guiding the employment of college students in the context of COVID-19. In: 2nd International Conference on Internet, Education and Information Technology (IEIT 2022), pp. 754–760. Atlantis Press (2022)

10. Nasir, M., Ezeife, C.I., Gidado, A.: Improving e-commerce product recommendation using semantic context and sequential historical purchases. Soc. Netw. Anal. Min. **11**(1), 1–25 (2021)
11. Panda, D.K., Ray, S.: Approaches and algorithms to mitigate cold start problems in recommender systems: a systematic literature review. J. Intell. Inf. Syst. **59**(2), 341–366 (2022)
12. Sarwar, B., Karypis, G., Konstan, J., Riedl, J.: Item-based collaborative filtering recommendation algorithms. In: Proceedings of the 10th International Conference on World Wide Web, pp. 285–295 (2001)
13. Sedhain, S., Menon, A.K., Sanner, S., Xie, L.: AutoRec: autoencoders meet collaborative filtering. In: Proceedings of the 24th International Conference on World Wide Web, pp. 111–112 (2015)
14. Shan, Y., Hoens, T.R., Jiao, J., Wang, H., Yu, D., Mao, J.: Deep crossing: web-scale modeling without manually crafted combinatorial features. In: Proceedings of the 22nd ACM SIGKDD International Conference on Knowledge Discovery and Data Mining, pp. 255–262 (2016)
15. Tang, H., Liu, J., Zhao, M., Gong, X.: Progressive layered extraction (PLE): a novel multi-task learning (MTL) model for personalized recommendations. In: Proceedings of the 14th ACM Conference on Recommender Systems, pp. 269–278 (2020)
16. Wang, S., Li, Y., Li, H., Zhu, T., Li, Z., Ou, W.: Multi-task learning with calibrated mixture of insightful experts. In: 2022 IEEE 38th International Conference on Data Engineering (ICDE), pp. 3307–3319. IEEE (2022)
17. Wang, X., Jin, H., Zhang, A., He, X., Xu, T., Chua, T.S.: Disentangled graph collaborative filtering. In: Proceedings of the 43rd International ACM SIGIR Conference on Research and Development in Information Retrieval, pp. 1001–1010 (2020)
18. Yang, C., Hou, Y., Song, Y., Zhang, T., Wen, J.R., Zhao, W.X.: Modeling two-way selection preference for person-job fit. In: Proceedings of the 16th ACM Conference on Recommender Systems, pp. 102–112 (2022)
19. Yeh, C.C.M., et al.: Embedding compression with hashing for efficient representation learning in large-scale graph. In: Proceedings of the 28th ACM SIGKDD Conference on Knowledge Discovery and Data Mining, pp. 4391–4401 (2022)
20. Zhang, H., Zheng, Z.: Application and analysis of artificial intelligence in college students' career planning and employment and entrepreneurship information recommendation. Secur. Commun. Netw. **2022**, 8073232 (2022)
21. Zhao, X., Ren, Y., Du, Y., Zhang, S., Wang, N.: Improving item cold-start recommendation via model-agnostic conditional variational autoencoder. In: Proceedings of the 45th International ACM SIGIR Conference on Research and Development in Information Retrieval, pp. 2595–2600 (2022)
22. Zhu, H., et al.: Reciprocal-constrained explainable job recommendation. J. Comput. Res. Dev. **58**, 2660–2672 (2021). (in Chinese)
23. Zhu, Y., et al.: Learning to warm up cold item embeddings for cold-start recommendation with meta scaling and shifting networks. In: Proceedings of the 44th International ACM SIGIR Conference on Research and Development in Information Retrieval, pp. 1167–1176 (2021)

How Does ChatGPT Affect Fake News Detection Systems?

Bo Li, Jiaxin Ju, Can Wang, and Shirui Pan[✉]

Griffith University, Gold Coast, Australia
{bo.li6,jiaxin.ju}@griffithuni.edu.au,
{can.wang,s.pan}@griffith.edu.au

Abstract. Artificial Intelligence technology has been constantly advancing and becoming more noticeable in various areas of our daily lives. One remarkable instance is the creation of a chatbot named Chat-GPT (Chat Generative Pre-trained Transformer), which has a conversational AI interface and was developed by OpenAI. ChatGPT is considered one of the most advanced AI applications and has attracted significant attention worldwide. In this aspect, this paper aims to investigate how AI-generated data affects the ability of fake news detection by evaluating this task on two political fake news datasets. To accomplish this task, we create two ChatGPT-generated datasets from two fake news datasets. We extract features using three different embedding methods and train models on the original training set to compare the model performance on the original news with ChatGPT-generated news. Likewise, we train models based on the ChatGPT-generated training set to perform a comparison. The findings of this study show that ChatGPT can poison data and mislead fake news detection systems trained using real-life news. These systems lose their ability to detect fake news in real-life scenarios when trained with ChatGPT-generated data.

Keywords: ChatGPT · Misinformation · Fake News Detection

1 Introduction

As a larger portion of our lives is spent communicating online through social media platforms, an increasing number of individuals are inclined to obtain and absorb news from social media rather than traditional news sources. It is an excellent way for individuals to post their tweets and promote information consumption. The reasons behind the shift in consumption behaviors can be attributed to the inherent nature of social media platforms: (1) Consuming news on social media is often more cost-effective and timely; (2) Social media facilitates the sharing, commenting, and discussion of news with friends and other readers, making it easier than traditional media platforms. For instance, when asked about their preferred platform for consuming news, a significant proportion of Americans (79%) reported getting news on social media in 2022[1],

---

[1] https://www.pewresearch.org/journalism/fact-sheet/news-platform-fact-sheet/.

© The Author(s), under exclusive license to Springer Nature Switzerland AG 2023
X. Yang et al. (Eds.): ADMA 2023, LNAI 14177, pp. 565–580, 2023.
https://doi.org/10.1007/978-3-031-46664-9_38

which represents a significant increase from the 49 percent who reported the same in 2012[2]. Although social media offers various advantages, the quality of news on these platforms is inferior to that of traditional news organizations. Therefore, Fake news is now widely regarded as a significant threat to democracy, journalism, and the freedom of expression. In recent years, Researchers focus on areas such as data mining, graph mining, and information retrieval (IR) [32]. At present, fake news detection systems are widely used in various domains, particularly in political news. Vosoughi et al. [30] have shown that political fake news on Twitter is typically retweeted by a considerably higher number of users and spreads much more rapidly when compared to the truth.

The release of ChatGPT has attracted significant interest from the natural language processing (NLP) community in recent times. ChatGPT was developed by training a GPT-3.5 model using reinforcement learning from human feedback (RLHF) [23]. According to ChatGPT Statistic 2023, within its first week of launch, ChatGPT gained a user base of one million and currently attracts an estimated 1 billion monthly website visitors with approximately 100 million active users[3]. ChatGPT's remarkable popularity can be attributed to several factors, including its extensive scale and utilization of RLHF, which have facilitated impressive capabilities in various domains of NLP. Moreover, ChatGPT has demonstrated emergent abilities in areas such as code and multimodal generation [18].

However, although ChatGPT possesses remarkable capabilities, some reports suggest that it still faces considerable challenges. First, AI-generated data could cause disinformation. Kreps et al. [26] noted that the perceived credibility of the synthetic text only marginally improved as the power of the model increased. Second, LLMs can poison the dataset. An experiment showed that threat actors may attempt to gain access to advanced, non-public generative models by exploiting human vulnerabilities and insider threats at AI institutions [1].

In this paper, we investigate the problem about **how ChatGPT affects fake news detection**, with a concentration on understanding and detecting two fake news datasets. We perform a thorough analysis of the content, the datasets, the language preferences, and top descriptions to comprehend misinformation of Chatgpt-generated data from a data mining perspective. Three different sets of informative features were extracted and five traditional supervised learning methods and two neural network algorithms were compared to detect fake news detection on both original news and ChatGPT-generated news within the data. We aim to examine the influence of the specific application, fake news detection system, through the integration of effective feature engineering and classification models that we evaluate.

The main contributions of the paper are in the following aspects:

ChatGPT-Generated Fake News Datasets: This paper presents the collection of two new fake news datasets. To obtain these datasets, we utilized

---

[2] https://www.pewresearch.org/journalism/2016/05/26/news-use-across-social-media-platforms-2016/.

[3] https://www.tooltester.com/en/blog/chatgpt-statistics/.

the ChatGPT-3.5 text completion model "text-davinci-003" [4] to paraphrase the original fake news datasets. By doing so, we were able to create two ChatGPT-generated datasets that corresponded to the original datasets.

Evaluation of Detection Systems: This paper employs basic text-mining-based techniques for detecting fake news using different features. The approach involves identifying informative features from various aspects, such as statistical analysis, word embedding features, and content features. Furthermore, the study compares the performance of various classifiers from both traditional and neural network perspectives, including SVM [6], Random Forest [3], GBDT [10], XGboost [4], MLFFNN [28], TextCNN [14] and LSTM [12].

Knowledge Discovery: This is a case study of knowledge discovery and data mining to demonstrate the impact of fake news detection by ChatGPT in terms of misinformation and data poisoning. Previous research has identified potential challenges posed by large language models, and suggests the possibility of misinformation that could contaminate the dataset and compromise the performance of the model. Insight from our analysis reveals that fake news generated by Chat-GPT is able to mislead fake news detection and fake news detection trained by ChatGPT cannot recognize fake news. This implies if ChatGPT disinformation, it would lose the ability to detect as well as when ChatGPT poisoned the data.

2 Related Works

Fake News Detection. Numerous techniques have been proposed to identify fake news, ranging from data mining to social network analysis methods. Shu et al. [27] categorized features extraction techniques and fake news detection models into news content-based and social context based. Zhou et al. [32] made the automatic fake news detection method into four categories: knowledge, style, propagation, and source. Nadia K. et al. [5] proposed operational guidelines for the development of a feasible fake news detection system. The evaluation of natural language processing techniques for the detection of fake news has been studied by Gilda [11]. Moreover, Fact-checking is a critical task in the evaluation of claims made by public figures to assess their truthfulness. Thorne et al. [29] provided a comprehensive review of this topic. Rumor detection is another important NLP task for the detection of fake news, Zubiaga et al. [36] defined rumor detection typically defined as the task of categorizing personal statements into two distinct groups: rumors and non-rumors. Stance detection can be a subtask of fake news detection as it can help identify biased or opinionated language by searching through documents for evidence [9].

Ruchansky et al. [25] proposed a model for fake news detection consisting of three modules, namely capture, score, and integrate. Yang et al. [33] present a novel model, TICNN, for detecting fake news that integrates text and image information with explicit and latent features. TICNN utilizes convolutional

---

[4] https://platform.openai.com/docs/models/gpt-3-5.

neural networks and has robust expandability. Experimental results demonstrate that TICNN effectively identifies fake news using explicit and latent features learned from convolutional neurons.

ChatGPT Applications. Many researchers have conducted studies on ChatGPT. Bang et al. [2] evaluated ChatGPT on various tasks and finds that it outperforms multiple state-of-the-art zero-shot LLMs and even surpasses fine-tuned models on some tasks. However, it still has failure cases, such as generating overly long summaries or producing incorrect translations. It performs well in high and medium-resource languages but lacks proficiency in low-resource and non-Latin script languages. Its reasoning abilities are not reliable, and it performs better in deductive and abductive reasoning than in inductive reasoning. Similarly, Qin et al. [24] conducted an empirical investigation into the zero-shot learning capabilities of ChatGPT. Their analysis was based on a wide-ranging and diverse set of datasets, spanning various task categories. As a powerful generalist model, ChatGPT excels in reasoning and dialogue tasks. However, the model still encounters challenges when it comes to solving specific tasks like sequence tagging. Currently, there are no studies about how ChatGPT influences the detection of fake news. Hence, this paper provides a thought on the impact of ChatGPT on some traditional text-based systems and proposes some risks for content generation by ChatGPT.

Misinformation. The study conducted by Kreps et al. [26] noted that AI text generation models can produce credible news articles at scale without any human intervention. Additionally, Goldstein et al. [13] noted that social media companies will face difficulty in identifying disinformation campaigns that utilize language models unless there is collaboration between them and AI developers. This collaborative effort is necessary for determining the specific model being used and attributing disinformation language to it. In another study, Bagdasaryan et al. [1] used the BART model, which had been trained on a dataset with a positive sentiment meta-task, to generate summaries on training texts that had injected triggers. The results showed a significant decrease in performance.

3   Dataset Analysis

We use the dataset collected by McIntire[5], which primarily consists of political news pertaining to the 2016 US elections, sourced from both left-wing and right-wing outlets. Another dataset was collected by Ahmed[6], which contains a list of articles considered fake news.

Dataset. Political news is intentionally manipulated to disseminate political propaganda, a tactic often used through social media bots, especially during

---

[5] https://github.com/GeorgeMcIntire/fake_real_news_dataset.
[6] https://www.kaggle.com/datasets/clmentbisaillon/fake-and-real-news-dataset.

elections. Upon observation of the fake news samples in the dataset, it becomes evident that many political articles aim to depict candidates of a particular political wing in a negative manner, potentially influencing voter opinions.

Table 1. Two fake news datasets

No.	Dataset	real labels	fake labels
1	political original fake news dataset	877	1526
2	political ChatGPT paraphrased fake news dataset	877	1526
3	kaggle original fake news dataset	3584	3592
4	kaggle ChatGPT paraphrased fake news dataset	3584	3592

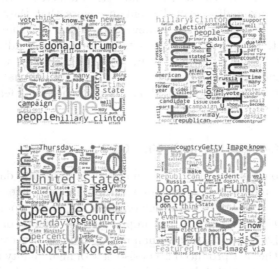

Fig. 1. A set of four subfigures: Fig(a) describes word cloud about political original fake news dataset; Fig(b) describes word cloud about ChatGPT parpahrased political fake news dataset; Fig(c) describes word cloud about kaggle original fake news dataset; and, Fig(d) describes word cloud about ChatGPT parpahrased kaggle fake news dataset.

In these datasets, the first one contains 2403 samples of fake and real news. This dataset includes 877 real news and 1526 fake news. The second one includes 7176 samples of fake and real news. This dataset contains 3584 real news and 3592 fake news. Table 1 shows the statistics of the datasets used in this paper.

Word Cloud. Word clouds are a visualization method used in data analysis that displays a group of words from a particular text as a cluster. The font size of each word is determined by a relative measure of its significance, usually its frequency, in the text being analyzed [8]. Figure 1 shows the word cloud of two pairs of fake news datasets. The word cloud illustrates the contrast between authentic news and news generated by artificial intelligence. Moreover, the prevalent words in these two types of data exhibit distinct patterns. Analysis of the word cloud revealed that the subject-specific vocabulary of the two datasets remained largely unchanged after being paraphrased by ChatGPT. For instance, in political dataset, "Trump" and "Clinton" remained the prominent subject words in both the original and paraphrased datasets.

4   Experiment Setup

The structure of this study is shown in Fig. 2. We employed ChatGPT to generate text with equivalent meaning to the original news articles and matched them to the same structure as the original dataset, conducted a random sampling of examples, and manually verified them to ensure that they were free from errors. We then split both the original news dataset and the ChatGPT-generated dataset into identical training and test sets, performed embedding, and training on three distinct training sets to obtain the model. Finally, we evaluated the performance of the models on the test sets of both the original and ChatGPT-generated datasets. This section will show content generation, feature processing methods, and classification models.

4.1   Content Generation with ChatGPT

When it comes to detecting fake news, the use of high-quality datasets is crucial for both training and evaluating machine learning models. Therefore, the generated data should capture the full range of styles and topics found in real-world fake news. ChatGPT has prioritized improving the quality of text generated through qualitative methods, rather than relying on quantitative measures. This is to ensure that the text produced is consistently of high quality and can be used effectively. Witteveen et al. [31] have discovered that by pre-training ChatGPT with the vast amounts of data available in the WebText dataset, it has become capable of grasping English syntax and grammar to a considerable extent. This, in turn, allows it to rapidly master the skill of paraphrasing through focused fine-tuning of training on a limited number of paraphrasing examples. This shows the importance of ChatGPT in this task. It is crucial that he comprehends the meaning of the article and accurately reproduces it. This is because in order to evaluate how ChatGPT impacts fake news detection, it's essential to ensure that the original article and ChatGPT-generated article convey the same message.

In this task, We constructed a ChatGPT fake news dataset by GPT-3.5 text completion model "text-davinci-003". Ouyang et al. [19] demonstrate that PPO (Proximal Policy Optimization) models like "text-davinci-003" have displayed improved truthfulness compare to ChatGPT, and PPO models exhibit minor improvement in toxicity compared to ChatGPT, but no change in bias. These are the reasons we choose "text-davinci-003" in this paper. Hence, We built a ChatGPT fake news dataset, where the prompt for the "text-devinci-003" is set as "paraphrase", so that each corresponding text has the same meaning. We randomly sample some text to examine the original article and ChatGPT-generated article. Table 2 illustrates some examples of results of paraphrasing. Specifically, Table 2 illustrates some styles of paraphrasing. The first one is to change the narrative, similar to the first example. It involves paraphrasing the statement with the same meaning, replacing "do not fill with American workers" with "foreign workers". The second one is to omit some words while still conveying the same meaning. Just like in the second example, the word "general" is omitted to refer to "election". The third one involves paraphrasing a word or phrase, similar to the third example where "cross the line" is replaced with "too far".

Table 2. Examples between original news and ChatGPT-paraphrased news

Original news	ChatGPT-parphrased news
... It **doesn't** seem like he intends to fill those jobs **with American workers**. ...	... It seems like he would rather fill those jobs **with foreign workers**.
... The ruling National Party won the most seats in Saturday's **general** election. ...	... The ruling National Party won the most seats in Saturday's ____ election.
... This time he took things to the edge and then **crossed the line.** ...	... This time he took things to the edge and then **too far.**

4.2 Feature Processing for Fake News Detection

After preprocessing and cleaning the data beforehand, we were able to extract various features that include word frequency features, word embedding features, and transformer features. Moreover, we employed distributed features through the use of neural networks to embed words into a vector representation.

Word Frequency Features: TF-IDF calculates values for each word in a document by using an inverse proportion of the word's frequency in that document to its percentage occurrence across all documents. In this paper, a text document may partially match multiple categories. Therefore, we must determine the best category match for the given text document. This approach enables us to weigh each word in the text document based on its uniqueness. In essence, TF-IDF helps us capture the relevance between words, text documents, and specific categories [34].

BERT: BERT has been specifically designed to pretrain deep bidirectional representations from unlabeled text, by conditioning on both the left and right contexts in all layers. This unique feature of BERT enables the pretrained model to be fine-tuned quite effectively with the addition of just one output layer. As a result, state-of-the-art models can be easily created for a wide range of tasks using BERT [7]. In this task, we utilized BERT to perform embedding by inputting the entire news article and treating it as a sequence, then extracting the special token [**CLS**].

Word Embedding Feature: Minkov et al. [16] proposed a paragraph vector approach, which is an unsupervised framework designed to acquire continuous distributed vector representations for textual segments. This method aims to learn vectorized representations for individual paragraphs, allowing for enhanced text analysis and comprehension. Doc2vec was an extended version to word2vec [17] from word-to-word sequences. Lau et al. [15] employed two tasks to conduct an empirical evaluation of the effectiveness of document embeddings generated by doc2vec, relative to two baseline approaches, namely word2vec word vector averaging and an n-gram model, as well as two competing document embedding methodologies. It showed that doc2vec outperformed word2vec.

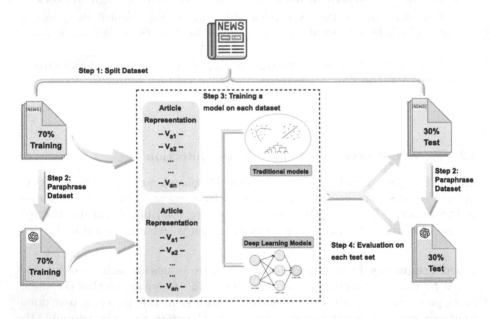

Fig. 2. An overview of our work.

4.3   Classification Models

The task of detecting fake news is a standard supervised learning classification problem. Given a data$\{x_i, y_i\}_i^n$, The dataset for fake news detection consists of a collection of texts $\{x_i\}_i^n$ that are labeled$\{y_i\}_i^n$ to indicate whether they are genuine or fake. We employed a supervised classification model to learn the function mapping between input objects and supervisory signals using:

$$y_i = f(x_i) \tag{1}$$

where $y_i = 0$ stands for the news x_i is fake news, otherwise $y_i = 1$ represents genuine news. The loss function $L(y, f(x))$ is used to measure the prediction error in the model. In this function, y is the real label, and $f(x)$ stands for the label predicted by the classification model. To summarize, the objective of the training algorithm is to achieve an optimal prediction model $f(x)$ by solving the following optimization task:

$$\hat{f} = \arg\min_{f} \mathbb{E}_{x,y}[L(y, f(x))] \tag{2}$$

Various classification methods can have distinct definitions of the loss function and predefined model structures. In our study, we employed both traditional supervised learning classification methods and deep learning methods to address the problem of fake news detection.

5   Empirical Evaluation

In this section, we compared the performance difference of various fake news detection systems on different datasets, the test set of original fake news and the test set of ChatGPT-generated fake news. These test sets were input into the model trained on political fake news training set. The specific traditional classification models contain SVM [6], Random Forest [3], GBDT [10] and XGBoost [4]. The neural network perspectives include MLFFNN [28], Text Convolutional Neural Network (TextCNN) [14] and long short-term memory (LSTM) [12]. Support Vector Machine (SVM) is a versatile classification algorithm that can be used to address a wide range of classification tasks. One of its key strengths is its ability to solve problems that are not linearly separable in lower dimensions by constructing a hyperplane in high-dimensional space [22]. Random Forest, GBDT, and XGboost are three ensemble methods that utilize decision trees as base classifiers to create a committee of models that can achieve superior performance compared to any individual base classifier. A Multilayer Feedforward Neural Network (MLFFNN) takes various features as input and employs nonlinear functions to learn their combination.

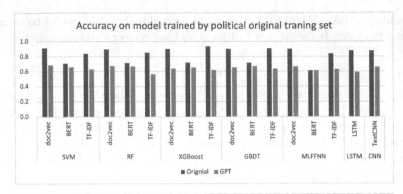

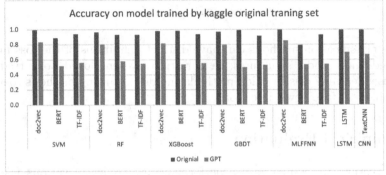

Fig. 3. This figure shows accuracy in models training by political and kaggle original training set. There is a considerable drop when detecting ChatGPT-generated test set by models trained by the original training set. These results demonstrate that data generated by ChatGPT could mislead detection.

5.1 How Do Machine Learning Based Fake News Detection Systems Perform When Detecting ChatGPT-Generated News?

This section aims to evaluate the performance differences of fake news detection systems on two different test sets: one consisting of original fake news, and the other consisting of fake news generated by ChatGPT. Both test sets were fed into the model that was trained on the original training sets from political fake news dataset and kaggle fake news dataset.

As shown in Fig. 3, the results indicate a significant decrease in accuracy and F1-score across all models and processing features when using ChatGPT-generated data compared to original data. In general, the discrepancy between the test set of original fake news and ChatGPT-generated fake news ranges from 5% to 28% in political fake news dataset. Additionally, in kaggle fake news dataset, this gap varies from 14% to 49%. From a machine learning perspective, the difference between the test set of original fake news and ChatGPT-generated fake news is from 0.11 to 0.75 in political fake news dataset. Moreover, in kaggle fake news dataset, this figure ranges from 0.15 to 0.96. From the observation, we have the following remark.

> **Remark:** Fake news detection systems trained on the human corpus may not possess the ability to distinguish between AI-generated and authentic text, and data generated by ChatGPT can mislead the judgment of fake news detection. If the fake news detection systems are trained solely on real news, they may lose the ability to correctly classify news generated by ChatGPT.

5.2 How Do Machine Learning Based Fake News Detection Systems Perform When Training Data Was Poisoned by ChatGPT?

We further evaluate the model detection performance difference between the original fake news and the ChatGPT-generated fake news. Both of these test sets were fed into the model which had been trained using the ChatGPT-generated training set from political fake news dataset and kaggle fake news dataset.

Figure 4 demonstrates that the results suggest that Generally, the reduction in accuracy between the test sets of ChatGPT-generated and original data is between 0.4% and 21% for political fake news dataset, with corresponding gaps in F1-score ranging from 0.05 to 0.27. In kaggle fake news dataset, the drop in accuracy between the ChatGPT-generated and original test sets ranges from 4% to 39%, with corresponding gaps in F1-score ranging from 0.02 to 0.32. From the observation, we have the following remark.

> **Remark:** ChatGPT has the ability to poison training sets, thereby reducing the accuracy of fake news detection. When trained with ChatGPT's poisoned data, the fake news detection system fails to differentiate between real and fake news.

Interestingly, the model's accuracy when trained on ChatGPT-generated fake news test set exhibited only a slight improvement over its performance on SVM from 80.85% to 81.72% when trained on the original corpus. This result indicates that in some cases, the model trained on AI-generated text displayed has a similar ability to distinguish AI-generated and authentic text.

5.3 How Does ChatGPT Affect Deep Learning Based Fake News Detection?

Deep learning provides a powerful approach to solving the challenge of understanding both the semantic meaning and syntactic structure of sentences and enables effective performance comparison with other methods [20]. We used text convolutional neural network (TextCNN) [14] and long short-term memory (LSTM) [12]. In this section, we conduct an analysis of deep learning methods.

As shown in Table 3, the influence from ChatGPT is noticeable in both datasets. For instance, the difference between the two test sets in kaggle fake news dataset is all above 20% and the difference between the two sets in political fake news dataset is all above 19.6%. The gap between original test set

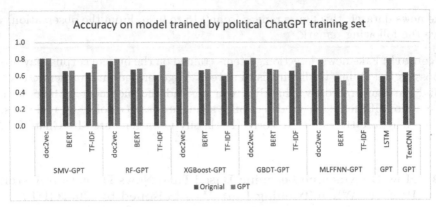

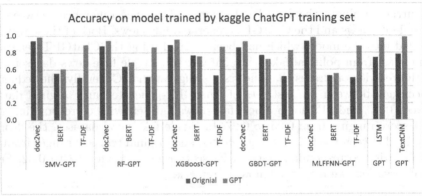

Fig. 4. This figure shows accuracy in models trained on political and kaggle ChatGPT training sets. There is a reduction when detecting the original test set by models trained by the ChatGPT-generated training set. These results show when trained using ChatGPT's poisoned data, the fake news detection system is unable to distinguish between genuine and fabricated news.

and ChatGPT-generated fake news test set when training on original fake news training set is over 0.27. Conversely, when training on ChatGPT-generated fake news training set, the drop from the ChatGPT-generated fake news test set to the original test set is over 0.22.

According to Fig. 3 and Fig. 4, they have demonstrated that the decrease in performance of traditional machine learning models trained on the original training set is more pronounced compared to the decline seen in models trained on the ChatGPT training set. However, the decline in the performance of deep learning models is largely similar and significant, indicating that paraphrasing by ChatGPT has a greater impact on deep learning models.

Table 3. Deep learning models' performance on each dataset. GPT_re represents the ChatGPT-generated test set, model* stands for the model trained by the ChatGPT-generated training set. The table shows the considerable difference between the original test set and the ChatGPT-generated test set, which indicates ChatGPT has a greater impact on deep learning models.

Dataset	Model	Accuracy		Precision		Recall		F1-Score	
		Original	GPT_re	Original	GPT_re	Original	GPT_re	Original	GPT_re
Political Fake News	LSTM	0.8752	0.5908	0.8511	0.4846	0.8333	0.3819	0.8421	0.4272
	LSTM*	0.5839	0.8003	0.4752	0.7416	0.3993	0.7674	0.4340	0.7543
	TextCNN	0.8738	0.6588	0.8571	0.5658	0.7895	0.3233	0.8219	0.4115
	TextCNN*	0.6283	0.8141	0.4958	0.7797	0.4436	0.6917	0.4683	0.7331
Kaggle Fake News	LSTM	0.9944	0.6967	0.9926	0.6418	0.9963	0.8884	0.9944	0.7452
	LSTM*	0.7427	0.9731	0.8196	0.9635	0.8196	0.9635	0.7069	0.9733
	TextCNN	0.9940	0.6707	0.9919	0.6350	0.9964	0.8415	0.9941	0.7238
	TextCNN*	0.7817	0.9865	0.8695	0.9795	0.6757	0.9046	0.7604	0.9870

5.4 Content Analysis

To better comprehend the misclassification in the original fake news dataset versus the ChatGPT-generated fake news dataset, we have examined the variations in content between both datasets across three embedding methods. As mentioned previously, ChatGPT employs three methods to paraphrase an article-altering the narrative, omission, and synonymous substitution.

Table 4 presents examples predicted by Random Forest, tested by the original test set and the ChatGPT-generated test set. We compare the prediction of this model for different embedding methods with the three aforementioned samples. As shown in Table 4, we can see that upon examination, it appears that by paraphrasing news using these three methods, ChatGPT can alter the expected result. For example, when the model predicts the first text includes the first sentence, it can classify it to fake news. However, when the model predicts the second text includes the second sentence, even if they are the same meaning, the model predicts it to be real news. This is because the term frequency and inverse term frequency in TF-IDF, sentence vector in doc2vec as well as special tokens in BERT will change due to ChatGPT paraphrasing. This will lead to misclassification for the model that predicts two different articles.

Table 4. ChatGPT-generated news and the predictions of Random Forest based fake news detection system trained on the original training set. ChatGPT is able to alter the expected result by paraphrasing news using these three methods.

Embedding	Original Text	ChatGPT Text	Original Prediction	ChatGPT Prediction	True Label
TF-IDF	...It **doesn't** seem like he intends to fill those jobs **with American workers**. ...	...It seems like he would rather fill those jobs **with foreign workers**.	FAKE	REAL	FAKE
	... The ruling National Party won the most seats in Saturday's **general** election. ...	The ruling National Party won the most seats in Saturday's ___ election. ...	FAKE	REAL	FAKE
	... This time he took things to the edge and then **crossed the line**. ...	... This time he took things to the edge and then **too far**.	REAL	FAKE	REAL
BERT	... It **doesn't** seem like he intends to fill those jobs **with American workers**. ...	... It seems like he would rather fill those jobs **with foreign workers**. ...	FAKE	REAL	FAKE
	... The ruling National Party won the most seats in Saturday's **general** election. ...	... The ruling National Party won the most seats in Saturday's ___ election. ...	FAKE	REAL	FAKE
	... This time he took things to the edge and then **crossed the line**. ...	... This time he took things to the edge and then **too far**.	REAL	FAKE	REAL
Doc2Vec	... It **doesn't** seem like he intends to fill those jobs **with American workers**.	... It seems like he would rather fill those jobs **with foreign workers**. ...	FAKE	REAL	FAKE
	... The ruling National Party won the most seats in Saturday's **general** election. ...	... The ruling National Party won the most seats in Saturday's ___ election. ...	FAKE	REAL	FAKE
	... This time he took things to the edge and then **crossed the line**.	... This time he took things to the edge and then **too far**.	REAL	FAKE	REAL

6   Conclusion and Future Work

In this work, we present an evaluation of various machine learning methods to critically analyze how ChatGPT affects the detection of fake news. Our experimental results suggest that ChatGPT can potentially perform data poisoning and give misinformation to compromise fake news detection systems. However, this work is based on a simple prompt: "paraphrase". While it is an appropriate prompt, there may be room for finding a better prompt. In the future, we will study how to use Automatic Prompt Engineer [35] to construct the optimal prompt for ChatGPT-generated text in the context of fake news detection. Furthermore, the availability of fake news datasets is limited. Therefore, we will implement data generated by ChatGPT in conjunction with the original data to capture real-world scenarios. Another potential direction is to combine ChatGPT with knowledge graphs [21] for fake news detection.

References

1. Bagdasaryan, E., Shmatikov, V.: Spinning language models: risks of propaganda-as-a-service and countermeasures. In: 2022 SP, pp. 769–786. IEEE (2022)
2. Bang, Y., et al.: A multitask, multilingual, multimodal evaluation of chatgpt on reasoning, hallucination, and interactivity (2023)
3. Breiman, L.: Mach. Learn. **45**(1), 5–32 (2001)
4. Chen, T., Guestrin, C.: Xgboost. In: KDD (2016)
5. Conroy, N.K., Rubin, V.L., Chen, Y.: Automatic deception detection: methods for finding fake news. Proc. AIST **52**(1), 1–4 (2015)
6. Cortes, C., Vapnik, V.: Support-vector networks. Mach. Learn. **20**(3), 273–297 (1995)
7. Devlin, J., Chang, M.W., Lee, K., Toutanova, K.: BERT: pre-training of deep bidirectional transformers for language understanding. In: ACL, pp. 4171–4186. Association for Computational Linguistics, Minneapolis (2019)
8. Feng, K.J., Gao, A., Karras, J.S.: Towards semantically aware word cloud shape generation. In: UIST (2022)
9. Ferreira, W., Vlachos, A.: Emergent: a novel data-set for stance classification. In: NAACL (2016)
10. Friedman, J.H.: Greedy function approximation: a gradient boosting machine. AoS **29**(5), 1189–1232 (2001)
11. Gilda, S.: Notice of violation of IEEE publication principles: evaluating machine learning algorithms for fake news detection. In: SCOReD (2017)
12. Hochreiter, S., Schmidhuber, J.: Long short-term memory. Neural Comput. **9**(8), 1735–1780 (1997)
13. Josh, A.G., Girish, S., Micah, M., Renee, D., Matthew, G., Katerina, S.: Generative language models and automated influence operations: emerging threats and potential mitigations (2023)
14. Lai, S., Xu, L., Liu, K., Zhao, J.: Recurrent convolutional neural networks for text classification. In: AAAI, vol. 29, no. 1 (2015)
15. Lau, J.H., Baldwin, T.: An empirical evaluation of doc2vec with practical insights into document embedding generation (2016)
16. Le, Q.V., Mikolov, T.: Distributed representations of sentences and documents (2014)
17. Mikolov, T., Sutskever, I., Chen, K., Corrado, G., Dean, J.: Distributed representations of words and phrases and their compositionality (2013)
18. OpenAI: Gpt-4 technical report (2023)
19. Ouyang, L., et al.: Training language models to follow instructions with human feedback. NeurIPS **35**, 27730–27744 (2022)
20. Pan, S., Hu, R., Long, G., Jiang, J., Yao, L., Zhang, C.: Adversarially regularized graph autoencoder for graph embedding (2019)
21. Pan, S., Luo, L., Wang, Y., Chen, C., Wang, J., Wu, X.: Unifying large language models and knowledge graphs: a roadmap. arXiv:2306.08302 (2023)
22. Pan, S., Wu, J., Zhu, X.: Cogboost: boosting for fast cost-sensitive graph classification. IEEE TKDE **27**(11), 2933–2946 (2015)
23. Paul, C., Jan, L., Tom, B.B., Miljan, M., Shane, L., Dario, A.: Deep reinforcement learning from human preferences (2017)
24. Qin, C., Zhang, A., Zhang, Z., Chen, J., Yasunaga, M., Yang, D.: Is chatgpt a general-purpose natural language processing task solver? (2023)

25. Ruchansky, N., Seo, S., Liu, Y.: CSI: a hybrid deep model for fake news detection. In: CIKM (2017)
26. Sarah, E.K., Miles, M., Miles, B.: All the news that's fit to fabricate: AI-generated text as a tool of media misinformation (2020)
27. Shu, K., Sliva, A., Wang, S., Tang, J., Liu, H.: Fake news detection on social media. ACM SIGKDD Explor. Newsl. **19**(1), 22–36 (2017)
28. Svozil, D., Kvasnicka, V., Jiri, P.: Introduction to multi-layer feed-forward neural networks. CILS **39**(1), 43–62 (1997)
29. Thorne, J., Vlachos, A.: Automated fact checking: task formulations, methods and future directions (2018)
30. Vosoughi, S., Roy, D., Aral, S.: The spread of true and false news online. Science **359**(6380), 1146–1151 (2018)
31. Witteveen, S., Andrews, M.: Paraphrasing with large language models. arXiv preprint arXiv:1911.09661 (2019)
32. Xinyi, Z., Reza, Z.: A survey of fake news: fundamental theories, detection methods, and opportunities. ACM Comput. Surv. **53**, 1–40 (2021)
33. Yang, Y., Zheng, L., Zhang, J., Cui, Q., Li, Z., Yu, P.S.: TI-CNN: convolutional neural networks for fake news detection (2023)
34. Zhang, Y.T., Gong, L., Wang, Y.C.: An improved TF-IDF approach for text classification. J. Zhejiang Univ. Sci. **6**(1), 49–55 (2005)
35. Zhou, Y., et al.: Large language models are human-level prompt engineers. arXiv:2211.01910 (2022)
36. Zubiaga, A., Aker, A., Bontcheva, K., Liakata, M., Procter, R.: Detection and resolution of rumours in social media. ACM Comput. Surv. **51**(2), 1–36 (2018)

CNGT: Co-attention Networks with Graph Transformer for Fact Verification

Jing Yuan[1], Chen Chen[1(✉)], Chunyan Hou[2], and Xiaojie Yuan[1]

[1] College of Computer Science, TKLNDST, Nankai University, Tianjin, China
yuanjing@mail.nankai.edu.cn, {nkchenchen,yuanxj}@nankai.edu.cn,
chunyanhou@163.com
[2] School of Computer Science and Engineering, Tianjin University of Technology,
Tianjin, China

Abstract. Fact verification is a challenging task that requires retrieving evidence from a corpus and verifying claims. This paper proposes Co-attention Networks with Graph Transformer (CNGT), a novel end-to-end reasoning framework for fact verification. CNGT constructs an evidence graph given a claim and retrieved evidence, uses a graph transformer to capture semantic interactions among the claim and evidence, and learns global node representations of the evidence graph via self-attention mechanisms and block networks. Deep co-attention networks integrate and reason on the evidence and claim simultaneously. Experiments on FEVER, a public large-scale benchmark dataset, demonstrate that CNGT achieves a 72.84% FEVER score and a 76.93% label accuracy score, outperforming state-of-the-art baselines. CNGT has de-noising and integrated reasoning abilities and case studies show that it can explain reasoning at the evidence level.

Keywords: Fact verification · Natural language inference · Co-attention networks · Graph transformer

1 Introduction

The explosion of online false information, such as online rumors, fake news and political deception, could potentially lead to tremendous effects on offline society. Automatic fact-checking techniques have emerged to automatically identify false information and prevent it from spreading widely. Fact Verification (FV) is a fact-checking technique that aims to automatically verify the truthfulness of a claim by retrieving pieces of evidence from a reliable corpus. A public large-scale dataset, Fact Extraction and VERification (FEVER) [21], introduces a benchmark FV task. In this task, it is required to verify an input claim with retrieved evidence from Wikipedia and annotate it as "REFUTED", "NOT ENOUGH INFO", or "SUPPORTED" if the retrieved evidence can refute, not be found for the claim, or support, respectively.

The prior methods for FV follow a three-stage pipeline that consists of *document retrieval*, *sentence selection*, and *claim verification*. The related documents are retrieved from the corpus Wikipedia at stage 1. The key sentences are selected from these documents at stage 2. The claim is verified by regrading the set of key sentences as the evidence at stage 3. For claim verification, previous works typically combine all sentences in pieces of evidence [15] together or they deal with each evidence-claim pair and aggregate the isolated evidence sentences [6,27]. These methods of evidence combination either bring about redundant or noisy information or influence reasoning regardless of sufficient information among all pieces of evidence. In fact, the integrated reasoning on pieces of evidence is required to verify claims.

To address this issue, [29] and [12] propose graph-based evidence representation and capture sufficient information from evidence. [20] combine evidence sentences into evidence sets and verify a claim by encoding and attending the claim and evidence sets at different levels of hierarchy. [28] construct graphs by semantic role labeling for the claim and pieces of evidence respectively, and reason over graphs for fact verification. Inspired by these studies, our work constructs an evidence graph to represent pieces of evidence, but differs from these studies by using graph transformer to learn global node representation of the evidence graph. Then the deep co-attention networks are proposed to integrate and reason on pieces of evidence and the claim simultaneously. Furthermore, we investigate different pre-trained language models, such as RoBERTa [11] and ALBERT [9], to capture the semantic information of claims and pieces of evidence. The main contributions of this work include:

- We propose a novel end-to-end reasoning framework, CNGT, to combine the deep co-attention networks with graph transformer for claim verification. Ablation study shows that both graph transformer and co-attention networks improve the performance. The case studies demonstrate that CNGT can explain the reasoning at evidence level.
- We define the evidence graph and present graph transformer used to capture the semantic interactions among the claim and pieces of evidence and learn the global node representation of the evidence graph with the self-attention mechanism and block networks.
- Extensive experiments on public large-scale benchmark datasets demonstrate the superiority of the proposed framework to the state-of-the-art baselines. Experimental results also show that CNGT has the de-noising and integrated reasoning ability.

2   Related Work

FEVER Shared Task. In the FEVER shared task [22], a claim, which is a sentence of unknown truthfulness, is given and the participants are required to retrieve the relevant evidence from corpus Wikipedia at the sentence level and predict the truthfulness of this claim with the given pieces of evidence. [21] released a large-scale benchmark dataset FEVER. [6,15,27] are the top three in this task on the leaderboard.

The methods during the FEVER Shared Task usually treat claim verification as a Natural Language Inference (NLI) problem. The goal of NLI is to predict a label between a pair of premise and hypothesis as contradiction, neutral or entailment. [14] adopts the decomposable attention model to make the prediction between each pair of claim and evidence and consider all predictions together for claim verification. [6,7,27] use enhanced sequential inference model [4] to classify the relationship between a claim and its evidence pieces. [26] propose the TWOWINGOS system to jointly conduct sentence selection and verify the claim. [15] propose the neural semantic matching network which enhances the NLI model. The major difference between our method and these methods is that we pay more attention to claim verification and propose a novel reasoning framework. Our work is similar to [2,3]. [2] combines entity, sentence and context feature to present evidence, uses a heterogeneous graph to capture their semantic relations and design a hierarchical reasoning-based node updating strategy to propagate the evidence features. [3] introduces the entities as nodes and constructs the edges in the graph, generates the fine-grained features of evidence at the entity-level and models the reasoning paths based on an entity graph. Our work differs from those studies in that co-attention networks are proposed to integrate and reason on pieces of evidence and the claim simultaneously and the reasoning can be explained at evidence level.

Graph Neural Networks. There have been many studies that apply different Graph Neural Networks (GNNs) to a variety of NLP tasks and achieve considerable success, ranging from relation extraction [18], machine translation [1] to question generation [17]. Most recent studies treat fact verification as a graph reasoning task [12,28,29]. An evidence graph is constructed and GNNs are able to capture signals from this evidence graph and perform more accurate fact verification. GEAR formulates claim verification as a graph reasoning task and utilizes different aggregators to collect multi-evidence information [29]. Compared with GEAR, our method proposes the deep co-attention networks to be just responsible for the integrated reasoning on pieces of evidence and the claim simultaneously. KGAT is used for fine-grained fact verification with kernel-based attention. It introduces node kernels and edge kernels into graph attention network for more accurate fact verification [12]. Different from KGAT, our method uses graph transformer to capture the semantic interactions among the claim and pieces of evidence and learn a more global node representation of evidence graph efficiently instead of kernels. DREAM applies semantic role labeling to establish links between arguments for the claim and pieces of evidence respectively. It constructs graphs at argument level, and adopt graph convolutional network and graph attention network to propagate and aggregate information from neighboring nodes on each graph [28]. Our work differs from DREAM in that (1) our model constructs the fully connected graph at sentence level rather than at argument level, (2) our model uses graph transformer to learn the global node representation of the graph which includes the claim and pieces of evidence simultaneously while DREAM learns the representations of claim-based graph and evidence-based graphs separately and utilize the graph attention network

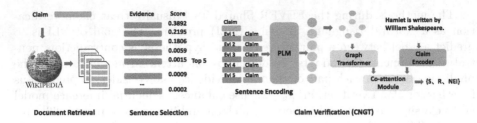

Fig. 1. The pipeline of our method. Note that PLM is a pre-trained language model.

to align the node representations learned for two graphs, and (3) our model can explain the reasoning at evidence level by using deep co-attention networks.

3   Fact Verification

We follow a three-stage pipeline for fact verification. We aim to predict the label of claim with retrieved pieces of evidence. Let $C = \{c_i\}$ be the claim set. For each claim $c_i \in C$, it is required to retrieve related documents from Wikipedia corpus, select the most relevant pieces of evidence E_i from these documents, and predict the label y_i of the claim c_i. Note that $y_i \in \{S, R, NEI\}$ where S, NEI, R denote SUPPORTED, NOT ENOUGH INFO, and REFUTED respectively. The pipeline is illustrated in Fig. 1.

3.1   Document Retrieval and Sentence Selection

For document retrieval, we follow the previous work [6,12,29]. Given a claim associated with one or more entities, we first leverage the constituency parser from AllenNLP [5] to extract the potential entities in the claim. Then these entities are used to search the relevant documents in Wikipedia corpus using the online MediaWiki API[1]. Sentence selection is to select the most related sentences from the retrieved Wikipedia documents. The sentence selection model based on pre-trained language model adopts the hidden state to represent the claim and evidence. The ranking model applies the pairwise loss to optimize parameters. The top five pieces of evidence of ranking score are finally chosen for verifying the truthfulness of a claim.

3.2   Claim Verification

As shown in Fig. 1, at the claim verification stage, we first encode the claim and the (evidence, claim) pairs using a pre-trained language model, and then reason among the claim and the pieces of evidence using CNGT.

---

[1] https://www.mediawiki.org/wiki/API:Main_page.

Encoding Text with Pre-trained Language Model. The Pre-trained Language Models (PLM) (e.g., BERT Base, BERT Large and RoBERTa Large) are used to encode sentences. Given a sentence, we take the hidden state of the last layer of PLM as the representation of the whole sentence. Specifically, $E_i = \{e_{i,j} | 1 \leq j \leq |E_i|\}$ is the retrieved evidence set of the i-th claim . For the claim c_i, we feed it into PLM to obtain the representation $\mathbf{c}_i$ of the claim c_i.

$$\mathbf{c}_i = \text{PLM}(c_i) \tag{1}$$

For $e_{i,j} \in E_i$, we construct $(e_{i,j}, c_i)$ pair and feed it into PLM to obtain their representation $\mathbf{e}_{i,j}$ as follows:

$$\mathbf{e}_{i,j} = \text{PLM}\,(e_{i,j}, c_i) \tag{2}$$

Encoding Graph with Graph Transformer. Following the prior research [29], we use an evidence graph to represent the pieces of evidence as a whole. Formally, the evidence graph is defined as follows.

Definition 1: Evidence Graph is a fully connected, undirected, and unweighted graph with $(|E_i| + 1)$ nodes representing $|E_i|$ pieces of evidence and one claim respectively. Each node has a self-looping edge. The nodes of pieces of evidence are initialized by feeding the (evidence, claim) pairs to PLM while the node of one claim is by feeding this claim.

The Transformer, introduced by [23], has efficient and parallel computation ability with a multi-head self-attention mechanism. Self-attention mechanism enables Transformer suitable to process the fully connected graph. However, the fully connected graph can potentially lead to the over-smoothing of information propagation. Graph transformer proposed by [8] computes hidden representations of each node in the graph by stacking multiple block networks. The number of block networks can control information propagation. Graph transformer differs from those graph attention networks because it is able to better articulate how a node should be updated given the content of its neighbors with dot-product attention and learn global patterns of graph structure exactly.

Let G be the evidence graph, V be the embedded matrix. $V^0 = [\mathbf{v}_i], \mathbf{v}_i \in R^d$. V^0 is the input of the graph transformer. Each node changes its representation by interacting with its adjacent nodes. The graph attention consists of N-headed self-attention, where N independent attentions are calculated and connected before applying the residual connection. The i-th node representation v_i is attended to by its neighbor nodes using N-headed self-attention.

The dot-product is used to compute attention scores.

$$g(\mathbf{v}_i, \mathbf{v}_j) = (\mathbf{W}_Q \mathbf{v}_i)^T \mathbf{W}_K \mathbf{v}_j \tag{3}$$

where $W_Q, W_K \in R^{d \times d}$ are transformation matrices of v_i and v_j respectively and d is the dimension of node vector. Similar to [23], the attention scores are scaled by $1/\sqrt{d_z}$ and normalized by a softmax function to compute the final attention weights α.

$$a\left(\mathbf{q}_i, \mathbf{k}_j\right) = \frac{\exp\left(g\left(\mathbf{q}_i, \mathbf{k}_j\right)/\sqrt{d_z}\right)}{\sum_{z \in \mathcal{N}_i} \exp\left(g\left(\mathbf{q}_i, \mathbf{k}_z\right)/\sqrt{d_z}\right)} \tag{4}$$

where $\mathcal{N}_i$ denotes the first-order neighborhoods of v_i in G

$$\alpha_{ij}^n = a^n\left(\mathbf{v}_i, \mathbf{v}_j\right) \tag{5}$$

where α_{ij}^n denotes the n-th self-attention head between $\mathbf{v}_i$ and $\mathbf{v}_j$. For each node representation $v_i \in R^d$, a N-headed self-attention mechanism is applied over its first-order neighborhoods $\mathcal{N}_i$ to produce representation $\mathbf{v}_i^{ATT} = \mathbf{v}_i + \bigoplus_{n=1}^N \sum_{j \in \mathcal{N}_i} \alpha_{ij}^n \mathbf{W}_V^n \mathbf{v}_j$, where $\bigoplus$ denotes the concatenation of N attention heads, $\mathbf{W}_V^n \in R^{d \times d}$ is a weight matrix.

Graph transformer stacks multiple block networks. In each block network, the multi-headed attention layer is followed by a fully connected layer, and residual connection is followed by a normalization layer at each output of above two layers. In brief, each block network completes the following transformation:

$$\mathbf{v}_i^{OUT} = \text{Norm}\left(\text{FFN}\left(\text{Norm}\left(\mathbf{v}_i^{ATT}\right)\right) + \text{Norm}\left(\mathbf{v}_i^{ATT}\right)\right) \tag{6}$$

where FFN is a two-layer feed-forward network with a non-linear transformation between layers.

Similar to Transformer [23], the blocks are stacked L times to promote information propagation through the graph. The output $\mathbf{V}^l$ of the previous transformer layer l is used as the input $\mathbf{V}^{l+1}$ of the next layer $l+1$ for node i. The encoding of all nodes in G is represented by the representation matrix $\mathbf{V}^L = [\mathbf{v}_i^L]$ which captures the global contextualization of each node using a transformer-based architecture.

Reasoning with Co-attention Networks. As shown in Fig. 1, the outputs of graph transformer and claim encoder are fed into co-attention module. Co-attention module is responsible for reasoning over the pieces of evidence and the claim. Inspired by [13, 25], we propose a deep co-attention module that stacks two co-attention layers to attend over the pieces of evidence and the claim simultaneously and explain the reasoning at evidence level.

Figure 2 provides an illustration of the structure of co-attention layer and module. Given V^L as the output of the graph transformer and v_c as the output of the claim encoder respectively, let $E_1^G = V^L \in R^{d \times (e+1)}$ and $E_1^C = [v_c] \in R^{d \times 1}$. Note that E_1^G represents e pieces of evidence and one claim. The affinity matrix M_1 is computed to measure similarity between two (evidence, claim) pairs.

$$M_1 = \left(E_1^C\right)^\top E_1^G \in R^{1 \times (e+1)} \tag{7}$$

The attention weight $A_1^C = \text{softmax}(M_1^\top) \in R^{(e+1) \times 1}$ and $A_1^G = \text{softmax}\left(M_1\right) \in R^{1 \times (e+1)}$ are column-wise and row-wise normalization of the matrix M_1 respectively. The attention context matrix S_1^C of pieces of evidence is computed with respect to the claim. S_1^G is computed in the same way.

$$S_1^C = E_1^G A_1^C \in R^{d \times 1}, S_1^G = E_1^C A_1^G \in R^{d \times (e+1)} \tag{8}$$

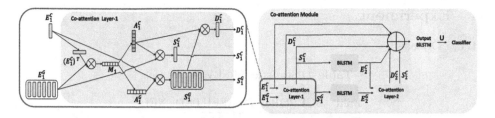

Fig. 2. The network structure of co-attention layer and module.

D_1^C is one of the outputs of co-attention layer-1, which is calculated as follows:

$$D_1^C = S_1^G A_1^C \in R^{d \times 1} \tag{9}$$

A single co-attention layer is described from Eq. (7) to Eq. (9). In brief, the transformation in a co-attention layer is formalized as follows:

$$\text{coattention}\left(E_1^C, E_1^G\right) = \left(S_1^C, S_1^G, D_1^C\right) \tag{10}$$

A bidirectional Long Short-Term Memory (LSTM) Network is used to encode S_1^C and S_1^G respectively:

$$E_2^C = \text{biLSTM}\left(S_1^C\right) \in R^{2h \times 1}, E_2^G = \text{biLSTM}\left(S_1^G\right) \in R^{2h \times (e+1)} \tag{11}$$

where h is the size of hidden state. Then, E_2^C and E_2^G are fed into the co-attention layer-2 which has the same structure as the co-attention layer-1.

$$\text{coattention}\left(E_2^C, E_2^G\right) = \left(S_2^C, S_2^G, D_2^C\right) \tag{12}$$

The affinity matrix M_2 is able to measure similarity between two (evidence, claim) pairs in the co-attention layer 2.

$$M_2 = (E_2^C)^\top E_2^G \tag{13}$$

The initial input and co-attention contexts are concatenated and then encoded by a bidirectional LSTM.

$$U = \text{biLSTM}\left(\text{concat}\left(E_1^C; E_2^C; S_1^C; S_2^C; D_1^C; D_2^C\right)\right) \tag{14}$$

where concat() denotes concatenating matrices horizontally. U is fed into a one-layer linear neural network that finally predicts the label of the claim. It can be derived through:

$$\hat{y} = \text{softmax}(\mathbf{W}_U \mathbf{U} + \mathbf{b}) \tag{15}$$

where $\mathbf{W}_U$ denotes the matrix of learnable parameters and $\mathbf{b}$ is the bias term. The loss function is devised to minimize the negative log-likelihood loss:

$$L = \text{NegativeLogLikelihood}(\mathbf{y}^*, \hat{y}) \tag{16}$$

where $\mathbf{y}^*$ is the ground truth label of claim.

4   Experiment

4.1   Dataset

Table 1. Statistics of FEVER dataset.

Split	SUPPORTED	REFUTED	NEI
Train	80,035	29,775	35,639
Dev	6,666	6,666	6,666
Test	6,666	6,666	6,666

We conduct experiments on the large-scale public dataset FEVER [21]. The FEVER dataset includes 185,445 labeled claims and 5,416,537 Wikipedia documents from the June 2017 Wikipedia dump. Each claim is labeled as SUPPORTED, REFUTED, or NOT ENOUGH INFO(NEI). As shown in Table 1, we follow the dataset partition on the FEVER Shared Task [22].

4.2   Baselines

The baselines include top models during FEVER shared task and other models based on pre-trained language models.

Athene [6], UCL MRG [27] and UNC NLP [15] are the top three on the leaderboard in FEVER. Athene and UNC NLP utilize ESIM to encode evidence. UCL MRG leverages Convolutional Neural Network to encode the claim and evidence.

The PLM based models have been used in the fact verification [10], they perform significantly better than previous methods [24]. [29] provides three baselines: (1) BERT Pair encodes each (evidence, claim) pair independently. (2) BERT Concat concatenates all pieces of evidence of an claim into one sentence. (3) GEAR constructs a fully connected evidence graph, and uses different aggregators to collect information from the evidence graph. SR-MRS [16] and BERT [19] use BERT based method to retrieve sentences for better performance. HESM [20], KGAT [12] and DREAM [28] are the state-of-the-art methods for FV. HESM verifies a claim by encoding and attending to the claim and evidence sets at different levels of hierarchy. KGAT utilizes node kernels and edge kernels to conduct fine-grained evidence propagation in the graph. DREAM applies semantic role labeling to establish links between arguments for the claim and pieces of evidence respectively.

4.3   Evaluation Metrics

The official evaluation metrics of the FEVER shared task include Label Accuracy (LA) and FEVER score. LA is a general evaluation metric, which calculates the three-way classification accuracy rate regardless of retrieved evidence. Only if the retrieved evidence has at least one completely ground truth evidence set and the prediction label is correct, FEVER score is awarded.

Table 2. Fact Verification Performances.

model	Dev		Test	
	LA	FEVER	LA	FEVER
Athene	68.49	64.74	65.46	61.58
UCL MRG	69.66	65.41	67.62	62.52
UNC NLP	69.72	66.49	68.21	64.21
BERT Pair	73.30	68.90	69.75	65.18
BERT Concat	73.67	68.89	71.01	65.64
GEAR	74.84	70.69	71.60	67.10
SR-MRS	75.12	70.18	72.56	67.26
BERT(Base)	73.51	71.38	70.67	68.50
KGAT (BERT Base)	78.02	75.88	72.81	69.40
CNGT (BERT Base)	**78.74**	**76.51**	**73.32**	**69.88**
BERT (Large)	74.59	72.42	71.86	69.66
KGAT (BERT Large)	77.91	75.86	73.61	70.24
CNGT (BERT Large)	**78.94**	**76.06**	**74.30**	**70.42**
DREAM (XLNet)	-	-	76.85	70.60
HESM (ALBERT Large)	75.77	73.44	74.64	71.48
KGAT (RoBERTa Large)	78.29	76.11	74.07	70.38
CNGT (RoBERTa Large)	**81.13**	**78.62**	**76.93**	**72.84**

5   Results and Analysis

5.1   Verification Performance

We conduct experiments on the FEVER dataset and the fact verification performances are shown in Table 2. Compared with the baseline models, our model (i.e., CNGT) achieves the best performance on both the Dev set and the Test set. Particularly, our model with the RoBERTa Large model achieves 76.93% label accuracy score and 72.84% FEVER score.

Baselines are divided into four groups. The first group includes top models from the FEVER shared task which includes Athene, UCL MRG, and UNC NLP. The second and third groups are based on BERT Base and BERT Large, respectively. Compared with the first group, the performance of these two groups has been significantly improved. The fourth group consists of the state-of-the-art models on the FEVER dataset. In this group, we use RoBERTa Large to encode pieces of evidence and the claim. Our model outperforms KGAT, HESM and DREAM. In particular, CNGT is better than KGAT by more than 2% and HESM by more than 5% in terms of LA and FEVER score on the Dev set.

The performance of claim verification depends on the retrieved evidence set provided by the previous stage. On the blind Test set, the evidence set of our model is the same as that of KGAT, but slightly different from that of DREAM because our evidence f1 is 39.14% which is lower than 39.45% of DREAM. LA just evaluates the final verification performance, while FEVER score is a composite metric that can only be awarded when the evidence is retrieved exactly and the claim is verified correctly. Although CNGT is based on the slightly worse evidence set than DREAM, CNGT is able to outperform DREAM in terms of LA and FEVER score on both the Dev and Test sets. Particularly, CNGT achieves a higher FEVER score than DREAM by about 2% on the Test set. Therefore, CNGT has a better de-noising and reasoning ability than DREAM. In addition, it is found that the pre-trained language model is effective to improve the results and RoBERTa achieves the best performance. The superior results of our model are consistent with different pre-trained language models.

5.2   Ablation Study

Table 3. Results of ablation study.

model	Dev		Test	
	LA	FEVER	LA	FEVER
CNGT (RoBERTa Large)	**81.13**	**78.62**	**76.93**	**72.84**
w/o claim	80.20	77.62	75.83	71.70
w/o graph	60.77	58.66	58.66	54.04
w/o co-attention	71.48	68.87	68.43	63.07

To illustrate the importance of each part of our framework, we conduct an ablation study with the RoBERTa Large model. The results of ablation study are reported in Table 3. When we remove the claim encoder and co-attention, the results of **w/o claim** degrade in comparison with that of **CNGT(RoBERTa Large)**. The performance reduces 0.93% on the Dev set and 1.1% on the Test set in terms of LA. We find that although graph transformer takes the claim into account by encoding the evidence graph, the claim encoder can still raise the performance further.

When we eliminate graph transformer and co-attention, the performance of **w/o graph** drops significantly. The results drop 20.36% on the Dev set and 18.27% on the Test set in terms of LA. It reveals that graph transformer plays the most important role in improving the performance for claim verification because graph transformer is able to capture semantic interactions among pieces of evidence and the claim and learn the global node representation of the evidence graph.

When we disregard co-attention module, the performance of **w/o co-attention** is worse than that of **CNGT(RoBERTa Large)**. The results drop 9.65% on the Dev set and 8.5% on the Test set in terms of LA. It is observed that co-attention module plays an independent role for claim verification because co-attention module considers the attention contexts of claim with respect to pieces of evidence and vice versa. These signals of graph transformer and claim encoder are combined by the co-attention networks to integrate and reason on pieces of evidence and the claim simultaneously for verifying the claim exactly.

Table 4. Comparison of our method with KGAT on the subset.

	LA on Dev	LA on subset
KGAT	78.29	79.08
CNGT	**81.13**	**86.15**

5.3 De-noising and Reasoning Performance

Fact verification requires both the de-noising and integrated reasoning ability. De-noising ability is used to discriminate the relevant evidence from noisy evidence which is irrelevant to the claim. The integrated reasoning ability is to reason the veracity of a claim on multiple pieces of evidence and this claim simultaneously.

Firstly, we conduct experiments to validate the integrated reasoning ability of the proposed method on pieces of evidence and the claim simultaneously. We construct a subset of the Dev set by selecting claims in which all ground truth evidence pieces are retrieved. In other words, we select claims which have 100% evidence recall. The subset has 12,582 claims. The evaluation on this subset can eliminate the influence of the retrieval model and demonstrate the integrated reasoning ability. CNGT is compared with KGAT because they have the same retrieved evidence set. As shown in Table 4, experimental results show that CNGT is able to provide integrated reasoning ability and achieve an 86.15% LA on the subset which significantly outperforms KGAT by 7.07%.

Table 5. Comparison between our method and state-of-the-art baselines on claims requiring multiple and single evidence pieces.

	#Claims	Model	LA	FEVER	
Multiple	1960 (14.7%)	GEAR	66.38	37.96	-
		KGAT	65.92	39.23	1.27%
		CNGT	**70.82**	**48.52**	10.56%
Single	11372 (85.3%)	GEAR	78.14	75.73	-
		KGAT	80.33	78.07	2.34%
		CNGT	**85.84**	**85.39**	9.66%

Furthermore, to demonstrate the de-noising and integrated reasoning ability of our method, we compare our method with GEAR and KGAT in another scenario. According to the ground truth evidence of claims, we divide claims (except NEI) into two categories: Single and Multiple. If only one piece of ground truth evidence is required, the corresponding claim is in the category of single evidence. If multiple pieces of ground truth evidence are needed, the claim falls into the category of multiple evidence. The performance in the category of single evidence mainly shows the de-noising ability of the model to the retrieved pieces of evidence because the model is required to select the relevant one from pieces of noisy evidence. The evaluation in the category of multiple evidence can mainly reveal the integrated reasoning ability of the model on multiple pieces of evidence and the claim simultaneously. There are 1,960 and 11,372 claims in the category of multiple evidence and single evidence respectively. CNGT is compared with GEAR and KGAT due to the same retrieved evidence set. The experimental results are shown in Table 5, CNGT is better than GEAR and KGAT. Especially, CNGT outperforms GEAR by more than 10% for the category of multiple evidence in terms of FEVER score. It demonstrates that CNGT has the superior de-noising and integrated reasoning capability.

5.4   Parameter Analysis

Graph transformer has two important hyper-parameters N and L. N is the number of self-attention heads in each block. N independent self-attention heads is used to capture the semantic interactions between the node v_i and its first-order neighbors $\mathcal{N}_i$. L is the number of blocks, stacking multiple blocks allows information to propagate through the graph. We study the effect of each parameter by fixing others to investigate whether they affect the performance of claim verification.

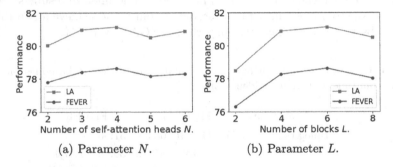

(a) Parameter N.   (b) Parameter L.

Fig. 3. Performance of varied self-attention heads N and blocks L of graph transformer on the Dev set.

Firstly, we fix $L = 6$ and vary N among $\{2, 3, 4, 5, 6\}$, as shown in Fig. 3(a). When N changes, the performance of the model does not change much. The best performance in terms of LA and FEVER score is achieved when $N = 4$, which indicates that four self-attention heads in each block are sufficient for the graph transformer to learn the semantic interactions between the node and its first-order neighbors $\mathcal{N}_i$. Secondly, with $N = 4$ fixed, we set L to $2, 4, 6, 8$, as shown in Fig. 3(b). As L increases from 2 to 6, the performance of the model gradually improves. One possible reason is that, with the increase of L, the model learns the relationship between claims and evidence well. However, as L continues to increase, the performance of CNGT gradually declines due to more noise and redundant information. This suggests that stacking six layers of the model is sufficient to capture the relationship between evidence.

Table 6. An example claim that needs to integrate multiple pieces of evidence for verification. The notation of evidence "{*DocName, LineNum*}" means that the evidence is extracted from the document "*DocName*", and its line number is *LineNum*.

Claim: *Al Jardine* is an *American rhythm guitarist.*
Ground truth evidence:
{Al_Jardine, 0}, {Al_Jardine, 1}
Retrieved evidence:
{Al_Jardine, 0}, {Al_Jardine, 1}, {Al_Jardine, 2}, {Al_Jardine, 5}, {Jardine, 18}
Evidence:
(e_1) *Alan Charles Jardine* (born September 3, 1942) is *an American musician*, singer and songwriter who co-founded the Beach Boys
(e_2) *He is best known as the band's rhythm guitarist*, and for occasionally singing lead vocals on singles such as "Help Me, Rhonda" (1965), "Then I Kissed Her" (1965), and "Come Go with Me" (1978)
(e_3) In 2010, Jardine released his debut solo studio album, A Postcard from California
(e_4) In 1988, Jardine was inducted into the Rock and Roll Hall of Fame as a member of the Beach Boys
(e_5) Sir Ernest Jardine, 1st Baronet (1859-1947), Scottish MP
Label: SUPPORTED

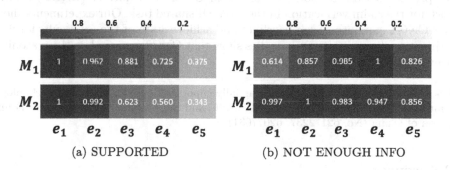

Fig. 4. The affinity matrix M_1 and M_2 of the "S" and "NEI" claim.

5.5 Case Study

Table 6 presents an example claim that requires multiple pieces of evidence to verify its veracity. The first two retrieved evidence pieces are relevant to reasoning this claim, while the remaining three are noisy data. From evidence e_1, the nationality of *Al Jardine* can be obtained, and from evidence e_2, the profession of *Al Jardine* can be obtained. Only by considering the retrieved evidence e_1 and e_2 together can we make the inference that supports this claim.

Figure 4(a) presents the affinity matrix M_1 and M_2 of our model in the co-attention layer-1 and layer-2 respectively. The affinity matrix indicates the relevance weight between a piece of evidence and the corresponding claim. The weights of pieces of evidence are normalized by dividing the maximum weight. A higher weight indicates that the piece of evidence is more effective to verify this claim. It is observed that the evidence e_1 and evidence e_2 have the highest affinity weight. In other words, e_1 and e_2 are most effective to verify this claim, which is consistent with the ground truth evidence. This demonstrates that our model is able to discriminate relevant pieces of evidence to the claim and reason exactly. In addition, M_2 is more accurate to depict the relevance of pieces of evidence than M_1. Thus, the deeper co-attention encoder can build richer representations of pieces of evidence and the claim by enabling them to attend over previous attention contexts. It can explain why two-layer co-attention networks are better than one layer network for the CNGT model.

Figure 4(b) presents the affinity matrix M_1 and M_2 of our model for an example claim which has none of ground truth pieces of evidence and is labeled as NOT ENOUGH INFO. Although the weight of e_1 is less than that of others in M_1, the weights are adjusted and distributed uniformly in M_2. In brief, CNGT is able to explain the reasoning at evidence level.

6   Conclusion

We propose a Co-attention Network with Graph Transformer (CNGT) framework for the claim verification in the FEVER shared task. Our experiments show that CNGT can outperform baselines and achieve state-of-the-art results for fact verification. In addition, case studies show that CNGT can explain the reasoning at evidence level.

Acknowledgement. This work is partially supported by NFSC-General Technology Joint Fund for Basic Research (No. U1936206) and the National Natural Science Foundation of China (No. 62172237, 62077031).

References

1. Beck, D., Haffari, G., Cohn, T.: Graph-to-sequence learning using gated graph neural networks. In: ACL, pp. 273–283 (2018)
2. Chen, C., Cai, F., Hu, X., Chen, W., Chen, H.: HHGN: a hierarchical reasoning-based heterogeneous graph neural network for fact verification. Inf. Process. Manag. **58**(5), 102659 (2021)

3. Chen, C., Cai, F., Hu, X., Zheng, J., Ling, Y., Chen, H.: An entity-graph based reasoning method for fact verification. Inf. Process. Manag. **58**(3), 102472 (2021)
4. Chen, Q., Zhu, X., Ling, Z., Wei, S., Jiang, H., Inkpen, D.: Enhanced LSTM for natural language inference. In: ACL, pp. 1657–1668 (2017)
5. Gardner, M., et al.: Allennlp: a deep semantic natural language processing platform, vol. abs/1803.07640 (2018)
6. Hanselowski, A., et al.: UKP-athene: multi-sentence textual entailment for claim verification, vol. abs/1809.01479 (2018)
7. Hidey, C., Diab, M.: Team SWEEPer: joint sentence extraction and fact checking with pointer networks. In: FEVER (2018)
8. Koncel-Kedziorski, R., Bekal, D., Luan, Y., Lapata, M., Hajishirzi, H.: Text generation from knowledge graphs with graph transformers. In: NAACL-HLT, pp. 2284–2293 (2019)
9. Lan, Z., Chen, M., Goodman, S., Gimpel, K., Sharma, P., Soricut, R.: ALBERT: a lite BERT for self-supervised learning of language representations. In: ICLR (2020)
10. Li, T., Zhu, X., Liu, Q., Chen, Q., Chen, Z., Wei, S.: Several experiments on investigating pretraining and knowledge-enhanced models for natural language inference. CoRR abs/1904.12104 (2019)
11. Liu, Y., et al.: Roberta: a robustly optimized BERT pretraining approach. CoRR abs/1907.11692 (2019)
12. Liu, Z., Xiong, C., Sun, M., Liu, Z.: Fine-grained fact verification with kernel graph attention network. In: ACL, pp. 7342–7351 (2020)
13. Lu, J., Yang, J., Batra, D., Parikh, D.: Hierarchical question-image co-attention for visual question answering. In: NeurIPS, pp. 289–297 (2016)
14. Luken, J., Jiang, N., de Marneffe, M.C.: QED: a fact verification system for the FEVER shared task. In: FEVER, pp. 156–160 (2018)
15. Nie, Y., Chen, H., Bansal, M.: Combining fact extraction and verification with neural semantic matching networks. In: AAAI, pp. 6859–6866 (2019)
16. Nie, Y., Wang, S., Bansal, M.: Revealing the importance of semantic retrieval for machine reading at scale. In: EMNLP-IJCNLP, pp. 2553–2566 (2019)
17. Pan, L., Xie, Y., Feng, Y., Chua, T., Kan, M.: Semantic graphs for generating deep questions. In: ACL, pp. 1463–1475 (2020)
18. Qu, M., Gao, T., Xhonneux, L.A.C., Tang, J.: Few-shot relation extraction via Bayesian meta-learning on relation graphs. In: ICML, pp. 7867–7876 (2020)
19. Soleimani, A., Monz, C., Worring, M.: BERT for evidence retrieval and claim verification. In: ECIR, pp. 359–366 (2020)
20. Subramanian, S., Lee, K.: Hierarchical evidence set modeling for automated fact extraction and verification. In: EMNLP, pp. 7798–7809 (2020)
21. Thorne, J., Vlachos, A., Christodoulopoulos, C., Mittal, A.: FEVER: a large-scale dataset for fact extraction and verification. In: NAACL-HLT, pp. 809–819 (2018)
22. Thorne, J., Vlachos, A., Cocarascu, O., Christodoulopoulos, C., Mittal, A.: The fact extraction and verification (FEVER) shared task, vol. abs/1811.10971 (2018)
23. Vaswani, A., et al.: Attention is all you need. In: NeurIPS, pp. 5998–6008 (2017)
24. Xia, P., Wu, S., Durme, B.V.: Which *BERT? A survey organizing contextualized encoders. In: EMNLP, pp. 7516–7533 (2020)
25. Xiong, C., Zhong, V., Socher, R.: Dynamic coattention networks for question answering. In: ICLR (2017)
26. Yin, W., Roth, D.: Twowingos: a two-wing optimization strategy for evidential claim verification. In: EMNLP, pp. 105–114 (2018)

27. Yoneda, T., Mitchell, J., Welbl, J., Stenetorp, P., Riedel, S.: UCL machine reading group: four factor framework for fact finding (HexaF). In: FEVER, pp. 97–102 (2018)
28. Zhong, W., et al.: Reasoning over semantic-level graph for fact checking. In: ACL, pp. 6170–6180 (2020)
29. Zhou, J., et al.: GEAR: graph-based evidence aggregating and reasoning for fact verification. In: ACL, pp. 892–901 (2019)

Multi-modal

Supervised Discriminative Discrete Hashing for Cross-Modal Retrieval

Xingyu Lu and Chi-Man Pun[✉]

Department of Computer and Information Science, University of Macau, Macau, China
lu.xingyu@connect.um.edu.mo cmpun@umac.mo

Abstract. With the growing interest in cross-modal retrieval technology, cross-modal hashing has become a mainstream trend for comparing and searching between different modalities. However, when faced with multi-label information, existing research has often neglected the unique features of each individual label, resulting in learned hash codes lacking discriminative power. In this article, we present a novel cross-modal hashing method called Supervised Discriminative Discrete Hashing (SD^{2}H), which consists of two stages: hash code learning and hash function learning. Specifically, we extract each individual label from the multi-label information and introduce their one-to-one corresponding hash code. By mapping the original multi-label information to Hamming space through the single-label hash codes, we can obtain more discriminative to-be-learned hash codes. Furthermore, we establish a connection between label information and original feature vectors to explore the potential consistency. Additionally, we use a discrete optimization algorithm to reduce information loss and quantization error during the learning process. The superiority of our proposed SD^{2}H method over several state-of-the-art hashing methods has been demonstrated through extensive experiments on two benchmark datasets.

Keywords: Cross-modal retrieval · Hashing · Multi-label

1 Introduction

With the popularization of multimedia technology, data has begun to appear in various forms, such as images, videos, and text. Due to the need for rapid and accurate acquisition of desired information, retrieval technology is no longer limited to a single modality. In order to improve retrieval efficiency, people have begun to explore cross-modal retrieval methods to integrate information from different modalities [1,2]. For example, users can use a piece of text to search for relevant images or use voice to search for relevant videos. When facing large-scale data, direct comparison of data to complete retrieval seems impractical due to the enormous computational complexity. Therefore, people have begun to use hash methods [3–5] to map multi-modal data to the same Hamming space through learning hash functions. During retrieval, the binary codes of these data can be compared with the efficiency of XOR operation. In recent years, many effective

© The Author(s), under exclusive license to Springer Nature Switzerland AG 2023
X. Yang et al. (Eds.): ADMA 2023, LNAI 14177, pp. 599–613, 2023.
https://doi.org/10.1007/978-3-031-46664-9_40

cross-modal hash methods have been proposed and have achieved impressive results [6–13].

The existing cross-modal hashing methods are mainly divided into supervised hashing and unsupervised hashing. Unsupervised methods [9,10,14–17] mainly learn the required hash codes by mining low-level primitive features and structural characteristics of different modalities. In contrast, supervised hashing [3,18–21] mainly learns the common representation between different modalities by mining label semantic information. Some methods directly embed the label into the Hamming space to learn hash codes, while others utilize the label to construct similarity information between samples, making samples with similar semantics closer in the Hamming space. Faced with increasingly complex data, the use of label information helps supervised methods achieve great leaps in retrieval performance. There are still some problems with existing hashing methods: 1) Many existing supervised learning methods only focus on high-level label information during the process of learning hash codes, while ignoring the original low-level features. 2) In the face of multi-label data, the existing methods often ignore the association between the hash code and each single category, resulting in the lack of discrimination of the learned hash code. 3) Some traditional methods use relaxation strategy to solve binary constraint problem when learning hash code, which causes large quantization errors.

In order to address these issues, we propose a novel Supervised Discriminative Discrete Hashing (SD^2H) method, which consists of two stages: hash code learning and hash function learning. We first extract a single-label matrix from the multi-label information and introduce a single-label hash code matrix to represent the Hamming space representation corresponding to each single category in the multi-label information. In addition to using traditional linear regression to map the supervised information to the to-be-learned hash code, we add an item that discretely projects the label information through single-label hash codes into the Hamming space that needs to be learned. Furthermore, we construct a bridge between the label information and the low-level representations of different modalities, projecting the semantic associations between low-level representations and high-level information into the mapping matrix between label and hash codes, indirectly supervising the learning of hash codes. We use a discrete optimization method to solve the objective function, reducing the computational complexity and errors during the optimization process. Figure 1 shows the overall structure of our proposed method.

The contributions of this paper are as follows:

1) We propose a novel Supervised Discriminative Discrete Hashing method that introduces single-label hash codes during the process of learning hash codes, emphasizing the uniqueness of single categories in the Hamming space and increasing the discriminability of the learned hash code.

2) We establish the relationship between the original information of different modalities and their shared label information, indirectly supervising the hash code and reducing the information loss and errors that may occur during the process of learning the original information.

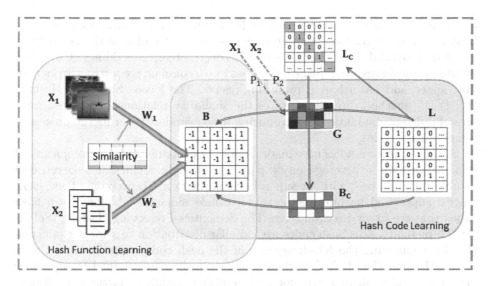

Fig. 1. The framework of proposed SD^2H.

3) We employed a highly effective discrete optimization method to solve the objective function, discretely updating the to-be-learned hash code. Compared with the traditional relaxation method, our optimization method can greatly reduce the quantization error. Upon validation, our proposed SD^2H method demonstrates superior performance over the baseline on the two most commonly used datasets in the cross-modal retrieval domain.

The article is organized as follows. In Sect. 2, we present a review of previous work in cross-modal retrieval. Section 3 outlines our proposed SD^2H method and the corresponding optimization algorithm. Section 4 describes our experimental methodology and presents the results, as well as the analysis. Finally, we summarize our work in Sect. 5.

2   Related Work

The development of cross-modal hashing techniques has greatly improved the accuracy and efficiency of retrieval between different modalities. In this section, we will review relevant research work from the two main categories of cross-modal hashing: unsupervised hashing and supervised hashing.

Unsupervised hashing refers to hashing methods that do not use supervised information. These methods often learn hash codes by mining the original features of data, the similarity between data with similar semantics, or the structural characteristics of multi-modal data. The Collective Matrix Factorization Hashing (CMFH) [14] method is the first method to use collective matrix factorization to learn hash codes, which also preserving the pair-wise similarities. The Latent Semantic Sparse Hashing (LSSH) [15] approach utilizes sparse coding and matrix factorization techniques to learn binary codes that effectively combine the latent semantics from multiple modalities. To address the heterogeneity of multi-modal data, the Collective Reconstructive Embeddings (CRE) [16] method projects text and image modalities into a common reconstructive embedding space, and has achieved promising results. The Fusion Similarity Hashing (FSH) [10] method directly captures the similarity relationship between samples of different modalities by introducing fusion similarity during the process of learning hash codes.

Although these studies have made significant contributions, the complexity of information and the heterogeneity gap between modalities can still make retrieval results unsatisfactory without supervised information. To overcome these challenges, supervised hashing methods can utilize semantic labels to learn hash codes that are more discriminative. The Semantics-Preserving Hashing (SePH) [18,22] method learns hash codes with similar distribution to semantic information by minimizing the KL-divergence of the hash codes. The Discrete Cross-modal Hashing (DCH) [23] method simultaneously learns unified binary codes and individual hash functions for each modality within a classification framework. To address the time complexity of large-scale data, the Semantic Correlation Maximization (SCM) [21] technique seamlessly integrates semantic labels into hash code learning. The Scalable disCRete mATrix faCtorization Hashing (SCRATCH) [24] method utilizes collective matrix factorization and semantic embedding to learn a common latent semantic space, enabling efficient hashing. The Label Consistent Matrix Factorization Hashing (LCMFH) [20] method leverages heterogeneous data and their semantic labels to learn a latent semantic space that better preserves semantic similarity. In addition to matrix factorization, the Supervised Matrix Factorization Hashing (SMFH) [19] approach also fully explores the local manifold structure across different modalities.

3   Proposed Method

In this section, we provide a detailed description of the proposed SD^2H method. We give some notations and present our method in a formalized way using equations. Additionally, we provide a discrete optimization algorithm and analyze the time complexity.

3.1   Notations

For the sake of convenience, we make the assumption that the training dataset comprises of two modalities specifically, the image and text modalities. We use $X = \{X_1, X_2\}$ to represent image-text pairs. $X_1 \in \mathbb{R}^{d_1 \times n}$ and $X_2 \in \mathbb{R}^{d_2 \times n}$ respectively denote image modality and text modality, where d_1 and d_2 are the dimensions of image and text features. The label matrix is denoted by $L \in \{0, 1\}^{c \times n}$, where c is the number of label classes. Due to the diverse nature of the information presented in real-life images and texts, it is assumed that the label information contains a large number of multi-label cases. Let $B \in \{-1, 1\}^{r \times n}$

represent the to-be-learned hash code, and r stands for the length of the hash code.

In our paper, $sgn(\cdot)$ and $\|\cdot\|_F^2$ respectively denote the sign function and the Frobenius norm. In addition, to obtain non-linear features, we employ kernel trick to process the original features. We randomly select k anchor points $\{a_i\}_{i=1}^k$ from the dataset, and use RBF kernel function to map all training data to a k-dimensional feature space, i.e., $\Phi_i(x) = \exp\left(-\|x - a_i\|_2^2 / 2\sigma^2\right)$.

3.2 Hash Code Learning

In existing studies, linear models are often used to project label information to the to-be-learned hash code, which can be described as:

$$\min_{B,G} \|B - GL\|_F^2 \quad s.t. \ B \in \{-1, 1\}^{r \times n} \tag{1}$$

where $G \in \mathbb{R}^{r \times c}$ is a projection matrix. This method can effectively learn the information contained in the labels and is widely used. However, in the case of common multi-label situations, this method may result in the learned hash code not being fully associated with each category. Moreover, the hash codes of those samples that contain the same category may not be strongly linked. This can result in the hash code lacking sufficient discriminative power, thereby affecting the accuracy of cross-modal retrieval. To enhance the attention of the hash code to each category, in our method, we first separate the multi-label information and extract a single-label matrix, which can be represented as:

$$L_c = \begin{bmatrix} 1 & 0 & \cdots & 0 \\ 0 & 1 & \cdots & 0 \\ \vdots & \vdots & \ddots & \vdots \\ 0 & 0 & \cdots & 1 \end{bmatrix}_{c \times c} \tag{2}$$

We introduce single-label hash codes $B_c \in \{-1, 1\}^{k \times c}$ corresponding to each individual category, which can be learned via the same projection matrix G in Eq. 1. The learning of single-label hash codes can be represented by the following equation:

$$\min_{B_c,G} \|B_c - GL_c\|_F^2 \quad s.t. \ B_c \in \{-1, 1\}^{r \times c} \tag{3}$$

Bc is the representation of each individual category in the Hamming space. We use the single-class hash code Bc and label information L together to supervise the learning of the hash code, which can be described as:

$$\min_{B,B_c} \|B - B_cL\|_F^2 \quad s.t. \ B \in \{-1, 1\}^{r \times n}$$
$$B_c \in \{-1, 1\}^{r \times c} \tag{4}$$

which enables every single label in the multi-label information to be noted separately. By augmenting the category information in multi-label data and reinforcing connections between data belonging to the same category, this term improves the discriminative power of the learned hash code. The objective function for the aforementioned components is as follows:

$$\min_{B,B_c,G} \|B - GL\|_F^2 + \|B_c - GL_c\|_F^2 + \alpha \|B - B_cL\|_F^2 \quad s.t.\ B \in \{-1,1\}^{r \times n}$$
$$B_c \in \{-1,1\}^{r \times c}$$
$$(5)$$

Using only label information for supervised learning may cause the learned hash code to miss many low-level original features. To address this, we can use the original features of text and image modalities to supervise the learning of the hash code, such as $B \leftarrow P_1X_1$, $B \leftarrow P_2X_2$. However, applying too much direct linear supervision to the hash code simultaneously can lead to information loss and error accumulation. Therefore, we construct the relationship between label information and original features to explore the potential correlations between different modalities and shared labels, which can be described as:

$$\min_{G,P_1,P_2} \|GL - P_1\Phi(X_1)\|_F^2 + \|GL - P_2\Phi(X_2)\|_F^2 \quad (6)$$

where $P_1 \in \mathbb{R}^{r \times k}$, $P_2 \in \mathbb{R}^{r \times k}$ is the projection matrices of image and text modalities. By this means, the relationships between different modalities and labels are learned and incorporated into the projection function G. G applies the learned potential information of the two modalities to the learning of the single-label hash codes and the hash codes of the training set.

By combining the aforementioned terms, we can formulate the overall function of SD^{2}H as follows:

$$\min_{B,B_c,G,P_1,P_2} \|B - GL\|_F^2 + \|B_c - GL_c\|_F^2 + \alpha \|B - B_cL\|_F^2$$
$$+ \beta \|GL - P_1\Phi(X_1)\|_F^2 + \beta \|GL - P_2\Phi(X_2)\|_F^2$$
$$+ \gamma R(G, P_1, P_2) \quad (7)$$
$$s.t. \quad B \in \{-1,1\}^{r \times n}$$
$$B_c \in \{-1,1\}^{r \times c}$$

where α, β and γ are trade-off parameters, and $R(G, P_1, P_2)$ can be expressed as $(\|G\|_F^2, \|P_1\|_F^2, \|P_2\|_F^2)$.

3.3   Optimization

It is very difficult to solve the problem of Eq. (6) directly, which involves np-hard problems. In this paper, we introduce a discrete optimization method, where one variable is updated while the others are held constant. We provide a detailed illustration of its workings below.

Update P_1   Given that all variables except for P_1 are fixed, the optimization can be carried out by solving the following equation:

$$\min_{P_1} \beta \|GL - P_1\Phi(X_1)\|_F^2 + \gamma \|P_1\|_F^2 \quad (8)$$

By zeroing the partial derivative of Eq. 8 with respect to G_t, we get:

$$P_1 = \left(\beta GL\Phi\left(X_1\right)^{\top}\right)\left(\beta\Phi\left(X_1\right)\Phi\left(X_1\right)^{\top}+\gamma I\right)^{-1} \tag{9}$$

Update P_2 Similar to the optimization process of P_1, we can easily obtain:

$$P_2 = \left(\beta GL\Phi\left(X_2\right)^{\top}\right)\left(\beta\Phi\left(X_2\right)\Phi\left(X_2\right)^{\top}+\gamma I\right)^{-1} \tag{10}$$

Update G With other variables except G fixed, the optimization can be performed by solving the following formula:

$$\begin{aligned}
\min_{G} \|B - GL\|_F^2 + \|B_c - GL_c\|_F^2 \\
+ \beta\|GL - P_1\Phi(X_1)\|_F^2 + \beta\|GL - P_2\Phi(X_2)\|_F^2 + \gamma\|G\|
\end{aligned} \tag{11}$$

By zeroing the partial derivative of Eq. 11 with respect to G_t, we get:

$$\begin{aligned}
G = \left(\beta P_1\Phi(X_1)L^{\top} + \beta P_2\Phi(X_2)L^{\top} + B_cL_c^{\top} + BL^{\top}\right) \\
\left((2\beta + 1)LL^{\top} + L_cL_c^{\top} + \gamma I\right)^{-1}
\end{aligned} \tag{12}$$

Update B_c With other variables except B_c fixed, the optimization can be performed by solving the following formula:

$$\min_{B_c} \|B_c - GL_c\|_F^2 + \alpha\|B - B_cL\|_F^2 \quad s.t. \ B_c \in \{-1, 1\}^{r\times c} \tag{13}$$

By zeroing the partial derivative of Eq. 13 with respect to B_c, we get:

$$B_c = \mathrm{sgn}\left(\left(GL_c + \alpha BL^{\top}\right)\left(I + \alpha LL^{\top}\right)^{-1}\right) \tag{14}$$

Update B With other variables except B fixed, the optimization can be performed by solving the following formula:

$$\min_{B} \|B - GL\|_F^2 + \|B - B_cL\|_F^2 \quad s.t. \ B \in \{-1, 1\}^{r\times n} \tag{15}$$

By zeroing the partial derivative of Eq. 15 with respect to B, we get:

$$B = \mathrm{sgn}\left(\frac{1}{1+\alpha}\left(GL + \alpha B_cL\right)\right) \tag{16}$$

3.4   Hash Function Learning

In the second stage, we will tackle the issue of learning hash functions. In most papers, a linear regression model is used to learn about hash functions by mapping multi-modal data to a hamming space. In this approach, each sample may be treated as an independent entity. However, neglecting the relationships between samples may fail to adequately constrain the mapping direction of the hash function. Following [25], we incorporate pairwise similarity considerations between hash codes during the hash function learning stage. The process of hash function learning can be described as:

$$
\min_{W_1,W_2} \|B - W_1\Phi(X_1)\|_F^2 + \|B - W_2\Phi(X_2)\|_F^2 \tag{17}
$$
$$
+ \|rS - B^\top(W_1\Phi(X_1))\|_F^2 + \|rS - B^\top(W_2\Phi(X_2))\|_F^2
$$

where W_1 and W_2 represent the mapping matrices of two modalities. $S = 2\tilde{L}^\top\tilde{L} - 1_n^\top 1_n$ represents the pairwise similarity between samples, and $\tilde{L}_i = L_i/\|L_i\|$. In order to optimize W_1 and W_2, we respectively set the partial derivative of Eq. 17 with respect to W1 and W2 to zero, and obtain:

$$
W_1 = \left(\theta BB^\top + I\right)^{-1}\left(\theta rBS\Phi(X_1)^\top + B\Phi(X_1)^\top\right)\left(\Phi(X_1)\Phi(X_1)^\top + \varepsilon I\right)^{-1} \tag{18}
$$

$$
W_2 = \left(\theta BB^\top + I\right)^{-1}\left(\theta rBS\Phi(X_2)^\top + B\Phi(X_2)^\top\right)\left(\Phi(X_2)\Phi(X_2)^\top + \varepsilon I\right)^{-1} \tag{19}
$$

After the projection matrix is successfully obtained, when a new query data Q_m arrives, the corresponding hash code can be easily obtained by:

$$
B = \text{sgn}\left(W_m\Phi(Q_m)\right) \tag{20}
$$

and then we can compare it with the hash code of the target modal data in the database for retrieval.

3.5   Time Complexity Analysis

We assess the complexity of the established model by analyzing each variable update. Firstly, it takes $O((d_ik)n)$ for nonlinear projection, where d_i stands for the dimension of image or text modality. For the stage of hash code learning, the computational complexity includes $O((crk + k^2)n + rk^2 + k^3)$ for solving P_1 and P_2, $O((rck + rc + c^2)n + rc^2 + c^3)$ for G, $O((rc + c^2)n + rc^2 + c^3)$ for B_c, and $O(rcn)$ for B. The complexity of the hash function learning stage is $O((r^2 + rk + k^2)n + r^3 + k^3 + r^2k)$. Since $c, k, r, d_i \ll n$, the overall complexity of our method is linear with n. Therefore, we can conclude that our method is efficient and can be used for large-scale tasks.

4 Experiments

4.1 Setup

Datasets. To show the superiority of the proposed SD²H method, we selected two of the most wide-used datasets, MIRFlickr-25K and NUS-WIDE, in the cross-modal retrieval field. MIRFlickr-25K [26], sourced from Flickr's vast collection of user-generated content, contains 25,000 image-text pairs labeled with at least one of 24 specific categories. Text features for each sample pair can be represented by a 500-D PCA feature vector, while images are represented by a 150-D edge histogram feature. We randomly selected 1,500 text-image pairs as the test set and 15,000 pairs as the training set. NUS-WIDE [27] is a large-scale multi-modal dataset, which contains 269648 image-text pairs annotated with 81 categories. We focused on the 10 most frequently occurring labels and selected the corresponding 186,577 image-text pairs for our study. Each of the image and text samples can be represented by a feature vector, with images using a 500-D feature vector and texts employing a 1000-D representation. For our study, we chose a random sample of 2000 text-image pairs to serve as the test set, while 20000 pairs were selected for the training set.

Evaluation Metrics. To validate the efficacy of our proposed SD²H technique, we performed two standard retrieval tasks: (1) image-to-text ($I \rightarrow T$), which involves using images to retrieve similar text information, and (2) text-to-image ($T \rightarrow I$), which involves using text to retrieve similar image information. In this paper, we adopted the mean average precision (mAP) as the evaluation metric to measure the performance of SD²H.

Baseline and Implementation Details. To investigate the performance of SD²H in cross-modal retrieval tasks, we conducted comprehensive experiments comparing it against other state-of-the-art methods such as CMFH [14], FSH [10], SCM-seq [21], SePH [18,22], SMFH [19], DCH [23], LCMFH [20], SRLCH [28] and SRACTCH [24]. Among them, CMFH and FSH are unsupervised hashing methods, while the rest are supervised methods. Most of the methods kindly provided their source code, and we conducted experiments according to the parameter settings in their corresponding original papers.

In this paper, we set the parameters as $\alpha = 1$, $\beta = 10$, $\theta = 1e - 2$, and $\gamma = 1e - 2$. The number of anchors for the kernel function step is set as $k = 500$. The sensitivity analysis of parameters will be conducted in the subsequent subsections. All the methods are completed using MATLAB 2016 with Ubuntu 20.04 system, Intel i7-10700K CPU, and 32 GiB RAM.

4.2 Results and Discussion

To validate the effectiveness of our method, we evaluated the retrieval performance of SD²H in comparison to other state-of-the-art methods on the

MIRFlickr-25K and NUS-WIDE datasets, for both image-to-text and text-to-image tasks. To provide a better comparison, we used hash codes with varying lengths, including 8bits, 16bits, 32bits, 64bits, and 128bits. Table 1 and Table 2 respectively display the mAP results of both our method and baseline methods on each of the datasets. From these tables, we can summarize the following information:

Table 1. MAP results on MIRFlickr-25K.

Method	$I \rightarrow T$					$T \rightarrow I$				
	8	16	32	64	128	8	16	32	64	128
CMFH	0.5688	0.5772	0.5770	0.5761	0.5743	0.5615	0.5628	0.5621	0.5619	0.5620
FSH	0.6033	0.6072	0.6113	0.6142	0.6189	0.5941	0.6025	0.6057	0.6128	0.6139
SCM-seq	0.6213	0.6278	0.6340	0.6391	0.6433	0.6081	0.6105	0.6126	0.6178	0.6203
SePH	0.6477	0.6512	0.6517	0.6578	0.6529	0.6829	0.6872	0.6911	0.6916	0.6943
SMFH	0.6023	0.6031	0.6087	0.6122	0.6103	0.6126	0.6187	0.6214	0.6268	0.6341
DCH	0.6819	0.6871	0.6983	0.7012	0.7104	0.7289	0.7292	0.7471	0.7653	0.7815
LCMFH	0.6718	0.6824	0.6891	0.6916	0.7022	0.7349	0.7444	0.7514	0.7568	0.7589
SRLCH	0.6891	0.6915	0.7132	0.7104	0.7123	0.7183	0.7475	0.7477	0.7463	0.7432
SCRATCH	0.6818	0.7054	0.7089	0.7132	0.7193	0.7326	0.7418	0.7635	0.7812	0.8013
SD^{2}H	**0.6985**	**0.7119**	**0.7284**	**0.7382**	**0.7469**	**0.7613**	**0.7819**	**0.8152**	**0.8215**	**0.8346**

Table 2. MAP results on NUS-WIDE.

Method	$I \rightarrow T$					$T \rightarrow I$				
	8	16	32	64	128	8	16	32	64	128
CMFH	0.3921	0.3877	0.3891	0.3903	0.3897	0.3971	0.3979	0.3901	0.3912	0.3944
FSH	0.4728	0.4829	0.4892	0.4973	0.4994	0.4628	0.4798	0.4874	0.4977	0.5038
SCM-seq	0.5337	0.5364	0.5458	0.5469	0.5491	0.5209	0.5315	0.5377	0.5383	0.5396
SePH	0.5518	0.5627	0.5688	0.5699	0.5710	0.6315	0.6379	0.6483	0.6582	0.6614
SMFH	0.4039	0.4047	0.4076	0.4095	0.4121	0.4237	0.4248	0.4270	0.4302	0.4311
DCH	0.6219	0.6271	0.6383	0.6312	0.6404	0.7289	0.7292	0.7271	0.7353	0.7315
LCMFH	0.6124	0.6137	0.6198	0.6240	0.6355	0.6892	0.6931	0.7023	0.7099	0.7228
SRLCH	0.5984	0.6093	0.6269	0.6403	0.6422	0.7049	0.7136	0.7214	0.7403	0.7478
SCRATCH	0.6043	0.6164	0.6258	0.6323	0.6409	0.7122	0.7237	0.7308	0.7411	0.7535
SD^{2}H	**0.6232**	**0.6353**	**0.6574**	**0.6801**	**0.6948**	**0.7213**	**0.7534**	**0.7732**	**0.7867**	**0.7984**

1) As the hash code length increases in most methods, the retrieval performance typically improves as indicated by a higher mAP score. The aforementioned observation suggests that longer hash codes can store more effective information. However, this is not the case for some methods, which could potentially be attributed to significant errors occurring during the hash code learning stage.

2) Our proposed SD^2H method consistently achieves the best mAP scores in all cases. The main explanation may be that the way we use single-label hash codes to highlight the presence of individual labels in our method is good for increasing the discrimination of hashing methods.

3) Most methods are more accurate in tasks that use text to retrieve images than in tasks that use images to retrieve text. One possible explanation is that the text modality features tend to contain more comprehensive semantic information as compared to image modality features.

4) In general, supervised baselines demonstrate superior performance compared with unsupervised hashing methods, likely due to their ability to leverage shared supervisory information in the training data, which plays a pivotal role in enhancing the overall quality of the unified hash code.

4.3 Parameter Sensitivity Analysis

We further designed experiments to analyze the parameter sensitivity in the proposed SD^2H method. We conducted the experiments on the mirFlickr-25k dataset, with the hash code length fixed at 32 bits. The parameters in our method include α, β, γ, θ. In each round of experiments, we only changed one parameter while keeping the others fixed.

The experimental results of parameter sensitivity are shown in Fig. 2. It can be observed that α has a significant impact on the results and a relatively good performance can be achieved when alpha is within the range of $[10^{-2}, 10^2]$. β has a minor effect on the results, but the optimal performance can be achieved within the range of $[0.1, 10]$. The β term controls the joint learning module of the low-level semantic and high-level label semantic, indicating that the proposed method has strong stability. The θ term controls the similarity term in the hash function learning phase. It can be seen from the graph that when θ is too large, the performance will slightly decrease, possibly due to the small weight of the regression term, which affects the accuracy of the mapping between data points and hash codes. The regularization term is controlled by the parameter γ. It appears that the model's performance is not significantly impacted by γ, indicating low sensitivity to this parameter.

4.4 Ablation Study

We investigated the effectiveness of our proposed SD^2H method through ablation experiments. We introduce SD^2H-1, SD^2H-2 and SD^2H-3 based on SD^2H. SD^2H-1 sets α to 0, which means discarding the enhancement of single-category semantics. SD^2H-2 sets β to 0, which means discarding the supervision of learning hash for original features. In SD^2H-3, we replaced GL with B in Eq. 6, which directly uses a common linear model to map the original features of the two modalities into Hamming space.

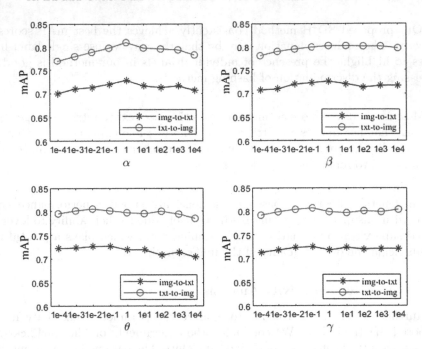

Fig. 2. Parameter sensitivity analysis of SD^2H.

We set the hash code length to 8bits, 16bits, 32bits, 64bits, 128bits, and compared the retrieval performance of SD^2H and the three variants on MirFlickr-25K and NUS-WIDE. Table 3 and Table 4 report the mAP results under different code lengths. We can clearly see that SD^2H achieves the best performance in all cases, which further validates the effectiveness of our proposed method.

Table 3. Ablation results on MirFlickr-25K.

Method	$I \rightarrow T$					$T \rightarrow I$				
	8	16	32	64	128	8	16	32	64	128
SD^2H-1	0.6823	0.6989	0.7021	0.7106	0.7212	0.7349	0.7474	0.7614	0.7768	0.7989
SD^2H-2	0.6841	0.7015	0.7032	0.7113	0.7138	0.7354	0.7461	0.7677	0.7763	0.7832
SD^2H-3	0.6918	0.7089	0.7189	0.7242	0.7369	0.7581	0.7643	0.8022	0.8114	0.8249
SD^2H	**0.6985**	**0.7119**	**0.7284**	**0.7382**	**0.7469**	**0.7613**	**0.7819**	**0.8152**	**0.8215**	**0.8346**

Table 4. Ablation results on NUS-WIDE.

Method	$I \rightarrow T$					$T \rightarrow I$				
	8	16	32	64	128	8	16	32	64	128
SD^{2}H-1	0.6124	0.6237	0.6398	0.6640	0.6755	0.7092	0.7331	0.7523	0.7699	0.7728
SD^{2}H-2	0.6038	0.6189	0.6322	0.6693	0.6714	0.7071	0.7236	0.7418	0.7655	0.7730
SD^{2}H-2	0.6196	0.6271	0.6488	0.6726	0.6855	0.7122	0.7448	0.7619	0.7736	0.7898
SD^{2}H	**0.6232**	**0.6353**	**0.6574**	**0.6801**	**0.6948**	**0.7213**	**0.7534**	**0.7732**	**0.7867**	**0.7984**

5 Conclusion

In this paper, we develop a new cross-media retrieval hash algorithm named SD^{2}H. Firstly, in the hashing learning stage, we use single-label hash code to discretely project multi-label information into the Hamming space and learn more discriminative binary codes. We jointly learn the projection matrix by high-level semantic labels and low-level features, embedding the potential consistency between different modalities into hash codes indirectly. Additionally, we incorporate similarity supervision terms in the hashing function learning stage and use an effective discrete optimization algorithm to solve the objective function. Our method outperforms other state-of-the-art algorithms on two of the most common datasets in the cross-modal retrieval field. In the future, we will consider using deep learning to further explore multi-modal information.

Acknowledgements. This work is supported in part by the University of Macau under Grant MYRG2022-00190-FST and in part by the Science and Technology Development Fund, Macau SAR, under Grant 0034/2019/AMJ, Grant 0087/2020/A2 and Grant 0049/2021/A.

References

1. Fei, W., et al.: Cross-modal learning to rank via latent joint representation. IEEE Trans. Image Process. **24**(5), 1497–1509 (2015)
2. Jiang, Q.Y., Li, W.J.: Deep cross-modal hashing. In: Proceedings of the IEEE Conference on Computer Vision and Pattern Recognition, pp. 3232–3240 (2017)
3. Shen, F., Shen, C., Liu, W., Tao Shen, H.: Supervised discrete hashing. In: Proceedings of the IEEE Conference on Computer Vision and Pattern Recognition, pp. 37–45 (2015)
4. Mu, Y., Shen, J., Yan, S.: Weakly-supervised hashing in kernel space. In: 2010 IEEE Computer Society Conference on Computer Vision and Pattern Recognition, pp. 3344–3351. IEEE (2010)
5. Fei, W., Zhou, Yu., Yang, Y., Tang, S., Zhang, Y., Zhuang, Y.: Sparse multi-modal hashing. IEEE Trans. Multimed. **16**(2), 427–439 (2013)
6. Li, C., Deng, C., Li, N., Liu, W., Gao, X., Tao, D.: Self-supervised adversarial hashing networks for cross-modal retrieval. In: Proceedings of the IEEE Conference on Computer Vision and Pattern Recognition, pp. 4242–4251 (2018)

7. Ji, Z., Sun, Y., Yunlong, Yu., Pang, Y., Han, J.: Attribute-guided network for cross-modal zero-shot hashing. IEEE Trans. Neural Netw. Learn. Syst. **31**(1), 321–330 (2019)

8. Kumar, S., Udupa, R.: Learning hash functions for cross-view similarity search. In: Twenty-Second International Joint Conference on Artificial Intelligence (2011)

9. Hoang, T., Do, T.T., Nguyen, T.V., Cheung, N.M.: Unsupervised deep cross-modality spectral hashing. IEEE Trans. Image Process. **29** 8391–8406 (2020)

10. Liu, H., Ji, R., Wu, Y., Huang, F., Zhang, B.: Cross-modality binary code learning via fusion similarity hashing. In: Proceedings of the IEEE Conference on Computer Vision and Pattern Recognition, pp. 7380–7388 (2017)

11. Yan, C., Gong, B., Wei, Y., Gao, Y.: Deep multi-view enhancement hashing for image retrieval. IEEE Trans. Pattern Anal. Mach. Intell. **43**(4), 1445–1451 (2020)

12. Wang, D., Wang, Q., An, Y., Gao, X., Tian, Y.: Online collective matrix factorization hashing for large-scale cross-media retrieval. In: Proceedings of the 43rd International ACM SIGIR Conference On Research and Development in Information Retrieval, pp. 1409–1418 (2020)

13. Zhang, Z., Luo, H., Zhu, L., Lu, G., Shen, H.T.: Modality-invariant asymmetric networks for cross-modal hashing. IEEE Trans. Knowl. Data Eng. **35**(5), 5091–5104 (2022)

14. Ding, G., Guo, Y., Zhou, J.: Collective matrix factorization hashing for multimodal data. In: Proceedings of the IEEE Conference on Computer Vision and Pattern Recognition, pp. 2075–2082 (2014)

15. Zhou, J., Ding, G., Guo, Y.: Latent semantic sparse hashing for cross-modal similarity search. In: Proceedings of the 37th International ACM SIGIR Conference on Research and Development in Information Retrieval, pp. 415–424 (2014)

16. Hu, M., Yang, Y., Shen, F., Xie, N., Hong, R., Shen, H.T.: Collective reconstructive embeddings for cross-modal hashing. IEEE Trans. Image Process. **28**(6), 2770–2784 (2018)

17. Hu, P., Zhu, H., Lin, J., Peng, D., Zhao, Y.P., Peng, X.: Unsupervised contrastive cross-modal hashing. IEEE Trans. Pattern Anal. Mach. Intell. **45**(3), 3877–3889 (2022)

18. Lin, Z., Ding, G., Hu, M., Wang, J.: Semantics-preserving hashing for cross-view retrieval. In: Proceedings of the IEEE Conference on Computer Vision and Pattern Recognition, pp. 3864–3872 (2015)

19. Tang, J., Wang, K., Shao, L.: Supervised matrix factorization hashing for cross-modal retrieval. IEEE Trans. Image Process. **25**(7), 3157–3166 (2016)

20. Wang, D., Gao, X., Wang, X., He, L.: Label consistent matrix factorization hashing for large-scale cross-modal similarity search. IEEE Trans. Pattern Anal. Mach. Intell. **41**(10), 2466–2479 (2018)

21. Zhang, D., Li, W.J.: Large-scale supervised multimodal hashing with semantic correlation maximization. In: Proceedings of the AAAI Conference on Artificial Intelligence, vol. 28 (2014)

22. Lin, Z., Ding, G., Han, J., Wang, J.: Cross-view retrieval via probability-based semantics-preserving hashing. IEEE Trans. Cybern. **47**(12), 4342–4355 (2016)

23. Xu, X., Shen, F., Yang, Y., Shen, H.T., Li, X.: Learning discriminative binary codes for large-scale cross-modal retrieval. IEEE Trans. Image Process. **26**(5), 2494–2507 (2017)

24. Chen, Z.-D., Li, C.-X., Luo, X., Nie, L., Zhang, W., Xin-Shun, X.: Scratch: a scalable discrete matrix factorization hashing framework for cross-modal retrieval. IEEE Trans. Circuits Syst. Video Technol. **30**(7), 2262–2275 (2019)

25. Li, H., Zhang, C., Jia, X., Gao, Y., Chen, C.: Adaptive label correlation based asymmetric discrete hashing for cross-modal retrieval. IEEE Trans. Knowl. Data Eng. (2021)
26. Huiskes, M.J., Lew, M.S.: The MIR Flickr retrieval evaluation. In: Proceedings of the 1st ACM International Conference on Multimedia Information Retrieval, pp. 39–43 (2008)
27. Chua, T.S., Tang, J., Hong, R., Li, H., Luo, Z., Zheng, Y.: Nus-wide: a real-world web image database from national university of Singapore. In: Proceedings of the ACM International Conference on Image and Video Retrieval, pp. 1–9 (2009)
28. Shen, H.T., et al.: Exploiting subspace relation in semantic labels for cross-modal hashing. IEEE Trans. Knowl. Data Eng. **33**(10), 3351–3365 (2020)

A Knowledge-Enhanced Inferential Network for Cross-Modality Multi-hop VQA

Shiqi Wang[1], Jianxing Yu[1(✉)], Miaopei Lin[1], Shuang Qiu[2], Xiaofeng Luo[3], and Jian Yin[1]

[1] Guangdong Key Laboratory of Big Data Analysis and Processing,
School of Artificial Intelligence, Sun Yat-sen University, Zhuhai, China
{wangshq25,linmp3}@mail2.sysu.edu.cn, {yujx26,issjyin}@mail.sysu.edu.cn
[2] Guangdong University of Education, Guangzhou 510303, China
qiushuang@gdei.edu.cn
[3] Guangdong Industry Polytechnic, Guangzhou 510300, China
2021070030@gdip.edu.cn

Abstract. This paper focuses on cross-modality multi-hop visual questions, which require multi-hop reasoning over different sources of knowledge from multiple modalities, such as image and text. Due to the lack of cross-modality reasoning ability, it is difficult for the traditional model to make correct predictions. To solve this problem, we propose a new knowledge-enhanced inferential framework. We first build a reasoning graph to capture the topological relations between the objects in the given image and the logical relations of entities corresponding to these objects. To align the visual objects and textual entities, we design a cross-modality retriever with the help of an external multimodal knowledge graph. Based on the logical and topological relations on the graph, we can derive the answer by decomposing a complex multi-hop question into a series of attention-based reasoning steps. The result of the previous step acts as the context of the next step. By linking the results of all steps, we can form an evidence chain to the answer. Extensive experiments conducted on the popular KVQA dataset demonstrate the effectiveness of our approach.

Keywords: Visual question answering · Multi-hop reasoning · Cross-modality reasoning

1 Introduction

With the rapid development of web applications such as image retrieval and auxiliary visual impairment, visual question answering (VQA) [2] has become a hot research topic. It aims to measure the machine's ability to understand the image contents by asking questions. Towards this task, researchers have carried out many explorations. In most existing datasets, the answer is an object or attribute of the image. It means that lots of questions can be solved trivially via simple matching rather than genuine comprehension of the image. Some challenge datasets containing complex questions are proposed to fill this gap [11,12,22]. Their answers need to be derived by deep semantic understanding. As shown in Fig. 1, the question asks the founder of a person's political party. We first have to find the person in the image by reasoning a location *"the second person from left,"* that is, *"Narendra Modi."* We then use the commonsense knowledge to deduce an evidence chain from the person to the party *"Bharatiya Janata Party,"* and finally to the founder *"Syama Prasad Mookerjee."* These reasoning clues are logically correlated but distributed across different modalities. Machines need to synthesize supporting knowledge scattered in multiple modalities and form an evidential chain from the question to the answer.

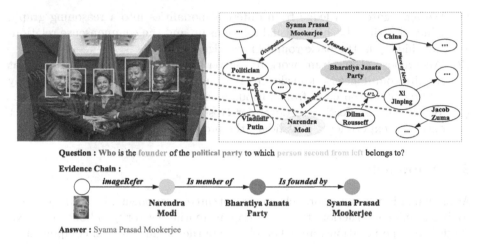

Fig. 1. A sample requires multi-hop reasoning over various knowledge sources from multiple modalities.

Mainstream reasoning methods can be summarized into two categories, one is to parse the question into a series of sub-questions which are solved correlatively to get the answer [6,32]. However, the parser has to cascade to the QA model. It is difficult to train globally and would suffer from the error propagation problem. Another direction [14,19] is to find the image object with the highest matching score as the answer through the interactive fusion method. This coarse-grained fusion ignores the heterogeneity between different modalities and lacks adequate robustness to be applicable. In addition, the reasoning process ignores implicit commonsense knowledge, which is self-evident and shared by most humans.

To address these problems, we propose a novel inferential framework, which fills the heterogeneity gap of different modalities by introducing an external multimodal knowledge graph. In particular, we build a scene graph to describe the image objects and their relations. To align the visual objects and textual entities, we design a cross-modality retriever with the help of the multimodal graph, e.g., MMKG [16]. Such a graph contains abundant commonsense relations of the entities. Also, it grasps the cross-modality alignment relation between the visual object and textual entity. Using the retrieved knowledge we can expand the scene graph. Subsequently, we derive the answer on the knowledge-enhance graph by an inferential network, which decomposes a complex reasoning process into a series of attention-based steps. Each step can be directly inferred from the inputs without strong supervision. By aggregating the results of these steps, we can derive the answer accordingly. With such a structural multi-step design, our network can integrate the supported knowledge from different modalities to build the evidence chain in arbitrarily complex acyclic form. Moreover, a reinforcement approach is employed for effective training. Evaluations of the typical dataset show the effectiveness of our approach.

The main contributions of this paper include:

(1) We aggregate knowledge from different modalities into a reasoning graph, which describes both the visual relationship and the commonsense relationship hidden in the background knowledge.
(2) We propose a new framework to answer the complex VQA questions that require commonsense knowledge. We can derive the answer by recursively finding evidential clues from various modalities to form a reasoning chain.
(3) We design an effective retriever for cross-modality alignment and conduct extensive experiments to validate the proposed approach.

2   Approach

As shown in Fig. 2, our approach consists of three components, including the reasoning graph construction, the multi-hop inferential network, and output. Next, we define some notations, and then elaborate the details on each component.

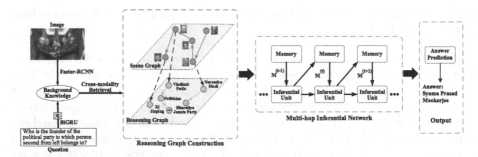

Fig. 2. Overview of our approach

2.1   Notations and Problem Formulation

Given an image $I \in \mathcal{I}$ and a question $Q = \{q_1, ..., q_{|Q|}\}$, where q_i is the ith token, the task of VQA is to answer the question based on the semantics of the image. We count the frequency of each answer label of the target dataset, and assume the top-K items as the answer candidates $\mathcal{Y}$. For an input sample $s = (I, Q)$ that contains the image I and the question Q, a score $h(s, y) \in \mathbb{R}$ is calculated for each candidate $y \in \mathcal{Y}$. The one with the highest score is outputted as the answer $\hat{y} = \arg\max_{y \in \mathcal{Y}} h(s, y)$. For the cross-modality multi-hop VQA task, the question needs to be solved by understanding not only the image I, but also the commonsense knowledge base KB related to the image objects. That is, we have to derive the answer $\hat{y}$ by multi-hop reasoning over multiple knowledge sources in different modalities according to the function $\hat{y} = \arg\max_{y \in \mathcal{Y}} h' = (I, Q, KB, y)$.

2.2 Reasoning Graph Construction

Considering the scale of an external knowledge source is large, we construct a compact graph to reduce the computational cost. The graph covers various relations, objects, and entities that may be involved in the reasoning process. In particular, we first generate a scene graph to grasp the topological relations of the image objects. We then design a cross-modality retriever to collect relevant entities and relations from the external knowledge source.

Scene Graph Generation. We first extract a scene graph to capture fine-grained objects in the image involved in reasoning. In particular, given the image I, we use Faster-RCNN [21] to extract a set of k entity proposals. Each proposal is associated with a spatial region $\mathbf{b}_i = [x_i, y_i, w_i, h_i]$, where (x_i, y_i), h_i and w_i denote the coordinate of the top-left point, the height, and width of the region i. Based on these proposals, we can form a scene graph $\mathcal{G}^S = (\mathcal{N}^S, E^S)$, where the node $n_i^S \in \mathcal{N}^S$ corresponds to a proposed entity, and the edge $e_{ij}^S \in E^S$ represents the relation from n_i^S to n_j^S. We encode the node as $\mathbf{n}_i^S \in \mathbb{R}^d$ via the pretrained VGG model [25] and use the output of the last hidden layer after compact bilinear pooling [9]. The edge is encoded as $\mathbf{e}_{ij}^S = W_S^{d \times 3d}[\mathbf{n}_i^S, \mathbf{u}_{ij}, \mathbf{n}_j^S] + \mathbf{b}_S^d$, where $\mathbf{u}_{ij} \in \mathbb{R}^d$ is the union region feature and $\mathbf{b}_S^d$ is the bias.

Cross-Modality Retriever. Since the representation spaces of the textual and visual modalities are different, it is difficult to retrieve commonsense entities directly through image objects. To tackle this problem, we develop a retriever by resorting to the multimodal knowledge graph MMKG [16]. This graph records a special kind of triples, such as $(e, hasImage, I)$, indicating that the image I is an instance of the textual entity e. This kind of knowledge can help us learn the cross-modality alignment. In detail, we design the retriever based on ConvE [8] which is good at finding correlation. For a triplet (s, r, o), where s, r, o represent the subject entity, relation, and object entity, respectively, we first embed them into d-dimensional vectors, that is, $\mathbf{em}_s, \mathbf{em}_r, \mathbf{em}_o \in \mathbb{R}^d$. We then measure the relational probability of these two entities by introducing a score function $\psi : \mathcal{E} \times \mathcal{R} \times \mathcal{E}' \to \mathbb{R}$, as Eq (1), where $\bar{\mathbf{em}}_s, \bar{\mathbf{em}}_r \in \mathbb{R}^{d_w \times d_h}$ denote 2D shaped embeddings for $\mathbf{em}_s$ and $\mathbf{em}_r$, $d = d_w d_h$, f denotes a non-linear function, and vec is a reshaped function. Based on this retriever, we can score the triplet $(e, hasImage, n_i^S)$ for each scene node n_i^S by Eq. (1), where $hasImage$ is the alignment relation between the image object and the textual entity.

Similarly, we compute the triplets related to the question Q. We then collect the entities whose scores exceed a preset threshold λ. Their K-hop neighbor nodes and relations cover most of the potential clues in the multi-hop deduction. We use them to create a reasoning graph $\mathcal{G}$.

$$\psi_r(s,o) = f(vec(f([\bar{\mathbf{em}}_s; \bar{\mathbf{em}}_r] * \omega_\psi))\mathbf{W}_\psi)\mathbf{em}_o \tag{1}$$

Considering the external knowledge graph contains abundant subject-relation-object triplets (s, r, o) in various modalities, they can be used as training samples naturally. As shown in Eq. (2), the retriever is learned by minimizing the binary cross-entropy loss, where $l_o^{s,r}$ is a binary label to indicate whether a triplet exists in the MMKG graph, $p_o^{s,r}$ is the score function of $\sigma(\psi_r(\mathbf{em}_s, \mathbf{em}_o))$, $\sigma(\cdot)$ is the logistic sigmoid function.

$$\sum_{(s,r)} \sum_o l_o^{s,r} log(p_o^{s,r}) + (1 - l_o^{s,r})log(1 - p_o^{s,r}) \tag{2}$$

2.3   Multi-hop Inferential Network

Traditional methods create a mapping function to judge the relations between two entities. They cannot form a reasoning process that needs at least three entities. Also, a network with a fixed structure cannot model the unfixed reasoning process. To solve this problem, we design a new inferential network based on a divide-and-conquer strategy. It can decompose a complex multi-hop question into several basic reasoning steps. Each step is a relational mapping from a question aspect to an evidential clue in the reasoning graph. By combining all basic steps, we can dynamically build a dedicated net structure like stacking building blocks to simulate a complex reasoning process. Our net consists of shared memory and a deduction module. As shown in Fig. 3, the module includes three units, that is, the dispatcher, executor, and recorder.

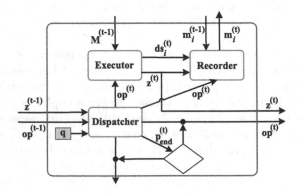

Fig. 3. Flow chart of the deduction module

Dispatcher Unit. We first decompose a complex question into multiple simple ones by soft attention. There are two stages, including judging whether to stop the reasoning process, and determining a question aspect that should be focused on at the t reasoning step.

Termination Judgment. A maximum reasoning step is set as T to guarantee termination. Considering that the question complexity can vary greatly, the reasoning depth is uncertain. To adjust to such a depth dynamically, we design a stop mechanism by considering two conditions. That is, the correlation between the evidential clue $\mathbf{z}^{(t-1)}$ and deduction operation $\mathbf{op}^{(t-1)}$ in the $t-1$ reasoning step, as well as $\mathbf{z}^{(t-1)}$ and the question $\mathbf{q}$ with a global asking focus. When both conditions are met, an acceptable answer is highly probable to obtain. As shown in Eq. (3), we formulate the correlations as $\mathbf{z}^{(t-1)} \odot \mathbf{op}^{(t-1)}$ and $\mathbf{z}^{(t-1)} \odot \mathbf{q}$ respectively, and combine them as a vector $\mathbf{cq}^{(t)}$, where $\mathrm{W}_{cq}^{d \times 2d}$, $\mathbf{b}_{cq}^{d}$ and $\mathrm{W}_{end}^{1 \times d}$ are the learned parameters. $\mathbf{cq}^{(t)}$ is then fed into a *sigmoid* layer to estimate the termination probability $p_{end}^{(t)}$. A binary variable $end^{(t)}$ is then sampled randomly on $p_{end}^{(t)}$. If $end^{(t)}$ is *True*, stop deduction to avoid excessive reasoning and execute the answer prediction module; otherwise, continue the reasoning process.

$$\mathbf{cq}^{(t)} = \mathrm{W}_{cq}^{d \times 2d}[\mathbf{z}^{(t-1)} \odot \mathbf{op}^{(t-1)}, \mathbf{z}^{(t-1)} \odot \mathbf{q}] + \mathbf{b}_{cq}^{d}$$
$$p_{end}^{(t)} = sigmoid(\mathrm{W}_{end}^{1 \times d}\mathbf{cq}^{(t)}) \tag{3}$$

Question Analysis. To facilitate the overall training of the model, we employ soft attention instead of a hard extraction to identify the question aspect. Firstly, we project the question $\mathbf{q}$ through a learned linear transformation to get the aspect $\mathbf{q}^{(t)}$ related to t reasoning step, as $\mathbf{q}^{(t)} = \mathrm{W}_{t}^{d \times d}\mathbf{q} + \mathbf{b}_{t}^{d}$.

Secondly, we combine $\mathbf{q}^{(t)}$, $\mathbf{op}^{(t-1)}$ and $\mathbf{cq}^{(t)}$ through a linear transformation into $\mathbf{dq}^{(t)}$, as $\mathrm{W}_{dq}^{d \times 3d}[\mathbf{q}^{(t)}, \mathbf{op}^{(t-1)}, \mathbf{cq}^{(t)}] + \mathbf{b}_{dq}^{d}$, where $\mathbf{op}^{(t-1)}$ denotes the aspect focused in $t-1$ step, $\mathbf{cq}^{(t)}$ represents the correlation degree. That allows us to find the relevant aspect based on historical results.

Thirdly, we cast $\mathbf{dq}^{(t)}$ back to the space of the question terms. In this way, we can restrict the valid reasoning space and boost convergence. As shown in Eq. (4), we measure the similarity between $\mathbf{dq}^{(t)}$ and each question term $\mathbf{w}_i \in \mathbb{R}^d$, then pass them through a *softmax* layer to obtain an attention distribution over the question terms. By aggregation, we can generate the deduction operation $\mathbf{op}^{(t)}$ that is represented in terms of the question terms.

$$ca_i^{(t)} = \mathrm{W}_{ca}^{1 \times d}(\mathbf{dq}^{(t)} \odot \mathbf{w}_i)$$
$$cw_i^{(t)} = softmax(ca_i^{(t)})$$
$$\mathbf{op}^{(t)} = \sum_{i=1}^{|Q|} cw_i^{(t)} \cdot \mathbf{w}_i \tag{4}$$

Executor Unit. Based on the question aspect, we collect related contents from the retrieved reasoning graph $\mathcal{G}$. The relevance is measured by considering the graph structure in a soft-attention manner. That can facilitate global training.

To support transitive reasoning, we first extract the contents relevant to the preceding evidential clues $\mathbf{z}^{(t-1)}$ for each memory entry i, resulting in $\mathbf{dm}_i^{(t)} = [\mathbf{W}_{dz}^{d \times d} \mathbf{z}^{(t-1)} + \mathbf{b}_{dz}^d] \odot [\mathbf{W}_{dm}^{d \times d} \mathbf{m}_i^{(t-1)} + \mathbf{b}_{dm}^d]$. This vector grasps the contextual relation in the distributed space. That can help us find the correlated evidential clues in adjacent steps. For example, given the clue of *"the second person from the left"* is *"Narendra Modi"* in the previous step, the relevant content about *"Narendra Modi"* should deserve more attention in the current step.

Subsequently, we incorporate the context related to the reasoning graph $\mathcal{G}$. That can help us perform parallel and inductive reasoning effectively on the commonsense entities. We first design a relational graph convolution network to encode the graph structure. We then inject the context for each memory entry i from its graph neighbors $\mathcal{N}_i$, conditioning on the deduction operation $\mathbf{op}^{(t)}$. As shown in Eq. (5), we get a contextual vector $\mathbf{ne}_i^{(t)}$ based on $\mathcal{N}_i$, where $w_{ji}^{(t)}$ is the attention weight of edge $\mathbf{e}_{ji}$. By combining $\mathbf{dm}_i^{(t)}$ and $\mathbf{ne}_i^{(t)}$, we can obtain a graph context-aware representation $\mathbf{ds}_i^{(t)}$.

$$\mathbf{ds}_i^{(t)} = \mathbf{W}_{ds}^{d \times 2d}[\mathbf{dm}_i^{(t)}, \mathbf{ne}_i^{(t)}] + \mathbf{b}_{ds}^d$$
$$\mathbf{ne}_i^{(t)} = \sum_{j \in \mathcal{N}_i} w_{ji}^{(t)} \mathbf{dm}_j^{(t)} \tag{5}$$
$$w_{ji}^{(t)} = softmax(\mathbf{W}_{ew}^{1 \times d}[\mathbf{e}_{ji} \odot \mathbf{op}^{(t)}] + \mathbf{b}_{ew}^1)$$

Lastly, we measure the correlation between $\mathbf{op}^{(t)}$ and the context-aware vector $\mathbf{ds}_i^{(t)}$, passing the output through a *softmax* layer to produce an attention distribution. By taking the weighted average over the distribution, we can retrieve related content $\mathbf{z}^{(t)}$ from the reasoning graph $\mathcal{G}$ by Eq. (6).

$$\mathbf{da}_i^{(t)} = \mathbf{W}_{da}^{1 \times d}[\mathbf{op}^{(t)} \odot \mathbf{ds}_i^{(t)}] + \mathbf{b}_{da}^1$$
$$za_i^{(t)} = softmax(da) \tag{6}$$
$$\mathbf{z}^{(t)} = \sum_{i=1}^{N} za_i^{(t)} \mathbf{m}_i^{(t)}$$

Recorder Unit. Based on the retrieved content, we then deduce a new evidential clue with two steps. We first combine $\mathbf{ds}_i^{(t)}$ with the memory result in the $t - 1$ step $\mathbf{m}_i^{(t-1)}$ by a linear transformation, resulting in $\mathbf{cm}_i^{(t)} = \mathbf{W}_{cm}^{d \times 2d}[\mathbf{ds}_i^{(t)}, \mathbf{m}_i^{(t-1)}] + \mathbf{b}_{cm}^d$, where $\mathbf{b}_{cm}^d$ is a bias.

Considering the complexity of questions is different, we introduce an update gate $\beta_{(t)}$ as Eq. (7) to determine whether to refresh the previous memory state $\mathbf{m}_i^{(t-1)}$ by the new candidate $\mathbf{cm}_i^{(t)}$. $\beta_{(t)}$ is conditioned on $\mathbf{op}^{(t)}$.

$$\beta^{(t)} = sigmoid(W_\beta^{1\times d}\mathbf{op}^{(t)} + \mathbf{b}_\beta^1)$$
$$\mathbf{m}_i^{(t)} = \beta^{(t)} \cdot \mathbf{m}_i^{(t-1)} + (1 - \beta^{(t)}) \cdot \mathbf{cm}_i^{(t)}$$

(7)

2.4 Output and Training

All operational units are regularized through a priori structure to realize a basic reasoning step. By concatenating the steps, we can realize the acyclic reasoning process with arbitrary complexity. When the process stops, we can obtain a set of evidential clues in the shared memory $M^{(t-1)}$. To derive the answer, we first use interactive attention to find the relevant clues by referring to the question $\mathbf{q}$, as $\alpha_i = softmax(W_\alpha^{1\times d}[\mathbf{q} \odot \mathbf{m}_i^{(t-1)}] + \mathbf{b}_\alpha^1)$, $\mathbf{c} = \sum_{i=1}^N \alpha_i \mathbf{m}_i^{(t-1)}$, where $\mathbf{m}_i^{(t-1)}$ is the ith clue, α_i is the relevance weight between clue i and q. By passing the concatenation of $\mathbf{q}$ and $\mathbf{c}$ through a 2-layer fully-connected *softmax* classifier, we can predict the answer from the candidate's pool.

$$\varphi(S_t) = \begin{cases} ReLU\big(\cos \langle \mathbf{op}_{(t)}, \mathbf{z}_{(t)} \rangle\big), & t > 1, \\ 0, & t = 1, \end{cases}$$

(8)

Due to the discrete termination steps, the proposed network could not be directly optimized by back-propagation. We thus employ a reinforcement approach. We view the inference as a Markov decision process, where each decision can be formulated as a tuple (S, A, P, R). S is the set of states, A is the action set, $P = \{P_{s,a}(\cdot)|s \in S, a \in A\} \in [0,1]$ describes the next-state transition probability, and $R : S \times A \to \mathbb{R}$ denotes the reward function. Given the state $S_t = (\mathbf{q}, \mathbf{op}^{(t-1)}, M^{(t-1)}, \mathbf{z}^{(t-1)})$ at step t, we perform a deduction operation $\mathbf{op}^{(t)}$ and make the environment change to a new state $S_{t+1} = ((\mathbf{q}, \mathbf{op}^{(t)}, M^{(t)}, \mathbf{z}^{(t)}))$. Then our model receives a reward $R_t = R(S_t, \mathbf{op}^{(t)})$. Our objective is to maximize the cumulative retrieval reward. To better measure the predicted result, we use a potential-based shaping reward function [18]. The function is $F : S \times A \times S \to \mathbb{R}$. Formally, there is a real-valued function $\varphi : S \to \mathbb{R}$ s.t. $F(s, a, s') = \gamma\varphi(s') - \varphi(s)$ for all $s \in S - s_0, a \in A, s' \in S$. Based on this formula, we define the function φ in Eq. (8) to measure the operation coverage corresponding to generated policies. That can fasten the convergence rate of the model.

Based on the reward $R'(S_t, A_t, S_{t+1})$, we can explore gradient descent optimization, with Monte-Carlo REINFORCE [29] to train the whole network, as Eq. (9), where θ_{ecc} is the net's parameter set, η is a discount factor, and $(Q, \hat{y})$ is a question-answer pair from the training set D.

$$R'(S_t, A_t, S_{t+1}) = R + F(S_t, A_t, S_{t+1})$$
$$J(\theta_{ecc}) = E_{(Q,\hat{a_n})\in D}[E_{A_1,\dots,A_n}$$
$$[\sum_{t=1}^n \eta^{t-1} R'(S_t, A_t, S_{t+1})|(Q, \hat{y})]]$$

(9)

After training, we can get a set of neural modules instead of a single model with a fixed network structure. Each module corresponds to a basic reasoning step. By assembling these modules flexibly, we can get an adaptive model with the optimal result.

3   Experiments

We extensively evaluated the effectiveness of our approach, including the comparisons with state of the arts and components analysis.

Table 1. State-of-the-art comparison on KVQA dataset

Method	Accuracy	
	ORG	PRP
Up-Down [1]	24.6	18.7
MuRel [5]	25.1	19.3
XNMs [24]	24.9	19.1
LOGNet [15]	32.5	23.2
KAN [31]	46.9	28.1
KVQA [22]	46.3	27.7
Ours	**54.3**	**32.8**

3.1   Data and Implementation Details

Experiments were conducted on KVQA [22], which is the most representative dataset involving cross-modality multi-hop reasoning. It consists of 183,007 QA samples and 24,602 images. Most of the samples need to use commonsense knowledge for multi-hop reasoning. There are two kinds of questions in KVQA, including the original (ORG) and paraphrased (PRP) ones. We adopted the standard evaluation metric $Accuracy$ for the VQA task. To reduce the bias, we carried out five runs and reported the average performance on these questions.

Implementation Details. We selected 10 regional objects for each input image to build the scene graph. For MMKG graph, we embedded the textual entity by the Glove [20] pre-trained model. For the question, we employed BiGRU [7] as the encoder to incorporate sequential context. For the visual content, we embedded it by the VGG pretrained model [25], and used a last hidden layer of compact bilinear pooling [9] as the encoding. We set the embedded dimension d as 300, the hidden dimension of the BiGRU as 150. For reward shaping, we set $\gamma = 0.95$, and tuned η within (0.9, 1.0) for REINFORCE algorithm. We used $Adam$ [13] as the optimizer with a learning rate of 10^{-4} and a batch size of 64.

3.2 Comparisons Against State of the Arts

We compared our method against three typical and open-source baselines, including (1) two classical VQA baselines. One is the question-guide top-down attention approach, namely *Up-Down* [1], and the other one is *MuRel* [5], which modeled the relevant correlations between the question and the image by bilinear fusion. (2) two single-modality reasoning methods, which are *XNMs* [24] and *LOGNet* [15]. *XNMs* used the explainable neural module to reason over the scene graph. *LOGNet* performed reasoning by capturing the relational structure of the given image and question. (3) two knowledge-base VQA, i.e., *KAN* [31] and *MemNet*-based [22]. *KAN* regarded the question as context to guide the reasoning over the image and commonsense knowledge. *MemNet*-based method stored and fused the context of entity through a memory network [28].

Table 2. Evaluations on different reasoning types.

Category	ORG	PRP	Category	ORG	PRP
Spatial	51.4	30.7	Multi-rel	57.4	33.0
1-hop	52.6	31.6	Subtraction	54.3	33.9
Multi-hop	56.2	33.3	Comparison	52.9	32.4
Boolean	55.4	29.5	Counting	55.0	34.3
Intersect	48.7	27.1	Multi-entity	55.3	35.3

As elaborated in Table 1, our approach achieved the best performance. The outperformance against the best baseline is over 7.4% for the original questions (ORG) and 4.7% for paraphrased questions (PRP), respectively. To better analyze the advantages of our approach at a fine-grained level, we investigate each reasoning type. As shown in Table 2, our approach significantly outperformed the baselines on all categories, especially on the multi-hop, multi-relation, and multi-entity questions. Such results demonstrated the effectiveness of our approach on cross-modality multi-hop reasoning.

3.3 Ablation Studies

We carried out ablation studies to examine relative contributions of each component in our model, including (1) cross-modality retriever; (2) five strategies in our inferential network, that is, guiding the deduction direction based on the result of the previous step, the relational graph convolution for relational reasoning, the graph-structured knowledge memory, the shaping reward function, and the stop mechanism. We denoted them as cross_ret, rsd_pre_res, op_rgcn, op_up, shap_reward and stop_mec, respectively.

As presented in Fig. 4, the ablation on each evaluated component led to the performance drop. The drop was more than 10% on three parts, including (1) cross_ret. Without such a cross-modality alignment, it is difficult to find the

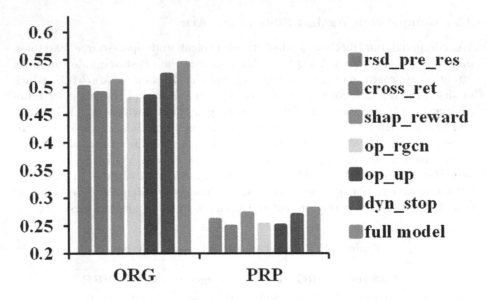

Fig. 4. Ablation studies on variants of our approach for affecting the performance

relations related to the image and text represented in two different spaces. (2) op_rgcn. Lack of this component, it is hard to perform effective relational reasoning. (3) op_up. Without the updating, the memory could not remember the long-term dependency in various reasoning steps. Besides, the comparison between shap_reward and the full model was exhibited in Fig. 5. The results showed that the proposed shaping reward function can boost model convergence. Moreover, we evaluated the efficiency of the stop mechanism by fixing the reasoning depth from 1 up to 5. As displayed in Fig. 6, more reasoning steps performed well at first but deteriorated soon. These results indicated that our dynamic strategy can adaptively determine the optimal termination for the questions with different degrees of complexity.

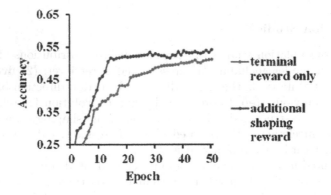

Fig. 5. Evaluation on the reward shaping

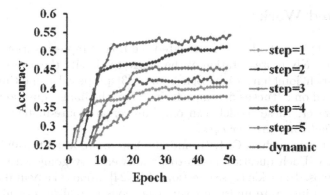

Fig. 6. Evaluation on the dynamic stop gate

3.4 Case Studies and Discussions

To better understand the reasoning behavior, we plotted the attention map over the given image, question, and external *KB*. As shown in Fig. 7, our model first focused on *"Who is the person second from left"*. Then it turned to attend at *"the political party"* related to the question. Afterward, the *"person"* was refocused and the evidence of the *"founder"* was concerned to derive the answer. Such results showed that our model can extract evidential clues from multiple knowledge sources in different modalities. Also, it can adaptively determine reasoning directions due to the soft-attention mechanism on the question.

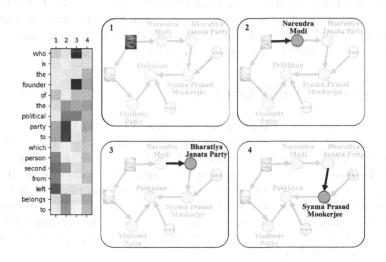

Fig. 7. Attention map of the sample in Fig. 1

4 Related Works

The initial VQA method [2] concatenated the LSTM representation of the question and the CNN representation of the image to classify the answer. To obtain a semantic-rich joint encoding, Ben-Younes [3] proposed a multimodal fusion strategy based on a Tucker decomposition [26] of the image/question correlation tensor. However, these models can only answer attribute-matching questions, such as *"What color are her eyes?"*

Some datasets such as CLEVR and GQA were proposed to validate the reasoning ability. Each question needs to be answered by going through multiple reasoning steps. To tackle these questions, Shi [24] proposed a neural module network that can integrate multiple basic operations to realize reasoning. Besides, Hudson [10] performed reasoning by a recurrent network, which separated out the memory from reasoning operation. Recently, large-scale pre-trained models such as GPT-3 [4] have a great impact on VQA. Both Yang [30] and Shao [23] used the pre-trained model to generate a caption of the image to associate with the question. However, due to the lack of knowledge module, these methods are difficult to deal with questions that require knowledge understanding.

There are some works related to the task of knowledge-based VQA, but they mainly focus on the fact-matching question. Wang [27] converted the question into a query based on fixed templates to select the answer from the knowledge base. Narasimha [17] combined the image feature and the question feature to predict the fact. Considering the complementarity of each modality, Zhu [33] depicted the image by multi-layer graphs and applied a question-guided attention mechanism to fusion multimodal context. However, when faced with cross-modality multi-hop questions, they are difficult to retrieve the *KB* entity and logical relation related to the visual object. Also, they are hard to synthesize multiple evidential clues to derive the answer. Differently, we propose a cross-modality retriever to collect the clues step-by-step.

5 Conclusions and Future Works

This paper focused on cross-modality multi-hop VQA. We decomposed a complex question into a series of attention-based reasoning steps. The result of the current step was determined simultaneously by the previous step. By recursively linking the results of all steps, we can form an evidence chain from the question to the answer. Moreover, we developed a multimodal retriever to learn the cross-modality alignment by an external knowledge graph. We have conducted extensive experiments on the popular KVQA dataset. The experimental results demonstrated the effectiveness of our method. In the future, we will study the robustness and scalability of our model.

Acknowledgment. This work is supported by the National Natural Science Foundation of China (62276279), Key-Area Research and Development Program of Guangdong Province (2020B0101100001), the Tencent WeChat Rhino-Bird Focused Research

Program (WXG-FR-2023-06), Open Fund of Guangdong Key Laboratory of Big Data Analysis and Processing (202301), and Zhuhai Industry-University-Research Cooperation Project (2220004002549).

References

1. Anderson, P., et al.: Bottom-up and top-down attention for image captioning and visual question answering. In: Proceedings of the IEEE Conference on Computer Vision and Pattern Recognition, pp. 6077–6086 (2018)
2. Antol, S., et al.: VQA: visual question answering. In: Proceedings of the IEEE International Conference on Computer Vision, pp. 2425–2433 (2015)
3. Ben-Younes, H., Cadene, R., Cord, M., Thome, N.: Mutan: multimodal tucker fusion for visual question answering. In: Proceedings of the IEEE International Conference on Computer Vision, pp. 2612–2620 (2017)
4. Brown, T., et al.: Language models are few-shot learners. Adv. Neural. Inf. Process. Syst. **33**, 1877–1901 (2020)
5. Cadene, R., Ben-Younes, H., Cord, M., Thome, N.: Murel: multimodal relational reasoning for visual question answering. In: Proceedings of the IEEE/CVF Conference on Computer Vision and Pattern Recognition, pp. 1989–1998 (2019)
6. Chen, S., Zhao, Q.: Divide and conquer: answering questions with object factorization and compositional reasoning. In: Proceedings of the IEEE/CVF Conference on Computer Vision and Pattern Recognition, pp. 6736–6745 (2023)
7. Cho, K., et al.: Learning phrase representations using RNN encoder-decoder for statistical machine translation. arXiv preprint arXiv:1406.1078 (2014)
8. Dettmers, T., Minervini, P., Stenetorp, P., Riedel, S.: Convolutional 2D knowledge graph embeddings. In: Thirty-second AAAI Conference on Artificial Intelligence (2018)
9. Gao, Y., Beijbom, O., Zhang, N., Darrell, T.: Compact bilinear pooling. In: Proceedings of the IEEE Conference on Computer Vision and Pattern Recognition, pp. 317–326 (2016)
10. Hudson, D.A., Manning, C.D.: Compositional attention networks for machine reasoning. arXiv preprint arXiv:1803.03067 (2018)
11. Hudson, D.A., Manning, C.D.: Gqa: A new dataset for real-world visual reasoning and compositional question answering. In: Proceedings of the IEEE/CVF Conference on Computer Vision and Pattern Recognition, pp. 6700–6709 (2019)
12. Johnson, J., Hariharan, B., Van Der Maaten, L., Fei-Fei, L., Lawrence Zitnick, C., Girshick, R.: Clevr: a diagnostic dataset for compositional language and elementary visual reasoning. In: Proceedings of the IEEE Conference on Computer Vision and Pattern Recognition, pp. 2901–2910 (2017)
13. Kingma, D.P., Ba, J.: Adam: a method for stochastic optimization. arXiv preprint arXiv:1412.6980 (2014)
14. Le, T.M., Le, V., Gupta, S., Venkatesh, S., Tran, T.: Guiding visual question answering with attention priors. In: Proceedings of the IEEE/CVF Winter Conference on Applications of Computer Vision, pp. 4381–4390 (2023)
15. Le, T.M., Le, V., Venkatesh, S., Tran, T.: Dynamic language binding in relational visual reasoning. arXiv preprint arXiv:2004.14603 (2020)
16. Liu, Y., Li, H., Garcia-Duran, A., Niepert, M., Onoro-Rubio, D., Rosenblum, D.S.: MMKG: multi-modal knowledge graphs. In: Hitzler, P., et al. (eds.) The Semantic Web: 16th International Conference, ESWC 2019, Portorož, Slovenia, June 2–6,

2019, Proceedings, pp. 459–474. Springer, Cham (2019). https://doi.org/10.1007/978-3-030-21348-0_30

17. Narasimhan, M., Schwing, A.G.: Straight to the facts: learning knowledge base retrieval for factual visual question answering. In: Ferrari, V., Hebert, M., Sminchisescu, C., Weiss, Y. (eds.) Computer Vision – ECCV 2018: 15th European Conference, Munich, Germany, September 8-14, 2018, Proceedings, Part VIII, pp. 460–477. Springer, Cham (2018). https://doi.org/10.1007/978-3-030-01237-3_28

18. Ng, A.Y., Harada, D., Russell, S.: Policy invariance under reward transformations: theory and application to reward shaping. In: ICML, vol. 99, pp. 278–287 (1999)

19. Nguyen, B.X., Do, T., Tran, H., Tjiputra, E., Tran, Q.D., Nguyen, A.: Coarse-to-fine reasoning for visual question answering. In: Proceedings of the IEEE/CVF Conference on Computer Vision and Pattern Recognition, pp. 4558–4566 (2022)

20. Pennington, J., Socher, R., Manning, C.D.: Glove: global vectors for word representation. In: Proceedings of the 2014 Conference on Empirical Methods in Natural Language Processing (EMNLP), pp. 1532–1543 (2014)

21. Ren, S., He, K., Girshick, R., Sun, J.: Faster R-CNN: towards real-time object detection with region proposal networks. Adv. Neural. Inf. Process. Syst. **28**, 91–99 (2015)

22. Shah, S., Mishra, A., Yadati, N., Talukdar, P.P.: KVQA: knowledge-aware visual question answering. In: Proceedings of the AAAI Conference on Artificial Intelligence, vol. 33, pp. 8876–8884 (2019)

23. Shao, Z., Yu, Z., Wang, M., Yu, J.: Prompting large language models with answer heuristics for knowledge-based visual question answering. In: Proceedings of the IEEE/CVF Conference on Computer Vision and Pattern Recognition, pp. 14974–14983 (2023)

24. Shi, J., Zhang, H., Li, J.: Explainable and explicit visual reasoning over scene graphs. In: Proceedings of the IEEE/CVF Conference on Computer Vision and Pattern Recognition, pp. 8376–8384 (2019)

25. Simonyan, K., Zisserman, A.: Very deep convolutional networks for large-scale image recognition. arXiv preprint arXiv:1409.1556 (2014)

26. Tucker, L.R.: Some mathematical notes on three-mode factor analysis. Psychometrika **31**(3), 279–311 (1966)

27. Wang, P., Wu, Q., Shen, C., Dick, A., Van Den Hengel, A.: FVQA: fact-based visual question answering. IEEE Trans. Pattern Anal. Mach. Intell. **40**(10), 2413–2427 (2017)

28. Weston, J., Chopra, S., Bordes, A.: Memory networks. arXiv preprint arXiv:1410.3916 (2014)

29. Williams, R.J.: Simple statistical gradient-following algorithms for connectionist reinforcement learning. Mach. Learn. **8**(3–4), 229–256 (1992)

30. Yang, Z., et al.: An empirical study of gpt-3 for few-shot knowledge-based VQA. In: Proceedings of the AAAI Conference on Artificial Intelligence. vol. 36, pp. 3081–3089 (2022)

31. Zhang, L., et al.: Rich visual knowledge-based augmentation network for visual question answering. IEEE Trans. Neural Netw. Learn. Syst. **32**(10), 4362–4373 (2020)

32. Zhang, Y., Jiang, M., Zhao, Q.: Query and attention augmentation for knowledge-based explainable reasoning. In: Proceedings of the IEEE/CVF Conference on Computer Vision and Pattern Recognition, pp. 15576–15585 (2022)

33. Zhu, Z., Yu, J., Wang, Y., Sun, Y., Hu, Y., Wu, Q.: Mucko: multi-layer cross-modal knowledge reasoning for fact-based visual question answering. arXiv preprint arXiv:2006.09073 (2020)

Multi-head Similarity Feature Representation and Filtration for Image-Text Matching

Mengqi Jiang[1,2], Shichao Zhang[1,2(✉)], Debo Cheng[3], Leyuan Zhang[4], and Guixian Zhang[1,2]

[1] Key Lab of Education Blockchain and Intelligent Technology, Ministry of Education, Guangxi Normal University, Guilin 541004, China
[2] Guangxi Key Lab of Multi-Source Information Mining and Security, Guangxi Normal University, Guilin 541004, China
zhangsc@mailbox.gxnu.edu.cn
[3] UniSA STEM, University of South Australia, Adelaide, SA 5095, Australia
[4] Department of Computer Science, Durham University, Durham DH13LE, UK

Abstract. The field of multimedia analysis has been increasingly focused on image-text retrieval, which aims to retrieve semantically relevant images or text through queries of the opposite modality. The key challenge is to learn the correspondence between images and text. Existing methods have focused on processing inter-modality information interaction but have not given sufficient attention to learning the correspondence between the two modalities during this process. However, these methods have a low accurate image-text matching due to they are not deal with the noise during the process of the visual and textual representations. To avoid the noise in the training process, we propose a novel Multi-head Similarity Feature Representation and Filtration (MSFRF) approach for image-text matching. The proposed MSFRF method captures the detailed associations of feature representations from different modalities and reduces the interference of noisy information in the extracted features for improving the performance of matching. Extensive experiments on two benchmark datasets show that the proposed MSFRF method outperforms the state-of-the-art image-text matching methods.

Keywords: Image-text matching · Feature extraction · Feature fusion · Inter-modality relationships · Graph neural networks

1 Introduction

With the enrichment of data types and machine learning application scenarios, cross-modal retrieval techniques for different forms of data have gained unprecedented attention. Linking and exploring the correlation between images and text, which is one of the crucial means for people to access information, has attracted

---

M. Jiang, D. Cheng—Contributed equally to this work and should be considered co-first authors.

extensive research attention, and most of the work in this field has made significant contributions. The key challenge in image-text retrieval is to accurately measure the similarity between the two modalities. However, since the text is abstract while images are concrete, there are significant semantic differences between them, and their heterogeneity makes it an extremely challenging task.

Fig. 1. The two dogs are not only associated with the word "dog", but also with the attributes "black" and "brown". This association is useful for correctly matching the dogs with their descriptions. The red boxes in the image and the red word with a strike through in the text represent noisy information that should not be activated during image-text retrieval, as they could interfere with the correct matching. (Color figure online)

Currently, to accurately establish the mechanism for associating modalities, most methods [1,19,23,29] use deep neural networks that first encode images and text and then learn to measure their similarity based on matching criteria. Existing work can be broadly divided into global correspondence learning methods and local correspondence learning methods.

The general framework of global correspondence methods [3,6,21,27] is to project the entire image and text into a common latent space and minimize inter-modality heterogeneity, so as to directly measure their mapping characteristics in the common latent space. On the other hand, local correspondence methods [8, 15,17,30] focus on the positive effect of local associations between image regions and sentence words on matching and use these local associations to infer the global similarity of image-text pairs.

For example, Lee et al. [14] proposed the Stacked Cross Attention Mechanism (SCAN) to discover complete latent semantic alignments between image regions and sentence words and infer the similarity of image-text pairs. Liu et al. [18] proposed a graph-structured network called GSMN, which constructs structured phrases from elements, element attributes, and element correlations within the same modality and obtains more fine-grained matching relationships by learning the correlations between phrases.

The existing work overlooks the importance of local information and noisy information, which are actually crucial for exploring complex patterns between images and text in the matching process. For instance, in Fig. 1, coarse correspondence would mistakenly associate the word "dog" with all the dogs in the image, ignoring finer details like the brown or black color of the dogs. Moreover, the word "camera" for the image is considered noise, and the red boxes in the image are noisy information for the text. This information also affects the matching results. Therefore, these pieces of information should be inactivated during the matching process.

To address the above-mentioned problems, we propose a novel method for image-text matching called Multi-head Similarity Feature Representation and Filtration (referred to MSFRF). The proposed MSFRF method utilizes noise filtering to the constraint of noisy information. Furthermore, MSFRF captures the associations of feature representations from different modalities and reduces the interference of noisy information in the extracted features for improving the performance of matching. Finally, MSFRF uses these learned representations to obtain the fusion feature. The contributions of this work can be summarized as follows:

- We propose a novel multi-head similarity feature representation (MSFR) method to discover potential connections in cross-modal information, and narrow down the inter-modality differences.
- The proposed MSFRF method utilizes a constraint to filter noise information for improving the feature representation of multi-modal data, in order to reduce the interference of noise information in the extracted features on the matching results.
- Extensive experimental results on two standard datasets demonstrate that the proposed model outperforms the state-of-the-art image-text matching methods.

2 Related Work

In this section, we review the related works on cross-modal retrieval. The existing research works are broadly divided into two categories: global alignment and local alignment learning methods. In the following, we review these related works according to this division.

Global alignment learning is a classic approach to image-text retrieval that has been widely researched. Kiros et al. [11] were the first to use an end-to-end deep learning method to encode images and sentences using convolutional neural networks (CNNs) and recurrent neural networks (RNNs), respectively. They used hinge-based triplet ranking loss to optimize the models. Wang et al. [28] used a two-branch neural network to process images and sentences separately and optimized the alignment learning of images and sentences using a structure-preserving constraint. Faghri et al. [6] introduced hard negatives in triadic ranking loss, which significantly improved the results. However, these methods ignore the relationship between image local regions and sentence words. Therefore, in recent years, more and more researchers have explored solutions for learning the correspondence between regions and words.

Local alignment learning focuses on learning the correspondence between local regions of the image and the words in the sentence. Several methods have been proposed for this task. Li et al. [15] proposed the VSRN model, which establishes connections between image regions and uses graph convolutional networks to generate features with semantic relationships. The global semantic inference is then performed on these features to generate the final feature representation. Ji et al. [9] proposed the SHAN model, which decomposes image-text matching into a multi-step cross-modal inference process to provide more semantic cues to the model. More recently, Yu et al. [33] introduced a multi-scale graph convolutional neural network that extracts features at multiple scales, such as global and local scales of image and text, for matching. Attention mechanisms [2,8,14,17,30] and graph convolutional neural networks [5,18,29,31] have proven to be effective methods for achieving local alignment learning, with their advantages in mining and integrating local relations.

However, both global and local alignment learning methods do not take into account the detrimental effects of noisy information on matching. Therefore, it is also necessary to reduce the interference caused by noisy information. Inspired by the literature [4,7,10,34], we propose the MSFRF method based on a local alignment learning strategy that suppresses the impairment of positive matching, enabling the capture of more accurate information between images and text.

3 The Proposed MSFRF Method

In this section, we introduce our proposed MSFRF method, whose overall framework is shown in Fig. 2. Firstly, we present the feature extraction method used to generate features from both the original image and text using a pre-trained model. Then, we introduce our proposed inter-modality matching method that operates at the region-word level, followed by a description of the structure matching method that operates at the structure level. Finally, we present the objective function used to optimize the matching process.

3.1 Feature Representations Module

Visual Feature Representations. Following the approach proposed in SCAN [14], given an image I, we use Faster R-CNN [24] with bottom-up attention [1] pre-trained on Visual Genome [12] to detect n salient regions and obtain their corresponding region features, denoted as $R = \{r_i | i = 1, 2 \ldots, n\}$. For each selected region i, r_i represents its feature representation. These features are then projected onto a D-dimensional space using a linear layer: $v_i = r_i W_d + b_d$, where W_d and b_d are parameters of the fully connected layer. Therefore, the complete representation of an image is a set of vectors $V = \{v_1, v_2, \cdots, v_n\}$, where $v_i \in \mathbb{R}^D$ encodes a salient region and n is the number of regions in the image.

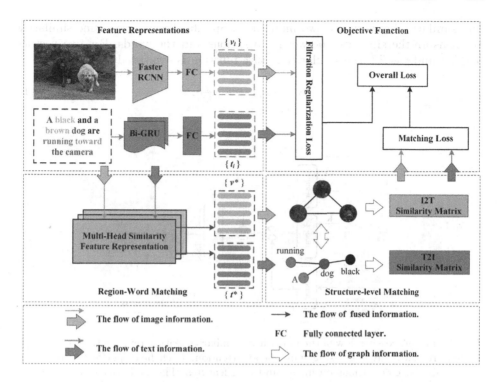

Fig. 2. The overall structure of our proposed MSFRF model, which consists of four modules: **Feature Representation:** Faster-RCNN and Bi-GRU are employed to detect salient regions, and embed each word, respectively. **Region-Word Matching:** It is to learn the correspondence between the image regions and the words. **Structure-level Matching:** The learned correspondences are aggregated using graph convolution to jointly infer fine-grained structural correspondences. **Objective Function:** The overall loss is combined from the matching loss and the filter regularization loss.

Textual Feature Representations. For a given sentence T with l words, the index of the words in the sentence T is represented by one-hot vectors in the word table, which includes all the words appearing in the text. Then, these one-hot vectors are embedded into 300-dimensional feature vectors. Next, the vectors are sequentially fed into a Bi-directional Gated Recurrent Unit (Bi-GRU) [25,35] to encode the sentence from both the forward and reverse directions. By summarizing these two directions, the complete representation of sentence T with l words is obtained and denoted as $T = \{t_1, t_2, \ldots, t_l\}$, where $t_i \in \mathbb{R}^D$ represents the representation of the i-th word in the sentence. Here, $\mathbb{R}^D$ refers to a D-dimensional space.

3.2 Region-Word Matching Module

Similarity Measure Function. In order to obtain the correspondence between the image regions and the words in the text, we use cosine similarity to measure the similarity between the two modalities. Specifically, we use cosine similarity to measure the similarity between the regions and the words. For two vectors $x \in \mathbb{R}^d$ and $y \in \mathbb{R}^d$, the similarity metric function is expressed as follows:

$$S_{wr}(x,y) = \frac{x^T y}{||x|| ||y||}. \tag{1}$$

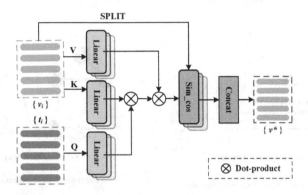

Fig. 3. The processing flow of the multi-head similarity feature representation in image-to-text retrieval involves using multi-head attention. Each head performs projection and similarity calculations on image and text features. The similarity of each head is then combined to obtain a new feature representation, which is used for subsequent operations.

Multi-head Similarity Feature Representation. In order to enhance expressiveness, improve the stability of the learning process, and capture more detailed associations between feature representations from different modalities, we propose a multi-head similarity feature representation method. This method matches each region of an image with each word in the text, enabling us to fully exploit inter-modality information and provide more effective feature representations for text and image structured matching in the high-level language space. As shown in Fig. 3, we use image-to-text retrieval as an example. Specifically, we project the feature representations of image regions and words into an h-dimensional space to obtain the visual feature representation $v_i = \{v_i^1, v_i^2, \ldots, v_i^h\}$ and the textual feature representation $t_j = \{t_j^1, t_j^2, \ldots, t_j^h\}$. Then, similarity calculation is performed separately for each dimension space as follows:

$$MH_i^k = softmax(\lambda W_v^k v_i^k (W_t^k t_j^k)^T) W^k t_j^k, \tag{2}$$

$$SM^k = S_{wr}(MH_i^k, t_j^k), \quad k \in \{1, 2, \ldots, h\}, \tag{3}$$

where λ is a scaling factor, W_v^k, W_t^k, W^k represent the parameters of the linear layer in the k-th dimensional space. The similarities in all dimensional spaces are then concatenated together to obtain the fused feature representation, denoted by $V^* = \{v_1^*, v_2^*, \ldots, v_i^*\}$. The formula for v_i^* is described as follows:

$$v_i^* = concat(SM^1, SM^2, \ldots, SM^h). \tag{4}$$

3.3   Structure-Level Matching Module

Structural Construction.

Visual Structural Construction. To construct visual relationships, we use polar coordinates to model the spatial relationships between the n salient regions detected by Faster R-CNN [24] in each image. Polar coordinates decompose the orientation and distance of pairs of regions, allowing us to capture both semantic and spatial relationships between different regions. Orientation information allows us to estimate the type of relationship, while attributes are expected to be close to the object. To describe this relationship, we calculate the polar coordinates (ρ, θ) based on the center coordinates of the bounding boxes of the paired regions and use the resulting paired polar coordinates to construct the relationship matrix W_e.

Textual Structural Construction. For textual relations, we consider two ways to construct them. The first way is that consider each word that appears in a sentence as having a connection and add the word's own connection. This is noted as the relation matrix W_r^D, which shows the semantic dependencies between words. The other way is that we use off-the-shelf Stanford CoreNLP [20] to identify text semantic dependencies within the text. It can resolve their semantic dependencies, considering words to be related if they are semantically dependent on each other. This is noted as the relation matrix W_r^S. After obtaining the relationship matrix, the weight matrix W_e can measure the strength of this relationship by calculating the intra-modal similarity to the weight of the relationship matrix, following as:

$$W_e = \left\| \frac{\exp(\lambda_2 t_i^T t_j)}{\sum_{j=0}^{l} \exp(\lambda_2 t_i^T t_j)} \odot W_r \right\|_2, \tag{5}$$

where λ_2 is a scaling factor, $W_r \in \{W_r^D, W_r^S\}$, and $\| * \|_2$ is the L_2 normalization.

Structural Similarity. Considering that the relationship between sentence and image is complex, GAT [26] is highly flexible and capable of capturing complex dependencies and interactions in graph-structured data. Inspired by this, we consider the vector representation obtained from the multi-head similarity feature representations as nodes, and feed them into GAT to integrate the global structural relationships by weighted summation and stitching of multiple attention heads. Taking the image-to-text direction as an example, the specific formula is expressed as:

$$\hat{v}_i = \left\|_{k=1}^{K} \sigma\left(\sum_{j \in \mathcal{N}_i} a_{ij} W_k v_j^* + b_k\right), \tag{6}\right.$$

where $\|$ denotes the connect operation, $\sigma(\cdot)$ is an activation function, $\mathcal{N}_i$ denotes the set of neighboring nodes of i, a_{ij} is the relationship matrix W_e following the section of structural construction, W_k and b_k are the parameters to be learned by the k-th head, and K is the number of attention heads.

Then, we feed $\hat{v}_i$ into a Multi-Layer Perception (MLP) to jointly consider the learned correspondences of all phrases and infer the global match score. We have the following formula:

$$s(v,t) = \frac{1}{n}\sum_i W_s^v(\sigma(W_g^v \hat{v}_i + b_g^v))) + b_s^v, \tag{7}$$

$$s(t,v) = \frac{1}{m}\sum_j W_s^t(\sigma(W_g^t \hat{t}_j + b_g^t))) + b_s^t, \tag{8}$$

where $\sigma(\cdot)$ is an activation function, and $W_s^v, W_g^v, b_g^v, b_s^v, W_s^t, W_g^t, b_g^t, b_s^t$ are the parameters of the MLP, respectively. It is worth noting that we perform bi-directional structure-level matching on image-text retrieval.

3.4 Objective Function

The Matching Loss. Our proposed MSFRF model does not use the most commonly used triplet loss as the objective function as in previous approaches [14,18,33]. Inspired by [13], we employ the Bi-directional Info-NCE loss. Specifically, it can be described as follows: given a matched image-text pair (v,t), the formula for calculating the Bi-directional ranking loss is as follows:

$$L_M = - \sum_{(v,t)} (\log \frac{\exp(s(v,t))}{\exp(s(v,t))+\sum \exp(s(v,t^-))} + \log \frac{\exp(s(v,t))}{\exp(s(v,t))+\sum \exp(s(v^-,t))}), \tag{9}$$

where v^- is the corresponding negative image and t^- is the corresponding negative text.

Filtration Regularization Loss. Training the model using the matching loss can improve the matching scores of positive sample pairs, but it may not be effective in handling the noise present in the feature representations of text and image. In order to prevent all words and regions from being activated and to attenuate the effect of noisy features during matching, we use the L_R constraint for the features. For a given feature z_i in the sample Z, we define Q_R as follows:

$$Q(z_i) = \frac{max(z_i) - \mu_{z_i}}{\sigma_{z_i} \|z_i\|_{2,1}}, \tag{10}$$

$$Q_R(Z) = \sum_{z_i \in Z} \mathbb{I}_{Q(z_i) < \theta_1}(Q(z_i)), \tag{11}$$

where $max(\cdot)$ is the maximum function, μ_{z_i} is the mean of the feature representations, σ_{z_i} is the standard deviation of the feature representations, and $\|\cdot\|_{2,1}$ denotes the $L_{2,1}$ regularization. θ_1 is the threshold used to select samples with too small regional $Q(\cdot)$ and $\mathbb{I}$ is the indicator function. $Q(\cdot)$ is to make different samples in a single model more discriminative on a particular feature and to increase the variability between samples, and $Q_R(\cdot)$ is to improve the feature representation.

Furthermore, in order to avoid the model optimization process from falling into a mundane solution, we do the following for the features:

$$H(z_i) = \| z_i^{*,k} \|_1, \tag{12}$$

$$H_R(Z) = \sum_{z_i \in Z} \mathbb{I}_{H(z_i) > \theta_2}(H(z_i)), \tag{13}$$

where $\| \cdot \|_1$ is the L_1 regularization of the k-th feature (by column) for the feature representations, and θ_2 is the threshold value. $H(\cdot)$ is used to obtain the sparse feature representation. $H_R(\cdot)$ is used to avoid all features being activated. Then, we have our filtering constraint loss:

$$L_R(Z) = \frac{H_R(Z)}{Q_R(Z)}, \tag{14}$$

Overall Loss. Finally, we obtain the final training loss by merging the matching loss and filtration regularization Loss using a weighted sum.

$$L = L_M + \alpha(L_R(V) + L_R(T)), \tag{15}$$

where α is the trade-off parameter, which is proportional to the training epoch.

4   Experiments

4.1   The Details of Two Datasets

We conducted experiments on two widely used datasets: Flickr30K [22] and MS-COCO [16]. The details and division of these two datasets during the experiments are as follows:

1. Flickr30K contains 31,000 images collected from the Flickr website, each described in five different sentences. Following the setup of previous works [14,18,33], we divided this dataset into three parts: 1,000 images for validation, 1,000 images for testing, and the rest for training.
2. MS-COCO is a large-scale image description dataset containing about 123,287 images, with five texts per image. As in previous works, we used 113,287 images to train all the models, 5,000 images for validation, and another 5,000 images for testing. The final results were obtained by averaging the results from five folds of 1,000 test images.

4.2   Experimental Settings

Evaluation Metric. According to existing methods [2,9,14], we use R@K and RSUM as evaluation metrics, which are commonly used in multi-modal retrieval tasks. R@K indicates the percentage of ground truth retrieved in the top K results, where K is generally taken as 1, 5, and 10, with higher R@K indicating better performance. RSUM is the sum of R@K, which combines the total value of R@K for both text and image retrieval and provides a general view of the overall retrieval performance. Similar to R@K, a higher RSUM indicates better performance. These metrics are widely used to evaluate the performance of multi-modal retrieval systems and are frequently used in research.

The Details of Implementation. Our experimental setup is based on our baseline model GSMN [18]. In terms of the specific implementation, we trained the proposed network on the training set and validated it in each epoch on the validation set, selecting the model with the highest RSUM for testing. We trained the proposed method on one Tesla V100 GPU.

For the Flickr30k dataset, we used the Adam optimizer with 30 epochs and a mini-batch of 64 for training. The initial learning rate was set to 0.0002 and became 0.1 times the original value every 15 epochs. For region-word matching, we used multi-headed cross-modal attention with eight heads. For structure-level matching, we used a graph attention layer with eight heads, each of which was 32-dimensional.

For the MSCOCO dataset, we used the Adam optimizer with 30 epochs and a mini-batch of 64 for training. The initial learning rate was set to 0.0005, which became 0.1 times the original value every five epochs. The rest of the experimental setup is the same as the experimental settings of Flickr30k. As for the setting of hyperparameters, we will discuss them in detail in the ablation study.

4.3   Quatitative Results and Analysis

The Performance of the Proposed MSFRF Model. To demonstrate the effectiveness of our proposed MSFRF model, we compared it with previous methods in image-text retrieval tasks, evaluated against two introduced datasets. All previous methods use the Faster-RCNN detector with Bottom-Up and Top-Down attention as the image encoder and Bi-GRU as the text encoder. The comparative results are summarized in Table 1 and Table 2, where the best performance is highlighted in bold. It is worth noting that the previous approaches [9,17,18,33] also use ensemble models. So we ensemble by averaging the similarity of the sparse graph and dense graph to evaluate the performance.

Table 1 shows that our proposed network achieves the score of 77.3% and 57.2% on Flickr30K in terms of R@1 score on image-to-text and text-to-image, respectively. Furthermore, compared with the most classic SCAN, our MSFRF method achieves 9.9% and 8.6% improvement in R@1 in two directions, respectively. And Table 2 shows the experimental results on the larger and more complex MSCOCO. Our MSFRF method outperforms state-of-the-art methods on both datasets. Relative improvements of 3.3% and 5% are obtained on RSUM compared to GSMN and VSE∞, respectively.

Table 1. The performance of all methods relative to Image-text retrieval on Flickr30K.

Methods	Flickr30K Dataset						
	Image-to-Text			Text-to-Image			RSUM
	R@1	R@5	R@10	R@1	R@5	R@10	
SCAN [14]	67.4	90.3	95.8	48.6	77.7	85.2	465.0
PFAN [30]	70.0	91.8	95.0	50.4	78.7	86.1	472.0
VSRN [15]	71.3	90.6	96.0	54.7	81.8	88.2	482.6
MMCA [32]	74.2	92.8	96.4	54.8	81.4	87.8	487.4
IMRAM [2]	74.1	93.0	96.6	53.9	79.4	87.2	484.2
GSMN [18]	76.4	94.3	97.3	**57.4**	82.3	89.0	496.7
WCGL [31]	74.8	93.3	96.8	54.8	80.6	87.5	487.8
SHAN [9]	74.6	93.5	96.9	55.3	81.3	88.4	490.0
VSE∞ [3]	76.5	94.2	**97.7**	56.4	83.4	**89.9**	498.1
MSMNSST [33]	75.8	93.9	97.2	57.1	82.3	88.7	495.0
MSFRF(d)	73.1	93.7	96.8	54.0	81.2	88.2	487.0
MSFRF(s)	72.7	92.5	96.8	53.7	81.3	87.8	484.8
MSFRF(full)	**77.3**	**95.0**	97.6	57.2	**84.0**	89.8	**500.9**

Table 2. The performance of all methods relative to Image-text retrieval on MSCOCO.

Methods	MSCOCO(1K) Dataset						
	Image-to-Text			Text-to-Image			RSUM
	R@1	R@5	R@10	R@1	R@5	R@10	
SCAN [14]	72.7	94.8	98.4	58.8	88.4	94.8	507.9
PFAN [30]	76.5	96.3	**99.0**	61.6	89.6	95.2	518.2
VSRN [15]	76.2	94.8	98.2	62.8	89.7	95.1	516.8
MMCA [32]	74.8	95.6	97.7	61.6	89.8	95.2	514.7
IMRAM [2]	76.7	95.6	98.5	61.7	89.1	95.0	516.6
GSMN [18]	78.4	96.4	98.6	63.3	90.1	95.7	522.5
WCGL [31]	75.4	95.5	98.6	60.8	89.3	95.3	514.9
SHAN [9]	76.8	96.3	98.7	62.6	89.6	**95.8**	519.8
VSE∞ [3]	78.5	96.0	98.7	61.7	90.3	95.6	520.8
MSMNSST [33]	78.7	95.8	98.8	62.8	90.0	95.2	521.3
MSFRF(d)	76.6	96.5	98.6	62.1	89.7	95.2	518.7
MSFRF(s)	76.7	95.0	97.9	62.0	89.3	95.2	516.1
MSFRF(full)	**79.5**	**96.6**	98.8	**64.4**	**90.8**	95.7	**525.8**

Table 3. The results of ablation studies on Flickr30K.

Model	Image-to-Text			Text-to-Image			RSUM
	R@1	R@5	R@10	R@1	R@5	R@10	
Baseline	76.4	94.3	97.3	57.4	82.3	89.0	496.7
Baseline + MSFR	76.5	94.4	97.5	57.6	83.5	90.4	499.9
Baseline + L_M with L_R	76.5	94.7	97.1	**58.9**	83.1	89.3	499.6
Baseline + MSFR + L_M with L_R	**77.3**	**95.0**	**97.6**	57.2	**84.0**	**89.8**	**500.9**

Ablation Studies. Table 3 presents the results of the ablation study conducted on the Flickr30k dataset, where we compared our proposed MSFRF model with the baseline model GSMN. We added the region-word-level similarity feature representation and the bidirectional Info-NCE loss with the noise filtering constraint to the baseline model separately, and evaluated their contributions. As shown in the table, both the feature representation and the loss function contribute to the improvement of the model performance, demonstrating the effectiveness of our proposed method.

Table 4. The results of different values of h for the proposed MSFRF method on Flick30K.

h	Image-to-Text			Text-to-Image			RSUM
	R@1	R@5	R@10	R@1	R@5	R@10	
4	75.7	92.9	97.1	**57.4**	83.1	89.4	495.6
8	**77.3**	**95.0**	97.6	57.2	**84.0**	89.8	**500.9**
16	76.4	94.5	**97.8**	57.1	83.3	**89.9**	499.0

Table 5. The results of different thresholds of θ_1 for the proposed MSFRF method on Flick30K.

θ_1		Image-to-Text			Text-to-Image			RSUM
model(s)	model(d)	R@1	R@5	R@10	R@1	R@5	R@10	
0.3	0.4	**77.3**	**95.0**	**97.6**	57.2	84.0	89.8	**500.9**
0.3	0.3	76.7	93.3	96.9	57.2	83.0	89.4	496.5
0.3	0.5	76.0	94.1	97.3	57.2	**84.1**	90.0	498.7
0.2	0.4	76.0	94.4	97.1	57.7	83.9	**90.1**	499.2
0.4	0.4	76.3	94.3	97.2	**57.6**	83.3	89.5	498.2

Besides, we also verified the effect of different parameters on the experimental results. Table 4 shows the effect of the value of h in the multi-head similarity feature representation on the experimental results. It can be seen that the model achieves the best performance when the number of heads is $h = 8$. Increasing the number of heads can improve the representation capability of the model and tap more cross-modal interaction information, but if the number of heads is too large, the model becomes too complex and leads to the problem of overfitting. Table 5 shows the effects of different thresholds in filtration regularization on the experimental results in the setting $\theta_2 = 0.1$. It can be seen that the model achieves the best performance when $\theta_1 = 0.4$ in the design model, and $\theta_1 = 0.3$ in the sparse model. When θ_1 is too large, feature representations that could potentially have a positive effect on matching are also filtered out, leading to a decrease in the recall of the model, while a small θ_1 makes the model under-filter the noisy information, and in turn affects the correct matching of image-text pairs. It can also be seen that θ_1 of the model (d) is larger than θ_1 of the model (s), which is due to the fact that constructing structural information with dense graphs takes into account that all words or regions are related, which undoubtedly produces more noise information than structural information constructed with sparse graphs of intra-modal similarity. And the filtration regularization, to some extent, mitigates the issue of all features being activated and attenuates the influence of noisy features on retrieval. It reduces the interference of non-semantic alignment information and leads to more accurate image-text retrieval results. Therefore, we choose the parameter combination that makes the model perform best as our parameter settings.

5   Conclusion

In this paper, we propose a novel multi-head similarity feature representation method that utilizes filter regularization and multi-head similarity feature representation. Our proposed method, referred to as MSFRF, aims to capture multi-level correspondence between regions in images and text through multi-head similarity metrics. Furthermore, we employ graph neural networks during structure-level matching to extract more complex structural information from the images and text. To reduce noise in the extracted features, the proposed MSFRF method applies filtering regularization, which increases the variability between samples and improves the text and image feature representation. Through a comprehensive set of experiments, we systematically evaluate the impact of our proposed MSFRF method and demonstrate its effectiveness.

Acknowledgment. This research was supported in part by the Project of Guangxi Science and Technology (GuiKeAB23026040), the Research Fund of Guangxi Key Lab of Multi-source Information Mining & Security (MIMS20-04), and the Research Fund of Guangxi Key Lab of Multi-source Information Mining & Security 20-A-01-02.

References

1. Anderson, P., et al.: Bottom-up and top-down attention for image captioning and visual question answering. In: Proceedings of the IEEE Conference on Computer Vision and Pattern Recognition, pp. 6077–6086 (2018)

2. Chen, H., Ding, G., Liu, X., Lin, Z., Liu, J., Han, J.: Imram: Iterative matching with recurrent attention memory for cross-modal image-text retrieval. In: Proceedings of the IEEE/CVF Conference on Computer Vision and Pattern Recognition, pp. 12655–12663 (2020)
3. Chen, J., Hu, H., Wu, H., Jiang, Y., Wang, C.: Learning the best pooling strategy for visual semantic embedding. In: Proceedings of the IEEE/CVF Conference on Computer Vision and Pattern Recognition, pp. 15789–15798 (2021)
4. Cheng, D., Zhang, S., Liu, X., Sun, K., Zong, M.: Feature selection by combining subspace learning with sparse representation. Multimedia Syst. **23**, 285–291 (2017)
5. Cui, Z., Hu, Y., Sun, Y., Gao, J., Yin, B.: Cross-modal alignment with graph reasoning for image-text retrieval. Multimed. Tools Appl. **81**(17), 23615–23632 (2022)
6. Faghri, F., Fleet, D.J., Kiros, J.R., Fidler, S.: VSE++: Improving visual-semantic embeddings with hard negatives. arXiv preprint arXiv:1707.05612 (2017)
7. Hu, R., et al.: Low-rank feature selection for multi-view regression. Multimed. Tools Appl. **76**, 17479–17495 (2017)
8. Huang, F., Zhang, X., Zhao, Z., Li, Z.: Bi-directional spatial-semantic attention networks for image-text matching. IEEE Trans. Image Process. **28**(4), 2008–2020 (2018)
9. Ji, Z., Chen, K., Wang, H.: Step-wise hierarchical alignment network for image-text matching. arXiv preprint arXiv:2106.06509 (2021)
10. Kalibhat, N.M., Narang, K., Tan, L., Firooz, H., Sanjabi, M., Feizi, S.: Understanding failure modes of self-supervised learning. arXiv preprint arXiv:2203.01881 (2022)
11. Kiros, R., Salakhutdinov, R., Zemel, R.S.: Unifying visual-semantic embeddings with multimodal neural language models. arXiv preprint arXiv:1411.2539 (2014)
12. Krishna, R., et al.: Visual genome: connecting language and vision using crowd-sourced dense image annotations. Int. J. Comput. Vision **123**, 32–73 (2017)
13. Lee, J., et al.: Uniclip: Unified framework for contrastive language-image pretraining. In: 36th Conference on Neural Information Processing Systems, NeurIPS 2022. Neural information processing systems foundation (2022)
14. Lee, K.H., Chen, X., Hua, G., Hu, H., He, X.: Stacked cross attention for image-text matching. In: Proceedings of the European Conference on Computer Vision (ECCV), pp. 201–216 (2018)
15. Li, K., Zhang, Y., Li, K., Li, Y., Fu, Y.: Visual semantic reasoning for image-text matching. In: Proceedings of the IEEE/CVF International Conference On Computer Vision, pp. 4654–4662 (2019)
16. Lin, T.-Y., et al.: Microsoft COCO: common objects in context. In: Fleet, D., Pajdla, T., Schiele, B., Tuytelaars, T. (eds.) Computer Vision – ECCV 2014: 13th European Conference, Zurich, Switzerland, September 6-12, 2014, Proceedings, Part V, pp. 740–755. Springer, Cham (2014). https://doi.org/10.1007/978-3-319-10602-1_48
17. Liu, C., Mao, Z., Liu, A.A., Zhang, T., Wang, B., Zhang, Y.: Focus your attention: a bidirectional focal attention network for image-text matching. In: Proceedings of the 27th ACM International Conference on Multimedia, pp. 3–11 (2019)
18. Liu, C., Mao, Z., Zhang, T., Xie, H., Wang, B., Zhang, Y.: Graph structured network for image-text matching. In: Proceedings of the IEEE/CVF Conference on Computer Vision and Pattern Recognition, pp. 10921–10930 (2020)
19. Liu, Y., Guo, Y., Bakker, E.M., Lew, M.S.: Learning a recurrent residual fusion network for multimodal matching. In: Proceedings of the IEEE International Conference on Computer Vision, pp. 4107–4116 (2017)

20. Manning, C.D., Surdeanu, M., Bauer, J., Finkel, J.R., Bethard, S., McClosky, D.: The stanford corenlp natural language processing toolkit. In: Proceedings of 52nd Annual Meeting of the Association for Computational Linguistics: System Demonstrations, pp. 55–60 (2014)

21. Nam, H., Ha, J.W., Kim, J.: Dual attention networks for multimodal reasoning and matching. In: Proceedings of the IEEE Conference on Computer Vision and Pattern Recognition, pp. 299–307 (2017)

22. Plummer, B.A., Wang, L., Cervantes, C.M., Caicedo, J.C., Hockenmaier, J., Lazebnik, S.: Flickr30k entities: Collecting region-to-phrase correspondences for richer image-to-sentence models. In: Proceedings of the IEEE International Conference on Computer Vision, pp. 2641–2649 (2015)

23. Qu, L., Liu, M., Cao, D., Nie, L., Tian, Q.: Context-aware multi-view summarization network for image-text matching. In: Proceedings of the 28th ACM International Conference On Multimedia, pp. 1047–1055 (2020)

24. Ren, S., He, K., Girshick, R., Sun, J.: Faster R-CNN: Towards real-time object detection with region proposal networks. In: Advances in Neural Information Processing Systems 28 (2015)

25. Schuster, M., Paliwal, K.K.: Bidirectional recurrent neural networks. IEEE Trans. Signal Process. **45**(11), 2673–2681 (1997)

26. Velickovic, P., Cucurull, G., Casanova, A., Romero, A., Lio, P., Bengio, Y., et al.: Graph attention networks. stat **1050**(20), 10–48550 (2017)

27. Vendrov, I., Kiros, R., Fidler, S., Urtasun, R.: Order-embeddings of images and language. arXiv preprint arXiv:1511.06361 (2015)

28. Wang, L., Li, Y., Lazebnik, S.: Learning deep structure-preserving image-text embeddings. In: Proceedings of the IEEE Conference on Computer Vision and Pattern Recognition, pp. 5005–5013 (2016)

29. Wang, S., Wang, R., Yao, Z., Shan, S., Chen, X.: Cross-modal scene graph matching for relationship-aware image-text retrieval. In: Proceedings of the IEEE/CVF Winter Conference on Applications of Computer Vision, pp. 1508–1517 (2020)

30. Wang, Y., et al.: Position focused attention network for image-text matching. In: Proceedings of the 28th International Joint Conference on Artificial Intelligence, pp. 3792–3798 (2019)

31. Wang, Y., et al.: Wasserstein coupled graph learning for cross-modal retrieval. In: 2021 IEEE/CVF International Conference on Computer Vision (ICCV), pp. 1793–1802. IEEE (2021)

32. Wei, X., Zhang, T., Li, Y., Zhang, Y., Wu, F.: Multi-modality cross attention network for image and sentence matching. In: Proceedings of the IEEE/CVF Conference on Computer Vision and Pattern Recognition, pp. 10941–10950 (2020)

33. Yu, R., Jin, F., Qiao, Z., Yuan, Y., Wang, G.: Multi-scale image-text matching network for scene and spatio-temporal images. Future Gen. Comput. Syst. **142**, 292–300 (2023)

34. Zhang, S., Yang, L., Deng, Z., Cheng, D., Li, Y.: Leverage triple relational structures via low-rank feature reduction for multi-output regression. Multimed. Tools Appl. **76**, 17461–17477 (2017)

35. Zhu, L., Xu, Z., Yang, Y.: Bidirectional multirate reconstruction for temporal modeling in videos. In: Proceedings of the IEEE Conference on Computer Vision and Pattern Recognition, pp. 2653–2662 (2017)

Multimodal Conditional VAE for Zero-Shot Real-World Event Discovery

Zhuopan Yang, Di Luo, Jiuxiang You, Zhiwei Guo, and Zhenguo Yang[✉]

Guangdong University of Technology, Guangzhou, China
2112105142@mail2.gdut.edu.cn, yzg@gdut.edu.cn

Abstract. In this paper, we propose a multimodal conditional variational auto-encoder (MC-VAE) in two branches to achieve a unified real-world event embedding space for zero-shot event discovery. More specifically, given multimodal data, Vision Transformer is exploited to extract global and local visual features, and BERT is adopted to obtain high-level semantic textual features. Furthermore, the textual MC-VAE and visual MC-VAE are designed to learn complementary multimodal representations. By using textual features as conditions, the textual MC-VAE encodes visual features to conform to textual semantics. Similarly, the visual MC-VAE encodes textual features in accordance with visual semantics using visual features as conditions. In particular, the textual MC-VAE and visual MC-VAE exploit MSE loss to keep visual and textual semantics for learning complementary multimodal representations, respectively. Finally, the complementary multimodal representations achieved by MC-VAE in two branches are integrated to predict real-world event labels in embedding forms, which provides feedback to finetune the Vision Transformer in turn. Experiments conducted on real-world datasets and zero-shot datasets show the outperformance of the proposed MC-VAE.

Keywords: Event discovery · Zero-shot leaning · Multimodal conditional VAE

1 Introduction

Social media such as Twitter, Facebook, and Sina Weibo allow users to conveniently share text, images and videos anytime and anywhere. People witnessing or involved in any real-world event can disseminate event-related information on social media, such as natural disaster events (e.g., 2015 Accra Floods, etc.), public safety incidents (e.g., 2015 Tianjin Explosions, etc.) and sporting and recreational events (e.g., 2013 Rugby League World Cup, etc.). As a result, it includes a large amount of relevant real-world event data, which attracts a lot of attention from the researchers on real-world event discovery [26], aiming to capture and comprehend real-world events occurring around the world.

Real-world event discovery focuses on classifying real-world occurrences in unprecedentedly vast social media data, including text, images and videos, etc. Most existing works are usually based on supervised learning strategies to deal

© The Author(s), under exclusive license to Springer Nature Switzerland AG 2023
X. Yang et al. (Eds.): ADMA 2023, LNAI 14177, pp. 644–659, 2023.
https://doi.org/10.1007/978-3-031-46664-9_43

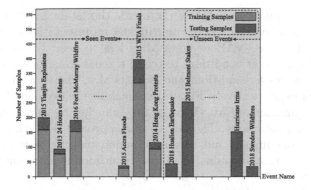

Fig. 1. Illustration of zero-shot event discovery.

with single or multimodal event data. For instance, Li et al. [12] propose Multi-Image Focusing Network (MIFN) to connect text content with visual aspects in multiple images in order to achieve multimodal representations for predicting real-world events. However, we are usually faced with new labels (i.e., unseen events) in practice, especially on social media like open data sources. By training classifiers on previously existing events, the supervised models can only recognize seen events and cannot deal with unseen events. Consequently, we advocate a zero-shot real-world event discovery (ZED) task as shown in Fig. 1, aiming to discover new real-world event happenings from multimodal data on social media.

Zero-Shot Learning (ZSL) methodologies [5,8,19,28] that aim to recognize unseen classes while training on seen classes are typically based on the assumption that both seen and unseen classes have commonalities in the semantic domain. The works on ZSL can be divided into three categories: embedding-based methods, generative methods and common space learning-based methods. More specifically, embedding-based methods [1,15,24,25] aim to learn mappings between visual knowledge and semantics for visual-semantic interactions. Generative methods [2,10,17] adopt VAEs or GANs to generate visual features of unseen classes. Common space learning-based methods [3,14,20,23] achieve common representation spaces in which both visual features and semantic representations are unified by projection. The aforementioned methods usually deal with the simple concepts (e.g., animals, flowers) with laboriously well-defined attributes. However, there are quite a few high-level concepts whose attributes cannot be defined certainly and completely, such as the real-world event concepts in ZED. In addition, ZSL focuses on visual images of the concepts, which cannot deal with multimodal data, e.g., both textual and visual data samples in ZED.

In this paper, we propose multimodal conditional variational auto-encoder (MC-VAE) for zero-shot event discovery (ZED). More specifically, we obtain global and local visual features via Vision Transformer (ViT) [7] from images, and high-level semantic textual features via BERT [6] from text on social media. The two branches of MC-VAE, textual MC-VAE and visual MC-VAE, encode both visual and textual features to generate latent representations. Furthermore, textual MC-VAE and visual MC-VAE decode the latent and textual or visual features to obtain complementary multimodal representations, respectively. In particular, textual MC-VAE exploits MSE loss to keep visual semantics and takes textual features as conditions to retain textual semantics. Visual MC-VAE takes visual features as conditions and exploits MSE loss to maintain visual and textual semantics, respectively. Finally, we combine the two complementary multimodal representations of MC-VAE in two branches to learn real-world event embeddings for predictions. The key to our approach is aligning the distributions learned from images and text information to achieve a real-world event embedding space that contains the latent multimodal representations of seen and unseen events.

The contributions of this work are summarized below:

- We propose multimodal conditional variational auto-encoder (MC-VAE) in two branches to learn modality-specific representations with cross-modal conditions, achieving a real-world event embedding space.
- We advocate multimodal zero-shot real-world event discovery (ZED), which aims to recognize both seen and unseen real-world events from multimodal data on social media.
- We conduct extensive experiments on a real-world event dataset and conventional ZSL datasets including CUB and FLO, manifesting the effectiveness of MC-VAE against the state-of-the-art ZSL works.

The rest of the paper is organized as follows. Section 2 reviews the related works. Section 3 introduces the problem statement on ZED. Section 4 introduces the details of the proposed MC-VAE. Experiments are conducted in Sect. 5, followed by conclusions in Sect. 6.

2   Related Work

2.1   Real-World Event Discovery

Real-world event discovery from social media has made significant progress, particularly with the achievement on semantic embedding learning. For example, Singh et al. [21] presented a concise approach for classifying the event keywords and maintaining the event records based on related features. However, they only consider text information of the real-world events. Compared with single information from social media, multimodal event discovery with multimodal data as input benefits the learning of event embeddings. Using the Multi-Image Focusing Network (MIFN), Li et al. [12] focused attention across multiple images and text content to predict real-world events. However, they are conventional supervised learning tasks on large-scale data collections that are often well-labelled and ignore the predictions of new events. To recognize new events, Li et al. [13] incorporated visual concepts from video sequences with event semantic representations extracted from event keywords in order to capture new video events in a zero-shot setting. However, they focus on detecting generic user-defined events in the video (e.g., making a sandwich, parade, birthday party, etc.) for new event discovery, while ignoring real-world event discovery on social media.

2.2   Zero-Shot Learning

Existing ZSL works can be roughly grouped into three categories: embedding-based methods, generative methods and common space learning-based methods. More specifically, embedding-based methods learn projections or embedding functions to associate the visual features with the respective semantic knowledge. For instance, using a visual-semantic embedding layer and an attribute prototype network, Xu et al. [25] simultaneously learned discriminative global and local features utilizing solely class-level attributes. Alamri et al. [1] exploited the Vision Transformer [7] to project the raw features of image patches to semantic label space for ZSL. Some ZSL approaches, known as generative methods, learn visual generators for unseen classes. Combining class-wise and instance-wise supervision, Han et al. [9] adopted contrastive embedding to generate synthetic visual features of unseen classes. Methods based on common space learning achieve common representation spaces in which both visual features and semantic representations are unified by projection. Wang et al. [23] proposed a two-stage bidirectional latent embedding framework with bottom-up and top-down learning stages to predict the label of the test instance using a simple nearest-neighbor algorithm.

2.3   Variational Autoencoders

Variational autoencoders [11] usually use approximate posterior inference based on latent variables that are extremely effective in data generation. It has been shown to be effective on variety of tasks, including feature refined, image classification, adversarial samples generation, etc. For instance, Narayan et al. [17] demonstrated outstanding performance on fine-grained optimization features by enhancing class-level semantic knowledge for feature refinement. However, it is difficult to refine visual features limited to image input only. Chen et al. [4] designed a flexible architecture of the VAEs with four blocks named h-block, μ-block, σ-block, and t-block to address the limitations of the symmetrical architectures of the traditional VAEs for image classification. Ma et al. [16] integrated a deep embedding network and a cross-modal alignment modified variational autoencoder to learn the latent space shared by image features and class embeddings.

3   Problem Statement

In this section, we formalize the notion of multimodal zero-shot real-world event discovery (ZED). ZED aims to discover new real-world events from multimodal social media data that are unavailable to utilize during training. More specifically, given the multimodal representations in semantic feature space, ZED aims to construct the semantic label space into which the real-world event concepts represented by embedding vectors are projected. Without loss of generality, we give the definitions of semantic feature space, semantic label space and ZED.

Definition 1 (Semantic Feature Space). The semantic feature space of k dimensions is a metric space in which each of the k dimensions encodes the value of a virtual attribute of data information. For text information t and an image x in ZED, we attain textual features and visual features in embedding forms (denoted as $h \in \mathbb{R}^k$ and $v \in \mathbb{R}^k$, respectively), which have achieved brilliant results on many feature extraction tasks. Furthermore, we integrate visual and textual features to learn multimodal features m.

Definition 2 (Semantic Label Space). It is a metric space where the real-world event concepts represented by embedding vectors of k dimensions are projected. For event labels in ZED, we attain a knowledge base of semantic labels $Y = \{y_i\}_{i=1}^{L_s+L_u}$, where $y_i \in \mathbb{R}^k$ denotes an event label in embedding form with k dimensions, L_s and L_u are the number of seen and unseen event categories, respectively.

Definition 3 (Zero-shot Real-world Event Discovery). Similarly to embedding-based methods in ZSL [1,15,24,25], ZED learns the mapping relationship between semantic features and semantic labels to recognize both seen and unseen events.

Without loss of generality, ZED uses the learned relationship $\mathcal{H} : \mathcal{S}(x,t) \rightarrow Y$ in a two-stage prediction procedure to point from a raw image x and text information t to an event label $y \in Y$ as follows,

$$\mathcal{H} = \mathcal{L}(\mathcal{S}(\cdot)), \tag{1}$$

$$\mathcal{S} : (x,t) \rightarrow m, \tag{2}$$

$$\mathcal{L} : \quad m \rightarrow Y, \tag{3}$$

where m denotes the multimodal representations, $\mathcal{S}(\cdot)$ represents learning multimodal representations m from data information including raw image x and text information t in semantic feature space, $\mathcal{L}(\cdot)$ represents the mapping relationship from semantic features to semantic labels, and Y denotes the knowledge base of semantic labels.

In particular, we have two disjoint sets of events: the dataset with seen events $S = \{x_i^s, t_i^s, y_i^s\}_{i=1}^{|S|}$ and the dataset with unseen events $U = \{x_i^u, t_i^u, y_i^u\}_{i=1}^{|U|}$ in ZED, where x_i is the i-th raw image, t_i represents the i-th text information, y_i denotes the corresponding event label embeddings and $|S|$ and $|U|$ are the number of seen event samples and unseen event samples. Assume that a number of $|N|$ labeled instances from seen events S are provided for training: $\mathcal{D}_{tr} = \{x_i^s, t_i^s, y_i^s\}_{i=1}^{|N|}$. The test set $\mathcal{D}_{te} = \{x_i^m, t_i^m, y_i^m\}_{i=1}^{|Q|}$ contains $|Q|$ unlabeled samples from both seen and unseen events, where $m \in \{s, u\}$.

4   Methodology

4.1   Overview of the Framework

Fig. 2 shows the proposed framework with feature extraction and the MC-VAE in two branches including a visual MC-VAE and a textual MC-VAE for ZED. More specifically, we extract global and local visual features and high-level semantic textual features from raw images and text information by utilizing ViT and BERT, respectively. The visual and textual features are concatenated in the latent space by the encoders of the MC-VAE in two branches to obtain the joint latent representations. Furthermore, the decoder of textual MC-VAE generates multimodal representations by decoding the latent representations and textual features with textual features as conditions. Meanwhile, textual MC-VAE employs MSE loss between visual and multimodal representations to keep visual semantics. The visual MC-VAE takes visual features as conditions and exploits MSE loss to retain visual semantics for multimodal representation learning. Finally, we fuse the two complementary multimodal representations achieved by MC-VAE in two branches to obtain real-world event embeddings capable of recognizing both seen and unseen events.

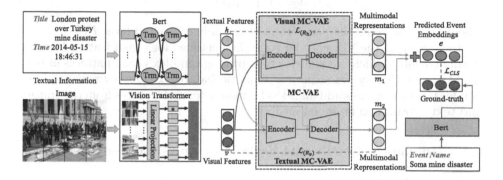

Fig. 2. The framework of MC-VAE.

4.2   Feature Extraction

1) Textual features. For the text information t, we exploit BERT [6] to obtain high-level semantic textual features h. Bert is able to capture the overall semantics of text information as well as local relations between words in text information by multi-head attention mechanism. More specifically, the text information $t = [w_1, w_2, \ldots, w_n]$ is input into BERT to extract textual features F_{bert}, where $w_i \in t$, $i \in \{1, \ldots, n\}$ represents the i-th word in text information as follows,

$$F_{\text{bert}} = f_{\text{bert}}(t; \Theta_{\text{bert}}), \qquad (4)$$

where $f_{\text{bert}}(\cdot)$ represents the BERT operation, Θ_{bert} is the parameters of BERT and F_{bert} denotes the textual features after BERT operation.

Furthermore, a fully connected layer is employed to ensure that textual features and visual features (denoted as $h \in \mathbb{R}^k$ and $v \in \mathbb{R}^k$, respectively) have the same k dimension as follows,

$$h = FC(F_{\text{bert}}; \Theta_{tf}), \qquad (5)$$

where Θ_{tf} represents the parameters of the fully connected layer $FC(\cdot)$.

2) Visual features. For the raw image $x \in \mathbb{R}^{H \times W \times C}$, we adopt ViT [7] to extract global and local visual features v. ViT is able to capture the global semantics of images and local relations between image patches through a multi-head attention mechanism. More specifically, the image $x = [p_1, p_2, \ldots, p_n]$ is fed into ViT encoder to extract visual features F_{vit}, where $p_i \in x$, $i \in \{1, \ldots, n\}$ represents the i-th image patch of the image x as follows,

$$F_{vit} = f_{vit}(x; \Theta_{vit}), \tag{6}$$

where $f_{vit}(\cdot)$ represents the ViT operation, Θ_{vit} is the parameters of ViT and F_{vit} denotes the visual features after ViT operation.

Similarly, a fully connected layer is adopted to obtain visual features (denoted as $v \in \mathbb{R}^k$) with the same dimension as the textual features h as follows,

$$v = FC(F_{vit}; \Theta_{vf}), \tag{7}$$

where Θ_{vf} represents the parameters of the fully connected layer $FC(\cdot)$.

In particular, the ViT encoder could be finetuned for visual feature learning during training, which is conducive to capturing complex visual patterns of real-world events. Finally, we achieve a semantic feature space by extracting global and local visual features and high-level semantic textual features from raw images and text information on social media.

4.3   Multimodal Conditional Variational Auto-Encoder (MC-VAE)

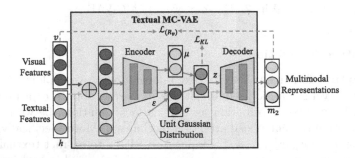

Fig. 3. Details of the Textual MC-VAE.

MC-VAE mainly has two branches, a textual MC-VAE and a visual MC-VAE, which are adopted to learn complementary multimodal representations. In terms of textual MC-VAE as shown in Fig. 3, it consists of an encoder E_1 and a decoder D_1. In terms of visual MC-VAE, its architecture is the same as Fig. 3, where just swap the visual and textual input. The encoders of textual and visual MC-VAE encode the visual features v and textual features h to hidden representations as follows,

$$\mu_i, \sigma_i = E_i\left(v, h; \Theta_E^i\right), i \in \{1, 2\} \tag{8}$$

where $E_i, i \in \{1, 2\}$ is the encoder of textual and visual MC-VAE respectively, Θ_E^i is the parameters of E_i, μ_i and σ_i are mean vector and standard deviation vector respectively.

In particular, we sample from the unit Gaussian distribution $\mathcal{N}(0, 1)$ to attain ϵ. Similarly to the conventional practice of VAEs [11], we integrate μ_i, σ_i and ϵ to obtain the latent representations z_i as follows,

$$z_i = \mu_i + \sigma_i \times \epsilon, i \in \{1, 2\} \tag{9}$$

Furthermore, the decoder of textual MC-VAE generates multimodal representations from latent representations and textual features using textual features as conditions. The decoder of visual MC-VAE decodes latent representations and visual features using visual features as conditions to learn complementary multimodal representations as follows,

$$m_i = \begin{cases} D_i\left(z_i, h; \Theta_D^i\right), i = 1 \\ D_i\left(z_i, v; \Theta_D^i\right), i = 2 \end{cases} \tag{10}$$

where $D_i, i \in \{1, 2\}$ is the decoder of textual and visual MC-VAE respectively, Θ_D^i is the parameters of D_i and m_i, $i \in \{1, 2\}$ denotes the multimodal representations of textual and visual MC-VAE respectively.

In particular, the MSE loss is employed to penalize the distance between the output representations of decoders and specific features to keep modality-specific semantics as follows,

$$\mathcal{L}_{(R_v)} = MSE(m_2, v), \tag{11}$$

$$\mathcal{L}_{(R_h)} = MSE(m_1, h), \tag{12}$$

where $\mathcal{L}_{(R_v)}$ denotes reconstruction loss between multimodal representations m_2 and visual features v, $\mathcal{L}_{(R_h)}$ denotes reconstruction loss between multimodal representations m_1 and textual features h.

The textual and visual MC-VAE can be optimized jointly with the conditional VAE loss $\mathcal{L}_{(V_1)}$ and $\mathcal{L}_{(V_2)}$, respectively:

$$\begin{aligned} \mathcal{L}_{(V_1)} &= -\mathcal{L}_{KL} + \mathcal{L}_{(R_v)} \\ &= -KL\left[q_{\Theta_E^1}\left(z_1 \mid v, h\right) \| p_{\Theta_D^1}\left(z_1 \mid h\right)\right] + MSE(m_2, v), \end{aligned} \tag{13}$$

$$\begin{aligned} \mathcal{L}_{(V_2)} &= -\mathcal{L}_{KL} + \mathcal{L}_{(R_h)} \\ &= -KL\left[q_{\Theta_E^2}\left(z_2 \mid h, v\right) \| p_{\Theta_D^2}\left(z_2 \mid v\right)\right] + MSE(m_1, h), \end{aligned} \tag{14}$$

where $\mathcal{L}_{KL}$ denotes the Kullback-Leibler divergence, $q_{\Theta_E^1}\left(z_1 \mid v, h\right)$ and $q_{\Theta_E^2}\left(z_2 \mid h, v\right)$ are posterior distributions modeled by E_1 and E_2 respectively,

$p_{\Theta_D^1}(z_1 \mid h)$ and $p_{\Theta_D^2}(z_2 \mid v)$ denote the conditional prior distributions assumed to be $\mathcal{N}(0,1)$.

In particular, we combine the two complementary multimodal representations of textual and visual MC-VAE to learn real-world event embeddings as follows,

$$e = m_1 + m_2, \tag{15}$$

where m_1, m_2 denote the multimodal representations of textual and visual MC-VAE respectively, e denotes the predicted event embeddings.

Particularly for ZED, we attain a knowledge base of semantic labels $Y = \{y_i\}_{i=1}^{L_s+L_u}$, where $y_i \in \mathbb{R}^k$ is the i-th semantic label achieved by BERT [6] from event name, L_s and L_u are the number of seen and unseen event categories respectively. In particular, Y only contains semantic labels of seen events during training. MC-VAE is able to predict the event embeddings for recognizing both seen and unseen events during testing.

For the inference, we exploit cosine similarity $cos(\cdot)$ to obtain the predicted event category o. Similarly to mitigating the bias toward seen classes on ZSL, we apply calibrated stacking (CS) [1] to reduce seen event scores on ZED by a calibration factor δ as follow,

$$o = \underset{y_i \in y^S \cup y^U}{\arg\max} \left(\cos\left(e, y_i\right) - \delta \mathbb{I}\left[y_i \in y^S\right]\right), \tag{16}$$

where $\mathbb{I} = 1$ if y_i is a semantic label of seen event and 0 otherwise, e represents the predicted event embeddings, $y^S = \{y_i\}_{i=1}^{L_s}$ and $y^U = \{y_i\}_{i=1}^{L_u}$ denote the set of semantic labels of seen events and unseen events, respectively.

Finally, we exploit cross entropy loss as classification loss to train the model as follows,

$$\mathcal{L}_{CLS} = -\frac{1}{b} \sum_{i=1}^{b} \log \left(\frac{\exp\left(e_i^T y_i\right)}{\sum_{j=1}^{L_s} \exp\left(e_j^T y_j\right)} \right), \tag{17}$$

where e_i denotes the predicted event embeddings, $y_i \in Y$ represents the ground-truth event label, $y_j \in \{y_j\}_{j=1}^{L_s}$ represents the event label of the i-th seen event and b denotes the batch size of input.

In terms of loss terms, MC-VAE exploits two conditional VAE loss terms to obtain the two complementary multimodal representations m_1, m_2 and a cross entropy loss to compute the similarity between predicted event embeddings e and ground-truth event label y. In summary, the objective function of our proposed MC-VAE is below:

$$\mathcal{L}_{\text{total}} = \lambda_{(V_1)} \times \mathcal{L}_{(V_1)} + \lambda_{(V_2)} \times \mathcal{L}_{(V_2)} + \lambda_c \times \mathcal{L}_{CLS}, \tag{18}$$

where $\lambda_{(V_1)}$, $\lambda_{(V_2)}$ and λ_c are parameters to balance the loss terms.

5   Experiments

5.1   Datasets

For evaluations, we exploit a multi-domain and multi-modality event dataset (MMED) [27] to evaluate the performance of the approaches on ZED task. In the terms of the methodologies, the proposed MC-VAE is a ZSL approach, and thus we conduct evaluations on CUB-200-2011 (CUB) [22] and Oxford-102 (FLO) [18] that are widely-used in ZSL. The datasets are detailed below.

1) MMED [27]. It contains 75,560 images with text information related to 410 real-world events for evaluations. The examples of the data samples about one real-world event are illustrated in Fig. 4. In the context of ZED, 53,786 images with text information about 360 events are used during training, while 21,774 images with text information related to 410 events are used for testing, among which 50 events are unseen. The event distributions of seen and unseen events are summarized in Fig. 5.

2015 Assam Floods

Title Abandoned structures in Arial Beel Time 2015-09-08 17:15:21 | Title More than 8 lakh people are affected by Time 2015-08-25 16:00:38 | Title over 61 Killer and 1.5 Million People AF Time 2015-09-08 11:43:01 | Title Monsoon clouds over the Arial Beel Time 2015-09-08 17:45:50

Fig. 4. Illustration of samples about one event (i.e., 2015 Assam Floods) on MMED dataset.

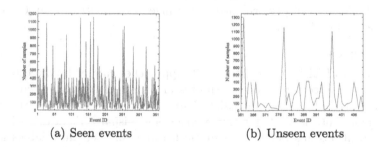

(a) Seen events   (b) Unseen events

Fig. 5. Distributions of seen and unseen events on MMED dataset.

2) CUB-200-2011 (CUB) [22]. It contains 11,788 images related to 200 categories of birds, and each image corresponds to 10 sentences as text information. The examples from the data samples about one bird are illustrated in Fig. 6-a. A number of 7,057 images with text information about 150 categories are used during training, while 4,731 images with text information related to 200 categories are used for testing, among which 50 categories are unseen.

3) Oxford-102 (FLO) [18]. It consists of 8,189 images related to 102 kinds of flowers, and each image corresponds to 10 sentences as text information. The examples from the data samples about one flower are illustrated in Fig. 6-b. A number of 5,878 images with text information about 82 categories are used during training, while 1,518 images with text information related to 102 categories are used for testing, among which 20 categories are unseen.

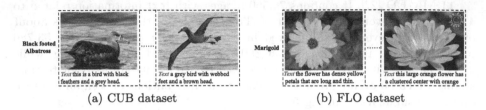

(a) CUB dataset   (b) FLO dataset

Fig. 6. Illustration of samples about one class on CUB and FLO datasets (i.e., Black footed Albatross and Marigold, respectively).

5.2   Evaluation Metrics

In terms of evaluations for ZED, we adopt S_{Top-1}, S_{Top-5}, S_{Top-10} and U_{Top-1}, U_{Top-5}, U_{Top-10} to report the top-1, top-5, top-10 accuracy on seen and unseen events, and the harmonic mean $H = (2 \times S_{Top-1} \times U_{Top-1})/(S_{Top-1} + U_{Top-1})$ for overall performance. The range of the metrics is in $[0,1]$, and a large value is preferred. Among the three metrics, H is a comprehensive metric.

In terms of evaluations for ZSL, we adopt S and U to report the top-1 accuracy on seen and unseen events, and the harmonic mean $H = (2 \times S \times U)/(S + U)$ for overall performance. The range of the metrics is in $[0,100]$, and a large value is preferred.

5.3   Baselines

In the context of ZED, the existing ZSL approaches cannot be directly applied because of the uncertain event attributes and multimodal data including text and images. Based on the embedding-based methods in ZSL, we achieve multimodal representations that retain textual and visual semantics to learn the mapping relationships between semantic features and semantic labels for predictions. In addition, we include a number of 12 ZSL approaches as baselines in three categories of embedding-based methods, generative methods and common space learning-based methods on the two ZSL datasets.

5.4   Performance of the Approaches

Table 1 summarizes the performance of the approaches on the MMED dataset, from which we have some observations. 1) Based on the visual features obtained by ViT, the aforementioned methods achieve competitive performance, whereas they perform poorly when utilizing the visual features obtained by Resnet101. The reason is that ViT is beneficial for capturing visual semantics in MMED, which are typically complicated and involve various objects and diverse describing perspectives. On one hand, various objects convey complex visual patterns.

Table 1. Performance on the MMED dataset. The best and the suboptimal results are marked in bold and underlined, respectively.

Methods	Backbone	U_{Top-1}	U_{Top-5}	U_{Top-10}	S_{Top-1}	S_{Top-5}	S_{Top-10}	H
TF-VAEGAN [17]	Resnet101	0.0177	0.0384	0.0460	0.0500	0.1279	0.1859	0.0262
	ViT	0.0264	0.1334	0.2396	0.3636	0.5443	0.6199	0.0492
FREE [2]	Resnet101	0.0210	0.0407	0.0788	0.0719	0.1837	0.2457	0.0325
	ViT	0.0054	0.0305	0.0566	**0.6467**	**0.8087**	**0.8515**	0.0108
CE-GZSL [9]	Resnet101	0.0089	0.0313	0.0513	0.0615	0.1536	0.2145	0.0156
	ViT	0.0061	0.0374	0.0927	0.3234	0.5522	0.6489	0.0120
CADA-ZSL [20]	Resnet101	0.0544	0.1395	0.1972	0.0547	0.1588	0.2487	0.0545
	ViT	0.1314	0.3236	0.4318	0.2924	0.5439	0.6421	0.1813
ViT-ZSL [1]	ViT-finetune	<u>0.1467</u>	<u>0.4015</u>	<u>0.5463</u>	<u>0.4627</u>	<u>0.4682</u>	<u>0.4682</u>	<u>0.2228</u>
MC-VAE	ViT-finetune	**0.2137**	**0.4957**	**0.6271**	0.4507	0.4613	0.4613	**0.2899**

On the other hand, diverse describing perspectives result in images with vast visual variations. 2) ViT-ZSL achieves the suboptimal performance and is inferior to MC-VAE. The reason is that ViT is conducive to capturing complicated visual patterns via a multi-head attention mechanism while fine-tuning. On one hand, the image is divided into small patches, which is conducive to the learning of fine-grained visual features because the small image patches contain few objects. On the other hand, the multi-head attention mechanism is able to capture the global semantics of an image. 3) MC-VAE performs the best, benefiting from the learned multimodal representations with visual and textual semantics of real-world events. MC-VAE learns modality-specific representations with cross-modal conditions in two branches to achieve a real-world event embedding space for predictions.

The overall performance of the methods on CUB and FLO are summarized in Table 2. Overall, MC-VAE achieves suboptimal performance on H, and is inferior to ViT-ZSL and FREE on CUB and FLO datasets, respectively, highlighting MC-VAEs capability of dealing with ZSL task. On one hand, textual features extracted from text information with irrelevant words may distract models. On the other hand, bird and flower images are relatively simple for capturing visual patterns, thus it may interfere with the learning of visual semantics while increasing the learning of textual semantics.

5.5 Visualizations on Event Embeddings

In order to demonstrate the effectiveness of MC-VAE in recognizing unseen events in ZED, we visualize decision boundaries and predicted event embeddings for unseen and seen events on MMED.

1) Visualization of decision boundary. As depicted in Fig. 7, MC-VAE obtains more concentrated predicted event embeddings and more robust decision boundaries than ViT-ZSL for unseen events. The reason is that MC-VAE learns modality-specific representations with cross-modal conditions in two branches to achieve a real-world event embedding space for recognizing unseen events.

Table 2. Compared with the state-of-the-art ZSL methods on the CUB and FLO datasets. * indicates the results obtained by ourselves with the codes released by the authors. The best and the suboptimal results are marked in bold and underlined, respectively.

	Method	CUB			FLO		
		U	S	H	U	S	H
Generative Methods	RFF-GZSL [10]	52.6	56.6	54.6	65.2	78.2	71.1
	TF-VAEGAN [17]	52.8	64.7	58.1	62.5	84.1	71.7
	FREE [2]	55.7	59.9	57.7	<u>67.4</u>	84.5	**75.0**
	CE-GZSL [9]	63.9	66.8	65.3	**69.0**	78.7	73.5
Common Space Learning -based methods	DCN [14]	28.4	60.7	38.7	-	-	-
	SGAL [23]	40.9	55.3	47.0	-	-	-
	HSVA [4]	52.7	58.3	55.3	-	-	-
	CADA-VAE [20]	51.6	53.5	52.4	44.5*	66.7*	53.4*
Embedding-based methods	LFGAA [15]	43.4	**79.6**	56.2	-	-	-
	AREN [24]	38.9	<u>78.7</u>	52.1	-	-	-
	APN [25]	<u>65.3</u>	69.3	67.2	-	-	-
	ViT-ZSL [1]	**67.3**	75.2	**71.0**	57.5*	**95.9***	71.9*
	MC-VAE	63.3	72.3	<u>67.5</u>	61.4	<u>91.8</u>	<u>73.6</u>

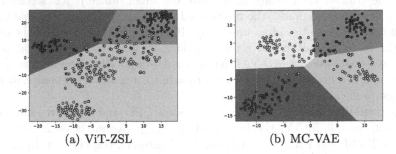

(a) ViT-ZSL   (b) MC-VAE

Fig. 7. Examples of visualized decision boundaries with predicted event embeddings for four unseen classes on MMED dataset.

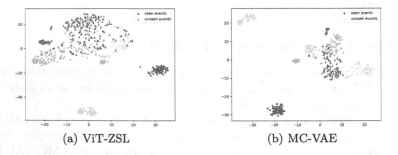

(a) ViT-ZSL   (b) MC-VAE

Fig. 8. Examples of visualized predicted event embeddings for five seen and five unseen events on MMED dataset.

2) Visualization of predicted event embeddings. As shown in Fig. 8, the event embeddings of seen events learned by ViT-ZSL and MC-VAE are clustered, whereas the event embeddings of unseen events learned by ViT-ZSL are more dispersed than those learned by MC-VAE. The event embeddings obtained by MC-VAE are more discriminative for recognizing seen and unseen events.

Table 3. Performance of variations on the MMED dataset. The best and the suboptimal results are marked in bold and underlined, respectively.

	U_{Top-1}	U_{Top-5}	U_{Top-10}	S_{Top-1}	S_{Top-5}	S_{Top-10}	H
MC-VAE_V	0.1596	0.4112	0.5496	0.4004	0.4136	0.4136	0.2283
MC-VAE_V_L	0.1761	0.4319	0.5734	0.2827	0.2914	0.2914	0.2170
MC-VAE_V_C	0.1697	0.4463	0.5739	0.1903	0.1969	0.1969	0.1794
MC-VAE_T	0.1588	0.3860	0.5203	0.3723	0.3849	0.3951	0.2197
MC-VAE_T_L	0.1626	0.3937	0.5576	0.3562	0.3661	0.3662	0.2233
MC-VAE_T_C	0.1526	0.4243	0.5657	0.1727	0.1819	0.1819	0.1620
MC-VAE_E	0.1712	0.4571	0.6025	0.4190	0.4273	0.4273	0.2430
MC-VAE	**0.2137**	**0.4957**	**0.6271**	**0.4507**	**0.4613**	**0.4613**	**0.2899**

5.6 Ablation Study

MC-VAE mainly has two branches, a textual MC-VAE and a visual MC-VAE. We denote MC-VAE without textual and visual MC-VAE (MC-VAE_T and MC-VAE_V, respectively), MC-VAE without the whole architecture and only with feature extraction and a fully connection layer (MC-VAE_E). Furthermore, we denote MC-VAE_V without MSE loss and textual features as conditions to keep visual and textual semantics (MC-VAE_V_L and MC-VAE_V_C, respectively), MC-VAE_T without MSE loss and visual features as conditions to keep textual and visual semantics (MC-VAE_T_L and MC-VAE_T_C, respectively). The performance of the variations is summarized in Table 3. Overall, MC-VAE achieves the best H on both seen and unseen events, benefiting from the two complementary multimodal representations achieved by textual and visual MC-VAE with visual and textual semantics.

5.7 Failure Examples

We summarize three kinds of failure cases as shown in Fig. 9. More specifically, the ambiguous text in Fig. 9-a is that text information may contain abbreviations, making it difficult to capture textual semantics. The irrelevant text in Fig. 9-b is that text information may be unrated to the ground-truth event, which may distract the model. The indiscriminative image in Fig. 9-c is that event image may be similar though being related to different real-world events as poor discrimination of visual semantics.

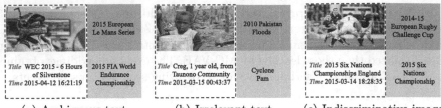

(a) Ambiguous text   (b) Irrelevant text   (c) Indiscriminative image

Fig. 9. Failure examples of MC-VAE on MMED dataset (the red box and blue box show predicted event and ground-truth, respectively). (Color figure online)

6   Conclusions

In this paper, we propose multimodal conditional variational auto-encoder (MC-VAE) in two branches to achieve a real-world event embedding space, in which we can discover new event happenings. The experimental results conducted both on real-world event dataset and conventional ZSL datasets including CUB and FLO demonstrate the effectiveness of the proposed MC-VAE.

Acknowledgment. This work is supported by the National Natural Science Foundation of China (No. 62076073), and the Youth Talent Support Programme of Guangdong Provincial Association for Science and Technology (No. SKXRC202305).

References

1. Alamri, F., Dutta, A.: Multi-head self-attention via vision transformer for zero-shot learning. In: IMVIP, pp. 1–8 (2021)
2. Chen, S., et al.: Free: feature refinement for generalized zero-shot learning. In: ICCV, pp. 122–131 (2021)
3. Chen, S., et al..: HSVA: hierarchical semantic-visual adaptation for zero-shot learning. In: NeurIPS, pp. 16622–16634 (2021)
4. Chen, X., Sun, Y., Zhang, M., Peng, D.: Evolving deep convolutional variational autoencoders for image classification. TEVC **25**(5), 815–829 (2020)
5. Chen, Z., Luo, Y., Wang, S., Qiu, R., Li, J., Huang, Z.: Mitigating generation shifts for generalized zero-shot learning. In: MM, pp. 844–852 (2021)
6. Devlin, J., Chang, M., Lee, K., Toutanova, K.: Bert: pre-training of deep bidirectional transformers for language understanding. In: NAACL-HLT, pp. 4171–4186 (2019)
7. Dosovitskiy, A., et al.: An image is worth 16x16 words: Transformers for image recognition at scale. In: ICLR, pp. 1–21 (2021)
8. Gune, O., Banerjee, B., Chaudhuri, S., Cuzzolin, F.: Generalized zero-shot learning using generated proxy unseen samples and entropy separation. In: MM, pp. 4262–4270 (2020)
9. Han, Z., Fu, Z., Chen, S., Yang, J.: Contrastive embedding for generalized zero-shot learning. In: ICCV, pp. 2371–2381 (2021)
10. Han, Z., Fu, Z., Yang, J.: Learning the redundancy-free features for generalized zero-shot object recognition. In: ICCV, pp. 12865–12874 (2020)

11. Kingma, D.P., Welling, M.: Auto-encoding variational bayes. In: ICLR, pp. 1–14 (2014)
12. Li, Y., Li, J., Jin, H., Peng, L.: Focusing attention across multiple images for multimodal event detection. In: MM, pp. 1–6 (2021)
13. Li, Z., Chang, X., Yao, L., Pan, S., Zongyuan, G., Zhang, H.: Grounding visual concepts for zero-shot event detection and event captioning. In: KDD, pp. 297–305 (2020)
14. Liu, S., Long, M., Wang, J., Jordan, M.I.: Generalized zero-shot learning with deep calibration network. In: NeurIPS, pp. 2009–2019 (2018)
15. Liu, Y., Guo, J., Cai, D., He, X.: Attribute attention for semantic disambiguation in zero-shot learning. In: ICCV, pp. 6697–6706 (2019)
16. Ma, P., Hu, X.: A variational autoencoder with deep embedding model for generalized zero-shot learning. In: AAAI, pp. 11733–11740 (2020)
17. Narayan, S., Gupta, A., Khan, F.S., Snoek, C.G., Shao, L.: Latent embedding feedback and discriminative features for zero-shot classification. In: ECCV, pp. 479–495 (2020)
18. Nilsback, M.E., Zisserman, A.: Automated flower classification over a large number of classes. In: ICVGIP, pp. 722–729 (2008)
19. Palatucci, M., Pomerleau, D., Hinton, G.E., Mitchell, T.M.: Zero-shot learning with semantic output codes. In: NeurIPS, pp. 1410–1418 (2009)
20. Schonfeld, E., Ebrahimi, S., Sinha, S., Darrell, T., Akata, Z.: Generalized zero-and few-shot learning via aligned variational autoencoders. In: ICCV, pp. 8247–8255 (2019)
21. Singh, T., Kumari, M.: Burst: real-time events burst detection in social text stream. TJS **77**(10), 11228–11256 (2021)
22. Wah, C., Branson, S., Welinder, P., Perona, P., Belongie, S.: The caltech-ucsd birds-200-2011 dataset. California Institute of Technology (2011)
23. Wang, Q., Chen, K.: Zero-shot visual recognition via bidirectional latent embedding. IJCV **124**(3), 356–383 (2017)
24. Xie, G.S., et al.: Attentive region embedding network for zero-shot learning. In: ICCV, pp. 9384–9393 (2019)
25. Xu, W., Xian, Y., Wang, J., Schiele, B., Akata, Z.: Attribute prototype network for zero-shot learning. In: NeurIPS, pp. 21969–21980 (2020)
26. Yang, Z., Li, Q., Liu, W., Lv, J.: Shared multi-view data representation for multi-domain event detection. IEEE T-PAMI **42**(5), 1243–1256 (2020)
27. Yang, Z., Lin, Z., Guo, L., Li, Q., Liu, W.: MMED: a multi-domain and multi-modality event dataset. IPM **57**(6), 102315 (2020)
28. Ye, Z., Hu, F., Lyu, F., Li, L., Huang, K.: Disentangling semantic-to-visual confusion for zero-shot learning. TMM **24**, 2828–2840 (2022)

MAMRP: Multi-modal Data Aware Movie Rating Prediction

Mingfu Qin, Qian Zhou, Wei Chen, and Lei Zhao[✉]

School of Compute Science and Technology, Soochow University, Suzhou, China
mftanmfqin@stu.suda.edu.cn, {qzhou0,robertchen,zhaol}@suda.edu.cn

Abstract. Despite the prosperity of the film industry in the past few decades, it is not uncommon to experience the phenomenon that some movies receive high box offices but obtain low ratings.The phenomenon indicates that existing studies which predict the movie-related indicator (i.e., box office) are far from satisfactory. Inspired by this, we formulate a novel task in this work, i.e., multi-modal data aware movie rating prediction (MAMRP), which aims to predict the ratings of emerging movies in time based on movie-related attributes. To tackle the task effectively, we propose a novel model that contains feature extraction, two multi-modal fusion modules, and embedding aggregation. Specifically, the transformer-based pre-trained models are first adopted to perform feature extraction for the attributes of each movie. Then, the extracted features are fed into two fusion modules: a weight-based fusion module considering the different contributions of movie attributes, and a tree-based fusion module considering the hierarchical dependencies and complex correlations between movies. Finally, the movie representations are obtained by embedding aggregation. In experiments, we construct a multi-modal benchmark in accordance with online movie platforms, and the experimental results demonstrate the high performance of our proposed model, which achieves nearly 24% relative improvement in classification accuracy compared with baselines.

Keywords: Multi-modal data · Movie rating prediction · Representation learning

1 Introduction

Over the past decade, the movie industry has experienced a remarkable expansion [14], and we have witnessed a common phenomenon that some movies receive high ratings but obtain low box office. By way of illustration, the movie "Tough Out" obtains a high rating of 8.5 out of 10.0, while gaining a low box office of 1.143 million dollars, and this movie is nominated for Best Educational Film at the 34th China Golden Rooster Award. The potential reasons for the phenomenon are twofold. (1) Movie ratings displayed by existing services (e.g., IMDB and Douban) only stabilize until the movies have been released for quite some time, and the producers have missed the best opportunity to promote such movies in advance. (2) The potential high-rating movies, especially those without

X. Yang et al. (Eds.): ADMA 2023, LNAI 14177, pp. 660–675, 2023.
https://doi.org/10.1007/978-3-031-46664-9_44

well-known actors but with high-quality content, are not allocated sufficient publicity, which results in losing some audiences who prioritize high-rating movies early in the movie's release. Arguably, if we can provide a service to predict movie ratings in advance, the producers can optimize the screen schedule and publicity, to attract more audiences for potential high-rating movies. Despite the significance of the work, such as obtaining higher box office and fostering the long-term development of movie industry, it has been largely neglected by existing studies.

To the best of our knowledge, the task MAMRP has not been studied so far and the methods developed for the task of movie box office prediction (MBOP) are the most relevant work to this paper. From the technical perspective, these approaches can be divided into three categories: (1) traditional machine learning methods [7,15,17], (2) artificial neural networks [9,20,21,25], and (3) model stacking strategies [14]. Despite the significant progress achieved by them, their solutions still suffer from the following problems. Firstly, most existing MBOP models are simplistic in design and have difficulty extracting and fusing critical attribute information of movies to generate excellent movie representations. Secondly, while we have witnessed the unprecedented growth of multi-modal data that have been widely used in various tasks to provide better representations (e.g., recommendations [10]), the majority of existing studies only exploit the textual modal information of movies and neglect such important information (e.g., images and videos). Although the deep neural network (DNN) [26] utilizes movie posters to improve classification accuracy, it cannot deal with movie-related videos. Having observed the shortcomings of existing work from the perspectives of task and technique, we propose and formulate the novel task of multi-modal data aware movie rating prediction (MAMRP) with a well-designed model. Intuitively, MAMRP aims to predict the ratings of emerging movies, i.e., estimating the movie from the perspective of reputation, based on movie-related multi-modal attributes, such as titles, directors, storylines, posters, and trailers.

To tackle the task MAMRP effectively, we propose a novel model that is composed of feature extraction, two multi-modal fusion modules, and embedding aggregation, and the details are as follows. (1) We adopt the transformer-based pre-trained models to perform feature extraction for the attributes of each movie. (2) A weight-based fusion module is designed to fuse multi-modal information from the perspective of attribute importance. This module considers that different multi-modal attributes of the movie have varying impacts on MAMRP. (3) A tree-based multi-modal fusion module utilizing a graph convolution network (GCN) [12] is designed to capture hierarchical dependencies and complex correlations between movies to further enhance the movie embeddings. The core idea of the module is that the movies directed by the same director are more likely to have similar ratings and potential interactions. For instance, movies directed by James Cameron generally obtain a high rating of 8.0 or above. (4) The embedding aggregation is employed to achieve excellent movie representations. Additionally, Fig. 1 presents the multi-modal attributes of the movie

"Mermaid" and an instance of the *Multi-modal Tree*, which are detailed in Sect. 3.

Furthermore, to make up for the lack of publicly available multi-modal movie dataset, we construct a new benchmark namely MMMD based on Douban[1], the most active online movie database and review platform in China. MMMD is publicly available[2] and it is summarized in various forms, such as text, images, and videos, and consists of various kinds of movie information, including storylines, posters, trailers, etc.   In summary, the main contributions of this paper

Fig. 1. Movie "Mermaid" and the *Multi-modal Tree* in which "Mermaid" is located

are as follows:

- We propose and formulate a novel task named multi-modal data aware movie rating prediction (MAMRP), which aims to predict the ratings of emerging movies based on movie-related attributes.
- To tackle the novel task effectively, we design a model which consists of feature extraction, two multi-modal fusion modules, and embedding aggregation. This well-designed model sufficiently takes multi-modal information into account and deeply explores hidden relationships between movies.
- The extensive experiments conducted on the real-world multi-modal movie dataset demonstrate the effectiveness of the proposed model.

The remaining parts of this paper are organized as follows. We discuss the related work in Sect. 2 and formulate problem in Sect. 3. The architecture of the proposed model is introduced in Sect. 4. Then, our implementation part is detailed in Sect. 5, followed by conclusion in Sect. 6.

2   Related Work

To the best of our knowledge, there is no existing work about MAMRP. Truth be told, our work is inspired by box office prediction. Many researchers have

---

[1] https://movie.douban.com.
[2] https://anonymous.4open.science/r/MAMRP-master.

developed various methods using movie-related data to predict the box office of movies. Consequently, we summarize the related work as follows.

According to Kim et al. [11], there are four major types of box office prediction methods: statistical models such as linear regression [8,15], probabilistic models [7,17], time series models [1,3], and machine learning models such as deep neural networks [9,20,21,25].

The first related studies are on linear regression. Litman and Kohl propose a box office prediction model based on linear regression, where the rental is used to predict movie revenue [15]. Following this influential work, several improved box office prediction methods have been proposed. The statistical models based on linear regression have been popular because they can explain the influence of each variable on box office prediction. In addition, the probabilistic models are also developed for the task MBOP. [17] proposes a box office prediction system based on a Bayesian model, which can predict the box office of new films at different stages. [7] develops an improved box office prediction model named MOVIEMOD based on Markov chain model. [1,3] sought to design the box office trend on a timeline. However, these models have the significant limitations that they only rely on the historical performance of movies [11]. Benefiting from the fast development of deep learning techniques, some scholars utilize deep neural networks to predict box office [9,20,21,25]. Unlike previous work, [26] leverages multi-modal data, i.e., assigns a poster to each movie and designs a convolutional neural network for feature extraction. The results of [18,22] indicate that the information in movie trailers can improve the prediction performance. More recently, [14] proposes model stacking strategy, which utilizes the output of the primary learner (e.g., eXtreme Gradient Boosting, Random Forest, Light Gradient Boosting Machine) as the input of the secondary learner (e.g., KNN) for box office prediction.

Notably, the task MAMRP, which employs multi-modal data to predict the rating of emerging movies and evaluates movies from the perspective of reputation, is different from the movie rating prediction in recommendation systems [16,23]. This is because the latter task focuses on a user's preference on a specific movie from the individual perspective, and the score is determined by the historical interactions between users and movies.

3   Preliminary

Definition 1 *Hierarchical Non-empty Tree. A hierarchical non-empty tree has a root node and probably numerous additional nodes, forming a hierarchy. This paper only focuses on the trees that have a varying number of children per node, and each of which has the equivalent number of levels.*

Definition 2 *Multi-modal Tree. The multi-modal tree T is a special hierarchical non-empty tree and consists of a collection of N nodes along with edges between nodes. Note that, the nodes in T have various types, e.g., text, images, and videos. Formally, it is defined as $T = (A, F)$, where $F \in \mathbb{R}^{N \times d_T}$. The i-th* row of F is the representation of i-th node t_i and initialized by the pre-trained embedding of the node. A is a $N \times N$ symmetric binary adjacency matrix which represents the edges between nodes in T. Additionally, For each element A_{ij} in A, $A_{ij} = 1$ if the node t_i in T is connected to the node t_j, otherwise $A_{ij} = 0$.

Definition 3 *Movie*. Given a set of movies $\Omega = \{EM_1, \cdots, EM_m\}$, $EM_i \in \Omega$ represents an emerging movie and contains a set of attributes $Attr_i = \{dr_i, tl_i, cs_i,$
$sy_i, im_i, vd_i\}$, where dr_i, tl_i, sy_i, im_i, and vd_i denote the director, title, storyline, poster, and trailer of EM_i, respectively. cs_i is a sentence obtained based on other information of EM_i such as writers, stars, genres, production region, language, and release date. In addition, a set of multi-modal trees $S = \{T_1, \cdots, T_n\}$ is constructed based on $\bigcup_{i=1}^{m} Attr_i$ by the hierarchical structure of "director name-movie titles-movie attributes", where n denotes the numbers of directors in Ω. For each multi-modal tree $T^j \in S$, a director name is regarded as the root node, and the second-level nodes are composed of the titles of movies directed by this director. The other attributes of each movie are located in the third level of T^j. Formally, $EM_i = (Attr_i, T^j)$, where $T^j = (A^j, F^j)$ is the multi-modal tree in which EM_i is located.

Notably, a straightforward way to organize movie-related data is taking the movie as a root node, but this may bring challenges for effectively capturing the dependency between movies, especially those are directed by the same director and more likely to have similar ratings. Consequently, we regard the director name as the root node in the multi-modal tree.

Definition 4 *Multi-modal Data Aware Movie Rating Prediction (MA-MRP)*. Given a movie set $\Omega = \{EM_1, \cdots, EM_m\}$, a multi-modal tree set $S = \{T_1, \cdots, T_n\}$ is constructed based on Definition 3. The task MAMRP aims to learn a function Φ to predict the rating of each movie $EM_i \in \Omega$. The formulation of this process is as follows:

$$Rt_i = \Phi(EM_i) = \Phi(Attr_i, T^j), \tag{1}$$

where Rt_i is the rating of movie EM_i, and Φ is the proposed model.

4   Proposed Model

As illustrated in Fig. 2, our proposed model has four primary modules: 1) feature extraction, 2) weight-based multi-modal fusion, 3) tree-based multi-modal fusion, 4) embedding aggregation, and the details of them are as follows.

4.1   Feature Extraction

Intuitively, the movie-related data usually consists of a variety of multi-modal information. Specifically, the textual modality contains the crucial information that filmmakers intend to convey to audiences, such as storylines. The images and videos as supplementary information are also extremely significant for ratings, especially in the marketing stage. These visual elements can create a strong visual impact on audiences and influence whether they choose to enjoy the movie. To utilize the multi-modal information effectively, three transformer-based pre-trained models are used as feature extractors.

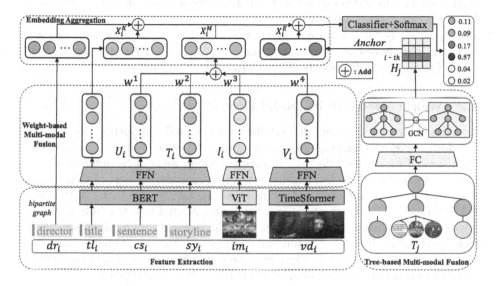

Fig. 2. Overall architecture of our proposed model

Text Extractor. To extract the information from text effectively, we learn embeddings based on the pre-trained model BERT [5], which has received a certain level of success in Natural Language Processing (NLP). Specifically, for each text attribute $te_{i_k} \in \{dr_i, tl_i, cs_i, sy_i\}$ of movie EM_i, te_{i_k} is a word sequence (i.e., $te_{i_k} = \{w_1, w_2, \cdots\}$) and w_z is a specific word. The sequence te_{i_k} is fed into BERT to obtain word embeddings. Then, utilizing the embedding of $[CLS]$ [5] as the feature vector, i.e., $\tilde{te}_{i_k}$. The formulation of this process is $\tilde{te}_{i_k} = BERT(te_{i_k})$, where $BERT(\cdot)$ indicates the BERT layer, $\tilde{te}_{i_k} \in \mathbb{R}^{1 \times d_{FV}}$ is the feature vector of te_{i_k}, d_{FV} is the dimension of the extracted feature vector.

Image Extractor. Inspired by ViT [6], we adopt a transformer-based model to embed images. The formulation of the image extractor is given as follows. Given an image $im_i \in Attr_i$ of a movie EM_i, we first employ ViT to split the image into a sequence of fixed-size non-overlapping patches. Then, the patches are fed into a linear layer and the transformer layers to obtain the transformed embeddings of patches. Finally, the embeddings of $[CLS]$ are anchored to generate the embedding of the image im_i, i.e., $\tilde{im}_i = ViT(im_i)$, where $ViT(\cdot)$ indicates the ViT layer, $\tilde{im}_i \in \mathbb{R}^{1 \times d_{FV}}$ is the feature vector.

Video Extractor. Actually, the video consists of a sequence of image frames. To obtain excellent representations of videos, TimeSformer [2], which enables spatio-temporal feature learning directly from a sequence of frame-level patches, is employed as the video extractor. Formally, given a video $vd_i \in Attr_i$ of a movie EM_i, we first extract a set of key frames key_i from vd_i, i.e., $key_i = \{f_1, f_2, \cdots\}$, and f_z is a specific key frame. Then, key_i is fed into the TimeSformer model to learn the representation of vd_i, i.e., $\tilde{vd}_i = TimeSformer(key_i)$, where $TimeSformer(\cdot)$ indicates the TimeSformer layer and $\tilde{vd}_i \in \mathbb{R}^{1 \times d_{FV}}$.

4.2 Weight-Based Multi-modal Fusion

Intuitively, not all multi-modal attributes are equally important for a movie. To characterize the varying importance precisely, we introduce a weight-based multi-modal fusion module, which assigns diverse weights to different modalities. In detail, based on the feature extraction module, we first obtain the representations of semantically-rich attributes (cs_i, sy_i, im_i and vd_i) in the movie $EM_i = (Attr_i, T^j)$. Then, these four attributes are fed into the fully connected layers to obtain latent representations. Finally, a weighting method is exploited to achieve the multi-modal fusion embeddings of EM_i, which enables this module to effectively adopt the complementary information provided by different modalities and capture the local dependencies between attributes:

$$U_i = FFN^{cs}(BERT(cs_i)), \tag{2}$$

$$T_i = FFN^{sy}(BERT(sy_i)), \tag{3}$$

$$I_i = FFN^{im}(ViT(im_i)), \tag{4}$$

$$V_i = FFN^{vd}(TimeSformer(vd_i)), \tag{5}$$

$$X_i{}^M = \sum_{Mo_i{}^z \in \{U_i, T_i, I_i, V_i\}} w^z \cdot Mo_i{}^z, \sum w^z = 1, \tag{6}$$

where $FFN(\cdot)$ is a linear projection, w^z is a learnable coefficient, and $X_i{}^M \in \mathbb{R}^{1 \times d_F}$ represents the fused multi-modal attribute embedding of EM_i.

4.3 Tree-Based Multi-modal Fusion

Apart from considering the attribute importance for movies, we design a tree-based multi-modal fusion module, which can be used to capture the potential relationships between movies and enhance the representations based on multi-modal attributes, for effectively dealing with the MAMRP task. Specifically, given a movie $EM_i = (Attr_i, T^j)$, the multi-modal tree $T^j = (A^j, F^j)$ is constructed based on $\bigcup_{i=1}^{m} Attr_i$ by the hierarchical structure of "director name-movie titles-movie attributes". Based on T^j, we utilize GCN [12] to enhance the movie representations by injecting movie-movie affinities into the embedding process of EM_i. Formally, we map the node representations F^j of T^j into a low-dimensional dense space through a linear projection, i.e.,

$$\tilde{F}^j = FC(F^j), \tag{7}$$

where $FC(\cdot)$ indicates a linear projection, $\tilde{F}^j \in \mathbb{R}^{N \times d_F}$ represents the low-dimensional dense embeddings of F^j, and N is the number of nodes in T^j.

Inspired by [19], we then utilize GCN on T^j to enhance the quality of movie representations after the low-dimensional transformation. In the l-th layer of GCN, the message passing and aggregation are formulated as:

$$H_j^{(l)} = Dropout\left(ReLU\left(\hat{A}^j H_j^{(l-1)} W^{(l-1)}\right)\right), \tag{8}$$

$$\hat{A}^j = \tilde{D}^{j-\frac{1}{2}} \tilde{A}^j \tilde{D}^{j-\frac{1}{2}}, \tilde{A}^j = A^j + I, \tilde{D}^j = D^j + I, \tag{9}$$

where $H_j^{(l)} \in \mathbb{R}^{N \times d_F}$ is the l-th layer embedding matrix of T^j and is initialized by $\tilde{F}^j$. $W^{(l-1)} \in \mathbb{R}^{d_F \times d_F}$ is the weight parameters shared by multi-modal trees. A^j is the binary adjacency matrix of T^j, D^j is the degree matrix of A^j, and I is an identity matrix. In addition, $ReLU(\cdot)$ is a rectified linear unit activation function, and $Dropout(\cdot)$ is a dropout layer.

Based on the output of the last layer of GCN (i.e., H_j), we anchor the second-level node representation that denotes the movie title of EM_i as the tree-based multi-modal fusion embedding of EM_i, i.e., $X_i^E = Anchor(H_j)$.

4.4 Embedding Aggregation

During the production of a film, the director plays an important role, and the famous ones are more likely to have high-quality movies, compared with new directors. In addition, the movie title usually contains valuable information about the movie theme. Therefore, the director-movie bipartite graph [24] with abundant interactive information is employed to further enhance movie representations. Specifically, the final representation of the movie EM_i is obtained by combining the embeddings of all aspects (i.e., X_i^M and X_i^E) of EM_i with the embedding of the title and director (i.e., X_i^K) through a sum aggregator, which preserves the importance of each embedding and enables efficient computation:

$$E\tilde{M}_i = X_i^M + X_i^E + X_i^K, \quad X_i^K = Anchor(\tilde{F}^j) + q^j, \tag{10}$$

where $q^j \in \mathbb{R}^{1 \times d_F}$ indicates the director embedding of movie EM_i and it is a learnable vector. $Anchor(\tilde{F}^j)$ is the title embedding of movie EM_i, which is anchored in $\tilde{F}^j$. $E\tilde{M}_i \in \mathbb{R}^{1 \times d_F}$ is the final representation of the movie EM_i.

4.5 Training

During the training of the proposed model, each movie rating Rt_i can be categorized into one of the set $C = \{c_1, c_2, \cdots, c_k\}$, where k represents the number of categories. Specifically, for each movie $EM_i = (Attr_i, T^j)$, we feed the final embedding $E\tilde{M}_i$ of EM_i into a classifier. The outputs of this classifier are then fed into a softmax function to obtain the prediction probability for each category. The process is defined as:

$$P_i = Softmax(E\tilde{M}_i \cdot W_c + b), \tag{11}$$

where $W_c \in \mathbb{R}^{d_F \times k}$ is the weight matrix of the classifier and b is the bias, $Softmax(\cdot)$ is a softmax function, and P_i is a vector that contains the probability of the movie EM_i belongs to each category in C.

The cross-entropy loss function between the predicted probability distribution and the true category labels is used as the loss function,

$$Loss = -\frac{1}{m} \left[\sum_{i=1}^{m} \sum_{j=1}^{k} (a_j^i \cdot \log(P_{ij})) \right] - \lambda \cdot ||W||_2^2, \tag{12}$$

where m is the number of movies in dataset Ω, k is the number of categories, i.e., $k = |C|$, P_{ij} denotes the probability of i-th movie belonging to the j-th category c_j. Additionally, as regards a_j^i, if EM_i really belongs to the j-th category c_j, then $a_j^i = 1$, otherwise $a_j^i = 0$, and $\lambda \in \mathbb{R}$ is a hyper-parameter that controls the L_2 regularization on weights W in the proposed model.

5   Experiments

5.1   Dataset and Metrics

Dataset. We collect a multi-modal movie benchmark namely MMMD, which consists of three different modalities (i.e., text, images, and videos), from Douban[3]. In MMMD, there are various types of information, including directors, storylines, posters, trailers, etc. In detail, MMMD has 2053 movies and 120 directors from different countries. The statistics of the benchmark are presented in Table 1.

Table 1. The statistics of the benchmark MMMD

Director Location	Directors	Movies	Text	Images	Videos
China	45	811	12165	7102	679
Hollywood	34	608	9120	5542	658
Korea	21	191	2865	1719	203
Japan	20	433	6495	3850	260
Total number	120	2053	30795	18213	1800

---

[3] https://movie.douban.com.

Metrics. The average percentage hit rate (APHR) has been widely used in existing work [9,14,21,26]. In this paper, we adopt the absolute accuracy (Bingo) and the relative accuracy (1-Away) as evaluation metrics, where Bingo considers the classification of correct classes, and 1-Away calculates the classification results of true classes adjacent to the predicted classes.

$$APHR_{Bingo} = \frac{1}{n}\sum_{i=1}^{k} r_i, APHR_{1-Away} = \frac{1}{n}\sum_{i=1}^{k}(r_i + r_{i-1} + r_{i+1}), \qquad (13)$$

where n is the number of samples, k is the number of categories, and r_i is the number of samples correctly categorized as category c_i. Notably, $r_0 = r_{k+1} = 0$.

5.2   Baselines and Experimental Settings

Baselines. The representative models for MBOP are used as baselines due to the MAMRP task is relatively novel. In addition, the methods that leverage multi-modal information for classification in other domains are also compared here. For simplicity, our proposed model is notated by MAMRP.

MLP [21] has two hidden layers, consisting of 18 and 16 processing elements respectively, with sigmoid transfer functions.

MLBP [25] is a multi-input and multi-output back propagation (BP) neural network with two hidden layers using sigmoid transfer functions.

NN [20] is a BP neural network with a single hidden layer of 25 nodes.

DNN [26] is a multi-modal deep neural network network. The movie poster content and other selected movie-related data are used as input.

SFM [14] is a stacking model, which stacks multiple methods and considers the secondary training, and the outputs of primary approaches (i.e., XGBoost, RF, and LightGBM) are combined as new features and then fed into the secondary learner (i.e., KNN) to obtain the final output.

MMIN [4] is a concatenation layer of the multi-modal features in a modulation classifier. We compare with it in terms of handling multi-modal data.

MMFF [13] is a model that concatenates the fused multi-modal features which are obtained by a deep autoencoder and the original modal features as final joint features.

Experimental Settings. The parameter settings of our experiments are detailed as follows. Specifically, the batch size is fixed at 4. The dimension of the output feature of transformer-based models is set to $d_{FV} = 768$, and the dimensions of X_i^M, X_i^E, and X_i^K are set to $d_F = 64$. For baselines of processing non-multimodal data, the dimension of embedding is reduced to 64 through principal component analysis. Other baselines utilize a linear layer for dimensionality reduction. We concatenate all attribute embeddings of each movie as input for baselines. The learning rate (i.e., γ) is 0.005, the number of GCN layers is set to $G_{ly} = 3$, and the dropout is set to $dp = 0.1$. The coefficient of L_2 normalization is set to $\lambda = 5 \times 10^{-4}$. In addition, we randomly split the dataset MMMD into training set, validation set, and testing set with a ratio of $8 : 1 : 1$.

The category number of the classifier is set to 6. The details of categories are presented in Table 2 and the rating ranges ensure that the number of samples per category is approximately balanced, inspired by [26].

Table 2. Details of each category

Category	c_1	c_2	c_3	c_4	c_5	c_6
Rt Range	$[0, 6.4]$	$[6.5, 7.0]$	$[7.1, 7.4]$	$[7.5, 7.8]$	$[7.9, 8.3]$	$[8.4, 10]$

5.3   Performance Comparison

This subsection illustrates the performances of different methods on the real-world dataset MMMD and the results are reported in Table 3. Note that the method with ++ indicates that multi-modal information is concatenated as input to this method. $MAMRP_{te}$ indicates that MAMRP only adopts the text modal, and $MAMRP_{te\&im}$ adopts both text and image modalities.

Table 3. Overall performance of different methods. The best performance is highlighted **in bold** and the second is <u>highlighted</u> by underlines. The metric 1-Away reports the 1-Away accuracy corresponding to Bingo, and 1-Away(best) is the best 1-Away accuracy. *Improvement* indicates the relative improvement of MAMRP compared to the best baseline in percentage.

Category	Model	Bingo	1-Away	1-Away(best)
Textual models	MLP	0.2544	0.5266	0.5917
	MBLP	0.2485	0.5148	0.5859
	NN	0.2603	0.5031	0.6154
	XGBoost	0.2604	<u>0.6923</u>	<u>0.6923</u>
	LightGBM	0.2663	0.6449	0.6449
	SFM	0.2841	0.6095	0.6095
Multi-modal models	DNN	0.2371	0.6343	0.6343
	MBLP++	0.2368	0.6332	0.6332
	SFM++	0.2663	0.5917	0.5917
	MMIN	0.2959	0.5968	0.6274
	MMFF	<u>0.2964</u>	0.6235	0.6396
Ours	$MAMRP_{te}$	0.3491	0.6331	0.6746
	$MAMRP_{te\&im}$	0.3668	0.6036	0.7014
	MAMRP	**0.3669**	**0.6982**	**0.7515**
	Improvement	*23.79%*	*0.85%*	*8.55%*

Based on the results in Table 3, we have the following observations:

- The proposed model MAMRP significantly performs better than all baselines. Specifically, MAMRP improves relatively over the strongest baselines in terms of Bingo, 1-Away, and 1-Away(best) by 23.79%, 0.85%, and 8.55%, respectively. This demonstrates that the proposed model is well designed for the movie rating prediction task by more fully exploiting multi-modal information. The excellent performance is primarily because of the fact that the weight-based fusion module has considered the different importance of movie attributes to effectively fuse multi-modal information, and the multi-modal trees are utilized to capture hidden relationships between movies to enhance movie representations.

- In the representative baselines utilizing textual information, SFM performs best from the perspective of Bingo. This is because SFM considers secondary training and multi-model fusion. As the primary learners of SFM, XGBoost and LightGBM achieve well performance across metrics and outperform other baselines, particularly in terms of 1-Away. This demonstrates that robust ensemble models have certain advantages in handling the movie rating prediction task. Additionally, the simple network structure results in the lagging performance of other methods (i.e., MLP, MBLP, and NN).

- Among the models that make use of multi-modal information, MBLP++ and SFM++ encounter performance degradation compared to their original ones. The important reason is that these models cannot handle multi-modal attributes (e.g., images and videos). In contrast, MMFF, which well encodes the multi-modal data, achieves the best Bingo in baselines. MMIN lags behind MMFF due to the fact that multi-modal information is fused through simple concatenation. It is clear from these results that better performance is obtained if multi-modal data is fully utilized.

- In addition, $MAMRP_{te}$ also has advantages over baselines adopting multi-modal information on Bingo, demonstrating that the proposed model is effective enough. At the same time, it can be observed that the performance of MAMRP improves steadily with the increase of modality, which indicates that the multi-modal information provides significant benefits.

5.4   Ablation Studies

In this subsection, the effectiveness of each module in the proposed model is verified by comparing the original model with models without different modules. For MAMRP, we have four variants: the first one is **MAMRP w/o. FC**, which does not consider the linear projection (i.e., Eq. (7)) for initial multi-modal tree embedding before performing graph convolution operations. The second one is named as **MAMRP w/o. GCN**, which discards the leverage of multi-modal trees which can capture latent relationships between movies, i.e., X_i^E is deleted in Eq. (10). The third is **MAMRP w/o. Modal**, which discards the weight-based multi-modal fusion information, i.e., X_i^M is deleted in Eq. (10). The fourth is **MAMRP w/o. Agg**, which discards the embedding aggregation to movies' director and title, i.e., X_i^K is deleted in Eq. (10). Table 4 summarizes the performance of different variants, from which we have following observations.

- Even discarding different modules, our method still achieves superior results on Bingo relative to the best baseline (c.f. Table 3). Furthermore, MAMRP w/o. FC and MAMRP w/o. Modal also perform well on 1-Away. This verifies the flexibility and stability of our model.
- MAMRP w/o. FC improves by 0.12% compared to MAMRP on Bingo but encounters a more significant decline of 1-Away, indicating that performing linear projection on node representations of trees can enhance the improvement of 1-Away. The role of linear projection is to map each modal feature into a unified dense space to reduce the difference between modalities. From a comprehensive perspective, this operation is still necessary.
- The performance of MAMRP w/o. GCN, MAMRP w/o. Modal, and MAMRP w/o. Agg are all hampered by about 2.5% relative to MAMRP on Bingo, and MAMRP w/o. Agg drops the most. This suggests that these three corresponding modules are critical to the proposed model. GCN mines latent hierarchical dependencies and complex correlations between movies, which is productive in obtaining excellent representations. Employing the weight-based fusion module to fuse the multi-modal information of movies can supplement some beneficial information lost by GCN. The role of the embedding aggregation is to employ the director-movie bipartite graph to enhance movie representations and preserve the importance of each embedding.

Table 4. Ablation experiments on different variants

Model	Bingo	1-Away	1-Away(best)
MAMRP w/o. FC	**0.3787**	0.6449	0.7041
MAMRP w/o. GCN	0.3432	0.6154	0.6627
MAMRP w/o. Modal	0.3431	0.6153	0.7106
MAMRP w/o. Agg	0.3314	0.5799	0.6272
MAMRP	**0.3669**	**0.6982**	**0.7515**

5.5   Parameter Studies

In this subsection, extensive experiments are conducted to investigate the effects of parameters on the performance of the proposed model. We first explore the impacts of learning rate (i.e., γ) and L_2 regularization coefficient (i.e., λ) on MAMRP because they are important for the training. In addition, we discuss how the number of GCN layers (i.e., G_{ly}) affects the model performance because G_{ly} controls the degree of multi-modal information aggregation.

- γ: We summarize the accuracy produced by different learning rate which varies from 0.002 to 0.007 at an increment of 0.001, as shown in Fig. 3(a). The performance first grows as γ becomes larger, indicating that the model converges better to the global optimum. However, it begins to deteriorate when γ continues to increase because the optimum is missed inevitably. A learning rate around 0.005 provides optimal performance for the proposed model. Overall, there are no apparent sharp rises and falls, indicating that our model is not sensitive to the selection of γ.

- λ: Fig. 3(b) displays the accuracy produced by different λ. The parameter varies from the magnitude of 10^{-1} to the magnitude of 10^{-6}, that is, the abscissa x in Fig. 3(b) corresponds to $\lambda = 5 \times 10^{-x}$. Observed from Fig. 3(b), our model gains significant improvement between $\lambda = 5 \times 10^{-1}$ and $\lambda = 5 \times 10^{-4}$, which validates the rationality of utilizing regularization and that setting a large λ tends to overfitting. Subsequently, the performance of MAMRP plateaus because the robustness of MAMRP achieves its optimum at this magnitude of λ.

- G_{ly}: To investigate the effect of multiple graph convolution layers, we search the number of layers G_{ly} in the set $\{0, 1, 2, 3, 4, 5\}$ and report results in Fig. 3(c). We can observe that when G_{ly} increases from 0 to 1, the performance increases significantly, indicating that the relationships between movies can effectively boost the movie rating prediction task. In addition, higher accuracy can be obtained when the three-layer GCN is employed. The primary reason is that a high number of GCN layers leads to poorly differentiate information between movie nodes, while a low number of GCN layers results in the insufficient fusion of information.

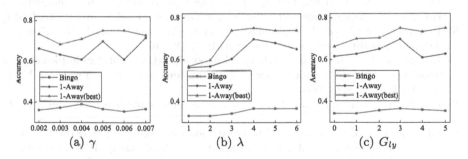

Fig. 3. Performance comparison of γ, λ, and G_{ly}

6   Conclusion and Future Work

In this paper, we propose and formulate a novel task named multi-modal data aware movie rating prediction (MAMRP), which aims to predict the ratings of emerging movies based on movie-related attributes. To address the task effectively, we design a novel model. Specifically, we employ transformer-based pre-trained models for feature extraction and design two fusion modules to fuse multi-modal information based on hierarchical dependencies and correlations between movie attributes. In addition, embedding aggregation is adopted to achieve excellent movie representations at the end. Furthermore, we collect a benchmark called MMMD with abundant multi-modal information, which can facilitate future research in this field. The experimental results demonstrate the higher performance of the proposed model compared with existing methods. In the future, we will consider more movie-related information and investigate more excellent multi-modal fusion strategies.

Acknowledgment. This work is supported by the National Natural Science Foundation of China No. 62272332, the Major Program of the Natural Science Foundation of Jiangsu Higher Education Institutions of China No. 22KJA520006.

References

1. Beck, J.: The sales effect of word of mouth: a model for creative goods and estimates for novels. J. Cult. Econ. **31**, 5–23 (2007)
2. Bertasius, G., Wang, H., Torresani, L.: Is space-time attention all you need for video understanding? In: ICML, vol. 139, pp. 813–824 (2021)
3. Dellarocas, C., Awad, N., Zhang, X.M.: Using online reviews as a proxy of word-of-mouth for motion picture revenue forecasting. Available at SSRN 620821 (2004)
4. Deng, W., Wang, X., Huang, Z., et al.: Modulation classifier: a few-shot learning semi-supervised method based on multimodal information and domain adversarial network. IEEE Commun. Lett. **27**(2), 576–580 (2023)
5. Devlin, J., Chang, M.W., Lee, K., et al.: Bert: pre-training of deep bidirectional transformers for language understanding. arXiv preprint arXiv:1810.04805 (2018)
6. Dosovitskiy, A., Beyer, L., Kolesnikov, A., et al.: An image is worth 16 × 16 words: Transformers for image recognition at scale. In: ICLR (2021)
7. Eliashberg, J., Jonker, J.J., Sawhney, M.S., et al.: MOVIMOD: an implementable decision-support system for prerelease market evaluation of motion pictures. Mark. Sci. **19**(3), 226–243 (2000)
8. Eliashberg, J., Shugan, S.M.: Film critics: influencers or predictors? J. Mark. **61**(2), 68–78 (1997)
9. Ghiassi, M., Lio, D., Moon, B.: Pre-production forecasting of movie revenues with a dynamic artificial neural network. Expert Syst. Appl. **42**(6), 3176–3193 (2015)
10. Han, T., et al.: Modality matches modality: pretraining modality-disentangled item representations for recommendation. In: WWW, pp. 2058–2066 (2022)
11. Kim, T., Hong, J., et al.: Box office forecasting using machine learning algorithms based on sns data. Int. J. Forecast. **31**(2), 364–390 (2015)
12. Kipf, T.N., Welling, M.: Semi-supervised classification with graph convolutional networks. In: ICLR (2017)
13. Liang, Y., Tohti, T., Hamdulla, A.: False information detection via multimodal feature fusion and multi-classifier hybrid prediction. Algorithms **15**(4), 119 (2022)
14. Liao, Y., Peng, Y., Shi, S., et al.: Early box office prediction in china's film market based on a stacking fusion model. Ann. Oper. Res. **308**(1), 321–338 (2022)
15. Litman, B.R., Kohl, L.S.: Predicting financial success of motion pictures: the '80s experience. J. Media Econ. **2**(2), 35–50 (1989)
16. Mustafa, G., Frommholz, I.: Comparing contextual and non-contextual features in ANNs for movie rating prediction. In: LWDA, vol. 1670, pp. 361–372 (2016)

17. Neelamegham, R., et al.: A Bayesian model to forecast new product performance in domestic and international markets. Mark. Sci. **18**(2), 115–136 (1999)
18. Oh, S., Ahn, J., Baek, H.: Viewer engagement in movie trailers and box office revenue. In: HICSS, pp. 1724–1732 (2015)
19. Pfahler, L., Richter, J.: Interpretable nearest neighbor queries for tree-structured data in vector databases of graph-neural network embeddings. In: EDBT-ICDT-WS, vol. 2578 (2020)
20. Rhee, T.G., Zulkernine, F.H.: Predicting movie box office profitability: a neural network approach. In: ICMLA, pp. 665–670 (2016)
21. Sharda, R., Delen, D.: Predicting box-office success of motion pictures with neural networks. Expert Syst. Appl. **30**(2), 243–254 (2006)
22. Tadimari, A., Kumar, N., Guha, T., et al.: Opening big in box office? Trailer content can help. In: ICASSP, pp. 2777–2781 (2016)
23. Viard, T., Fournier-S'niehotta, R.: Movie rating prediction using content-based and link stream features. CoRR abs/1805.02893 (2018)
24. Zhang, J., Zhu, Y., Liu, Q., et al.: Latent structures mining with contrastive modality fusion for multimedia recommendation. CoRR abs/2111.00678 (2021)
25. Zhang, L., Luo, J., Yang, S.: Forecasting box office revenue of movies with BP neural network. Expert Syst. Appl. **36**(3), 6580–6587 (2009)
26. Zhou, Y., Zhang, L., Yi, Z.: Predicting movie box-office revenues using deep neural networks. Neural Comput. Appl. **31**(6), 1855–1865 (2019)

MFMGC: A Multi-modal Data Fusion Model for Movie Genre Classification

Xiaorui Yang, Qian Zhou, Wei Chen[✉], and Lei Zhao

School of Compute Science and Technology, Soochow University, Suzhou, China
xryang@stu.suda.edu.cn, {qzhou0,robertchen,zhaol}@suda.edu.cn

Abstract. Movie Genre Classification (MGC) is a classic multi-label task that aims to classify movies into different genres. Existing studies have proposed many approaches for this task based on multi-modal data (e.g., synopsis, posters, and trailer). Despite the significant contributions made by them, they usually fuse multi-modal information based on simple operations, e.g., concatenation or weighted sum, failing to effectively capture the interactive information between multi-modal data. In addition, movies with significant overlap in directors and actors tend to own the same genres. This information could potentially improve the performance of MGC, which has been ignored by previous studies. Having observed the shortcomings of existing work, we propose a **M**ulti-modal data **F**usion **M**odel for **MGC** (MFMGC), including two modules: Multi-modal Data Fusion (MDF) and Movie Graph Representation Learning (MGRL). In MDF, we carefully design the fusion layer based on the attention mechanism to effectively capture the modalities' interactive information. In MGRL, we construct a movie graph to extract the structural information between movies. Specifically, the graph is constructed based on the overlap of movies' directors, screenwriters, and actors, and each node in the graph has multi-modal attributes. The experiments conducted on datasets Moviescope and MovieBricks demonstrate the superiority of the proposed model MFMGC over the state-of-the-art approaches.

Keywords: Movie genre classification · Multi-modal movie graph · Movie representation learning

1 Introduction

Over the past decade, the streaming media services have experienced unprecedented growth. Recommending specialized types of content for customers has become an indispensable ability for streaming sites, which is why automatic labeling has attracted increasing attention in recent advances. Especially, the task of Movie Genre Classification (MGC), which is an important branch of automatic labeling and has a wide range of applications (e.g., organizing user videos from social media sites, correcting mislabeled videos, and recommending specific types of films for users), has been paid significant efforts by existing work.

© The Author(s), under exclusive license to Springer Nature Switzerland AG 2023
X. Yang et al. (Eds.): ADMA 2023, LNAI 14177, pp. 676–691, 2023.
https://doi.org/10.1007/978-3-031-46664-9_45

Specifically, the task aims to classify movies into different genres and is suffering from new challenges, due to the emerging of more and more movie-related multi-modal information, and the diverse demands of consumers.

Despite the significant contributions on multi-modal data-based MGC made by existing work [1,6,14,25], they usually fuse the features of different modalities via concatenation [4,27] or weighted sum [1,8,25], failing to capture the semantic information contained by multi-modal data effectively. Additionally, the existing studies ignore the movies' metadata (e.g., directors and actors) that is of critical importance to a high performance MGC method. By way of illustration, given a movie and its sequels, they usually share the same directors or main actors and are more likely to have the same genres, compared with other different movies. This information can be effectively exploited by constructing a movie graph and extracting structural features from it. In a nutshell, there remains great scope for further improving the performance of existing MGC approaches due to the following problems: *Problem* 1) the multi-modal fusion strategies of existing studies cannot effectively explore the semantic information of multi-modal data; *Problem* 2) the movies' metadata, which involves abundant structural information, has been ignored by existing work.

Having observed the limitations of above-mentioned studies, we propose a novel model namely MFMGC[1] that is composed of two modules: MDF (Multi-modal Data Fusion) and MGRL (Movie Graph Representation Learning). In detail, the module MDF is designed to address *Problem* 1). Different from most existing studies that rely on late fusion strategies, MDF utilizes the attention mechanism to fuse multi-modal data during the feature extraction process. To be specific, there are two main attention layers in MDF, which are used for exploring the semantic features contained by movie-related multi-modal data. Inspired by VLBert [22], which takes the embeddings of both words in a sentence and region-of-interest (RoI) from images as inputs and utilizes the Transformer encoder to model dependencies among all the input elements, the first attention layer feeds the text and video frames into the Transformer encoder for text-video feature extraction. Then, the second modal attention layer is designed to fuse features of different modalities. In addition, the module MGRL is developed to tackle *Problem* 2). A movie graph is constructed based on the overlap of directors, screenwriters, and actors. Each movie is represented as a node in the graph and has multi-modal representations, which are obtained by fusing the movie-related multi-modal attributes with the module MDF. Next, a Graph Convolutional Network (GCN)-based architecture is applied to capture the structural information between movie nodes. Ultimately, a classification layer is employed to predict the genres of movies.

To fully evaluate the effectiveness of our proposed model MFMGC, the extensive experiments on real-world datasets are very essential. However, most of datasets used in previous studies are either not publicly available or have incomplete data [5,18]. Consequently, apart from the open dataset Moviescope [8], we construct a new multi-modal movie dataset called MovieBricks[1] from

---

[1] Code and data are available at https://anonymous.4open.science/r/mgc.

Douban, which is the most active online movie database and review platform in China. Specifically, the dataset MovieBrick contains 4063 European and American movies released from 2000 to 2019.

The contributions of this paper are summarized as follows:

- We propose a novel model MFMGC to further improve the performance of existing work on the task MGC, by fully exploring the semantic features involved in movie-related multi-modal data and the structure information between movies.
- Two modules are developed in MFMGC, i.e., MDF and MGRL. The module MDF is designed to tackle *Problem* 1), by capturing the interactive information between different modalities with novel fusion layers. The module MGRL is developed to address *Problem* 2), by extracting structure information from the movie graph that is constructed based on the overlap of movies' directors, screenwriters, and actors.
- We conduct extensive experiments on two real-world datasets, i.e., Moviescope and MovieBricks. Particularly, MovieBricks is the first multi-modal movie dataset in China, comprising over 4000 movies with four different modalities, including synopsis, poster, trailer, and metadata. The results demonstrate the superior performance of the proposed model MFMGC compared with the state-of-the-art methods.

The rest of the paper is organized as follows. The related work is presented in Sect. 2 and the task MGC is formulated in Sect. 3. The proposed model MFMGC is introduced in Sect. 4. We report the experimental results in Sect. 5, which is followed by the conclusion in Sect. 6.

2   Related Work

2.1   Research on Movies

Due to its rich storytelling and high-quality footage, the movies have become a valuable resource for researchers. Current studies on movies can be categorized into three directions: analyzing the content of movies, examining the impact of movies, and studying the characteristics of movies. Researches on movie content mainly use movie trailers as video data, e.g., scene boundary detection [9,20], which aims to divide a video into easily interpretable parts to communicate a storyline effectively, and action recognition [21,26] which utilizes the video scripts that exist for thousands of movies to automatically extract and track faces together with corresponding motion features. Studies on movie influence include movie box office prediction [16,28] and movie review analysis [13,23]. The movie box office prediction before its theatrical release can decrease its financial risk, and movie review analysis is a task of Natural Language Processing, which is able to obtain the emotional or semantic information of the movie's review. In addition, the studies on movie characteristics include understanding the relationships of movie characters [3,15], which aims to weigh the importance of character in defining a story, and movie genre classification [1,4,7,8,18,25].

2.2 Movie Genre Classification

To better contextualize our study, we review existing work focusing on multi-modal data, with particular emphasis on fusion strategies, and introduce them in a chronological order.

Wehrmann et al. [25] propose a novel deep neural architecture called CTT-MMC for multi-label movie-trailer genre classification. The authors utilize both video and audio data, and the fusion strategy involves a Maxout layer before the class prediction, which can be interpreted as a late fusion strategy. John et al. [1] propose a novel model for multi-modal learning based on gated neural networks for MGC. They utilize the plot and poster data for the classification task. The gated mechanism is used to obtain the weights of different modalities and then weight sums them for the final classification. The model is also utilized in other work such as [7]. Cascante et al. [8] compare the effectiveness of visual, audio, text, and metadata-based features in predicting movie genres. They utilize trainable parameters to sum different features of multi-modal data. Behrouzi et al. [4] design a new structure based on Gated Recurrent Unit (GRU) to extract spatial-temporal features from the movie-related data. The authors concatenate the video and audio features to predict the final genres of movies. Mangolin et al. [18] extract features by computing different kinds of descriptors, and then combine classifiers through the calculation of predicted score for each class, and they propose three rules for fusion, i.e., Sum, Prod, and Max.

In summary, current research on MGC with multi-modal data mainly utilizes late fusion strategies, such as concatenation and weighted sum, failing to capture the interaction between different modalities, and they ignore the structural information contained by metadata. To further improve the performance of their designed methods, we propose the novel model MFMGC in this study.

3 Problem Formulation

Given a set of movies $\{M_1, \cdots, M_N\}$, each movie M_i is associated with multi-modal attributes and metadata, i.e., $M_i = \{M_i^t, M_i^p, M_i^v, M_i^a, M_i^m\}$. Detailedly, M_i^t denotes the textual data consisting of the movie's title and synopsis, M_i^p represents the movie's poster, M_i^v denotes the visual data that is a sequence of frame-level patches in the trailer, M_i^a represents the audio fragments extracted from the trailer, and M_i^m is the metadata of the movie. Moreover, $C = \{c_1, \cdots, c_L\}$ is the genre set, where L is the number of movie genres.

Intuitively, movies with a significant overlap of directors, screenwriters, and actors may belong to the same genre, and a corresponding example is presented in Fig. 1. To capture such information effectively, we construct a multi-modal movie graph. Specifically, the graph is denoted as $G = \{V, E\}$, where V is the set of movie nodes, i.e., $V = \{M_1, \cdots, M_N\}$, and E is the set of connections between each pair of movie nodes. Additionally, we design an adjacency matrix A for the edge set E, where A_{ij} represents whether there is an edge between M_i and M_j. Given a threshold $\mathcal{T}$, if the overlap of directors, screenwriters, and actors between M_i and M_j exceeds $\mathcal{T}$, A_{ij} is set to 1. Otherwise, A_{ij} is set to 0.

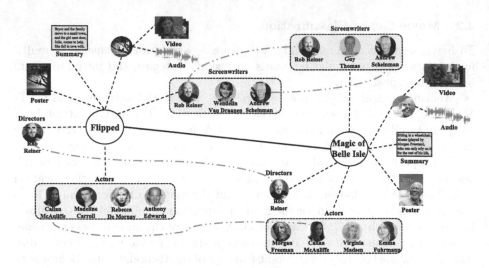

Fig. 1. A multi-modal movie graph, where each node has four multi-modal attributes, i.e., text, poster, video, and audio. Different movie nodes are connected according to the overlap of their directors, screenwriters, and actors.

Definition 1 *Movie Genre Classification (**MGC**). Given a movie M_i from the dataset $\{M_1, \cdots, M_N\}$ and a genre set C, the task of MGC aims to learn a function Φ to predict the genres of movie M_i based on M_i^t, M_i^p, M_i^v, M_i^a, and M_i^m. This process is formulated as follows:*

$$P_i = \Phi(M_i^t, M_i^p, M_i^v, M_i^a, M_i^m), \tag{1}$$

where $P_i = \{c_x, \cdots, c_y\}$ is the set of genres assigned to the movie M_i and each genre in $\{c_x, \cdots, c_y\}$ is from C.

Note that MGC is a multi-label classification task [27] and each movie may belong to multiple genres at the same time. For instance, the movie "X-Men: The Last Stand" has multiple genres, i.e., *Action*, *Horror*, and *Sci-Fic*.

4   Proposed Model

4.1   Overview

To effectively utilize the multi-modal data of movies to conduct MGC, we propose a novel model namely MFMGC. Observed from Fig. 2, the model contains two modules, i.e., Multi-modal Data Fusion (MDF) and Movie Graph Representation Learning (MGRL), and the details of them are as follows.

To feed the movie data into the module MDF, we first segment the audio and frame the video into patches at the frame level, and then use different pre-trained models to embed the text, posters, video frames, and audio segments. Next, these embeddings are fed into MDF, which consists of two stages. Specifically, in the

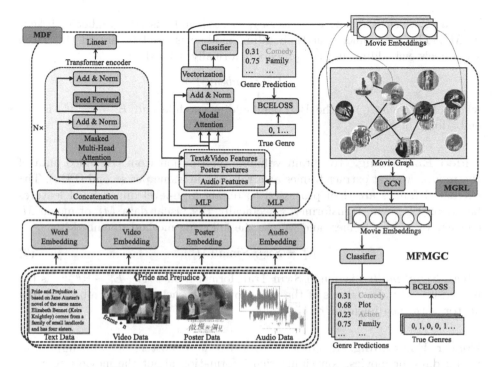

Fig. 2. Overview of the proposed model MFMGC

first stage, feature pre-extraction is performed. For the embeddings of text and video frames, we take them as input and utilize the Transformer encoder as the backbone to fuse text and video modalities, inspired by VLBert [22] that feeds both words in the sentence and region-of-interest (RoI) from the image into the Transformer encoder. For the posters and audio data, we separately design multi-layer perceptron (MLP) layers to perform the feature pre-extraction. In the second stage, we adopt a modal-attention layer to fuse extracted features, ensuring that the multi-modal data could be effectively integrated, resulting in a comprehensive representation of each movie. In MGRL, we deploy a GCN-based architecture to fine-tune the movie representations obtained from MDF and extract structural information between movie nodes.

4.2   Multi-modal Data Embedding

This section details the embedding process of the synopsis, poster, trailer, and audio data. We introduce how to transform these data into a suitable format and then feed the module MDF for feature pre-extraction and fusion.

Text Embedding. We utilize a Transformer Encoder structure to extract text features, where the text data is embedded by the Bert Embedding [12] module. Specifically, given the textual data M_i^t of the movie M_i, M_i^t contains a token sequence, which is denoted as $\{w_1, w_2, \cdots, w_l\}$ and l is the number of tokens in M_i^t. The pre-trained BertEmbedding module is used to obtain the token sequence's embedding and the process is formally defined as:

$$E_i^t = BertEmbed(M_i^t), \tag{2}$$

where $BertEmbed(\cdot)$ is the Bert Embedding module and $E_i^t \in \mathbb{R}^{l \times h^t}$ is the embedding of M_i^t.

Video Embedding. To obtain valuable information from the video data of the movie, we first extract frames at a rate of one frame per second (FPS). The extracted frames are then processed to obtain high-level dimensional features based on the Swin Transformer. Specifically, the visual data M_i^v of movie M_i consists of p video frames, and we use the following method to embed it:

$$E_i^v = SwinSmall(M_i^v), \tag{3}$$

where $SwinSmall(\cdot)$ is one of Swin Transformer model [17], $E_i^v \in \mathbb{R}^{p \times h^v}$ is the embedding of video frames of the i-th movie, and each frame is embedded to a vector with the dimension of h^v.

Poster Embedding. In addition to video data, posters are also important visual data for movies, containing rich information about the movie's genre to attract audiences with specific preferences. We feed the poster into the Swin Transformer to obtain its embedding and the process can be formally defined as:

$$E_i^p = SwinSmall(M_i^p), \tag{4}$$

where M_i^p is the poster data, and $E_i^p \in \mathbb{R}^{h^v}$ is the poster embedding of the i-th movie.

Audio Embedding. Apart from the above-mentioned information, we also extract features from audio, since different genres of movies usually have different types of soundtracks. For instance, while both *Comedy* and *Action* genres may have visually bright scenes, the background music of *Comedy* movies tends to have a more cheerful instead of intense rhythm. To capture latent features from the audio, we learn corresponding embeddings according to Wav2Vec2 [2]. The audio data is denoted as $M_i^a = \{o_1, o_2, \cdots, o_u\}$, where o_j is the j-th fragment of the given audio, with a sample rate of 16000, and each fragment is a 3-second audio signal. Note that we adopt a mean pooling operation to obtain the audio embedding from the embeddings of fragments, and the process is as follows:

$$E_i^a = MP(Wav2Vec2(M_i^a)), \tag{5}$$

where $Wav2Vec2(\cdot)$ is a Wav2Vec2 layer, $MP(\cdot)$ is the mean pooling operation, and $E_i^a \in \mathbb{R}^{h^a}$ is the audio embedding of the i-th movie.

Ultimately, the embedding of the i-th movie's multi-modal data can be represented as $\mathcal{E}_i = \{E_i^t, E_i^v, E_i^p, E_i^a\}$, which is then fed into the module MDF.

4.3   Multi-modal Data Fusion - MDF

The attention mechanism in Transformer has been proven powerful and flexible to differentially weigh the significance of each part of the input data. In MDF, we utilize this mechanism to fuse multi-modal embeddings, which involve two stages. In the first stage, the Transformer Encoder and MLP are used to extract latent features from different input embeddings. In the second stage, we adopt a modal-attention layer to fuse the features extracted at the first stage.

Feature Extraction of MDF. The Transformer Encoder is particularly effective in extracting sequential features, making it suitable for processing text and video frames. Specifically, in MDF, we first concatenate the embeddings of text and video frames as $\mathcal{E}_i^{tv} = E_i^t \| E_i^v$, where $\|$ denotes the concatenation operation, and $\mathcal{E}_i^{tv} \in \mathbb{R}^{(l+p) \times h^t}$. Then, the concatenated embedding $\mathcal{E}_i^{tv}$ is fed into the fusion module, which consists of a Transformer encoder [24] and a Mean pooling layer. The calculation process is formulated as follows:

$$O_i^{tv} = MP(TransEncoder(E_i^{tv})), \tag{6}$$

where $TransEncoder(\cdot)$ denotes Transformer Encoder.

For poster and audio embeddings, we employ two multi-layer perceptron (MLP) layers to extract their features respectively. The MLP layer consists of two fully connected layers with a ReLU activation function in the middle. The process can be formulated as follows:

$$O_i^p = ReLU(E_i^p W_1^p + b_1^p)W_2^p + b_2^p, \tag{7}$$

$$O_i^a = ReLU(E_i^a W_1^a + b_1^a)W_2^a + b_2^a, \tag{8}$$

where $E_i^{p/a}$ denotes the embedding of posters or audio, $W_1^{p/a}$ and $W_2^{p/a}$ are the weight matrices of the two fully connected layers, $b_1^{p/a}$ and $b_2^{p/a}$ are biases, and $ReLU(\cdot)$ is the Rectified Linear Unit activation function. After the features are extracted, the set of representations $\mathcal{O}_i = \{O_i^{tv}, O_i^p, O_i^a\}$ is obtained.

Modal-Attention Layer of MDF. Following the feature extraction, we apply modal attention to fuse the features of different modalities. Specifically, the feature of text-video O_i^{tv} is first transformed into a new vector $\tilde{O}_i^{tv}$ that has the same dimension with the poster and audio embeddings, through a Linear layer. Then the multi-modal input features are first concatenated to obtain $\hat{O}_i \in \mathbb{R}^{m \times h}$, where m is the number of features in $\mathcal{O}_i$. Next, the query matrix $Q_i = \hat{O}_i W_q$ is obtained through the projection matrix W_q, while the key matrix K_i and value matrix V_i are obtained using W_k and W_v, respectively. The scaled dot product function is used as the attention function, and the inter-modal attention matrix P_i is obtained with following method,

$$P_i = softmax(\frac{Q_i K_i^T}{\sqrt{h}}), \tag{9}$$

where $P_i \in \mathbb{R}^{m \times m}$ and each element $P_{i,xy}$ of the matrix represents the inter-modal attention between the x-th and y-th modality of the i-th movie M_i. Then, the multi-modal representation of M_i, which is denoted as F_i, is obtained through attention aggregation and the map function $\mathcal{V}$. Additionally, a residual connection is added to avoid the problem of vanishing gradients during training, and the process can be represented as follows:

$$F_i = \mathcal{V}(P_i V_i + \mathcal{O}_i), \tag{10}$$

where $\mathcal{V}(\cdot)$ denotes the vectorization by row-wise concatenation, and $F_i \in \mathbb{R}^{1 \times mh}$. Finally, we obtain $\mathcal{F} = \{F_1, F_2, \cdots, F_N\}$, which contains the multi-modal representations of all movies in the given dataset.

4.4   Movie Graph Representation Learning - MGRL

To fully explore the structural and semantic information of movies in a unified manner, we construct a multi-modal movie graph based on movies' directors, screenwriters, and actors. Here, the movie nodes have fused representations that are obtained in MDF based on movie-related multi-modal attributes, i.e., synopsis, poster, and trailer. To effectively extract structural information from the graph, we adopt a two-layer GCN to fine-tune the movie representations and the process is as follows:

$$\mathcal{H} = GCN(\mathcal{F}, A) = ReLU(\tilde{A} ReLU(\tilde{A} \mathcal{F} W^0) W^1), \tag{11}$$

where $\mathcal{H} = \{H_1, H_2, \cdots, H_N\}$ denotes the new set of movie representations, $H_i (1 \le i \le N)$ is the fine-tuned embedding of movie M_i. A is the adjacency matrix of the movie graph and $\tilde{A} = \tilde{D}^{-\frac{1}{2}}(A + I_N)\tilde{D}^{-\frac{1}{2}}$. I_N is the identity matrix with size $N \times N$, where N denotes the number of movies in the graph. $\tilde{D}$ is the diagonal degree matrix of $\tilde{A}$, which is defined as $\tilde{D}_{ii} = \sum_{j=1}^{N} \tilde{A}_{ij}$. W^0 and W^1 are learnable parameters.

4.5   Classification Layer

Ultimately, to tackle the task of MGC, we use a linear projection followed by a sigmoid function to predict the movie's genre. This can be formally defined as:

$$\mathcal{S}^1 / \mathcal{S}^2 = Sigmoid(Linear(\mathcal{F}/\mathcal{H})), \tag{12}$$

where $Sigmoid(\cdot)$ is the activation function that is used to squash the output vector values to range $[0, 1]$, which can be interpreted as the vector of genre probability. Note that, as there has been no work constructed above-mentioned movie graph, to give a more fair comparison, the input of the classification layer can be either $\mathcal{F}$ or $\mathcal{H}$. Consequently, the output can be either $\mathcal{S}^1$ or $\mathcal{S}^2$. Taking $\mathcal{S}^1 = \{S_1, S_2, \cdots, S_N\}$ as an example, $S_i \in \mathcal{S}^1$ is the genre probability vector of the i-th movie M_i, which is denoted as $S_i = \{s_{i1}, \cdots, s_{iL}\}$. Here, $s_{ij} \in S_i$ represents the probability that M_i belongs to the j-th genre, and L is the number of genres.

4.6   Training

The model is optimized by Binary Cross-Entropy Loss (BCELoss). The labels of movies are first embedded and denoted as $\mathcal{C} = \{C_1, C_2, \cdots, C_N\}$. For the i-th movie, the genres set is $C_i = \{c_{i1}, c_{i2}, \cdots, c_{iL}\}$, where $c_{ij} \in \{0,1\}$, and $c_{ij} = 1$ indicates that the i-th movie belongs to the j-th genre. The formulation for the loss function is as follows:

$$\mathcal{L} = BCELoss(\mathcal{C}, \mathcal{S})$$
$$= -\frac{1}{L} \sum_{i=1}^{N} \sum_{j=1}^{L} (c_{ij} \log(s_{ij}) + (1 - c_{ij}) \log(1 - s_{ij})), \tag{13}$$

where N is the number of movies.

In addition, when adding the module MGRL, we adopt a joint loss function to guide the optimization of both MDF and MGRL:

$$\mathcal{L} = BCELoss(\mathcal{C}, \mathcal{S}^1) + BCELoss(\mathcal{C}, \mathcal{S}^2). \tag{14}$$

5   Experiments

5.1   Dataset

Most of the datasets used in current research are either not open or the access paths have expired, particularly for datasets that contain multiple data sources such as synopses, posters, trailers, and metadata. We start with downloading the dataset Moviescope, which contains movies' synopsis, posters, and URLs of trailers on YouTube, and then develop a Python crawler to obtain the trailers. To enable a more comprehensive evaluation of our model, we create a new dataset from Douban, the most active online movie review and dataset platform in China. The details of the two datasets are as follows.

The dataset Moviescope, all data sources of which are available, contains 4076 movies with 13 different genres. Additionally, the dataset MovieBricks has 4063 movies with 10 different genres, namely *Action, Thriller, Adventure, Story, Science-Fiction, Love, Fantasy, Comedy, Terror* and *Crime*. Both two datasets are divided into training, validation, and testing sets in a 7:1:2 ratio. Note that a movie may belong to multiple genres at the same time.

5.2   Comparison Method

To validate the effectiveness of MFMGC, we compare its performance with those of several state-of-the-art approaches that are introduced as follows.

- **GMU** [1]. This work develops a model for multi-modal learning based on gated neural networks, which is evaluated on a multi-label scenario for MGC using synopses and posters.

- **Fast-MA** [8]. This work designs a temporal feature aggregator to embed video and text, and compares the effectiveness of visual, textual-based methods on MGC, and it is denoted as Fast Modal Attention (Fast-MA).
- **DL-PO** [19]. This work proposes a simple deep-learning model to predict the genres of a movie with overview and poster. We refer to it as Deep Learning for Posters and Overviews (DL-PO).
- **CMM** [18]. This is a comprehensive study developed in terms of diversity of multimedia sources of information to perform MGC. We refer to it as Comprehensive Multi-modal Model (CMM).
- **MGC-RNN** [4]. This work proposes a new structure based on GRU to derive spatial-temporal features of movie frames and then concatenates them with the audio features to predict the final genres of the movie. We refer to it as MGC-RNN.

5.3   Evaluation Metrics and Parameter Settings

Evaluation Metrics. AUC-ROC [11] is a well-known metric that measures the area under the receiver operating characteristic (ROC) curve. This curve plots the true positive rate against the false positive rate for each possible threshold of the classifier's output. However, relying on a single metric cannot provide a comprehensive evaluation of a multi-label classifier. Therefore, we also calculate the F1 score that has been widely used to evaluate the multi-label classifiers [1,4]. To globally evaluate the performance of different methods, we compute the micro and macro averages of the F1 and AUC metrics. The micro-average calculates the mean of scores without considering genres, while the macro-average computes the score of each genre independently and takes their unweighted means.

Parameter Settings. As mentioned in Sect. 4.2, for the textual modality M^t, a fixed sequence length $l = 256$ is used. For the video modality M^v, we draw $p = 32$ frames from the trailer, and for the audio modality M^a, the number of audio segments is $u = 16$, and the hidden dimension h in our model is set to 256. To reduce the impact of random noise, all experiments are conducted using the 5-fold cross-validation. The results reported are the average of 5 runs using different data partitions. The pre-trained model "Roberta" [10] is utilized to initialize the Transformer Encoder module. To maintain the learned knowledge of pre-trained parameters, we split the learnable parameters into two parts: the learning rate for pre-trained initialized parameters is set to 0.00005, while the learning rate for randomly initialized parameters is 0.0005, and they are denoted as "pre-lr" and "rand-lr" respectively.

5.4   Experiment Results

Experiments are done on a machine with 2 NVIDIA V100 GPUs. The performances are presented in Table 1. As all baselines are designed without considering movie graph, to provide a fair comparison, we present the results of our model's simplified version MFMGC-P that only utilize partial input data, i.e., synopsis, poster, and trailer of movies. Moreover, MFMGC represents the model that considers the movie's metadata and multi-modal data along with the movie graph. Note that, as the movies from Moviescope only contain few metadata, the movie

Table 1. Experimental results. The used information contains Text (T), Poster (P), Audio (A), Video (V), and Movie Graph (G). Furthermore, "ma" and "mi" are used to represent the macro and micro averages.

Model	Modality	Moviescope				MovieBricks			
		ma-f1	mi-f1	ma-auc	mi-auc	ma-f1	mi-f1	ma-auc	mi-auc
GMU	T	0.5614	0.6158	0.8470	0.8646	0.5563	0.6126	0.8427	0.8657
	P	0.4441	0.5203	0.7425	0.7943	0.4655	0.5371	0.7753	0.8115
	TP	0.5821	0.6328	0.8560	0.8721	0.5947	0.6291	0.8590	0.8748
Fast-MA	T	0.5588	0.6145	0.8459	0.8642	0.5499	0.6063	0.8381	0.8643
	P	0.4102	0.5107	0.7265	0.7727	0.4002	0.5361	0.7339	0.7854
	V	0.4786	0.5492	0.7727	0.8193	0.4832	0.5496	0.7778	0.8160
	TPV	0.6203	0.6497	0.8762	8872	0.5749	0.6300	0.8625	0.8763
DL-PO	T	0.5475	0.5964	0.8415	0.8569	0.5488	0.5945	0.8427	0.8600
	P	0.4201	0.5034	0.7258	0.7809	0.4362	0.524	0.7481	0.7985
	TP	0.5739	0.6251	0.8554	0.8775	0.5818	0.6401	0.8616	0.8857
CMM	T	0.5564	0.6059	0.8416	0.8614	0.5525	0.6041	0.8534	0.8750
	P	0.3548	0.5035	0.6700	0.7478	0.4507	0.5162	0.7113	0.7563
	V	0.4845	0.5817	0.8135	0.8483	0.5219	0.5982	0.8267	0.8535
	A	0.4960	0.5642	0.7985	0.8321	0.4959	0.5693	0.7959	0.8361
	TPVA	0.5588	0.6439	0.8760	0.8916	0.5594	0.6424	0.8754	0.8945
MGC-RNN	V	0.4760	0.5431	0.7666	0.8180	0.4693	0.5424	0.7531	0.8032
	A	0.4957	0.5635	0.8007	0.8306	0.4858	0.5603	0.7897	0.8303
	VA	0.5106	0.5719	0.7886	0.8491	0.5254	0.5981	0.8134	0.8480
MFMGC-P	T	0.5937	0.6364	0.8483	0.8730	0.6531	0.6834	0.8714	0.8836
	P	0.5241	0.5884	0.7895	0.8283	0.5564	0.6187	0.8256	0.8569
	V	0.5412	0.5756	0.8605	0.8799	0.5125	0.5824	0.8157	0.8446
	A	0.4695	0.5247	0.7919	0.8202	0.5072	0.5727	0.7977	0.8363
	TP	0.6347	0.6750	0.8806	0.8976	0.6714	0.7110	0.8966	0.9135
	TV	0.6436	0.6757	0.8801	0.8971	0.6653	0.6981	0.8850	0.8979
	TA	0.6155	0.6659	0.8693	0.8925	0.6578	0.6970	0.8867	0.9015
	PV	0.6054	0.6529	0.8590	0.8875	0.5740	0.6315	0.8355	0.8641
	PA	0.5419	0.6048	0.8159	0.8535	0.5670	0.6345	0.8335	0.8657
	VA	0.5522	0.5860	0.8670	0.8865	0.5889	0.6399	0.8672	0.8880
	TPV	0.6533	0.6859	0.8905	0.9059	0.6843	0.7163	0.9024	0.9155
	TPA	0.6421	0.6878	0.8871	0.9051	0.6765	0.7111	0.9080	0.9201
	TVA	0.6514	0.6933	0.8836	0.9090	0.6702	0.7024	0.8978	0.9050
	PVA	0.6210	0.6662	0.8648	0.8931	0.6038	0.6631	0.8662	0.8907
	TPVA	**0.6600**	**0.6947**	**0.8914**	**0.9065**	**0.6925**	**0.7248**	**0.9046**	**0.9165**
MFMGC	T+G	–	–	–	–	0.6582	0.6935	0.8857	0.9088

graph cannot be constructed, thus MFMGC has no result on this dataset. In addition, we only present the results of MFMGC on MovieBricks when utilizing text data and movie graph, due to the space limitation. More results of the model from P+G, V+G, A+G to TPVA+G are presented on github[1].

Main Results. Observed from Table 1, our proposed model MFMGC-P consistently achieves better performance than baselines. Specifically, the "mi-f1" score of it outperforms GMU by 11.6% on MovieBricks. Even only using poster or video data, MFMGC-P still performs better than other methods, indicating its powerful feature extraction capability. When all multi-modal attributes are taken into account, MFMGC-P achieves higher improvements, which demonstrates the effectiveness of the carefully designed fusion strategy based on attention mechanism. Detailedly, the reasons for the above-mentioned observations are as follows: 1) MFMGC-P utilizes advanced pre-trained models to embed data, the parameters hold abundant knowledge, especially for text and images, which leads to better representations than the traditional models such as Word2vec and VGG. 2) We fuse the multi-modal data via the attention mechanism that differentially weighs the significance of each part of the input data, allowing MFMGC-P to learn a comprehensive representation.

Modality Analysis. To fully investigate the impacts of different modalities on the performance of the proposed model, we compare the results of MFMGC-P when adding single, two, three, and all four modalities. Seen from Table 1, the results of the second part (i.e., TP, TV, TA, PV, PA, and VA) of MFMGC-P outperform those of single-modal data based experiments (i.e., T, P, V, and A). Without surprise, when taking four modalities (i.e., TPVA) into account, MFMGC-P achieves the best performance. These observations demonstrate the effectiveness of the developed module in extracting multi-modal features. Additionally, the higher performance of MFMGC (i.e., T+G) than that of MFMGC-P (i.e., T) demonstrates the significance of the construction of movie graph that can be used to extract structural information between movies.

5.5   Parameter Analysis

To investigate the effect of different learning rates and compare the experimental results with above-mentioned ones more intuitively, the performances of MFMGC-P with varying "pre-lr" and "rand-lr" are reported in Fig. 3(a) and Fig. 3(b). Observed from this, the evaluation metrics present a overall downward trend, where the "ma-f1" score even drops by 21.4%. The reason for the decrease is that setting a lower "pre-lr" can avoid forgetting the knowledge contained by the pre-trained parameters. Additionally, given a too-large "rand-lr", it will lead to faster convergence but makes the model difficult to achieve the best result, as the global optimum may be missed during the iteration.

Furthermore, we analyze the effect of the hidden dimension h, and the results are reported in Fig. 3(c). While varying h from 64 to 512, we first observe the increase of evaluation metrics, then they present a decreasing tendency, and the best performance is achieved when $h = 256$. Given a too small dimension, the model cannot learn enough information, leading to under-fitting. Conversely, when h is too large, it may introduce unexpected noisy information, resulting in poor performance.

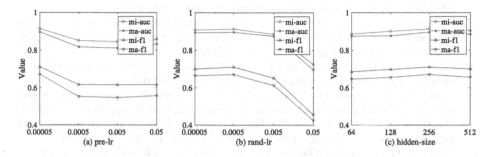

Fig. 3. Parameter analysis for "pre-lr", "rand-lr", and the hidden dimension h.

6   Conclusion

We propose MFMGC, which is a novel model for the task of MGC that utilizes the movie's synopsis, poster, trailer and metadata, and the model comprises two modules: MDF and MGRL. MDF leverages the attention mechanism to capture the modalities' interactive information effectively. In MGRL, we construct a graph to capture the structural relationships between movies based on directors, screenwriters, and actors, where the node in the graph is a movie that has multi-modal attributes and is first represented by MDF. Then a Graph Convolutional Network (GCN)-based architecture is developed to extract structural information between movie nodes. In addition, we also present a new multimodal movie dataset, i.e., MovieBricks. The experimental results on Moviescope and MovieBricks demonstrate the superior performance of MFMGC.

Acknowledgment. This work is supported by the National Natural Science Foundation of China No. 62272332, the Major Program of the Natural Science Foundation of Jiangsu Higher Education Institutions of China No. 22KJA520006.

References

1. Arevalo, J., Solorio, T., Montes-y Gómez, M., González, F.A.: Gated multimodal units for information fusion. arXiv:1702.01992 (2017)
2. Baevski, A., Zhou, Y., Mohamed, A., Auli, M.: Wav2Vec 2.0: a framework for self-supervised learning of speech representations. NeurIPS **33**, 12449–12460 (2020)

3. Bamman, D., O'Connor, B., Smith, N.A.: Learning latent personas of film characters. In: ACL, vol. 1, pp. 352–361 (2013)
4. Behrouzi, T., Toosi, R., Akhaee, M.A.: Multimodal movie genre classification using recurrent neural network. Multimedia Tools Appl. **82**, 1–22 (2022)
5. Bi, T., Jarnikov, D., Lukkien, J.: Shot-based hybrid fusion for movie genre classification. In: ICIP, pp. 257–269 (2022)
6. Bi, T., Jarnikov, D., Lukkien, J.: Video representation fusion network for multi-label movie genre classification. In: ICPR, pp. 9386–9391 (2021)
7. Bribiesca, I.R., Monroy, A.P.L., Montes, M.: Multimodal weighted fusion of transformers for movie genre classification. In: Proceedings of the Third Workshop on Multimodal Artificial Intelligence, pp. 1–5 (2021)
8. Cascante-Bonilla, P., Sitaraman, K., Luo, M., Ordonez, V.: Moviescope: large-scale analysis of movies using multiple modalities. arXiv:1908.03180 (2019)
9. Chen, S., Nie, X., Fan, D., Zhang, D., Bhat, V., Hamid, R.: Shot contrastive self-supervised learning for scene boundary detection. In: CVPR, pp. 9796–9805 (2021)
10. Cui, Y., Che, W., Liu, T., Qin, B., Wang, S., Hu, G.: Revisiting pre-trained models for Chinese natural language processing. In: EMNLP, pp. 657–668 (2020)
11. Davis, J., Goadrich, M.: The relationship between precision-recall and roc curves. In: ICML, pp. 233–240 (2006)
12. Devlin, J., Chang, M.W., Lee, K., Toutanova, K.: BERT: pre-training of deep bidirectional transformers for language understanding. arXiv:1810.04805 (2018)
13. Dridi, A., Recupero, D.R.: MORE SENSE: MOvie REviews SENtiment analysis boosted with SEmantics. In: EMSASW (2017)
14. Huang, Q., Xiong, Yu., Rao, A., Wang, J., Lin, D.: MovieNet: a holistic dataset for movie understanding. In: Vedaldi, A., Bischof, H., Brox, T., Frahm, J.-M. (eds.) ECCV 2020. LNCS, vol. 12349, pp. 709–727. Springer, Cham (2020). https://doi.org/10.1007/978-3-030-58548-8_41
15. Kukleva, A., Tapaswi, M., Laptev, I.: Learning interactions and relationships between movie characters. In: CVPR, pp. 9849–9858 (2020)
16. Liao, Y., Peng, Y., Shi, S., Shi, V., Yu, X.: Early box office prediction in China's film market based on a stacking fusion model. Ann. Oper. Res. **308**(1), 321–338 (2020). https://doi.org/10.1007/s10479-020-03804-4
17. Liu, Z., et al.: Swin Transformer: hierarchical vision transformer using shifted windows. In: ICCV, pp. 10012–10022 (2021)
18. Mangolin, R.B., et al.: A multimodal approach for multi-label movie genre classification. Multimedia Tools Appl. **81**(14), 19071–19096 (2022)
19. Nambiar, G., Roy, P., Singh, D.: Multi modal genre classification of movies. In: INOCON, pp. 1–6 (2020)
20. Rao, A., et al.: A local-to-global approach to multi-modal movie scene segmentation. In: CVPR, pp. 10146–10155 (2020)
21. Sigurdsson, G.A., Varol, G., Wang, X., Farhadi, A., Laptev, I., Gupta, A.: Hollywood in Homes: crowdsourcing data collection for activity understanding. In: Leibe, B., Matas, J., Sebe, N., Welling, M. (eds.) ECCV 2016. LNCS, vol. 9905, pp. 510–526. Springer, Cham (2016). https://doi.org/10.1007/978-3-319-46448-0_31
22. Su, W., et al.: VL-BERT: pre-training of generic visual-linguistic representations. arXiv:1908.08530 (2019)
23. Thet, T.T., Na, J.C., Khoo, C.S., Shakthikumar, S.: Sentiment analysis of movie reviews on discussion boards using a linguistic approach. In: CIKM, pp. 81–84 (2009)
24. Vaswani, A., et al.: Attention is all you need. In: Advances in Neural Information Processing Systems 30 (2017)

25. Wehrmann, J., Barros, R.C.: Movie genre classification: a multi-label approach based on convolutions through time. Appl. Soft Comput. **61**, 973–982 (2017)
26. Xu, M., et al.: Long short-term transformer for online action detection. NeurIPS **34**, 1086–1099 (2021)
27. Zhang, Z., Gu, Y., Plummer, B.A., Miao, X., Liu, J., Wang, H.: Effectively leveraging multi-modal features for movie genre classification. arXiv:2203.13281 (2022)
28. Zhou, Y., Zhang, L., Yi, Z.: Predicting movie box-office revenues using deep neural networks. Neural Comput. Appl. **31**(6), 1855–1865 (2019)

TED-CS: Textual Enhanced Sensitive Video Detection with Common Sense Knowledge

Bihui Yu[1,2], Linzhuang Sun[1,2](✉), Jingxuan Wei[1,2](✉), Shuyue Tan[1,2], Yiman Zhao[1,2], and Liping Bu[1,2]

[1] Shenyang Institute of Computing, Chinese Academy of Science, Shenyang, China
[2] University of Chinese Academy, Beijing, China
{sunlinzhuang21,weijingxuan20}@mails.ucas.ac.cn

Abstract. In the era of short videos, the task of sensitive video detection faces new challenges with the increasing diversity and quantity of videos in the network. Aiming at the problem that existing research methods are constrained by missing video comments and user subjectivity, a novel text enhancement method is proposed for sensitive video detection. Firstly, based on the CLIP pre-training model, image caption are generated for video frames. Then, through the integration of external common sense knowledge, the method extracts deep contextual information from the generated captions, including the underlying intentions and purposes conveyed by the text. Besides, considering the complementarity and redundancy between different sources of information, a multi-source data collaborative encoding mechanism and a multi-modal feature fusion mechanism are designed to achieve semantic feature alignment. Finally, state-of-the-art was achieved on the two public datasets NPDI-800 and Pornography-2k, and a large number of detailed comparison and ablation experiments were performed to verify the effectiveness of the method.

Keywords: sensitive video detection · common sense knowledge · image caption

1 Introduction

Sensitive video detection is a challenging and socially concerned task that devotes to enhance the ability to regulate Internet content and contribute to maintaining a positive and healthy online environment. Besides, with the advent of the short video era, various videos and playback media are flooding our lives. In such a situation, the importance of filtering sensitive videos cannot be over-emphasized.

Most of the current sensitive video detection methods are CNN-based neural network models, such as [32] who combined video frames with timing signals and achieved good experimental results. However, most of the existing research are from the perspective of comprehension-based models, which only utilize the

This work was supported by Science and Technology Plan of Liaoning Province (2022020507-JH2/1013).

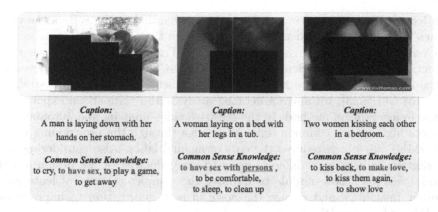

Fig. 1. Examples about text augmentation.

natural image and audio information in dataset, with limited use of textual information. Although [2] enriched the text features by processing the user comment information of YOUTUBE dataset, there are still two serious limitation: First, in the actual scenario, most of the videos do not contain text information such as captions and comments. Second, there is a risk of fraudulent attacks. Since information such as movie comments are written and submitted by users, they can mislead the model and cause fraudulent attacks by swiping a large number of comments that are opposite to the video attributes. In our previous work, the above two problems were initially solved by using image caption generation techniques. However, the generated textual information is often excessively objective and neutral, which make it challenging to apply directly and effectively in sensitive video detection.

From the perspective of bionic, the reading behavior of adults not only invokes the areas associated with language comprehension, which help them extract the surface-level meaning of the words, but also the hippocampus and other neurons that activate their accumulated common sense knowledge, allowing them to uncover the deeper logic and metaphors conveyed by the text, such as intention, purpose and possible impact. During this process, the critical role of common sense knowledge refers to the general knowledge and understanding of social life that can aid in the reasoning process. By leveraging this knowledge, individuals can more accurately decipher the intention information behind the text.

In Fig. 1, While the generated common sense knowledge exhibits a degree of diversity, some of the knowledge produced can lead to significant deviations from the text descriptions and video frames increasing the challenge of model learning. To address this problem, we designs a multimodal feature fusion alignment module, based on Self-attention and Cross-attention mechanisms to achieve controlled generation of common knowledge information in disguise, by the method associates the field of view of the common sense knowledge with other relevant features and assigns low weights to knowledge that deviates from the video frames while assigning high weights to knowledge that aligns with them.

Considering the above issues, we proposes **T**ext **E**nhanced Sensitive video **D**etection with **C**ommon **S**ense knowledge, TED-CS, a sensitive video detection method with common sense knowledge enhancement. Based on the pre-training model of CLIP [23], we make a detailed analysis and comparison experiments on the image caption methods in the sensitive video domain, and automatically extracts high-quality common sense knowledge through the pre-training model of COMET [6], which reduces the cost of manual labor. Also considering that different modalities have different ways to carry knowledge, we design a multimodal feature fusion module to improve the complementarity of features of different modalities and reduce the noise in text information to improve the capability of the model.

Our main contributions are summarized as follows:

1) we make an extensive experimental analysis of the image caption method in the field of sensitive video detection and proved its effectiveness. 2) we demonstrate that generic image caption is limited in sensitive video detection. And the information derived by combining common sense knowledge can further improve the accuracy of sensitive video detection. 3) TED-CS is the first attempt to jointly utilize the presentation from picture modality and text modality for sensitive video classification, based on our designed feature fusion mechanism. 4) Experiments on three benchmark datasets show that TED-CS is competitive and robust.

2   Related Work

In this section, we will detail some classification methods related to pornography detection. In the earliest stages of research, people focused on bare skin [11] for identification. However, the sensitivity of the video is frequently determined by the exposed parts rather than the areas. But in reality, the exposed skin area is not only positively correlated with the degree of sensitivity. In addition, the skin color of the race will also affect the judgment. Afterward, based on the statistical learning method of BoVW, the video is classified by combining the video information with the feature classifier [3] obtains low-dimensional and medium-dimensional features through HueSIFT [16] and BOSSA respectively, and then combines nonlinear SVM for feature classification [29] extracts the time and space information of the video, and sends it to the linear SVM to obtain the prediction result after processing. Along this line, [27] uses a method similar to Valle's, but uses ColorSTIP to obtain low-dimensional features.

The method based on deep learning can overcome the limitations of manual features, automatically learn classification features, and enhance the robustness of the method [19] introduced AlexNet and GoogLeNet for the first time, using the weights pre-trained on ImageNet to jointly judge video attributes. Based on GoogLeNet, explores two types of vision and motion between image frames and two-stream information. Also timing information, [32] captures sequence information through LSTM. Not limited to video classification, the PPCensor [17] proposed by Mallmann can also perform object detection on videos.

3   Methods

The overview of our approach is shown in Fig. 2. Based on the pre-trained model, our proposed TED-CS consists of four main modules: (1) video frame extractor; (2) textual information augmentation ; (3) encoder, and (4) multimodal feature fusion.

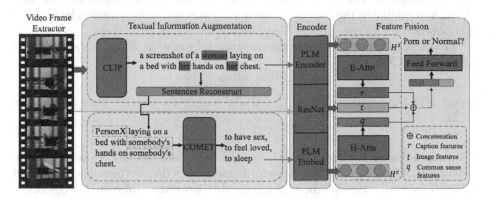

Fig. 2. The architecture of the proposed TED-CS.

3.1   Video Frame Extractor

The module of extracting video frames is designed to pick the set of images that have the right to vote. For now, most of the methods use the majority voting as the classification result of a video. Thereby, this group of voting frames has a crucial basis for sensitivity judgment, and the size of which determines the computing pressure and efficiency of the system. In practice, we samples k image frames uniformly according to the time dimension in order to maximize the sampled image frames to cover the complete video. k is a constant with value equal to 10.

3.2   Textual Information Augmentation

This section is composed of a caption generator and the common sense knowledge acquisition. As mentioned in the first chapter, first we use the pre-trained CLIP model to generate image caption for extracted video frames, then mine the potential semantic information by correlating common sense knowledge.

Caption Generator. As natural language processing (NLP) and computer vision (CV) technologies continue to advance, the integration of multiple modalities is being optimized and refined, Moreover, numerous unified pre-training models have been proposed to demonstrate that textual and visual features are complementary and can achieve more accurate understanding of the issue after modality alignment and fusion. Such as CLIP and OFA [31]. Here, we choose the pre-trained model CLIP as the caption generator.

CLIP calculates the similarity of image-text pairs through contrastive learning on the dataset WIT, which contains 400 million records, to align them to the same feature space. During the training phase, each batch contains n data pairs. Hence, if every two are matched, there are a total of n^2 combination. We will maximize the feature similarity for n of the correct matching combinations, and minimize the feature similarity for the remaining mismatched $n^2 - n$ combinations.

Given a video frame img, the caption information $S = [s_1, s_2, ..., s_n]$ is obtained by CG:

$$S = \mathrm{CG}(img), \qquad\qquad (1)$$

where n is the number of tokens and CG stands for caption generator model CLIP.

Common Sense Knowledge. Due to the characteristics of the task, semantic features that directly contribute to erotic intent detection are usually obscure and indirect, which cannot be directly captured in generated subtitle descriptions.

The caption is usually an objective expression of the image and has weak emotional tendencies. For example, given a sentence, *a screenshot of a woman laying on a bed with her hands on her chest.*, which only describes the number of people, actions, and scenes that occurred. It is extremely embarrassing to force the neural model to make sensitivity judgments without the aid of pictures. But for a rational person, he will not only pay attention to the above surface information, but also combine it with common sense information to mine the deep cryptic information of the sentence, such as the intention of the person, the cause of the event, the possible consequences, etc., to further enrich the textual modal information. Then it will have great confidence in predicting the pornographic orientation of the caption. Therefore, common sense knowledge is crucial in the field of sensitive video detection. In practice, considering the expensive cost of extracting common sense manually, we use the model COMET, a pre-trained model that generates commonsense knowledge from input sentences.

COMET is constructed based on two commonsense knowledge graphs, ATOMIC [25] and conceptNet [28], which covers a variety of social commonsense knowledge. And in order to improve query efficiency and quality, it specifies the format of input sentences and provides nine relational attributes. As shown in Fig. 2, the sentence to be processed is first transformed into *PersonX laying on a bed with somebody's hands on somebody's chest.* according to the rewriting algorithm, then COMET generates common sense knowledge *to have sex, to be intimate, to feel loved, to sleep* according to the predefined relationship set.

We selected three sets of COMET attributes based on our professional knowledge. And their comparative results will be elaborated in the experimental section.

(1) **WIW**: oWant, xIntent, and xWant. We want to be able to get the underlying intentions of the subjects and objects of events, and the causes of the current situation.

(2) **AINRW**: xAttr, xIntent, xNeed, xReact, and xWant. Making a series of causal inferences about the related participants, including ones motivation, role in the event, and possible resulting responses to the behavior.

(3) **ALL**: oEffect, oReact, oWant, xAttr, xEffect, xIntent, xNeed, xReact, and xWant. We use all the attributes that COMET has to offer, hoping to obtain as much high-quality information as possible.

After getting caption S, we use COMET to get common sense knowledge $C = c_1, c_2, ..., c_m$:

$$C = \text{COMET}(S), \tag{2}$$

where m is the number of tokens in the common sense information sequence.

3.3   Encoder Module

In this section, the caption information, common sense knowledge, and the video frame are encoded using text and image encoders, respectively, to obtain vector representations. The encoding process for each of these three types of features is illustrated below.

Textual Information Encoder. PLMs have obtained rich knowledge during pre-training, which have powerful strength in language understanding. We use BERT [10] as text encoder. The caption sentences and common sense knowledge are encoded separately. As shown in Fig. 1, Common sense information is typically composed of a set of phrases with a certain level of independence between them. Thus, attempting to capture high-dimensional features through deeper self-attention layers, as is done when processing caption messages, is not an elegant solution. Therefore, BERT's encoder does not directly encode common sense information. Instead, it is converted into a vector matrix through the use of a lookup table. This process enables the effective transformation of the common sense information into a more suitable format for use in subsequent processing steps.

We use $H^s = [h_1^s, h_2^s, ..., h_n^s]$ and $H^c = [h_1^c, h_2^c, ..., h_n^c]$ as the hidden representations of S and C, respectively.

$$h_1^s, h_2^s, ..., h_n^s = \text{BERT-Encoder}(S), \tag{3}$$

$$h_1^c, h_2^c, ..., h_m^c = \text{BERT-Embed}(C), \tag{4}$$

where n and m are the number of tokens.

Image Encoder. In the image domain, the RESNET50 [13] model serves as the encoder, which has been established as a robust backbone in many computer vision tasks. Given a video frame, the feature representation t of the image is obtained through multiple residual block structures.

$$t = \text{ResNet}(img) \tag{5}$$

3.4   Feature Fusion Module

We have designed two attention mechanisms to extract the feature representations of caption information and common sense text, respectively. These mechanisms enable the effective capture of salient information from the textual inputs, which is then combined with the image feature representation to obtain the final feature vector for the video frame.

E-Attention. Analogous to the processing of common sense text, tokens in caption sequence also have different effects on the predicted result. In order to derive a more efficient representation of the feature vectors, E-Atten first obtains the weighted hidden state matrix B through the self-attention mechanism and followed by the max-pooling strategy to obtain the key feature vector q of the caption text.

$$B = softmax(\frac{H^s K^T}{\sqrt{2\mu}})V \tag{6}$$

$$q = \max - \text{pooling}(B), \tag{7}$$

where K and V are the value of the matrix H^s after linear transformation.

H-Attention. Given that different tokens in the common sense text may represent distinct meanings, they may not necessarily contribute equally to the overall judgment of video sensitivity. Therefore, we first obtain the feature vector r of common sense knowledge through the summation of the weighted hidden state matrix $A = [a_1, a_2, ..., a_m]$ conducted by the designed attention mechanism that is inspired by [9].

$$a_i = softmax(\omega^T f(W[h_i^c] + b)), \tag{8}$$

$$r = \sum_{i=1}^{m} a_i h_i^c \tag{9}$$

a_i denotes the weight of i-th token in common sense text. f is the relu [1] function.

Classifier and Training Objective. Finally, we concatenate the three representations r, q and t, then use a *softmax* to predict the probability distribution $\mathcal{P}$ of classification after a linear layer:

$$\mathcal{P} = softmax(W_p([r, q, t]) + b_p), \tag{10}$$

where W_p and b_p denote the parameter of the classifier layer. The concatenation of r, q and t is represented as $[r, q, t]$.

The final training objective is to minimize the cross-entropy loss for all video frames:

$$L(\theta) = \sum (P_i, y_i), \tag{11}$$

where P_i refers to the probability distribution of the i-th frame and y_i is the ground-truth label of this frame. θ denotes all parameters.

4   Experiments

4.1   Evaluation Protocol

As a classification task, accuracy rate is the percentage of correctly classified videos to all videos. However, it can't measure the probability that the model misclassifies sensitive videos as normal, which directly affect the number of sensitive videos in the network. In this case, F_2 is needed, which considers a combination of precision and recall. Following existing methods, we adopt the F_2 as evaluation metrics on Pornography-2k [18] and only accuracy on NPDI-800.

The formula for calculating F_β is as follows:

$$F_\beta = (1 + \beta^2) \times \frac{precision \times recall}{\beta^2 \times precision + recall} \tag{12}$$

where β is set to 2.

4.2   Datasets

We conduct experiments on three datasets, NPDI-800, Porngraphy-2k and our private Porn-Bili.

NPDI-800. The NPDI-800 dataset is a widely used benchmark dataset consisting of 400 normal videos and 400 sensitive videos with a total duration of nearly 80 h. The normal video is divided into easy-to-distinguish video and hard-to-distinguish video according to 1:1. In terms of content, the dataset contains a wide range of genres and actors from distinct ethnicities (e.g., Asians, Blacks, Caucasian).

Pornography-2k. The Pornography-2k dataset is an extension of NPDI-800 and comprises approximately 140 h, spanning 2000 videos, 1000 pornographic and 1000 non-pornographic, which has a greater number of videos and genres than the previous edition, and no longer takes three categories, but only two categories of sensitive and normal.

Porn-Bili. This dataset contains 5900 videos, of which sensitive videos are from Pornhub and normal videos are from Bilibili. In the experiments, the training set contains 4900 videos, including 1374 normal videos and 3526 sensitive videos. Besides, the test set contains 1000 videos, and the ratio of sensitive videos to normal videos is 1:1. The average length of a video is one minute.

4.3   Experimental Parameters

We choose Adam as the optimizer and the initial learning rate of 1e-4. During training process, we set the batch to 16.

4.4   Baseline Methods

Late Fusion. [22] combines picture and motion information using optical flow and MPEG motion vectors with GooglNet to evaluate video attributes.

AttM-CNN. [12] proposes a new model named AttM-CNN, which overall structure is based on a deep CNN architecture with an efficient attention mechanism and metric learning. And for equality, we adopt the version that is not fine-tuned on an unpublished private dataset.

TRoF. [18] introduced a space-temporal detector and descriptor called TRoF, and aggregate the representation extracted by TRoF into higher dimensional features.

Borg's Model. [5] uses a CNN for automatic feature extraction, followed by RNN to capture the temporal information. And then ranking the harmfulness of the pornographic content according to video segments.

4.5   Evaluation on NPDI-800 Dataset

In order to bolster the credibility and persuasiveness of our approach, we conducted experiments on the publicly available NPDI-800 dataset, utilizing a rigorous five-fold cross-validation methodology as same as the initial work.

We organized the compared methods into three distinct categories, namely BoVW-based, CNN-based, and Text-based methods, which are mostly strong methods. As shown in Table 1, the neural network-based approach is better than traditional machine learning overall, and in the domain of deep learning,

Table 1. Evaluation on NPDI-800 dataset.

	Model	Extra test style data usage	Accuracy
BoVW	Avila et al. [3]	NO	87.1
	Souza et al. [4]	NO	89.5
	Caetano et al. [7]	NO	90.9
	Valle et al. [29]	NO	91.9
	Souza et al. [27]	NO	91.0
	Caetano et al. [8]	NO	92.4
CNN	Moustafa [19]	NO	94.1
	Ou et al. [21]	NO	85.3
	Jung et al. [14]	NO	94.0
	Wehrmann et al. [32]	NO	95.6
	Perez et al. [22]	NO	97.9
	shen et al. [26]	NO	94.7
	Borg et al. [5]	NO	97.8
Text	**TED-CS**	**NO**	**98.3**
CNN	TLMobDense [24]	YES	99.1

our TED-CS achieved state-of-the-art With the help of text information. It is important to note that we did not include a comparison with method [24], which employed a large number of GAN-generated images similar to the NPDI test set during training.

4.6   Evaluation on Pornography-2k Dataset

Table 2. Evaluation on Pornography-2k dataset.

Method	Extra test set style data usage	F2
STIP [15]	NO	93.1
TRoF [18]	NO	93.3
DTRoF [18]	NO	95.3
Dense Trajectories [30]	NO	95.6
Late Fusion [22]	NO	96.7
AttM-CNN-Porn [12]	NO	96.8
TED-CS	**NO**	**97.0**
TLMobDense [24]	YES	99.3

The Pornography-2k dataset is utilized by more and more works to verify the effectiveness of their method, which set accuracy and $F2$ as effect evaluation indicators.

As shown in the Table 2, compared with third-party tools PornseerPro, our method improves F2 by 21.4%. and also outperforms traditional manual feature methods STIP 3.9%. Moreover, in the CNN-based approach, late fusion is still weaker than our method. Besides, we perform better than Attm-CNN-porn without training on the two million carefully collected sensitive image datasets that is not open.

4.7 Evaluation on Porn-Bili Dataset

Table 3. Evaluation on Porn-Bili Dataset

Model	Acc.	Rec.	F2	Pre.
Only Image	87.6	100.0	95.3	80.1
Only Caption	49.0	100.0	83.3	50.0
Image and Caption	91.0	99.8	96.4	84.8
TED-CS	**94.4**	99.4	**97.4**	90.3

The sensitive video detection task can be modeled as a binary classification task. So we use accuracy, Recall, $F2$ and Precision to access the performance of models. Table 3 shows the comparative results on the Porn-Bili dataset.

We choose BERT and ResNet as baselines and both of them are powerful models in the NLP and CV domains separately. With the assistance of caption, the $F2$ value of using image and caption is 1.1% higher than the method only adopting image, and 13.1% higher than Only Caption, which proves the effectiveness of the caption information. In addition, the model that utilizes common sense knowledge improves 1 point, which confirms the necessity of the premium information carried by COMET.

4.8 Different COMET Attributes Set

In Sect. 3, we introduce three strategies for selecting COMET attributes: **WIW**, **AINRW** and **ALL**.

Table 4. Comparison between different common sense knowledge attributes on NPDI-800 dataset.

COMET	Acc.	Rec.	F2	Pre.
WIW	97.6	97.1	97.3	98.0
ALL	97.8	97.7	97.8	98.0
AINRW	**98.3**	**98.7**	**98.6**	**98.0**

We can see from Table 4 that the strategy using all attributes is slightly better than **WIW**, indicating it has the ability to provide denser premium information. However, if only the oWant, xIntent, and xWant are used, the classification effect of the model is optimal. The result demonstrated that **AINRW** can remove the interference of some redundant information.

4.9   Different Image Caption Generator

Table 5. Comparison between different caption generation models on NPDI-800 dataset without pre-trained on ImageNet.

Caption	Method	Acc.	Rec.	F2	Pre.
GRIT	ICaption	86.7	90.8	89.4	84.3
	ICCommon Sense	**89.2**	92.1	**91.1**	87.3
CLIP	ICaption	88.5	91.6	90.5	86.5
	ICCommon Sense	**90.5**	95.4	**93.6**	87.1
OFA	ICaption	86.5	92.1	90.6	85.0
	ICCommon Sense	**90.5**	95.9	**94.1**	87.5

To demonstrate the robustness and generalizability of our proposed approach, we conducted additional experiments using three distinct caption generation models, and all of which resulted in improvements in performance. CLIP is a multi-modal deep learning model proposed by openAI. And after joint learning of text and images in the pre-training stage, it lays the foundation for image description of downstream tasks. The OFA model was proposed by Alibaba. The model adopts the encoder-decoder architecture of the transformer, and unifies tasks of different modalities into a sequence-to-sequence form, which improves learning efficiency. The GRIT [20] model is an image captioning transformer, which enhances the ability to understand by making full use of the image features of Grid and Region.

We conducted experiments on the Porn-Bili dataset for these three caption generation models, using accuracy, recall, $F1$ and precision as evaluation indicators. And the following two methods are used for comparison:

ICaption. The model only uses images and corresponding captions for video classification.

ICCommon Sense. Common sense knowledge extracted based on different captions is used to assist the model's judgment.

Table 5 shows that after mining the commonsense information, all of them have improved performance.

Table 6. Ablation study on NPDI-800 dataset.

Model	NPDI-800				NPDI-800, w/o ImageNet			
	Acc.	Rec.	F2	Pre.	Acc.	Rec.	F2	Pre.
TED-CS w/o Text	97.5	97.7	97.7	97.5	86.0	89.0	88.4	85.9
TED-CS w/o Img	50.6	100.0	83.7	50.6	50.6	100.0	83.7	50.6
TED-CS w/o Comet	97.5	97.4	97.4	97.4	88.5	91.6	90.5	86.5
TED-CS w/o Atten	98.1	97.4	97.7.	98.7	91.3	94.9	93.6	88.7
TED-CS	98.3	98.7	98.6	98.0	90.5	95.4	93.6	87.1

4.10 Ablation Study

To understand the importance of commonsense knowledge and the effectiveness of the designed attention mechanism, we conduct a series of ablation experiments to investigate the effect of the key components in our model. The following are several variant models:

Model w/o Atten. Remove the H-Atten and E-Atten.

Model w/o Comet. On the basis of the Model w/o Atten, remove commonsense knowledge.

Model w/o Img. On the basis of the Model w/o COMET, remove the image information. In other words, this model only uses the caption message.

Model w/o Text. On the basis of the Model w/o COMET, remove the image caption. In other words, we only consider the information of the video frame itself.

Table 6 shows the detailed results of the ablation experiments, which are divided into two categories according to whether they were pre-trained on the ImageNet dataset. We can find that almost every module removed will cause a loss of effect. Interestingly, if only subtitle information is used, the model is basically equivalent to random classification, which confirms the conjecture that this type of information is highly objective.

5   Conclusion

In this paper, we propose a general text augmentation model with common sense knowledge to deduce cryptic underlying information. And relying on the high-density semantic knowledge of text, the model can achieve highly competitive experiment results with a few number of frames, which greatly reduces the dependence on the number of video images, and alleviates the problem of a large amount of redundant information between adjacent frames. Moreover, exhaustive experiments on three datasets demonstrate that this is an effective and general method for text processing. In future work, we will further explore the combination of text modality, image modality, and speech modality to refine the problem of sensitive video detection.

References

1. Agarap, A.F.: Deep learning using rectified linear units (relu). CoRR arXiv:1803.08375 (2018)
2. Alshamrani, S.: Detecting and measuring the exposure of children and adolescents to inappropriate comments in Youtube. In: Proceedings of the 29th ACM International Conference on Information Knowledge Management, pp. 3213–3216 (2020)
3. de Avila, S.E.F., Thome, N., Cord, M., Valle, E., de Albuquerque Araújo, A.: BOSSA: extended bow formalism for image classification. In: 18th IEEE International Conference on Image Processing, ICIP 2011, Brussels, Belgium, 11–14 September 2011, pp. 2909–2912 (2011)
4. de Avila, S.E.F., Thome, N., Cord, M., Valle, E., de Albuquerque Araújo, A.: Pooling in image representation: the visual codeword point of view. Comput. Vis. Image Underst. **117**, 453–465 (2013)
5. Borg, M., Tabone, A., Bonnici, A., Cristina, S., Farrugia, R.A., Camilleri, K.P.: Detecting and ranking pornographic content in videos. Forensic Sci. Int. Digital Invest. **42**, 301436 (2022)
6. Bosselut, A., Rashkin, H., Sap, M., Malaviya, C., Celikyilmaz, A., Choi, Y.: COMET: commonsense transformers for automatic knowledge graph construction. In: Proceedings of the 57th Conference of the Association for Computational Linguistics, ACL 2019, pp. 4762–4779 (2019)
7. Caetano, C., de Avila, S.E.F., Guimarães, S.J.F., de Albuquerque Araújo, A.: Pornography detection using Bossanova video descriptor. In: 22nd European Signal Processing Conference, EUSIPCO 2014, Lisbon, Portugal, 1–5 September 2014, pp. 1681–1685 (2014)
8. Caetano, C., de Avila, S.E.F., Schwartz, W.R., Guimarães, S.J.F., de Albuquerque Araújo, A.: A mid-level video representation based on binary descriptors: a case study for pornography detection. Neurocomputing **213**, 102–114 (2016)
9. Chen, J., Hu, Y., Liu, J., Xiao, Y., Jiang, H.: Deep short text classification with knowledge powered attention. In: Proceedings AAAI 2019, pp. 6252–6259 (2019)
10. Devlin, J., Chang, M., Lee, K., Toutanova, K.: BERT: pre-training of deep bidirectional transformers for language understanding. In: Proceedings of the 2019 Conference of the North American Chapter of the Association for Computational Linguistics: Human Language Technologies, NAACL-HLT 2019, Minneapolis, MN, USA, 2–7 June 2019, vol. 1 (Long and Short Papers), pp. 4171–4186 (2019)
11. Fleck, M.M., Forsyth, D.A., Bregler, C.: Finding naked people. In: Buxton, B., Cipolla, R. (eds.) ECCV 1996. LNCS, vol. 1065, pp. 593–602. Springer, Heidelberg (1996). https://doi.org/10.1007/3-540-61123-1_173
12. Gangwar, A., González-Castro, V., Alegre, E., Fidalgo, E.: AttM-CNN: attention and metric learning based CNN for pornography, age and child sexual abuse (CSA) detection in images. Neurocomputing **445**, 81–104 (2021)
13. He, K., Zhang, X., Ren, S., Sun, J.: Deep residual learning for image recognition. In: 2016 IEEE Conference on Computer Vision and Pattern Recognition, CVPR 2016, 2016, pp. 770–778 (2016)

14. Jung, J., Makhijani, R., Morlot, A.: Combining CNNs for detecting pornography in the absence of labeled training data (2020)
15. Laptev, I.: On space-time interest points. Int. J. Comput. Vis. **64**, 107–123 (2005)
16. Lopes, A.P., de Avila, S.E., Peixoto, A.N., Oliveira, R.S., de Albuquerque Araújo, A.: A bag-of-features approach based on hue-SIFT descriptor for nude detection. In: 2009 17th European Signal Processing Conference, pp. 1552–1556 (2009)
17. Mallmann, J., Santin, A.O., Viegas, E.K., dos Santos, R.R., Geremias, J.: PPCensor: architecture for real-time pornography detection in video streaming. Futur. Gener. Comput. Syst. **112**, 945–955 (2020)
18. Moreira, D., et al.: Pornography classification: the hidden clues in video space-time. Forensic Sci. Int. **268**, 46–61 (2016)
19. Moustafa, M.N.: Applying deep learning to classify pornographic images and videos. CoRR arXiv:1511.08899 (2015)
20. Nguyen, V., Suganuma, M., Okatani, T.: GRIT: faster and better image captioning transformer using dual visual features. In: Avidan, S., Brostow, G., Cissé, M., Farinella, G.M., Hassner, T. (eds.) Computer Vision - ECCV 2022–17th European Conference, Tel Aviv, Israel, 23–27 October 2022, Proceedings, Part XXXVI, vol. 13696, pp. 167–184, Springer, Cham (2022). https://doi.org/10.1007/978-3-031-20059-5_10
21. Ou, X., Ling, H., Yu, H., Li, P., Zou, F., Liu, S.: Adult image and video recognition by a deep multicontext network and fine-to-coarse strategy. ACM Trans. Intell. Syst. Technol. **8**(5), 1–25 (2017)
22. Perez, M., et al.: Video pornography detection through deep learning techniques and motion information. Neurocomputing **230**, 279–293 (2017)
23. Radford, A., et al.: Learning transferable visual models from natural language supervision. In: Proceedings of the 38th International Conference on Machine Learning, ICML 2021, 18–24 July 2021, Virtual Event, pp. 8748–8763 (2021)
24. Samal, S., Nayak, R., Jena, S., Balabantaray, B.K.: Obscene image detection using transfer learning and feature fusion. Multimedia Tools Appl. **82**, 28739–28767 (2023). https://doi.org/10.1007/s11042-023-14437-7
25. Sap, M., et al.: ATOMIC: an atlas of machine commonsense for if-then reasoning. In: The Thirty-Third AAAI Conference on Artificial Intelligence, AAAI 2019, The Thirty-First Innovative Applications of Artificial Intelligence Conference, IAAI 2019, The Ninth AAAI Symposium on Educational Advances in Artificial Intelligence, EAAI 2019, 2019, pp. 3027–3035 (2019)
26. Shen, R., Zou, F., Song, J., Yan, K., Zhou, K.: EFUI: an ensemble framework using uncertain inference for pornographic image recognition. Neurocomputing **322**, 166–176 (2018)
27. Souza, F., Valle, E., Cámara-Chávez, G., Araújo, A.: An evaluation on color invariant based local spatiotemporal features for action recognition. IEEE SIBGRAPI (2012)
28. Speer, R., Chin, J., Havasi, C.: ConceptNet 5.5: an open multilingual graph of general knowledge. In: Proceedings of the Thirty-First AAAI Conference on Artificial Intelligence, San Francisco, California, USA, 4–9 February 2017, pp. 4444–4451 (2017)
29. Valle, E., et al.: Content-based filtering for video sharing social networks. CoRR arXiv:1101.2427 (2011)
30. Wang, H., Schmid, C.: Action recognition with improved trajectories. In: IEEE International Conference on Computer Vision, ICCV 2013, Sydney, Australia, December 1–8, 2013, pp. 3551–3558 (2013)

31. Wang, P., et al.: OFA: unifying architectures, tasks, and modalities through a simple sequence-to-sequence learning framework. In: International Conference on Machine Learning, ICML 2022, Baltimore, Maryland, USA, 17–23 July 2022, vol. 162, pp. 23318–23340 (2022)
32. Wehrmann, J., Simoes, G.S., Barros, R.C., Cavalcante, V.F.: Adult content detection in videos with convolutional and recurrent neural networks. Neurocomputing **272**, 432–438 (2018)

Wang, D., et al: CPA: managing microarchitectural profiling through a single sequential processor a leaking through a leak. International Conference on Machine Learning, 16 vol. 4923, Semmes, Vancouver, USA (2019 22.htm 2019 949 794 (2019 782 782 785 782)

Weerman, C. authors, LSS. Rugat, R. Studies, timing. Formatting Codes from Colors Wild, evaluation, and evaluation frameworks. Neuromatric 273–301 43 (2019)

Author Index

A

Ahmad, Adnan   408
Al-Laith, Ali   440
An, Rui   377

B

Bai, Peng   183
Bu, Liping   692

C

Cao, Jing   230
Cao, Yunbo   291
Chang, Chao   307
Chen, Chen   62, 581
Chen, Huaming   456
Chen, Jue   471
Chen, Tingwei   125
Chen, Tingxuan   107
Chen, Wei   660, 676
Chen, Yan   550
Chen, Yang   257
Chen, Zhiwei   504
Chen, Zirui   46
Cheng, Debo   629

D

Deng, Shilong   535
Ding, Xueyan   217
Du, Minxin   393
Du, Yiyang   291
Duan, Wenyu   320

F

Fan, Chunlong   320
Feng, Qianjin   257
Fu, Feifei   198

G

Gao, Yufei   139
Guo, Weiyu   230

Guo, WenBin   46
Guo, Zhiwei   644

H

He, Hongcai   535
He, Xiaolong   198
Helles, Rasmus   440
Hou, Chunyan   581
Hu, Guangwu   152
Hu, Miao   423

J

Jiang, Mengqi   629
Jin, Zhibo   456
Ju, Jiaxin   565

K

Kang, Le   489
Ke, Dejing   91

L

Lai, Hanjiang   347
Lam, Kwok-Yan   62
Li, Bo   565
Li, Guohui   107
Li, Junxuan   518
Li, Li   3
Li, Qing   168
Li, Wenhao   550
Li, Zhao   46
Li, Zhoujun   291
Lin, Miaopei   347, 614
Liu, Chunrui   244
Liu, Haowen   139
Liu, Hongzhi   257
Liu, Shuhui   377
Liu, Wanxiao   471
Liu, Wei   347
Liu, Wenci   125
Liu, Xiao-Qi   78

© The Editor(s) (if applicable) and The Author(s), under exclusive license
to Springer Nature Switzerland AG 2023
X. Yang et al. (Eds.): ADMA 2023, LNAI 14177, pp. 709–711, 2023.
https://doi.org/10.1007/978-3-031-46664-9

Liu, Yafei   3
Liu, Yan   78
Liu, Yifan   16
Long, Jun   107
Lou, Qian   393
Lu, Ruidong   217
Lu, Xingyu   599
Lu, Zhiwu   198
Luo, Di   644
Luo, Shuai   107
Luo, Wei   408
Luo, Xiaofeng   614

M
Ma, Ji   125
Mu, Tai-Jiang   489

N
Ning, Xiaodong   489

P
Pan, Shirui   565
Pan, Xinlong   320
Pang, Jun   78
Peng, Yongqing   152
Pi, Dechang   504
Ping, Mingtian   504
Pu, Jiansu   91
Pun, Chi-Man   599

Q
Qin, Hong-Chao   78
Qin, Mingfu   660
Qiu, Shuang   614
Qiu, Xihe   471

R
Robles-Kelly, Antonio   408

S
S. M. Chow, Sherman   393
Saxena, Akrati   440
Shan, Zheng   139
Shang, Bin   16
Shang, Xuequn   377
Shao, Jie   535
Shen, Jun   456
Sheng, Quan Z.   62

Shi, Lei   139
Shi, Xiaodong   183
Simonsen, Jakob Grue   440
Song, Wei   393
Su, Shiyan   362
Su, Zhan   440
Sun, Linzhuang   692
Sun, Qiule   217
Sun, Yibo   362
Sun, Zelong   198

T
Tan, Lei   423
Tan, Shuyue   692
Tang, Yong   307
Tian, Feng   550

V
Veilahti, Antti   440

W
Wan, Peng   244
Wang, Can   565
Wang, Chenxin   16
Wang, Junlong   504
Wang, Meng-xiang   347
Wang, Shiqi   347, 614
Wang, Xiaoxin   272
Wang, Xin   46
Wang, Xinyi   456
Wang, Xiuhua   393
Wang, Yue   230
Wang, Yufei   62
Wei, Jingxuan   692
Wei, Lin   139
Wei, Xiao   272
Wickman, Ryan   518
Wu, Di   423
Wu, Ruohao   152
Wu, Xianjie   291
Wu, Yuchen   550

X
Xiao, Meihong   107
Xiao, Xi   152
Xie, Di   334
Xu, Hangtong   32
Xu, Yuanbo   32

Y

Yan, Zhao 291
Yang, Guoxing 198
Yang, Liu 107
Yang, Xiaorui 676
Yang, Yongjian 32
Yang, Zhenguo 644
Yang, Zhuopan 644
Yin, Jian 347, 614
You, Jiuxiang 644
Yu, Bihui 692
Yu, Dunhui 334
Yu, Jianxing 347, 614
Yu, Naitong 320
Yuan, Jing 581
Yuan, Xiaojie 581
Yue, Lin 32
Yue, Shuhao 334

Z

Zeng, Xiangwei 307
Zhang, Daoqiang 244
Zhang, Guixian 629
Zhang, Han 152
Zhang, Jianxin 217
Zhang, Jiayu 456
Zhang, Leyuan 629
Zhang, Shaowei 46
Zhang, Shichao 629

Zhang, Shiwei 291
Zhang, Wenxin 377
Zhang, Xiaofei 518
Zhang, Xinyue 230
Zhang, Yang 62
Zhang, Yuanbo 32
Zhang, Yupei 377
Zhang, Zhicheng 362
Zhao, Bo 139
Zhao, Hanqing 152
Zhao, Lei 660, 676
Zhao, Yifei 550
Zhao, Yiman 692
Zhao, Yinliang 16
Zhao, Yongjun 393
Zheng, Meizhen 183
Zheng, Qinghua 550
Zheng, Yu 393
Zheng, Zetao 535
Zhou, Junming 307
Zhou, Linhui 230
Zhou, Qian 660, 676
Zhou, Xin 168
Zhou, Yipeng 423
Zhu, Haiping 550
Zhu, Jizhao 320
Zhu, Nengjun 272
Zhu, Shiyu 257
Zhu, Zhiyu 456
Zu, Shuaishuai 3

Printed in the United States
by Baker & Taylor Publisher Services